HANDBOOK
OF
COMPUTABILITY THEORY

STUDIES IN LOGIC

AND

THE FOUNDATIONS OF MATHEMATICS

VOLUME 140

ELSEVIER

AMSTERDAM • LAUSANNE • NEW YORK • OXFORD • SHANNON • SINGAPORE • TOKYO

HANDBOOK OF COMPUTABILITY THEORY

Edited by

Edward R. GRIFFOR
Communication Advisors, Inc., Southfield, MI 48109, USA

N·H

1999

ELSEVIER

AMSTERDAM • LAUSANNE • NEW YORK • OXFORD • SHANNON • SINGAPORE • TOKYO

ELSEVIER SCIENCE B.V.
Sara Burgerhartstraat 25
P.O. Box 211, 1000 AE Amsterdam, The Netherlands

First edition 1999

Library of Congress Cataloging-in-Publication Data

Handbook of computability theory / edited by E.R. Griffor.
p. cm. -- (Studies in logic and the foundations of mathematics : v. 140)
Includes bibliographical references and indexes.
ISBN 0-444-89882-4 (alk. paper)
1. Computable functions. I. Griffor, Edward R. II. Series.
QA9.59.H36 1999
511.3--dc21 99-25568
CIP

ISBN: 0 444 89882 4

⊗ The paper used in this publication meets the requirements of ANSI/NISO Z39.48-1992 (Permanence of Paper).

Printed in The Netherlands

Preface

Background

In the *Handbook of Mathematical Logic* an entire part was devoted to *recursion theory*. That area has seen so much activity since that time so as to justify devoting an entire volume to that development. We have tried to avoid unnecessary overlap with that part. The elementary theory and several other topics like the theory of inductive definability are not included here. During the last fifteen plus years many long standing open questions have been answered, often using entirely new methods. At the same time a new era in the development of computability has begun. This new era involves a fundamental interaction between the efforts of mathematicians to understand the *concepts of computability in an era where scientific and everyday experience are dominated by computing* and the efforts of computer scientists to develop their activity into a science.

The chapters of this volume all have their own level of presentation. The topics have been chosen based on the active research interest associated with them. Since the interest in some topics is older than that in others, some presentations contain fundamental definitions and basic results while others relate very little of the elementary theory behind them and aim directly toward an exposition of advanced results. Presentations of the latter sort are in some cases restricted to a short survey of recent results (due to the complexity of the methods and proofs themselves). Hence the variation in level of presentation from chapter to chapter only reflects the conceptual situation itself. One example of this is the collective efforts to develop an acceptable theory of computation on the real numbers. The last two decades has seen at least two new definitions of effective operations on the real numbers.

Contents and Presentation

Both the notions of a decidable set and that of a computable function can be explained in terms of intuitive and informal notions. Using finiteness as a primitive, a relation between input and output is referred to as computable if it can be described as a *finitary procedure*, where input and output have *finitary representations*.

Thus the explanations of what computable and decidable mean can be given in terms of other informal notions like that of an algorithm.

Computability theory refers to the study of different formalizations of the informal notions of computable function or algorithmic procedure. In the classical case of functions whose inputs and outputs are natural numbers, examples of such formalizations are Turing machines and Kleene's generating schemes for the partial recursive functions. Turing machines' own likeness to a physical machine gives credence to the claim that they witness the computability of the operations they carry out on their inputs. In particular, we acknowledge their being finitary procedures and that the representation of their input strings on a tape is an example of a finitary representation. Kleene's generating schemes represent conceptually a radically different approach, namely, that of inductively defining the result of applying an operation to an input (if it converges or has a value at all on that input). The definition explains roughly that a formal expression, as an operation on its specified arguments, is defined or takes a value just in case certain of its constituent parts are defined and, in that case, the value is given in a trivial way from those of its constituents. Each one of those parts can in turn be regarded as subject to the same sort of analysis and so on. What is developed in this way is a *computation tree* associated with the originally considered operation and inputs. An analysis of these trees can be used to see the computability of *scheme-computable* functions or, alternatively, in order to prove that the same functions are Kleene and Turing computable.

Many of the topics taken up later in this volume are instances of *generalized computability theory*. What is more general about the computations considered is in many cases the inputs and outputs allowed. They may even be infinite giving rise to infinitely branching computation trees. *Set recursion* is an example of this where the inputs and outputs may be arbitrary sets. Classically the terms set and function were used in computability theory to refer to sets of natural numbers and functions on the natural numbers. One sees in the chapter on the theory of *numerations* that even this approach is quite general. There it is explained how computation on essentially large variety of structures can be framed within the classical theory. Some of the examples of generalized computability contained are computable functions on the real numbers, computable functionals of higher type and computable functions on sets.

Conclusion

For an adequate treatment of the elementary results of computability theory, the reader can consult standard references like H. Rogers and, more recently, P. Odifreddi. These references contain central results of the classical theory that remain challenges to our understanding. For example, Kleene's recursion or fixed points theorems express fundamental notions about a *stored program computing device* as it was described in H. Enderton's chapter in the *Handbook of Mathematical Logic*. They explain the meaning or consequences of the fact that one machine may take the description of another and run it, interpret it or apply it in the course of carrying out the computation of its own function.

Finally, an in-depth discussion of the terminology *computability* vs. *recursion theory* is contained in one of the first chapters of this handbook and I shall not anticipate it here. I would add however one remark. I have chosen the title *Handbook of Computability Theory* for this volume for reasons independent of the more profound reasons discussed there. They have more to do with a wish to *broaden the debate of the key issue in the study of computability*. My sense was and remains that this can only serve to strengthen our intuitions about computation and the science of computability.

Acknowledgement

Our goal in preparing this handbook has been to selectively reflect the research activity of the various directions that make up computability theory through contributions presenting the dominant themes of the period since the publication of the *Handbook of Mathematical Logic*. During this period some topics, like the study of the Turing degrees, have been worked on intensively and have seen may of their open problems and questions answered. Other topics included here saw their birth during the same period. We have not excluded other areas of study for lack of importance, but rather to stay within the scope of our original purpose. I would like to thank Robert Soare, Barry Cooper and John Tucker for their suggestions and the staff at Elsevier Science Publishers for their support.

Edward R. Griffor
Grosse Pointe Park, MI USA

List of Contributors

Ambos-Spies, K., *Universität Heidelberg, Heidelberg* (Ch. 18).
Cenzer, D., *University of Florida, Gainesville, FL* (Ch. 2).
Chong, C.T., *National University of Singapore, Singapore* (Ch. 9).
Clote, P., *Ludwig-Maximilians-Universität München, München* (Ch. 17).
Cooper, S.B., *University of Leeds, Leeds, England, UK* (Ch. 4).
Ershov, Y.L., *Novosibirsk State University, Novosibirsk* (Ch. 14).
Friedman, S.D., *Massachusetts Institute of Technology, Massachusetts, MA* (Ch. 9).
Hinman, P.G., *University of Michigan, Ann Arbor, MI* (Ch. 11).
Millar, T.S., *University of Wisconsin–Madison, 480 Lincoln Drive, WI* (Ch. 15).
Normann, D., *University of Oslo, Oslo* (Ch. 8).
Odifreddi, P., *University of Torino, Torino* (Ch. 3).
Pour-El, M.B., *University of Minnesota, Minneapolis, MN* (Ch. 13).
Sacks, G.E., *Harvard University, Cambridge, MA* (Ch. 10).
Schwichtenberg, H., *University of Munich, Munich* (Ch. 16).
Shore, R.A., *Cornell University, Ithaca, NY* (Ch. 6).
Slaman, T.A., *The University of California at Berkeley, Berkeley, CA* (Ch. 5).
Soare, R.I., *University of Chicago, Chicago, IL* (Chs. 1, 7).
Stoltenberg-Hansen, V., *Uppsala University, Uppsala* (Ch. 12).
Tucker, J.V., *University of Wales Swansea, Swansea, Wales, UK* (Ch. 12).

Contents

Preface v
List of Contributors ix

Part 1: Fundamentals of Computability Theory

1. The history and concept of computability 3
   R.I. Soare
2. Π^0_1 classes in recursion theory 37
   D. Cenzer

Part 2: Reducibilities and Degrees

3. Reducibilities 89
   P. Odifreddi
4. Local degree theory 121
   S.B. Cooper
5. The global structure of the Turing degrees 155
   T.A. Slaman
6. The recursively enumerable degrees 169
   R.A. Shore
7. An overview of the computably enumerable sets 199
   R.I. Soare

Part 3: Generalized Computability Theory

8. The continuous functionals 251
   D. Normann
9. Ordinal recursion theory 277
   C.T. Chong and S.D. Friedman
10. E-recursion 301
    G.E. Sacks
11. Recursion on abstract structures 315
    P.G. Hinman

Part 4: Mathematics and Computability Theory

12. Computable rings and fields 363
V. Stoltenberg-Hansen and J.V. Tucker
13. The structure of computability in analysis and physical theory: An extension of Church's thesis 449
M.B. Pour-El
14. Theory of numberings 473
Y.L. Ershov

Part 5: Logic and Computability Theory

15. Pure recursive model theory 507
T.S. Millar
16. Classifying recursive functions 533
H. Schwichtenberg

Part 6: Computer Science and Computability Theory

17. Computation models and function algebras 589
P. Clote
18. Polynomial time reducibilities and degrees 683
K. Ambos-Spies

Author Index 707

Subject Index 715

Part 1
Fundamentals of Computability Theory

CHAPTER 1

The History and Concept of Computability

Robert I. Soare*

Department of Mathematics, University of Chicago, Chicago, IL 60637-1546, USA
E-mail: soare@math.uchicago.edu or soare@cs.uchicago.edu
Anonymous ftp: cs.uchicago.edu: ftp/pub/users/soare
World Wide Web: http://www.cs.uchicago.edu/~soare

Contents

1. Introduction 5
2. A brief history of computability 6
   2.1. The concepts of computability and recursion 6
   2.2. The origin of recursion 7
   2.3. The origin of computable functions 8
   2.4. General recursive functions 8
   2.5. The flaw in Church's thesis 9
3. Turing's contributions to computability 11
   3.1. Turing's idealized human computor 11
   3.2. Accepting Turing's Thesis 13
   3.3. The Church–Turing Thesis as a definition 15
   3.4. Other theses became definitions 16
   3.5. Register machines 17
4. Later developments in computability 18
   4.1. Kleene's normal form and his μ-recursive functions 18

*The author was partially supported by National Science Foundation Grant DMS 94-00825. Much of this material was presented in Soare [1996]. It is presented in a revised and shortened form here with the permission of the Association for Symbolic Logic. Helpful suggestions, comments, and criticisms on preliminary drafts of the former paper are acknowledged by name in Soare [1996].

This chapter is dedicated to Paul Snowden Russell, former President of the Harris Bank, and long time trustee of the University of Chicago. At Russell's memorial service in 1950 the University of Chicago President and distinguished educator, Robert Maynard Hutchins, said of Russell, "He was tough minded and realistic, ... realistic in the sense that he was against cant and hypocrisy. He called things by their right names." He "was faithful to his youth." Without the Paul Snowden Russell Distinguished Service Professorship awarded to him in 1994, Soare might never have begun two weeks later the now successful movement to return the terminology, concepts, and paradigm of the subject from "recursion" to "computability," might never have "*called things by their right names*," and this volume might have had its original 1994 proposed title, *The Handbook of Recursion Theory*, instead of its final title, *The Handbook of Computability Theory*.

HANDBOOK OF COMPUTABILITY THEORY
Edited by E.R. Griffor

4.2. Computably enumerable sets and Post 19
4.3. History of relative computability 21
4.4. Higher order computability 22
4.5. How the terms became fixed 24
4.6. Current usage of the concepts and terms 25
5. Mathematical, scientific, and general English usage 28
6. Themes and goals of computability theory 29
7. Analysis 30
References 31

1. Introduction

We consider the informal concept of a "computable" or "effectively calculable" function on natural numbers and two of the formalisms used to define it, "*computability*" and "(*general*) *recursiveness*." We consider their origin, exact technical definition, concepts, history, how they became fixed in their present roles, and how they were first and are now used. All functions are on the nonnegative integers, $\omega = \{0, 1, 2, \ldots\}$, and all sets will be subsets of ω. The central concept of the field of computability theory is the notion of an "effectively calculable" or "computable" function.

DEFINITION 1.1. A function is "*computable*" (also called "*effectively calculable*" or simply "*calculable*") if it can be calculated by a finite mechanical procedure. (For a more precise description see Section 3.1.)

DEFINITION 1.2.
(i) A function is *Turing computable* if it is definable by a Turing machine, as defined by Turing [1936]. (See Kleene [1952a] or Soare [1987].)
(ii) A set A is *computably enumerable* (c.e.) if A is $\emptyset$ or is the range of a Turing computable function.
(iii) A function f is *recursive* if it is *general recursive*, as defined by Gödel [1934]. (See also Kleene's variant [1936, 1943] and [1952a, p. 274].)
(iv) A set A is *recursively enumerable* (*r.e.*) if A is $\emptyset$ or is the range of a general recursive function.[1]

Later we shall say more of these *formal* definitions and their meanings. For the moment we regard these terms strictly with their *intensional* meaning as above, and we do not *extensionally* identify them with each other or with other formal notions known to be mathematically equivalent (such as λ-definability, μ-recursiveness, or Post normal systems described later), nor with the *informal* notions of computable or effectively calculable under Church's Thesis or Turing's Thesis.

The subject of computability theory was accidentally named "*recursive function theory*" or simply "*recursion theory*" in the 1930's but has recently acquired the more descriptive name of "Computability Theory," which is also historically more accurate based on the work of Gödel and Turing, the inventors of the two concepts.

In this paper we examine the meaning, origin, and history of the concepts "recursive" and "computable" with an eye toward reexamining how we use them in practice. The ultimate aim is to ask: "*What is the subject really about*?" For example, is it about computability, recursions, definability, or something else?

---

[1] This terminology is the same as that introduced in the 1930's and used since then, except for the term "computably enumerable," recently introduced, because Turing and Gödel did not explicitly introduce a term for these corresponding *sets*, but just for the computable *functions*. Post [1944] explicitly added the empty set as an r.e. set (see Davis [1965, p. 308]), which Church and Kleene had omitted.

In Section 2 we review the origin and history of each concept and the formal definitions. In Section 3 we consider the Church–Turing Thesis that the intuitively computable functions coincide with the formally computable ones, and consider using the thesis as a definition. In Section 4 we trace the historical development of certain parts of the subject after the 1930's and show how the present practices were adopted. General English usage is discussed in Section 5, and a conclusion and analysis is given in Section 6.

We cite references in the usual convention by author and year, e.g., [Post, 1944] or simply Post [1944]. To save space we omit from our bibliography some references which appear in Soare [1987], van Heijenoort [1967], or Kleene [1952a], and we cite them as there by year.

2. A brief history of computability

2.1. *The concepts of computability and recursion*

A *computation* is a process whereby we proceed from initially given objects, called *inputs*, according to a fixed set of rules, called a *program*, *procedure*, or *algorithm*, through a series of *steps* and arrive at the end of these steps with a final result, called the *output*. The algorithm, as a set of rules proceeding from inputs to output, must be precise and definite, with each successive step clearly determined. The concept of *computability* concerns those objects which may be specified in principle by computations, and includes relative computability (computability from an oracle as explained in Section 4.3) which studies the relationship between two objects which holds when one is computable relative to the other. For the Gödel–Church–Turing case of computability on ω (called *ω-computability theory*) the inputs, outputs, the program, and computation will all be finite objects, but in Kleene's higher order computability such as computability on constructive ordinals, or higher types, computations may be more general objects such as finite path trees (well-founded trees), and the inputs may be infinite objects such as type 1 objects, namely functions from ω to ω. In α-computability theory (computability on admissible ordinals) the inputs and outputs are likewise suitably generalized.

The concept of *recursion* stems from the verb "recur," "to return to a place or status." The primary mathematical meaning of recursion (Section 5) has always been "definition by induction" (i.e. by recursion), namely defining a function f at an argument x using its own previously defined values (say $f(y)$ for $y < x$), and also using "simpler" functions g (usually previously defined). The advantage of the Herbrand–Gödel definition of a (general) recursive function (Section 2.4) was that it encompassed recursion on an arbitrary number of arguments, and many felt it "included all possible recursions."

The Kleene Fixed Point Theorem gave a still more powerful form of this "reflexive program call" permissible in programs. Let $\{P_n\}_{n\in\omega}$ be an effective listing of all

(Turing) programs and let φ_n be the computable partial function computed by P_n. The Kleene Fixed Point Theorem (Recursion Theorem) asserts that for every Turing computable total function $f(x)$ there is a fixed point n such that $\varphi_{f(n)} = \varphi_n$. This gives the following recursive call as described in Soare [1987, pp. 36–38]. Using the Kleene s-m-n-theorem we can define a computable function $f(x)$ by specifying $\varphi_{f(x)} : \ldots x \ldots$, according to some program $P_{f(x)}$, which may mention x and may even call program P_x. Taking a fixed point n for $f(x)$ we have $\varphi_n = \varphi_{f(n)} : \ldots n \ldots$, so that the program P_n for computing φ_n can in effect "call itself" (or more precisely call a program which computes the same function) during the execution of the program. We call this a "*reflexive program call*." Platek's thesis [1966] in higher types stresses the role of fixed points of certain functionals and is often cited as an example of a more general type of recursion. (See Section 4.4.)

The *concept of recursion* used here includes: (1) induction and the notion of *reflexive program call* (including primitive recursion and also Kleene's Recursion Theorem); (2) the notion of a *fixed point* for some function, and the more general Platek style fixed points in higher types (see Section 4.4); (3) other phenomena related to (1) and (2) specified for certain situations and structures. However, the concept of recursion does *not* include the notion of "computable" or "algorithmic" or decidable as described in the first paragraph, even though it was often used from 1935 to 1995 with these meanings before the present terminology became established.

2.2. *The origin of recursion*

Well before the nineteenth century mathematicians used the principle of defining a function by induction. Dedekind [1888] proved, using accepted axioms, that such a definition defines a *unique* function, and he applied it to the definition of the functions $m + n$, $m \times n$, and m^n. Based on this work of Dedekind, Peano [1889] and [1891] wrote the familiar five axioms for the positive integers. As a companion to his fifth axiom, mathematical induction, Peano used definition by induction, which has been called *primitive* recursion (since Péter [1934] and Kleene [1936]), namely

$$\text{Scheme (V)} \quad \begin{cases} f(0, \vec{y}) = h(\vec{y}), \\ f(x+1, \vec{y}) = g(x, f(x, \vec{y}), \vec{y}) \end{cases} \tag{1}$$

where g and h are previously defined functions, and $\vec{y}$ denotes a (possibly empty) sequence, $y_1, \ldots, y_n$, of additional variables (parameters). This is Scheme (V) in the well-known five schemata used to define the class of primitive recursive functions, see Soare [1987, pp. 8–9], or Kleene [1952a, p. 219]. The other schemata are: (I) successor $\lambda x\,[x + 1]$; (II) constant functions $\lambda x_1 \ldots x_n\,[k]$; (III) projections $\lambda x_1 \ldots x_n\,[x_i]$; and (IV) composition $f(\vec{x}) = h(g_1(\vec{x}), \ldots, g_m(\vec{x}))$. The concept of recursion played an important role in the foundations of mathematics and in the work of Skolem [1923], Hilbert [1926], Gödel [1931], and Péter [1934].

2.3. *The origin of computable functions*

Mathematicians have studied calculation and algorithms since the time of the Babylonians. Kleene [1988, p. 19] wrote, "The recognition of algorithms goes back at least to Euclid (c. 330 B.C.)." For example there is Euclid's famous greatest common divisor algorithm. The name "algorithm" comes from the name of the ninth century Arabian mathematician Al-Khowarizmi. Along with the development of theoretical mathematical algorithms there developed an interest in actual calculating machines. In 1642 the French mathematician and scientist, Blaise Pascal, invented an adding machine which may be the first digital calculator. In 1671 the German mathematician and philosopher, Gottfried Wilhelm Leibniz, co-inventor with Newton of the calculus, invented a machine that performed multiplication. Leibniz's machine, called a *stepped reckoner* could not only add and multiply, it could divide, and extract square roots, by a series of repeated additions, and is used even today. His stepped gear wheel still appears in a few twentieth century devices. Leibniz' main contribution was the demonstration of the superiority of the binary over the decimal representation for mechanical computers. In the work of Leibniz the symbolic representation of problems was combined with a search for their algorithmic solutions. Sieg [1994, p. 73] wrote that Leibniz "viewed algorithmic solutions of mathematical and logic problems as paradigms of problem solving in general. Remember that he recommended to disputants in *any* field to sit down at a table, take pens in their hands, and say 'Calculemus'!" Leibniz searched for a universal language (*lingua characteristica*) and a calculus of reasoning ("*calculus ratiocinator*") with which to facilitate his program. Around 1834 Babbage invented the idea of an "Analytic Engine," which could be programmed to perform long and tedious calculations, and formulated what Gandy [1988, p. 58] called "Babbage's Thesis," that "the whole of the development and operations of analysis are now capable of being executed by machinery." Gandy [1988, p. 57] pointed out that considering Babbage's Analytic Engine as a register machine, his proposed operations define precisely the Turing computable functions.

2.4. *General recursive functions*

These two trends of recursion and computability were brought together in the 1930's by Gödel, Church, Kleene, Turing, and others partly in response to questions raised earlier by Hilbert. At the end of the nineteenth century Hilbert [1899] gave an axiomatization of geometry and showed that the question of the consistency of geometry reduced to that for the real-number system, and that in turn to arithmetic by results of Dedekind (at least in a second order system). Hilbert [1904] proposed proving the consistency of arithmetic by what became known by [1928] as his *finitist program.* He proposed using the finiteness of mathematical proofs in order to establish that contradictions could not be derived. This tended to reduce proofs to manipulation of finite strings of symbols devoid of intuitive meaning which stimulated the development of mechanical processes to accomplish this. Closely related was the *Entscheidungsproblem*,[2] the decision problem for first order logic, which emerged in the early 1920's in lectures by Hilbert and was described in Hilbert and Ackermann [1928]. It was to give a decision procedure [*Entscheidungsverfahren*] "that allows one to decide the validity[3] (respectively satisfiability) of a given logical expression by a finite number of operations" (Hilbert and Ackermann [1928, pp. 72–73]). Hilbert characterized this as the fundamental problem of mathematical logic.

Gödel [1931] proved his *first* incompleteness theorem which (stated in modern terms and with an improvement by Rosser) asserts roughly that any consistent extension T of elementary number theory is incomplete. By arithmetizing the proof Gödel obtained his *second* incompleteness theorem which asserts that such a T cannot prove its own consistency (see Kleene [1952a, §42]), which was a setback for Hilbert's program (see Gödel [1958]).

In his proof Gödel [1931] (see Gödel [1986, p. 158]) used the notion of a primitive recursive function (which he called "recursive" [*eine rekursive Funktion*]) because these functions were easily representable in Gödel's formal system **P** for arithmetic, and were sufficient to enable him to "Gödel number" all the syntactic objects so that he could obtain self-reference and thereby incompleteness. Gödel realized, however, that the primitive recursive functions did not include *all* effectively calculable functions,[4] and in [1934] he proposed a wider class of functions based on an earlier suggestion of Herbrand. Gödel called these the *general recursive functions*. Herbrand had written Gödel a letter on April 7, 1931 (see Gödel [1986, p. 368] and Sieg [1994, p. 81]), in which he wrote, "If φ denotes an unknown function, and $\psi_1, \ldots, \psi_k$ are known functions, and if the ψ's and φ are substituted in one another in the most general fashions and certain pairs of resulting expressions are equated, then if the resulting set of functional equations has one and only one solution for φ, φ is a recursive function." Gödel made two restrictions on this definition to make it *effective*, first that the left-hand sides of the functional equations be in standard form with φ being the outermost symbol, and second that for each set of natural numbers $n_1, \ldots, n_j$ there exists a unique m such that $\varphi(n_1, \ldots, n_j) = m$ is a derived equation. Kleene [1936, 1943, 1952a] introduced variants of Gödel's two rules which give an equivalent formulation of the Herbrand–Gödel definition.

2.5. *The flaw in Church's thesis*

In 1930 Church had been studying a class of effectively calculable functions called *λ-definable functions*. Church's student, Kleene, showed by 1933 that a large class

---

2 Kleene [1987b, p. 49] states, "The *Entscheidungsproblem* for various formal systems had been posed by Schröder [1895], Löwenheim [1915], and Hilbert [1918]."

3 Here "valid" means "true in the standard structure," not the modern sense of valid as true in all structures.

4 Ackermann in [1928] had produced a function defined by double recursion which was not primitive recursive.

of number theoretic functions were λ-definable. On the strength of this evidence, Church proposed to Gödel around March, 1934 (see Davis [1965, pp. 8–9]) that the notion of "effectively calculable" be identified with "λ-definable," a suggestion which Gödel rejected as "thoroughly unsatisfactory."

Following this encounter with Gödel, Church changed formal definitions from "λ-definable" to "recursive," his abbreviation for Herbrand–Gödel general recursive, and Church presented on April 19, 1935, to the American Mathematical Society his famous proposition published in [Church, 1936a] and known (since Kleene [1952a]) as *Church's Thesis* which asserts that the effectively calculable functions should be identified with the recursive functions. This is apparently the first published appearance of the term "recursive" to mean "general recursive." On the basis of this Thesis, Church [1936a] announced the unsolvability of Hilbert's *Entscheidungsproblem*.

Gödel, however, remained unconvinced of the validity of Church's Thesis through its publication [Church, 1936a]. This is all the more significant, first, because Gödel had originated the first formalism, that of the general recursive functions, and the one upon which Church based his Thesis. Second, much of the evidence for Church's Thesis rested on the coincidence of these formal classes, and this was based largely on Kleene's use of arithmetization, the method that Gödel *himself* had introduced so dramatically in [1931]. The reasons why Gödel did not accept or invent the thesis himself are explained in Davis [1982] and below.

The flaw in Church's argument [1936a, §7] for his thesis was this. Church began by defining an "effectively calculable" function to be one for which "there exists an algorithm for the calculation of its values." Church analyzed the informal notion of the calculation of a value $f(n) = m$ according to a step-by-step approach (so called by Gandy [1988, p. 77]) from two points of view, first by an application of an algorithm, and second as the derivation in some formal system, because, as he pointed out, Gödel had shown that the steps in his formal system P were primitive recursive. Following Davis [1958, p. 64] or Shoenfield [1967, pp. 120–121] it is reasonable to suppose that the calculation of f proceeds by writing expressions on a sheet of paper, and that the expressions have been given code numbers, $c_0, c_1, \ldots, c_n$. Define $\langle c_0, c_1, \ldots, c_n \rangle = p_0^{c_0} \cdot p_1^{c_1} \cdots p_n^{c_n}$, where p_n denotes the nth prime number. We say that the calculation is *stepwise recursive* if there is a partial recursive function ψ such that $\psi(\langle c_0, \ldots, c_i \rangle) = c_{i+1}$ for all i, $0 \leqslant i < n$.

If the basic steps are stepwise recursive, then it follows easily by the Kleene Normal Form Theorem (see Section 4.1) which Kleene had proved and communicated to Gödel before November, 1935 (see Kleene [1987b, p. 57]), that the entire process is recursive. The fatal weakness in Church's argument was the core assumption that the atomic steps were stepwise recursive, something he did not justify. Gandy [1988, p. 79] and especially Sieg [1994, pp. 80, 87] in their excellent analyses brought out this weakness in Church's argument. Sieg (p. 80) wrote, "...this core does not provide a convincing analysis: steps taken in a calculus must be of a restricted character and they are assumed, for example by Church, without argument to be recursive." Sieg (p. 78) wrote, "It is precisely here that we encounter the major stumbling block for Church's analysis, and that stumbling block was quite clearly seen by Church," who wrote that without this assumption it is difficult to see how the notion of a system of logic can be given any exact meaning at all. It is exactly this stumbling block which Turing overcame by a totally new approach.

3. Turing's contributions to computability

In the spring of 1935 a twenty-two year old student at Cambridge University, who had just given an independent proof of the Central Limit Theorem (see [Zabell, 1995]), heard the lectures of Professor M.H.A. Newman on Gödel's paper and on the Hilbert *Entscheidungsproblem*. Turing worked on the problem for the remainder of 1935 and submitted his solution to the incredulous Newman on April 15, 1936. Turing's monumental paper [1936] was distinguished because: (1) Turing analyzed an idealized *human* computing agent (a "*computor*") which brought together the intuitive conceptions of a "function produced by a mechanical procedure" which had been evolving for more than two millennia from Euclid to Leibniz to Babbage and Hilbert; (2) Turing specified a remarkably simple formal device (*Turing machine*) and proved the equivalence of (1) and (2); (3) Turing proved the unsolvability of Hilbert's *Entscheidungsproblem* which established mathematicians had been studying intently for some time; (4) Turing proposed a *universal* Turing machine, one which carried within it the capacity to duplicate any other, an idea which was later to have great impact on the development of high speed digital computers and considerable theoretical importance. Gödel enthusiastically accepted Turing's Thesis and his analysis, and thereafter Gödel always gave credit to Turing (not to Church or to himself) for the definition of mechanical computability and computable function.

3.1. *Turing's idealized human computor*

In 1935 Turing and everyone else used the term "*computer*" for an idealized *human* calculating with extra material such as pencil and paper, or a desk calculator, a meaning very different from the use of the word today. (Even ten years later in his 1946 report [Turing, 1986, p. 20] on the Automatic Computing Engine (A.C.E.) Turing used the term "computer" to refer to a human with paper, as in [Turing, 1986, p. 106]. Turing wrote that A.C.E. can do any job of a (human) computer in one ten-thousandth of the time.) To avoid confusion we shall follow Gandy [1988] and Sieg [1994] and use the term "*computor*" to mean such an *idealized human* calculating in a purely mechanical fashion, and the term "*computer*" for a *machine*, either an idealized machine like a Turing machine or register machine, or for a physical device like a high speed digital computer. The analysis in this subsection was not completely clear in Turing [1936], and is due almost entirely to Sieg [1994] and [1995], who built upon Gandy [1988].

To analyze what it means for a function to be "calculable by an algorithm" or a "mechanical procedure," Turing put certain conditions on the calculation. Turing

[1939, §9, pp. 249–254] assumed that the computation was being done by the computor "writing certain symbols on paper," and that the paper was one dimensional and divided into squares. He also proposed a set of *states* (of mind) of the computor. First, Turing required three *finiteness* conditions: (F1) the number of symbols; (F2) the number of squares observed at any one moment; (F3) the number of states. Turing proposed a number of simple operations "so elementary that it is not easy to imagine them further subdivided." Turing allowed the computor to observe a set of squares and in one atomic operation to: change the set of squares being observed or print on an observed square; and change its state in accordance with the following *neighborhood* conditions: (C1) the computor can change the symbol only in an *observed* square and then at most *one* symbol; (C2) the computor can move to a different set of observed squares but only within a certain bounded distance L of an observed square; (C3) the atomic operation must depend only on the current state and the symbols in the observed squares. Turing also imposed a determinacy condition (D) that from the state and observed symbols there was at most one atomic operation which could be performed, but this is unnecessary, since it is now well-known how to simulate a nondeterministic process by a deterministic one.[5]

From the precise description of his computor, Turing then formally defined his familiar *automatic machine*, now known as a *Turing machine*, a finite state machine with a two-way infinite tape, whose squares contained symbols from a finite alphabet, with a read/write head which scans one square at a time, and a finite set of instructions (Turing program), see Soare [1987, p. 12]. Turing called a function defined by a Turing machine a "*computable function.*" Using Turing's analysis we can now repair the weakness in Church's argument. (See Sieg [1994, p. 95].) To show that an effectively calculable function is recursive, take the algorithm which calculates it, find the technical description of the corresponding computor, then the associated Turing machine, and then the associated recursive function, from the equivalence of the latter two classes.

DEFINITION 3.1. A function is *computorable* if it can be calculated by an idealized human computor as defined above.

Turing then proved *Turing's Theorem*: *Any computorable function is Turing computable.* Although not proved in a formal system, Turing's proof is as rigorous as many in mathematics. Gandy [1988, p. 82] observed, "Turing's analysis does much more than provide an argument for" Turing's Thesis, "*it proves a theorem.*"[6] Fur-

[5] Hodges [1983, p. 96] suggests that Turing's computor may have grown out of his analysis of a typewriter: "Alan had dreamt of inventing typewriters as a boy; Mrs. Turing had a typewriter; and he could well have begun by asking himself what was meant by calling a typewriter 'mechanical.' " Turing's computor does resemble a kind of erasing typewriter with an infinite carriage but with a finite program.

[6] Gandy actually wrote "Church's thesis" not "Turing's thesis" as written here, but surely Gandy meant the latter, at least intensionally, because Turing did not prove anything in [1936] or anywhere else about general recursive functions.

thermore, as Gandy [1988, pp. 83–84] pointed out, "Turing's analysis makes no reference whatsoever to calculating machines. Turing machines appear as a result, a codification, of his analysis of calculations by humans." *Turing's Thesis* [1936, §9] is that every intuitively computable function is computable by a Turing machine. By Turing's Theorem, Turing's Thesis reduces to the following thesis (called by Sieg [1994] Turing's "Central Thesis").

TURING'S THESIS (TT-COMPUTOR). [7] If a function is informally computable (i.e. definable by a finite mechanical procedure or algorithm) then it is *computorable* (i.e. computed by a Turing idealized human computor).

Thus, the relationship between: (1) effectively calculable; (2) computorable; and (3) Turing computable is that: (1) $\Longrightarrow$ (2) by TT-Computor, and also (2) $\Longrightarrow$ (3) by Turing's Theorem, as Sieg [1994] and [1995] has also pointed out. Subsequent work has been done to show that more functions fall into one of these two classes. For example, Sieg and Byrnes [1995] generalized the concept of computorable and thus weakened TT-computor and strengthened Turing's Theorem. Similarly, Gandy [1980] analyzed discrete deterministic mechanical devices (DDMD machines) proving them to be Turing computable, a variant of TT known as TT-DDMD. However, these results and other subsequent work do not affect the original Turing Thesis TT-Computor which we regard not so much as a thesis but rather as a *definition* of the two thousand year old notion of an algorithmic function. It is now seen to encompass all modern high speed digital computers as well.

3.2. *Accepting Turing's Thesis*

If we review the conceptions of algorithms and mechanical procedures over the last two millennia from the Euclidean algorithm, Pascal's and Leibniz' conceptions of calculating, Babbage's analytic engine, Hilbert's and Gödel's computation of a function in a formal system, and many others included in the concept of computation described in Section 2.1, we see that they all fit within the computor model. Indeed, we claim that the common conception of mechanical procedure and algorithm envisioned over this period is exactly what Turing's computor captures.

This may be viewed as roughly analogous to Euclidean geometry or Newtonian physics capturing a large part of everyday geometry or physics, but not necessarily all *conceivable* parts. Here, Turing has captured the notion of a function computable by a mechanical procedure, and as yet there is no evidence for any kind of computability which is *not* included under this concept. If it existed, such evidence would not affect

[7] Here we follow Gandy [1980] in using appended words or letters to pinpoint the exact version of Turing's Thesis proposed. For example, Gandy [1980, p. 124] wrote of "*Theorem T.* What can be calculated by an abstract human being working in a routine way is computable," and distinguishes it from, "*Thesis M.* What can be calculated by a machine is computable." Gandy goes on to propose his own Thesis P about discrete deterministic mechanical devices (DDMD).

Turing's thesis about mechanical computability any more than hyperbolic geometry or Einsteinian physics refutes the laws of Euclidean geometry or Newtonian physics. Each simply describes a different part of the universe.

Turing machines and Turing's analysis were enthusiastically accepted by the founders of the subject, Gödel, Church, and Kleene, as the correct definition of computability. In [193?][8] paper (see [Gödel, 1995, p. 168]) Gödel wrote regarding the formal definitions of computability, "That this really is the correct definition of mechanical computability was established beyond any doubt by Turing." Gödel left no doubt that he regarded Turing's approach as superior to all other previous definitions (including his own recursive functions) when he wrote in [1964] (see [Davis, 1965, p. 72, footnote]), speaking of Turing machines, that, "As for previous equivalent definitions of computability, which, however, are much less suitable for our purpose, see A. Church [1936a, pp. 256–358]." (Gödel's reference is to Church's Thesis Section 9 which we have just analyzed in Section 2.5.) Kleene wrote [1981b, p. 49], "Turing's computability is intrinsically persuasive" but "λ-definability is not intrinsically persuasive" and "general recursiveness scarcely so (its author Gödel being at the time not at all persuaded)." Church, in his review [1937a] of Turing [1936], wrote that of the three different notions: computability by a Turing machine, general recursiveness of Herbrand–Gödel–Kleene, and λ-definability, "The first has the advantage of making the identification with effectiveness in the ordinary (not explicitly defined) sense evident immediately – i.e. without the necessity of proving preliminary theorems." Most people today accept Turing's Thesis. Sieg [1994, p. 96] wrote, "Thus, Turing's clarification of effective calculability as *calculability by a mechanical computor* should be accepted."

Some have cast doubt on Turing's Thesis on the grounds that there might be physical or biological processes which may produce, say, the characteristic function of the halting problem. It is possible that these may exist (although there is presently no evidence) but if so, this will have absolutely *no effect* on Turing's Thesis because they will not be algorithmic or mechanical procedures as required in Section 2.1 and in Turing's Thesis. Although suggesting the possibility of noncomputational *mental* processes [Gödel, 1995, p. 310], Gödel was unequivocal in his support of Turing's Thesis TT-Computor. Regarding the possibility of other nonmechanical procedures, Gödel [1964] wrote,

> Note that the question of whether there exist *non-mechanical* procedures not equivalent with any algorithm, has nothing whatsoever to do with the adequacy of the definition of "formal system" and of "mechanical procedure."
> – *Gödel* [1964] (see [Davis, 1965, p. 72])

[8] In referencing this paper [193?] of Gödel we follow the bibliographic referencing and numbering in his collected papers [Gödel, 1995, p. 156] where the editors use "[193?]" and explain, "This article is taken from handwritten notes in English, evidently for a lecture, found in the *Nachlass* in a spiral notebook." Although the date of the piece is not known, some conjectures about this will be discussed later.

3.3. *The Church–Turing Thesis as a definition*

When Church [1936a] first proposed Church's Thesis, he thought of it as a *definition*, not as a *thesis*. Church (see [Davis, 1965, p. 90]) wrote, "The purpose of the present paper is to propose a definition of effective calculability." Similarly, Turing [1936] did not use the term "definition," but he spoke (see [Davis, 1965, p. 135]) of showing "that all computable numbers are [Turing] 'computable'," and clearly regarded it as the definition of computable. Gödel stated on several occasions that the correct *definition* of computability had unquestionably been achieved by Turing.

> ... one has for the first time succeeded in giving an absolute definition of an interesting epistemological notion, i.e. one not depending on the formalism chosen. ... For the concept of computability, however, although it is merely a special kind of demonstrability or decidability, the situation is different. By a kind of miracle it is not necessary to distinguish orders, and the diagonal procedure does not lead outside the defined notion.
> – *Gödel:* [1946] *Princeton Bicentennial* [Gödel, 1946, p. 84]

> The greatest improvement was made possible through the precise definition of the concept of finite procedure, ... This concept, ... is equivalent to the concept of a "computable function of integers" ... The most satisfactory way, in my opinion, is that of reducing the concept of finite procedure to that of a machine with a finite number of parts, as has been done by the British mathematician Turing.
> – *Gödel: Gibbs lecture* [1951] [Gödel, 1995, pp. 304–305]

> But I was completely convinced only by Turing's paper.
> – *Gödel: letter to Kreisel of May 1, 1968* [Sieg, 1994, p. 88]

The theses of Church and Turing were not even called "theses" at all until Kleene [1943, p. 60] referred to Church's "definition" as "Thesis I," and then in [1952a] Kleene referred to "Church's Thesis" and "Turing's Thesis." What is even more curious is that the phrase "Church's Thesis" came to denote also "Turing's Thesis" and perhaps others as well, thereby blurring all intensional distinctions. (This, of course, stems partly from the Recursion Convention in Section 4.6 that "recursive" denotes "computable," because under this convention Turing's Thesis follows from Church's Thesis, whereas in reality the reverse was true as seen in Section 3.1.) There are many examples of this in the literature. For example, Gandy's [1980] paper is entitled, "*Church's thesis and principles for mechanisms*". However, Gandy's paper is entirely about *Turing's* Thesis and whether or not certain intuitively defined classes are *Turing* computable (i.e. *mechanistic*), not whether or not they are *recursive*. The hypotheses are stated in terms of variants of Turing's Thesis (TT) such TT-H, TT-M, TT-DDMD. They are presented both informally and formally entirely in the language of *machines*. Gandy's main result is that what can be calculated by a discrete deterministic mechanical device (DDMD) is Turing computable.

In contrast, the distinction of *intensional* meaning which distinguishes between Church's Thesis and Turing's Thesis is preserved by others, for example Sieg [1994] and Tamburrini [1997], and here. Here we also use the phrase "*Church–Turing Thesis (CTT)*" to refer to the amalgamation of the two theses (these and others) where we identify all the informal concepts of Definition 1.1 with one another and we identify all the formal concepts of Definition 1.2, and their mathematical equivalents, with one another and suppress their intensional meanings.

We now propose that Turing's Thesis be used as a *definition* of a computable function as Turing and Gödel suggested. Other theses in the past have dealt with very problematic topics but have eventually become definitions as we now discuss.

3.4. *Other theses became definitions*

One senior logician objected to this proposed definition because he said we should view the Church–Turing Thesis as certainly correct, but as "a one of a kind, without any true analogue in mathematics. I think we recursion theorists should be proud of this, and not (as you seem to suggest) replace it by a change of our definitions." There is no reason why we cannot use Turing's Thesis as a definition of computability and still maintain awe and pride at a fundamental discovery. As to the uniqueness of this discovery in the history of mathematics, it is informative to consider the history of other "theses." In the early 1800's mathematicians were trying to make precise the intuitive notion of a continuous function, namely one with no breaks. What we might call the "Cauchy–Weierstrass Thesis" asserts that a function is intuitively continuous iff it satisfies the usual formal δ-ε-definition found in elementary calculus books. Similarly, what we might call the "Curve Thesis" asserts that the intuitive notion of the length of a continuous curve in 2-space is captured by the usual definition as the limit of sums of approximating line segments. The "Area Thesis" asserts that the area of an appropriate continuous surface in 3-space is that given by the usual definition of the limit of the sum of the areas of appropriate approximating rectangles. These are no longer called *theses*, rather they are simply taken as *definitions* of the underlying intuitive concepts.

The same senior logician argued that these analogies are misleading because "Only a moment's thought is needed to see that Weierstrass' definition is a correct formulation of the intuitive notion of continuity. However, it takes a lot of thought to convince oneself that every function which can be computed by an algorithm can be computed by a Turing machine."

This impression of the simplicity in verifying the other theses and the belief in the unique historical place of the Church–Turing Thesis in formally capturing a difficult intuitive notion seems to ignore the history of the other "theses." What is problematic to one generation seems the obvious definition to another. Kline [1972, p. 354] wrote,

> "Up to about 1650 no one believed that the length of a curve could equal exactly the length of a line. In fact, in the second book of *La Geometrie*, Descartes says the relation between curved lines and straight lines is not nor ever can be known.

> But Robertval found the length of an arch of a cycloid. The architect Christopher Wren (1632–1723) rectified the cycloid ... Fermat, too, calculated some lengths of curves. These men usually found the sum of the segments, then let the number of segments become infinite as each got smaller."

Kline asserts (p. 355) that finding the lengths of curves was one of "the four major problems that motivated the work on the calculus." Regarding the Area Thesis Kline remarked (p. 355) that during the same period Huygens "was the first to give results on the areas of surfaces beyond that of the sphere."

A second distinguished senior logician stated that for him the Curve Thesis is *more* difficult to accept than Turing's Thesis and explained his reasons with references from Kline. With the Curve Thesis there is no upper bound closing downward toward the length of the curve, but merely the *lower* bound of the sum of the lengths of line segments, which increases with ever finer subdivisions. Likewise, for the Area Thesis (unlike the area under a curve in 2-space) there is no *upper* bound to the area, but merely the sum of the areas of finitely many rectangles approaching the correct value from *below*. He notes that *in all three cases,* the length of a curve in 2-space, the area of a surface in 3-space, and the set of all computable functions, *there is no upper bound,* just a lower bound. Furthermore, in both the Curve Thesis and Turing's Thesis one breaks the demonstration into smaller and smaller pieces until it becomes evident.[9]"

3.5. *Register machines*

Closely related to Turing machines is the formalism proposed much later of *register machines* by Shepherdson and Sturgis [1963]. (See also Cutland [1980], or Shoenfield [1991].) These have the advantage of more closely resembling modern digital computers which manipulate data and instructions stored in various "registers" rather than having to go back and forth through the data on a single tape. In the version of Cutland [1980, p. 9] the register machine contains an infinite number of registers $\{R_n\}_{n \in \omega}$, each of which contains an integer, r_n. The program P is a finite set of instructions built up from the four basic instructions: zero $Z(n)$ (replace r_n by 0), successor $S(n)$ (replace r_n by $r_n + 1$), transfer $T(m, n)$ (replace r_n by r_m), and jump instructions $J(m, n, q)$ (if $r_m = r_n$ go to the qth instruction of P, and otherwise go to the next instruction of P). It is easily shown that Turing machines can compute the same class of functions as register machines.

---

[9] The second logician pointed out that Kline goes on to say that during the second half of the 17th century various curves were rectified (using essentially the modern definition of arc length). Kline (p. 107) describes how other axioms involve the lengths of concave curves and surfaces. Kline tells how Archimedes deals axiomatically with arc length, and describes how Archimedes gave what can be construed as a proof of the Curve Thesis for certain curves since his axiom gives a way of handling an upper-bound on the length. The second logician suggested that this is analogous to Gandy's proof of TT-DDMD in Section 3.1, and stated "there is apparently no Gandy-like proof of the Curve Thesis for arbitrary rectilinear curves; in that case arclength is a definition. On the other hand there is a Gandy-like proof of Turing's Thesis for the case of TT-DDMD (namely Gandy's).

4. Later developments in computability

4.1. *Kleene's normal form and his μ-recursive functions*

From 1931 to 1934 Kleene tested many operations on functions to see whether they preserve λ-definability. Among these was the least number operator, "the least y such that," which, since [Kleene, 1938], has been denoted by "μy." Kleene proved that if $R(x, y)$ is a λ-definable relation then so is the partial function $\psi(x) = (\mu y)R(x, y)$.

Kleene [1936a, 1943] used this to prove his Normal Form Theorem which asserts that there is a primitive recursive predicate $T(e, x, y)$ and a primitive recursive function $U(y)$ such that for any general recursive function $\varphi(x)$, there is an index e (corresponding to the system E of equations defining φ) such that

$$\varphi(x) = U\big(\mu y\, T(e, x, y)\big). \tag{2}$$

(This is the [Kleene, 1943] version. The [Kleene, 1936a] version had a U with an additional parameter.) Kleene's Normal Form Theorem establishes that every general recursive (partial) function is μ-recursive, and conversely. Since the application of μ often leads to only *partial* functions Kleene [1938] introduced the *partial recursive functions* (i.e. *computable partial functions*). The Normal Form Theorem also holds if we replace total by *partial* recursive functions. Define the class $\mathcal{C}$ to be the smallest class of partial functions closed under the five schemata for primitive recursion (see Section 2.2) and, in addition, the following schema,

$$\text{Scheme (VI) (Unbounded Search)} \quad \varphi(x) = (\mu y)\big[g(x, y) = 0\big],$$

where $g(x, y) \in \mathcal{C}$ and $g(x, y)$ is total. Scheme (VI) is also sometimes called "minimalization," or the "least number operator." Kleene [1952a] referred to this $\mathcal{C}$ as the class of *(partial) μ-recursive* functions (reserving the term "*recursive*" for Herbrand–Gödel recursive), and used the term (partial) μ-recursive in later papers such as [1959] and [1963].

The (partial) μ-recursive functions constitute a robust class and one which plays a very important role in the subject. For example, by Gödel numbering the configurations of a computation we can easily prove a normal form theorem for the Turing computable functions (see Soare [1987, p. 15]). The μ-recursive functions are a mathematically definable class of functions almost independent of syntax and formalism. They have sometimes been used as the *definition* of a (partial) recursive function (see Table 1 in Section 4.6), but when used precisely and by Kleene, the formal meaning of "recursive" has been "defined by a Herbrand–Gödel system of equations."

It has sometimes been erroneously written that an advantage of the formalism of recursive functions or μ-recursive functions is that one can precisely write down a proof in either one of the two, but that this would have been infeasible using Turing machines or λ-definable functions. It is true that the latter two are unsuitable for writing proofs, but the former two are not much more suitable. Kleene [1981a, p. 62] wrote, "Under Herbrand–Gödel general recursiveness and my partial recursiveness adapted from it one works with systems E of equations that can be very unwieldy." The general recursive formalism has almost *never* been used for writing papers, so the writers are probably using "recursive" to refer to "*μ-recursive,*" which was earlier used by Kleene for writing his proofs from [1936a] to [1963] and even later, and by some followers like Sacks in his book [1963] on degrees. These expositions were extremely difficult to read (not unlike machine code) and were virtually completely abandoned by the mid 1960's in favor of the style of Rogers' book [1967] which has prevailed in subsequent texts (given in the table in Section 4.6). This style is to use rigorous proofs but written in the usual informal mathematical style and usually based on the formalism of Turing machines or the closely related register machines to define necessary items such as the number of steps of the computation, the "use function" measuring the number of oracle squares scanned during a computation, and so on.

4.2. *Computably enumerable sets and Post*

Since they were motivated by formalizing algorithms and possible decision procedures in connection with Hilbert's *Entscheidungsproblem*, the first formalizations of computability were designed to define a computable *function*. However, it had been recognized that effectiveness also occurs with *generating* objects, such as sets of formulas. Church, in his paper (see [Davis, 1965, p. 96]) on Church's Thesis, introduced the term "*recursively enumerable set*" for a set which is the range of a recursive function as in Definition 1.2. This is apparently the first appearance of the term "recursively enumerable" in the literature and the first appearance of "recursively" as an adverb meaning "effectively" or "computably."

Church goes on to prove in Section 6 and Section 8 various theorems and corollaries about recursively enumerable sets of well-formed formulas. Church also used the term "*effectively enumerable*" for the *informal* concept of a recursively enumerable set but used the latter for *both* the informal and formal concepts.

In the same year Kleene [1936] mentioned (see [Davis, 1965, p. 238]) a "*recursive enumeration*" and noted that there is no recursive enumeration of Herbrand–Gödel systems of equations which gives only the systems which define the (total) recursive functions. By a "recursive enumeration" Kleene states that he means "a recursive sequence (i.e. the successive values of a recursive function of one variable)." Effectively enumerable or recursively enumerable sets were not mentioned much thereafter until Post's paper [1943] on normal (production) systems which led to generated sets and then his famous [1944] paper which inaugurated the modern study of computably enumerable sets.

In the same year as Turing [1936], Post [1936] independently of Turing (but not independently of the work by Church and Kleene in Princeton) defined a "*finite combinatory process*" which closely resembles a Turing machine. From this it is often and erroneously written (Kleene [1987b, p. 56] and [1981a, p. 61]) that Post's contribution here was "essentially the same" as Turing's, but in fact it was much less. Post did not attempt to prove that his formalism coincided with any other such as general recursiveness but merely expressed the expectation that this would turn out to be true, while Turing proved the Turing computable functions equivalent to the λ-definable ones. Post gave no hint of a universal Turing machine. Most important, Post gave no analysis as did Turing in Section 3.1 above of why the intuitively computable functions are computable in his formal system. Post offers only as a "working hypothesis" that his contemplated "wider and wider formulations" are all "logically reducible to formulation 1." Lastly, Post, of course, did not prove the unsolvability of the *Entscheidungsproblem* because at the time Post was not aware of Turing's paper [1936], and Post believed that Church had settled the *Entscheidungsproblem*. (Post may have been aware of the flaw in Church's Thesis discussed in Section 2.5, and perhaps this is why he objected to the use of the term "definition.")

Later, Post [1941] and [1943] introduced a *second* and unrelated formalism called a *production* system and (in a restricted form) a *normal* system, which he explained again in [1944]. Post's (normal) canonical system is a *generational* system, rather than a *computational* system as in general recursive functions or Turing computable functions, and led Post to concentrate on *effectively enumerable sets* rather than computable functions. He showed that every recursively enumerable set is a normal set (one derived in his normal canonical system) and therefore normal sets are formally equivalent to recursively enumerable sets. Post, like Church and Turing, gave a thesis [1944, p. 201] but stated in terms of generated sets and production systems, which asserted that "any generated set is a normal set."

Post used the terms "effectively enumerable set" and "generated set" almost interchangeably, particularly for sets of positive integers. Post [1944, p. 285] (like Church [1936a]) defined a set of positive integers to be *recursively enumerable* if it is the range of a recursive function and then stated, "The corresponding intuitive concept is that of an *effectively enumerable* set of positive integers." Post [1944, p. 286] explained his informal concept of a "generated set" of positive integers this way,

> "Suffice it to say that each element of the set is at some time written down, and earmarked as belonging to the set, as a result of predetermined effective processes. It is understood that once an element is placed in the set, it stays there."

Post then (p. 286) restated his thesis from [1943] that *"every generated set of positive integers is recursively enumerable,"* [the italics are Post's] and he remarked that "this may be resolved into the two statements: every generated set is effectively enumerable, every effectively enumerable set of positive integers is recursively enumerable." Post continued, "their converses are immediately seen to be true."

Hence, this amounts to an assertion of the identification (at least extensionally) of the three concepts. Post accepted the Church–Turing Thesis even though he was reluctant to call it a definition, as Church and Turing would have done. Post [1944, p. 307, footnote 4] calls attention to Kleene's first use [1943, p. 201] of the word "thesis" in this context, but remarks "We still feel that, ultimately, "Law" will best describe the situation," and Post refers to his [1936] where this term was first proposed. This suggests that Post perhaps thought of the thesis as a kind of natural law like the laws of Newtonian physics.

In his famous and very influential paper [1944], Post continued with the intuitive concepts of "effectively enumerable" and "generated set," which he explains again at some length. The *formalism* Post used was that of his own normal (production) system, i.e. "normal set." He used the term "recursively enumerable set" (Church's term from [1936a]) as a name for *both* his *informal* "effectively enumerable set" and for his *formal* version, "normal set." However, the *concept* or formal *definition* of "recursive" does not enter Post's paper at all, only the *terms* "recursive" and "recursively enumerable." Post's use of the term "recursively enumerable" is one of several ambiguities in the subject (ambiguous at least from an *intensional* viewpoint).

In spite of this ambiguity, Post's entrance on the scene was fortunate for recursively enumerable sets and for the entire subject. Previously, the papers in the subject had been written in the very technical formalism of μ-recursive functions (see the last paragraph of Section 4.1), with little intuition. Recursively enumerable sets had attracted very little attention since their debut in [Post, 1936]. Post's papers brought excitement, intuitive appeal, and an informal style of proof, much closer to ordinary mathematical proofs, and represented the real birth of the subject of recursively enumerable sets [1943] and [1944] and degrees of unsolvability [1948]. The results and machinery they generated (Post's problem, Friedberg–Muchnik priority method) not only heavily influenced computability on ω but also provided a goal for higher excursions such as meta-recursion theory, α-recursion theory, recursion in higher types, E-recursion theory, and others. These papers of Post stimulated the entire subject for decades, but unfortunately they simultaneously helped to fix the use of the terms "recursive" and "recursively enumerable" to acquire the additional meanings, "computable" and "computably enumerable."

4.3. *History of relative computability*

The problem of *computability* of a set A *relative* to a set B is that of giving an algorithm for answering every question of the form "Is $x \in A$" by a computation which asks at most finitely many questions of the form "Is $y_1 \in B$?," ..."Is $y_k \in B$?" The first formal definition of relative computability (also called "relative reducibility") was given by Turing [1939, §4], in terms of an "*oracle Turing machine*." This is best visualized as a Turing machine with an extra infinite "oracle tape" on which is written the characteristic function of B (see Soare [1987, p. 47]). Other formal definitions were later given by Kleene [1943] and [1952a] of a function φ being *general recursive in* a function ψ if the latter is simply added to the equations E defining φ. Post [1948] formulated another definition by modifying his definition [1943] of a canonical (production) system. These three definitions can be proved to be equivalent. (See Kleene [1952a].) Using Turing reducibility (denoted $A \leqslant_{\mathrm{T}} B$), we say that two sets A and B have the same *information content* or have the same *Turing degree* if $A \leqslant_T B$ and $B \leqslant_T A$. Post [1948] introduced this extremely influential concept of Turing degree, also called *degree of unsolvability*. Kleene and Post [1954] laid the foundation for the abstract structure of the degrees, where there has been much research ever since.

It is interesting that all but one of the texts from Table 1 in Section 4.6 use Turing machines or their variant, register machines, to define $A \leqslant_T B$, but they apply the term "recursive in" (rather than "computable in") to the result. For example, Shoenfield uses register machines, and his entire apparatus is machine based, as is all his terminology to formulate the definition. He speaks (p. 40) of an "oracle" for a function "F" in the sense of Turing, asking for a value "we have computed" to be used in "the rest of the computation," "the use of an algorithm," the "notion of a program computing a function for this machine," and the "Φ-machine" (oracle machine) being "obtained from the basic machine by adding all F-instructions for all F in Φ." Yet after all this definitional background which heavily uses both the *formalism* and the *concepts* of *machine computation* but *none* of the formalism or concepts of recursion, Shoenfield concludes with the formal definition, "A function is *recursive relative to* Φ if it is computed by some program for the Φ-machine." This is typical of most of these references and is another instance where a *concept* like computability is used to *define* a function, but then a different *name* like "recursion" is assigned afterward even though the concept of recursion is not used in the definition.

> If we replace recursive by computable in results in recursion theory, we often obtain a statement which is evident, or at least more evident than the original result.
> – *Shoenfield* [1995, p. 15]

4.4. *Higher order computability*

Kleene opened the frontiers of computability on higher type objects in a series of papers first on constructive ordinals and hierarchies of number-theoretical predicates[10] and later on computability in higher types. Although Kleene calls the functions here by the *word* "recursive" he often used *concepts* of computability to define, explain, and prove theorems about them. For example, in [1955b] Kleene wrote,

> "By *general recursive* functions (predicates) we mean ones whose values can be computed (decided) by ideal computing machines not limited in their space for storing information. A theory of such machines was given by Turing [1936] and in less detail by Post [1936]."

It is on computability on higher types that the concept of recursion comes into one of its more splendid realizations. An object of type 0 is a number; an object of type

[10] From this work grew later the very beautiful subject of descriptive set theory, although when he began Kleene was unaware of the work in classical descriptive set theory from the early 1900's.

$n+1$ is a mapping from the set of objects of type n into ω. Thus, an object of type 1 is a *real* (i.e. identified with a function α from ω to ω). A well-known type 2 object is $\mathbf{E}$ where $\mathbf{E}(\alpha) = 0$ if $(\exists x)[\alpha(x) = 0]$, and $\mathbf{E}(\alpha) = 1$ otherwise.

In order to formally define computable functions of higher type, Kleene [1959] used a *schemata-based* definition very much like that for the μ-recursive functions in Section 4.1. Kleene began by giving (p. 3) *schemata* (S1) to (S8) which closely resemble the previous primitive recursive schemata (I)–(V) of Section 2.2. After proving various properties about these primitive recursive functions of higher type, Kleene addressed the general recursive case (§3, p. 10). Kleene began by talking about Turing oracles and "computations being carried out by a preassigned procedure." To obtain the partial recursive functions Kleene added an additional schema (S9) (p. 13) which is a kind of enumeration schema and, together with (S1) to (S8), forms a huge induction. If instead of schema (S9) one adds a schema (S10) which closely resembles the unbounded search schema (VI) of Section 4.1 then Kleene obtained the "*partial* μ*-recursive*" functions which are a strictly smaller class than the partial recursive functions (see Kleene §8.4), unlike the ω case where the two classes coincide.

In a later paper Kleene defined his schema (S11): $\varphi(\theta; \overrightarrow{a}) = \psi(\varphi, \theta; \overrightarrow{a})$, and he declared "This schema gives an absolutely general form of recursion." Later, in his Ph.D. dissertation, Platek developed very elegant abstract form of recursion in higher types. For example, if H is finitary operation with certain properties and and

$$F_{n+1} = H(F_n, x)$$

then $F = \bigcup_{n \in \omega} F_n$ is a fixed point, but is not a recursion on any argument. Some people have cited this work by Kleene, Platek, and others to prove that in higher types recursion plays the main role and computability plays very little role if any, but this is not accurate.

Consider Kleene's papers [1959] and [1963] laying the foundations for higher types. Although Kleene used the *name* recursive for his higher type functions, Kleene used the concept of computability to explain them and to carry out his proofs. From the moment Kleene introduced the general recursive case on p. 10 of [1959] he used the *concept* and terminology of computability, including: "computation," "oracle," "preassigned procedure," "mechanical character," and many more. Words like these, particularly "terminating" or "nonterminating" "computations" occur on average several times per page throughout the rest of the article. For example, Kleene showed "how the inductive definition of $\{z\}(\overrightarrow{a}) \simeq w$ provides a computation process," The "stages" of a computation can be arranged in a tree (p. 22), and termination or nontermination of a computation along a certain branch of the tree (p. 32) is crucial to the overall computation. Kleene went on in [1962a] and [1962b] to develop what he called "Turing-machine computable functionals of finite types."

Dag Normann is the author of an authoritative text [1980] on the subject of recursion on countable functionals. Normann gave a lecture at Oberwolfach in January,

1996, a main theme of which was that the subject of higher types has much more to do with computability than with recursion.

It is fair to say that the subject of higher types represents a very interesting and beautiful new arena where *both* the concepts of recursion and computability play a key role. Kleene's work and Platek's have raised the pure concept of recursion to new heights with unexpected discoveries of new kinds of fixed points. At the same time motivation and methods have often been those associated with the concept of computability, suitably generalized. Indeed is there any area of recursion theory which has been opened *merely* to study the concepts of self-reference, fixed points, reflexive call and other aspects of recursion alone with *no* intent of studying the effective or computable content of the new area?

4.5. *How the terms became fixed*

If both Turing and Gödel, the inventors of the two formal definitions and the two names, preferred the terminology "computable" for this class of functions, how did the word "recursive" become preferred for it and for the subject? When Turing's [1939] paper appeared, he had already been recruited by the British government as a cryptanalyst on September 4, 1939 [Hodges, 1983, p. 161], three days after Britain was plunged into World War II. Turing played the major role [Hodges, 1983] in 1940 in breaking the German cipher, Enigma. After the war Turing worked on the design of high speed digital computers, first, at the British National Physical Laboratory from 1945 to 1948 and then at the Computing Machine Laboratory in Manchester from 1948 until his death in 1954. Turing wrote a report in 1946 (see [Turing, 1986]) on the design of A.C.E., a high speed digital computer (partly inspired by his universal Turing machine). Gödel moved to set theory and proved his famous results about the consistency of the axiom of choice and the generalized continuum hypothesis which appeared in 1938 and 1939. He returned to computability with his well-known *Dialectica* paper [1958] in which he speaks of "computable functions of finite type" (see [Gödel, 1990, p. 245]). Gödel made many statements expressing his preference for "computable" over "recursive" (see the quotes here from his collected works [Gödel, 1986, 1990, 1995]), but neither Turing nor Gödel had much influence on the terminology of the subject after 1939.

The present terminology came from Church and Kleene. They had worked in the λ-definable functions until 1935 when they changed to recursive functions because it was more in the mathematical mainstream and had more audience appeal, as explained by Kleene [1987b]. They had both committed themselves to the new "recursive" terminology before they ever heard of Turing or his results. Furthermore, using "computable" in 1935 would not have increased audience appeal because a "computer" meant, even as late as 1946, a human being calculating with paper. Ironically, the personal computer revolution of the late 1970's which brought the technology, concept and terminology of computability to tens of millions arrived just as Kleene was retiring.

After 1938 Church had little influence on the subject or its terminology, although he did produce in the late 1940's and 1950's a number of students who later became quite prominent. Kleene, with his steady stream of papers giving fundamental tools like the hierarchies, normal form theorems, and recursion theorem (fixed point theorem) and opening new areas, dominated the subject from the late 1930's until at least the late 1950's, and his papers and book [1952a] set the standard for the results and terminology, such as "recursive," "recursively enumerable," and "Church's Thesis." Post [1944] changed from his own terminology to that of Church and Kleene in his use of "recursive" and "recursively enumerable." The enormous popularity and influence of Post's paper and of Post's Problem firmly and widely established the Church–Kleene terminology. After the solution to Post's Problem by Friedberg and Muchnik in 1956–1957 and the introduction of their priority method, the field greatly expanded, and there was no single dominant figure, but the existing terminology had been established and has continued to the present day.

4.6. *Current usage of the concepts and terms*

There is a current tendency in the subject to work in one formalism (usually that of Turing computable functions) but then to *name* the results using the terminology of *recursive* functions not *computable* functions. For example, consider from an *intensional viewpoint* the following quote from Putnam's recent review [1995, p. 371] of Roger Penrose's new book [1994][11] about "a noncomputational ingredient in our conscious thinking."

> "First Penrose provides the reader with a proof of a form of the Gödel Theorem due to Alan Turing, the father of the modern digital computer and the creator of the mathematical subject *recursion theory*, which analyzes what computers can and cannot in principle accomplish."
> – *Hilary Putnam, review of Penrose* [1994]

There is very good reason to agree with Putnam[12] on his two assertions about Turing. However, Turing certainly *never* used the term "recursion theory" or "recursive function theory" for the subject. Turing mentioned the term "recursive function"

[11] Physicist Penrose, like most scientists, never mentions the term "recursive," but he has an extensive discussion of Turing and Turing machines covering a whole chapter. Penrose (p. 66) writes, "by a computation (or algorithm) I indeed mean the action of some Turing machine, i.e. in effect, just the operation of a computer according to some computer program." This is a good example of the acceptance in the scientific world of Turing's Thesis. (See §3).

[12] Putnam's *own article* [1995] is an excellent example of the modern use of computer related concepts and terms rather than recursive functions to describe computational processes. Putnam's review (like Penrose's book) is written *entirely* using the *Turing machine model*, speaking of "machines," "programs," "output," "lines of code," a "debugged" program, "Turing-machine action," and "programs which output theorems." Putnam uses words like: "computer," "computational," "machine," and "program" over three dozen times, while "recursion" is mentioned only once (namely, in the quote above) and "primitive recursive" only once.

only very briefly in [1937b] and [1939, Section 2] to say that these functions were mathematically equivalent to his Turing computable functions, and then Turing *dismissed* general recursive functions with the phrase, "we will not be much concerned here with this particular definition." Turing certainly never used "recursive" to mean "computable," and Turing did not refer to "recursive functions" again. Clearly from a *strictly intensional* viewpoint, the term "*recursion theory*" does not analyze anything about what *computers* can or cannot accomplish at all (contrary to Putnam's assertion); it deals with the properties of Herbrand–Gödel general recursive functions, the concepts of induction, recursion, reflexive program calls, and fixed points.

Gödel, who had invented [1934] the formal definition of general recursive function, abandoned it almost completely after seeing Turing machines [1936] and Turing's demonstration of Turing's Thesis. After 1936 Gödel rarely spoke of recursive functions, and never used the term "recursive" to mean "computable" or "decidable." Gödel often asserted later that Turing's was the correct definition of the notion of mechanical computability, and spoke often of the concept of computability [1946, p. 84], [1951] (see [Gödel, 1995, pp. 304–305]), [193?] (see [Gödel, 1995, p.168]), (see [Sieg, 1994, p. 88]), [1936] (see [Davis, 1965, p. 82]), [1964] (see [Davis, 1965, p. 71–72]).

Both Turing and Gödel, even later in life, rejected "recursive" as a name for the subject and often for their results. At his lecture [1949] on verifying program correctness, Turing used the term "induction variable," to which Prof. Hartree objected that the term should be "recursive variable" to distinguish it from the sense of mathematical induction. Turing [1949, p. 141] rejected the suggestion. In the three volumes of his collected works, [1986, 1990, 1995], Gödel *never* used the term "recursive function theory" to name the subject; when others did Gödel reacted sharply negatively, as related by Martin Davis.

> In a discussion with Gödel at the Institute for Advanced Study in Princeton about 1952–54, Martin Davis casually used the term "recursive function theory" as it was used then. Davis related, "To my surprise, Gödel reacted sharply, saying that the term in question should be used with reference to the kind of work Rosza Peter did."

In spite of the strong preference for "computable" by Turing and Gödel, the founders of the two formalisms and concepts, the name "recursive" instead of "computable" has been associated with almost all objects of the subject since the late 1930's. In spite of the computer revolution of the last few decades which Turing's work did so much to spawn, and which has given new connections between the subject and many outside areas in the scientific community, logicians have been slow to change the terminology and concepts of "recursive" to "computable." For various historical reasons there gradually emerged from 1936 to 1960 the following unspoken convention to use "recursive" as an all encompassing term for the concepts and for the name of the subject.

The *Recursion Convention* is to: (1) use the terms of the general recursive formalism (i.e. "recursive," "recursively enumerable," "recursive in") to describe results

Table 1

Book	Definition of computable	Definition of relative computability	Name used for function defined
Kleene [1952a]	general recursive	general recursive	recursive
Rogers [1967]	Turing machines	Turing machines	recursive
Cutland [1980]	register machines	register machines	computable
Lerman [1983]	μ-recursive	Turing machines	recursive
Soare [1987]	Turing machines	Turing machines	recursive
Odifreddi [1989]	μ-recursive	μ-recursive	recursive
Shoenfield [1991]	register machines	register machines	recursive

about the subject, even if the proofs are based on the concepts and formalism of Turing computability; (2) use the term "*Church's Thesis*" to denote the amalgamation of the several theses, including theses by Church, Turing, and Post, in Section 2 and Section 3, even though Church's demonstration of his thesis (that all effectively calculable functions are general recursive) was flawed (Section 2.5) and was rejected by Gödel in its original form, and even though Turing gave an "unquestionably adequate" [Gödel's words] demonstration (Section 3.1) of Turing's Thesis (that all intuitively computable functions are Turing machine computable); (3) name the subject, and any new excursions such as to higher recursion, using the language of recursion, even if the concept of computability plays a very strong role there.

The Recursion Convention has been followed for over fifty years. Consider the following table of the basic *texts* on the subject, the *formalisms* that they use to define computable functions and relative computability, and the *names* that they assign afterward.

Here Turing computable and (general) recursive are as in Definition 1.2, register machines are in Section 3.5, and μ-recursive in Section 4.1. Of these texts no modern book (i.e. after 1965) uses general recursive functions as the formalism for defining computable functions; two use μ-recursive functions (which is not the same as general recursive and was not used by Kleene to define "recursive") for ordinary computability, but then one (Lerman) changes to Turing machines for the more complicated case of relative computability, while the other (Odifreddi) stays with the μ-recursive definition of relative computability, but then gives a nonstandard proof of results about relative Turing computability, such as the Friedberg–Muchnik solution to Post's Problem. And yet all the authors (omitting Cutland who is writing a more elementary text for a general audience including computer scientists) use the *name* "recursive" for both the intuitive concept and the formally defined object, whether they have used a computability style definition or not.

The Recursion Convention has brought "recursive" to have at least four different meanings as discussed in Section 5. This leads to some ambiguity. When a speaker uses the word "recursive" before a general audience, does he mean "defined by induction," "related to fixed points and reflexive program calls," or does he mean "computable?"

> The first rule of good taste in writing is to use words whose meaning will not be misunderstood; and if a reader does not know the meaning of the words, it is infinitely better that he should know he does not know it.
> – *Charles Sanders Peirce*, Ethics of Terminology, [1960, p. 131]

Worse still, the Convention leads to *imprecise thinking* about the basic concepts of the subject; the term "recursion" is often used when the concept of "computability" is meant. (By the term "recursive function" does the writer mean "inductively defined function" or "computable function?") Furthermore, ambiguous and little recognized terms and imprecise thinking lead to *poor communication* both within the subject and to outsiders, which leads to isolation and lack of progress within the subject, since progress in science depends on the collaboration of many minds.

5. Mathematical, scientific, and general English usage

The term "computable" appears as early as 1646 in English usage according to the *Oxford English Dictionary* (*O.E.D.*) [O.E.D., 1989]. O.E.D. and *Webster's Third International Dictionary* [Webster, 1993] give the definition of "computable" as roughly synonymous with "calculable," capable of being ascertained or determined by a mathematical process especially of some intricacy. The meaning of "calculate" is somewhat more general including "to figure out," "to design or adapt for a purpose," "to judge to be probable," while "compute" means more "to determine by a mathematical process," or "to determine or calculate by means of a computer."

When Dedekind [1888] proved that a definition by recursion uniquely defines a function, he called it "definition by induction." Hilbert [1904] used the term "*rekurrent(e)*," and in [1923] he used "*Rekursion*". The term "*recursive*" was apparently first used in English by Ramsey (see Gandy [1988, p. 73]). Skolem in [1923] showed that many number-theoretic functions are primitive recursive, and he used "*rekurrierend.*" In [1926] Hilbert expanded the use of the term to include transfinite types and essentially transfinite recursion. Ackermann [1928] considered functions which can be defined using primitive recursion at all finite types. He gave a definition of a particular function using double nested recursion and showed that it was not primitive recursive. R. Péter [1934] and [1951] examined primitive recursive functions and special recursive functions (where recursion on more than one variable is allowed).

The current meanings of "recursive" derive from the verb "*recur*" which means to return to a place or status, or the concept of "*definition by recursion*", like Scheme (V) in (1), for which Webster gives the meaning: a definition of a function permitting values of the function to be calculated systematically in a finite number of steps, especially a mathematical definition in which the first case is given and the nth case is defined in terms of one or more previous cases, especially the immediately preceding one. Thus, the term "*recursive*" is presently used in the subject in at least *four* different ways which we now summarize as a definition for future reference.

DEFINITION 5.1. The current meanings of *recursive* and *recursion* are these:

(i) *recursion* is used with meanings derived from the verb "recur," as in the dictionary definition of "recursion" above;

(ii) *recursion* is used in the sense of "definition by recursion" (i.e. definition by induction) as defined in equation (1) of Section 2.1 and in the dictionary entry of definition by recursion above;

(iii) following Kleene [1936] and Church [1936a] the term "*recursive*" denotes "general recursive" and any of its mathematically equivalent *formal* variants, such as "Turing computable," "λ-definable," "specified by a Post [1944] normal system," or Kleene's "μ-recursive".

(iv) "*recursive*" is used to mean any of the *informal* variants of Definition 1.1 such as "(intuitively) computable," "effectively calculable," "defined by a mechanical process," or "specified by an algorithm."

Most dictionaries give meaning (i) and usually (ii). Most people outside the subject including computer scientists, mathematicians, and scientists understand "recursive" as (ii) if they know it at all. Of the dictionaries only O.E.D. gives meaning (iii) and then only in the fine print, without a definition, but with reference to Kleene [1936] and [1952a], an entry written by Gandy. *None* of these dictionaries gives meaning (iv) that "recursive" means "computable" or "decidable." This is a meaning understood by very few outside the subject.

6. Themes and goals of computability theory

Many believe that the present subject of recursion theory would benefit from: (1) the pruning of some more technical and specialized topics while retaining most of the present research content; (2) a broadening of horizons and problems to others areas in logic, mathematics, computer science, and science in general in interaction with computability; (3) a better communication of present and future results in both (1) and (2) in terms of some of the basic concepts below to the larger scientific community.[13] Before presenting his lecture or paper, the author should ask himself, "What light does this shed on the basic themes and goals of the subject such as computability, enumerability, information content, relation to other branches of logic and mathematics?" As Harvey Friedman has suggested, every morning one should wake up and reflect on the conceptual and foundational significance of one's work. This reevaluation process should be carried out *regardless of names* for the subject.

The following items should be considered a mixture of concepts, goals, themes, and connections with other areas: computability; enumerability; relative computability (Turing reducibility, $A \leqslant_T B$); information content, normally measured by Turing degree; computational complexity and computing with bounds on resources of

[13] Of course, a desire by its proponents to improve a field is not an indication that it is in greater need of improvement than other fields, but rather indicates the intention to strengthen it still further.

space and time; polynomial hierarchy questions; definability; invariance and automorphisms; elementary theory; relationship of computability, enumerability, and information content to algebraic structures; relationship of computability to model theory, and set theory, and proof theory, for example: models of arithmetic; provably computable functions, reverse mathematics and levels in the arithmetic hierarchy; relationship of computability to topology, to algebra and combinatorics, to analysis, e.g., to descriptive set theory; relationship to number theory (e.g., Hilbert's 10th Problem); relationship of computability to computer science; Kleene arithmetic hierarchy and the Meyer–Stockmeyer hierarchy for polynomial reducibility, structures in complexity; relationship of computability to other fields, e.g., biology, quantum physics, economics, etc.

7. Analysis

Both of the concepts of recursion and computability have played a crucial role in the development of the subject and will continue to do so.

The term and concept of "computable" is associated with the notion of computation (Section 2.1), algorithm (Section 2.3), and with the functions defined by (or sets enumerated by) Turing machines (Section 3.1) or register machines (Section 3.4), and also with relative Turing computability (Section 4.3).

The term and concept of "recursive" is associated with: definition by recursion (induction) (Section 2.2), general recursive functions in the sense in Herbrand–Gödel (Section 2.4), fixed points as in the Kleene Recursion Theorem or more generally Kleene's schema (S11), which Kleene believed included all possible recursions, and Kleene's μ-recursive functions (Section 4.1).

Researchers in the subject have recently changed the the name of the subject from "Recursion Theory" to "Computability Theory" in order to make clear this distinction. Thus, the term "recursive" no longer carries the additional meaning of "computable" or "decidable," as it once did. This reinforces the original meaning of "recursive" and induction as understood by Dedekind [1888], Peano [1889] and [1891], Hilbert [1904] and [1926], Skolem [1923], Gödel [1931] and [1934], and Péter [1934] and [1951], and by most modern computer scientists, mathematicians, and physical scientists, and as expanded to fixed points, the recursion theorem, and to other kinds of recursion by Kleene, Platek, and others.

Presently, if functions are defined, or sets are enumerated, or relative computability is defined using Turing machines, register machines, or variants of these (as in the texts in Table 1 of Section 4.6 or in the Putnam [1995] review), then the name "computable" rather than "recursive" will be attached to the result, as in Cutland [1980], Davis [1958], as well as [Boolos and Jeffrey, 1974], the subtitle of [Soare, 1987] and others.

Thus, the terms "recursive" and "computable" have reacquired their traditional and original meanings, and those understood by most outsiders (Section 5). This is in accord with the usage and opinion of the founders of the two concepts and terms,

Turing and Gödel, both of whom used "computability" in this sense and both of whom rejected the use of "recursive" to mean computable (Section 4.6).

This will improve communication with many researchers outside the field. It will also give a new scientific precision to our discussions within the subject and will remove various ambiguities mentioned above. (For example, by a "recursive function" do we mean computable one or one defined by induction?) It will enable us to speak with greater clarity and precision about our own subject from r.e. sets (á la Post Section 4.2) to computability in higher types where the relative role of the two concepts has always been controversial, even to the experts (see Section 4.4).

Researchers will now also distinguish between the *intensional* meaning of Church's Thesis (that all effectively calculable functions are general recursive) versus that of Turing's Thesis (that all intuitively computable functions are computable by a Turing machine). When one writes a paper dealing with which classes of functions are Turing computable (i.e. mechanistic), as in Gandy [1980] and in many other places, one now refers to "*Turing's Thesis*" (as in Sieg [1994] and Tamburrini [1997]) not to "Church's Thesis."

Philosopher Charles S. Peirce described the importance of language for science this way.

> "the woof and warp of all thought and all research is symbols, and the life of thought and science is the life inherent in symbols; so that it is wrong to say that a good language is *important* to good thought, merely; for it is of the essence of it."
> – *Charles Sanders Peirce*, The Ethics of Terminology, Volume II, *Elements of Logic*, in: [Peirce, 1960, p. 129].

References

G. BOOLOS AND R. JEFFREY

[1974] *Computability and Logic*, Cambridge Univ. Press, Cambridge, England.

A. CHURCH

[1935] An unsolvable problem of elementary number theory, Preliminary report (abstract), *Bull. Amer. Math. Soc.*, 41, pp. 332–333.

[1936a] An unsolvable problem of elementary number theory, *Amer. J. Math.*, 58, pp. 345–363.

[1936b] A note on the Entscheidungsproblem, *J. Symbolic Logic*, 1, pp. 40–41. Correction, pp. 101–102.

[1937a] Review of Turing [1936], *J. Symbolic Logic*, 2(1), pp. 42–43.

[1937b] Review of Post [1936], *J. Symbolic Logic*, 2(1), p. 43.

[1938] The constructive second number class, *Bull. Amer. Math. Soc.*, 44, pp. 224–232.

A. CHURCH AND S. C. KLEENE

[1936] Formal definitions in the theory of ordinal numbers, *Fund. Math.*, 28, pp. 11–21.

N. CUTLAND

[1980] *Computability: An Introduction to Recursive Function Theory,* Cambridge Univ. Press, Cambridge, England.

M. DAVIS

[1958] *Computability and Unsolvability,* McGraw-Hill, New York; reprinted in 1982 by Dover Publications.

[1965] *The Undecidable. Basic Papers on Undecidable Propositions, Unsolvable Problems, and Computable Functions*, Raven Press, Hewlett, New York.
[1982] Why Gödel did not have Church's Thesis, *Information and Control*, 54, pp. 3–24.
[1988] Mathematical logic and the origin of modern computers, in: *Herken* [1988], pp. 149–174.

R. DEDEKIND
[1872] Stetigkeit und irrational Zahlen, Braunschweig (5th ed.), 1927. Also in: *Dedekind Gesammelte mathematische Werke*, Vol. III, Braunschweig (Vieweg & Sohn) 1932, pp. 315–334. English transl. by Wooster Woodruff Beman entitled *Continuity and irrational numbers*, pp. 1–24 of *Essays on the Theory of Numbers*, Chicago, Open Court, 1901, 115 pp.
[1888] Was sind und was sollen die Zahlen?, Braunschweig (6th ed.), 1930. Also in *Dedekind Gesammelte mathematische Werke*, Vol. III, pp. 335–391. English transl. by Wooster Woodruff Beman, *The nature and meaning of numbers*, loc. cit., pp. 31–105. (English transl. in Dedekind, *Essays on the Theory of Numbers*, Chicago, Open Court, 1901, 29–115.)

R. L. EPSTEIN AND W. CARNIELLI
[1989] *Computability: Computable Functions, Logic, and the Foundations of Mathematics*, Brooks/Cole Advanced Books and Software, Pacific Grove, CA.

S. FEFERMAN
[1988] Turing in the land of $O(z)$, in: *Herken* [1988], pp. 113–147.
[1992] Turing's "Oracle": From absolute to relative computability – and back, in: *The Space of Mathematics*, J. Echeverria et al., eds., Walter de Gruyter, Berlin, pp. 314–348.

M. FITTING
[1987] *Computability Theory, Semantics, and Logic Programming*, Oxford Univ. Press, Oxford.

R. GANDY
[1980] Church's thesis and principles for mechanisms, in: *The Kleene Symposium*, North-Holland, Amsterdam, pp. 123–148.
[1988] The confluence of ideas in 1936, in: *Herken* [1988], pp. 55–111.

K. GÖDEL
[1931] Über formal unentscheidbare sätze der Principia Mathematica und verwandter systeme. I, *Monatsh. Math. Phys.*, 38, pp. 173–178. (English transl. in *Davis* [1965], pp. 4–38, and in *van Heijenoort* [1967], pp. 592–616.
[1934] *On undecidable propositions of formal mathematical systems*, Notes by S. C. Kleene and J. B. Rosser on lectures at the Institute for Advanced Study, Princeton, NJ (1934), 30 pp. (Reprinted in *Davis* [1965], pp. 39–74.)
[1936] On the length of proofs, in: *Gödel* [1986], pp. 397–399; reprinted in *Davis* [1965], pp. 82–83, with a *Remark* added in proof [of the original German publication].
[193?] Undecidable diophantine propositions, in: *Gödel* [1995], pp. 156–175.
[1946] Remarks before the Princeton bicentennial conference of problems in mathematics, Reprinted in: *Davis* [1965], pp. 84–88.
[1951] Some basic theorems on the foundations of mathematics and their implications, in: *Gödel* [1995], pp. 304–323. (This was the Gibbs Lecture delivered by Gödel on December 26, 1951 to the Amer. Math. Soc.)
[1958] Über eine bisher noch nicht benützе Erweiterung des finiten Standpunktes, *Dialectica*, 12, pp. 280–287. (German and English transl. in *Gödel* [1986], pp. 240–251, with introductory note by A. S. Troelstra, pp. 217–241.)
[1964] Postscriptum to Gödel [1931], written in 1946, printed in: *Davis* [1965], pp. 71–73.
[1972] Some remarks on the undecidability results (written in 1972), in: *Gödel* [1990], pp. 305–306.
[1986] *Collected Works Volume I: Publications 1929–1936*, S. Feferman et al., eds., Oxford Univ. Press, Oxford.
[1990] *Collected Works Volume II: Publications 1938–1974*, S. Feferman et al., eds., Oxford Univ. Press, Oxford.

[1995] *Collected Works Volume III: Unpublished Essays and Lectures*, S. Feferman et al., eds., Oxford Univ. Press, Oxford.

L. HARRINGTON AND R. I. SOARE

[1996] Definability, automorphisms, and dynamic properties of computably enumerable sets, *Bulletin of Symbolic Logic*, 2, pp. 199–213.

R. HERKEN

[1988] *The Universal Turing Machine: A Half-Century Survey*, Oxford Univ. Press, Oxford.

D. HILBERT

[1899] *Grundlagen der Geometrie,* 7th ed., Tuebner-Verlag, Leipzig, Berlin, 1930.

[1904] Über die Grundlagen der Logik und der Arithmetik, in: *Verhandlungen des Dritten Internationalen Mathematiker-Kongresses in Heidelberg vom 8. bis 13. August 1904*, pp. 174–185, Teubner, Leipzig, 1905. Reprinted in *van Heijenoort* [1967], pp. 129–138.

[1918] Axiomatisches Denken, *Math. Ann.*, 78, pp. 405–415.

[1926] Über das Unendliche, *Math. Ann.*, 95, pp. 161–190. (English transl. in *van Heijenoort* [1967], pp. 367–392.)

[1928] Die Grundlagen der Mathematik, *Abhandlungen aus dem mathematischen Seminar der Hamburgischen Universität, Die Grundlagen der Mathematik*, 6, pp. 65–85. Reprinted in *van Heijenoort* [1967], pp. 464–479.

D. HILBERT AND W. ACKERMANN

[1928] *Grundzüge der theoretischen Logik*, Springer, Berlin (English transl. of 1938 edition, Chelsea, New York, 1950).

D. HILBERT AND P. BERNAYS

[1934] *Grundlagen der Mathematik* I (1934), II (1939), 2nd ed., I (1968), II (1970), Springer, Berlin.

A. HODGES

[1983] *Alan Turing: The Enigma*, Burnett Books and Hutchinson, London, and Simon and Schuster, New York.

S. C. KLEENE

[1936a] General recursive functions of natural numbers, *Math. Ann.*, 112, pp. 727–742.

[1936b] λ-definability and recursiveness, *Duke Math. J.*, 2, pp. 340–353.

[1936c] A note on recursive functions, *Bull. Amer. Math. Soc.*, 42, pp. 544–546.

[1938] On notation for ordinal numbers, *J. Symbolic Logic*, 3, pp. 150–155.

[1943] Recursive predicates and quantifiers, *Trans. Amer. Math. Soc.*, 53, pp. 41–73.

[1944] On the forms of the predicates in the theory of constructive ordinals, *Amer. J. Math.*, 66, pp. 41–58.

[1952a] *Introduction to Metamathematics*, Van Nostrand, New York. 9th reprint 1988, Walters-Noordhoff Publishing Co., Gröningen and North-Holland, Amsterdam.

[1952b] Recursive functions and intuitionistic mathematics, in: *Proceedings of the International Congress of Mathematicians, Cambridge, MA, USA, Aug. 30–Sept. 6, 1950*, Amer. Math. Soc., Providence, RI, Vol. 1, pp. 679–685.

[1955a] Arithmetical predicates and function quantifiers, *Trans. Amer. Math. Soc.*, 79, pp. 312–340.

[1955b] On the forms of the predicates in the theory of constructive ordinals (second paper), *Amer J. Math.*, 77, pp. 405–428.

[1955c] Hierarchies of number-theoretical predicates, *Bull. Amer. Math. Soc.*, 61, pp. 193–213.

[1959] Recursive functionals and quantifiers of finite type I, *Trans. Amer. Math. Soc.*, 91, pp. 1–52.

[1962a] Turing-machine computable functionals of finite types I, in: *Logic, Methodology, and Philosophy of Science: Proceedings of the 1960 International Congress*, Stanford Univ. Press, pp. 38–45.

[1962b] Turing-machine computable functionals of finite types II, *Proc. London Math. Soc.*, 12, no. 3, pp. 245–258.

[1963] Recursive functionals and quantifiers of finite type II, *Trans. Amer. Math. Soc.*, 108, pp. 106–142.
[1981a] Origins of recursive function theory, *Annals of the History of Computing*, 3, pp. 52–67.
[1981b] The theory of recursive functions, approaching its centennial, *Bull. Amer. Math. Soc. (N.S.)*, 5, pp. 43–61.
[1981c] Algorithms in various contexts, in: *Proc. Sympos. Algorithms in Modern Mathematics and Computer Science (dedicated to Al-Khowarizimi)* (Urgench, Khorezm Region, Uzbek SSR, 1979), Springer, Berlin.
[1987a] Reflections on Church's Thesis, *Notre Dame J. Formal Logic,* 28, pp. 490–498.
[1987b] Gödel's impression on students of logic in the 1930's, in: *Gödel Remembered,* P. Weingartner and L. Schmetterer, eds., Bibliopolis, Naples, pp. 49–64.
[1988] Turing's analysis of computability, and major applications of it, in: *Herken* [1988], pp. 17–54.

S. C. KLEENE AND E. L. POST
[1954] The upper semi-lattice of degrees of recursive unsolvability, *Ann. of Math.*, 59, pp. 379–407.

M. KLINE
[1972] *Mathematical Thought from Ancient to Modern Times*, Oxford Univ. Press, Oxford.

M. LERMAN
[1983] *Degrees of Unsolvability: Local and Global Theory*, Springer, Heidelberg.

L. LÖWENHEIM
[1915] Über Möglichkeiten im Relativkalkül, *Math. Ann.,* 76, pp. 447–470.

D. NORMANN
[1980] *Recursion on Countable Functionals*, Lecture Notes in Mathematics, Vol. 811, Springer, Heidelberg.

P. ODIFREDDI
[1989] *Classical Recursion Theory*, North-Holland, Amsterdam.

O.E.D.
[1989] *Oxford English Dictionary Second Edition*, J. A. Simpson and E. S. C. Weiner, eds., Clarendon Press, Oxford, 24 volumes.

G. PEANO
[1889] *Arithmetices Principi, Nova Methodo Exposita*, Turin, Bocca, xvi+20 pp. English transl. in *van Heijenoort* [1967], pp. 83–97.
[1891] Sul concetto di numero, *Rivista di Matematica*, 1, pp. 87–102, 256–267.

C. S. PEIRCE
[1960] Book II. Speculative Grammar, in: *Collected Papers of Charles Sanders Peirce, Volume II: Elements of Logic*, C. Hartshorne and P. Weiss, eds., The Belknap Press of Harvard Univ. Press, Cambridge, MA and London, England.

R. PENROSE
[1994] *Shadows of the Mind*, Oxford Univ. Press, Oxford.

R. PÉTER
[1934] Über den Zussammenhang der verschiedenen Begriffe der rekursiven Funktion, *Math. Ann.*, 110, pp. 612–632.
[1951] *Rekursive Funktionen*, Akadémiai Kiadó (Akademische Verlag), Budapest, 206 pp. *Recursive Functions*, 3rd rev. ed., Academic Press, New York, 1967, 300 pp.

R. PLATEK
[1966] *Foundations of Recursion Theory*, Ph.D. Thesis, Stanford University, Stanford, CA.

E. L. POST

[1936] Finite combinatory processes – formulation I, *J. Symbolic Logic*, 1, pp. 103–105. Reprinted in: *Davis* [1965], pp. 288–291.

[1941] Absolutely unsolvable problems and relatively undecidable propositions: Account of an anticipation. (Submitted for publication in 1941.) Printed in: *Davis* [1965], pp. 340–433.

[1943] Formal reductions of the general combinatorial decision problem, *Amer. J. Math.*, 65, pp. 197–215.

[1944] Recursively enumerable sets of positive integers and their decision problems, *Bull. Amer. Math. Soc.*, 50, pp. 284–316. Reprinted in Davis [1965], pp. 304–337.

[1947] Recursive unsolvability of a problem of Thue, *J. Symbolic Logic*, 12, pp. 1–11. Reprinted in Davis [1965], pp. 292–303.

[1948] Degrees of recursive unsolvability: preliminary report (abstract), *Bull. Amer. Math. Soc.*, 54, pp. 641–642.

H. PUTNAM

[1995] Review of Penrose [1994], *Bull. Amer. Math. Soc.*, 32, pp. 370–373.

H. ROGERS, JR.

[1967] *Theory of Recursive Functions and Effective Computability*, McGraw-Hill, New York, 482 pp.

G. E. SACKS

[1963] Degrees of unsolvability, in: *Ann. of Math. Stud.*, 55, Princeton Univ. Press, Princeton, NJ.

[1990] *Higher Recursion Theory*, Springer, Heidelberg.

E. SCHRÖDER

[1895] *Algebra und Logik der Relative, Part 1*, Leipzig, 1895, 400 pp.

J. R. SHOENFIELD

[1967] *Mathematical Logic*, Addison-Wesley, Reading, MA, 344 pp.

[1991] *Recursion Theory*, Lecture Notes in Logic, Springer, Heidelberg.

[1995] The mathematical work of S. C. Kleene, *Bull. A.S.L.*, 1, pp. 8–43.

W. SIEG

[1994] Mechanical procedures and mathematical experience, in: *Mathematics and Mind*, A. George, ed., Oxford Univ. Press.

W. SIEG AND J. BYRNES

[1995] *K-graph machines: generalizing Turing's machines and arguments*, Preprint.

T. SKOLEM

[1923] Begründung der elementaren Arithmetik durch die rekurrierende Denkweise ohne Anwendung scheinbare Veränderlichen mit unendlichen Ausdehnungsbereich, *Skrifter utgit av Videnskapsselskapet i Kristiania, I. Mathematisk-Naturvidenskabelig Klasse*, 6, 38 pp. (English transl. in *van Heijenoort* [1967], pp. 302–333.)

R. I. SOARE

[1981] Constructions in the recursively enumerable degrees, in: *Recursion Theory and Computational Complexity, Proceedings of Centro Internazionale Matematico Estivo (C.I.M.E.), June 14–23, 1979, in Bressanone, Italy*, G. Lolli, ed., Liguori Editore, Naples, Italy.

[1987] *Recursively Enumerable Sets and Degrees: A Study of Computable Functions and Computably Generated Sets*, Springer, Heidelberg.

[1996] Computability and recursion, *Bulletin of Symbolic Logic*, 2, pp. 284–321.

[1997] Computability and enumerability, in: *Logic and Scientific Methods*, Vol. 1 of the Proceedings of the Tenth International Congress of Logic, Methodology, and Philosophy of Science, August, 1995, pp. 221–237.

[1999] An overview of the computably enumerable sets, in: *Handbook of Computability Theory*, E. R. Griffor, ed., Elsevier, Amsterdam, pp. 199–248.

G. TAMBURRINI

[1997] Mechanistic theories in cognitive science: The import of Turing's Thesis, in: *Logic and Scientific Methods*, Vol. 1 of the Proceedings of the Tenth International Congress of Logic, Methodology, and Philosophy of Science, August, 1995, pp. 239–257.

A. M. TURING

[1936] On computable numbers, with an application to the Entscheidungsproblem, *Proc. London Math. Soc. Ser. 2*, 42 (Parts 3 and 4), pp. 230–265.[14]

[1937a] A correction, *ibid.*, 43, pp. 544–546.

[1937b] Computability and λ-definability, *J. Symbolic Logic*, 2, pp. 153–163.

[1939] Systems of logic based on ordinals, *Proc. London Math. Soc.*, 45 (3), pp. 161–228; reprinted in *Davis* [1965], pp. 154–222.

[1948] Intelligent machinery, in: *Machine Intelligence*, 5, pp. 3–23. (Written in September, 1947 and submitted to the National Physical Laboratory in 1948.)

[1949] Text of a lecture by Turing on June 24, 1949, in: F. L. Morris and C. B. Jones, "An early program proof by Alan Turing," *Annals of the History of Computing*, 6, pp. 139–143.

[1950a] Computing machinery and intelligence, *Mind*, 59, pp. 433–460.

[1950b] The word problem in semi-groups with cancellation, *Ann. of Math.*, 52, pp. 491–505.

[1954] Solvable and unsolvable problems, *Science News*, 31, pp. 7–23.

[1986] Lecture to the London Mathematical Society on 20 February 1947, in: *A. M. Turing's ACE Report of 1946 and Other Papers*, B. E. Carpenter and R. W. Doran, eds., Cambridge Univ. Press, pp. 106–124.

J. VAN HEIJENOORT

[1967] *From Frege to Gödel, A Sourcebook in Mathematical Logic, 1879–1931,* Harvard Univ. Press, Cambridge, MA.

WEBSTER

[1993] *Webster's Third New International Dictionary of the English Language* (*unabridged*), Ph. B. Gove, ed., Merriam-Webster Inc. Publishers, Springfield, MA, 2, 662 pp.

S. L. ZABELL

[1995] Alan Turing and the Central Limit Theorem, *Amer. Math. Monthly*, 102 (6), pp. 483–494.

[14] Many papers, Kleene [1943, p. 73], [1987a, 1987b], Davis [1965, p. 72], Post [1943, p. 200], and others, mistakenly refer to this paper as "[Turing, 1937]," perhaps because the volume 42 is 1936–37 covering 1936 and part of 1937, or perhaps because of the two page minor correction [1937a]. Others, such as Kleene [1952a, 1981a, 1981b], Kleene and Post [1954, p. 407], Gandy [1980], Cutland [1980], and others, correctly refer to it as "[1936]," or sometimes " [1936–37]." The journal states that Turing's manuscript was "Received 28 May, 1936–Read 12 November, 1936." It appeared in two sections, the first section of pp. 230–240 in Volume 42, Part 3, issued on November 30, 1936, and the second section of pp. 241–265 in Volume 42, Part 4, issued December 23, 1936. No part of Turing's paper appeared in [1937a], but the two page minor correction [1937a] did. Determining the correct date of publication of Turing's work is important to place it chronologically in comparison with Church [1936a], Post [1936], and Kleene [1936].

CHAPTER 2

Π_1^0 Classes in Computability Theory

Douglas Cenzer

Department of Mathematics, University of Florida, Gainesville, FL 32611, USA

Contents

1. Introduction 38
   1.1. Background 38
   1.2. Preliminaries 40
2. Π_1^0 sets and classes 44
3. Basis and anti-basis results 51
4. Cantor–Bendixson rank 61
5. Minimal and thin classes 69
6. Applications 74
   6.1. Logical theories 74
   6.2. Graph-coloring problems 79
   6.3. Other applications 82
References 82

HANDBOOK OF COMPUTABILITY THEORY
Edited by E.R. Griffor

1. Introduction

1.1. *Background*

This article is a survey of recent results on Π^0_1 classes and their applications. It includes some new results, in particular Theorem 5.4 which shows the existence of a r.b. Π^0_1 class with no member of high degree. The focus will be on the study of *recursively bounded* (*r.b.*) Π^0_1 classes as a branch of recursion theory. In particular the notion of a Π^0_1 class may be viewed as the generalization of the notion of a Π^0_1 set, which is simply the complement of a Σ^0_1, or recursively enumerable set. In our usage, we denote by "set" a set of the natural numbers. For us a "class" denotes a subset of the space ω^ω of functions from ω to ω or of the subspspace $\{0, 1\}^\omega$ of 0,1-valued functions. Now a subset A of ω may be identified with its characteristic function $\chi_A \in \{0, 1\}^\omega$, so that a class in $\{0, 1\}^\omega$ will be called a "class of sets".

Following the standard Levy hierarchy of logic, we say that $X \subset \omega^\omega$ is a Π^0_1 class if there is a Δ_0 formula $\phi(n, y)$ such that

$$x \in X \iff (\forall n)\phi(n, x),$$

where the language is that of first-order arithmetic together with free function variables and new atomic formulae $x(m) = n$ associated with each function variable x. There is an alternative representation for classes $X \subset \{0, 1\}^\omega$ of sets $A \subset \omega$, where the new atomic formulae have the form "$n \in A$".

A Π^0_1 class may be described topologically as an effectively closed subset of the product space ω^ω. The study of Π^0_1 classes and their members provides an important link between recursion theory and descriptive set theory.

The key problem here is the connection between the topological structure of the Π^0_1 class P and the degree-theoretic complexity of the members of P. This problem has been studied by many recursion theorists, going back to the Kleene basis theorem [Kleene, 1943], which showed that every Π^0_1 class contains a member which is recursive in some Σ^1_1 set and the Kreisel–Shoenfield basis theorem [Shoenfield, 1958], which showed that every r.b. Π^0_1 class contains a member of degree $< 0'$. Two fundamental papers in this area are Jockusch and Soare [1972a, 1972b]. They show, among other things, that there is a Π^0_1 class with no recursive members and such that any two members have mutually incomparable Turing degree.

Kreisel [1959] first noticed that the degree of a member x of a Π^0_1 class is related to the Cantor–Bendixson rank of x in P, when he showed that every member of a countable Π^0_1 class P is hyperarithmetic and in particular that P has a recursive member. This relationship has been developed in detail in recent work on countable Π^0_1 classes and Cantor–Bendixson rank by Cenzer, Clote, Smith, Soare and Wainer [1986], Cenzer and Smith [1989] and more recently by Cholak and Downey [1993] as well as work by Cenzer, Downey, Jockusch and Shore [1993] on countable thin Π^0_1 classes.

Of particular interest for recursion theory is the connection between a *retraceable* Π_1^0 set subset A of ω and the Π_1^0 class $P(A)$ of *initial* subsets of A. The notion of a *co-maximal* Π_1^0 set also plays an important role and has a natural version for Π_1^0 classes, that of a *minimal* Π_1^0 class P. Here P is minimal if every Π_1^0 subclass Q of P is either finite or is cofinite in P.

Π_1^0 classes are also important in the application of recursion theory to numerous branches of mathematics, including combinatorics, algebra and analysis. For example, the set of k-colorings of a recursive graph may be represented as a Π_1^0 class, and, conversely, Remmel [1986] showed that any r.b. Π_1^0 class may be represented by the set of k-colorings of a recursive graph for $k \geqslant 3$. It follows that there exist 3-colorable recursive graphs with no recursive 3-coloring, whereas any 3-colorable recursive graph with only countably many 3-colorings has a recursive 3-coloring.

The class $S(A, B)$ of separating sets for a pair of disjoint r.e. sets A, B is always a r.b. Π_1^0 class.

The first area where Π_1^0 classes were applied was to theories of arithmetic. Shoenfield [1960] showed that the set of complete consistent extensions of a decidable first-order theory is always a Π_1^0 class. Ehrenfeucht [1961] showed that, conversely, every Π_1^0 class is represented by the set of complete, consistent extensions of some first-order theory.

We will limit ourselves to a sketch of the application of Π_1^0 classes to logical theories and graph colorings and a brief mention of some other applications. The forthcoming survey paper of Cenzer and Remmel [ta2] contains a development of the applications of Π_1^0 classes in recursive mathematics. The topics of index sets for Π_1^0 classes, developed in [Cenzer and Remmel, 1998] and of feasible Π_1^0 classes, developed in [Cenzer and Remmel, 1992], will also be covered in [Cenzer and Remmel, ta2].

A more direct connection between Π_1^0 sets and logic is the representation of Π_1^0 sets (and, more generally, relations) by formulas in the language of arithmetic. The fundamental point here is that while any Σ_1^0 sentence of arithmetic is provable in any reasonable theory (say, Robinson's arithmetic), Π_1^0 sentences may be true but unprovable. For example, the standard unprovable sentence $CON(P)$, which asserts the consistency of Peano Arithmetic, is logically equivalent to a Π_1^0 sentence. A Π_1^0 sentence $(\forall n)\phi(n)$ makes infinitely many assertions.

Another aspect of this essential non-finiteness of Π_1^0 relations makes them a link between recursive and elementary or hyperarithmetic relations. The fundamental result of Spector [1961] is that any Π_1^1 subset of ω is 1–1 reducible to the closure of a monotone Π_1^0 inductive definition $\theta : \mathcal{P}(\omega) \to \mathcal{P}(\omega)$. Any such inductive definition is naturally represented by a Π_1^0 class. Note that the repeated application of the Cantor–Bendixson derivative to a Π_1^0 class P which eventually reduces P to its perfect kernel K may be viewed as a monotone Π_1^0 inductive definition with closure the complement of K.

1.2. *Preliminaries*

Some definitions are needed. The set $\{0, 1, 2, \ldots\}$ of natural numbers is denoted by ω. Let Σ be a (usually finite) alphabet. Then for a natural number n, Σ^n denotes the set of strings $\sigma = (\sigma(0), \sigma(1), \ldots, \sigma(n-1))$ of n letters from Σ; the length n of σ is denoted by $|\sigma|$. The empty string has length 0 and will be denoted by $\emptyset$. Σ^* (or sometimes $\Sigma^{<\omega}$) denotes the set $\bigcup_{n\in\omega} \Sigma^n$ and Σ^ω denotes the set of infinite sequences. The empty string has length 0 and will be denoted by $\emptyset$. Strings may be coded by natural numbers in the usual fashion. First let $[x, y]$ denote the standard pairing function $\frac{1}{2}(x^2+2xy+y^2+3x+y)$ and in general $[x_0, x_1, \ldots, x_n] = [[x_0, \ldots, x_{n-1}]x_n]$. Then we can code strings of arbitrary length $n > 0$ by

$$\langle\sigma\rangle = \big[n, [\sigma(0), \sigma(1), \ldots, \sigma(n-1)]\big]$$

and also $\langle\emptyset\rangle = 1$. A constant string σ of length n will be denoted k^n. For $m < |\sigma|$, $\sigma\lceil m$ is the string $(\sigma(0), \ldots, \sigma(m-1))$; σ is an *initial segment* of τ (written $\sigma \prec \tau$) if $\sigma = \tau\lceil m$ for some m. The *concatenation* $\sigma^\frown\tau$ (or sometimes just $\sigma\tau$) is defined by

$$\sigma^\frown\tau = \big(\sigma(0), \sigma(1), \ldots, \sigma(m-1), \tau(0), \tau(1), \ldots, \tau(n-1)\big),$$

where $|\sigma| = m$ and $|\tau| = n$; in particular we write $\sigma^\frown a$ for $\sigma^\frown(a)$ and $a^\frown\sigma$ for $(a)^\frown\sigma$. For any $x \in \Sigma^*$ and any finite n, the *initial segment* $x\lceil n$ of x is $(x(0), \ldots, x(n-1))$. We write $\sigma \prec x$ if $\sigma = x\lceil n$ for some n. For any $\sigma \in \Sigma^n$ and any $x \in \Sigma^*$, we have

$$\sigma^\frown x = (\sigma(0), \ldots, \sigma(n-1), x(0), x(1), \ldots).$$

For a sequence $a_0 < a_1 < \cdots < a_n$, we denote by $\lfloor a_0, \ldots, a_n \rfloor$ the string $\sigma \in \{0, 1\}^{a_n}$ such that $\sigma(k) = 1$ if and only if $k = a_i$ for some $i < n$. Thus $\lfloor a_0, a_1, \ldots, a_n \rfloor = 0^{a_0}10^{a_1-a_0-1}1\cdots0^{a_{n-1}-a_{n-2}-1}10^{a_n-a_{n-1}-1}$.

For any $x, y \in \omega^\omega$, the *join* $x \oplus y = z$, where $z(2n) = x(n)$ and $z(2n+1) = y(n)$. For two classes P and Q, the join $P \oplus Q = \{x \oplus y\colon x \in P \,\&\, y \in Q\}$.

A *tree* T over Σ is a set of finite strings from Σ^* which is closed under initial segments. We say that $\tau \in T$ is an *immediate successor* of a string $\sigma \in T$ if $\tau = \sigma^\frown a$ for some $a \in \Sigma$. Since our alphabet will always be countable and effective, we may assume that $T \subset \omega^{<\omega}$. Such a tree is said to be *ω-branching*, since each node has potentially a countably infinite number of immediate successors. If each node of T has finitely many immediate successors, then T is said to be *finite branching*. A recursive tree T is said to be *highly recursive* if there is a partial recursive function f such that, for any $\sigma \in T$, σ has at exactly $f(\sigma)$ immediate successors in T. For any tree T, an *infinite path* through T is a sequence $(x(0), x(1), \ldots)$ such that $x\lceil n \in T$ for all n. Let $[T]$ be the set of infinite paths through T. A subset X of ω^ω is a *Π^0_1 class* if $X = [T]$ for some recursive tree T; if the tree T is finite branching, we will say that

$P = [T]$ is a *bounded* Π_1^0 *class*, and if T is highly recursive, then $X = [T]$ is said to be a *recursively bounded* (*r.b.*) Π_1^0 class. We observe that T is bounded (respectively r.b.) if and only if there exists a function (respectively a recursive function) f such that $\sigma(n) < f(n)$ for all n and all $\sigma \in T$ with $\sigma > n$.

The topology on ω^ω is determined by a basis of intervals $I(\sigma) = \{x\colon \sigma \prec x\}$. Notice that each interval is also a closed set and is therefore said to be *clopen* and that the clopen subsets of the Cantor space $\{0, 1\}^\omega$ are just the finite unions of intervals. A subset P of ω^ω is closed if and only if $P = [T]$ for some tree T. This justifies the description of a Π_1^0 class as an effectively closed subset of ω^ω. It is easy to see that a Π_1^0 class P is compact if and only if P is bounded. The standard Lebesgue measure on $\{0, 1\}^\omega$ is determined by letting the measure of $I(\sigma)$ equal $2^{-|\sigma|}$.

We refer the reader to Odifreddi [1989] for the basic definitions of recursion theory. Let ϕ_i be the i-th partial recursive function. Given a string $\sigma \in \{0, 1\}^*$, we write $\phi_{i,s}(\sigma)\downarrow$ if the computation of $\phi_i(\sigma)$ gives an output in s or fewer steps. We write $\phi_e(\sigma)\downarrow$ if $(\exists s)(\phi_{e,s}(\sigma)\downarrow)$ and $\phi_e(\sigma)\uparrow$ if not $\phi_e(\sigma)\downarrow$. Given two sets A and B, we write $A \leqslant_T B$ if A is Turing reducible to B (that is, $A = \phi_e^B$ for some e) and we write $A \equiv_T B$ if both $A \leqslant_T B$ and $B \leqslant_T A$; the *Turing degree* of a set A is the equivalence class of A under $\equiv_T$. We let $\mathbf{0}$ denote the Turing degree of a recursive set and we let $\mathbf{0}'$ denote the Turing degree of the jump of a recursive set. A set A is said to be *truth-table reducible to* B (written $A \leqslant_{tt} B$) if there exists a recursive relation R and a recursive function f such that for any n, $n \in A \iff R(\langle B\lceil f(n)\rangle)$.

The notation "Π_1^0" indicates that a Π_1^0 class may be represented in arithmetic by a formula having one universal quantifier, ranging over natural numbers. The following lemma makes the connection between trees and quantified relations precise.

LEMMA 1.1. *For any class* $P \subset \omega^\omega$, *the following are equivalent*:
(a) $P = [T]$ *for some recursive tree* $T \subset \omega^{<\omega}$;
(b) $P = [T]$ *for some primitive recursive tree* T;
(c) $P = \{x\colon (\forall n)R(n, x)\}$, *for some recursive relation* R;
(d) $P = [T]$ *for some* Π_1^0 *tree* $T \subset \omega^{<\omega}$.

PROOF. [(a) $\to$ (b)]: Suppose that $P = [S]$, where S is a recursive tree and let ϕ_e be a total $\{0, 1\}$-valued recursive function such that $\sigma \in S$ if and only if $\phi_e(\sigma) = 1$. Define the primitive recursive tree T by $\tau \in T \iff (\forall n < |\tau|)\neg\phi_{e,|\tau|}(\tau\lceil n) = 0$. Clearly $T \subset S$, so that $[T] \subset [S]$. Suppose now that $x \notin [T]$. Then for some n, $x\lceil n \notin T$. Thus we have some m such that $\phi_{e,m}(x\lceil n) = 0$. Then for any $k > \max\{m, n\}$, we clearly have $x\lceil k \notin S$. It follows that $x \notin [S]$.

[(b) $\to$ (c)]: Suppose that $P = [T]$, where T is a primitive recursive tree. Define the relation R by $R(n, x) \iff x\lceil n \in T$. Then we have

$$x \in [T] \iff (\forall n)x\lceil n \in T \iff (\forall n)R(n, x).$$

[(c) $\to$ (d)]: Suppose that $x \in P \iff (\forall n)R(n, x)$ where R is a recursive relation, that is, there is a recursive functional $\phi = \phi_e$ such that $R(n, x) \iff \phi(n, x) = 1$

and $\neg R(x) \iff \phi(n, x) = 0$. Define the tree T by

$$\sigma \in T \iff \big(\forall k, c < |\sigma|\big) \neg \varphi_{e,c}(k, \sigma) = 0.$$

It is clear that $P = [T]$.

[(d) $\to$ (a)]: Suppose that the tree T is a Π_1^0 subset of $\omega^{<\omega}$, so that there is a recursive relation R such that $\sigma \in T \iff (\forall n) R(n, \sigma)$. Define the recursive tree $S \supset T$ by

$$\sigma \in S \iff (\forall m, n \leqslant |\sigma|) R(m, \sigma\lceil n).$$

It is easily verified that $[T] = [S]$. □

An important corollary of Lemma 1.1 is that the Π_1^0 classes can be effectively enumerated.

LEMMA 1.2. *There is a uniformly recursive sequence T_e of primitive recursive trees such that, for every Π_1^0 class P, $P = [T_e]$ for some e.*

PROOF. Let $\pi_0, \pi_1, \ldots$ be a recursive enumeration of the primitive recursive functions which map ω to $\{0, 1\}$. Define the e-th tree by

$$\sigma \in T_e \iff (\forall \tau \preceq \sigma) \pi_e(\langle \tau \rangle) = 1.$$

It is easy to see that each T_e is a tree and that if T is a primitive recursive tree with characteristic function π_e, then $T = T_e$. The fact that every Π_1^0 class is equal to one of the $[T_e]$ follows from Lemma 1.1. □

The set $\mathrm{Ext}(T)$ of extendible nodes of a tree is defined by

$$\sigma \in \mathrm{Ext}(T) \iff (\exists x)[x \in [T] \;\&\; \sigma \prec x]. \tag{1.1}$$

For a finite branching tree T, $\sigma \in \mathrm{Ext}(T) \iff (\forall n)(\exists \tau)[|\tau| = n \;\&\; \sigma^\frown\tau \in T]$.

A node $\sigma \in T$ is said to be a *dead end* if $\sigma \notin \mathrm{Ext}(T)$, that is, if σ has no infinite extension in $[T]$.

It is often easier to work with binary trees than with highly recursive trees. Thus we give the following lemma.

LEMMA 1.3. *Every recursively bounded Π_1^0 class P is recursively homeomorphic to a Π_1^0 class $Q \subset \{0, 1\}^\omega$.*

PROOF. Let $T \subset \omega^{<\omega}$ be a recursive tree, let $P = [T]$, and let f be a recursive function such that for all $\tau \in T$ and all $n < |\tau|$, $\tau(n) < f(n)$.

Define a recursive map k from ω^ω to $\{0,1\}^\omega$ such that $k(x)$ has the same Turing degree as x for all $x \in \omega^\omega$, by

$$k(x) = (0^{x(0)}{}^\frown 10^{x(1)}{}^\frown 1 \ldots). \tag{1.2}$$

For each $\tau \in \omega^{<\omega}$ with $|\tau| = n$, let

$$k(\tau) = (0^{\tau(0)}{}^\frown 10^{\tau(1)}{}^\frown 1 \ldots 0^{\tau(n-1)}{}^\frown 1). \tag{1.3}$$

Now define the tree $S \subset \{0,1\}^{<\omega}$ by

$$S = \{k(\tau)^\frown 0^i \colon \tau \in T \ \&\ i < f(|\tau|)\}. \tag{1.4}$$

It is then clear that k is a recursive homeomorphism from P onto $Q = [S]$. □

The Cantor–Bendixson (C-B) derivative $D(P)$ of a compact subset P of ω^ω is the set of nonisolated points of P. Thus a point $x \in P$ is not in $D(P)$ if and only if there is some open set U containing x which contains no other point of P. Equivalently, $x \notin D(P)$ if and only if there is some clopen set U such that $U \cap P = \{x\}$. Another useful observation is that, for any compact set P, $D(P)$ is empty if and only if P is finite.

The iterated Cantor–Bendixson derivative $D^\alpha(P)$ of a closed set P is defined for all ordinals α by the following transfinite induction.

$D^0(P) = P$; $D^{\alpha+1}(P) = D(D^\alpha(P))$ for any α; $D^\lambda(P) = \bigcap_{\alpha<\lambda} D^\alpha(P)$ for any limit ordinal λ.

The Cantor–Bendixson rank of a closed set P is the least ordinal α such that $D^{\alpha+1}(P) = D^\alpha(P)$. A set A has Cantor–Bendixson rank α if α is the least ordinal such that, for some Π^0_1 class P, $A \in D^\alpha(P) \setminus D^{\alpha+1}(P)$.

An ordinal α is said to be a *recursive ordinal* if there is a recursive well-ordering of ω of order type α. The least nonrecursive ordinal is denoted by $\omega_1^{\text{C-K}}$, and was introduced by Church and Kleene [1937].

The natural, or Hessenberg sum $\alpha \oplus \beta$ of two ordinals α and β, may be defined as follows. Let $\alpha = \omega^{\gamma_1}a_1 + \omega^{\gamma_2}a_2 + \cdots + \omega^{\gamma_k}a_k$ and $\beta = \omega^{\gamma_1}b_1 + \omega^{\gamma_2}b_2 + \cdots + \omega^{\gamma_k}b_k$ be the Cantor normal forms of α and β, where we have inserted $a_i = 0$ and $b_j = 0$ to obtain expressions with the same powers of ω. Then

$$\alpha \oplus \beta = \omega^{\gamma_1}(a_1 + b_1) + \omega^{\gamma_2}(a_2 + b_2) + \cdots + \omega^{\gamma_k}(a_k + b_k).$$

Thus we treat ordinals as polynomials over ω with natural number coefficients. This natural addition is commutative. For any ordinals α and β, $\alpha + \beta \leqslant \alpha \oplus \beta$. See Kuratowski and Mostowski [1968, p. 253] for details.

A fundamental idea here is that the complexity of an element x of a Π^0_1 class P is related to the Cantor–Bendixson rank of x in P. The basic result is the following.

LEMMA 1.4. *For any* $x \in \{0,1\}^\omega$, *the following are equivalent*:

(a) *x is recursive*;
(b) $\{x\}$ *is a* Π_1^0 *class*;
(c) *x has Cantor–Bendixson rank* 0.

PROOF. Suppose first that x is recursive. Then $\{x\} = [T]$, where

$$\sigma \in T \iff (\forall i < |\sigma|)\big(\sigma(i) = x(i)\big).$$

Next suppose that $\{x\}$ is a Π_1^0 class. Then the rank of x in $\{x\}$ is 0, so that the C-B rank of x is 0. Next suppose that x has C-B rank 0 and let $P = [T]$ be a Π_1^0 class such that x is isolated in P, where T is a recursive tree. Then for each sufficiently large n, $x\lceil n+1$ is the unique path of length n which has an extension in P. Thus we may compute $x\lceil n+1$ (and therefore compute $x(n)$) by searching for the least m such that all strings $\sigma \in T$ of length m have the same initial segment $\sigma\lceil n$. □

We remark that Lemma 1.4 and its proof can be relativized to recursion in B for any set B.

As usual, a *first-order language* $\mathcal{L}$ is given by a set $\{R_i\}_{i\in S}$ of relation symbols, a set $\{f_j\}_{j\in T}$ of function symbols, and a set $\{c_i\}_{i\in U}$ of constant symbols, together with functions $m(i)$ and $n(i)$ such that R_i is an $m(i)$-ary relation symbol and f_i is an $n(i)$-ary function symbol. We assume here that S, T and U are subsets of ω. The language also includes variables and both existential and universal quantifiers using these variables. The set of terms of $\mathcal{L}$ and the set Sent($\mathcal{L}$) of sentences of $\mathcal{L}$ are defined as usual by induction. A propositional language is given by a set of 0-ary relation symbols, or propositional variables. The reader is referred to Shoenfield [1967] for details.

A *model* $\mathcal{A} = (A, \{R_i^{\mathcal{A}}\}_{i\in S}, \{f_i^{\mathcal{A}}\}_{i\in T}, \{c_i^{\mathcal{A}}\}_{i\in U})$, or *structure* for the language $\mathcal{L}$ is given by a set A together with interpretations of the relation, function and constant symbols.

A structure $\mathcal{A}$ with universe A is said to be a *recursive structure* if

(i) A is a recursive subset of ω.
(ii) For each $i \in S$, $R_i^{\mathcal{A}}$ is a recursive subset of $A^{m(i)}$.
(iii) For each $j \in T$, $f_j^{\mathcal{A}}$ is a recursive function from $A^{n(i)}$ into A.
(iv) If $S = \omega$, then there is a recursive relation R such that, for all i and all $(a_0, \dots, a_{m(i)-1})$, $R_i^{\mathcal{A}}(a_0, \dots, a_{m(i)-1}) \iff R(i, \langle a_0, \dots, a_{m(i)-1}\rangle)$.
(v) If $T = \omega$, then there is a partial recursive function f such that, for all i and all $(a_0, \dots, a_{n(i)-1})$, $f_i^{\mathcal{A}}(a_0, \dots, a_{n(i)-1}) = f(i, \langle a_0, \dots, a_{n(i)-1}\rangle)$.

2. Π_1^0 sets and classes

There are numerous connections between Π_1^0 and r.e. subsets of ω and Π_1^0 classes. We will consider in particular the power set $\mathcal{P}(C)$ of a Π_1^0 set C, the class $S(A, B)$

of separating sets for a pair of r.e. sets A, B, and the class $I(C)$ of initial subsets of a Π_1^0 sets C.

The most basic example here is that, for any Π_1^0 set C, the power set $\mathcal{P}(C)$ is a Π_1^0 class. More generally, consider the class of separating sets. If A and B are infinite disjoint *r.e.* sets, then C is a *separating set* for A and B if $A \subset C$ and $B \cap C = \emptyset$. The class of separating sets for A and B is denoted by $S(A, B)$; of course, $C \in S(A, B)$ if and only if $C \in P(\omega \setminus B)$ and $\omega \setminus C \in P(\omega \setminus A)$. The sets A and B are said to be *recursively inseparable* if there is no recursive separating set C for A and B. This concept was introduced by Kleene [1950], where recursively inseparable r.e. sets were constructed. Shoenfield [1958] showed that every nonrecursive r.e. degree contains a pair of recursively inseparable sets. Shoenfield [1960] observed that the class $S(A, B)$ of separating sets for A and B is a r.b. Π_1^0 class. Observe that $S(A, B)$ is finite if and only if $A \cup B$ is cofinite, in which case A and B are both recursive and every separating set is also recursive. Otherwise, $S(A, B)$ is a perfect set and thus has the cardinality of the continuum. In either case, both the r.e. set A and the co-r.e. set $\omega \setminus B$ are of course separating sets for A and B.

For any infinite set A, the *principal function* p_A enumerates the elements $a_0 < a_1 < \cdots$ in increasing order. A is said to be *hyperimmune* if, for any recursive function f, there is an n such that $a_n > f(n)$ and is said to be *retraceable* if there is a partial recursive function ϕ such that $\phi(a_{n+1}) = a_n$ for all n. Retraceable sets were introduced by Dekker and Myhill [1985], who proved that any retraceable nonrecursive Π_1^0 set A is hyperimmune. For Π_1^0 sets A, a stronger characterization can be given.

THEOREM 2.1. *The following are equivalent for any infinite Π_1^0 set A:*

(a) *A is retraceable.*

(b) *There is a total recursive function Φ such that, for all n, $\Phi(a_{n+1}) = a_n$ and, for all y, $\{x\colon \Phi(x) = y\}$ is finite.*

(c) *There is a total recursive function Ψ such that, for all n, $\Psi(a_n) = n$ and $\{x\colon \Psi(x) = n\}$ is finite.*

PROOF. Let $A = \{a_0 < a_1 < \cdots\}$ be an infinite Π_1^0 set and let A be the decreasing intersection of uniformly recursive sets A_s.

(a) $\rightarrow$ (b): Let ϕ be a partial recursive retracing function for A. Assume, without loss of generality, that $\phi(a_0) = a_0$. Then for any x, we define $\Phi(x)$ as follows. Look for the least s such that either $x \notin A_s$, or such that $\phi_s(x) = y$ converges and, for all z with $x < z < y$, $z \notin A_s$. In the former case, we let $\Phi(x) = x$ and in the latter case, we let $\Phi(x) = \phi(x)$. Note that if $x \notin A$, then the former case will obtain and if $x \in A$, then the latter case will obtain, so that Φ is total and is a retracing function for A. It follows from the definition that for every x, $\Phi(x) \leqslant x$ and there are no elements of A between $\Phi(x)$ and x. Now for any y, let a be the least such that $a > y$ and $a \in A$. Then $\Phi(x) = y$ implies that $x \leqslant a$, so that $\{x\colon \Phi(x) = y\}$ is finite, as desired.

(b) $\rightarrow$ (c): Let Φ be given as described. Then we define $\Psi(x)$ to be length n of the chain $x > \Phi(x) > \Phi(\Phi(x)) > \cdots > \Phi^n(x) = a_0$, if there is such an n-chain, and

$\Psi(x) = x$ if $\Phi^{i+1}(x) = \Phi^i(x)$ for some i. Thus for $a_n \in A$, we obtain $\Psi(a_n) = n$. To complete the proof, we show by induction that, for each n, there are only finitely many n-chains $x > \Phi(x) > \cdots > \Phi^n(x) = a_0$ of length n. For $n = 1$, this follows from the assumption that $\Phi(x) = a_0$ for only finitely many x. Suppose now that there are only finitely many such n-chains of length n. Then any $n+1$-chain must extend one of these and, by our assumption, there are only finitely extensions of each chain. Thus there can be only finitely many $n+1$-chains.

(c) → (a): Let Ψ be given as described. Then for $a = a_{n+1} \in A$, $\Psi(a) = n+1$ and the retracing function $\phi(a_{n+1}) = a_n$ may now be computed by searching for the least s such that exactly $n+1$ elements of A_s are less than a and taking a_n to be the largest of those elements. □

We say that an infinite set $A = \{a_0 < a_1 < \cdots\}$ is *second-retraceable* if there is a total recursive function Φ such that, for any $m < n$, $\Phi(a_m, a_n) = m$. In general, A is k-retraceable if there is a total recursive Φ such that $\Phi(a_{m_1}, a_{m_2}, \ldots, a_{m_k}) = m_1$ for any $m_1 < m_2 < \cdots < m_k$. Of course, any k-retraceable set is also $k+1$-retraceable.

A subset F of the set $\{a_0 < a_1 < \cdots\}$ is said to be an *initial subset* of A if $a_{n+1} \in F$ implies $a_n \in F$ for all n. Thus the initial subsets of A are A together with the finite sets $\{a_0, \ldots, a_{n-1}\}$ for each n. Let $I_1(A)$ denote the class of initial subsets of A. In general, the k-initial subsets $I_k(A)$ are the subsets F of A such that for any elements $a < b_1 < b_2 < \cdots b_k$ of A, if $b_1, \ldots, b_k \in F$, then $a \in F$.

THEOREM 2.2. *For each finite k, the set* $A = \{a_0 < a_1 < \cdots\}$ *is* Π_1^0 *and k-retraceable if and only if the class* $I_k(A)$ *of k-initial subsets of A is a* Π_1^0 *class.*

PROOF. Suppose first that A is k-retraceable via the function Φ and that A is a Π_1^0 set. Let A_s denote the recursive approximation to the set A at stage s, so that $A = \bigcap_s A_s$. Now define the recursive tree T as follows.

$$\lfloor b_0, b_1, \ldots, b_m, s \rfloor \in T \iff \text{for all } n \leqslant m$$
$$\text{(i) } (\forall i \leqslant n)(b_i \in A^s) \text{ and}$$
$$\text{(ii) if } n \geqslant k, \text{ then } \Phi(b_{n-k+1}, b_{n-k+2}, \ldots, b_{n-1}, b_n) = n - k + 1.$$

It is easy to check that $[T] = I_k(A)$, so that $I_k(A)$ is a Π_1^0 class.

Now suppose that $I_k(A)$ is a Π_1^0 class and let T be a recursive tree so that $I_k(A) = [T]$. We will explain how to compute a k-retracing function Φ. Given $b_1 = a_{m_1} < b_2 = a_{m_2} < \cdots < b_k = a_{m_k}$, observe that there is only one possible string $\sigma = \lfloor a_0, a_1, \ldots, a_{m_1-1}, b_1, b_2, \ldots, b_k \rfloor^\frown 1$ of the form $\lfloor c_0, c_1, \ldots, c_r, b_1, b_2, \ldots, b_k \rfloor^\frown 1$ which has an extension in T; $\Phi(b_1, \ldots, b_k) = m_1$ is then easily computed from σ. To find σ, we just search through all strings of length $m > b_k$ until we find m large enough so that all strings τ in T of length m and with $\tau \lceil b_k + 1$ of the desired form, start with the same initial segment (σ) of length $b_k + 1$.

To see that A is a Π_1^0 set, recall that $\mathrm{Ext}(T)$ is Π_1^0 and observe that

$$a \in A \iff (\exists\sigma)[|\sigma| = a+1 \ \&\ \sigma \in \mathrm{Ext}(T)\ \&\ \sigma(a) = 1].$$

□

We can now give a quick proof that any retraceable non-recursive Π_1^0 set is hyperimmune.

THEOREM 2.3 (Dekker–Myhill). *If $A = \{a_0 < a_1 < \cdots\}$ is a retraceable non-recursive Π_1^0 set, then A is hyperimmune.*

PROOF. By Theorem 2.2, $I(A)$ is a Π_1^0 class. Now suppose by way of contradiction that there were a recursive function f which dominated p_A, that is, $f(n) > a_n$ for all n. Then the set $\{A\}$ would be the intersection of $I(A)$ with the following Π_1^0 class:

$$\{B : (\forall n)(\mathrm{card}(B \cap \{0, 1, \ldots, f(n)\}) \geqslant n\}.$$

Thus $\{A\}$ would be a Π_1^0 class, so that A would be recursive by Lemma 1.4. This contradiction demonstrates the result. □

For any k-retraceable Π_1^0 set, the Π_1^0 class $I_k(A)$ provides an example of a class with C-B rank k.

THEOREM 2.4. *For any set A, $D^k(I_k(A)) = \{A\}$.*

PROOF. This follows by induction since $D(I(A)) = \{A\}$ and, for each k, $D(I_{k+1}(A)) = I_k(A)$. □

It follows that if A is k-retraceable, then A has rank k in $I_k(P)$ and thus has rank $\leqslant k$.

We next give a result which shows how to define a retraceable Π_1^0 set by Π_1^0-recursion.

THEOREM 2.5. *Suppose that the set $A = \{a_0 < a_1 < \cdots\}$ is defined recursively by a Π_1^0 relation $Q(x, y)$ such that, for all n and x, $x = a_n \iff Q(x, \langle a_0, \ldots, a_{n-1}\rangle)$. Then A is a Π_1^0 set and is retraceable.*

PROOF. Define the Π_1^0 relation $R(n, x)$ by

$$R(n, x) \iff (\exists x_0 < \cdots < x_{n-1} < x_n = x)(\forall i < n)Q(x_i, \langle x_0, \ldots, x_{i-1}\rangle).$$

Then the set A is Π_1^0 since

$$a \in A \iff (\exists n \leqslant a)R(n, a).$$

Define the uniformly recursive relations $R_s(n, x)$ as in the definition of R above with Q_s in place of Q.

The counting function Ψ such that $\Psi(a_n) = n$ can be defined by the fact that n is the unique y such that $R(y, a_n)$. Since, given $a \in A$, there is a unique $n \leqslant a$ such that $a = a_n$, $\Psi(a) = n$ may be computed by searching for an s large enough so that $R_s(n, a)$ holds for only one number $n \leqslant a$. □

This result can now be applied to give a quick proof of the following theorem of Dekker and Myhill [1985] (Theorem T3).

THEOREM 2.6 (Dekker–Myhill). *Every r.e. set B is Turing equivalent to a retraceable Π_1^0 set A.*

PROOF. Let the r.e. set B be the union of uniformly recursive sets B_s and define the set A by Π_1^0 recursion as follows. There are two cases in the definition of a_n. If $n \notin B$, then $a_n = a_{n-1} + 1$ and if $n \in B$, then a_n is the least $s > a_{n-1}$ such that $n \in B_s$. It is clear that this is a Π_1^0-recursion, so that A is a Π_1^0 retraceable set. The definition also shows that A is recursive in B. On the other hand, for any n, we have $n \in B \iff n \in B_{a_n}$, so that B is recursive in A. □

It follows from Theorems 2.4 and 2.6 that every non-zero r.e. degree contains a set A of C-B rank one. A slightly better result can be obtained. This was shown by Cenzer, Downey, Jockusch and Shore [1993], hereafter abbreviated as C-D-J-S.

THEOREM 2.7 (C-D-J-S). *Every r.e. non-recursive set B is Turing equivalent to a hypersimple r.e. set E of rank one; furthermore there is a recursive tree U with no dead ends such that $D([U]) = \{E\}$.*

PROOF. Let $A = a_0 < a_1 < \cdots$ be the Π_1^0 retraceable set defined in Theorem 2.6 and let A be the intersection of the uniformly recursive, decreasing sequence A_s. Define the recursive tree S to be a slight extension of the tree T defined in Theorem 2.2. That is, we let Φ be the retracing function given by Theorem 2.1 for A so that $\Phi(a_{n+1}) = a_n$ and so that $\{x \colon \Phi(x) = y\}$ is finite for each y, and define

$$\lfloor c_0, \ldots, c_n, c_{n+1} \rfloor \in S \iff$$
$$c_0 = a_0 \mathbin{\&} (\forall i \leqslant n)[c_i \in A_{c_n} \mathbin{\&} (i > 0 \to \Phi(c_i) = c_{i-1})].$$

S has no dead ends because for any string $\sigma \in S$, it is clear that $\sigma^\frown 0 \in S$.

Let $P = [S]$. We claim that $D(P) = \{A\}$. Since S extends the tree T of Theorem 2.2, it follows that $I(A) \subset P$ and $A \in D(P)$. We claim that any other set $C \in P$ is isolated. First note that if $C = \{c_0 < c_1 < \cdots\} \in P$ is infinite, then by the definition of T, we have $c_i \in A_{c_t}$ for all t, so that each $c_i \in A$. We also have $\Phi(c_{i+1}) = c_i$ for each i, so that, since $c_0 = a_0$, it follows by induction that $c_i = a_i$ for all i and thus $C = A$. Next suppose that $C = \{c_0 < c_1 < \cdots < c_n\} \in P$. Then any $\sigma = \lfloor c_0, c_1, \ldots, c_n, x, y \rfloor \in T$ must have $\Phi(x) = c_n$. Thus there is some upper bound u on all such x. It follows that C is isolated as the unique extension in P of $\lfloor c_0, c_1, \ldots, c_n, u \rfloor$.

To obtain the r.e. set E, we note that the complement function $F(C) = \omega \setminus C$ is a recursive homeomorphism of $\{0,1\}^\omega$ to itself, so that the r.e. set $E = \omega \setminus A$ has rank one in the Π_1^0 class of complements $\{F(C)\colon C \in P\}$. Finally, any retraceable nonrecursive Π_1^0 set is hyperimmune by Theorem 2.3, so that E is hypersimple. □

The 2-retraceable sets may be applied to give the following, which is Theorem 3.2 of Cenzer, Downey, Jockusch and Shore [1993].

THEOREM 2.8 (C-D-J-S). *There is a maximal set A with Cantor–Bendixson rank at most two.*

PROOF. The Friedberg construction (see Soare [1987]) of a co-maximal set $A = \{a_0 < a_1 < \cdots\}$ is modified to yield a co-maximal set A which is second-retraceable. (This suffices by Theorems 2.2 and 2.4.) Specifically, we meet the comaximality requirements in standard fashion and, in addition, ensure that whenever $a_e^s \neq a_e^{s+1}$, then $a_{e+1}^{s+1} \geqslant s+1$. This easily implies that A is second-retraceable with $\Phi(a,b)$ defined to be the unique n with $a = a_n^b$.

The details are left to the reader. □

Of course a co-maximal set Π_1^0 set $A = \{a_0 < a_1 < \cdots\}$ cannot be retraceable, since $B = \{a_{2n}\colon n < \omega\}$ is an infinite Π_1^0 subset of A with $A \setminus B$ also infinite. A concept related to retraceability was used in Theorem 3.3 of Cenzer, Downey, Jockusch and Shore [1993] to show that the rank of A in Theorem 2.8 must be exactly 2. An infinite set A is called *hyperhyperimmune* (*h.h.i.*) if there is no uniformly r.e. sequence $\{U_e\}$ of pairwise disjoint finite sets all intersecting A. Then every co-maximal Π_1^0 set is h.h.i.

THEOREM 2.9 (C-D-J-S). *If A is a Σ_2^0 h.h.i. set, then there is no Π_1^0 class $P \subseteq 2^\omega$ with $D(P) = \{A\}$.*

PROOF. We sketch the part of the proof which involves Π_1^0 classes and sets. Let $A = \{a_0 < a_1 < \cdots\}$ and assume that $D(P) = \{A\}$ and let T be a recursive tree such that $P = [T]$. Then it can be shown that there is an r.e. relation $R(n,x)$ such that

(i) $R(n+1,x) \to R(n,x)$ for all n and x;

(ii) for each $a_n \in A$, $R(i,a_n)$ implies $i \leqslant n$;

(iii) for each n, $R(n,a)$ for all but finitely many $a \in A$.

Observe that for a retraceable Π_1^0 set A, we can let $R(n,x) \iff \Psi(x) \geqslant n$ to obtain a recursive relation satisfying these properties. Then $\Psi(x) = \max_n R(n,x)$. Note that if A is a recursive set, then we may easily obtain such a relation R by $R(i,x) \iff x \geqslant a_i$. Thus we may assume that A is nonrecursive and therefore infinite.

To define $R(n, x)$ in general, first let $n(\sigma, x) = \mathrm{card}(\{m < x\colon \sigma(m) = 1\}$. Then we let $R(n, x)$ if and only if, for every σ with $|\sigma| = a + 1$, if $\sigma(a) = 1$ and $\sigma \in \mathrm{Ext}(T)$, then $n(\sigma, x) \geqslant n$. Thus

$$\max_n R(n, x) = \min\bigl\{n(\sigma, x)\colon \sigma(a) = 1 \;\&\; \sigma \in \mathrm{Ext}(T)\bigr\}.$$

Thus $R(0, x)$ for all x and also $R(n+1, x) \to R(n, x)$. Now for $a = a_n \in A$, $\sigma = \lfloor a_0, a_1, \ldots, a_n, a_n + 1\rfloor \in \mathrm{Ext}(T)$ is a witness that $R(k, x)$ can only hold for $i \leqslant n$.

Finally, suppose that for some n, there are infinitely many $a \in A$ such that $\neg R(n, a)$. For each such a, let $\sigma_a \in \mathrm{Ext}(T)$ such that $\sigma(a) = 1$ and such that $n(\sigma_a) < n$, and let $\sigma_a \prec B_a \in P$. It follows by compactness that P contains a set B which is a limit point of the B_a's, and $B \neq A$ since $\mathrm{card}(B) \leqslant n$.

The remainder of the proof consists of showing that the existence of such a relation $R(n, x)$ for a Σ_2^0 set A precludes A being h.h.i. For details, see Cenzer, Downey, Jockusch and Shore [1993]. □

We conclude this section with another connection between Π_1^0 classes and Π_1^0 sets, via the notion of a *Dedekind cut*. Effective Dedekind cuts were studied by R. Soare [1969].

The elements of ω^ω may be linearly ordered by the lexicographic ordering, where $x \leqslant_L y$ if and only if $x(n) < y(n)$, where n is the least such that $x(n) \neq y(n)$. The elements of $\omega^{<\omega}$ are similarly ordered and may be embedded into ω^ω by taking σ to $\sigma^\frown 0^\omega$. Then we define the *Dedekind cut* $L(x) = \{\sigma \in \omega^{<\omega}\colon \sigma \leqslant_L x\}$. The interval $[x, y]$ is defined to be $\{z\colon x \leqslant_L z \leqslant_L y\}$; $[x, \infty] = \{z\colon x \leqslant_L z\}$.

THEOREM 2.10.

(a) *For any* $x < y$ *in* ω^ω, *the interval* $[x, y]$ *is a* Π_1^0 *class if and only if* $L(x)$ *is a* Σ_1^0 *set and* $L(y)$ *is a* Π_1^0 *set.*

(b) $L(x)$ *has the same Turing degree as* x.

(c) *For* $x = \chi_A \in \{0, 1\}^\omega$, *if* A *is a* Σ_1^0 *(respectively* Π_1^0 *set), then* $L(x)$ *is a* Σ_1^0 *(respectively* Π_1^0*) set.*

PROOF. (a) We begin with $x < y \in \omega^\omega$. We claim that $[x, y]$ is a Π_1^0 class if and only if both $[x, \infty]$ and $[0^\omega, y]$ are Π_1^0 classes. The *if* direction follows from the fact that $[x, y] = [x, \infty] \cap [0, y]$. For the other direction, choose σ and n such that $x\lceil n \leqslant_L \sigma < y\lceil n$. Then the result follows from the fact that $[x, \infty] = [x, y] \cup [\sigma^\frown 1^\omega, \infty]$ and $[0, y] = [0, \sigma^\frown 0^\omega] \cup [x, y]$.

It suffices to show that $[x, \infty)$ is a Π_1^0 class if and only if $L(x)$ is a Σ_1^0 set and that $[0^\omega, y]$ is a Π_1^0 class if and only if $L(y)$ is a Π_1^0 set. We will demonstrate the former of these.

Observe that if $x = \tau^\frown 0^\omega$ for some τ, then of course $[x, \infty]$ is a Π_1^0 class and $L(x)$ is a Σ_1^0 set. Thus we assume that x has infinitely many "ones".

Now suppose that $[x, \infty]$ is a Π_1^0 class, that is, $[x, \infty] = [T]$ for some recursive tree T. Then $\sigma \in L(x)$ if and only if $\sigma^\frown 0^\omega \notin [T]$, which is if and only if $(\exists n)\sigma^\frown 0^n \notin T$. Next suppose that $L(x)$ is a Σ_1^0 set, that is, there is a recursive relation R such that for all σ, $\sigma \leqslant_L x \iff (\exists n)R(n, \sigma)$. Then we have

$$z \in [x, \infty] \iff (\forall m)(z\lceil m \notin L(x)) \iff (\forall m)(\forall n)\neg R(n, z\lceil m),$$

which shows that $[x, \infty]$ is a Π_1^0 class.

The argument that $L(y)$ is a Π_1^0 set if and only if $[0^\omega, y]$ is a Π_1^0 class is similar.

(b) We first see that $L(x)$ is recursive in x, since $\sigma \in L(x) \iff \sigma \leqslant x\lceil|\sigma|$. Next we see that x is recursive in $L(x)$, since for each n, $x(n+1)$ is the least a such that $x\lceil n^\frown a \in L(x)$ & $x\lceil n^\frown a+1 \notin L(x)$.

(c) Suppose that x is the characteristic function of a Π_1^0 set, so that x is the limit of a decreasing sequence $x_0, x_1, \ldots$. Then $\sigma \in L(x) \iff (\forall n)(\sigma \leqslant_L x_n\lceil n)$. Similarly, if x is the characteristic function of a Σ_1^0 set, so that x is the limit of an increasing sequence x_n, then $\sigma \in L(x) \iff (\exists n)(\sigma \leqslant_L x_n\lceil n)$. □

We note that Theorem 2.10 may be relativized to show in particular that $L(x)$ is Δ_2^0 if and only if x is Δ_2^0.

Since a bounded Π_1^0 class is compact, it of course possesses a maximum and a minimum element.

THEOREM 2.11. *For any $x \in \{0,1\}^\omega$, x is the maximum element of some Π_1^0 class if and only if $L(x)$ is a Π_1^0 set and x is the minimum element of some Π_1^0 class if and only if $L(x)$ is a Σ_1^0 set.*

PROOF. Suppose first that $L(x)$ is a Σ_1^0 set. Then it follows from Theorem 2.10 that $[x, \infty]$ is a Π_1^0 class with minimum element x. Similarly if $L(y)$ is Π_1^0, then y is the maximum element of $[0, y]$.

Next suppose that the Π_1^0 class P has a maximum element $x \in \{0,1\}^\omega$. We may assume without loss of generality that $P \subset \{0,1\}^\omega$ and let $P = [T]$, where $T \subset \{0,1\}^{<\omega}$ is recursive. Then $\sigma \leqslant_L x \iff (\forall n)(\exists \tau \in \{0,1\}^n)(\tau \in T \mathbin{\&} \sigma \prec \tau)$.

The argument for a minimum element is similar. □

REMARK. Theorems 2.10 and 2.11 imply that every r.b. Π_1^0 class contains a member of r.e. degree.

3. Basis and anti-basis results

The class $\Gamma \subset \omega^\omega$ is said to be a *basis* for a family Θ of subclasses of ω^ω if every nonempty class from Θ has a member from Γ. For example, the class Δ_0^0 of recursive functions is a basis for the family of open subclasses of ω^ω. This is an example of a "basis theorem". On the other hand, the class of recursive functions is not a basis for the family of closed subclasses of ω^ω since every singleton is a closed class. This is an example of an "anti-basis theorem".

The fundamental result here is König's Lemma, which states that any infinite, finite branching tree has an infinite branch; the effective version is the Kleene Basis Theorem.

THEOREM 3.1 (Kleene). *For any tree T such that the Π^0_1 class $P = [T]$ is nonempty, P contains a member which is recursive in* $\mathrm{Ext}(T)$.

PROOF. The infinite path x through T can be defined recursively by letting $x(0)$ be the least n such that $(n) \in \mathrm{Ext}(T)$ and, for each k, letting $x(k+1)$ be the least n such that $(x(0), \ldots, x(k), n) \in \mathrm{Ext}(T)$. □

THEOREM 3.2. *For any recursive tree $T \subset \omega^{<\omega}$:*
(a) $\mathrm{Ext}(T)$ *is a Σ^1_1 set;*
(b) *if T is finite branching, then* $\mathrm{Ext}(T)$ *is a Π^0_2 set;*
(c) *if T is highly recursive, then* $\mathrm{Ext}(T)$ *is a Π^0_1 set.*

PROOF. (a) In general,

$$\sigma \in \mathrm{Ext}(T) \iff (\exists x)(\forall n > |\sigma|)[x\lceil n \in T \ \&\ \sigma \prec x\lceil n)].$$

(b) If T is finite branching, then König's Lemma implies that

$$\sigma \in \mathrm{Ext}(T) \iff (\forall n > |\sigma|)(\exists \tau)[|\tau| = n \ \&\ \sigma \prec \tau \ \&\ \tau \in T).$$

(c) Finally, suppose that T is highly recursive and let f be a recursive function such that $\sigma(i) < f(i)$ for all $\sigma \in T$ and all $i < |\sigma|$. Then the quantifier "$\exists\tau$" in (b) is bounded, since $\tau(i) < f(i)$ for all $i < n$. □

Combining Theorems 3.1 and 3.2, we get the following.

THEOREM 3.3. *For any nonempty Π^0_1 class $P \subset \omega^\omega$:*
(a) *P has a member recursive in some Σ^1_1 set;*
(b) *if P is bounded, then P has a member recursive in* $\mathbf{0}''$;
(c) *if P is recursively bounded, then P has a member recursive in* $\mathbf{0}'$;
(d) *if $P = [T]$, where T is recursive and has no dead ends, then P has a recursive member.*

Part (a) of this theorem is the Kleene basis theorem [Kleene, 1943] and part (c) is the Kreisel basis theorem [Kreisel, 1953].

Jockusch and Soare [1972a] obtained several important refinements of Theorem 3.3(c). They showed that every nonempty r.b. Π^0_1 class contains members a and b

such that any function recursive in both a and b is recursive; if a and b are nonrecursive, then they form a *minimal pair*. They also showed in [1972b] that any *r.b.* Π_1^0 class contains a hyperimmune member a such that $\mathbf{a}'' = \mathbf{0}''$. Note also that Shoenfield [1960] showed first that any nonempty r.b. Π_1^0 class P contains a member a which has degree $\mathbf{a} < \mathbf{0}'$. We state the minimal pair result in order to apply it in Section 6.

THEOREM 3.4. *Every nonempty r.b.* Π_1^0 *class with no recursive member contains a minimal pair.*

Perhaps the most cited result in the theory of Π_1^0 classes is the Low Basis Theorem of Jockusch and Soare [1972a]. We will use the notion of a *generic* real to prove this theorem. An element $x \in \{0, 1\}^*$ is said to be 1-*generic* if, for every Π_1^0 class P, there exists n such that either $I(x\lceil n) \subset P$ or $I(x\lceil n) \cap P = \emptyset$.

The existence of a generic real is obtained by the method known as *forcing with* Π_1^0 *classes*. See Section III.6 of Hinman [1978] for a full development. A subset D of $\{0, 1\}^{<\omega}$ is said to be *dense* if, for any σ, there exists a $\tau \succ \sigma$ such that $\tau \in D$. Now let $\{D_i\colon i \in I\}$ be a family of dense sets. The element x of $\{0, 1\}^{\omega}$ is said to be *generic* for this family if, for each i, $x\lceil n \in D_i$ for some n.

The standard forcing theorem shows that any countable family of dense sets possesses a generic set. If $\{D_i\colon i < \omega\}$ of dense sets is uniformly recursive in $\mathbf{0}'$, then we can construct a generic $x \leqslant_T 0'$ by letting σ_0 be the least $\sigma \in D_0$ and for each n, letting σ_{n+1} be the least proper extension of σ_n such that $\sigma_{n+1} \in D_{n+1}$, where the strings σ are ordered according to their codes $\langle\sigma\rangle$.

Recall that by Lemma 1.2, we may enumerate the Π_1^0 classes P_e by letting $P_e = [T_e]$, where T_e is the e-th primitive recursive tree. Then the standard family of dense sets is now

$$D_i = \big\{\sigma\colon \sigma \notin T_i \ \vee\ (\forall \tau \succeq \sigma)\tau \in T_i\big\}.$$

Observe each D_i is dense and that this sequence of sets is uniformly Π_1^0. The element $x \in \{0, 1\}^{\omega}$ is 1-generic if it is generic for this sequence of dense sets. Then the remarks above imply the existence of a 1-generic real. The crucial property of a 1-generic real is given by the following well-known fact.

THEOREM 3.5. *For any* 1*-generic* $x \in \{0, 1\}^{<\omega}$, *if* $x \leqslant \mathbf{0}'$, *then* $x' \leqslant x \oplus \mathbf{0}'$.

PROOF. Let x be 1-generic. For each e, let the Π_1^0 class $Q_e = \{y\colon \phi_e^y(e)\uparrow\}$, so that $Q_e = [U_e]$, where $\sigma \in U_e \iff \phi_e^{\sigma}(e)\uparrow$. Thus $e \in x' \iff x \notin Q_e$. If $e \in x'$, then of course there is some n such that $x\lceil n \notin U_e$. Since x is 1-generic, if $e \notin x'$, then there is some n such that $(\forall \tau \succeq x\lceil n)\tau \in U_e$. Let $f(e)$ be the least n such that either $x\lceil n \notin U_e$ or $(\forall \tau \succeq x\lceil n)\tau \in U_e$. Then f is recursive in $x \oplus \mathbf{0}'$. But then we have $e \in x'$ if and only if $x\lceil f(e) \notin U_e$, so that x' is also recursive in $x \oplus \mathbf{0}'$. □

It follows that if a 1-generic real x is recursive in $\mathbf{0}'$, then $x' = \mathbf{0}'$, that is, x is low. For the low basis theorem, a modification of this argument is used.

THEOREM 3.6 (Low Basis Theorem (Jockusch and Soare)).

(i) *Every nonempty r.b. Π^0_1 class P contains a member of low degree.*

(ii) *There is a low degree* **a** *such that every nonempty r.b. Π^0_1 class contains a member of degree $\leqslant$* **a**.

PROOF.

(i) By Lemma 1.3, we may assume that $P \subset \{0,1\}^\omega$. Let $P = [T]$ and, as above, let $\sigma \in U_e \iff \phi^\sigma_e(e)\uparrow$. We will define, recursively in $\mathbf{0}'$, a sequence $T = S_0 \supset S_1 \supset S_2 \supset \cdots$ of infinite subtrees of T and show that any member x of $\bigcap_e [S_e]$ has low degree. There are two cases in the definition of S_{e+1}.

(a) If $S_e \cap U_e$ is finite, then $S_{e+1} = S_e$.

(b) If $S_e \cap U_e$ is infinite, then $S_{e+1} = S_e \cap U_e$.

Observe that this construction is recursive in $\mathbf{0}'$, in that there is a function $f \leqslant \mathbf{0}'$ such that $S_e = U_{f(e)}$ for each e. This is because the determination of whether $S_e \cap U_e$ is finite can be made using a $\mathbf{0}'$ oracle. Since each S_e is infinite by the construction, it follows that each $[S_e]$ is nonempty, so that $\bigcap_e [S_e]$ is nonempty. Now suppose that $x \in \bigcap_e [S_e]$. Then, for any e, we have $e \in x'$ if and only if $x \notin [U_e]$, and it follows from the construction that $x \notin [U_e]$ if and only if $S_e \cap U_e$ is finite. It follows by the observation above that $x' \leqslant_T 0'$.

(ii) It suffices to prove the result for binary classes by Lemma 1.3. Let $P_e = [T_e]$ be an effective enumeration of the binary Π^0_1 classes and let p_e be the e-th prime number. Let the tree T be the amalgamation of the nonempty Π^0_1 classes, in the following sense. Let $\sigma \in T$ if, for each e such that P_e is nonempty and each k such that $p^k_e < |\sigma|$, $(\sigma(p_e), \sigma(p^2_e), \ldots, \sigma(p^k_e)) \in T_e$. Since we can test recursively in $\mathbf{0}'$ whether P_e is nonempty, the tree T is recursive in $\mathbf{0}'$. Thus the construction above can be carried out to produce a member x of $P = [T]$ of low degree. Then any nonempty class P_e has a member $(x(p_e), x(p^2_e), \ldots)$ recursive in x. □

The same technique can be used to prove other basis results. For example, a generalization shows that if the r.b. class P contains no recursive member, then for any degree $\mathbf{b}$, P has a member A of degree $\mathbf{a}$ such that $\mathbf{a} \oplus \mathbf{0}' = \mathbf{a}' = \mathbf{b}'$. It follows (for $\mathbf{b} = \mathbf{0}'$) that any Π^0_1 class P has a member A of degree $\mathbf{a}$ such that $\mathbf{a} \oplus \mathbf{0}' = \mathbf{a}' = \mathbf{0}''$. Now a degree $\mathbf{a}$ is said to be *high* if $\mathbf{a} \leqslant \mathbf{0}'$ and $\mathbf{a}' = \mathbf{0}''$. A natural question is whether every r.b. Π^0_1 class contains a member of high degree. This question will be answered in the negative by Theorem 5.4.

The following result is from Jockusch and Soare [1972a].

THEOREM 3.7 (Jockusch and Soare). *Every nonempty r.b. Π^0_1 class P contains a member of hyperimmune-free degree.*

PROOF. We sketch the proof indicated in Soare [1987, p. 109]. Let $P = [T]$, where T is an infinite recursive binary tree. A is hyperimmune-free if every function f recursive in A is majorized by some recursive function. Thus we want to find $A \in P$ such that ϕ^A_e, whenever total, is majorized by a recursive function. We define the decreasing sequence S_e of recursive subtrees of T beginning with $S_0 = T$. Then for each e and i, let $U_e^i = \{\sigma \in S_e\colon \phi_e^\sigma(i)\uparrow\}$. There are again two cases in the definition of S_{e+1}.

(i) If U_e^i is finite for every i, let $S_{e+1} = S_e$.

(ii) If U_e^i is infinite for some i, choose such an i and let $S_{e+1} = S_e \cap U_e^i$.

Suppose now that $A \in \bigcap_e S_e$ and that $f = \phi_e^A$ is total. If the second case applied in the definition of S_{e+1}, then $\phi_e^A(i)\uparrow$, so that ϕ_e^A is not total. If the first case applied, then to majorize $\phi_e^A(i)$, we look for n such that $\phi_e^\sigma(i)$ converges for every string $\sigma \in S_{e+1}$ of length n—then $\phi_e^A(i) \leqslant \max\{\phi_e^\sigma(i)\colon \sigma \in T \;\&\; |\sigma| = n\}$. □

For countable Π_1^0 classes, the results of Theorem 3.3 were significantly improved by Kreisel [1959].

THEOREM 3.8 (Kreisel). *Let P be a Π_1^0 class.*

(a) *Any isolated member of P is hyperarithmetic; if P is finite, then every member of P is hyperarithmetic.*

(b) *Suppose that P is bounded. Then any isolated member of P is recursive in $0'$; if P is finite, then every member of P is recursive in $0'$.*

(c) *Suppose P is recursively bounded. Then any isolated member of P is recursive; if P is finite, then every member of P is recursive.*

PROOF. As in the proof of Lemma 1.4, we see that if x is isolated in P, then $\{x\} = [T]$ for some recursive tree T. It follows that $\mathrm{Ext}(T) = \{x\lceil n\colon n < \omega\}$, so that x is recursive in $\mathrm{Ext}(T)$. Now consider the three cases:

(a) For an (unbounded) tree T, $\mathrm{Ext}(T)$ is Σ_1^1 by Theorem 3.2 and we have

$$\sigma \in \mathrm{Ext}(T) \iff (\forall \tau \in \omega^{|\sigma|}(\tau \neq \sigma \to \tau \notin \mathrm{Ext}(T)),$$

so that $\mathrm{Ext}(T)$ is also Π_1^1. It follows that $\mathrm{Ext}(T)$ and hence x are hyperarithmetic.

(b) For a finite branching tree T, $\mathrm{Ext}(T)$ is Π_2^0 by Theorem 3.2. Now it follows from Konig's Lemma that for each n, there is some $k \geqslant n$ such that every sequence in T of length k is an extension of $x\lceil n$. Thus we have

$$\sigma \in \mathrm{Ext}(T) \iff (\exists k \geqslant |\sigma|)(\forall \tau \in \omega^k)(\tau \in T \to \sigma \prec \tau),$$

so that $\mathrm{Ext}(T)$ is also Σ_2^0. It follows that $\mathrm{Ext}(T)$ and hence x is recursive in $0'$.

(c) In this case, since $\{x\}$ is a r.b. Π_1^0 class, x is recursive by Lemmas 1.3 and 1.4.

Suppose now that P is finite. The conclusion in each case follows from the fact that every member of P will be isolated. □

Now it is a classical result of descriptive analysis that any perfect set in a Polish space must have cardinality of the continuum (see Moschovakis [1980, p. 16]), so that any countable closed set must have an isolated point. Thus we obtain the following corollary.

THEOREM 3.9 (Kreisel). *Let P be a countable Π_1^0 class.*

(a) *P has a hyperarithmetic member.*

(b) *If P is bounded, then P has a member recursive in* $0'$.

(c) *If P is recursively bounded, then P has a recursive member.*

Next we consider some *anti-basis* results. It follows from the construction of recursively inseparable r.e. sets by Kleene [1950] that there is a Π_1^0 class of sets with no recursive member. We give an improvement of this result due to Jockusch [1974]. Recall that a set A is *immune* if it has no infinite recursive subset; A is said to be *bi-immune* if both A and $\omega \setminus A$ are immune. Since any infinite r.e. set has an infinite recursive subset, it follows that the difference of two r.e. sets cannot be bi-immune and then by induction that no Boolean combination of r.e. sets can be bi-immune. Let D_n denote the n'th finite set, as in Soare [1987, p. 33].

THEOREM 3.10 (Jockusch). *There is a nonempty Π_1^0 class of sets containing only bi-immune sets.*

PROOF. Let W_e be the eth r.e. set and let D_n be the n-th finite set. Let ψ be a partial recursive function such that, whenever $|W_e| \geqslant e+3$, then $\psi(e)$ is defined and $D_{\psi(e)} \subset W_e$ and $|D_{\psi(e)}| = e+3$. Define the Π_1^0 class $P = \bigcap_e P_e$, where $A \in P_e$ if and only if, if $\psi(e)$ is defined, then $A \cap D_{\psi(e)} \neq \emptyset$ and $(\omega \setminus A) \cap D_{\psi(e)} \neq \emptyset$. Any element A of P is clearly bi-immune. To see that P is nonempty, note that for each e, $\{0, 1\}^\omega \setminus P_e$ has measure $\leqslant 2^{-e-2}$. (For $A \notin P_e$, either all $e+3$ elements of $D_{\psi(e)}$ are in A or all $e+3$ elements are not in A, which allows only 2 of the 2^{e+3} possibilities.) It follows that $\{0, 1\}^\omega \setminus P$ has measure $\leqslant \sum_e 2^{-e-2} = \frac{1}{2}$, so that $P \neq \emptyset$. □

THEOREM 3.11 (Jockusch). *There is a nonempty r.b. Π_1^0 class with no member a Boolean combination of r.e. sets.*

PROOF. This is immediate from Theorem 3.7 and the fact that no Boolean combination of r.e. sets is bi-immune. □

Recall that the set A is *effectively immune* if there is a recursive function g such that for any e, if $W_e \subset A$, then $|W_e| \leqslant g(e)$. For the Π_1^0 class P constructed in the proof of Theorem 3.10, it is clear that any set $A \in P$ is effectively bi-immune via the function $g(e) = e+3$.

The following corollary is immediate.

THEOREM 3.12. *There is a nonempty r.b. Π_1^0 class with no member of r.e. degree strictly below* $\mathbf{0}'$.

PROOF. Let P be the Π_1^0 class defined in the proof of Theorem 3.10. The remarks above show that every member of P is effectively immune. Now if P had a member

A of r.e. degree, then it follows from an exercise in Soare [1987, p. 87] that A would have degree $\mathbf{0}'$. □

Next we consider Π_1^0 classes with no recursive members, which were studied by Jockusch and Soare [1972a, 1972b]. We will give a number of results here without proof. It was shown in [Jockusch and Soare, 1972a] that if the Π_1^0 class P has no recursive members, then there is some non-zero r.e. degree $\mathbf{a}$ such that P has no members of degree $\leqslant \mathbf{a}$.

It follows from Lemma 1.4 that if P has no recursive members, then it has no isolated members, so that P is perfect and therefore contains continuum many elements. This was improved by Jockusch and Soare [1972a] as follows.

THEOREM 3.13 (Jockusch–Soare). *For any r.b. Π_1^0 class P with no recursive members and any countable set $\{\mathbf{a_i}\colon i < \omega\}$ of nonrecursive degrees, P has continuum many mutually incomparable members x such that the degree of x is incomparable with each $\mathbf{a}_i$; in particular, P has members a, b such that $\mathbf{a} \wedge \mathbf{b} = \mathbf{0}$.*

For any Π_1^0 class P, let $\mathcal{D}(P) = \{x\colon (\exists y \in P)(x \equiv_T y)\}$. For later use, we will say that two classes P and Q are *degree-isomorphic* if $\mathcal{D}(P) = \mathcal{D}(Q)$.

A r.b. Π_1^0 class P is constructed in [Jockusch and Soare, 1972a] such that $\mathcal{D}(P)$ has Lebesgue measure 1. On the other hand, they also show that, even for an arbitrary (possibly unbounded) Π_1^0 class P, the set $\mathcal{D}(P)$ is meager.

There are several constructions of r.b. classes with no recursive members in [Jockusch and Soare, 1972a, 1972b, 1971].

THEOREM 3.14 (Jockusch–Soare). (a) *There is a nonempty r.b. Π_1^0 class P such that for any degree $\mathbf{a}$ of a member of P and any r.e. degree $\mathbf{b} \geqslant \mathbf{a}$, $\mathbf{b} = \mathbf{0}'$.*

(b) *For any r.e. degree $\mathbf{c}$, there is a r.b. Π_1^0 class P such that the r.e. degrees of members of P are exactly those $\geqslant \mathbf{c}$.*

(c) *For any degree $\mathbf{a}$, there is a r.b. Π_1^0 class P with no recursive members and with no members of degree $\mathbf{a}$.*

(d) *There is a nonempty r.b. Π_1^0 class P such that any two members of P are Turing incomparable.*

Kucera [1986] constructed a nonempty Π_1^0 class P such that, for any degree $\mathbf{a} \leqslant \mathbf{0}'$ of a member of P, there is a nonrecursive r.e. degree $\mathbf{b}$ with $\mathbf{b} \leqslant \mathbf{a}$.

Recently, Groszek and Slaman [ta3] constructed a Π_1^0 class P such that for every $x \in P$, there is a minimal degree below $\mathbf{x}$.

We next consider some notions of measure and randomness associated with binary Π_1^0 classes.

First observe that if r is the measure of Π_1^0 class P, then $L(r)$ is a Π_1^0 set, since if $P = [T]$, then r can be approximated from above by the measure of the clopen sets $P_n = \bigcup\{I(\sigma)\colon |\sigma| = n \;\&\; \sigma \in T\}$.

THEOREM 3.15. *Any Π^0_1 class with positive, recursive measure has a recursive member.*

PROOF. Let P have recursive measure $r > 0$. It follows that the measure of $P \cap I(\sigma)$ is recursive uniformly in σ, since we can approximate the measure from below by subtracting the measure of the complement from r. Thus we can recursively select paths $\sigma = (x(0), \ldots, x(n))$ of length n such that $P \cap I((x(0), \ldots, x(n)))$ always has measure $\geqslant r/2^n$. The infinite path x will be a recursive member of P. □

The following result is given on p. 110 of Hinman [1978].

THEOREM 3.16. *There is a Π^0_1 class with positive measure containing no recursive member (and therefore having nonrecursive measure).*

The notions of category are often considered analogous to that of measure. Recall that a class A is *dense* in an interval I if the closure of A contains I, that A is nowhere dense if it is not dense in any interval and that A is meager, or first category, if it is the countable union of nowhere dense sets. The complement of a meager set is said to be co-meager. It is easy to see that a closed set is non-meager if and only if it includes an interval. Thus any non-meager closed set has a recursive member. In fact, it is shown in Hinman [1978, p. 130], that every non-meager Σ^0_3 class contains a recursive real.

We now consider three related notions of randomness. We have already encountered the notion of a 1-generic real. The following result summarizes a few properties of the set of 1-generic reals.

THEOREM 3.17.
(i) *The class of* 1*-generic reals is comeager.*
(ii) *There are continuum many* 1*-generic reals in each interval.*
(iii) *Every non-meager class contains a* 1*-generic real.*
(iv) *The class of* 1*-generic reals has measure* 0.
(v) *The* 1*-generic reals are not a basis for the Π^0_1 classes of positive measure.*

PROOF. (i) A real x is 1-generic if, for each Π^0_1 class P_e, either $x \notin P_e$ or x is in the interior of P_e, that is, x is not in the boundary of P_e. Thus the class of non-1-generic reals is the union of the boundaries of the P_e, each of which is nowhere dense.

(ii) and (iii) are immediate from (i).

(iv) It is easy to construct a Π^0_1 class P having a boundary with arbitrarily large measure. For example, let U_n be the class of reals x such that $x(i) = 0$ for all i with $n(n+1) \leqslant i < (n+1)(n+2)$, so that U_n has measure 4^{-n-1}. Now let

$$P = \{0, 1\}^\omega \setminus \bigcup_n U_n.$$

Then P is a Π^0_1 class with no interior and with measure 2/3.

(v) The Π_1^0 class constructed above with measure 2/3 has no 1-generic members. □

Let us define a *random* real to be any $x \in \{0, 1\}^\omega$ such that x does not belong to any Π_1^0 class of measure 0. It follows from this definition that the class of random reals is simply the intersection of all Σ_1^0 classes of measure 1. Thus the set of random reals itself has measure one and therefore **any** set of positive measure must contain a random real. It is also clear that the set of random reals is co-meager, since any Π_1^0 class of measure 0 is nowhere dense. Recall that if a 1-generic real x belongs to a Π_1^0 class P, then P includes some interval and thus has positive measure. Thus any 1-generic real is random.

P. Martin-Löf [1966] introduced the notion of a 1-*random* real. The real x is said to be 1-random if for any recursive function f such that $\mu(P_{f(n)}) > 1 - 2^{-n}$ for each n, there is some n such that $x \in P_{f(n)}$. It is not hard to see that any 1-random real is random. Martin-Löf proved that there is a universal, increasing sequence P_{e_n} of Π_1^0 classes such that $\mu(P_{e_n}) > 1 - 2^{-n}$ and such that $\bigcup_n P_{e_n}$ is precisely the class of 1-random reals. Thus the class of 1-random reals also has measure 1. The degree **a** of a 1-random set will be called a 1-*random degree*. Recall that a function $f : \omega \to \omega$ is said to be *fixed-point-free* if there is no e such that $\phi_e = \phi_{f(e)}$; the degree of a fixed-point-free function is said to be a *fixed-point-free degree*. The following results are taken from Kucera [1985].

THEOREM 3.18 (Kucera).
(i) *Any* Π_1^0 *class P of positive measure has members of every* 1*-random degree.*
(ii) *Every* 1*-random degree is a fixed-point-free degree.*
(iii) *Every degree* **a** *above* $\mathbf{0}'$ *is a* 1*-random degree.*

The results of this section can all be relativized to classes $\Pi_1^{0^A}$, that is, classes $P = [T]$ where T is a tree recursive in A. For example, the relativized low basis theorem says that for any tree T which is recursive in A and bounded by a function recursive in A, if $[T]$ is nonempty, then it has a member x such that $x' \leqslant_T A'$. The basis problem for Π_1^{0A} classes can be discussed using the notation $\mathbf{a} \ll \mathbf{b}$ to mean that every nonempty Π_1^{0A} class of sets contains a member of degree $\leqslant \mathbf{b}$.

This notion is discussed by Simpson [1977], where the following results are given. The first theorem is elementary.

THEOREM 3.19.
(i) $\mathbf{a} \ll \mathbf{b}$ *implies* $\mathbf{a} < \mathbf{b}$.
(ii) $\mathbf{a} \leqslant \mathbf{b} \ll \mathbf{c} \leqslant \mathbf{d}$ *implies* $\mathbf{a} \ll \mathbf{d}$.
(iii) $\mathbf{a} \ll \mathbf{b}$ *and* $\mathbf{b} \ll \mathbf{c}$ *implies* $\mathbf{a} \ll \mathbf{c}$.
(iv) $\mathbf{a} \ll \mathbf{a}'$ *for all* $\mathbf{a}$.

THEOREM 3.20 (Jockusch and Soare [1972a]).

(i) *If* $\mathbf{a} \ll \mathbf{b}$, *then every countable partially ordered set is embeddable in* $\{\mathbf{d}\colon \mathbf{a} < \mathbf{d} < \mathbf{b}\}$.

(ii) *If* $\mathbf{a} \ll \mathbf{b}$ *and* $\mathbf{b}$ *is r.e. in* $\mathbf{a}$, *then* $\mathbf{a}' = \mathbf{b}$.

(iii) *For all* $\mathbf{a}$, *there exists* $\mathbf{b}$ *such that* $\mathbf{a} \ll \mathbf{b}$ *and* $\mathbf{a}' = \mathbf{b}'$.

PROOF. (i) This follows from the relativized version of Theorem 3.14.

(ii) This follows from the relativized version of Theorem 3.14(a).

(iii) This follows from a uniform version of the low basis theorem. □

Simpson also states in [Simpson, 1977] the following two results. First, that if $\mathbf{a} \ll \mathbf{b}$, then $\mathbf{a} \ll \mathbf{c} \ll \mathbf{b}$ for some c. Second, that for all $\mathbf{a}$ and all $\mathbf{b} \geqslant \mathbf{a}$, there exists $\mathbf{c} \gg \mathbf{a}$ such that $\mathbf{a} = \mathbf{b} \wedge \mathbf{c}$.

We give one more result on the ordering "$\ll$" from the Simpson article which is useful for applications of Π^0_1 classes.

THEOREM 3.21. *For any degree* $\mathbf{b}$, *the following are equivalent*

(i) $\mathbf{b} \gg \mathbf{0}$;

(ii) $\mathbf{b}$ *is the degree of a complete extension of Peano Arithmetic*;

(iii) $\mathbf{b}$ *is the degree of a set which separates some effectively inseparable, disjoint pair of r.e. sets.*

PROOF. The implication from (ii) to (i) is the Scott Basis Theorem [Scott, 1962].

The reverse implication from (i) to (ii) is due to Solovay (unpublished). Suppose that $\mathbf{b} \gg \mathbf{0}$. Since the complete extensions of Peano Arithmetic form a Π^0_1 class, there exists a complete extension T of Peano arithmetic of degree $< \mathbf{b}$. The result follows from the upward closure of the degrees of complete extensions of Peano Arithmetic. A proof of this upward closure is given on p. 511 of Odifreddi [1989].

The implication from (ii) to (iii) is simply the observation that any complete extension of Peano Arithmetic separates the set of theorems from the set of negations of theorems.

The implication from (iii) to (ii) follows from Theorem 6.3 below, that any Π^0_1 class of sets may be represented as the class of complete extensions of some theory T, together with the result of Jockusch and Soare [1972a] that the class of complete extensions of any two effectively inseparable theories are degree-isomorphic. □

More generally, one can consider the following notion. A nonempty set S of subsets of ω is said to be a *Scott Set* if it satisfies the following three conditions.

(i) If A and B are in S, then their disjoint union $A \oplus B$ is in S.

(ii) If $B \in S$ and $A \leqslant_T B$, then $A \in S$.

(iii) If $A \in S$, then there exists $B \in S$ with $A \ll B$.

For example, if M is a model of Peano Arithmetic, the Scott set of M, $\{A \cap \omega : A \in \mathcal{M}\}$ is easily shown to satisfy the conditions above. On the other hand, it can be shown that any countable Scott set is the Scott set of some model of Peano Arithmetic.

4. Cantor–Bendixson rank

In this section, we develop the notion of the Cantor–Bendixson (C-B) rank to give a finer analysis of the complexity of the members of a Π^0_1 class. The classical Cantor–Bendixson theorem states that any closed subset P of the Cantor space $\{0, 1\}^\omega$ is the union of a perfect set $K(P)$, called the *perfect kernel* of P, with a countable set S and that $K(P) = \bigcap_\alpha D^\alpha(P)$. For $A \in S$, the C-B rank $\mathrm{rk}_P(A)$ of A in P is the least ordinal α such that $A \in D^\alpha(P) \setminus D^{\alpha+1}(P)$; the C-B rank $\mathrm{rk}(P)$ is the least α such that $D^{\alpha+1}(P) = D^\alpha(P)$ (which $= K(P)$). The set A is *ranked* if there is a Π^0_1 class P such that $A \in P \setminus K(P)$, and the C-B rank $\mathrm{rk}(A)$ is the least α such that $\mathrm{rk}_P(A) = \alpha$ for some Π^0_1 class P. Kreisel [1959] used the Boundedness Principle of Spector to show that $\mathrm{rk}(P) \leqslant \omega_1^{\mathrm{C-K}}$ for any Π^0_1 class P, so that any ranked point A has $\mathrm{rk}(A) < \omega_1^{\mathrm{C-K}}$. For the sake of completeness, let $\mathrm{rk}_P(A) = \mathrm{rk}(P)$ if $A \in K(P)$. This means that the rank function will define a prewellordering on P.

We begin with a series of lemmas from [Cenzer and Smith, 1989].

LEMMA 4.1. *For any compact subset Q of $\{0, 1\}^\omega$, any clopen set K and any ordinal α,*

(a) $D^\alpha(K \cap Q) = K \cap D^\alpha(Q)$.

(b) $\mathrm{rk}_{K\cap Q}(A) = \mathrm{rk}_Q(A)$ *for any A.*

LEMMA 4.2. *For any set A and any recursive ordinal α,* $\mathrm{rk}(A) \leqslant \alpha$ *if and only if there is some Π^0_1 class P such that $D^\alpha(P) = \{A\}$.*

PROOF. The "if" direction is immediate from the definition of rank. Suppose now that $\mathrm{rk}(A) \leqslant \alpha$, so that $A \in D^\alpha(Q) \setminus D^{\alpha+1}(Q)$ for some Π^0_1 class Q. Thus A is isolated in $D^{\alpha(Q)}$, so that for some interval I, $I \cap D^\alpha(Q) = \{A\}$. Let $P = I \cap Q$. It follows from Lemma 4.1 that

$$D^\alpha(P) = D^\alpha(I \cap Q) = I \cap D^\alpha(Q) = \{A\}.$$

□

LEMMA 4.3.

(a) *Let Φ be a continuous map from $\{0, 1\}^\omega$ into $\{0, 1\}^\omega$ and let P, Q be compact sets such that $\Phi(P) = Q$. Then for any $y \in Q$,* $\mathrm{rk}_Q(y) \leqslant \max\{\mathrm{rk}_P(x)\colon x \in P \;\&\; \Phi(x) = y\}$.

(b) *For any sets A, B, if $A \leqslant_{tt} B$ and B is ranked, then A is ranked and* $\mathrm{rk}(A) \leqslant \mathrm{rk}(B)$.

(c) *For any sets A, B, if $A \equiv_{tt} B$ and B is ranked, then A is ranked and* $\mathrm{rk}(A) = \mathrm{rk}(B)$.

PROOF. Let $\mathrm{rk}_Q(y) = \beta$, so that $y \in D^\beta(Q) \setminus D^{\beta+1}(Q)$. It is shown in Lemma 1.2 of [Cenzer and Smith, 1989] that $D^\alpha(\Phi(P)) \subset \Phi(D^\alpha(P))$ for any α. Since $y \in$

$D^{\beta}(Q)$, it follows that $y = \Phi(x)$ for some $x \in D^{\beta}(P)$, so that $\mathrm{rk}_Q(y) = \beta \leqslant \mathrm{rk}_P(x)$, where it is possible that x is not ranked in P.

(b) By a theorem of Nerode and Trakhtenbrot [1957], if $A \leqslant_{tt} B$, then there is a recursive function $\Phi : \{0, 1\}^{\omega} \to \{0, 1\}^{\omega}$ such that $\Phi(B) = A$. Now let $\mathrm{rk}(B) = \alpha$ and, by Lemma 4.2, let P be a Π^0_1 class such that $D^{\alpha}(P) = \{B\}$, so that $\mathrm{rk}_P(B) = \max\{\mathrm{rk}_P(x)\colon x \in P\}$, and let $Q = \Phi(P)$. It follows from (a) that $\mathrm{rk}(A) \leqslant \mathrm{rk}_Q(A) \leqslant \mathrm{rk}_P(B) = \mathrm{rk}(B)$.

(c) This is immediate from (b). □

The following improvement of Lemma 1.4 of [Cenzer and Smith, 1989] is due to J. Owings, and C. Laskowski [1997]. Recall that $\alpha \oplus \beta$ is the Hessenberg sum of two ordinals, that $A \oplus B$ is the disjoint union of two sets and that $P \oplus Q = \{A \oplus B\colon A \in P \;\&\; B \in Q\}$ for two classes P and Q of sets.

THEOREM 4.1 (Owings). *For any sets $A, B \in \{0, 1\}^{\omega}$ and any compact $P, Q \subset \{0, 1\}^{\omega}$, $\mathrm{rk}_{P\oplus Q}(A \oplus B) = \mathrm{rk}_P(A) \oplus \mathrm{rk}_Q(B)$.*

PROOF. We first show by induction on $\mathrm{rk}_P(A) \oplus \mathrm{rk}_Q(B)$ that $\mathrm{rk}_{P\oplus Q}(x \oplus y) \leqslant \mathrm{rk}_P(A) \oplus \mathrm{rk}_Q(B)$. If $\mathrm{rk}_P(A) \oplus \mathrm{rk}_Q(B) = 0$, then A is isolated in P and B is isolated in Q, so that there are open intervals I and J such that $P \cap I = \{A\}$ and $Q \cap J = \{B\}$. It follows that $(P \oplus Q) \cap (I \oplus J) = \{A \oplus B\}$, so that $\mathrm{rk}_{P\oplus Q}(A \oplus B) = 0$. Now let $\mathrm{rk}_P(A) = \alpha$ and $\mathrm{rk}_Q(B) = \beta$ and suppose the inequality holds for all x, y such that $\mathrm{rk}_P(x) \oplus \mathrm{rk}_Q(y) < \alpha \oplus \beta$. By intersecting with open intervals, as above, we may assume that $D^{\alpha}(P) = \{A\}$ and that $D^{\beta}(Q) = \{B\}$. It suffices to show that $\mathrm{rk}_{P\oplus Q}(x \oplus y) < \alpha \oplus \beta$ for all $x \oplus y \neq A \oplus B$ in $P \oplus Q$. But if $x \oplus y \neq A \oplus B$, then either $x \neq A$ or $y \neq B$, so that either $\mathrm{rk}_P(x) < \alpha$ or $\mathrm{rk}_Q(y) < \beta$. In either case, $\mathrm{rk}_P(x) \oplus \mathrm{rk}_Q(y) < \alpha \oplus \beta$, so that $\mathrm{rk}_{P\oplus Q}(x \oplus y) < \alpha + \beta$.

For the reverse inequality, we prove by induction on $\alpha \oplus \beta$ that

$$D^{\alpha}(P) \oplus D^{\beta}(Q) \subset D^{\alpha\oplus\beta}(P \oplus Q).$$

For $\alpha \oplus \beta = 0$, this is obvious. We also need the case where $\alpha \oplus \beta = 1$. Suppose without loss of generality that $\alpha = 1$ and $\beta = 0$ and suppose that $A \in D(P)$ and $B \in Q$. Then for any interval $I \subset \{0, 1\}^{\omega}$, there is some $A' \neq A$ in $P \cap I$. Then for any basic open set $I \oplus J \in \{0, 1\}^{\omega} \oplus \{0, 1\}^{\omega}$, there is an element $A' \oplus B \neq A \oplus B$ in $(P \oplus Q) \cap (I \oplus J)$. Thus $A \oplus B \in D(P \oplus Q)$. This shows that

$$D(P) \oplus Q \subset D(P \oplus Q).$$

Now suppose the inclusion holds for all ordinals σ, τ with $\sigma \oplus \tau < \alpha \oplus \beta$. There are two cases.

Case 1. If $\alpha \oplus \beta$ is a limit ordinal, then α and β are both limit ordinals and $\alpha \oplus \beta = \sup\{\sigma \oplus \tau\colon \sigma < \alpha \ \&\ \tau < \beta\}$. Thus

$$\begin{aligned} D^{\alpha\oplus\beta}(P \oplus Q) &= \bigcap_{\gamma<\alpha\oplus\beta} D^{\gamma}(P \oplus Q) = \bigcap_{\sigma<\alpha,\,\tau<\beta} D^{\sigma\oplus\tau}(P \oplus Q) \\ &= \bigcap_{\sigma<\alpha,\,\tau<\beta} D^{\sigma}(P) \oplus D^{\tau}(Q) = D^{\alpha}(P) \oplus D^{\beta}(Q). \end{aligned}$$

Case 2. If $\alpha \oplus \beta$ is a successor, then either α is a successor or β is a successor – without loss of generality say that $\alpha = \gamma + 1$, so that $\alpha \oplus \beta = (\gamma \oplus \beta) + 1$. Then

$$\begin{aligned} D^{\alpha}(P) \oplus D^{\beta}(Q) &= D(D^{\gamma}(P)) \oplus D^{\beta}(Q) \subset D(D^{\gamma}(P) \oplus D^{\beta}(Q)) \\ &\subset D(D^{\gamma\oplus\beta}(P \oplus Q)) = D^{\alpha\oplus\beta}(P \oplus Q). \end{aligned}$$

□

Thus we have the following corollary.

THEOREM 4.2. *For any sets A and B,* $\max\{\mathrm{rk}(A), \mathrm{rk}(B)\} \leqslant \mathrm{rk}(A \oplus B) \leqslant \mathrm{rk}(A) \oplus \mathrm{rk}(B)$.

PROOF. The first inequality follows from Lemma 4.3(b), since both A and B are $\leqslant_{tt} A \oplus B$. The second inequality follows from Theorem 4.1. □

Now we turn to the main goal of the section, which is a comparison of the rank of a set with its hyperarithmetic complexity. The primary focus will be on sets of either low rank or low complexity.

THEOREM 4.3. *If* $\mathrm{rk}(A) = \lambda + n$, *where λ is either 0 or a limit ordinal and n is finite, then* $x \leqslant_T 0^{(\lambda+2n)}$.

PROOF. Let T be a recursive binary tree such that $D^{\lambda+n}([T]) = \{A\}$. We define by transfinite recursion a tree $d^{\lambda+n}(T)$ such that $D^{\lambda+n}([T]) = [d^{\lambda+n}(T)]$. Define

$$d(T) = \{\sigma \in T\colon (\exists\tau)[\sigma \prec \tau \ \&\ \tau^\frown 0 \in \mathrm{Ext}(T) \ \&\ \tau^\frown 1 \in \mathrm{Ext}(T)]\}.$$

Then $d(T) \leqslant_T \mathrm{Ext}(T) \leqslant_T T^{(2)}$. We iterate this derivative by

$$d^0(T) = T,\ d^{\alpha+1}(T) = d(d^{\alpha}(T)) \text{ and } d^{\lambda}(T) = \bigcap_{\alpha<\lambda} d^{\alpha}(T) \text{ for limit } \lambda.$$

It is proved by transfinite induction, using a system of notations for the ordinal λ as in [Rogers, 1967, p. 211], that $d^{\lambda}(T)$ is recursive in $0^{(\lambda+1)}$ and that, since $d^{\lambda}(T)$ has no dead ends, that $d^{\lambda+1}(T)$ is recursive in $0^{(\lambda+2)}$ and thus, $d^{\lambda+n}(T)$ is recursive in $0^{(\lambda+2n)}$ for all finite $n > 0$. Since A is the only element of $d^{\lambda+n}(T)$, it follows immediately that A is also recursive in $0^{(\lambda+2n)}$ for $n > 0$. If T has no dead ends, then $d^n(T)$ and A are both $\leqslant_T 0^{(2n-1)}$. For $n = 0$, we observe that, for each k, there is unique string σ of length k in $d^\lambda(T)$ (that is, $\sigma = A\lceil k$). Thus, by compactness, there must be some $\alpha < \lambda$ such that σ is the unique string of length k in $d^\alpha(T)$. Thus we can compute $A(k)$ by using the system of notations for $0^{(\lambda)}$ to compute such an α. □

Thus in particular any set of rank one must be recursive in $0''$. We now analyze the rank one sets further. Recall from Theorem 2.7 that every r.e. set is Turing equivalent to a hypersimple r.e. set of rank one. This result has the following improvement from [Cenzer and Smith, 1989].

THEOREM 4.4 (Cenzer–Smith). *For any nonrecursive degree $\mathbf{b} \leqslant \mathbf{0}'$, there is a hyperimmune set B with degree $\mathbf{b}$ of rank one; furthermore, there is a recursive tree T with no dead ends such that $D([T]) = \{B\}$.*

PROOF. Let A be a set of degree $\mathbf{b}$. By the limit lemma, there is a recursive function f such that, for all e,

$$A(e) = \lim_n f(n, e).$$

Let $n(0)$ be the least $n > 0$ such that $f(n, 0) = A(0)$ and, for any e, let $n(e+1)$ be the least $n > n(e)$ such that, for all $i < e+2$, $f(n, i) = A(i)$. Then $n(0) < n(1) < \cdots$ is a *modulus* for the set A, so that $B = \{n(0), n(1), \ldots\}$ has degree $\mathbf{b}$.

We define a Π^0_1 class P with $\mathrm{rk}_P(B) = 1$.

The (possibly finite) set $C = \{m(0) < m(1) < \cdots\}$ is in P if and only if $0 < m(0)$ and for all e, i, and m:

(i) $(0 < m < m(0)\ \&\ C \neq 0) \rightarrow f(0, m) \neq f(0, m(0))$;

(ii) $e < i < \mathrm{card}(C) \rightarrow f(m(i), e) = f(m(e), e)$;

(iii) $(e+1 < \mathrm{card}(C)\ \&\ m(e) < m < m(e+1)) \rightarrow (\exists j < e+2)(f(m(e+1), j) \neq f(m, j))$.

It is clear that P may be defined by a tree T without dead ends and in fact closed under extension by 0. Also, P contains all initial subsets of B and in fact is closed under initial subsets, so that $\mathrm{rk}_P(B) \geqslant 1$.

If $C = \{m(0) < m(1) < \cdots\}$ is any infinite set in P, then it follows from (ii) that $f(m(e), e) = f(n(e), e) = A(e)$ for all e. But it then follows from (i) or, by induction, from (iii) that $m(e) = n(e)$ for all e, so that $C = B$.

If $C = \{m(0) < m(1) < \cdots < m(k)\}$ is any finite set in P, then it follows from (i) and (iii) that there are at most two extensions $C \cup \{m\}$ of C in P (one with $f(m, k+1) = 0$ and one with $f(m, k+1) = 1$), so that C is isolated in P.

Thus B is the only non-isolated element of P and therefore $\mathrm{rk}_P(B) = 1$.

To see that B is hyperimmune, suppose by way of contradiction that h were a recursive function with $h(e) > n(e)$ for all e. Then we could define a Π^0_1 subclass Q of P by adding the restriction

(iv) $(\forall e)[\mathrm{card}(C \cap \{0, 1, \ldots, h(e) - 1\}) > e]$. □

We shall see later that not every Σ_2^0 degree contains a ranked set. Thus we have a more complicated result for degrees below $\mathbf{0}''$. This is a variant of the proof given in Cenzer, Clote, Smith, Soare and Wainer [1986], hereafter abbreviated as C-C-S-S-W.

THEOREM 4.5 (C-C-S-S-W). *For any degree* $\mathbf{b}$ *such that* $\mathbf{0}' \leqslant \mathbf{b} \leqslant \mathbf{0}''$, *there is a set* B *with degree* $\mathbf{b}$ *of rank one.*

PROOF. We sketch the proof, which is a modification of the proof of Theorem 4.4. Let A be a set of degree $\mathbf{b}$ and let f be a recursive function such that, for all e,

$$A(e) = \lim_{p\to\infty} \lim_{n\to\infty} F(p, n, e).$$

For each p and e, let $F_p(e) = \lim_{n\to\infty} F(p, n, e)$.

We define "outer modulus" values $p = p_k$ such that $F_p(e) = A(e)$ for all $e < k$ and "inner modulus" values $n = n_k$ such that $F(p, n, e) = A(e)$ for all $e < k$.

Let p_1 be the least $p > 0$ such that $F_p(0) = A(0)$ and let n_1 be the least $n > p_1$ such that, for all $m \geqslant n$, $F(p, m, 0) = A(0)$. Now inductively define p_{e+1} to the least $p > n_e$ such that $F_p(i) = A(i)$ for all $i \leqslant e + 1$. Define n_{e+1} to be the least $n > p_{e+1}$ such that

(1) for all $i \leqslant e + 1$ and all $m \geqslant n$, $F(p_{e+1}, m, i) = A(i)$, and

(2) for all p with $n_e < p < p_{e+1}$, there is an $i \leqslant e + 1$ such that, for all $m \geqslant n$, $F(p, m, i) \neq A(i)$.

Thus our choice of n_{e+1} not only verifies that $F_{p_{e+1}}(i) = A(i)$ for all $i \leqslant e + 1$, it also verifies that p_{e+1} is the least $p > n_e$ with this property.

Let $B = \{n_1 < n_1 + p_1 < n_1 + p_1 + n_2 < n_1 + p_1 + n_2 + p_2 < \cdots\}$. It is clear from the definition that A is recursive in B and that B is recursive in $0' \oplus A$, which is recursive in A by the hypothesis.

Elements of the recursive tree T with $D([T]) = \{B\}$ will be initial segments of strings of the form $\sigma = \lfloor n(1), n(1) + p(1), \ldots, n(1) + p(1) + \cdots + n(k) + p(k), n(1) + p(1) + \cdots + n(k) + p(k) + 1\rfloor$. The string σ is said to be *consistent* if it satisfies the following conditions:

(i) For all i, j, k and m, if $i \leqslant j \leqslant k$ and $n(j) \leqslant m \leqslant n(k)$, then $F(p(j), m, i) = F(p(i), n(i), i)$.

(ii) For all j, k and p, if $j \leqslant k$ and $n(j - 1) < p < p(j)$, then there is an $i \leqslant j$ such that, for all m with $n(i) \leqslant m \leqslant n(k)$, $F(p, m, i) \neq F(p(i), n(i), i)$.

The string $\sigma' = \lfloor n(1), n(1) + p(1), \ldots, n(1) + p(1) + \cdots + n(k) + p(k), n(1) + p(1) + \cdots + n(k) + p(k) + n(k + 1)\rfloor$ is consistent if σ is consistent and the conditions are satisfied when $n(k)$ is replaced by $n(k + 1)$. The string $\sigma'' = \lfloor n(1), n(1) + p(1), \ldots, n(1) + p(1) + \cdots + n(k) + p(k) + n(k + 1), n(1) + p(1) + \cdots + n(k) + p(k) + n(k + 1) + p(k + 1)\rfloor$ is consistent if σ' is consistent and $p(k + 1) < n(k + 1)$.

Finally, a consistent string τ with $\sigma \prec \tau \prec \sigma''$ is put into the tree T if each n_i is minimal, that is,

(iii) For all i, k and n, if $i \leqslant k$ and $n(i-1) < n < n(i)$, then $\lfloor n(1), n(1) + p(1), \ldots, n(1) + p(1) + \cdots + n(i-1) + p(i-1), n(1) + p(1) + \cdots + n(i-1) + p(i-1) + n, n(1) + p(1) + \cdots + n(i-1) + p(i-1) + n + p_i, n(1) + p(1) + \cdots + n(i-1) + p(i-1) + n + p(i) + 1 \rfloor$ is not consistent.

It can be seen as in the proof of Theorem 4.4 that B is the unique infinite set in $[T]$ and that every other set in $[T]$ is isolated, and that $B \in D(P)$ since, for each k, $B_k = \{n_1 < n_1 + p_1 < \cdots < n_1 + p_1 + n_2 + p_2 + \cdots + n_k + p_k\}$ is in $[T]$. □

This result can be extended to higher levels of the hyperarithmetic hierarchy. The following is an improvement from Cenzer and Remmel [1998] of a result from Cenzer, Clote, Smith, Soare and Wainer [1986]. The proof is only sketched.

THEOREM 4.6. *For any recursive ordinal λ which is either* 0 *or a limit and any finite n:*

(a) *there is a* $B \equiv_T 0^{(\lambda+2n)}$ *with* $\mathrm{rk}(B) = \lambda + n$;

(b) *for any degree* $\mathbf{a}$ *such that* $\mathbf{0}^{(\lambda+2n+1)} \leqslant \mathbf{a} \leqslant \mathbf{0}^{(\lambda+2n+2)}$, *there is a B of degree* $\mathbf{a}$ *with* $\mathrm{rk}(B) = \lambda + n + 1$.

PROOF. A recursive system of notations is used to define a uniformly recursive family of trees having rank α for each $\alpha \leqslant \lambda + n + 1$. For successor ordinals $\beta + 1$, the $\beta + 1$st tree is built from the βth tree by a "Jump Lemma" based on Theorem 4.5. For limit ordinals λ, the λth tree is built from the preceding trees by a "Stitching Lemma". We omit the details. □

Next we briefly consider sets which cannot be ranked. The first result is from Cenzer and Smith [1989].

THEOREM 4.7 (Cenzer–Smith). *For any hyperimmune set A, there is a* $C \equiv_T A$ *which is not ranked.*

PROOF. Let $A = \{f(0) < f(1) < \cdots\}$ be hyperimmune. Let $[T_0], [T_1], \ldots$ enumerate the Π_1^0 classes as in Lemma 1.2. We first define $B \leqslant_T A$ so that $[T_i]$ is uncountable whenever $B \in [T_i]$, which implies that B is not ranked. The characteristic function of B is the limit of a sequence of strings σ_n of length $f(n)$ which is recursive in A. Let $\sigma_0 = 0$. Given σ_n, σ_{n+1} is defined in two cases.

Case 1. There is some $i \leqslant f(n)$ and some σ of length $f(n)$ such that

(i) $\sigma_n \in T_i$,

(ii) $\sigma \notin T_i$,

(iii) $\sigma \lceil n = \sigma_n \lceil n$,

(iv) for any $j < i$, if $\sigma_n \notin T_i$, then $\sigma \notin T_j$.

Then we let i be the least for which there is a σ satisfying the conditions and we let σ_{n+1} be the (lexicographically) least corresponding to that i.

Case 2. If there is no such i, then $\sigma_{n+1} = \sigma_n^\frown 0^{f(n+1)-f(n)}$.

Let B have characteristic function $\bigcup_n \sigma_n$. It is clear that $B \equiv_T A$. The proof that $B \in [T_i]$ implies $[T_i]$ uncountable is by induction. Suppose true for all $i < j$, suppose that $B \in [T_i]$, and choose m large enough that $B \notin [T_i]$ implies $B\lceil m \notin T_i$ for all $i < j$. Then for any $n > f(m)$, any extension $\sigma \notin T_i$ of $B\lceil n$ of length $f(n)$ will satisfy the conditions of Case 1. Thus the shortest extension σ of $B\lceil n$ not in T_i must have length $> f(n)$. If $[T_i]$ were countable, then every σ would have an extension not in T_i, so that we could define a function $h(n)$ to be the least k such that any string σ of length n has an extension of length k which is not in T_i. Then for $n > f(m)$, $h(n) > f(n)$, contradicting the assumption that A is hyperimmune.

It follows that B is unranked. Finally, let $C = A \oplus B$. Then $C \equiv_T A$ and $B \leqslant_{tt} C$, so that C is unranked by Lemma 4.3(b). □

Of course this implies that every r.e. degree contains an unranked set. We say that **a** is *completely unranked* if every set A of degree **a** is unranked. Jockusch and Shore [1984] construct a Σ^0_2 degree which is completely unranked. Downey observed that since sets with the same truth-table degree have the same rank and since all sets in a hyperimmune-free degree have the same truth-table degree (see Odifreddi [1989, p. 589]), the construction of an unranked set of hyperimmune-free degree will provide a completely unranked degree. This led to the following improvement in [Downey, 1991] of the Jockusch–Shore result.

THEOREM 4.8 (Downey). *There is a hyperimmune-free degree which is completely unranked.*

On the other hand, Downey [1991] also showed that there exists a completely ranked degree below $0''$, again using a hyperimmune-free degree **a**, so that every A of degree $\leqslant$ **a** is in fact $\leqslant_{tt}$ **a**.

THEOREM 4.9 (Downey). *There exists a degree* $\mathbf{a} \leqslant \mathbf{0}''$ *such that every set A of degree* $\leqslant$ **a** *is ranked.*

Cenzer and Smith [1989] consider the problem of sets below $\mathbf{0}'$ but with high rank. They showed that for every recursive ordinal α, there is a Δ^0_2 set A of rank α. This was improved by Cholak and Downey [1993].

THEOREM 4.10 (Cholak–Downey). *For each recursive ordinal α, there is an r.e. set of rank α.*

We close this section with some interesting recent results of J. Owings [1997] and G. Martin [1993].

A set A is said to be a *Cantor singleton* if it is the unique nonrecursive member of some Π^0_1 class. Theorem 2.2 implies that every nonrecursive Π^0_1 retraceable set is a Cantor singleton. G. Martin [1993] improved this result by showing that any nonrecursive set which is the union of a recursive set with a Π^0_1 retraceable set is a Cantor singleton. We note that the ranked sets constructed in Theorems 2.7, 4.4 and 4.5 are all Cantor singletons, being in fact the unique infinite sets in the Π^0_1 classes constructed. A Π^0_1 class P is said to be *rank-faithful* if $\text{rk}_P(x) = \text{rk}(x)$ for all $x \in P$. The Π^0_1 classes constructed in Theorems 2.7, 4.4 and 4.5 are clearly rank-faithful.

G. Martin [1993] improved the result of Cholak and Downey as follows.

THEOREM 4.11 (G. Martin). *For each recursive ordinal α and every non-zero r.e. degree* **a**, *there are r.e. sets A and B of degree* **a** *and rank α such that A is a Cantor singleton and B belongs to a rank-faithful Π^0_1 class.*

Owings [1997] improved Lemma 4.3 for Cantor singletons as follows.

THEOREM 4.12 (Owings). *For any sets A, B, if B is a Cantor singleton and $A \leqslant_{tt} B$, then A is a Cantor singleton.*

PROOF. Let A be the unique nonrecursive element of the Π^0_1 class P and let Φ be a recursive function as in the proof of Lemma 4.3 such that $\Phi(B) = A$. Then A is the unique nonrecursive element in $\Phi(P)$. □

THEOREM 4.13 (Owings). (a) *If $A \oplus B$ is a Cantor singleton, then either B is recursive or A is recursive in B, so that the degree of $A \oplus B$ is either the degree of A or the degree of B.*

(b) *If* $\text{rk}(B) = \text{rk}(A \oplus B)$, *then A is recursive in B.*

PROOF. (a) Suppose that $A \oplus B$ is the unique nonrecursive element of P and that B is nonrecursive. Let $Q = P \cap (\{0,1\}^\omega \oplus \{B\}$ and observe that $B \leqslant_{tt} C$ for every $C \in Q$, so that any $C \in Q$ is nonrecursive. Thus $A \oplus B$ is the unique element of Q and is therefore recursive in B by the relativized version of Lemma 1.4.

(b) Let $\alpha = \text{rk}(A \oplus B)$ and let P be a Π^0_1 class such that $D^\alpha(P) = \{A \oplus B\}$. As in (a), let $Q = P \cap (\{0,1\}^\omega \oplus \{B\})$. It follows from Lemma 4.3(b) that $\text{rk}(C) \geqslant \alpha$ for all $C \in Q$, so that in fact $Q = \{A \oplus B\}$. Then $A \oplus B$ is recursive in B as above. □

Recall that a set A is *autoreducible* if there is a recursive functional F such that, for all n, $A(n) = F(n, A \setminus \{n\})$.

THEOREM 4.14 (Owings). *Every ranked set A is autoreducible.*

PROOF. Let $\text{rk}(A) = \alpha$ and let $P = [T]$ be a Π^0_1 class such that $D^\alpha(P) = \{A\}$. Note that for any n, $A \setminus \{n\}$ and $A \cup \{n\}$ are both $\equiv_{tt} A$ and thus also have rank α. It follows that only one of them can belong to P. Thus, given n and $A \setminus \{n\}$, we search for the least k such that every $\sigma \in T$ of length k consistent with $A \setminus \{n\}$, except possibly at n, has the same value $\sigma(n)$. Then $A(n) = \sigma(n)$. □

5. Minimal and thin classes

The notions of *thin* and of *minimal* Π_1^0 classes belong to the study of the lattice of Π_1^0 classes. The concept of a thin Π_1^0 class arose from the study of theories of arithmetic. Martin and Pour-El [1970] constructed an axiomatizable, essentially undecidable theory T each extension of which was principal. The class of complete extensions of such a theory may be interpreted as a (perfect) *thin* Π_1^0 class P, where P is said to be thin if, for every Π_1^0 subclass Q of P, there is a clopen set U such that $Q = U \cap P$. The notion of thinness was first made explicit by Downey [1982]. Perfect thin classes were also constructed by Simpson and are related to superminimal profinite groups by the work of R. Smith [1981].

An infinite Π_1^0 class C is said to be *minimal* if every Π_1^0 subclass Q of C is either finite or cofinite in C. Thus the notion of a minimal Π_1^0 class is the analog of the notion of a co-maximal Π_1^0 subset of ω. In particular, if C is a co-maximal set, then the class of subsets of C containing either one or no elements is an example of a minimal Π_1^0 class which is not thin.

THEOREM 5.1. *Any thin Π_1^0 class P is rank-faithful.*

PROOF. Suppose that $A \in P$ and that $\mathrm{rk}(A) = \alpha$. Let $D^\alpha(Q) = \{A\}$. Then $A \in P \cap Q$ and $P \cap Q = P \cap U$ for some clopen set U. It follows from Lemma 4.1 that $\mathrm{rk}_P(A) = \mathrm{rk}_{P\cap Q}(A) \leqslant \mathrm{rk}_Q(A)$. Equality follows since $\mathrm{rk}_Q(A)$ is minimal by assumption. □

The connection between thin and minimal classes is given by the following.

THEOREM 5.2 (C-D-J-S). *The following are equivalent for any Π_1^0 class P.*
(a) *P is thin and $D(P)$ is a singleton.*
(b) *P is minimal and has a non-recursive member.*

PROOF. (a) $\to$ (b): Suppose that P is thin and that $D(P) = \{A\}$. Then A is non-recursive by Theorem 5.1. Let Q be a Π_1^0 class such that $Q \subset P$. Then $Q = U \cap P$ for some clopen U, so that $P \setminus Q$ is also a Π_1^0 class. If both Q and $P \setminus Q$ were infinite, then both would contain limit points, contradicting the assumption that P has only one limit point.

(b) $\to$ (a): Suppose that P is minimal and has a nonrecursive member A. Then $A \in D(P)$ by Lemma 1.4. For any $B \neq A$ in P, let U be an interval such that $A \in U$ and $B \notin U$. Then $U \cap P$ is infinite, and therefore $P \setminus U$ must be finite, which implies that $B \notin D(P)$. Therefore $D(P) = \{A\}$. Now let Q be any Π_1^0 subclass of P. For any $B \neq A$ in P, let $U(B)$ be an interval such that $U(B) \cap P = \{B\}$. Since P is minimal, there are two cases.

Case 1: Q is finite. Then $Q = P \cap \bigcup_{B\in Q} U(B)$.

Case 2: $P \setminus Q$ is finite. Then $Q = P \cap [2^\omega \setminus \bigcup_{B\in P\setminus Q} U(B)]$. □

We next give several constructions of thin Π^0_1 classes.

THEOREM 5.3 (Martin–Pour-El). *There exists a perfect thin Π^0_1 class with no recursive member.*

PROOF. Let $P_e = [T_e]$ be the e-th Π^0_1 class as in Lemma 1.2 and let ϕ_e be the e-th partial recursive function from ω into $\{0, 1\}$. We will construct a recursive tree S with corresponding Π^0_1 class P and a homeomorphism F from $\{0, 1\}^\omega$ onto P. F will be constructed by means of a map $f : \{0, 1\}^{<\omega} \to T$ such that $\sigma \prec \tau \iff f(\sigma) \prec f(\tau)$; then for $x \in \{0, 1\}^\omega$, $F(x) = \bigcup_n f(x\lceil n)$.

To ensure that P has no recursive member, we construct f so that

(1_e): For each e, if ϕ_e is total, then there exists m such that for each $\sigma \in \{0, 1\}^{2e+1}$, $f(\sigma)$ is inconsistent with ϕ_e.

To ensure that P is thin, we construct f so that

(2_e): For each e and for each $\sigma \in \{0, 1\}^{2e+2}$, if $f(\sigma) \in T_e$, then $(\forall\tau)(\sigma \prec \tau \to f(\tau) \in T_e)$.

To see that this makes P thin, let $U = \bigcup\{I(f(\sigma)) \colon |\sigma| = 2e + 2 \ \&\ f(\sigma) \in T_e\}$ and observe that if $P_e \subset P$, then $P_e = P \cap U$.

The map f is defined in uniformly recursive stages f_s, beginning with f_0 as the identity function. At odd stages of the construction, we address the first type of requirement and at even stages we address the second type of requirement.

(Stage $2s+1$): Look for $e < s$ and $\sigma \in \{0, 1\}^{2e+1}$ and $\tau \succ \sigma$ with $|\tau| \leqslant 2s+1$ such that $f_{2s}(\sigma) \in T_e$, but $f_{2s}(\tau) \notin T_e$. If such e, σ and τ exist, then we take the least such e and the lexicographically least σ and τ for that e. Then we let $f_{2s+1}(\sigma) = f_{2s}(\tau)$ and in general, for any ρ we let

$$f_{2s+1}(\sigma^\frown\rho) = f_{2s}(\tau^\frown\rho) \quad \text{and}$$
$$f_{2s+1}(\rho) = f_{2s}(\rho) \quad \text{for } \rho \text{ incomparable with } \sigma.$$

If no such e, σ and τ exist, then we just let $f_{2s+1} = f_{2s}$.

(Stage $2s + 2$): Look for $e \leqslant s$, $\sigma \in \{0, 1\}^{2e+2}$ and $\tau \succ \sigma$ with $|\tau| \leqslant 2s + 2$ such that $\phi_{e,s}(a)\downarrow$ for all $a < |f(\tau)|$ and such that $f(\sigma)(a) = \phi_e(a)$ for all $a < |f(\sigma)|$ but $f(\tau)$ is inconsistent with ϕ_e. If such e, σ and τ exist, then we take the least such e and the lexicographically least σ and τ for that e.

Then we let $f_{2s+2}(\sigma) = f_{2s+1}(\tau)$ and in general, for any ρ we let

$$f_{2s+2}(\sigma^\frown\rho) = f_{2s+1}(\tau^\frown\rho) \quad \text{and}$$
$$f_{2s+1}(\rho) = f_{2s+1}(\rho) \quad \text{for } \rho \text{ incomparable with } \sigma.$$

If no such e, σ and τ exist, then we just let $f_{2s+2} = f_{2s+1}$.

It is easy to see by induction on $|\sigma|$ that for each σ, $f_s(\sigma)$ converges to a limit $f(\sigma)$, which makes $f \leqslant_T 0'$. Then we see by induction on e that the requirements (1_e) and (2_e) are satisfied. □

The proof of Theorem 5.3 in fact proves something more. It follows from the remarks after Theorem 3.6 (with $\mathbf{b} = \mathbf{0}'$) that any Π^0_1 class P has a member A of degree $\mathbf{a}$ such that $\mathbf{a} \oplus \mathbf{0}' = \mathbf{a}' = \mathbf{0}''$. A natural question is whether every r.b. Π^0_1 class contains a member of high degree. The following new result answers this question in the negative.

THEOREM 5.4. *There exists a perfect, thin, r.b. Π^0_1 class P with no recursive members such that if $\mathbf{a}$ is the degree of a member of P, then $\mathbf{a}' \leqslant \mathbf{a} \oplus \mathbf{0}'$.*

PROOF. Let P be the Π^0_1 class constructed in Theorem 5.3 and let f be the function defined therein. Then f is the limit of a uniformly recursive sequence of functions and is therefore recursive in $\mathbf{0}'$ by the Limit Lemma. As in the proof of Theorem 3.5, let $U_e = \{\sigma\colon \phi^\sigma_e(e)\uparrow\}$. It follows from the S-m-n theorem that there is a recursive function ϕ such that $U_e = T_{\phi(e)}$ for each e. Then for any element $A = f(x)$ of P and any e, we have

$$e \notin A' \iff (\exists \sigma \prec A)\sigma \in U_e \iff x\lceil f(2e+2) \in T_{\phi(e)}.$$

This shows that A' is recursive in $A \oplus \mathbf{0}'$. □

THEOREM 5.5 (C-D-J-S). *For any recursive ordinal α, there is a thin Π^0_1 class P_α with Cantor–Bendixson rank α. Furthermore, we may take P_α as the set of paths through a recursive tree with no dead ends.*

PROOF. We first sketch the proof for $\alpha = 1$. We construct a sequence $\tau_e \in \{0,1\}^{<\omega}$ such that $\tau_e{}^\frown 1 \prec \tau_{e+1}$ for all e, a set $A = \cup_e \tau_e$ and a Π^0_1 class $P = [T]$ such that

(1) $D(P) = \{A\}$, and

(2) for any e, if $A \in [T_e]$ then $P \cap I(\tau_e) \subset [T_e]$.

These conditions imply that A is nonrecursive, since if A were recursive, then $\{A\} = [T_e]$ for some e, so that by (2), $P \cap I(\tau_e) = \{A\}$, contradicting (1).

These conditions also imply that P is minimal (and therefore thin by Theorem 5.2). To see this, suppose that $[T_e] \subset P$. If $A \notin [T_e]$, then $[T_e]$ has no limit point and is therefore finite. If $A \in [T_e]$, then $P \setminus [T_e] \subset P \setminus I(\tau_e)$ by (2) has no limit point and is finite.

The construction is in stages, so that at stage s we have a tree T^s and strings τ^s_e. At stage $s+1$, we simply look for $e \leqslant s$ such that some $\tau \succ \tau^s_e$ is in $T^s \setminus T_e$ and let $\tau^{s+1}_e = \tau$ for the least such e and τ. For $i < e$, let $\tau^{s+1}_i = \tau^s_i$ and let $\tau^{s+1}_{e+i} = 0\tau{}^\frown 1^i$. We leave the details to the reader.

The general construction for a recursive ordinal α is accomplished using a recursive system of notations for α and a uniformly recursive family of trees of rank up to α is in the proof of Theorem 4.6. The details are omitted. □

Next we consider the possible degrees of members of thin Π^0_1 classes.

THEOREM 5.6 (C-D-J-S). *There is a Π^0_1 set A of degree $\mathbf{0}'$ and a minimal, thin Π^0_1 class P such that $D(P) = \{A\}$.*

PROOF. Let $B = \mathbf{0}'$ be the union of uniformly recursive sets B^s. Let $T_0, T_1, \ldots$ be an effective enumeration of the primitive recursive trees on $2^{<\omega}$. We will define a Π^0_1 retraceable set $A = \{a_0 < a_1 < \cdots\}$ and a corresponding Π^0_1 class $P = I(A)$ of initial subsets of A, by Theorem 2.2, such that

(1) For any e, $e \in B \iff e \in B^{a_e}$.

(2) For any Π^0_1 class $P_e = [T_e]$, if $A \in P_e$, then $A_n \in P_e$ for all $n \geqslant e$.

By property (1), $\mathbf{0}'$ is recursive in A, so that, since A is Π^0_1, A has degree $\mathbf{0}'$. It then follows from (2) as in the proof of Theorem 5.5 that P is minimal and thin.

The sequence $a_0 < a_1 < \cdots$ is defined by Π^0_1 recursion in the style of Theorem 2.5 by making a_n the least a which satisfies the following:

(i) For all $m < n$, $a_m < a$.

(ii) $n \in B \rightarrow n \in B^a$.

(iii) For all $m < n$, either $\langle a_0, \ldots, a_{n-1}, a\rangle \notin T_m$ or $(\forall x)(\langle a_0, \ldots, a_{n-1}, x\rangle \in T_m)$.

(iv) For all $x < a$, either

 (a) $x \leqslant a_{n-1}$ or

 (b) $n \in B^a \setminus B^x$ or

 (c) for some $m < n$, $\langle a_0, a_1, \ldots, a_{n-1}, x\rangle \in T_m$ & $\langle a_0, a_1, \ldots, a_i, a\rangle \notin T_m$.

The details are left to the reader. □

The following result is Theorem 2.13 of Cenzer, Downey, Jockusch and Shore [1993, p. 102]. The full hypothesis of this theorem was not stated in Cenzer et al. [1993], so we take this opportunity to correct that omission.

THEOREM 5.7 (C-D-J-S). *Let T be a recursive tree and P a Π^0_1 class such that $P = [T]$. Then for any set $A \in P$,*

(a) *If $P \subset \mathcal{P}(A)$, then $A \leqslant_T \mathrm{Ext}(T)$.*

(b) *If A is a Π^0_1 set and P is thin, then $A \leqslant_T \mathrm{Ext}(T)$.*

(c) *If P is thin and T has no dead ends and A is either r.e. or co-r.e., then A is recursive.*

PROOF. (a) To test whether $n \in A$, simply see if there is a $\sigma \in \mathrm{Ext}(T)$ of length $n+1$ such that $\sigma(n) = 1$.

(b) Note that $\mathcal{P}(A)$ is a Π^0_1 class, so that $Q = P \cap \mathcal{P}(A)$ is a Π^0_1 subclass of P and is nonempty since $A \in Q$. Since P is thin, we must have $Q = P \cap U$ for some clopen $U = I(\sigma_0) \cup \cdots \cup I(\sigma_k)$. If we now define

$$T_Q = \{\sigma \in T\colon \sigma \text{ is compatible with } \sigma_i, \text{ for some } i \leqslant k\},$$

then it is clear that $Q = [T_Q]$ and that

$$\mathrm{Ext}(T_q) = \{\sigma \in \mathrm{Ext}(T)\colon \sigma \text{ is compatible with } \sigma_i, \text{ for some } i \leqslant k\},$$

so that $\mathrm{Ext}(T_Q)$ is recursive in $\mathrm{Ext}(T)$. Now $Q \subset \mathcal{P}(A)$, so that by (a) we have $A \leqslant_T \mathrm{Ext}(T_Q) \leqslant_T \mathrm{Ext}(T)$.

(c) It is immediate from (b) that if A is Π^0_1 set, then A is recursive. If A is an r.e. set, then $\omega \setminus A$ is Π^0_1 and belongs to the thin Π^0_1 class $\{\omega \setminus X\colon X \in P\}$. □

THEOREM 5.8 (C-D-J-S). *Let T be a recursive tree such that $P = [T]$ is a thin Π^0_1 class and let $A \in P$. Then*

(a) *$A' \leqslant_T A \oplus \mathbf{0}''$ (so that it is not possible that $A \geqslant_T \mathbf{0}''$).*

(b) *If T has no dead ends, then $A' \leqslant_T A \oplus \mathbf{0}'$ (so that it is not possible that $A \geqslant_T \mathbf{0}'$).*

PROOF. (a) Let $P = [T]$ be thin and suppose $A \in P$. For each e, let $Q_e = \{C\colon \phi^C_e(e)\uparrow\}$. Then Q_e is a Π^0_1 class, so there is a clopen set $U(e)$ such that $P \cap Q_e = P \cap U(e)$. Thus if $\phi^A_e(e)\uparrow$, then there is some $\sigma = A\lceil n$ such that σ forces $\phi^A_e(e)\uparrow$, that is, for any $B \in P$, if $\sigma \prec B$ then $\phi^B_e(e)\uparrow$. Now define the Π^0_2 relation $R(e,\sigma)$ which says that σ forces $\phi^B_e(e)\uparrow$, by

$$R(e,\sigma) \iff (\forall \tau \succ \sigma)[(\tau \in T \mathbin{\&} \phi^\tau_e(e)\downarrow) \rightarrow \tau \notin \mathrm{Ext}(T)].$$

Then we can compute from A together with $\mathbf{0}''$, whether $e \in A'$ by searching for the least n such that, for $\sigma = x\lceil n$, either $\phi^\sigma_e(e)\downarrow$, in which case $e \in A'$, or $R(e,\sigma)$, in which case $e \notin A'$.

(b) Observe that if $\mathrm{Ext}(T)$ is recursive, then the relation R defined above will be recursive in $\mathbf{0}'$. □

It follows from (b) and the proof of Theorem 4.3 above that if A has rank one in a thin Π^0_1 class $P = [T]$, where T has no dead ends, then A has *low* degree $\mathbf{a}$, that is, $\mathbf{a}' = \mathbf{0}'$.

Part (a) of this theorem is best possible in the sense that, as shown in Theorem 2.18 of Cenzer, Downey, Jockusch and Shore [1993], there is a minimal thin Π^0_1 class P and a set A such that $D(P) = \{A\}$ and $A \oplus \mathbf{0}' \equiv_T \mathbf{0}''$.

We conclude this section by stating without proof several further results from Cenzer, Downey, Jockusch and Shore [1993].

THEOREM 5.9 (C-D-J-S). *Between any two distinct r.e. degrees $\mathbf{b} < \mathbf{c}$, there is a degree $\mathbf{a}$, a set A of degree $\mathbf{a}$ and a minimal, thin Π^0_1 class P with $D(P) = \{A\}$.*

There is a family of r.e. degrees which contain members of thin Π^0_1 classes. In particular, it follows from Theorem 4.9 of Downey, Jockusch and Stob [1990] that all array nonrecursive (*a.n.r.*) degrees and hence all non-low$_2$ degrees contain members of thin Π^0_1 classes.

Theorem 5.8 tells us that no set of degree $\mathbf{0}''$ can even belong to a thin Π^0_1 class. Two further results give lower degrees which also contain no members of thin classes.

THEOREM 5.10 (C-D-J-S). (a) *There is an r.e. degree* **a** *such that no set B of degree* **a** *belongs to any thin Π_1^0 class.*

(b) *There is a minimal degree* $\mathbf{a} < \mathbf{0}'$ *such that no set A of degree* **a** *is a member of any thin Π_1^0 class.*

In contrast, we have the following improvement of Theorem 4.9.

THEOREM 5.11 (C-D-J-S). *There is a non-recursive set $A \leqslant_T \mathbf{0}''$ such that every non-recursive set $B \leqslant_T A$ is a rank* 1 *member of a minimal, thin Π_1^0 class.*

Finally, there is another connection with maximal r.e. sets.

THEOREM 5.12 (C-D-J-S). *There is a maximal set A which is not a member of any thin Π_1^0 class.*

6. Applications

There are applications of Π_1^0 classes in a wide variety of mathematical areas, including logic, combinatorics, game theory and analysis. In each case, the set of solutions to a given recursively presented problem, such as the set of k-colorings of a recursive graph, is shown to be a Π_1^0 class. Then we can conclude, by Theorem 3.3, that any k-colorable recursive graph must have a coloring which is recursive in $\mathbf{0}'$. Conversely, it can be shown that any r.b. Π_1^0 class may be represented as the set of k-colorings of some recursive graph. Thus we can conclude, by Theorem 3.12 that there is a k-colorable recursive graph with no recursive k-coloring. In some cases, only the weaker result that every Π_1^0 class of separating sets may be represented as the set of solutions to the given recursively presented problem. This still implies the existence of a recursively presented problem with no recursive solution, by the result of Kleene [1950]. In this section we outline some applications to logic, graph theory, and the theory of partial orderings. For a more complete list of applications and a more detailed presentation, see Cenzer and Remmel [ta2].

6.1. *Logical theories*

Here we consider the problem of finding a complete consistent extension of a first-order logical theory. A recursively presented instance of a logical theory will be an axiomatizable theory, that is a theory with a recursively enumerable set of axioms. We will show that the set of solutions to a given effective problem can always be represented by a r.b. Π_1^0 class and that any Π_1^0 class can be represented by such a set.

Let $\mathcal{L}$ be an effective first-order language, as described in Section 1 and let Sent($\mathcal{L}$) be the set of sentences of $\mathcal{L}$. For any subset Γ of Sent($\mathcal{L}$), the set Con(Γ) of consequences of Γ is the closure of Γ under logical deduction and the set Ref(Γ) of refutations of Γ is the set of negations of the consequences of Γ. A subset Γ of Sent($\mathcal{L}$) is a first order logical theory if Γ is closed under logical deduction. Σ is said to be a set of axioms for Γ if $\Gamma = \text{Con}(\Sigma)$ and Γ is *axiomatizable* if Γ has a recursive set of axioms. It is not hard to see that Γ is axiomatizable if and only if Γ is recursively enumerable. A theory is said to be *decidable* if it is recursive. It is easy to see that any complete axiomatizable theory is decidable. The classical result here is that any consistent theory has an extension to a complete consistent theory. A theory is said to be *essentially undecidable theory* if it has no decidable complete consistent extension. The classic example is the theory of Peano Arithmetic, which Church and Rosser [1936] showed to be essentially undecidable. Shoenfield [1960] showed that in general, the family of complete, consistent extensions of an axiomatizable first order theory can be represented by a Π^0_1 class.

THEOREM 6.1 (Shoenfield). *For any r.e. theory Γ of an effective language $\mathcal{L}$, both the class of consistent extensions of Γ and the class of complete consistent extensions of Γ can be represented as Π^0_1 classes. Furthermore, if Γ is a decidable theory, then these classes can be represented by recursive trees with no dead ends.*

PROOF. Let $\mathcal{L}$ be an effective first-order language and let $S = \text{Sent}(\mathcal{L})$ have an effective enumeration as $\gamma_0, \gamma_1, \ldots$. Then the sentence γ_i may be identified with its index i, so that a theory Γ is represented by the set $\{i\colon \gamma_i \in \Gamma\}$, and a class of theories is represented by a class in $\{0, 1\}^\omega$. Let $\Gamma \models_s \gamma_i$ be the recursive relation on s, i which means that there is a proof of γ from Γ of length at most s. Then the class $P(\Gamma)$ of complete consistent extensions of Γ may be represented by the set of infinite paths through the recursive tree T defined so that for any $\sigma = (\sigma(0), \ldots, \sigma(n-1))$, σ is in T if and only if the following conditions hold.

(i) For any $i < n$, if $\Gamma \models_n \gamma_i$, then $\sigma(i) = 1$.

(ii) For any $i, j < n$, if $\Gamma \models_n \gamma_i \to \gamma_j$ and $\sigma(i) = 1$, then $\sigma(j) = 1$.

(iii) For any $i, j, k < n$, if $\gamma_k = \gamma_i \mathbin{\&} \gamma_j$, $\sigma(i) = 1$ and $\sigma(j) = 1$, then $\sigma(k) = 1$.

(iv) For any $i, j < n$, if $\sigma(i) = 1$ and $\gamma_j = \neg\gamma_i$, then $\sigma(j) = 0$.

(v) For any $i, j < n$, if $\gamma_j = \neg\gamma_i$, then either $\sigma(i) = 1$ or $\sigma(j) = 1$.

Let x be an infinite path through T and let $\Delta = \{\gamma_i\colon x(i) = 1\}$.

Condition (i) ensures that $\Gamma \subset \Delta$, while conditions (i), (ii), and (iii) ensure that Δ is a theory. Condition (iv) ensures that Δ is consistent and condition (v) ensures that Δ is complete. Thus to represent the class of consistent extensions of Γ, simply omit the final clause.

If Γ is decidable, then in each case we can modify the clauses given above as follows to get a tree S with no dead ends which has the same class of infinite paths. First, combine the first three clauses into the statement:

(i′): For any $k < n$, if $\Gamma \models \wedge\{\gamma_i\colon i < n \mathbin{\&} \sigma(i) = 1\} \to \gamma_k$, then $\sigma(k) = 1$.

Next, replace clause (iv) with

(iv′) It is not the case that $\Gamma \models [\wedge\{\gamma_i\colon i < n \mathbin{\&} \sigma(i) = 1\} \to (\gamma_0 \mathbin{\&} \neg\gamma_0)]$.

It follows that for any $\sigma \in S$,

$$\Gamma \cup \{\gamma_i\colon i < |\sigma| \;\&\; \sigma(i) = 1\} \cup \{\neg\gamma_i\colon i < n \;\&\; \sigma(i) = 0\}$$

is consistent and therefore has an extension to a complete consistent theory $\Gamma(\sigma)$ which will be represented by an extension of σ. Thus S has no dead ends. □

We note that the a recursively enumerable Boolean algebra $L(\Gamma)$ (the *Lindenbaum algebra*) may be associated with the theory Γ and the class of complete consistent extensions of Γ may be viewed as the class of maximal filters of $L(\Gamma)$. This connection is discussed further in Cenzer and Remmel [ta2].

We can now derive a number of immediate corollaries.

THEOREM 6.2. (a) *For any consistent, axiomatizable first-order theory Γ:*

(i) *Γ has a complete consistent extension in some r.e. degree.*

(ii) *Γ has complete consistent extensions Γ_1 and Γ_2 such that any function recursive in both Γ_1 and Γ_2 is recursive.*

(iii) *If Γ has only countably many complete consistent extensions, then Γ has a decidable complete consistent extension.*

(iv) *If Γ has only finitely many complete consistent extensions, then every complete consistent extension is decidable.*

(v) *If Γ has no decidable complete consistent extensions, then for any countable sequence of nonzero degrees $\{\mathbf{a}_i\}$, Γ has continuum many complete consistent extensions Σ which are mutually Turing incomparable and such that the degree of Σ is incomparable with each $\mathbf{a}_i$.*

(b) *Any consistent, decidable first order theory Γ has a complete, consistent, decidable extension.*

PROOF. (a) (i) This follows from Theorem 2.11.

(ii) This follows from Theorem 3.4.

(iii) This follows from Theorem 3.9.

(iv) This follows from Theorem 3.8.

(v) This follows from Theorem 3.13.

(b) This follows from Theorem 3.3(d). □

Next we turn to the other direction of our correspondence.

THEOREM 6.3. *Any r.b. Π^0_1 class P can be represented by the set of complete, consistent extensions of an axiomatizable theory Γ. If $P = [T]$ where T is a recursive tree with no dead ends, then Γ may be taken to be decidable.*

PROOF. We give the proof due to Ehrenfeucht [1961]. Let the language $\mathcal{L}$ consist of a countable sequence $A_0, A_1, \ldots$ of 0-place relations symbols, that is, propositional variables. For any $x \in \{0, 1\}^\omega$, we can define a complete consistent theory $\Delta(x)$ for

$\mathcal{L}$ as the theory generated by $\{C_i\colon i \in \omega\}$, where $C_i = A_i$ if $x(i) = 1$ and $C_i = \neg A_i$ if $x(i) = 0$. It is clear that every complete consistent theory of $\mathcal{L}$ is one of these. Thus for any Π_1^0 class $P \subset \{0, 1\}^\omega$, we want a theory Γ such that $\Delta(P) = \{\Delta(x)\colon x \in P\}$ is the set of complete, consistent extensions of Γ. For each finite sequence $\sigma = (\sigma(0), \dots, \sigma(n-1))$, let $\phi_\sigma = C_0 \wedge C_1 \wedge \cdots \wedge C_{n-1}$, where $C_i = A_i$ if $\sigma(i) = 1$ and $C_i = \neg A_i$ if $\sigma(i) = 0$. Let the binary tree T be given such that $P = [T]$ and define the theory $\Gamma(T)$ to be consist of all $\phi_\sigma \to A_n$ such that $\sigma \in T$ and $\sigma^\frown 0 \notin T$ and all $\phi_\sigma \to \neg A_n$ such that $\sigma \in T$ and $\sigma^\frown 1 \notin T$, where $|\sigma| = n$. We claim that $\Delta(P)$ is in fact equal to the set of complete consistent extensions of $\Gamma(T)$. Suppose first that $x \in P$ and let $\{C_i\colon i \in \omega\} = \Delta(x)$. Now any $\gamma \in \Gamma(T)$ is of the form $\phi_\sigma \to \pm A_i$ for some $\sigma \in T$; say that $|\sigma = n$. There are several cases. If $\sigma \neq x\lceil n$, then $\Delta(x) \models \neg\phi_\sigma$, so that we always have $\Delta(x) \models \phi_\sigma \to \pm A_n$. Thus we may suppose that $\sigma = x\lceil n$. If $\sigma^\frown 0 \notin T$, then of course $x(n) = 0$, so that $C_n = A_n \in \Delta(x)$ and therefore $\Delta(x) \models \phi_\sigma \to A_n$. Similarly, if $\sigma^\frown 1 \notin T$, then $\Delta(x) \models \phi_\sigma \to \neg A_n$. Thus $\Delta(x)$ is a complete consistent extension of $\Gamma(T)$. On the other hand, let Δ be a complete consistent extension of $\Gamma(T)$. Then, for each i, we have either $\Delta \models A_i$ or $\Delta \models \neg A_i$; let $C_i = A_i$ if $A_i \in \Delta$ and $C_i = \neg A_i$ otherwise. Define $x \in \{0, 1\}^\omega$ so that $x(i) = 1$ if and only if $\Delta \models A_i$. Then clearly $\Delta = \Delta(x)$. It remains to be shown that $x \in P$. If $x \notin P$, then there is some n such that $\sigma = x\lceil n + 1 \notin T$ and $x\lceil n \in T$. Then $\phi_\sigma = C_0 \wedge \cdots \wedge C_{n-1}$, so that $\Delta \models \phi_\sigma$, and $\phi_\sigma \to \neg C_i \in \Gamma(T)$, so that Δ is not consistent with $\Gamma(T)$. This contradiction proves that $x \in P$.

Now suppose that $P = [T]$, where T has no dead ends and let $\gamma = \gamma(A_0, \dots, A_{n-1})$ of the language $\mathcal{L}$ be given. We claim that $\Gamma(T) \models \gamma$ if and only if

$$\phi_\sigma \models \gamma \quad \text{for all } \sigma \in T \text{ with } |\sigma| = n.$$

This claim clearly implies that $\Gamma(T)$ is decidable.

Recall that $\Gamma(T) \models \gamma$ if and only if $\Delta(x) \models \gamma$ for every $x \in [T]$.

We argue by the contrapositive. Suppose first that $\Gamma(T) \not\models \gamma$. Then there is some $x \in [T]$ such that $\Delta(x) \models \neg\gamma$. Since γ depends only on $A_0, \dots, A_{n-1}$, it follows that $\phi_\tau \models \neg\gamma$, where $\tau = x\lceil n \in T$. Thus $\phi_\tau \models \gamma$ is clearly false, making the condition false. Suppose next that the condition is false. Then $\phi_\tau \models \gamma$ is false for some fixed $\tau \in T$, which means that $\phi_\tau \models \neg\gamma$ (since γ depends only on $A_0, \dots, A_{n-1}$). Since T has no dead ends, there is some $x \in P$ such that $\tau \prec x$ and therefore $\Delta(x) \models \neg\gamma$. It follows from the remark above that $\Gamma(T) \not\models \gamma$. □

Two other versions of this proof are of interest. Jockusch and Soare [1972a] gave a representation of an arbitrary Π_1^0 class as the complete extensions of an axiomatizable theory with a single binary relation and Hanf [1975] gave a representation as the complete extensions of a finitely axiomatizable theory.

This representation theorem has a number of immediate corollaries.

THEOREM 6.4.

(i) *There is a consistent axiomatizable first-order theory Γ which has no recursive consistent complete extension.*

(ii) *There is a consistent axiomatizable first-order theory Γ such that any two distinct complete consistent extensions of Γ are Turing incomparable, where distinct means having infinite symmetric difference* (*in the Lindenbaum algebra*).

(iii) *There is a consistent axiomatizable first-order theory Γ such that if* $\mathbf{a}$ *is the degree of any complete consistent extension of Γ and* $\mathbf{b}$ *is a r.e. degree with* $\mathbf{a} \leqslant \mathbf{b}$, *then* $\mathbf{b} = \mathbf{0}'$.

(iv) *If* $\mathbf{c}$ *is any r.e. degree, then there exists a consistent axiomatizable first-order theory Γ such that the set of r.e. degrees which contain complete consistent extensions of Γ equals the set of r.e. degrees* $\geqslant \mathbf{c}$.

PROOF. (i) This follows from Theorem 3.12.
(ii) This follows from Theorem 3.14(d).
(iii) This follows from Theorem 3.14(a).
(iv) This follows from Theorem 3.14(b). □

Before presenting further corollaries dealing with isolated points and thin classes, we consider the implications of these concepts for logical theories. For any $\sigma \in \{0,1\}^n$, let ϕ_σ be the conjunction of $\{A_i\colon \sigma(i) = 1\} \cup \{\neg A_i\colon \sigma(i = 0\}$. Then the interval $I(\sigma)$ represents the class of all theories Δ with $\phi_\sigma \in \Delta$. Now suppose that Δ is isolated in the Π_1^0 class $P = P(\Gamma)$ of complete consistent extensions of Γ, so that $P \cap I(\sigma) = \{A\}$ for some σ. Then Δ is clearly generated by $\Gamma \cup \{A\}$. Thus any isolated theory in $P(\Gamma)$ is finitely generated.

Now let Σ be some axiomatizable extension of Γ, so that $P(\Sigma) \subset P(\Gamma)$ and suppose that $P(\Sigma) = P(\Gamma) \cap U$, for some clopen set U. Then U is the union of finitely many intervals, so that by the same reasoning as above, there are finitely many sentences $\gamma_1, \ldots, \gamma_k$ such that $\Delta \in P(\Sigma)$ if and only if $\Delta \in P(\Gamma)$ and $\gamma_i \in \Delta$ for some i. Thus Σ is generated by $\Gamma \cup \{\gamma_1 \vee \gamma_2 \vee \cdots \vee \gamma_k\}$. It follows that if $P(\Gamma)$ is thin, then every consistent axiomatizable extension of Γ is finitely generated. Such a theory Γ is said to be a *Martin–Pour-El theory*.

If $P(\Gamma)$ is minimal, then for any axiomatizable extension Σ of Γ, either $P(\Sigma)$ or $P(\Gamma) \setminus P(\Sigma)$ must be finite.

Martin and Pour-El [1970] constructed an axiomatizable, essentially undecidable theory T such that each axiomatizable extension of T is a finite extension of T. The collection P of complete, consistent extensions of this theory T is the first known example of a thin Π_1^0 class.

THEOREM 6.5.

(i) *For any degree* $\mathbf{a} < \mathbf{0}'$, *there exists a consistent first-order decidable theory Γ with a complete consistent extension Δ of degree* $\mathbf{a}$ *such that Δ is the unique undecidable complete consistent extension of Γ and such that any other complete consistent extension of Γ is finitely generated.*

(ii) *For any degree* $\mathbf{a} < \mathbf{0}''$ *such that* $\mathbf{a}$ *is comparable with* $\mathbf{0}'$, *there exists a consistent first-order axiomatizable theory Γ with a complete consistent extension Δ of* *degree* **a** *such that* Δ *is the unique undecidable complete consistent extension of* Γ *and such that any other complete consistent extension of* Γ *is finitely generated.*

PROOF. (i) This follows from Theorem 4.7.
(ii) This follows from Theorem 4.8. □

THEOREM 6.6.
(i) (Martin–Pour-El) *There is an axiomatizable, essentially undecidable Martin–Pour-El Theory.*
(ii) *There is a decidable Martin–Pour-El theory* Γ *with a complete consistent extension* Δ *such that* Δ *is the unique undecidable complete consistent extension of* Γ*, such that any other complete consistent extension of* Γ *is finitely generated, and such that for any axiomatizable extension* Σ *of* Γ*, either there are only finitely many complete consistent extensions of* Σ *or else all but finitely many complete consistent extensions of* Γ *extend* Σ.
(iii) *There is a Martin-Pour-El theory* Γ *with a complete consistent extension* Δ *of degree* $\mathbf{0}'$ *such that* Δ *is the unique undecidable complete consistent extension of* Γ*, such that any other complete consistent extension of* Γ *is finitely generated, and such that for any axiomatizable extension* Σ *of* Γ*, either there are only finitely many complete consistent extensions of* Σ *or else all but finitely many complete consistent extensions of* Γ *extend* Σ.

PROOF. (i) This follows from Theorem 5.3.
(ii) This follows from Theorem 5.4.
(iii) This follows from Theorem 5.7. □

6.2. *Graph-coloring problems*

A recursive graph $G = (V, E)$ consists of a recursive subset V of ω (the vertices) together with a recursive subset E of $V \times V$ (the edges). We only consider *undirected* graphs, meaning that $(u, v) \in E \iff (v, u) \in E$. The *degree* of a vertex u of G is the cardinality of $\{v\colon (u, v) \in E\}$. A k*-coloring* of the graph G is a map g from V into $\{1, 2, \ldots, k\}$ such that $g(u) \neq g(v)$ whenever $(u, v) \in E$. $G = (V, E)$ is said to be *highly recursive* if there is a partial recursive function $f : V \to \omega$ such that, for each $v \in V$, $f(v)$ is the degree of v. Bean initiated the study of recursive graphs and colorings in [Bean, 1976], where he constructs a k-colorable highly recursive graph with no recursive k-coloring and a 3-colorable recursive graph with no recursive k-coloring for any k. He also proves the following representation theorem.

THEOREM 6.7 (Bean). *For any highly recursive graph G and any finite k, the set of k-colorings of G can be represented by a r.b.* Π_1^0 *class.*

PROOF. Let the vertices of G be enumerated as $v_0, v_1, \ldots$. Then a k-coloring g of G may be represented by a path $x \in \{1, 2, \ldots, k\}^\omega$ such that $x(i) \neq x(j)$

whenever $(v_i, v_j) \in E$. Define a recursive tree $T \subset \{1, 2, \ldots, k\}^{<\omega}$ by allowing $\sigma \in \{1, 2, \ldots, k\}^n \in T$ as long as $\sigma(i) \neq \sigma(j)$ whenever $(v_i, v_j) \in E$. Then T is clearly a highly recursive tree and $P = [T]$ represents the set of k-colorings of G. □

We can derive a number of immediate corollaries as above for logical theories.

THEOREM 6.8. *For any finite k and any k-colorable, highly recursive graph G:*

(i) *G has a k-coloring in some r. e. degree.*

(ii) *If G has no recursive k-coloring, then G has a minimal pair of k-colorings.*

(iii) *If G has only countably many k-colorings, then G has a recursive k-coloring.*

(iv) *If G has only finitely many k-colorings, then every k-coloring of G is recursive.*

(v) *If G has no recursive k-coloring, then then for any countable sequence of nonzero degrees* $\{\mathbf{a}_i\}$, *G has continuum many complete k-colorings g which are mutually Turing incomparable and such that the degree of g is incomparable with each* $\mathbf{a}_i$.

Remmel [1986] showed that, for any $k \geqslant 3$, any r.b. Π_1^0 class P may be represented by the set of k-colorings of some highly recursive graph. Here the representation is finite-to-1, since the colors of any k-coloring may be permuted to obtain $k!$ equivalent colorings.

We omit the proof.

THEOREM 6.9 (Remmel). *For any* $k \geqslant 3$, *any r.b.* Π_1^0 *class P can be represented by the set of k-colorings of a highly recursive graph G. If* $P = [T]$ *where T is a recursive tree with no dead ends, then G may be taken to be decidable.*

The representation Theorem 6.3 has a number of immediate corollaries.

THEOREM 6.10. *For any* $k \geqslant 3$,

(i) *There is a k-colorable, highly recursive graph which has no recursive k-coloring.*

(ii) *There is a k-colorable, highly recursive graph G such that any two distinct k-colorings of G are Turing incomparable, where distinct means up to a permutation of the colors.*

(iii) *There is a k-colorable, highly recursive graph G such that if* $\mathbf{a}$ *is the degree of any k-coloring of G and* $\mathbf{b}$ *is a r.e. degree with* $\mathbf{a} < \mathbf{b}$, *then* $\mathbf{b} = \mathbf{0}'$.

(iv) *If* $\mathbf{c}$ *is any r.e. degree, then there exists a k-colorable, highly recursive graph G such that the set of r.e. degrees which contain k-colorings of G equals the set of r.e. degrees* $\geqslant \mathbf{c}$.

If P is the set of k-colorings of a recursive graph G, then it is clear that an interval represents the k-colorings which extend a certain k-coloring of a finite (initial) subgraph. Thus an isolated k-coloring is the unique extension of some k-coloring of a finite subgraph.

If U is a relative clopen subset of the class $P(G)$ of k-colorings of G, then there is some finite n and a finite set of k-colorings of $\{v_0, \ldots, v_{n-1}\}$ such that U consists of all extensions of those finite colorings. In particular, if h is any recursive k-coloring of a recursive subgraph H of G, then the set Q of extensions of h is a Π_1^0 subclass of P; if this class is relatively clopen in $P(G)$, then there is actually some finite subgraph F of H and a list $f_1, \ldots, f_n$ of k-colorings of F such that Q is the set of all extensions of $f_1, \ldots, f_n$.

Let us say that a coloring g is *finitely determined* if there is some coloring f of a finite subgraph F of G such that g is the unique extension of f to G. Let us say that the k-coloring problem for G is *thin* if, for any recursive k-coloring h of a recursive subgraph H of G, there is a finite subgraph F of G and a list $f_1, \ldots, f_n$ of k-colorings of F such that, for any k-coloring g of G, g extends h if and only if g extends one of $f_1, \ldots, f_n$. If $P(G)$ is minimal, then for any k-coloring h of a recursive subgraph H of G, either h has only finitely many extensions to a k-coloring of G or else all but finitely many k-colorings of G extend h. Let us say that the k-coloring problem for G is decidable if there is an algorithm which decides, for any k-coloring f of a finite subgraph F of G, whether f is extendable to a k-coloring g of G. We have the following additional corollaries of Theorem 6.9.

THEOREM 6.11.

(i) *For any degree* $\mathbf{a} < \mathbf{0}'$, *there exists a k-colorable, highly recursive graph G with a decidable k-coloring problem, and a k-coloring g of G of degree* $\mathbf{a}$ *such that g is the unique nonrecursive k-coloring of G and such that any other k-coloring of G is finitely determined.*

(ii) *For any degree* $\mathbf{a} < \mathbf{0}''$ *such that* $\mathbf{a}$ *is comparable with* $\mathbf{0}'$, *there exists a k-colorable, recursive graph G with a k-coloring g of degree* $\mathbf{a}$ *such that g is the unique nonrecursive k-coloring of G and such that any other k-coloring of G is finitely determined.*

THEOREM 6.12.

(i) *There is a k-colorable, highly recursive graph G with no recursive k-coloring and with a thin k-coloring problem.*

(ii) *There is a k-colorable, highly recursive graph G with a decidable, thin k-coloring problem and a k-coloring g such that g is the unique nonrecursive k-coloring of G, such that any other k-coloring of G is finitely determined, and such that for any recursive coloring h of a recursive subgraph H of G, either there are only finitely many extensions of h to a k-coloring of G, or else all but finitely k-colorings of G extend h.*

(iii) *There is a k-colorable, highly recursive graph G with a thin k-coloring problem and a k-coloring g of degree* $\mathbf{0}'$ *such that g is the unique nonrecursive coloring of G, such that any other k-coloring of G is finitely determined, and such that for any recursive coloring h of a recursive subgraph H of G, either there are only finitely many extensions of h to a k-coloring of G, or else all but finitely k-colorings of G extend h.*

6.3. *Other applications*

We list here a number of problems for which the set of solutions may be represented as a Π_1^0 class.

(1) The marriage problem of Philip and Marshall Hall and other related matching problems, as studied by Manaster and Rosenstein [1972, 1973] and Remmel [1981].

(2) The Hamiltonian circuit problem, studied by D. Harel [1991].

(3) Three problems associated with partially ordered sets (posets).

(a) The problem of covering a partially ordered set (poset) with chains, studied by Kierstead [1981].

(b) The problem of covering a partially ordered set (poset) with antichains, studied by Schmerl [1986].

(c) The problem of expressing a poset as the intersection of linear orderings, studied by Kierstead, McNulty and Trotter [1984].

(4) The set of maximal or of prime ideals in a Boolean algebra or more generally in a commutative ring with unity, studied by Friedman, Simpson and Smith [1983].

(5) The problem of ordering a formally real field, studied by Metakides and Nerode [1979].

(6) Problems from recursive analysis, studied by many including Lacombe [1957, 1958], Nerode and Huang [1985], Cenzer [1993] and Cenzer and Remmel [ta1].

(a) The problem of finding the zeroes of a continuous function.

(b) The problem of finding the extrema of a continuous function.

(c) The problem of finding the fixed points of a continuous function.

(7) The problem of finding a winning strategy for a closed Gale–Stewart game, studied by Cenzer and Remmel [1992].

These problems and others are studied by Cenzer and Remmel [ta2] using the methods developed here for the analysis of Π_1^0 classes.

Two important open problems here are whether the set of maximal ideals of a recursive commutative ring with unity can represent an arbitrary r.b. Π_1^0 class and similarly whether the set of fixed points of a recursive continuous function on the square can represent an arbitrary r.b. Π_1^0 class.

References

D. Bean

[1976] Effective coloration, *J. Symbolic Logic*, 41, pp. 469–480.

D. Cenzer

[1993] Effective dynamics, in: *Logical Methods in honor of Anil Nerode's Sixtieth Birthday*, J. Crossley, J. Remmel, R. Shore and M. Sweedler, eds., Birkhäuser, pp. 162–177.

D. Cenzer, P. Clote, R. Smith, R. Soare and S. Wainer

[1986] Members of countable Π_1^0 classes, *Ann. Pure Appl. Logic*, 31, pp. 145–163.

D. Cenzer, R. Downey, C. Jockusch and R. Shore

[1993] Countable thin Π_1^0 classes, *Ann. Pure Appl. Logic*, 59, pp. 79–139.

D. CENZER AND J. REMMEL

[1992] Recursively presented games and strategies, *Math. Social Sciences*, 24, pp. 117–139.

[1998] Index sets for Π_1^0 classes, *Ann. Pure Appl. Logic*, 43, pp. 3–61.

[ta1] Index sets in computable analysis, *Theoretical Computer Science*, to appear.

[ta2] Π_1^0 classes in mathematics, in: *Recursive Mathematics*, Y. Ershov, S. Goncharov, A. Nerode, J. Remmel, eds., to appear.

D. CENZER AND R. SMITH

[1989] The ranked points of a Π_1^0 set, *J. Symbolic Logic*, 54, pp. 975–991.

P. CHOLAK AND R. DOWNEY

[1993] On the Cantor–Bendixson rank of recursively enumerable sets, *J. Symbolic Logic*, 58, pp. 629–640.

A. CHURCH AND S. KLEENE

[1937] Formal definitions in the theory of ordinal numbers, *Fund. Math.*, 28, pp. 11–21.

J. DEKKER AND J. MYHILL

[1985] Retraceable sets, *Canad. J. Math.*, 10, pp. 357–373.

R. DOWNEY

[1982] *Abstract Dependence, Recursion Theory and the Lattice of Recursively Enumerable Filters*, Dissertation, Monash University.

[1991] On Π_1^0 classes and their ranked points, *Notre Dame J. Formal Logic*, 32, pp. 499–512.

R. DOWNEY, C. JOCKUSCH AND M. STOB

[1990] Array non-recursive sets and multiple permitting arguments, in: *Recursion Theory Week (Proceedings Oberwolfach 1989)*, K. Ambos-Spies, G. Muller and G. Sacks, eds., Lecture Notes in Mathematics, Vol. 1432, Springer, pp. 141–173.

A. EHRENFEUCHT

[1961] Separable theories, *Bull. Acad. Polon. Sci. Sér. Sci. Math. Astronom. Phys.*, 9, pp. 17–19.

H. FRIEDMAN, S. SIMPSON AND R. SMITH

[1983] Countable algebra and set existence theorems, *Ann. Pure and Appl. Logic*, 25, pp. 141–181.

M. GROSZEK AND T. SLAMAN

[ta3] A recursive tree in which every path computes a minimal degree, *Proc. ASL Summer School, Haifa*, to appear.

W. HANF

[1975] The Boolean algebra of logic, *Bull. Amer. Math. Soc.*, 81, pp. 587–589.

D. HAREL

[1991] Hamiltonian paths in infinite graphs, *Israel J. Math.*, 76, pp. 317–336.

P. HINMAN

[1978] *Recursion-Theoretic Hierarchies*, Perspectives in Mathematical Logic, Springer, Berlin.

C. JOCKUSCH

[1974] Π_1^0 classes and Boolean combinations of recursively enumerable sets, *J. Symbolic Logic*, 39, pp. 95–96.

C. JOCKUSCH AND R. SHORE

[1984] Pseudo-jump operators, II, *J. Symbolic Logic*, 49, pp. 1205–1236.

C. JOCKUSCH AND R. SOARE

[1971] A minimal pair of Π_1^0 classes, *J. Symbolic Logic*, 36, pp. 66–78.

[1972a] Π_1^0 classes and degrees of theories, *Trans. Amer. Math. Soc.*, 173, pp. 33–56.

[1972b] Degrees of members of Π_1^0 classes, *Pacific J. Math.*, 40, pp. 605–616.

H. KIERSTEAD

[1981] An effective version of Dilworth's theorem, *Trans. Amer. Math. Soc.*, 268, pp. 63–77.

H. KIERSTEAD, G. MCNULTY AND W. TROTTER

[1984] Recursive dimension for partially ordered sets, *Order*, 1, pp. 67–82.

S. KLEENE

[1943] Recursive predicates and quantifiers, *Trans. Amer. Math. Soc.*, 53, pp. 41–73.

[1950] A symmetric form of Gödel's theorem, *Ind. Math.*, 12, pp. 244–246.

G. KREISEL

[1953] A variant to Hilbert's theory of the foundations of arithmetic, *British J. Philosophy of Science*, 4, pp. 107–129.

[1959] Analysis of the Cantor–Bendixson theorem by means of the analytic hierarchy, *Bull. Acad. Polon. des Sciences, Ser. Math., Astronom. et Phys.*, 7, pp. 621–626.

A. KUCERA

[1985] Measure, Π_1^0 classes and complete extensions of PA, in: *Recursion Theory Week (Proceedings Oberwolfach 1984)*, H.-D. Ebbinghaus, G. Muller and G. Sacks, eds., Lecture Notes in Mathematics, Vol. 1141, Springer, Berlin, pp. 245–259.

[1986] An alternative, priority free, solution to Post's problem, in: *Lecture Notes in Computer Science*, Vol. 233, pp. 493–500.

K. KURATOWSKI AND A. MOSTOWSKI

[1968] *Set Theory*, Studies in Logic, Vol. 86, North-Holland.

D. LACOMBE

[1957] Les ensembles recursivement ouverts ou fermes, et leurs applications a l'analyse recursive, *Comptes Rendus Hebdomaires des Sciences*, 245, pp. 1040–1043.

[1958] Les ensembles recursivement ouverts ou fermes, et leurs applications a l'analyse recursive, *Comptes Rendus Hebdomaires des Sciences*, 246, pp. 405–418.

A. MANASTER AND J. ROSENSTEIN

[1972] Effective matchmaking, *Proc. London Math. Soc.*, 25, pp. 615–654.

[1973] Effective matchmaking and k-chromatic graphs, *Proc. Amer. Math. Soc.*, 39, pp. 371–378.

D. MARTIN AND M. POUR-EL

[1970] Axiomatizable theories with few axiomatizable extensions, *J. Symbolic Logic*, 35, pp. 205–209.

G. MARTIN

[1993] *Cantor Singletons, Rank-Faithful Trees, and Other Topics in Recursion Theory*, Dissertation, University of Maryland.

P. MARTIN-LÖF

[1966] The definition of random sequences, *Information and Control*, 9, pp. 602–619.

G. METAKIDES AND A. NERODE

[1979] Effective content of field theory, *Ann. Math. Logic*, 17, pp. 289–320.

Y. MOSCHOVAKIS

[1980] *Descriptive Set Theory*, Studies in Logic, Vol. 100, North-Holland, Amsterdam.

A. NERODE

[1957] General topology and partial recursive functionals, in: *Summaries of Talks Presented at the Summer Institute for Symbolic Logic*, Cornell University, pp. 247–251.

A. NERODE AND W. HUANG

[1985] Applications of pure recursion theory to recursive analysis, *Acta Sinica*, 28.

P. ODIFREDDI

[1989] *Classical Recursion Theory*, North-Holland, Amsterdam.

J. OWINGS

[1997] Rank, join and Cantor singletons, *Archives of Mathematical Logic*, 36, pp. 313–320.

J. REMMEL

[1981] On the effectiveness of the Schroder–Bernstein theorem, *Proc. Amer. Math. Soc.*, 83, pp. 379–386.

[1986] Graph colorings and recursively bounded Π_1^0 classes, *Ann. Pure Appl. Logic*, 32, pp. 185–194.

H. ROGERS

[1967] *Theory of Recursive Functions and Effective Computability*, McGraw-Hill, New York.

B. ROSSER

[1936] Extensions of some theorems of Gödel and Church, *J. Symbolic*, 1, pp. 87–91.

J. SCHMERL

[1986] Jump number and width, *Order*, 3, pp. 227–234.

D. SCOTT

[1962] Algebras of sets binumerable in complete extensions of arithmetic, in: *Recursive Function Theory*, J. Dekker, ed., Amer. Math. Soc. Proc. Symposia in Pure Math., Vol. 5, pp. 117–121.

J. SHOENFIELD

[1958] Degrees of formal systems, *J. Symbolic*, 23, pp. 389–392.

[1960] Degrees of models, *J. Symbolic Logic*, 25, pp. 233–237.

[1967] *Mathematical Logic*, Addison-Wesley.

S. SIMPSON

[1977] Degrees of unsolvability, in: *Handbook of Mathematical Logic*, J. Barwise, ed., Studies in Logic, Vol. 90, North-Holland, Amsterdam, pp. 631–652.

R. SMITH

[1981] Effective aspects of profinite groups, *J. Symbolic Logic*, 46, pp. 851–863.

R. SOARE

[1969] Recursion theory and Dedekind cuts, *Trans. Amer. Math. Soc.*, 139, pp. 271–294.

[1987] *Recursively Enumerable Sets and Degrees*, Springer, Berlin.

C. SPECTOR

[1961] Inductively defined sets of natural numbers, in: *Infinitistic Methods* (*Proceedings of Warsaw Symposium*), Pergamon Press, pp. 301–384.

Part 2
Reducibilities and Degrees

CHAPTER 3

Reducibilities

Piergiorgio Odifreddi

University of Torino, Italy

Contents

0.1. From Algebra ... 90
0.2. ... to Recursion Theory 90
1. Many-one degrees 91
1.1. Algebraic global structure 93
1.2. Algebraic local structure 95
1.3. Metamathematical properties 96
2. Truth-table degrees 98
2.1. Truth-table reducibilities 99
2.2. Global structure 102
2.3. Local structure 104
3. Enumeration degrees 105
3.1. Global structure 106
3.2. Local structure 107
4. Progress on the problems of "Strong Reducibilities" 108
5. More problems 112
5.1. Truth-table degrees 112
5.2. Recursively enumerable truth-table degrees 113
5.3. Enumeration degrees 114
References 115

HANDBOOK OF COMPUTABILITY THEORY
Edited by E.R. Griffor

A good deal of Recursion Theory classifies sets of natural numbers by means of reflexive and transitive relations (called *reducibilities*), which generate equivalence relations and induce partial orderings on the corresponding equivalence classes (called *degrees*). To provide a feeling for what is done in the subject, we first give an example that does not require any recursion theoretical notion.

0.1. *From Algebra ...*

Let us consider the following relation on real numbers:

$$\alpha \leqslant_a \beta \Leftrightarrow \alpha \text{ is algebraically dependent on } \beta \text{ over the rationals.}$$

$\leqslant_a$ is a reflexive and transitive relation, which generates an equivalence relation $\equiv_a$ as follows:

$$\alpha \equiv_a \beta \Longleftrightarrow \alpha \leqslant_a \beta \text{ and } \beta \leqslant_a \alpha.$$

$\equiv_a$ partitions the reals into equivalence classes called *degrees*. Since $\leqslant_a$ is well-defined on the equivalence classes, it induces a partial ordering $\leqslant$ on the degrees, whose structure is completely determined by the following observations.

- *There are* $2^{\aleph_0}$ *degrees.*
  There are $2^{\aleph_0}$ reals, and each degree contains at most countably many of them (because, given α, there are at most countably many reals algebraic in α).
- *There is a least degree* **0**.
  An algebraic number is algebraic in any real, and any real algebraic in an algebraic real is algebraic. Thus the degree consisting of all algebraic numbers is below any other degree.
- *Every non-zero degree is minimal*, in the sense that there is no other degree between it and **0**.
  Take a non-algebraic real β (i.e. a representative of a non-zero degree), and any other real $\alpha \leqslant_a \beta$. Since β is non-algebraic, the (algebraic closure of the) field $Q[\beta]$ has transcendence degree 1. Since α is algebraically dependent on β, $Q[\alpha] \subseteq Q[\beta]$, and $Q[\alpha]$ must have transcendence degree $\leqslant 1$. If it has degree 0, then α is algebraic and thus is in the degree **0**; if it has degree 1, then $Q[\alpha] = Q[\beta]$, and thus α is in the same degree as β.

In other words, the degrees form a partially ordered structure consisting of a least degree, and of a continuum of pairwise incomparable degrees above it.

0.2. *... to Recursion Theory*

The example just given exhibits a trend quite typical of a substantial part of Recursion Theory: given a reflexive and transitive relation $\leqslant_r$ on the set of reals, one steps to the equivalence relation $\equiv_r$ generated by it, and partitions the reals into r-degrees (usually indicated by boldface letters such as $\boldsymbol{a}, \boldsymbol{b}, \boldsymbol{c}, \ldots$); then one studies the structure $\mathcal{D}_r$ of the r-degrees under the partial ordering $\leqslant$ induced by $\leqslant_r$, with the goal of characterizing it completely. As in the example, there are always $2^{\aleph_0}$ r-degrees, as well as a least r-degree $\mathbf{0}_r$.

A minor difference with the example above is that in Recursion Theory one actually considers not reals but sets of natural numbers (it is well-known that they are related, although different).

A major difference is that neither the final picture of the structure of degrees nor the method of proof of the example above are mathematically very exciting, while the structures of degrees of Recursion Theory are algebraically highly non-trivial, and require a good deal of mathematical sophistication in proofs.

Reducibilities and degrees are not only studied for their own sake, but have also provided a useful tool for the classification of problems (see, for example, the use made of them in Chapter 15). Moreover, they also arise in Higher Recursion Theory and Set Theory (see Chapters 9 and 10), Effective Mathematics (see Chapter 12), and Complexity Theory (see Chapter 18). Thus one can rightly say that we are dealing here with one of the central and most typical features of the subject.

The most popular notion of degree in Recursion Theory is induced by a reducibility introduced by Turing [1939], and appropriately called Turing reducibility ($\leqslant_T$). Its appeal is not only due to its intrinsic interest, as the most general notion of relativized computation, but also to the mathematical challenges that it presents, both in terms of results and proofs: this is why three chapters of this Handbook (Chapters 4–6) are needed to appropriately deal with it.

This chapter is devoted to an overview of the most important arithmetical reducibilities and degrees other than Turing's. They have been selected because of their intrinsic interest, testified not only by their natural definitions and connections with other areas, but also by the quantity of results obtained for them.

For a detailed study of these, as well as for a discussion of many other reducibilities and degrees, the reader is referred to the two volumes of *Classical Recursion Theory* (Odifreddi [1989] and [1999]). For additional and complementary surveys one should instead see Cooper [1990b], Shore [1993], and Nies [1993, 1997].

1. Many-one degrees

A look at undecidability proofs of (first-order) formal systems shows that they fall under two basic types:

- direct proofs, obtained by representing in the given formal system all the r.e. sets, and deducing undecidability from the existence of an r.e. non-recursive set;
- indirect proofs, obtained by effectively interpreting in the given formal system another formal system, whose undecidability has already been established.

An effective interpretation of a formal system $\mathcal{F}_1$ into another one $\mathcal{F}_2$ is usually obtained by an effective translation of formulas that preserves provability, i.e. by an effective function t such that

$$\vdash_{\mathcal{F}_1} \alpha \Longleftrightarrow \vdash_{\mathcal{F}_2} t(\alpha).$$

The method of arithmetization introduced by Gödel [1931] to prove its incompleteness result allows one to identify formulas with numbers, formal systems (or, better, their sets of theorems) with r.e. sets, and effective functions with recursive ones. The notion of translation of formal systems thus motivates the following:

DEFINITION 1.1 (*Post* [1944]). A set A is many-one reducible to a set B $(A \leqslant_m B)$ if there is a recursive function f such that, for all x,

$$x \in A \Longleftrightarrow f(x) \in B.$$

A is many-one equivalent to B $(A \equiv_m B)$ if $A \leqslant_m B$ and $B \leqslant_m A$.

Formal systems proved undecidable by the first type of undecidability proofs are universal, in the sense that all other formal systems are interpretable in them: they correspond to *m-complete r.e. sets*, to which all other r.e. sets are m-reducible.

Formal systems proved undecidable by interpreting in them a universal formal system are still universal: this corresponds to the fact that r.e. sets to which an m-complete r.e. set is m-reducible are still m-complete.

If all undecidable formal systems could be proved to be so by the two types of proofs they would all be universal, and formal systems could be neatly classified into decidable and universal: this corresponds to the problem of whether all non-recursive r.e. sets are m-complete, and was posed and solved negatively by Post [1944].

The negative solution was based on the observation, already noted explicitly by Turing [1939], that Gödel's way of producing formally undecidable sentences in universal formal systems could be iterated, thus effectively generating an infinite set of such formulas: this corresponds to the fact that the complement of an r.e. m-complete set has an infinite r.e. subset.

The solution to Post's problem can thus be obtained by constructing a *simple set*, i.e. an r.e. set whose complement is infinite, but does not have any infinite r.e. subset (notice that a coinfinite recursive set cannot be simple, since its complement is itself an infinite r.e. set). A simple set was constructed by Post, by a straightforward diagonalization against all infinite r.e. sets.

To deduce from this the existence of an undecidable formal system whose undecidability cannot be proved by the two methods discussed above one would still need to show that all r.e. m-degrees represent formal systems, but this is not so. Indeed, for any first-order formal system $\mathcal{F}$ there is an effective translation of pairs of formulas that preserves provability:

$$\vdash_{\mathcal{F}} \alpha \text{ and } \vdash_{\mathcal{F}} \beta \Longleftrightarrow \vdash_{\mathcal{F}} \alpha \wedge \beta,$$

but Fisher [1963] proved that there are r.e. sets A such that no recursive function f of pairs of elements exists that preserves membership, i.e. such that:

$$x \in A \text{ and } y \in A \Longleftrightarrow f(x, y) \in A.$$

To represent formal systems one thus needs a weakening of the notion of m-reducibility, introduced by Post and discussed in the next section.

These developments however started two programs for the study of the r.e. sets: on the one hand, a set-theoretical classification in terms of properties such as simplicity (which is the subject of Chapter 7); on the other hand, an algebraic-theoretical classification in terms of degrees (which is the subject of Chapters 3 and 6).

Moreover, once degrees were introduced, a degree-theoretical classification could be sought for all (and not only the r.e.) sets: this program was launched in Kleene and Post [1954], and is the subject of Chapters 3 and 5.

1.1. *Algebraic global structure*

Our main goal is to characterize the structure $\mathcal{D}_m$ of m-degrees as a partially ordered structure, w.r.t. the partial order $\leqslant$ induced by $\leqslant_m$.

$\mathcal{D}_m$ has a least element $\mathbf{0}_m$, i.e. the m-degree containing the non-trivial recursive sets (by convention one does not consider $\emptyset$ and ω, whose m-degrees $\{\emptyset\}$ and $\{\omega\}$ would be incomparable, and below any other m-degree).

Since there are only countably many m-reductions (i.e. recursive functions), $\mathcal{D}_m$ has $2^{\aleph_0}$ elements, each of them with at most countably many predecessors.

$\mathcal{D}_m$ is an uppersemilattice, in the sense that it admits a l.u.b. operation $\cup$, induced by the following operation (of disjoint union) on sets:

$$A \oplus B = \{2x \colon x \in A\} \cup \{2x + 1 \colon x \in B\}.$$

The main property distinguishing the m-degrees from practically all other notions of degrees used in Recursion Theory (with the sole exception of r.e. wtt-degrees) is distributivity, which amounts to the following property: if $\boldsymbol{a} \leqslant \boldsymbol{b} \cup \boldsymbol{c}$ then there are $\boldsymbol{b}_1 \leqslant \boldsymbol{b}$ and $\boldsymbol{c}_1 \leqslant \boldsymbol{c}$ such that $\boldsymbol{a} = \boldsymbol{b}_1 \cup \boldsymbol{c}_1$.[1]

$\mathcal{D}_m$ is distributive because if

$$x \in A \Longleftrightarrow f(x) \in B \oplus C$$

then $A \equiv_m B_1 \oplus C_1$ with $B_1 \leqslant_m B$ and $C_1 \leqslant_m C$, where

$$B_1 = \left\{x \colon f(x) \text{ even and } \frac{f(x)}{2} \in B\right\}$$

[1] More generally, an uppersemilattice $(A, \sqsubseteq, \sqcup)$ is distributive if, whenever $a \sqsubseteq b \sqcup c$, there are $b_1 \sqsubseteq b$ and $c_1 \sqsubseteq c$ such that $a = b_1 \sqcup c_1$. When the uppersemilattice is a lattice, i.e. it also admits a g.l.b. operation $\sqcap$, this reduces to the usual distributivity, i.e. to the law $a \sqcap (b \sqcup c) = (a \sqcap b) \sqcup (a \sqcap c)$, since then it is enough to let $b_1 = a \sqcap b$ and $c_1 = a \sqcap c$.

and

$$C_1 = \left\{x\colon\ f(x) \text{ odd and } \frac{f(x)-1}{2} \in C\right\}.$$

THEOREM 1.2 (Lachlan [1970]). *The principal ideals of $\mathcal{D}_m$ are exactly the direct limits (preserving* 0, 1 *and l.u.b.'s) of ascending sequences of finite distributive lattices.*

SKETCH OF PROOF. In a distributive uppersemilattice every finite subset closed under l.u.b.'s is embedded as an uppersemilattice (i.e. with l.u.b.'s preserved) in a finite distributive sublattice. Since a principal ideal is countable, the necessity of the condition follows.

For sufficiency, one has to build sets A such that the m-degrees below the m-degree $\boldsymbol{a}$ of A are isomorphic to given uppersemilattices. The basic tool is the observation that the map Φ defined as follows:

$$x \in \Phi(\mathcal{W}_e) \Longleftrightarrow f(x) \in A,$$

where f is any recursive function with range $\mathcal{W}_e$, induces a homomorphism of uppersemilattices between the r.e. sets modulo the finite sets (ordered by inclusion), and the m-degrees $\leqslant \boldsymbol{a}$. Thus, instead of controlling directly the m-degrees below A one controls the m-degrees of $\Phi(\mathcal{W}_e)$, for every r.e. set $\mathcal{W}_e$. □

This completely characterizes the principal ideals of $\mathcal{D}_m$, and shows in particular that: the finite principal ideals are exactly the finite distributive lattices; every countable distributive lattice is isomorphic to an ideal; and $\mathcal{D}_m$ is not a lattice.

THEOREM 1.3 (Ershov [1975], Paliutin [1975]). *The ideals of $\mathcal{D}_m$ are exactly the distributive uppersemilattices with least element, countable predecessor property, and at most $2^{\aleph_0}$ elements.*

SKETCH OF PROOF. Uncountable ideals cannot be approximated by finite pieces in countably many steps, and thus one approaches them by extending given countable pieces, in at most $2^{\aleph_0}$ steps (this is possible because of the countable predecessor property).

The main part of the proof is to show that every countable ideal $\mathcal{I}$ of $\mathcal{D}_m$ can be extended to an ideal $\mathcal{L}$ isomorphic to L, for any countable distributive uppersemilattice L having an ideal I isomorphic to $\mathcal{I}$.

To ensure one-oneness of the embedding without any appeal to the Continuum Hypothesis, one needs to show that there actually is a continuum of ideals isomorphic to L, and such that their parts isomorphic to $L - I$ are pairwise disjoint. □

This completely characterizes the ideals of $\mathcal{D}_m$, and shows in particular that: the well-ordered ideals are exactly the ordinals $\leqslant \omega_1$; and every countable ideal is the intersection of two principal ideals.

THEOREM 1.4 (Characterization of $\mathcal{D}_m$ (Ershov [1975], Paliutin [1975])). *Up to isomorphism, $\mathcal{D}_m$ is the only structure with the following properties:*
- *it is a distributive uppersemilattice with least element,*
- *every element has at most countably many predecessors,*
- *it has $2^{\aleph_0}$ elements,*
- *every principal ideal $\mathcal{I}$ with fewer than $2^{\aleph_0}$ elements can be extended to an ideal $\mathcal{L}$ isomorphic to L, for any distributive uppersemilattice L with less than $2^{\aleph_0}$ elements, and having an ideal I isomorphic to $\mathcal{I}$.*

The previous result provides an alternative, recursion-theoretical characterization of the continuum, and is the only known example of absolute characterization (i.e. proved in *ZFC*) of a degree structure in Recursion Theory. In particular, it easily provides answers to all global algebraic questions one could ask about the m-degrees.

THEOREM 1.5 (Global properties of $\mathcal{D}_m$).
- Homogeneity. *Any two principal filters of $\mathcal{D}_m$ are isomorphic.*
- Definability. $\mathbf{0}_m$ *is the only definable m-degree, and every non-empty definable set of m-degrees different from $\{\mathbf{0}_m\}$ has $2^{\aleph_0}$ elements.*
- Automorphisms. *$\mathcal{D}_m$ admits $2^{2^{\aleph_0}}$ automorphisms.*

All these results are best possible: principal ideals are not always isomorphic; there are definable sets with $2^{\aleph_0}$ elements (e.g. the set of minimal m-degrees); and there are only $2^{2^{\aleph_0}}$ functions from $\mathcal{D}_m$ to $\mathcal{D}_m$ (since $\mathcal{D}_m$ has $2^{\aleph_0}$ elements).

1.2. *Algebraic local structure*

We turn now to a study of the structure $\mathcal{R}_m$ of the r.e. m-degrees, i.e. of the m-degrees containing r.e. sets.

Since a set m-reducible to an r.e. set is still r.e., an r.e. m-degree contains only r.e. sets, and $\mathcal{R}_m$ is an ideal of $\mathcal{D}_m$. In particular, $\mathbf{0}_m$ is still the least element.

Moreover, the existence of m-complete r.e. sets implies the existence of a greatest r.e. m-degree $\mathbf{0}'_m$, and $\mathcal{R}_m$ is actually a principal ideal of $\mathcal{D}_m$. In particular, $\mathcal{R}_m$ is still a distributive uppersemilattice.

Since there are only countably many r.e. sets, $\mathcal{R}_m$ has countably many elements.

The asymmetry between a set and its complement, which is a basic feature of r.e. sets, is preserved by m-reducibility: this allows the structure of $\mathcal{R}_m$ to be a nice reflection of the structure of $\mathcal{D}_m$, an exceptional case among all usual reducibilities and degrees. The trend of global results about $\mathcal{R}_m$ is thus that they are effective versions of global results about $\mathcal{D}_m$, obtained by constructivizing their proofs. To be able to state them precisely, we need the following notions.

An *effective lattice* $(L, \sqsubseteq, \sqcup, \sqcap)$ is a (countable) lattice admitting a (not necessarily effective) enumeration $\{a_x\}_{x\in\omega}$ of its elements such that $a_x \sqsubseteq a_y$ is a Π^0_2 relation of x and y, and there are recursive functions f and g such that

$$a_x \sqcup a_y = a_{f(x,y)} \quad \text{and} \quad a_x \sqcap a_y = a_{g(x,y)}.$$

An *effective distributive uppersemilattice* is a direct limit (preserving 0, 1 and l.u.b.'s) of an ascending r.e. sequence of finite, uniformly effective distributive lattices.

THEOREM 1.6 (Lachlan [1972a]). *The principal ideals of $\mathcal{R}_m$ are, up to isomorphism, exactly the effective distributive uppersemilattices.*

This completely characterizes the principal ideals of $\mathcal{R}_m$, and shows in particular that: the finite principal ideals are exactly the finite distributive lattices; the well-ordered ideals are exactly the ordinals $\leqslant \omega_1^{ck}$ (the first non-recursive ordinal); and $\mathcal{R}_m$ is not a lattice.

THEOREM 1.7 (Characterization of $\mathcal{R}_m$ (Denisov [1978])). *Up to isomorphism, $\mathcal{R}_m$ is the only structure with the following properties*:
- *it is an effective distributive uppersemilattice with least and greatest element,*
- *every principal ideal $\mathcal{I}$ not containing the greatest element can be extended to an ideal $\mathcal{L}$ effectively isomorphic to L, for any effective distributive uppersemilattice L having an ideal I effectively isomorphic to $\mathcal{I}$.*

The previous result is the only example of a characterization of an r.e. degree structure in Recursion Theory, and it easily provides answers to all global algebraic questions one could ask about the r.e. m-degrees.

THEOREM 1.8 (Global properties of $\mathcal{R}_m$).
- Homogeneity. *Any two non-trivial principal filters of $\mathcal{R}_m$ (i.e. generated by an m-degree $< \mathbf{0}'_m$) are isomorphic.*
- Definability. *$\mathbf{0}_m$ and $\mathbf{0}'_m$ are the only definable r.e. m-degrees, and every non-empty definable set of r.e. m-degrees different from $\{\mathbf{0}_m\}$ and $\{\mathbf{0}'_m\}$ is countably infinite.*
- Automorphisms. *$\mathcal{R}_m$ admits $2^{\aleph_0}$ automorphisms.*

All these results are best possible: principal ideals are not always isomorphic; there are infinite coinfinite definable sets (e.g., the set of minimal r.e. m-degrees); and there are only $2^{\aleph_0}$ functions from $\mathcal{R}_m$ to $\mathcal{R}_m$ (since $\mathcal{R}_m$ is countable).

1.3. *Metamathematical properties*

The properties of (r.e.) m-degrees studied in the previous subsections are of algebraic nature, but one can also take a logical point of view.

THEOREM 1.9 (Nerode and Shore [1980]). *The first-order theory of $\mathcal{D}_m$ has the same complexity as the theory of second-order arithmetic.*

SKETCH OF PROOF. A translation of second-order arithmetic into the first-order theory of $\mathcal{D}_m$ (the opposite translation being trivial) is made through the following steps: a translation of second-order arithmetic into second-order logic on countable sets (trivial); a translation of second-order logic on countable sets into the theory of countable distributive lattices with quantification over ideals, by coding relations by graphs, and graphs by ideals of a distributive lattice; a translation of the theory of countable distributive lattices with quantification over ideals into the first-order theory of $\mathcal{D}_m$, by using the facts that every countable distributive lattice is isomorphic to an ideal of $\mathcal{D}_m$, and that every countable ideal is the intersection of two principal ideals (this latter fact is used to translate a quantification over ideals into a quantification over pairs of m-degrees, and thus second-order quantification into a first-order one). □

It follows in particular that the first-order theory of $\mathcal{D}_m$ is undecidable, and not axiomatizable. This was originally proved by Lachlan [1970], and requires only the fact that every finite distributive lattice is isomorphic to an ideal.

THEOREM 1.10 (Degtev [1979], Nies [1996]). *The two-quantifier theory of $\mathcal{D}_m$ is decidable, and the three-quantifier theory is undecidable.*

SKETCH OF PROOF. A two-quantifier sentence of the form $\forall\exists$ in the language of partial orderings says that whenever certain elements satisfy a given fixed configuration, there are other elements satisfying another fixed configuration: in other words, it says that an embedding of a given partial ordering can be extended to an embedding of another given partial ordering (in the full $\forall\exists$ theory, one actually has a disjunction of possible extensions). To decide the two-quantifier theory of $\mathcal{D}_m$ is thus enough to decide which extensions of finite embeddings can be realized in the m-degrees: the structure of distributive uppersemilattices and the possibility of having g.l.b.'s impose certain obstructions, whose avoidance is shown to be sufficient by an extension of embeddings theorem, and necessary by the characterization of finite ideals (obtained from Theorem 1.2). □

Turning to r.e. m-degrees, one has appropriate local versions of the previous results.

THEOREM 1.11 (Nies [1995]). *The first-order theory of $\mathcal{R}_m$ has the same complexity as the theory of first-order arithmetic.*

SKETCH OF PROOF. A translation of first-order arithmetic into the first-order theory of $\mathcal{R}_m$ (the opposite translation being trivial) relies on the following definability result: for each n, any countable Σ_n^0 relation bounded below $\mathbf{0}'_m$ is definable from finitely many parameters, in a uniform way.

A standard model of arithmetic bounded below $\mathbf{0}'_m$ exists in $\mathcal{R}_m$, since every countable partial ordering can be so embedded. By choosing the set of r.e. m-degrees representing numbers in a sufficiently simple way (e.g., as an r.e. sequence), such a model turns out to be definable using the previous definability result. For each model coded in the same way (and satisfying some basic axioms of arithmetic) the standard part is of arithmetic complexity: by quantifying over the parameters defining sets of the same complexity, one can then express in a first-order way the fact that the coded model is standard, thus showing that the standard model is definable in $\mathcal{R}_m$. □

It follows in particular that the first-order theory of $\mathcal{R}_m$ is undecidable, and not axiomatizable. This was originally proved by Lachlan [1972a], and requires only the fact that every finite distributive lattice is isomorphic to an ideal.

THEOREM 1.12 (Denisov [1978], Nies [1996]). *The two-quantifier theory of $\mathcal{R}_m$ is decidable, and the three-quantifier theory is undecidable.*

2. Truth-table degrees

The inadequacy of m-reducibility for the study of formal systems was exposed by the fact that there are r.e. sets such that the conjunction of two statements about them is not m-reducible to a single statement (see Fischer's result quoted above). One can thus try to obtain a reducibility appropriate to the study of formal systems by generalizing m-reducibility, in such a way as to make it at least invariant w.r.t. propositional manipulations.

To state a precise definition, let $\{\sigma_n\}_{n\in\omega}$ be an effective enumeration of all the propositional formulas, built from the atomic ones '$m \in X$', for $m \in \omega$. These are also called *truth-table conditions*, since they can be arranged in truth-tables. Given a set B, $B \models \sigma_n$ means that B satisfies σ_n, i.e. that the propositional formula σ_n becomes true when the variable X in the atomic formulas is interpreted as B.

DEFINITION 2.1 (*Post* [1944]). A is tt-reducible to B ($A \leqslant_{tt} B$) if there is a recursive function f such that, for all x,

$$x \in A \Longleftrightarrow B \models \sigma_{f(x)}.$$

A is tt-equivalent to B ($A \equiv_{tt} B$) if $A \leqslant_{tt} B$ and $B \leqslant_{tt} A$.

From the point of view of studying formal systems, tt-reducibility turns out to be appropriate: Feferman [1957] (and Hanf [1965]) proved that every r.e. tt-degree contains a (finitely axiomatizable) first-order formal system. More generally, tt-reducibility is appropriate also for the study of arbitrary logical theories,[2] since every tt-degree contains a first-order logical theory.

One can then go back to the original problem of classifying methods to prove undecidability, by allowing not only many-one translations of a formal system into another, but truth-table ones. Formal systems can now be universal in the generalized sense that all other formal systems are interpretable in them by means of truth-table translations: they correspond to *tt-complete r.e. sets*, to which all other r.e. sets are tt-reducible.

Formal systems proved undecidable by interpreting in them a universal formal system (in the generalized sense) are still universal: this corresponds to the fact that r.e. sets to which a tt-complete r.e. set is tt-reducible are still tt-complete.

If all undecidable formal systems could be proved to be so by the two types of proof previously discussed they would all be universal, and formal systems could still be neatly classified into decidable and universal: this corresponds to the problem of whether all non-recursive r.e. sets are tt-complete, and was again posed and solved negatively by Post [1944].

For a negative solution simple sets are not enough, since Post proved that there are simple tt-complete sets. He generalized the notion to *hypersimple sets*, by requiring not only that there is no infinite r.e. set contained in the complement (i.e. no infinite r.e. set of singletons all intersecting the complement), but that there is no infinite r.e. set of disjoint finite set (described explicitly, by means of their elements) all intersecting the complement.

Hypersimple sets are obviously not recursive (being simple), and Post proved that they are also not tt-complete. Moreover, by a non-trivial construction that anticipates some of the features of the *priority method*, he also showed that hypersimple sets exist, thus proving that there are undecidable formal systems that cannot be proved undecidable by interpreting in them a universal formal system (even in the generalized sense of truth-table translations).

2.1. *Truth-table reducibilities*

Since every propositional formula can be written in disjunctive or conjunctive normal form, using only the complete set of connectives $\{\neg, \wedge, \vee\}$, a number of possible variations of tt-reducibility immediately comes to mind, e.g. by restricting that set to:

- $\{\wedge, \vee\}$: positive reducibility ($\leqslant_p$),
- $\{\wedge\}$: conjunctive reducibility ($\leqslant_c$),
- $\{\vee\}$: disjunctive reducibility ($\leqslant_d$),
- $\{\neg\}$: bounded truth-table reducibility with norm 1 ($\leqslant_{btt(1)}$).

---

[2] The difference between formal systems and arbitrary logical theories is that the former are axiomatizable, and thus their sets of theorems are r.e., while the latter are not necessarily axiomatizable, and thus their sets of theorems are arbitrary sets.

Propositional connectives provide canonical ways of looking at Boolean functions, but they leave out a natural one: sum modulo 2, which gives rise to a further variation of tt-reducibility:

- linear reducibility ($\leqslant_l$)reducibility.

The next result justifies the list just given.

THEOREM 2.2 (Bulitko [1980], Selivanov [1982]). *The only truth-table reducibilities on non-trivial sets (different from $\emptyset$ and ω) are the following seven: $\leqslant_m$, $\leqslant_{btt(1)}$, $\leqslant_c$, $\leqslant_d$, $\leqslant_p$, $\leqslant_l$ and $\leqslant_{tt}$.*

On the non-trivial r.e. sets the truth-table reducibilities reduce to six, since $\leqslant_{btt(1)}$ and $\leqslant_m$ then coincide.

THEOREM 2.3 (Post [1944], Shoenfield [1957], Jockusch [1968], Degtev [1973, 1981, 1982], Bulitko [1980]). *On the positive side, the following implications hold on the r.e. sets, and no other does even on the complete sets*:

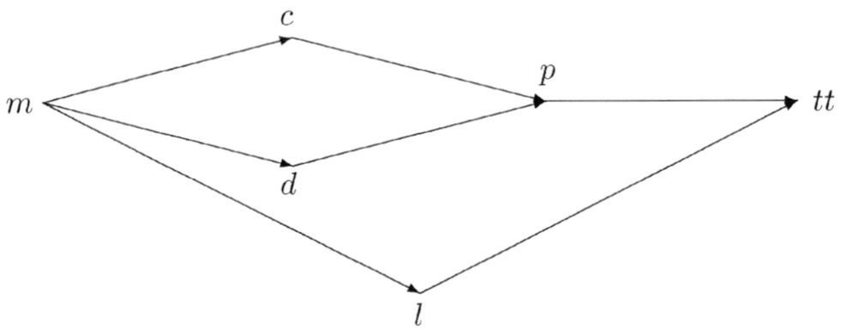

On the negative side, there are non-recursive r.e. sets on which all the reducibilities collapse; more precisely, there is an r.e. tt-degree that consists of only one r.e. m-degree.

The degrees induced by the six reducibilities provide different pictures of the world of r.e. sets.

THEOREM 2.4 (Degtev [1979, 1982, 1983a, 1983b]). *The structures of the r.e. m-, c-, d-, p-, l- and tt-degrees are pairwise not elementarily equivalent.*

SKETCH OF PROOF. The sentence

$$(\forall \boldsymbol{a} \neq \boldsymbol{0})(\exists \boldsymbol{b} \not\geqslant \boldsymbol{a})(\exists \boldsymbol{c} \not\geqslant \boldsymbol{a})(\boldsymbol{a} \leqslant \boldsymbol{b} \cup \boldsymbol{c})$$

is true for the r.e. d-, p-, l- and tt-degrees, but false for the r.e. m- and c-degrees.

The sentence expressing distributivity for uppersemilattices, i.e.

$$(\forall \boldsymbol{a})(\forall \boldsymbol{b})(\forall \boldsymbol{c})\big[\boldsymbol{a} \leqslant \boldsymbol{b} \cup \boldsymbol{c} \to (\exists \boldsymbol{b}_1 \leqslant \boldsymbol{b})(\exists \boldsymbol{c}_1 \leqslant \boldsymbol{c})(\boldsymbol{a} = \boldsymbol{b}_1 \cup \boldsymbol{c}_1)\big]$$

is true for the r.e. m-degrees but false for the r.e. c-degrees.

The sentence

$$(\exists \boldsymbol{a} \neq \mathbf{0})(\forall \boldsymbol{b})(\forall \boldsymbol{c})(\boldsymbol{c} \leqslant \boldsymbol{a} \cup \boldsymbol{b} \rightarrow \boldsymbol{c} \leqslant \boldsymbol{b} \vee \boldsymbol{a} \leqslant \boldsymbol{b} \cup \boldsymbol{c})$$

is true for the r.e. l-degrees but false for the r.e. d-, p- and tt-degrees.

The sentence

$$(\forall \boldsymbol{a} \neq \mathbf{0})(\exists \boldsymbol{b} \not\geqslant \boldsymbol{a})(\exists \boldsymbol{c} \not\geqslant \boldsymbol{a})(\boldsymbol{a} \leqslant \boldsymbol{b} \cup \boldsymbol{c} \wedge \boldsymbol{b} \leqslant \boldsymbol{a} \cup \boldsymbol{c})$$

is true for the r.e. tt-degrees but false for the r.e. d- and p-degrees.

Finally, the sentence

$$(\forall \boldsymbol{a} \neq \mathbf{0})(\exists \boldsymbol{b} \neq \mathbf{0})\big[\boldsymbol{b} \leqslant \boldsymbol{a} \wedge \\ (\forall \boldsymbol{c})(\forall \boldsymbol{d})(\boldsymbol{b} \leqslant \boldsymbol{c} \cup \boldsymbol{d} \wedge \boldsymbol{c} \leqslant \boldsymbol{b} \cup \boldsymbol{d}) \rightarrow (\exists \boldsymbol{e} \leqslant \boldsymbol{b}, \boldsymbol{c})(\boldsymbol{b} = \boldsymbol{d} \cup \boldsymbol{d})\big]$$

is true for the r.e. d-degrees but false for the r.e. p-degrees. □

On arbitrary sets, one has instead the following slightly more complicated picture.

THEOREM 2.5 (Jockusch [1968], Degtev [1982]). *On the positive side, the following implications hold on arbitrary sets, and no other does*:

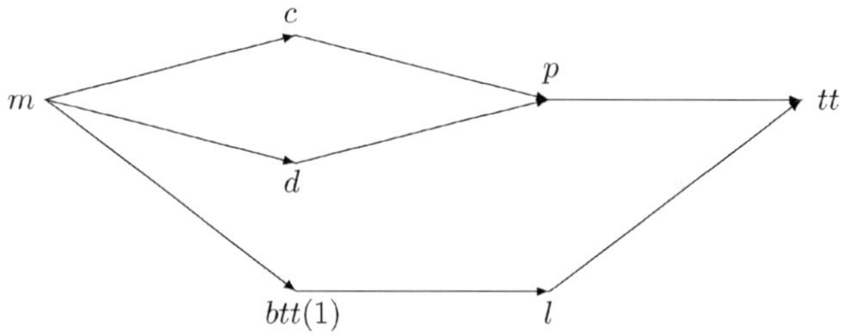

Moreover, p- and tt-reducibility never collapse in a non-trivial way, in the sense that each non-recursive tt-degree contains at least two p-degrees.

On the negative side, the following reducibilities can collapse in a non-trivial way: m- and p-; m- and l-; l- and tt-. More precisely, there are: non-recursive p-degrees consisting of only one m-degree; non-recursive l-degrees consisting of only one m-degree; and non-recursive tt-degrees consisting of only one l-degree.

SKETCH OF PROOF. The collapse of p- and m-reducibility is achieved by considering a *semirecursive set*, i.e. a set A for which there is a recursive function f of two variables such that:

- $f(x, y) = x$ or $f(x, y) = y$,
- if $x \in A$ or $y \in A$ then $f(x, y) \in A$.

If $B \leqslant_p A$ then $B \leqslant_m A$, and B is semirecursive too; thus the p-degree of a semirecursive set consists of a single m-degree.

The separation of tt- and p-reducibility is achieved by associating to every set A its *tt-cylinder* A^{tt}, defined as

$$A^{tt} = \{x\colon A \models \sigma_x\}.$$

Then $A \equiv_{tt} A^{tt}$, and it is enough to show that the tt-degree of A contains a semirecursive set B, and that if $A^{tt} \leqslant_p B$ then A is recursive.

The collapse of l- and m-reducibility is achieved by considering a set A for which there is a recursive function f of two variables such that:

$$f(x, y) \in A \Longleftrightarrow (x \in A \wedge y \notin A) \vee (x \notin A \wedge y \in A).$$

If $B \leqslant_l A$ then $B \leqslant_m A$; thus if A has minimal m-degree, its l-degree consists of a single m-degree. □

THEOREM 2.6 (Degtev [1979, 1985], Mal'cev [1985]). *The structures of m- and $btt(1)$-degrees are isomorphic, and so are the structures of c- and d-degrees.*

The structures of m-, c-, p-, l- and tt-degrees are pairwise not elementarily equivalent, with the only possible exception of p- and tt-degrees.

The isomorphism of c- and d-degrees is trivial, since

$$A \leqslant_c B \Longleftrightarrow \overline{A} \leqslant_d \overline{B}.$$

2.2. *Global structure*

Our main goal would now be to characterize the structure $\mathcal{D}_{tt}$ of tt-degrees as a partially ordered structure, w.r.t. the partial ordering $\leqslant$ induced by $\leqslant_{tt}$.

Like $\mathcal{D}_m$, $\mathcal{D}_{tt}$ is an uppersemilattice with least element $\mathbf{0}_{tt}$ (containing the recursive sets), countable predecessor property, and $2^{\aleph_0}$ elements.

Unlike $\mathcal{D}_m$, $\mathcal{D}_{tt}$ is not however a distributive uppersemilattice, and this makes its study more complicated (in particular, no complete characterization of the structure is known yet).

THEOREM 2.7 (Nerode and Shore [1980], Abraham and Shore [1986]). *The ideals of $\mathcal{D}_{tt}$ with at most $\aleph_1$ elements are exactly the uppersemilattices with least element, and countable predecessor property.*

SKETCH OF PROOF. A T-degree $\boldsymbol{a} \neq \mathbf{0}$ is called *hyperimmune-free*[3] if for every $A \in \boldsymbol{a}$ and $f \leqslant_T A$, f is majorized by a recursive function, i.e. there is a recursive function g such that $(\forall x)(f(x) \leqslant g(x))$.

If A has hyperimmune-free T-degree and $B \leqslant_T A$ then there is a recursive function majorizing the number of steps needed for the T-reduction to converge, and thus one actually has $B \leqslant_{tt} A$. In particular, the T-degrees below a hyperimmune-free T-degree are all tt-degrees.

Since it is a trivial matter to modify the constructions of ideals of T-degrees by making all the involved T-degrees hyperimmune-free (actually, this is automatically ensured by the usual constructions), one gets for tt-degrees the same ideals as for the T-degrees. □

This completely characterizes the countable ideals (as the uppersemilattices with least element), the principal ideals (which are countable by the countable predecessor property), the linearly ordered ideals (since a linear ordering with countable predecessor property can have at most $\aleph_1$ elements), and the well-ordered ideals (which are exactly the ordinals $\leqslant \omega_1$).

The result also completely characterizes the ideals of $\mathcal{D}_{tt}$ if the Continuum Hypothesis holds, and it is the best one can do without additional set-theoretical assumptions, since Groszek and Slaman [1983] have shown that it is consistent with ZFC that there are uncountable uppersemilattices with countable predecessor property which are not embeddable in $\mathcal{D}_{tt}$ even as uppersemilattices (let alone as ideals).

Practically nothing is known about extensions of ideals, except for the fact that even the simplest extension (i.e. adding a new top element) cannot be done in general. This is a consequence of the proof of the next result, which provides the only known answer to the usual global algebraic questions asked about degrees.

THEOREM 2.8 (Failure of homogeneity (Mohrherr [1984])). *The principal filter generated by $\mathbf{0}'_{tt}$ is not elementarily equivalent to $\mathcal{D}_{tt}$.*

SKETCH OF PROOF. By Theorem 2.7 there are minimal tt-degrees, but the tt-degrees above $\mathbf{0}'_{tt}$ are dense. □

We turn now to logical properties.

THEOREM 2.9 (Nerode and Shore [1980]). *The first-order theory of $\mathcal{D}_{tt}$ has the same complexity as the theory of second-order arithmetic.*

SKETCH OF PROOF. As in Theorem 1.9, it is enough to show that every countable distributive lattice is isomorphic to an ideal of $\mathcal{D}_{tt}$ (this follows from Theorem 2.7),

[3] The name comes from the fact that such a degree does not contain *hyperimmune sets*, i.e. sets having the property defining the complement of a hypersimple set (this is actually an equivalent alternative definition of the notion).

and that every countable ideal is the intersection of two principal ideals (this can easily be proved directly). □

It follows in particular that the first-order theory of $\mathcal{D}_{tt}$ is undecidable, and not axiomatizable.

THEOREM 2.10 (Shore [1978], Lerman [1983], Schmerl). *The two-quantifier theory of $\mathcal{D}_{tt}$ is decidable, and the three-quantifier theory is undecidable.*

SKETCH OF PROOF. Decidability follows in a way similar to Theorem 1.10. Undecidability is obtained by interpreting the theory of finite graphs (whose two-quantifier theory is strongly undecidable) into the theory of finite lattices, and by using the characterization of finite ideals (obtained from Theorem 2.7). □

2.3. *Local structure*

We turn now to a study of the structure $\mathcal{R}_{tt}$ of the r.e. tt-degrees, i.e. of the tt-degrees containing r.e. sets.

Like $\mathcal{R}_m$, $\mathcal{R}_{tt}$ has a least element $\mathbf{0}_{tt}$ (containing the recursive sets) and a greatest element $\mathbf{0}'_{tt}$ (containing the tt-complete sets), and is a countable uppersemilattice.

Unlike $\mathcal{R}_m$, $\mathcal{R}_{tt}$ is neither an ideal of $\mathcal{D}_{tt}$, nor a distributive uppersemilattice.

THEOREM 2.11 (Haught and Shore [1990a], Harrington and Haught [199?]). *On the negative side, every non-trivial finite ideal of $\mathcal{R}_{tt}$ has a least element above $\mathbf{0}_{tt}$.*

On the positive side, every finite dual partition lattice[4] *with an added least element, as well as every finite Boolean algebra with an added least element, is isomorphic to an ideal of $\mathcal{R}_{tt}$.*

This is a far cry from a complete characterization of even the finite ideals of $\mathcal{R}_{tt}$, but it is sufficient to prove that $\mathcal{R}_{tt}$ is undecidable and not axiomatizable.

THEOREM 2.12 (Nies and Shore [1995]). *The first-order theory of $\mathcal{R}_{tt}$ has the same complexity as the theory of first-order arithmetic.*

SKETCH OF PROOF. A translation of first-order arithmetic into the first-order theory of $\mathcal{R}_{tt}$ (the opposite translation being trivial) is made through the definition of a standard model of first-order arithmetic in an infinite dual partition lattice, with an added least element. The results about $\mathcal{R}_{tt}$ needed for the translation are the construction of a sufficiently simple ideal (precisely, of a Σ^0_3 ideal of T-incomplete tt-degrees) isomorphic to such a lattice (to define the model), and the fact that all sufficiently

[4] A dual partition lattice is the lattice of equivalence relations on n elements under reverse inclusion, for some n.

simple ideals are intersections of two principal ideals (to pick up the standard part).
□

A characterization of the quantifier level at which undecidability first appears is not known. A decision procedure for the one-quantifier theory, and partial results for the decidability of the two-quantifier theory, have been obtained by Fejer and Shore [1985]. The undecidability of the four-quantifier theory has been proved by Nies [1996].

3. Enumeration degrees

The reducibilities introduced in the previous two sections were inspired by the problem of reducing the decidability of a formal system (or, more generally, of a logical theory) to the decidability of another. In this section we turn to the different problem of reducing the axiomatizability of a logical theory to the axiomatizability of another.

Since in compact logics (such as the first-order one) only finitely many axioms are needed for the deduction of a theorem, and the deduction process is effective, we formalize the relative axiomatizability as follows, where $\{D_n\}_{n\in\omega}$ is an effective enumeration of all finite sets.

DEFINITION 3.1 (*Friedberg and Rogers* [1959]). A set A is enumeration reducible to a set B ($A \leqslant_e B$) if there is an r.e. relation R such that, for all x,

$$x \in A \Longleftrightarrow (\exists y)(R(x, y) \wedge D_y \subseteq B).$$

A is enumeration equivalent to B ($A \equiv_e B$) if $A \leqslant_e B$ and $B \leqslant_e A$.

Notice that only positive information about B can be used to show that $A \leqslant_e B$: thus, while $\leqslant_p$ implies $\leqslant_e$, this fails in general for $\leqslant_{tt}$ (more precisely, there are sets A such that $A \not\leqslant_e \overline{A}$).

Logical theories whose axiomatizability is reduced to that of an axiomatizable logical theory (i.e. a formal system) are axiomatizable: this corresponds to the fact that a set e-reducible to an r.e. set is r.e. More generally: the set of consequences of a logical theory is e-reducible to it, and a set e-reducible to a given set is r.e. in it.

From the point of view of studying logical theories, e-reducibility turns out to be appropriate: Feferman [1957] proved that every e-degree contains a first-order logical theory (because the proof actually works for p-degrees).

An additional justification for the notion of e-reducibility comes from the study not of logical theories, but of partial functions: by comparing graphs of partial functions (coded as sets) by means of $\leqslant_e$, one gets a natural way of comparing partial functions, which agrees with $\leqslant_T$ on total functions.

Also from the point of view of studying computability of partial functions, e-reducibility turns out to be appropriate: Plotkin [1972] and Scott [1975] have proved that the e-reductions provide an effective model of λ-calculus.

3.1. *Global structure*

Our main goal would now be to characterize the structure $\mathcal{D}_e$ of e-degrees as a partially ordered structure, w.r.t. the partial order $\leqslant$ induced by $\leqslant_e$.

As usual, $\mathcal{D}_e$ is an uppersemilattice with least element $\mathbf{0}_e$ (containing the r.e. sets), countable predecessor property, and $2^{\aleph_0}$ elements.

The next result provides an elementary difference with $\mathcal{D}_m$, $\mathcal{D}_{tt}$ and $\mathcal{D}_T$ (as well as most other degree structures).

THEOREM 3.2 (Gutteridge [1971]). *Every element of $\mathcal{D}_e$ has infinitely many predecessors. In particular, there is no finite ideal.*

SKETCH OF PROOF. One first shows that a set of minimal e-degree must be Δ_2^0 (by constructing an e-reduction Ψ with the property that if $\Psi(A)$ is either r.e. or e-equivalent to A then $A \in \Delta_2^0$). Then one shows that no $\Delta_2^0 - \Sigma_1^0$ set has minimal e-degree (by a priority argument). □

Although very little is known about (countable) ideals, the next result provides an answer to one of the usual global algebraic questions.

THEOREM 3.3 (Failure of homogeneity (Cooper)). *There is a principal filter not elementarily equivalent to $\mathcal{D}_e$.*

SKETCH OF PROOF. There is an e-degree with a minimal cover: thus in the principal filter generated by it there are minimal elements. □

We turn now to logical properties.

THEOREM 3.4 (Slaman and Woodin [1997]). *The first-order theory of $\mathcal{D}_e$ has the same complexity as the theory of second-order arithmetic.*

SKETCH OF PROOF. The approach of Theorem 1.9 does not apply here, because it uses the existence of simple initial segment embeddings. One thus uses an alternative approach similar to Theorem 1.11, by defining every countable relation from finitely many parameters in $\mathcal{D}_e$ in a uniform way, and using this (together with the fact, easily proved directly, that every countable ideal is the intersection of two principal ideals) to define a standard model of arithmetic in $\mathcal{D}_e$. □

It follows in particular that the first-order theory of $\mathcal{D}_e$ is undecidable, and not axiomatizable.

Since the e-degrees can be thought of as degrees of partial functions, a global substructure of $\mathcal{D}_e$ is the structure $\mathcal{T}_e$ of e-degrees of graphs of total functions (called *total e-degrees*).

THEOREM 3.5 (Medvedev [1955], Myhill [1961, Sorbi]). *$\mathcal{T}_e$ is properly embedded in $\mathcal{D}_e$ and generates it, but is not an ideal.*

Moreover, $\mathcal{T}_e$ is isomorphic to $\mathcal{D}_T$.

SKETCH OF PROOF. On the one hand, for any set A the e-degree of A' (the Turing jump of A) is total; in particular, so is the e-degree of $\emptyset''$. On the other hand, there is a set $B \in \Delta_2^0$ (hence such that $B \leqslant_e \emptyset''$) whose e-degree is not total.

The isomorphism between total e-degrees and T-degrees is provided by the fact that $\leqslant_e$ coincides with $\leqslant_T$ on total functions. □

Cooper [199?] has announced that *there are non-trivial automorphisms of $\mathcal{D}_e$*, and that *$\mathcal{T}_e$ is not definable in $\mathcal{D}_e$*.

3.2. *Local structure*

Since the r.e. sets all belong to the same e-degree $\mathbf{0}_e$, one has to look higher up in the arithmetical hierarchy to find a class of sets whose e-degrees are not trivial.

Since the r.e. sets are the Σ_1^0 ones, we can look at the next level, and consider the structure $\mathcal{R}_e$ of the Σ_2^0 e-degrees, i.e. of the ones containing Σ_2^0 sets.

As usual, $\mathcal{R}_e$ has a least element $\mathbf{0}_e$ (containing the r.e. sets) and a greatest element $\mathbf{0}_e'$ (containing the complete Σ_2^0 sets), and is a countable uppersemilattice.

Like $\mathcal{R}_m$, and unlike $\mathcal{R}_{tt}$, $\mathcal{R}_e$ is an ideal of $\mathcal{D}_e$.

The next result provides an elementary difference between $\mathcal{D}_e$ and both $\mathcal{R}_m$ and $\mathcal{R}_{tt}$.

THEOREM 3.6 (Cooper [1984]). *$\mathcal{R}_e$ is dense. In particular, it has no finite ideals.*

SKETCH OF PROOF. The proof that there are no Δ_2^0 sets of minimal e-degree (see Theorem 3.2) uses only the possibility of approximating the Δ_2^0 in a nice way, which is shared by the Σ_2^0 sets too: thus there are no Σ_2^0 sets whose e-degree is a minimal cover. □

$\mathcal{R}_e$ shares with $\mathcal{R}_T$ many other properties (for example, by McEvoy and Cooper [1985], it is not distributive), but it is not elementarily equivalent to it.

THEOREM 3.7 (Ahmad [1991, 199?], Ahmad and Lachlan [199?]). *The Splitting Theorem fails, and the Diamond Theorem holds, in $\mathcal{R}_e$.*

As for the logical properties, the only known result is the following.

THEOREM 3.8 (Slaman and Woodin [1997]). *The first-order theory of $\mathcal{R}_e$ is undecidable.*

4. Progress on the problems of "Strong Reducibilities"

Our 1981 paper "Strong Reducibilities" (Odifreddi [1981]) has become a widely quoted reference in the subject, and the 26 open problems there stated have produced a good deal of work toward their solution. Without attempting to place them in a broader context, for which the reader is referred both to the original paper and to our two volumes, we now report on their present status, and mark by a * the ones that have been completely solved.

PROBLEM 1. *Characterize the T-degrees containing η-maximal (η-hyperhypersimple) semirecursive sets.*

Miller [1981] proves that η-maximal semirecursive sets are low_2, and that there are both low_1 and non-low_1 ones; moreover, not every low_1 T-degree contains η-maximal semirecursive sets. On the other hand, there are high_1 η-hyperhypersimple semirecursive sets.

PROBLEM 2*. *Does every T-degree contain an m-degree consisting of only one 1-degree?*

Degtev [1979] shows that every T-degree not below $\mathbf{0}'$ does, and Downey [1993] shows the same for every non-recursive T-degree below $\mathbf{0}'$.

PROBLEM 3. *Does every r.e. tt-degree contain an r.e. m-degree consisting of only one 1-degree?*

Jockusch [1969] shows that every r.e. T-degree contains an r.e. m-degree consisting of only one 1-degree (see also Soare [1969]), and that every r.e. tt-degree contains an m-degree consisting of only one 1-degree.

PROBLEM 4. *Find a property P such that the m-degree of A consists of a single 1-degree iff A has the property P.*

Ershov [1971] shows that being perfect is a sufficient condition (a set A is perfect if it is η-closed, for some non-trivial r.e. equivalence relation η whose only recursive η-closed sets are $\emptyset$ and ω), but Denisov [1974] shows that it is not necessary.

PROBLEM 5. *If an m-degree has more than one 1-degree, does it contain an infinite antichain of 1-degrees (i.e. a set of mutually incomparable 1-degrees)?*

Young [1966] shows that if an m-degree has more than one 1-degree, then it contains an infinite chain of 1-degrees.

PROBLEM 6. *Does there exist an r.e. m-degree with infinitely many minimal 1-degrees?*

Degtev [1976] shows that an r.e. m-degree may contain any finite number of minimal 1-degrees.

PROBLEM 7*. *Does every non-recursive tt-degree contain an infinite antichain of m-degrees?*

Jockusch [1968, 1969] shows that every non-recursive tt-degree contains an infinite chain of m-degrees, as well as a pair of incomparable m-degrees. Stephan [199?] shows that every non-recursive tt-degree contains an infinite chain of m-degrees.

PROBLEM 8. *What is the situation for maximal and minimal m-degrees inside a given tt-degree?*

Rogers [1967] and Jockusch have shown that every tt-degree contains a greatest m-degree, but no tt-degree contains a least m-degree.

PROBLEM 9. *Find a property such that the tt-degree of an r.e. set A contains only one r.e. m-degree iff A has P.*

Degtev [1973] shows that having a rigid complement is a sufficient condition (rigidity means that any two subsets are pm-equivalent if and only if their symmetric difference is finite, where pm-reducibility is defined as m-reducibility, but using partial functions in place of total functions).

PROBLEM 10*. *Does every r.e. tt-degree contain either only one or infinitely many r.e. m-degrees?*

Downey [1989] shows that there are r.e. tt-degrees with an arbitrarily large finite number of r.e. m-degrees. Cholak and Downey [1994] improve on this by showing that there are r.e. tt-degrees with any finite number of r.e. m-degrees; more generally, that the structure of r.e. m-degrees within an r.e. tt-degree can be any given finite distributive lattice.

PROBLEM 11*. *Does every r.e. tt-degree have a greatest r.e. m-degree?*

Downey [1988b] shows that there are r.e. tt-degrees without a greatest r.e. m-degree.

PROBLEM 12*. *Is $\mathbf{0}'$ the only r.e. non-recursive T-degree with greatest r.e. tt-degree?*

On the positive side, Downey and Jockusch [1987] show that there even are r.e. non-recursive T-degrees with greatest r.e. 1-degree (i.e. not only greatest r.e. tt-degree). On the negative side, Jockusch [1972] shows that if a non-recursive r.e.

T-degree has a greatest r.e. tt-degree then it is low_2, and Downey and Jockusch [1987] add that it is not low_1.

Related to this, Downey and Shore [1995] show that the r.e. T-degrees which are bounded by an incomplete r.e. T-degree with a greatest r.e. tt-degree are exactly the low_2 ones.

PROBLEM 13*. *Is there an (r.e.) T-degree with a least (r.e.) tt-degree?*

On the positive side, Kobzev [1978] shows that there even are r.e. T-degrees with a least tt-degree (i.e. not only a least r.e. tt-degree). On the negative side, Downey and Jockusch [1987] show that $\mathbf{0}'$ does not have a least r.e. tt-degree, and Downey and Shore [1995] and Nies have shown that if an r.e. T-degree has a least r.e. tt-degree then it is low_2.

PROBLEM 14*. *Is every simple set A with the following two properties hyperhypersimple?*

- *every m-degree of an r.e. superset of A is below the m-degree of A;*
- *every m-degree below the m-degree of A is represented by an r.e. superset of A.*

Degtev [1989] shows that the two properties are satisfied by some simple, not hypersimple set.

PROBLEM 15*. *Classify $\{x\colon \mathcal{W}_x \equiv_1 A\}$ for A r.e. infinite and coinfinite.*

Herrmann [1986] shows that such a set is Σ^0_3-complete.

PROBLEM 16. *Are the theories of the orderings of tt-degrees below $\mathbf{0}'_{tt}$ and of the r.e. tt-degrees elementarily equivalent?*

Nies and Shore [1995] show that the first-order theories of both structures have the same complexity as the theory of first-order arithmetic, so they cannot be distinguished on this basis.

PROBLEM 17*. *Does every pair of r.e. sets have g.l.b. in the r.e. tt-degrees iff it has g.l.b. in the tt-degrees (and, if so, do the two g.l.b.'s coincide)?*

Fejer and Shore [1988] show that any of the five theoretical possibilities for the two g.l.b.'s (in the tt-degrees, and in the r.e. tt-degrees) can be achieved, namely: they exist and coincide; they exist and do not coincide; only one of them exists; none of them exists.

PROBLEM 18. *Characterize the countable initial segments of the 1-degrees.*

Lachlan [1969] shows that all initial segments of the m-degrees can also be realized as initial segments of the 1-degrees. He also shows that even a characterization of the finite initial segments is not trivial, since: every finite initial segment is a lattice; some finite initial segment is not distributive; and not all finite lattices are isomorphic to initial segments.

PROBLEM 19*. *Are the orderings of T-degrees and tt-degrees distinct (not elementarily equivalent)?*

Shore [1982] shows that the theories of T-degrees and tt-degrees are not elementarily equivalent. Mohrherr [1984] exhibits a natural sentence true in the tt-degrees, and false in the T-degrees (more precisely: there is a dense principal filter).

PROBLEM 20. *Characterize the finite (countable) initial segments of the r.e. tt-degrees.*

On the positive side, Haught and Shore [1990a] show that the finite dual partition lattices with an added least element are isomorphic to initial segments of the r.e. tt-degrees. On the negative side, Harrington and Haught [199?] show that any nontrivial finite initial segment of the r.e. tt-degrees must have a least element above $\mathbf{0}_{tt}$, and thus not every finite lattice is embeddable.

PROBLEM 21. *Characterize the countable initial segments of the r.e. 1-degrees.*

Lachlan [1969] shows that the principal finite ideals are exactly the finite distributive lattices.

PROBLEM 22*. *Are the theories of the orderings of r.e. tt-degrees and T-degrees decidable?*

Haught and Shore [1990a] show that the theory of r.e. tt-degrees (and, similarly, of tt-degrees below $\mathbf{0}'_{tt}$) is undecidable. Nies and Shore [1995] show that the theory has the same degree as first-order arithmetic.

Similar results hold for the theory of r.e. T-degrees, using of course different methods, have been proved by Harrington and Shelah [1982], and Harrington and Slaman (unpublished), respectively.

PROBLEM 23. *Is the ordering of r.e. btt-degrees elementarily equivalent to the ordering of r.e. tt-degrees? Is the ordering of btt-degrees elementarily equivalent to the ordering of tt-degrees (m-degrees)?*

Marchenkov [1977] had already solved the first half of the problem, by showing that the theory of r.e. btt-degrees is not elementarily equivalent to the theory of r.e. tt-degrees (in the second case, but not in the first, there is an incomplete degree bounding all minimal degrees).

PROBLEM 24*. *Does the complete r.e. wtt-degree contain infinitely many r.e. tt-degrees?*

Kobzev [1976], improving on Lachlan [1975], had already shown that every non-zero r.e. *wtt*-degree contains infinitely many r.e. *tt*-degrees.

PROBLEM 25. *Characterize the initial segments of the (r.e.) btt-degrees.*

Nies [1992] shows that the finite dual partition lattices with an added least element are isomorphic to initial segments of the r.e. *btt*-degrees.

PROBLEM 26*. *Are the theories of the orderings of r.e. btt-degrees and wtt-degrees decidable?*

Nies [1992] shows that the theory of r.e. *btt*-degrees is undecidable. Ambos-Spies, Nies and Shore [1992] show that the theory of r.e. *wtt*-degrees is undecidable, Ambos-Spies, Fejer, Lempp and Lerman [1996] that the two-quantifier theory is decidable, and Lempp and Nies [1995] that the four-quantifier theory is undecidable.

Related to this, Haught and Shore [1990b] show that the theory of *wtt*-degrees below $\mathbf{0}'_{wtt}$ is undecidable.

5. More problems

14 of the 26 problems of "Strong Reducibilities" remain to be solved, although good progress has been made on practically all of them. We add here some new problems, inspired by recent developments in the areas dealt with in the present chapter. For simplicity, we continue the previous enumeration.

5.1. *Truth-table degrees*

PROBLEM 27. *Are there non-trivial definable elements in $\mathcal{D}_{tt}$?*

For $\mathcal{D}_m$ the answer is negative, by Theorem 1.5. For $\mathcal{D}_T$ the answer is positive, for every sufficiently high degree (e.g., above $\mathbf{0}''$, by Slaman and Woodin [199?]) which is definable in second-order arithmetic.

PROBLEM 28. *Is every automorphism of $\mathcal{D}_{tt}$ the identity on a principal filter?*

For $\mathcal{D}_m$ the answer is negative, by Theorem 1.5. For $\mathcal{D}_T$ the answer is positive (e.g., for the principal filter generated by $\mathbf{0}''$, by Slaman and Woodin [199?]).

PROBLEM 29. *Are there non-trivial automorphisms of $\mathcal{D}_{tt}$, and if so how many?*

For $\mathcal{D}_m$ the answer is $2^{2^{\aleph_0}}$. For $\mathcal{D}_T$ Slaman and Woodin [199?] have proved that there are at most $\aleph_0$ automorphisms, and Cooper [1997] has announced that there are exactly $\aleph_0$. For hyperdegrees, there is no non-trivial automorphism.

It thus looks as if the number of automorphisms reflects the strength of the reducibility, and one may conjecture that that there are $2^{\aleph_0}$ automorphisms of $\mathcal{D}_{tt}$.

PROBLEM 30. *Are $\mathcal{D}_{tt}$ and $\mathcal{D}_p$ elementarily equivalent?*

By Theorem 2.6, this is the only case left open for truth-table reducibilities. The most obvious guess, namely to look at minimal degrees, does not work: Cooper [1990a] has proved that there are minimal p-degrees.

PROBLEM 31. *Are $\mathcal{D}_{tt}$ and $\mathcal{D}_{wtt}$ elementarily equivalent?*

All known proofs of results for $\mathcal{D}_{tt}$ automatically work for $\mathcal{D}_{wtt}$ too: while on the one hand this provides a good picture of $\mathcal{D}_{wtt}$, on the other hand it obviously leaves open the previous problem.

Shore [1982] shows that $\mathcal{D}_{tt}$ and $\mathcal{D}_{wtt}$ have isomorphic principal filters. Thus, even if the answer to the problem were negative, the two theories would still resemble each other more than any other pair among 1-, m-, tt-, wtt-, and T-degrees (none of which admits isomorphic principal filters).

5.2. *Recursively enumerable truth-table degrees*

Problem 20 asks for a characterization of the (finite) initial segments of $\mathcal{R}_{tt}$.

PROBLEM 32. *Characterize the quantifier level at which undecidability occurs for $\mathcal{R}_{tt}$.*

For $\mathcal{R}_m$ the answer is three, by Theorem 1.12. For $\mathcal{R}_T$ it is known that the one-quantifier theory is decidable and that the three-quantifier theory is undecidable (Lempp, Nies and Slaman [199?]), although partial results have been obtained toward the decidability of the two-quantifier theory (see Chapter A.6).

PROBLEM 33. *Are there elementarily non-equivalent non-trivial principal filters of $\mathcal{R}_{tt}$?*

For $\mathcal{R}_m$ the answer is negative, by Theorem 1.8. For $\mathcal{R}_T$ the answer is positive (e.g., there are both branching and non-branching degrees, by Lachlan [1966]).

PROBLEM 34. *Are there non-trivial definable elements in $\mathcal{R}_{tt}$?*

For $\mathcal{R}_m$ the answer is negative, by Theorem 1.8. For $\mathcal{R}_T$ the answer is also not known.

PROBLEM 35. *Are there non-trivial automorphisms of $\mathcal{R}_{tt}$?*

For $\mathcal{R}_m$ the answer is positive, by Theorem 1.8. Cooper [1997] has announced that the answer is positive for $\mathcal{R}_T$ too.

Problem 16 asks whether $\mathcal{R}_{tt}$ is elementarily equivalent to the theory of the ordering of tt-degrees below $\mathbf{0}'_{tt}$. For $\mathcal{R}_m$ the problem does not arise, since the m-degrees below $\mathbf{0}'_m$ coincide with the r.e. m-degrees. $\mathcal{R}_T$ is not elementarily equivalent to the T-degrees below $\mathbf{0}'_T$ (e.g., because there are minimal T-degrees below $\mathbf{0}'_T$, by Sacks [1961]).

PROBLEM 36. *Is $\mathcal{R}_{tt}$ definable in $\mathcal{D}_{tt}$?*

$\mathcal{R}_m$ is not definable in $\mathcal{D}_m$ by Theorem 1.5, and $\mathcal{R}_T$ is definable in $\mathcal{D}_T$ by Cooper [1990a].

PROBLEM 37. *Are the following realized as structures of r.e. m-degrees within an r.e. tt-degree?*

- *every finite distributive uppersemilattice*
- *the linear ordering of type ω.*

By Downey [1989], some finite distributive uppersemilattice which is not a lattice is realized; by Cholak and Downey [1994], all finite distributive lattices are realized.

5.3. *Enumeration degrees*

PROBLEM 38. *Characterize the countable ideals of $\mathcal{D}_e$.*

PROBLEM 39. *Are there non-trivial definable elements in $\mathcal{D}_e$?*

PROBLEM 40. *Are there non-trivial automorphisms of $\mathcal{D}_e$?*

$\mathcal{D}_e$ is undecidable by Theorem 3.4, and the next question asks for a logical refinement.

PROBLEM 41. *Characterize the quantifier level at which undecidability occurs for $\mathcal{D}_e$.*

We turn now to the local structure. $\mathcal{R}_e$ is undecidable by Theorem 3.8, and the next question asks for the usual logical refinements.

PROBLEM 42. *Characterize the quantifier level at which undecidability occurs for $\mathcal{R}_e$.*

PROBLEM 43. *Is the complexity of $\mathcal{R}_e$ the same as first-order arithmetic?*

We end with two questions about mutual relationships of substructures.

PROBLEM 44. *Is $\mathcal{R}_e$ definable in $\mathcal{D}_e$?*

PROBLEM 45. *Are the theories of Σ^0_n and Σ^0_m e-degrees not elementarily equivalent, for different $n, m \geqslant 2$?*

Acknowledgments

I wish to thank Barry Cooper, Carl Jockusch, André Nies, and Andrea Sorbi for their useful comments on a first draft of this chapter, and for their friendship.

References

U. ABRAHAM AND R.A. SHORE

[1986] Initial segments of the degrees of size $\aleph_1$, *Israel J. Math.*, 53, pp. 1–51.

S. AHMAD

[1991] Embedding the diamond in the Σ^0_2 enumeration degrees, *J. Symbolic Logic*, 55, pp. 195–212.

S. AHMAD AND A.H. LACHLAN

[199?] Some special pairs of Σ^0_2 *e*-degrees, to appear.

K. AMBOS-SPIES, P. FEJER, S. LEMPP AND M. LERMAN

[1996] Decidability of the two-quantifier theory of the recursively enumerable weak truth-table degrees and other distributive uppersemilattices, *J. Symbolic Logic*, 61, pp. 880–905.

K. AMBOS-SPIES, A. NIES AND R. A. SHORE

[1992] The theory of the recursively enumerable weak truth-table degrees is undecidable, *J. Symbolic Logic*, 57, pp. 864–874.

V. K. BULITKO

[1980] Reducibility by Zhegalkin linear tables, *Siberian Math. J.*, 21, pp. 23–31; transl. 21, pp. 332–339.

P. CHOLAK AND R. DOWNEY

[1994] Recursively enumerable *m*- and *tt*-degrees. III: realizing all finite distributive lattices, *Bull. London Math. Soc.*, 50, pp. 440–453.

S. B. COOPER

[1982] Partial degrees and the density problem, I, *J. Symbolic Logic*, 47, pp. 854–859.

[1984] Partial degrees and the density problem, II: the enumeration degrees of the Σ^0_2 sets are dense, *J. Symbolic Logic*, 49, pp. 503–511.

[1990a] The jump is definable in the structure of the degrees of unsolvability, *Bull. Amer. Math. Soc.*, 23, pp. 151–158.

[1990b] Enumeration reducibility, non-deterministic computations and relative computability of partial functions, in: *Lecture Notes in Mathematics*, Vol. 1432, pp. 57–110.

[1997] Beyond Gödel's theorem: the failure to capture information content, in: *Complexity, Logic and Recursion Theory*, Sorbi, ed., Dekker, pp. 93–122.

[199?] Hartley Rogers' 1965 agenda, to appear.

A. N. DEGTEV

[1973] tt- and m-degrees, *Algebra i Logika*, 12, pp. 143–161; transl. 12, pp. 78–89.

[1976] Partially ordered sets of 1-degrees contained in recursively enumerable m-degrees, *Algebra i Logika*, 15, pp. 249–266; transl. 15, pp. 153–164.

[1979] Some results on uppersemilattices and m-degrees, *Algebra i Logika*, 18, pp. 664–679; transl. 18, pp. 420–430.

[1981] Relationships among complete sets, *Sov. Math.*, 228, pp. 50–55; transl. 228, pp. 53–61.

[1982] Comparison of linear reducibility with other reducibilities truth-table like, *Algebra i Logika*, 21, pp. 511–529; transl. 21, pp. 339–353.

[1983a] Relationships between truth-table like degrees, *Algebra i Logika*, 22, pp. 35–52; transl. 22, pp. 26–39.

[1983b] Relationships between truth-table like reducibilities, *Algebra i Logika*, 22, pp. 243–259; transl. 22, pp. 173–185.

[1985] On the uppersemilattice of disjunctive and linear degrees, *Mat. Zametki*, 38, pp. 310–316; transl. 38, pp. 681–684.

[1989] On a question of P. Odifreddi, *Siberian Math. J.*, 30, pp. 185–187.

S. D. DENISOV

[1974] Three theorems on elementary theories and tt-reducibilities, *Algebra i Logika*, 13, pp. 5–8; transl. 13, pp. 1–2.

[1978] The structure of the uppersemilattice of recursively enumerable m-degrees and related questions, *Algebra i Logika*, 17, pp. 643–683; transl. 17, pp. 418–443.

R. DOWNEY

[1988a] Recursively enumerable m- and tt-degrees. II: the distribution of singular degrees, *Arch. Math. Logic*, 27, pp. 135–147.

[1988b] Two theorems on truth-table degrees, *Proc. Amer. Math. Soc.*, 103, pp. 281–287.

[1989] Recursively enumerable m- and tt-degrees. I: the quantity of m-degrees, *J. Symbolic Logic*, 54, pp. 553–567.

[1993] On irreducible m-degrees, *Rend. Sem. Math. Univ. Pol. Torino*, 51, pp. 109–112.

R. DOWNEY AND C. JOCKUSCH

[1987] T-degrees, jump classes, and strong reducibilities, *Trans. Amer. Math. Soc.*, 301, pp. 103–136.

R. DOWNEY AND R. A. SHORE

[1995] Degree-theoretic definitions of the low_2 recursively enumerable degrees, *J. Symbolic Logic*, 60, pp. 727–756.

Y. L. ERSHOV

[1971] Positive equivalences, *Algebra i Logika*, 10, pp. 620–650; transl. 10, pp. 378–394.

[1975] The uppersemilattice of enumerations of a finite set, *Algebra i Logika*, 14, pp. 258–284; transl. 14, pp. 159–175.

S. FEFERMAN

[1957] Degrees of unsolvability associated with classes of formalized theories, *J. Symbolic Logic*, 22, pp. 161–175.

P. A. FEJER AND R. A. SHORE

[1985] Embeddings and extensions of embeddings in the r.e. tt- and wtt-degrees, in: *Lecture Notes in Mathematics*, Vol. 1141, pp. 121–140.

[1988] Infima of recursively enumerable truth-table degrees, *Notre Dame J. Formal Logic*, 29, pp. 420–437.

P. C. FISCHER

[1963] A note on bounded truth-table reducibility, *Proc. Amer. Math. Soc.*, 14, pp. 875–877.

R. M. FRIEDBERG AND H. ROGERS

[1959] Reducibility and completeness for sets of integers, *Z. Math. Logik Grundl. Math.*, 5, pp. 117–125.

K. GÖDEL

[1931] Über formal unentscheidbare Sätze der Principia Mathematica und verwandter Systeme I, *Monash. Math. Phys.*, 38, pp. 173–198.

M. J. GROSZEK AND T. A. SLAMAN

[1983] Independence results on the global structure of the Turing degrees, *Trans. Amer. Math. Soc.*, 277, pp. 579–588.

L. GUTTERIDGE

[1971] *Some Results on Enumeration Reducibility*, PhD Thesis, Simon Frazer University.

W. HANF

[1965] Model-theoretic methods in the study of elementary logic, in: *The Theory of Models*, Addison et al., eds., North-Holland, Amsterdam, pp. 132–145.

L. HARRINGTON AND C. HAUGHT

[199?] Limitations on initial segment embeddings in the r.e. tt-degrees, to appear.

L. HARRINGTON AND S. SHELAH

[1982] The undecidability of the recursively enumerable degrees, *Bull. Amer. Math. Soc.*, 6, pp. 79–80.

C. HAUGHT AND R. A. SHORE

[1990a] Undecidability and initial segments of the (r.e.) tt-degrees, *J. Symbolic Logic*, 55, pp. 987–1006.

[1990b] Undecidability and initial segments of the wtt-degrees below $\mathbf{0}'$, in: *Lecture Notes in Mathematics*, Vol. 1432, pp. 223–244.

E. HERRMANN

[1986] The index set $\{e\colon \mathcal{W}_e \equiv_1 X\}$, *J. Symbolic Logic*, 51, pp. 110–116.

C. JOCKUSCH

[1968] Semirecursive sets and positive reducibility, *Trans. Amer. Math. Soc.*, 131, pp. 420–436.

[1969] Relationships between reducibilities, *Trans. Amer. Math. Soc.*, 142, pp. 229–237.

[1972] Degrees in which the recursive sets are uniformly recursive, *Canad. J. Math.*, 24, pp. 1092–1099.

S. C. KLEENE AND E. L. POST

[1954] The uppersemilattice of degrees of recursive unsolvability, *Ann. Math.*, 59, pp. 379–407.

G. N. KOBZEV

[1976] Relationships between recursively enumerable tt- and w-degrees, *Bull. Acad. Sci. Georg.*, 84, pp. 585–587.

[1978] On the tt-degrees of recursively enumerable Turing degrees, *Mat. Sb.*, 106, pp. 507–514.

A. H. LACHLAN

[1966] Lower bounds for pairs of recursively enumerable degrees, *Proc. London Math. Soc.*, 16, pp. 537–569.

[1969] Initial segments of one-one degrees, *Pacific J. Math.*, 29, pp. 351–366.

[1970] Initial segments of many-one degrees, *Canad. J. Math.*, 22, pp. 75–85.

[1972a] Two theorems on many-one degrees of recursively enumerable sets, *Algebra i Logika*, 11, pp. 216–229; transl. 11, pp. 127–132.

[1972b] Recursively enumerable many-one degrees, *Algebra i Logika*, 11, pp. 326–358; transl. 11, pp. 186–202.

[1975] wtt-complete sets are not necessarily tt-complete, *Proc. Amer. Math. Soc.*, 48, pp. 429–434.

R. E. LADNER AND L. P. SASSO

[1975] The weak truth-table degrees of recursively enumerable sets, *Ann. Math. Logic*, 8, pp. 429–448.

S. LEMPP AND A. NIES

[1995] Undecidability of the four-quantifier theory for the recursively enumerable Turing and wtt-degrees, *J. Symb. Log.*, 60, pp. 1118–1135.

S. LEMPP, A. NIES AND T. SLAMAN

[199?] The Π_3-theory of r.e. Turing degrees is undecidable, *Trans. Amer. Math. Soc.*

M. LERMAN

[1983] *The Degrees of Unsolvability*, Springer, Berlin.

A. A. MAL'CEV

[1985] The structure of the uppersemilattice of btt_1-degrees, *Siberian Math. J.*, 26, pp. 132–139; transl. 26, pp. 264–270.

S. S. MARCHENKOV

[1977] On recursively enumerable minimal btt-degrees, *Mat. Sb.*, 103, pp. 550–562; transl. 32, pp. 477–487.

K. MCEVOY AND S. B. COOPER

[1985] On minimal pairs of enumeration degrees, *J. Symbolic Logic*, 50, pp. 983–1001.

Y. T. MEDVEDEV

[1955] Degrees of difficulty of the mass problem, *Dokl. Akad. Nauk SSSR*, 104, pp. 501–504.

D. P. MILLER

[1981] *The Relationships Between the Structure and Degrees of Recursively Enumerable Sets*, PhD Thesis, University of Chicago.

J. MOHRHERR

[1984] Density of final segment of the truth-table degrees, *Pacific J. Math.*, 115, pp. 409–419.

J. MYHILL

[1961] Note on degrees of partial functions, *Proc. Amer. Math. Soc.*, 12, pp. 519–521.

A. NERODE AND R. A. SHORE

[1980] Second-order logic and first-order theories of reducibility orderings, in: *The Kleene Symposium*, Barwise et al., eds., North-Holland, Amsterdam, pp. 181–200.

A. NIES

[1992] *Definability and Undecidability in Recursion-Theoretic Semilattices*, PhD Thesis, Universität Heidelberg.

[1993] Interpreting true arithmetic in degree structures, in: *Lecture Notes in Comput. Sci.*, Vol. 713, pp. 255–263.

[1995] The last question on recursively enumerable many-one degrees, *Algebra i Logika*, 33, pp. 550–563.

[1996] Undecidable fragments of elementary theories, *Algebra Universalis*, 35, pp. 8–33.

[1997] On a uniformity in degree structures, in: *Complexity, Logic and Recursion Theory*, Sorbi, ed., Dekker, 1997, pp. 261–276.

A. NIES AND R. A. SHORE

[1995] Interpreting true arithmetic in the theory of the r.e. truth-table degrees, *Ann. Pure Appl. Logic*, 75, pp. 269–311.

P. G. ODIFREDDI

[1981] Strong reducibilities, *Bull. Amer. Math. Soc.*, 4, pp. 37–86.

[1985] The structure of m-degrees, in: *Lecture Notes in Mathematics*, Vol. 1141, pp. 315–332.

[1989] *Classical Recursion Theory*, North-Holland, Amsterdam.

[1999] *Classical Recursion Theory*, Vol. II, North-Holland, Amsterdam.

E. PALIUTIN

[1975] Addendum to the paper of Ershov [1975], *Algebra i Logika*, 14, pp. 284–287; transl. 14, pp. 176–178.

G. D. PLOTKIN

[1972] *A Set-Theoretical Definition of Application*, Mimeographed notes.

E. L. POST

[1944] Recursively enumerable sets of positive integers and their decision problem, *Bull. Amer. Math. Soc.*, 50, pp. 284–316.

H. ROGERS

[1967] *Theory of Recursive Functions and Effective Computability*, McGraw-Hill, New York.

G. E. SACKS

[1961] A minimal degree below $\mathbf{0}'$, *Bull. Amer. Math. Soc.*, 67, pp. 416–419.

D. SCOTT

[1975] Lambda calculus and recursion theory, in: *Proceedings of the Third Scandinavian Symposium*, Kanger, ed., North-Holland, Amsterdam, pp. 154–193.

V. L. SELIVANOV

[1982] On one class of reducibilities in the theory of recursive functions, *Probl. Math. Cyb.*, 18, pp. 83–100.

J. R. SHOENFIELD

[1957] Quasicreative sets, *Proc. Amer. Math. Soc.*, 8, pp. 964–967.

R. A. SHORE

[1978] On the $\forall\exists$ sentences of α-recursion, in: *Generalized Recursion Theory*, Fenstad et al., eds., North-Holland, Amsterdam, pp. 331–353.

[1982] The theories of the truth-table and Turing degrees are not elementarily equivalent, in: *Logic Colloquium '80*, Val Dalen et al., eds., North-Holland, Amsterdam, pp. 231–237.

[1993] The theories of the T, tt and wtt r.e. degrees: undecidability and beyond, *Notas de Logica Matematica*, 38, pp. 61–70.

T. SLAMAN AND H. WOODIN

[1997] Definability in the enumeration degrees, *Arch. Math. Logic*, 36, pp. 255–267.

[199?] Definability in degree structures, to appear.

F. STEPHAN

[199?] On the structure inside truth-table degrees, to appear.

R. I. SOARE

[1969] Recursion theory and Dedekind cuts, *Trans. Amer. Math. Soc.*, 140, pp. 271–294.

A. TURING

[1939] Systems of logic based on ordinals, *Proc. London Math. Soc.*, 45, pp. 161–228.

P. R. YOUNG

[1966] Linear orderings under one-one reducibility, *J. Symbolic Logic*, 31, pp. 70–85.

CHAPTER 4

Local Degree Theory*

S. Barry Cooper

University of Leeds, Leeds LS2 9JT, England, UK

Contents

1. Logic, hierarchies and approximations 123
2. Decidability and forcing below $0'$ 124
3. Deconstructing constructions: 1-generic degrees 126
4. Structure, jump and definability 128
5. Definability in cones 131
6. Degree and information content 133
7. The Ershov hierarchy for $\mathcal{D}(\leqslant \mathbf{0}')$ 137
8. Automorphisms and undefinability 140
9. Enumeration and Turing reducibilities: The local theory 142
References 145

*Preparation of this paper partially supported by E.P.S.R.C. research grants nos. GR/H91213 and GR/H02165, and by EC Human Capital and Mobility network 'Complexity, Logic and Recursion Theory'.

HANDBOOK OF COMPUTABILITY THEORY
Edited by E.R. Griffor

Relative computability is one of a handful of truly fundamental mathematical relations. However, it gives rise to structures and techniques of such complexity (and consequent challenge to the specialist) that quite basic prerequisites to theoretical sophistication are still unavailable. The richness of natural definable relations taken for granted in earlier mathematical structures, with its corresponding depth and beauty of infrastructure, has eluded computability theorists to the point where (for some) pathology dominates and the primary interest seems to lie in techniques as much as theorems. Nowhere has this been more true than in the local theory of the Turing degrees. We describe below how recent developments in the area have put things in a new perspective.

The structure of the arithmetical Turing degrees, and in particular of $\mathcal{D}(\leqslant \mathbf{0}')$, provided a refined context for the pathbreaking 1930's results of Gödel, Turing, Church and others—work which startlingly revealed how the non-computable impinges on everyday mathematical practice. Many intrinsically interesting phenomena which give depth and content to the general theory (recursively enumerable and n.r.e. degrees, PA degrees, 1-generics, etc.) inhabit local degree theory. $\mathcal{D}(\leqslant \mathbf{0}')$ has acted as a technical resource for the remarkable recent progress with long-standing global questions, and indirectly benefitted developments in model theory, set theory and logic generally. But (with one important exception) there do not exist local definitions even for those classes of degrees known to be nontrivially definable in the global structure $\mathcal{D}$, and the interplay between continuity/decidability and definability/undecidability seems to be a particularly subtle one. In contrast to the r.e. Turing degrees, with its apparent continuity of basic phenomena, here we find the ingredients for undecidability and definability (such as initial segments) in abundance, but absolute definability results seem almost as difficult to arrive at. The bi-interpretability conjectures of Slaman, Woodin and Harrington attempt a general characterisation of the definable relations on $\mathcal{D}$, both at the global and local levels, by extending the known parallels between the structures of the integers and of the Turing degrees. In the meantime, while the exact nature of Turing definability continues to be peculiar to computability theory, the local theory looks likely to play a crucial role in providing answers in which definability creatively enhances the role of natural classes of incomputable objects. This will enable further development at the local level of the theoretical sophistication originally assumed to be a principal identifying characteristic of the global theory.

It will be impossible to do full justice to all the work in such a complicated area (even allowing for the fact that the r.e. degrees and strong reducibilities are specifically dealt with elsewhere in this volume). The underlying theme of definability inevitably entails selectivity. Even within the bounds chosen, one cannot compensate for the fact that this is an area in which the long-term significance of results and techniques may be very unexpected. Also, we avoid all but the most superficial discussion of techniques. But in the belief that computability using partial information (inevitably allowing non-deterministic elements) is an important extension of Turing reducibility, we include a final section on $\mathcal{D}(\leqslant \mathbf{0}')$ in the context of the Σ^0_2 enumeration degrees.

For basic terminology and notation see Soare [1987] or Odifreddi [1989]. However, we follow Soare [1994] in making explicit many intended mental translations of "recursive" as "computable", systematically adopting as standard abbreviations and notation such as 'c.e.' for computably enumerable or '$\mathcal{E}$' for the class of all computably enumerable degrees. We are indebted to a number of previous accounts, in particular Odifreddi [ta], Lerman [1983], Epstein [1979] and Posner [1980]. For historical background see also Cooper [1974].

1. Logic, hierarchies and approximations

In the 1930's, Gödel [1931, 1934], Turing [1936], Church [1936] and others discovered the undecidability of a range of decision problems basic to mathematics. The notion of relative (Turing) computability which grew out of this work can be used to unite these superficially diverse examples (locating them in the Turing degree $\mathbf{0}'$), and to provide a natural fine structure theory for the wide range of non-canonical incomputable objects intrinsic to specific mathematical practice.

The local theory derives theoretically from this early work. Closing up under Turing reducibility $\leqslant_T$ leads to $\mathcal{D}(\leqslant \mathbf{0}')$, the set of degrees below $\mathbf{0}'$. Abstracting from Gödel's Theorem to the Turing jump and jump inversion provides a hierarchy of jump classes below $\mathbf{0}'$. (See Soare [1987] or Odifreddi [1989] for basic definitions.) Further analysis of quantifier forms leads to Post's Theorem [1948], and in particular the identification of $\mathcal{D}(\leqslant \mathbf{0}')$ as a classification of the Δ^0_2 sets, and as a context for the Σ^0_1 (= the computably enumerable) degrees $\mathcal{E}$ (a class given particular significance by the theorems of Feferman [1957], Hanf [1965] and Matiasevich [1970]).

Particular Δ^0_2 presentations arise as generalisations of the notion of "computably enumerable". Where enumerations and axiomatic theories fail as a framework for more complex computational/learning situations, we may approximate sets convergently via the Limit Lemma (see Soare [1987], p. 57). Of course, all this holds relative to any given oracle.

Just as, by Gödel's Theorem, $\mathbf{0}'$ is more important than $\mathbf{0}$ (the degree of the computable sets) in any attempt to qualify the search for truth in a general sense, so the Δ^0_2 degrees are *more important* than the c.e. degrees. Gödel's Theorem tells us that axiomatic theories, despite modelling limited data generating environments, are not powerful enough to fully reflect the way in which knowledge is accumulated in real life. But the set of true sentences of Peano arithmetic, say, is clearly out of reach (in $\mathbf{0}^{(\omega)}$ it turns out). Any attempt to transcend these limitations inevitably leads to a process of effective approximation, and the approximating complete theory is of Δ^0_2 degree. (In practice, one must use information acquired via consistent extensions of PA, at best relative to an oracle representing auxiliary empirical input, with no absolute guarantee of truth.) We will see in Section 6 (in a theorem distinguishing properly Δ^0_2 degrees in the same way that Gödel's Theorem distinguishes $\mathbf{0}'$) that any such degree other than $\mathbf{0}'$ is *necessarily* of non-c.e. degree.

Just as $\mathcal{D}(\leqslant \mathbf{0}')$ arises naturally from fundamental logical considerations, further development inevitably leads to a rich infrastructure. Turing computability applied to models of first order arithmetic give rise to the PA degrees (see Section 6). Inversion of iterated jumps leads to the high/low hierarchy:

DEFINITION 1.1 (*Cooper* [1972c], *Soare* [1974]). The high/low hierarchy is defined by

$$\mathbf{High}_n = \{\mathbf{a} \leqslant \mathbf{0}' \mid \mathbf{a}^{(n)} = \mathbf{0}^{(n+1)}\}, \qquad \mathbf{Low}_n = \{\mathbf{a} \leqslant \mathbf{0}' \mid \mathbf{a}^{(n)} = \mathbf{0}^{(n)}\},$$

for each $n \geqslant 1$.

The Boolean closure of $\mathcal{E}$ (the class of all computably enumerable sets) gives Ershov's hierarchy [1968a, 1968b] and the n.c.e. hierarchy of degrees below $\mathbf{0}'$ (Section 7). While the use of Cohen forcing [1963] as a presentational device in computability theory (originating independently with Gandy and Sacks, formalised in arithmetic by Feferman [1965] and refined by Hinman [1969]), yield the 1-generic (and more generally n-generic) degrees (Section 3 below).

2. Decidability and forcing below $\mathbf{0}'$

The local theory initially developed via effectivisation of (what has come to be recognised as) recursion theoretic forcing techniques (see Lerman [1983]). Local results derived messily from global ones via an analysis of constructive content of proofs for $\mathcal{D}$. The broad technical framework was imported, and the local theory encouraged a better understanding of its workings. It was not until later that a specifically local approach, with more in common with the theory of the computably enumerable degrees, made use of full approximation and priority. Recent years have seen a minor renaissance in the global-to-local approach, with improved coding techniques (see Slaman and Woodin [1986] or Odifreddi and Shore [1989]) being extensively applied to questions related to local definability and decidability.

Forcing with strings (finite functions defined on initial segments of ω) wedded to an appropriate bounding principle (see Lerman [1983]) underpins the seminal work of Kleene and Post [1954] (see also Section 3 below). In particular:

THEOREM 2.1. *There is a countable set* $\mathbf{S}$ *of independent degrees below* $\mathbf{0}'$ (*where a set* $\mathbf{S} \subset \mathcal{D}(\leqslant \mathbf{0}')$ *is independent if no* $\mathbf{a} \in \mathbf{S}$ *is bounded by a finite join in* $\mathbf{S} - \{\mathbf{a}\}$).

Spector [1956] used appropriate finite joins from such an $\mathbf{S}$ to show any finite poset embeddable in $\mathcal{D}(\leqslant \mathbf{0}')$, and hence:

COROLLARY 2.2. *The existential theory of* $\mathcal{D}(\leqslant \mathbf{0}')$ *is decidable.*

Current coding techniques are based on Spector's work [1956] concerning exact pairs.

DEFINITION 2.3. We say $\mathbf{a}_0, \mathbf{a}_1$ is an *exact pair* for a countable ideal $\mathbf{I}$ of $\mathcal{D}$ if $\mathbf{I} = \mathcal{D}(\leqslant \mathbf{a}_0, \mathbf{a}_1)$ $(= \mathcal{D}(\leqslant \mathbf{a}_0) \cap \mathcal{D}(\leqslant \mathbf{a}_1))$.

A counting argument involving ideals generated by subsets of an independent set $\mathbf{S}$ below $\mathbf{0}'$ show there to be $2^{\aleph_0}$ ideals below $\mathbf{0}'$ not having exact pairs below $\mathbf{0}'$ (in fact, by Yates [1970a], only Σ^0_4 ideals can have exact pairs below $\mathbf{0}'$). But:

THEOREM 2.4 (Spector [1956]). *Exact pairs below* $\mathbf{0}'$ *exist for uniformly low ideals* $\mathbf{I}$ (*that is, for* $\mathbf{I}$ *generated by some* $\{\deg(A_i)\}_{i\in\omega}$ *with* $\bigoplus_{i\in\omega} A_i$ *low*).

Since the set $\mathbf{S}$ of Theorem 2.1 can be chosen to be uniformly low, and the ideal finitely generated by an independent set has no greatest element, Theorem 2.4 gives:

COROLLARY 2.5 (Spector 1956). $\mathcal{D}(\leqslant \mathbf{0}')$ *is not a lattice.*

For further progress towards characterising those ideals with exact pairs below $\mathbf{0}'$ see Nerode and Shore [1980b] and Shore [1981]. There are also a number of questions concerning minimal upper bounds below $\mathbf{0}'$.

Shoenfield [1959] proved a jump inversion theorem for $\mathcal{D}(\leqslant \mathbf{0}')$ (every $\mathbf{c}$ c.e. in and above $\mathbf{0}'$ – for short $\mathbf{0}'$-CEA – is the jump of a degree below $\mathbf{0}'$), and showed that $\mathcal{D}(\leqslant \mathbf{0}')$ properly extends $\mathcal{E}$. These results were soon superseded by the Sacks Jump Inversion Theorem [1963c] (every $\mathbf{0}'$-CEA $\mathbf{c}$ is the jump of a c.e. degree) and Sacks' construction [1961] of a minimal degree below $\mathbf{0}'$ (replacing Spector's Σ^0_2 oracle [1956] by one which is Σ^0_1).

It was Shoenfield's short paper [1966] which first presented an initial segment construction in the now familiar tree framework, while the forcing content of such constructions was explicated and further developed by Sacks [1971]. More complicated initial segment results became an important investigative tool, both at the global and local levels, and formed a key ingredient in a number of major results relating to decidability, definability and homogeneity. Even now that recursion theoretically simpler codings have taken over much of this role, the theory of initial segments remains one of the foremost (and despite its complexity, most technically attractive) achievements of the subject and (especially at the local level) still confronts us with interesting questions.

The proofs of all the remaining (un)decidability results of this section initially depended on the coding power of initial segment embeddings below $\mathbf{0}'$, and those of Theorems 2.8 and 2.9 still do. They can all be proved using:

THEOREM 2.6 (Local Embedding Theorem) (Lerman [1983]). *Every* $\mathbf{0}^{(2)}$ *presentable (and in particular, finite) upper semi-lattice with least element is embeddable as an initial segment of* $\mathcal{D}(\leqslant \mathbf{0}')$.

The progress from the unbounded embeddings of Lachlan and Lebeuf [1976] to this remarkable result is far from straightforward. Epstein [1979] was the first to diagnose the need for full approximation (in the manner of Cooper [1973]) in even quite simple local embeddings (arising from the uniformity of the trees involved), and constructed an embedding of $\omega + 1$ with enough context below $\mathbf{0}'$ to obtain a local version of Simpson [1977]. This was sufficient to prove the undecidability and non-axiomatisability of the first order theory of $\mathcal{D}(\leqslant \mathbf{0}')$. In fact:

THEOREM 2.7 (Shore [1981]). *The first order theory of* $\mathcal{D}(\leqslant \mathbf{0}')$ *has degree* $\mathbf{0}^{(\omega)}$ (*the degree of first order arithmetic*).

(See Odifreddi and Shore [1989] for a proof using codings free of initial segments.)

Lerman and Shore [1988] use the embeddability of finite upper semi-lattices (u.s.l.'s), together with the reducibility of two-quantifier statements about $\mathcal{D}(\leqslant \mathbf{0}')$ to questions about extensions of embeddings, to get the following improvement of Theorem 2.2.

THEOREM 2.8. *The* Σ_2 *theory of* $\mathcal{D}(\leqslant \mathbf{0}')$ *is decidable.*

On the other hand, Lerman [1983] is able to use Theorem 2.6 to get the local version of Schmerl's undecidability result for the Σ_3 theory of $\mathcal{D}$.

THEOREM 2.9. *The* Σ_3 *theory of* $\mathcal{D}(\leqslant \mathbf{0}')$ *is undecidable.*

Technically, the main problem in fully characterising the initial segments of $\mathcal{D}(\leqslant \mathbf{0}')$ arises from the fact that it is the $\mathbf{0}^{(4)}$ presentable u.s.l.'s (such as $\mathcal{D}(\leqslant \mathbf{0}')$ itself) which are candidates for embedding as initial segments. However, unless they are $\mathbf{0}^{(3)}$ presentable they cannot be embedded in the low$_2$ degrees, which (by Jockusch and Posner [1978]) puts immediate limitations on such candidates (for instance they cannot be lattices) and suggests those remaining are not embeddable using the techniques of Theorem 2.6. So answers to the following questions are basic to any such general characterisation.

QUESTION 2.10. Does there exist a characterisation of the class of principal ideals $\mathcal{D}(\leqslant \mathbf{a})$ of $\mathcal{D}(\leqslant \mathbf{0}')$ with $\mathbf{a}^{(2)} > \mathbf{0}^{(2)}$?

QUESTION 2.11. Is every $\mathbf{0}^{(3)}$ presentable u.s.l. with least element embeddable as an initial segment of $\mathcal{D}(\leqslant \mathbf{0}')$?

3. Deconstructing constructions: 1-generic degrees

In computability theory, 1-genericity (like Baire category, measure and Banach–Mazur games) is an elegant presentational device, but with local applications. It expands the scope of finite extension arguments, and in the local context is useful in abstracting from finite injury constructions of Δ^0_2 sets. 1-generics can be woven into more complicated priority constructions, providing richness of structure in a uniform way. For this, one needs to know exactly what structure 1-genericity delivers.

DEFINITION 3.1 (*Jockusch* [1977]). (1) A set A is n-generic if and only if for every Σ^0_n set $S \subseteq 2^{<\omega}$, *either*

(i) $\exists n[A \upharpoonright n \in S]$, *or*

(ii) $\exists n(\forall \sigma \supset A \upharpoonright n)\,[\sigma \notin S]$.

(2) A degree $\mathbf{a}$ is *n-generic* if and only if it contains an n-generic set.

It is easy to show (Jockusch [1980]) that for all $n \geqslant 1$ n-generics satisfy $\mathbf{a}^{(n)} = \mathbf{a} \cup \mathbf{0}^{(n)}$, and exist below $\mathbf{0}^{(n)}$ (but not below $\mathbf{0}^{(n-1)}$). In particular, 1-generics (but not 2-generics) exist below $\mathbf{0}'$ (and below any c.e. $\mathbf{a} > \mathbf{0}$), and these are low.

Although the forcing power of an n-generic $\mathbf{a}$ is not restricted to $\mathcal{D}(\leqslant \mathbf{a})$, most applications of n-genericity are. Jockusch produces the expected analogues of Corollaries 2.2 and 2.5 above:

(1) If $\mathbf{a}$ is 1-generic, any finite poset is embeddable below $\mathbf{a}$. Hence the existential theory of $\mathcal{D}(\leqslant \mathbf{a})$ is decidable (and independent of $\mathbf{a}$).

(2) If $\mathbf{a}$ is 1-generic, $\mathcal{D}(\leqslant \mathbf{a})$ is not a lattice.

Every 1-generic bounds (and is even the join of) a minimal pair.

Jockusch [1980] also shows that although no 1-generic can bound a nonzero c.e. degree, any 1-generic $\mathbf{a}$ is $\mathbf{b}$-CEA for some $\mathbf{b} < \mathbf{a}$, so that the richness of structure below nonzero c.e. degrees is reproduced in appropriate intervals topped by 1-generics. In fact (Kumabe [1991]), *any* n-generic $\mathbf{a}$ is CEA some n-generic degree $< \mathbf{a}$.

In the local context, the precise limits of the power of 1-genericity are not obvious. Much interest has focused on the extent to which 1-generics below $\mathbf{0}'$ emulate 2-generics. Jockusch [1980] showed that for $n \geqslant 2$ there are no minimally n-generic degrees, and that no 2-generic bounds a minimal degree.

The analogous results for 1-generics (due to Chong and Jockusch [1984]) derive from a local version of Martin's Category Theorem [1967], and require detailed knowledge of the anatomy of Turing reductions and considerable technical ingenuity. The situation is clarified by the following surprising theorem:

THEOREM 3.2 (Haught [1986]). *The 1-generic degrees below $\mathbf{0}'$ (together with $\mathbf{0}$) are closed downwards.*

This is far from true of the 1-generics in general, or even for the 2-generics. Chong and Downey [1989] and Kumabe [1990] have shown the existence of 1-generics below $\mathbf{0}^{(2)}$ which bound minimal degrees. Kumabe [1993] showed that for each $n \geqslant 2$ the n-generic degrees are not dense.

An interesting question left open by Jockusch [1980] concerns degrees $\mathbf{a}$ with the *cupping property* (that is, such that $\mathbf{a}$ is nontrivially cuppable to any $\mathbf{c} > \mathbf{a}$). Jockusch proved that every 2-generic has the cupping property, observed that Yates' [1976] abundance theory gives a low degree with the cupping property, and asked:

Does every 1-generic degree have the cupping property? Kumabe [ta1] gave a negative answer by constructing a 1-generic **a** with a *strong minimal cover* **b** (that is, $\mathcal{D}(\leqslant \mathbf{a}) = \mathcal{D}(< \mathbf{b})$), which is even below $\mathbf{0}'$. (This also follows indirectly from Cooper's [1971] construction of a nonzero c.e. degree with a strong minimal cover.)

4. Structure, jump and definability

Essential ingredients of recent definability proofs are results and techniques concerned with local structure, which have taken on a role quite unforseen by their originators. A primary example is the coming together of Posner–Robinson cupping [1981], Lachlan nonsplitting [1975] and techniques from the theory of the d.c.e. (or d.r.e.) degrees in defining the Turing jump and the relation of 'computably enumerable in'. Meanwhile, the diversity of links found between jump and structure below $\mathbf{0}'$ opens up the possibility of finding natural (i.e. logically simple) definitions of levels of the high/low hierarchy in $\mathcal{D}(\leqslant \mathbf{0}')$.

In contrast to the specifically local approach, general coding methods have been imported from the global theory with impressive results (see Odifreddi and Shore [1989]). The basic lemma is:

THEOREM 4.1 (Slaman and Woodin [1986]). *Every uniformly low countable antichain is uniformly definable in $\mathcal{D}(\leqslant \mathbf{0}')$ from finitely many parameters.*

The proof uses the uniform bound **c** for the antichain and two other parameters **a** and **b** to define the antichain as the set of minimal solutions of $\mathbf{x} \neq (\mathbf{x} \cup \mathbf{a}) \cap (\mathbf{x} \cup \mathbf{b})$ below **c**. **a** and **b** are constructed using an oracle for $\mathbf{0}'$ in a (fairly sophisticated) development of the proof of Theorem 2.4.

Theorem 4.1 enables definitions below $\mathbf{0}'$ (using parameters) for any relation on degrees which is definable from a finite collection of uniformly low sets. For instance:

COROLLARY 4.2 (Definability of $\mathcal{E}$ using parameters) (Slaman and Woodin [1986]). *The set $\mathcal{E}$ of computably enumerable degrees is definable in $\mathcal{D}(\leqslant \mathbf{0}')$ from finitely many parameters.*

This uses Welch's [1981] observation that the Sacks [1963b] splitting of K into two low sets A and B, say, can be dissected to give uniformly low sets of degrees $U, V \leqslant \deg(A), \deg(B)$, respectively, which pairwise generate $\mathcal{E} - \{\mathbf{0}\}$ under join.

Using codings via initial segments, it is possible to get the absolute definability of $\mathbf{High}_3$, $\mathbf{Low}_3$, and hence:

THEOREM 4.3 (Shore Definability Theorem) (Shore [1988]). *$\mathbf{High}_n$ and $\mathbf{Low}_n$ are definable in $\mathcal{D}(\leqslant \mathbf{0}')$ for every $n \geqslant 3$.*

One can use the definability of $\mathbf{Low}_3$ (or of $\mathcal{E}$), with Shore's translation between $\mathcal{D}(\leqslant \mathbf{0}')$ and first order arithmetic, to show that the theory of $\mathcal{D}[\mathbf{a}, \mathbf{a}']$ is not independent of $\mathbf{a}$ (Shore [1981]): If $\mathcal{D}[\mathbf{a}, \mathbf{a}'] \equiv \mathcal{D}(\leqslant \mathbf{0}')$ then $\mathbf{a}^{(3)} = \mathbf{0}^{(3)}$. It would be interesting to find some natural difference between $\mathcal{D}[\mathbf{0}, \mathbf{0}']$ and $\mathcal{D}[\mathbf{0}', \mathbf{0}'']$, for instance. More generally:

QUESTION 4.4. Characterise the $\mathbf{a}$-independent fragment of the theory of $\mathcal{D}[\mathbf{a}, \mathbf{a}']$.

The limitation $n \geqslant 3$ in Theorem 4.3 is intrinsic to results proved purely via degree theoretic codings, although Slaman and Woodin [ta] have demonstrated the power of ingenious hybrid arguments. Of course, an extension of Shore's Theorem 2.7 sufficient for bi-interpretability between the standard model of first order arithmetic and $\mathcal{D}(\leqslant \mathbf{0}')$ (see Slaman [1991] and Section 8 below), would invest $\mathcal{D}(\leqslant \mathbf{0}')$ with enough of the familiar properties of ω to immediately give definability of all arithmetically describable sets. But at present there seems to be a point after which general definability depends on knowledge of local structure.[1]

The Posner–Robinson Cupping Theorem mentioned above, and Shoenfield's Capping Theorem [1966], provided important steps in proving:

THEOREM 4.5 (Complementation Theorem) (Posner and Robinson [1981], Posner [1981]). *$\mathcal{D}(\leqslant \mathbf{0}')$ is complemented.*

Slaman and Steel [1989] succeeded in removing the non-uniformity in the original proof (which treated the low_2 and non-low_2 cases differently). Seetapun and Slaman [1992] have shown (extending Cooper [1972a], Epstein [1975] and Posner [1977]) that the complement can be chosen to be minimal.

The earliest connections between jump and structure below $\mathbf{0}'$ grew out of questions concerning the distribution of minimal degrees below $\mathbf{0}'$. Cooper [1973], using Δ^0_2 approximations in a similar way to that of computable enumerations, and a notion of high permitting got via an extension of Robinson's [1968] characterisation of the c.e. degrees with jump $\mathbf{0}^{(2)}$, showed that no minimal degree below $\mathbf{0}'$ has jump $\mathbf{0}^{(2)}$. This contrasted with the Cooper Jump Inversion Theorem (every $\mathbf{a} \geqslant \mathbf{0}'$ is the jump of *some* minimal degree). Jockusch and Posner [1978] observed that the full power of high permitting was not needed here, and organised their improved result within two newly available general frameworks: those of n-genericity and of the generalised high/low hierarchy.

DEFINITION 4.6 (*Jockusch and Posner* [1978]). If $n \geqslant 1$, we say $\mathbf{a}$ is *generalised* $\textit{high}_n$ *(generalised* $\textit{low}_n$*)*, written $\mathbf{a} \in \mathbf{GH}_n$ ($\mathbf{a} \in \mathbf{GL}_n$, respectively), if and only if $\mathbf{a}^{(n)} = (\mathbf{a} \cup \mathbf{0}')^{(n)}$ ($\mathbf{a}^{(n)} = (\mathbf{a} \cup \mathbf{0}')^{(n-1)}$, respectively).

1-genericity can be used to substantiate the intuition that theorems about minimal degrees are really theorems about *any* structural property closely associated (in

[1] These remarks must now be qualified in the light of recent work of Nies, Shore and Slaman, improving Shore's earlier results by a factor of 1 in most cases.

the sense of Chong [1979]) with tree constructions. Recasting results in terms of the generalised high/low hierarchy (which of course agrees with the usual high/low hierarchy below $\mathbf{0}'$) helps explain the apparent non-uniformity of local and global phenomena (for example, the failure of the local version of the Cooper Jump Inversion Theorem).

THEOREM 4.7 (Jockusch and Posner [1978]). *Every degree is generalised low*$_2$ *or has a* 1*-generic predecessor. In particular, every minimal degree is generalised low*$_2$.

The proof exploited an earlier insight of Jockusch [1977]: Finite injury Δ_2^0 approximation arguments may often be recast as shorter (though not always more informative) $\mathbf{0}'$ oracle constructions. Fejer [1989] has found another characteristic of the cones below non-low$_2$ degrees: Any finite lattice can be embedded (preserving meets, joins and top element) below a non-low$_2$ (in fact non-$\mathbf{GL}_2$) degree.

In the other direction, Sasso [1974] showed that there are non-low minimal degrees below $\mathbf{0}'$. It turns out that the *jumps* of the minimal degrees below $\mathbf{0}'$ can be characterised as the *almost* Δ_2^0 *degrees*, a proper ideal of the $\mathbf{0}'$-CEA degrees low over $\mathbf{0}'$ (see Downey, Lempp and Shore [1996] and Cooper [1996a]).

There is a definable set of degrees which more closely approximates the class of high degrees.

THEOREM 4.8. (1) (Jockusch [1977], Cooper [1973]) *Every* $\mathbf{a} \in \mathbf{GH}_1$ *bounds a minimal degree, but*

(2) (Lerman [1986], Jockusch [1980]) *There is a* high$_2$ *degree which bounds no minimal degree.*

The proof of part (2) is interesting in that both the global techniques for obtaining atomless ideals are inapplicable: the proof of Martin [1967] using Baire category is closely related to 2-generic degrees (at best $\leqslant \mathbf{0}^{(2)}$), while Hugill's [1969] derivation via specific initial segments is subject to Theorem 4.7.

$\mathcal{D}(\leqslant \mathbf{0}')$ provides a rich environment for the computably enumerable degrees. Sasso [1970] (extending Yates [1967]) showed that there is a minimal degree $\mathbf{m}$ below $\mathbf{0}'$ incomparable with each c.e. $\mathbf{a} \neq \mathbf{0}$ or $\mathbf{0}'$. It is not known if $\mathbf{m}$ can *complement* each such $\mathbf{a}$, although Li Angsheng [ta] and Cooper and Seetapun have proved a dual of Sasso's theorem: There is a degree which nontrivially cups every c.e. $\mathbf{a} > \mathbf{0}$ to $\mathbf{0}'$.

Local techniques come into their own in defining $\mathcal{E}$ and the relation of "c.e. in" (see Cooper [1994]).

DEFINITION 4.9. (1) If $\mathbf{b} \not\leqslant \mathbf{c} < \mathbf{d}$, we say that $\mathbf{d}$ is *splittable in* $\mathcal{D}(\geqslant \mathbf{c}) - \mathcal{D}(\geqslant \mathbf{b})$ if and only if there exist $\mathbf{d}_0, \mathbf{d}_1 \in \mathcal{D}(\geqslant \mathbf{c}) - \mathcal{D}(\geqslant \mathbf{b})$ such that $\mathbf{d}_0 \cup \mathbf{d}_1 = \mathbf{d}$.

(2) Let

$$\begin{aligned} \mathcal{C}^{\mathbf{a}} = \big\{ &\mathbf{d} \mid (\forall \mathbf{c} \geqslant \mathbf{a})(\forall \mathbf{b} \not\leqslant \mathbf{c}) \\ &[\mathbf{d} \leqslant \mathbf{c} \vee \mathbf{c} \cup \mathbf{d} \text{ is splittable in } \mathcal{D}(\geqslant \mathbf{c}) - \mathcal{D}(\geqslant \mathbf{b})]\big\}. \end{aligned}$$

We notice that $\mathcal{C}^{\mathbf{a}}$ is obviously definable from $\mathbf{a}$ in $\mathcal{D}$.

By relativising the proof of the Sacks Splitting Theorem (see Soare [1987], p. 124) we get the set of degrees c.e. in $\mathbf{a} \subseteq \mathcal{C}^{\mathbf{a}}$. A full approximation construction relative to an oracle for $\mathbf{a}$, containing elements of the Lachlan Nonsplitting Theorem [1975] for the c.e. degrees, gives:

THEOREM 4.10 (Definability of 'computably enumerable in'). $\mathcal{C}^{\mathbf{a}} = \{\mathbf{d} \mid \mathbf{d}$ *is c.e. in* $\mathbf{a}\}$.

Since the proof of ($\mathbf{d}$ not c.e. $\Rightarrow \mathbf{d} \notin \mathcal{C}^{\mathbf{a}}$) is actually carried out in $\mathcal{D}(\leqslant \mathbf{a}')$, the same definition gives the absolute version of Theorem 4.2.

COROLLARY 4.11 (Definability of $\mathcal{E}$). *$\mathcal{E}$ is definable in $\mathcal{D}(\leqslant \mathbf{0}')$.*

Particular interest attaches to nontrivially definable singletons of $\mathcal{D}(\leqslant \mathbf{0}')$. A relativisable construction of a definable c.e. singleton would provide a positive solution to the longstanding:

QUESTION 4.12 (*The uniform solution to Post's problem*) (*Sacks* [1967]). Is there a degree invariant solution to Post's problem? That is, is there an index e such that for any $A, B \subseteq \omega$ we have

$$A <_T W_e^A <_T A' \quad \text{and} \quad A \equiv_T B \Rightarrow W_e^A \equiv_T W_e^B?$$

Sacks' question is not just of interest to computability theorists, but is closely connected to two very general conjectures in set theory (see Slaman [1994]). We note that a proof of bi-interpretability would not automatically give a degree invariant solution to Post's problem. The resulting definition of an intermediate c.e. degree $\mathbf{a}$ would just describe (degree theoretically) the index e for a representative W_e of $\mathbf{a}$.

5. Definability in cones

How far does the theory of $\mathcal{D}(\leqslant \mathbf{0}')$ resemble that for one of its principal ideals $\mathcal{D}(\leqslant \mathbf{a})$? The forcing results of Section 2 have analogues in $\mathcal{D}(\leqslant \mathbf{a})$ for a very general class of degrees $\mathbf{a}$. On the other hand a number of striking distinctions between such theories have emerged via full approximation techniques. These results are potentially useful tools in defining particular classes of computably enumerable and Δ_2^0 degrees. Moreover, a by-product of definability in lower cones may be a better understanding of such phenomena for $\mathcal{D}(\leqslant \mathbf{0}')$. Definability in lower cones has a special significance, since (by Ambos-Spies [1983]), every $\mathcal{D}(\leqslant \mathbf{a}$ c.e.) ($\mathbf{a} \neq \mathbf{0}$) is an *automorphism base* for $\mathcal{D}(\leqslant \mathbf{0}')$ (see Section 8 below).

Since any nonzero c.e. or non-low$_2$ Δ^0_2 degree $\mathbf{a}$ has a 1-generic predecessor, we immediately get the rich basic structure for the corresponding $\mathcal{D}(\leqslant \mathbf{a})$ (e.g., embeddings, decidability of existential theory, minimal pairs, non-c.e. elements, not a lattice, etc.) originally discovered by Sacks [1963a] and Yates [1970a].

Yates [1970b] used the first full approximation tree argument (confined uneasily within the global-to-local approach) to prove the important:

THEOREM 5.1 (Yates Minimal Degree Theorem). *Below any nonzero c.e. degree* $\mathbf{a}$ *there exists a minimal degree* $\mathbf{m}$.

Further development of full approximation tree constructions (Cooper [1973] and Epstein [1975]) eventually led to very strong initial segment embedding results for $\mathcal{D}(\leqslant \mathbf{a}$ c.e.), paralleling Theorems 2.6, 2.7 above:

THEOREM 5.2 (Lerman [1983]). *If* $\mathbf{a}$ *is either nonzero c.e. or high, then every* $\mathbf{0}'$*-presentable upper semi-lattice (with least element) is embeddable as an initial segment of* $\mathcal{D}(\leqslant \mathbf{a})$.

If $\mathbf{a} \leqslant \mathbf{0}'$ then $\mathcal{D}(\leqslant \mathbf{a})$ is $\mathbf{a}^{(3)} \leqslant \mathbf{0}^{(4)}$ presentable, so once again there is considerable scope for improvement here.

As for Theorem 2.7 above, Shore [1981] converts the structure provided by Theorem 5.2 into definability.

THEOREM 5.3 (Shore [1981]). *Let* $\mathbf{a}$ *be nonzero c.e. or high. Then the theory of* $\mathcal{D}(\leqslant \mathbf{a})$ *is* $\equiv_1$ *the theory of first order Peano arithmetic, and hence has degree* $\mathbf{0}^{(\omega)}$. *(But if* $\mathcal{D}(\leqslant \mathbf{a}) \equiv \mathcal{D}(\leqslant \mathbf{0}')$ *then* $\mathbf{a}^{(3)} = \mathbf{0}^{(4)}$.)

QUESTION 5.4. Is the $\forall\exists$-theory of $\mathcal{D}(\leqslant \mathbf{a})$ decidable for every c.e. or high degree $\mathbf{a}$?

Epstein's [1979] conjecture that every nonzero c.e. $\mathbf{b} < \mathbf{a}$ c.e. has a minimal complement $\mathbf{m} < \mathbf{a}$ leads to significant differences in definability and discontinuous structure in lower cones. Epstein [1981] confirmed the conjecture for $\mathbf{a}$ high c.e. and Seetapun and Slaman [1992] constructed a minimal complement for any $\mathbf{b} \in \mathcal{D}(\mathbf{0}, \mathbf{0}')$. But for the general case, Cooper and Epstein [1987] pointed to a more interesting situation:

- If $\mathbf{b} < \mathbf{a}$ is low, then one can find a minimal $\mathbf{m} < \mathbf{a}$ with $\mathbf{m} \mid \mathbf{b}$. But there exists a $\mathcal{D}(\leqslant \mathbf{a}$ c.e.) in which complementation fails. Moreover by Cooper [1986], $\mathbf{a}$ can be chosen to be high, so (answering Q11 of Epstein [1979]) not every $\mathcal{D}(\leqslant \mathbf{a}$ high) is elementarily equivalent to $\mathcal{D}(\leqslant \mathbf{0}')$.

In fact:

- (Slaman, private communication.) There exists a $\mathcal{D}(\leqslant \mathbf{a}$ c.e.) in which *no* $\mathbf{b} \in \mathcal{D}(\mathbf{0}, \mathbf{a})$ is complemented.

- (Cooper [1989], Slaman and Steel [1989].) There exists a $\mathcal{D}(\leqslant \mathbf{a}$ c.e.) in which cupping of some c.e. $\mathbf{b} > \mathbf{0}$ fails. (For an interesting alternative proof of this see Downey [1987].) And:
- (Cooper [1986].) There exists a $\mathcal{D}(\leqslant \mathbf{a}$ c.e.) in which capping of some $\mathbf{b} < \mathbf{a}$ fails.

The above result of Cooper and Slaman, and Steel is the key to a nontrivial definition of a c.e. singleton in some $\mathcal{D}(\leqslant \mathbf{a}$ c.e.). (A nontrivially definable c.e. $\mathbf{a} > \mathbf{0}$ in *some* lower cone of $\mathcal{D}(\leqslant \mathbf{0}')$ is provided by Cooper's [1971] construction of a strong minimal cover for such an $\mathbf{a}$.)

THEOREM 5.5 (See Cooper [1996b]). *There exist c.e. degrees* $\mathbf{0} < \mathbf{a} < \mathbf{c}$ *such that* $\mathbf{a}$ *is definable as the greatest degree in* $(\mathbf{0}, \mathbf{c})$ *not cuppable to* $\mathbf{c}$ *in* $\mathcal{D}(< \mathbf{c})$.

Very little is known about absolute definability in segments of $\mathcal{D}(\leqslant \mathbf{0}')$. For instance:

QUESTION 5.6. Is $\mathcal{E}(\leqslant \mathbf{a})$ definable in $\mathcal{D}(\leqslant \mathbf{a})$ for each c.e. $\mathbf{a}$?

QUESTION 5.7. Does every $\mathcal{D}(\leqslant \mathbf{a})$ with $\mathbf{a}$ c.e. or non-low$_2$ have a nontrivially definable singleton?

QUESTION 5.8. Characterise the definable relations of $\mathcal{D}(\leqslant \mathbf{a})$ for $\mathbf{a} \leqslant \mathbf{0}'$.

Finally, we note that jump inversion provides another limitation on the extent to which lower cones emulate $\mathcal{D}(\leqslant \mathbf{0}')$. By Soare and Stob [1982], if $\mathbf{a} < \mathbf{0}'$ then not every $\mathbf{0}'$-CEA degree is obtainable as the jump of some $\mathbf{x} \leqslant \mathbf{a}$.

6. Degree and information content

Gödel's theorem tells us that in a sufficiently strong language, complete information cannot be effectively generated. The underlying explanation for this can be found in the exact relationship between the computably enumerable sets and particular Π^0_1 classes.

A more artificial polarising of structure of information originates with Post's program [1944], which may be described as an attempt to differentiate degree via an analysis of information content (and in particular to find a property of c.e. *sets* which guarantees *degree* strictly between $\mathbf{0}$ and $\mathbf{0}'$). Just as productive sets arise out of the recursion theoretic analogue of first order Peano arithmetic (Myhill [1955]), the various immunity properties are strongly counter-related to such logical origins.

Both approaches realise and extend Post's program in subtle and contrasting ways.

For the purposes of the following definition, a *computable tree* is a computable ideal of $\omega^{<\omega}$ (with the ordering $\subseteq$ on strings).

DEFINITION 6.1. $\mathcal{A} \neq \emptyset$ is a Π_1^0 *class* if and only if there exists a computable tree T such that

$$f \in \mathcal{A} \Leftrightarrow \forall x \in \omega[f \upharpoonright x \in T].$$

That is, a Π_1^0 class $\mathcal{A}$ is the set of infinite branches of a computable tree T, where we write $\mathcal{A} = [T]$.

Examples of nonempty Π_1^0 classes are:

(1) (Shoenfield [1960]) The class of sets separating two disjoint c.e. sets,

(2) (Shoenfield [1960]) The set of all consistent (or complete) extensions of a consistently axiomatised first order theory, and

(3) (Jockusch) The class DNC of all *diagonally noncomputable functions*, that is those functions f for which $(\forall i) f(i) \neq \varphi_i(i)$.

Our special interest is in the degrees of complete extensions of first order Peano arithmetic (that is, the class **PA** of *PA degrees*), which Solovay has shown to be the same as the set of degrees of *consistent* extensions of Peano arithmetic. A useful observation of Jockusch is the coincidence of the *fixed point free* degrees **FPF** (that is, the degrees containing functions $f \in \omega^\omega$ for which $(\forall i)(W_i \neq W_{f(i)})$) with the DNC degrees **DNC**, with **PA** coinciding with the degrees of 0–1 valued DNC functions. **PA** and **FPF** are closed upwards. Kucera [1989] shows that the inclusion $\mathbf{PA} \subset \mathbf{FPF}$ is proper, even below $\mathbf{0}'$ (in fact, there is a jump inversion theorem for degrees in **FPF**–**PA**).

The basic relationship of **PA** with other important degree classes is best described in the general context of the (computably bounded, or equivalently, 0–1 valued) Π_1^0 classes.

DEFINITION 6.2. $\mathcal{B} \subseteq 2^\omega$ is a *basis* for (0–1 valued) Π_1^0 classes if and only if every Π_1^0 class has a member in $\mathcal{B}$. A set of degrees is a *basis* for Π_1^0 classes if and only if the union of the set is.

Then:

THEOREM 6.3. *The following are bases for* 0–1 *valued* Π_1^0 *classes*:

(1) (Scott Basis Theorem) (Scott [1962]) $\mathcal{D}(\leqslant \mathbf{a})$ *for any given* $\mathbf{a} \in \mathbf{PA}$.

(2) (Jockusch and Soare [1972a, 1972b]) $\mathcal{E}$, *but* (*Kleene* [1952]) *not* $\{\mathbf{0}\}$.

(3) (Low Basis Theorem) (Jockusch and Soare [1972b]) *The low degrees.*

(4) (Jockusch and Soare [1972b]) *The hyperimmune free degrees below* $\mathbf{0}^{(2)}$. (See Definition 6.7 below.)

(1) holds since we can prove enough in first order Peano arithmetic to be able to trace an infinite path through any given computable tree. (2) depends on the observation that the sector of strings to the left of the leftmost infinite path of a computable tree is c.e., and Turing equivalent to the leftmost path. The other results involve forcing with Π_1^0 classes, not at all compatible with the standard methods of

Sections 2 and 3 above. Locally the most useful of these results is the Low Basis Theorem, which of course provides us with low PA degrees. The local relevance of part (2) above is limited by the following generalisation of the Fixed Point Theorem:

THEOREM 6.4 (Arslanov Completeness Criterion) (Arslanov [1981]). *The only fixed point free c.e. degree is* $\mathbf{0}'$. *In particular,* $\mathbf{PA} \cap \mathcal{E} = \{\mathbf{0}'\}$.

Arslanov's completeness criterion for c.e. degrees has been extended by Jockusch, Lerman, Soare and Solovay [1989] to all finite levels of the n.c.e. (and even n-CEA) hierarchy.

On the other hand:

THEOREM 6.5 (Kucera [1986]). *Every fixed point free (and in particular PA) degree below* $\mathbf{0}'$ *has a nonzero c.e. predecessor.*

Kucera is able to use the Low Basis Theorem to get a low 0–1 valued DNC function, and hence a low $\mathbf{a} \in \mathbf{FPF}$. Since Jockusch and Soare's proof only uses a $\mathbf{0}'$ oracle, Kucera can apply his theorem to obtain a priority free solution to Post's problem. See Kucera [1989] for a range of applications to other results previously needing finite (and even infinite) injury priority constructions.

Another intriguing non-standard construction is due to Jockusch and Simpson [1980]. They first construct a 0–1 valued Π^0_1 class $\mathcal{P}$ with no computable members such that any two branches form a tt-minimal pair (in fact Slaman has obtained the analogous result for Turing reducibility by a somewhat more complicated argument). Then observing that every member of $\mathcal{P}$ of hyperimmune free degree is also of minimal degree, part (4) of Theorem 6.3 gives an alternative construction of a minimal degree (below $\mathbf{0}^{(2)}$).

This is related to the following problem left open by Jockusch and Simpson:

QUESTION 6.6. Does there exist a 0–1 valued Π^0_1 class each member of which bounds a minimal degree?[2]

The Scott Basis Theorem seems to support a positive answer, showing the ideal below any PA degree to be rich in structure. For instance, an application to $\mathcal{P}$ above shows that every hyperimmune free PA degree has a minimal predecessor, while Jockusch and Soare [1972b] use it to embed any countable poset below any given PA degree. On the other hand, Kumabe [ta2] has recently shown the existence of minimal FPF degrees.

The historical role of the immunity properties in computability theory (starting with Post [1944]) is analogous to that of structural discontinuity in regard to Turing definability. Locally, there are close relationships with the jump classes, indirectly providing important connections between information content and structure. For other aspects of this extensive and complex topic see Odifreddi [1989].

[2] Groszek and Slaman have recently announced a positive solution to Question 6.6.

The following definitions (apart from (4)) derive from Post [1944]. They aim to impose increasingly severe sparseness conditions on a set.

DEFINITION 6.7. Let A be an infinite subset of ω. Then:

(1) A is *immune* if and only if A contains no infinite c.e. subsets.

(2) A is *hyperimmune* if and only if there is no computable array $\{D_{f(i)}\}_{i\in\omega}$ of mutually disjoint finite sets, all intersecting with A. (Equivalently, the principal function of A is dominated by no computable function.)

(3) A is *(strongly) hyperhyperimmune* if and only if there is no computable array $\{W_{f(i)}\}_{i\in\omega}$ of mutually disjoint (finite) c.e. sets, all intersecting with A.

(4) (Myhill [1956]) A is *cohesive* if and only if there is no c.e. W for which $W \cap A$ and $\overline{W} \cap A$ are both infinite.

(A degree $\mathbf{a}$ is said to be *immune* etc. if and only if it contains an immune etc. set. The *hyperimmune free* degrees are those which are not hyperimmune.)

Of course, every degree $> \mathbf{0}$ is immune (any noncomputable path through $2^{<\omega}$ is an immune set of strings). For upward closure of the other associated degree classes see Jockusch [1973]. Locally, the properties are extremely well-behaved.

THEOREM 6.8. (1) (Miller and Martin [1968]) *Every nonzero degree below* $\mathbf{0}'$ *is hyperimmune.*

(2) (Jockusch [1969], Cooper [1972b]) *Below* $\mathbf{0}'$ *the (strongly) hyperhyperimmune and cohesive degrees coincide, and are exactly the high degrees.*

However, even below $\mathbf{0}^{(2)}$ the situation is more complicated. For instance, Miller and Martin [1968] used a tree argument to construct a hyperimmune free degree $< \mathbf{0}^{(2)}$, and Jockusch and Stephan [1993], using interesting connections between immunity properties and Π^0_1 classes relative to $\mathbf{0}'$, have found a cohesive degree $\mathbf{a}$ with $\mathbf{a}^{(2)} = \mathbf{0}^{(2)}$. Jockusch and Stephan (in a particularly nice sequence of results) actually succeed in finding non-equivalent characterisations of the jumps of the strongly hyperhyperimmune degrees (as precisely the degrees which are DNC relative to $\mathbf{0}'$) and of those of the cohesive degrees. So even the *jumps* do not coincide in general.

QUESTION 6.9. Characterise the cohesive degrees. Is there a cohesive, non-high minimal degree?

Returning to the original motivation for Post-inspired immunity properties, Marchenkov [1976] verified the hoped-for connection with Turing incompleteness (using Degtev's proof [1973] of the existence of a noncomputable, semirecursive, η-maximal set). Recently Harrington and Soare [1991] have found a property of c.e. sets which both fulfills Post's program and is definable in the lattice of c.e. sets.

See Odifreddi [1989] for other interesting properties which potentially relate local definability to information content.

7. The Ershov hierarchy for $\mathcal{D}(\leqslant \mathbf{0}')$

There are two related hierarchies starting with $\mathcal{E}$ which seek to classify the levels at which differences in the local theory appear. The n.c.e. (or n.r.e.) hierarchy is the finer and and more useful of these below $\mathbf{0}'$. Through its links with the n-CEA (= n-REA) hierarchy (see Jockusch and Shore [1983], [1984]) important applications of local structure to the global theory have been found, for example (Cooper [ta1]) in defining the Turing jump.

DEFINITION 7.1. (1) (Putnam [1965], Gold [1965]) Let $A \in \Delta^0_2$ have standard approximating sequence $\{A^s\}_{s\in\omega}$.

We say A is *n.c.e.* if and only if A^s changes value at most n times, each $x \in \omega$. (1.c.e. being just c.e., of course, and the 2.c.e. sets – or d.c.e. sets – being the differences $W_i - W_j$ of c.e. sets.)

We say A is *ω-c.e.* if and only if there is a computable bound on the number of such changes.

We write $\mathbf{D}_n$ $(n \geqslant 2)$ for the nth level of the corresponding *n.c.e. hierarchy* of degrees below $\mathbf{0}'$.

(2) (Jockusch and Shore [1984]) The *n-CEA hierarchy* of degrees is got from the hierarchy of sets in which $\emptyset$ is 0-CEA, and A is $(n+1)$-CEA if and only if A is B-CEA for some n-CEA B.

The n.c.e. hierarchy of sets is a natural extension of the c.e. sets in that it classifies the set of all finite Boolean combinations of computably enumerable sets and their complements. And just as the c.e. sets are the sets many-one reducible to a given creative set K, the n.c.e. sets are those $\leqslant_{\mathrm{btt}}$ (bounded truth-table reducible to) K (via truth-tables asking $\leqslant n$ membership questions of K). The ω-c.e. sets are those $\leqslant_{\mathrm{tt}}$ (truth-table reducible to) K (so there are Δ^0_2 degrees which are not even ω-c.e.). Ershov [1968a, 1968b, 1970] showed that the n.c.e. hierarchy could be extended exhaustively (but not uniquely due to its dependence on a notation system for the computable ordinals) into the transfinite. See Epstein, Haas and Kramer [1981] for a useful introduction to the α-c.e. degrees (α a computable ordinal). See Selivanov [ta] for a general study of the links between numerations, index sets and hierarchies.

The n.c.e. hierarchy of degrees does not collapse (Cooper [1971]), and by a similar argument (Jockusch and Shore [1984]) nor does the n-CEA hierarchy below $\mathbf{0}'$.

In relation to definability, the two most interesting topics are the structure of the d.c.e. (and more generally the n.c.e.) degrees, and the relationship of the n.c.e. hierarchy to $\mathcal{E}$ and the n-CEA hierarchy.

Lachlan observed that if A is d.c.e. then A is CEA the c.e. set $\{\langle x, s\rangle \mid x \in A^s - A\}$, so every d.c.e. degree is 2-CEA. In general $\mathbf{D}_n \subset n$-CEA, so $\mathbf{D}_n$ structurally relates to $\mathbf{0}$ much as $\mathcal{E}$ does. For instance, each $\mathbf{D}_n$ has no minimal degrees (although $\mathbf{D}_\omega$ does), but does have minimal pairs, noncappable degrees (so unlike $\mathcal{D}(\leqslant \mathbf{0}')$ is not complemented), and degrees which do not bound minimal pairs.

The intuition that $\mathbf{D}_n$ is elementarily equivalent for each $n \geqslant 1$ is reinforced by:

THEOREM 7.2 (Splitting theorem for $\mathbf{D}_n$) (Sacks [1963b], Cooper [1992]). *Every n.c.e.* $\mathbf{a}$ *can be nontrivially split in* $\mathbf{D}_n$, *for each* $n \geqslant 1$.

Here it is the capacity to change ones mind about which side of the splitting will accept traces for x entering $A \in \mathbf{a}$ which makes it possible to avoid infinite injury and maintain a Sacks splitting strategy at levels $n \geqslant 2$. It is not trivial to verify a computable bound for the set of traces for each x (necessary in avoiding disruption of restraints by infinitary outcomes).

Extending Sacks density we have:

THEOREM 7.3 (Weak density theorem) (Sacks [1964], Cooper, Lempp and Watson [1989]). *If* $\mathbf{a} < \mathbf{b}$ *are c.e., the n.c.e. hierarchy on the interval* $\mathcal{D}(\mathbf{a}, \mathbf{b})$ *does not collapse at any level* $n \geqslant 1$.

Arslanov [1990] and Ishmukametov [1985] have also investigated structural properties of the properly n.c.e. degrees.

A $\mathbf{0}^{(3)}$ priority argument shows that full density fails badly at the 2 level.

THEOREM 7.4 (Nondensity theorem for $\mathbf{D}_2$) (Cooper, Harrington, Lachlan, Lempp and Soare [1991]). *There exists a maximal d.c.e.* $\mathbf{a} < \mathbf{0}'$.

On the other hand, full density and splitting can be combined in the low_2 n.c.e. degrees.

THEOREM 7.5 (Low_2 density and splitting) (Shore and Slaman [1990], Cooper [1991]). *For* $n \geqslant 1$, *any* low_2 *n.c.e.* $\mathbf{a}$ *is splittable in* $\mathbf{D}_n$ *over any n.c.e.* $\mathbf{b} < \mathbf{a}$.

If one includes cone avoidance, one can obtain strong nonsplitting results for d.c.e. topped intervals, even in the Δ_2^0 degrees. Cooper [ta1] uses one such result to produce a 2-CEA operator (in the style of Jockusch and Shore [1984]) to define the Turing jump in $\mathcal{D}$.

In intervals with c.e. endpoints, Arslanov (private communication) has shown that $\mathbf{0}'$ is always splittable in the d.c.e. degrees above an c.e. $\mathbf{a}$, while Ding and Qian [ta2] has succeeded in replacing $\mathbf{0}'$ by any nonzero $\mathbf{b} > \mathbf{a}$. Cooper and Yi [ta] establish one remaining situation in which full density is possible in the d.c.e. degrees: if the bottom of the interval is c.e., then it properly contains a d.c.e. degree. (Geoffrey LaForte has observed that this result even holds with the top of the interval n.c.e.)

However, by Cooper and Yi [ta] (and indirectly Kaddah [1993]) the c.e. degrees are not dense in such intervals.

DEFINITION 7.6 (*Cooper and Yi* [ta]). A d.c.e. degree $\mathbf{d}$ is *isolated* if and only if there is a c.e. $\mathbf{a} < \mathbf{d}$ with $\mathcal{E} \cap \mathcal{D}(\mathbf{a}, \mathbf{d}] = \emptyset$. (The degree $\mathbf{a}$ is said to be an *isolating* c.e. degree.)

Then:

THEOREM 7.7 (LaForte [1995], Ding and Qian [1996]). *The isolated degrees (with their associated isolating degrees) are dense in the c.e. degrees.*

Also extending Cooper and Yi [ta]:

THEOREM 7.8 (Arslanov, Lempp and Shore [1996a]). *The nonisolated degrees are dense in the c.e. degrees.*

The first elementary difference between $\mathcal{E}$ and $\mathbf{D}_2$ was:

THEOREM 7.9 (Cupping for $\mathbf{D}_n$) (Arslanov [1985]). *Every c.e. (and hence n.c.e.) $\mathbf{a} > \mathbf{0}$ can be cupped to $\mathbf{0}'$ by a d.c.e. degree $\mathbf{b} < \mathbf{0}'$.*

Cooper, Lempp and Watson [1989 showed that we can replace $\mathbf{0}'$ by $\mathbf{h}$ high c.e., although in Section 5 we saw that there is no general cupping, even in $\mathcal{D}(\leqslant \mathbf{0}')$, to c.e. degrees $\mathbf{a} < \mathbf{0}'$.

Another striking elementary difference between the c.e. and the d.c.e. degrees is provided by Lachlan's nondiamond theorem [1966] and:

THEOREM 7.10 (Downey Diamond Theorem [1989]). *There exist incomparable d.c.e. degrees $\mathbf{a}, \mathbf{b}$ with $\mathbf{a} \cap \mathbf{b} = \mathbf{0}$ and $\mathbf{a} \cup \mathbf{b} = \mathbf{0}'$.*

Again, by Jiang [1993], $\mathbf{0}'$ can be replaced by any high c.e. $\mathbf{h}$ here. However, Yi [ta] has found a surprising elementary difference between $\mathbf{D}_2$ and $\mathbf{D}_2(\leqslant \mathbf{h})$, some high c.e. $\mathbf{h}$: We cannot replace $\mathbf{0}'$ by any high c.e. $\mathbf{h}$ in the nondensity theorem for $\mathbf{D}_2$.

An interesting conjecture relating to the major subdegree problem for the c.e. degrees is provided by Li: There is a high c.e. $\mathbf{a} < \mathbf{0}'$ such that for any c.e. $\mathbf{u} < \mathbf{a}$ there exists a d.c.e. $\mathbf{b} < \mathbf{0}'$ for which $\mathbf{b} \cup \mathbf{a} = \mathbf{0}'$ and $\mathbf{b} \cup \mathbf{u} \neq \mathbf{0}'$. (That is, there is a c.e. $\mathbf{a}$ with no c.e. major subdegree in the d.c.e. degrees.)

The exact relationship between the n.c.e. hierarchy and the n-CEA hierarchy is quite complex (see Arslanov, Lempp and Shore [1996b]). Jockusch and Shore [1984] showed that the 2-CEA degrees are cofinal in the α-c.e. hierarchy for any system of notations for the computable ordinals α.

There are particular problems, for instance, in analysing how particular c.e. degrees $\mathbf{a}$ contribute $\mathbf{a}$-CEA degrees to the different levels of the n.c.e. hierarchy (in intervals below $\mathbf{0}'$ or more generally). The contribution can be negligible: Arslanov, Lempp and Shore [1996b] have shown that there is a c.e. $\mathbf{a} \neq \mathbf{0}$ or $\mathbf{0}'$ (which cannot be high) for which $\mathbf{a}$-CEA=$\mathcal{E}(\geqslant \mathbf{a})$. A nice corollary of this is the existence of nonisolating c.e. degrees other than $\mathbf{0}$ and $\mathbf{0}'$ (see also Ding and Qian [ta1]).

The main open problem in the area is:

QUESTION 7.11 (*Downey's conjecture*). The first order theory of $\mathbf{D}_n$ is the same for every $n \geqslant 2$.

Downey and Shore have investigated possible differences in the lattices embeddable in distinct $\mathbf{D}_n$'s, but now conjecture that all finite lattices are immediately embeddable in $\mathbf{D}_2$. The question may well be related to:

QUESTION 7.12. Is $\mathcal{E}$, or $\mathbf{D}_n$, uniformly definable in $\mathbf{D}_{n+1}$ for each $n \geqslant 1$?

(The methods of Cooper [1994] can be adapted to define $\mathcal{E}$ in each n-CEA degree class, $n \geqslant 2$.)

Finally, there are questions of decidability. It seems likely that coding techniques can be used to characterise the degree of of the first order theory of $\mathbf{D}_n$, for each $n \geqslant 1$, as $\mathbf{0}^{(\omega)}$. But:

QUESTION 7.13. Is the Σ_2 theory of $\mathbf{D}_n$ decidable for each $n \geqslant 1$? Is the Σ_3 theory of $\mathbf{D}_n$ undecidable for each $n \geqslant 2$?

Characterising the finite final segments of $\mathbf{D}_2$ (extending Theorem 7.4) would be one approach to these questions.

8. Automorphisms and undefinability

The basic question as to whether there are any nontrivial automorphisms of the Turing degrees (otherwise we say $\mathcal{D}$ is *rigid*) was first raised by Sacks [1966] and Rogers [1967]. Rogers' interest in properties *invariant* under all automorphisms reflected the evidence at that time that there were no nontrivial obstacles to relativisation in computability theory. Since definability in a degree structure entails invariance under all its automorphisms, questions of rigidity are fundamental to any full understanding of definability.

Lerman's [1977] notion of an *automorphism base* (that is, a subset of the structure on which the action of any automorphism determines its global action) enables a remarkable reduction of the global theory of the Turing degrees to questions of local structure.

THEOREM 8.1 (Slaman and Woodin [ta]). *$\mathcal{E}$ and $\mathcal{D}(\leqslant \mathbf{0}')$ are automorphism bases for $\mathcal{D}$.*

So, by the definability of $\mathcal{E}$ and of the jump, rigidity of $\mathcal{D}$ would follow from that for $\mathcal{E}$ or $\mathcal{D}(\leqslant \mathbf{0}')$. In fact, by Ambos-Spies [1983], every nontrivial lower cone of $\mathcal{E}$ is an automorphism base for $\mathcal{E}$, and hence for $\mathcal{D}(\leqslant \mathbf{0}')$ and $\mathcal{D}$.

There are many local automorphism bases other than $\mathcal{E}$, providing indirect evidence for the rigidity of $\mathcal{D}(\leqslant \mathbf{0}')$.

THEOREM 8.2 (Jockusch and Posner [1981]). *The following classes of degrees generate $\mathcal{D}(\leqslant \mathbf{0}')$, and so are automorphism bases for $\mathcal{D}(\leqslant \mathbf{0}')$:*

(1) *Every atomic jump class* $\mathbf{J}(\mathbf{a}) = \{\mathbf{x} \leqslant \mathbf{0}' \mid \mathbf{x}' = \mathbf{a}'\}$ *with* $\mathbf{a} \leqslant \mathbf{0}'$. (*So in particular, every level of the high/low hierarchy is an automorphism base.*)

(2) *The* 1-*generic degrees below* $\mathbf{0}'$.

(3) *The minimal degrees below* $\mathbf{0}'$.

(4) *If* $\mathbf{a} > \mathbf{0}$, *the low degrees cupping* $\mathbf{a}$ *to* $\mathbf{0}'$.

(5) (*Posner*) *If* $\mathbf{a} < \mathbf{0}'$, *the set of degrees capping* $\mathbf{a}$.

(6) *If* $\mathbf{a} > \mathbf{0}$, $\mathcal{D}(\leqslant \mathbf{0}') - \mathcal{D}(\leqslant \mathbf{a})$.

Further indirect evidence for rigidity below $\mathbf{0}'$ is the compelling attractiveness of local versions of the *bi-interpretability conjecture*, arising from work of Harrington and Slaman. (See Slaman [1991] for a precise statement of the original conjecture of Woodin and Slaman.) According to this, the definability of the computably enumerable degrees is no longer a special consequence of local degree theory, but follows in a general way from the definability in $\mathcal{D}(\leqslant \mathbf{0}')$ of all such arithmetically describable relations. It involves an extension of Shore's Theorem 2.7 sufficient for bi-interpretability between the standard model of first order arithmetic and $\mathcal{D}(\leqslant \mathbf{0}')$. Roughly speaking, a bi-interpretation between ω and $\mathcal{D}(\leqslant \mathbf{0}')$ consists of:

(1) A coding of ω into $\mathcal{D}(\leqslant \mathbf{0}')$ (involving specifying a collection of degrees, and relations on those degrees which represent addition and multiplication),

(2) A mapping of each $x \in \mathcal{D}(\leqslant \mathbf{0}')$ to an index e for a representative Φ_e^K of $\mathbf{x}$, and

(3) A uniform degree-theoretic definition of the relationship between any $\mathbf{x}$ and the code of the index e for the representative Φ_e^K of $\mathbf{x}$.

This would invest $\mathcal{D}(\leqslant \mathbf{0}')$ with some well-known properties of ω (such as rigidity, definability of arithmetically describable sets – in particular, of all individual members of the domain).

Previously, the only counter-evidence was the lack of progress in proving facts basic to the local bi-interpretability scenario. These include local definability of relations (such as low) known to be definable in $\mathcal{D}$, automorphism bases consisting of upper cones of $\mathcal{E}$, and (in line with the analogous result of Slaman and Woodin [ta] for $\mathcal{D}$) a class of local automorphisms which is at most countable. Recently Lempp, Lerman and Shore [ta] constructed a nontrivial isomorphism between segments of c.e. weak truth table degrees. And a strategy for constructing nontrivial automorphisms of the Turing degrees is described in Cooper [1997, ta2]. As one would expect from the fact (Ambos-Spies [1983]) that the low promptly simple degrees form an automorphism base, the witnesses to nontriviality emerge via finite injury.

The rigidity of $\mathcal{D}(\leqslant \mathbf{0}')$ would have decided a whole range of fundamental open problems, and set a precedent for other degree structures. Nonrigidity not only leaves most such questions unanswered, but brings them into greater prominence. A complex situation opens out in which it makes sense to consider the *extent* to which a given relation is definable or invariant under automorphisms of $\mathcal{D}(\leqslant \mathbf{0}')$.

For instance:

QUESTION 8.3. Is every relation over $\mathcal{D}(\leqslant \mathbf{0}')$ which is definable in $\mathcal{D}$ also definable in $\mathcal{D}(\leqslant \mathbf{0}')$?

QUESTION 8.4. Is every relation on $\mathcal{D}(\leqslant \mathbf{0}')$ which is invariant under all automorphisms of $\mathcal{D}(\leqslant \mathbf{0}')$ also definable?

QUESTION 8.5. Is $\mathbf{D}_n$ definable in $\mathcal{D}(\leqslant \mathbf{0}')$ for any $n \geqslant 2$?

QUESTION 8.6. Are the low, high or low_2 degrees definable in $\mathcal{D}(\leqslant \mathbf{0}')$?[3]

(All the jump classes are definable in $\mathcal{D}$ by Cooper [ta1], all those from the three level upwards are definable in $\mathcal{D}(\leqslant \mathbf{0}')$ by Shore [1988], and Nies, Shore and Slaman have recently shown the latter together with $\mathbf{High}_2$ to be definable in $\mathcal{E}$.)

QUESTION 8.7. Are the 1-generic degrees definable in $\mathcal{D}(\leqslant \mathbf{0}')$?

QUESTION 8.8. Are the PA degrees below $\mathbf{0}'$ definable in $\mathcal{D}(\leqslant \mathbf{0}')$?

9. Enumeration and Turing reducibilities: The local theory

Enumeration reducibility accepts even partially defined Φ_i^K as objects computable from K. Computability relative to partial information must admit nondeterministic elements, and this is what e-reducibility provides, within a notationally concise theoretical framework. In any case, computability relative to auxiliary data accessed via enumerations rather than oracles is more applicable to computationally complex situations. It has close connections with the Scott graph model for lambda calculus (see Scott [1975a, 1975b]). It gives rise to a degree structure (the *enumeration degrees*, or *e-degrees*) which extends and enriches the Turing degrees. The local theory is particularly relevant and technically attractive, and we briefly review recent work in the area. See Cooper [1990] for a fuller introduction.

We first recall (Friedberg and Rogers [1959]) that $A \leqslant_e B$ (A is *enumeration reducible to* B) if and only if there is a uniform algorithm for obtaining an enumeration of the members of A from any given enumeration of B. A nice alternative characterisation due to Selman [1971] is:

$$A \leqslant_e B \Leftrightarrow \forall X (B \text{ c.e. in } X \Rightarrow A \text{ c.e. in } X).$$

The structure of the *enumeration degrees* $\mathcal{D}_e$ is derived from $\leqslant_e$ in the usual way. Identifying (possibly partial) functions with their graphs, $\leqslant_e$ and $\leqslant_T$ agree on the total functions, so there is a *natural embedding* (Myhill [1961]) of the Turing degrees

[3] In fact, Nies, Shore and Slaman have recently proved all jump classes except low to be locally definable.

into the e-degrees whereby we can write $\mathcal{D} \subset \mathcal{D}_e$ (where a Turing degree in $\mathcal{D}_e$ is called a *total* e-degree).

The *jump* on $\mathcal{D}_e$ (see Cooper [1984] and McEvoy [1985]) agrees with that on $\mathcal{D}$. The natural embedding takes $\{\mathbf{0}^{(n)}\}_{n\in\omega}$ to $\{\mathbf{0}_e^{(n)}\}_{n\in\omega}$, with Post's Theorem replaced by: (Cooper [1984], McEvoy [1985]) $A \in \mathbf{a}_e \leqslant \mathbf{0}_e^{(n)} \Leftrightarrow A \in \Sigma^0_{n+1}$ (each $n \geqslant 0$). So $\mathbf{0}_e$ consists of all c.e. sets, and the e-degrees $\leqslant \mathbf{0}'_e$ consist of exactly the Σ^0_2 sets. The high/low hierarchy is defined exactly as in the Turing degrees.

As for $\mathcal{D}$, a certain amount of local structure for $\mathcal{D}_e$ is obtainable via simple forcing (the forcing conditions are finite ω^*-valued strings, where $\omega^* = \omega \cup \{\uparrow\}$ and $\uparrow$ stands for 'undefined'). See Case [1971] and Copestake [1988] for more on the theory of generic and n-generic degrees. Sample local results are:

THEOREM 9.1 (Copestake [1988]). (1) *There exist n-generic (but not $(n+1)$-generic) e-degrees below $\mathbf{0}_e^{(n)}$ for each $n \geqslant 1$.*

(2) *Every* 1*-generic e-degree is quasi-minimal* (*that is, has no total predecessor other than* $\mathbf{0}_e$).

(3) *Every* 2*-generic e-degree bounds a minimal pair, but* (Cooper and Copestake) *not every* 1*-generic degree does.*

(4) *If* $\mathbf{a}$ *is* 1*-generic then every c.e. poset can be embedded below* $\mathbf{a}$ (*in fact, in the* 1*-generic e-degrees below* $\mathbf{a}$).

(5) (Copestake [1990]) *There exists a* 1*-generic e-degree* $< \mathbf{0}'_e$ *which is properly* Σ^0_2, *and hence not low.*

A surprising fact (Copestake [1988]) is that the e-degrees of 1-generic *sets* are distinct from the 1-generic e-degrees, but that every 1-generic e-degree bounds a set 1-generic e-degree.

QUESTION 9.2. Is the set of e-degrees of 1-generic sets closed downwards in $\mathcal{D}_e(\leqslant \mathbf{0}'_e)$?

Slaman and Woodin [1997] have applied coding techniques to the e-degrees (for instance, every countable relation on $\mathcal{D}_e$ is uniformly definable from parameters in $\mathcal{D}_e$, so that the first order theory of $\mathcal{D}_e$ is computably isomorphic to the second order theory of arithmetic), but full local versions await the answers to questions concerning local structure. For instance:

QUESTION 9.3. To what extent do exact pairs exist below $\mathbf{0}'_e$?

A major difference between the theories of $\mathcal{D}$ and $\mathcal{D}_e$ is that relativisation relative to nontotal degrees fails, and upper cones do not play the same important role in $\mathcal{D}_e$. There are no nice applications of determinacy. Although there are jump inversion theorems for $\mathcal{D}_e$ and $\mathcal{D}_e(\leqslant \mathbf{0}'_e)$ (due to McEvoy [1985]), the jumps are always total.

The immune and hyperimmune e-degrees are closed upwards (Rozinas [1978]), but:

QUESTION 9.4. Are the e-degrees of the cohesive or hyperhyperimmune sets closed upwards (below $\mathbf{0}'_e$)?

There are immune-free e-degrees below $\mathbf{0}_e^{(2)}$ (Rozinas [1978]), but every nonzero e-degree below $\mathbf{0}'_e$ contains a hyperimmune set.

The local theories of $\mathcal{D}$ and $\mathcal{D}_e$ are very different both in result and technique. Nowhere does this emerge more dramatically than in:

THEOREM 9.5 (Gutteridge's Theorem [1971]). (1) *No total e-degree has a minimal cover (so there are no minimal e-degrees).*

(2) *Every e-degree $\mathbf{a}_e$ has at most countably many minimal covers, all below $\mathbf{a}'_e$.*

The proof of part (2) (see Odifreddi [ta]) is of great technical interest. The proof that there are no minimal Δ_2^0 e-degrees can be extended to give:

THEOREM 9.6 (Density of $\mathcal{D}_e(\leqslant \mathbf{0}'_e)$) (Lachlan and Shore [1992]). *The $\mathbf{a}_e$-CEA e-degrees are dense, each $\mathbf{a}_e$. In particular* (Cooper [1984]), *$\mathcal{D}_e(\leqslant \mathbf{0}'_e)$ is dense.*

This result is the best possible in that:

THEOREM 9.7 (Nondensity of $\mathcal{D}_e(\leqslant \mathbf{0}_e^{(2)})$) (Cooper [1990]). *$\mathcal{D}_e$ is not dense. In fact* (Calhoun and Slaman [1996]), *there is a Π_2^0 e-degree which is a minimal cover in $\mathcal{D}_e(\leqslant \mathbf{0}_e^{(2)})$.*

In the real world, auxiliary information is not always available on demand, and enumeration reducibility models this situation. The resulting reducibility is inevitably more combinatorially complex, as are local constructions in the e-degrees. (There commonly appears to be an extra quantifier involved in answering a question in the context of the e-degrees.) The problems have been overcome in a number of cases.

THEOREM 9.8 (Cooper and Copestake [1988]). *There exists an e-degree below $\mathbf{0}'_e$ incomparable with all the Δ_2^0 e-degrees (other than $\mathbf{0}_e$ and $\mathbf{0}'_e$).*

The natural embedding of the c.e. Turing degrees into $\mathcal{D}_e(\leqslant \mathbf{0}'_e)$ is the set of Π_1^0 e-degrees. (In general, the e-degrees of the n.c.e. sets are total.)

THEOREM 9.9 (McEvoy and Cooper [1985]). *Any lattice embedding in the low c.e. degrees is also a lattice embedding in $\mathcal{D}_e(\leqslant \mathbf{0}'_e)$.*

So minimal pairs and the embedding results of Lachlan [1972] for the c.e. degrees give similar results for $\mathcal{D}_e(\leqslant \mathbf{0}'_e)$ (so, for instance, $\mathcal{D}_e(\leqslant \mathbf{0}'_e)$ is not distributive).

But unlike $\mathcal{D}(\leqslant \mathbf{0}')$, $\mathcal{D}_e(\leqslant \mathbf{0}'_e)$ is not complemented.

THEOREM 9.10 (NonCapping Theorem) (Cooper and Sorbi [1996]). *There exist noncappable e-degrees in* $\mathcal{D}_e(\leqslant \mathbf{0}'_e)$.

THEOREM 9.11 (NonCupping Theorem) (Cooper, Sorbi and Yi [1996]). *There exist noncuppable e-degrees in* $\mathcal{D}_e(\leqslant \mathbf{0}'_e)$.

However (Cooper and Sorbi [ta], Cooper, Sorbi and Yi [1996]), noncappable and noncuppable e-degrees are not Δ^0_2.

Despite Gutteridge's theorem, there may still be a useful theory of initial segments, with possible applications in analysing fragments of the theory of $\mathcal{D}_e(\leqslant \mathbf{0}'_e)$.

QUESTION 9.12. Characterise the decidable fragment of the first-order theory of $\mathcal{D}_e(\leqslant \mathbf{0}'_e)$.

QUESTION 9.13. Characterise the possible order-types of initial segments of $\mathcal{D}_e(\leqslant \mathbf{0}'_e)$.

Ahmad [1989] has independently shown that splitting fails in $\mathcal{D}_e(\leqslant \mathbf{0}'_e)$ (in the low e-degrees, in fact). She also shows:

THEOREM 9.14 (Diamond Theorem) (Ahmad [1991]). *There exist incomparable e-degrees* $\mathbf{a}$, $\mathbf{b}$ *with* $\mathbf{a} \cup \mathbf{b} = \mathbf{0}'_e$ *and* $\mathbf{a} \cap \mathbf{b} = \mathbf{0}_e$.

Hence $\mathcal{D}_e(\leqslant \mathbf{0}'_e) \not\equiv \mathcal{E}$ (by the Lachlan Nondiamond Theorem for $\mathcal{E}$).

Although even most global questions, apart from homogeneity are still open, current techniques may be sufficient to answer:

QUESTION 9.15. Characterise the degree of the first-order theory of $\mathcal{D}_e(\leqslant \mathbf{0}'_e)$.

(See Slaman and Woodin [1997] for an effective version of their coding lemma, sufficient to give undecidability of the first order theory of $\mathcal{D}_e(\leqslant \mathbf{0}'_e)$.)

QUESTION 9.16. Characterise the automorphisms of $\mathcal{D}_e(\leqslant \mathbf{0}'_e)$.

QUESTION 9.17. Is $\mathcal{D}(\leqslant \mathbf{0}')$ definable in $\mathcal{D}_e(\leqslant \mathbf{0}'_e)$?

QUESTION 9.18. Characterise the finite lattices embeddable in $\mathcal{D}_e(\leqslant \mathbf{0}'_e)$.

References

S. AHMAD

[1989] Some results on the structure of the Σ_2 enumeration degrees, *Recursive Function Theory Newsletter*, 38, item 373 (abstract).

[1991] Embedding the diamond in the Σ_2 enumeration degrees, *J. Symbolic Logic*, 50, pp. 195–212.

K. AMBOS-SPIES

[1983] Automorphism bases for the r.e. degrees (abstract), in: *Extended Abstracts of Short Talks of the 1982 Summer Institute on Recursion Theory Held at Cornell University*, I. Kalantari, ed., special publication of *Recursive Function Theory Newsletter*, pp. 3–4.

M. M. ARSLANOV

[1981] On some generalisations of the theorem on fixed points, *Izv. Vyssh. Uchebn. Zaved. Mat.*, 228, pp. 9–16 (Russian); *Sov. Math. (Izv. VUZ)*, 25, pp. 1–10 (English translation).

[1985] Structural properties of the degrees below $\mathbf{0}'$, *Dokl. Akad. Nauk SSSR, N.S.*, 283, pp. 270–273 (Russian).

[1990] On the structure of degrees below $\mathbf{0}'$, in: *Recursion Theory Week, Oberwolfach 1989*, K. Ambos-Spies, G. H. Muller, G. E. Sacks, eds., Lecture Notes in Mathematics, Vol. 1432, Springer, Heidelberg.

M. M. ARSLANOV, S. LEMPP AND R. A. SHORE

[1996a] On isolating r.e. and isolated d-r.e. degrees, in: *Computability, Enumerability, Unsolvability: Directions in Recursion Theory*, S. B. Cooper, T. A. Slaman and S. S. Wainer, eds., London Mathematical Society Lecture Note Series 224, Cambridge University Press, Cambridge, pp. 61–80.

[1996b] Interpolating d-r.e. and REA degrees between r.e. degrees, *Ann. Pure Appl. Logic*, 78, pp. 29–56.

W. CALHOUN AND T. A. SLAMAN

[1996] The Π^0_2 e-degrees are not dense, *J. Symbolic Logic*, 61, pp. 1364–1379.

J. CASE

[1971] Enumeration reducibility and partial degrees, *Ann. Math. Logic*, 2, pp. 419–439.

C. T. CHONG

[1979] Generic sets and minimal α-degrees, *Trans. Amer. Math. Soc.*, 254, pp. 157–169.

C. T. CHONG AND R. G. DOWNEY

[1989] On degrees bounding minimal degrees, *Math. Proc. Cambridge Philos. Soc.*, 105, pp. 211–222.

C. T. CHONG AND C. G. JOCKUSCH, JR.

[1984] Minimal degrees and 1-generic sets below $\mathbf{0}'$, in: *Computation and Proof Theory, Proceedings of the Logic Colloquium held in Aachen, July 1983*, Part II, M. M. Richter et al., eds., Lecture Notes in Mathematics, Vol. 1104, Springer, Heidelberg, pp. 63–77.

A. CHURCH

[1936] A note on the Entscheidungsproblem, *J. Symbolic Logic*, 1, pp. 40–41 and 101–102.

P. J. COHEN

[1963] The independence of the continuum hypothesis I, *Proc. Natl. Acad. Sci. USA*, 50, pp. 1143–1148.

S. B. COOPER

[1971] *Degrees of Unsolvability*, Ph. D. Thesis, University of Leicester.

[1972a] Degrees of unsolvability complementary between recursively enumerable degrees, Part I, *Ann. Math. Logic*, 4, pp. 31–73.

[1972b] Jump equivalence of the Δ^0_2 hyperhyperimmune sets, *J. Symbolic Logic*, 37, pp. 598–600.

[1972c] *Distinguishing the arithmetical hierarchy*, Preprint, Berkeley, October 1972.

[1973] Minimal degrees and the jump operator, *J. Symbolic Logic*, 38, pp. 249–271.

[1974] An annotated bibliography for the structure of the degrees below $\mathbf{0}'$ with special reference to that of the recursively enumerable degrees, *Recursive Function Theory Newsletter*, 5, pp. 1–15.

[1984] Partial degrees and the density problem. Part 2: The enumeration degrees of the Σ_2 sets are dense, *J. Symbolic Logic*, 49, pp. 503–513.

[1986] Some negative results on minimal degrees below $\mathbf{0}'$, *Recursive Function Theory Newsletter*, 34, item 353 (abstract).
[1989] The strong anti-cupping property for recursively enumerable degrees, *J. Symbolic Logic*, 54, pp. 527–539.
[1990] Enumeration reducibility, nondeterministic computations and relative computability of partial functions, in: *Recursion Theory Week, Proceedings Oberwolfach 1989*, K. Ambos-Spies, G. H. Müller and G. E. Sacks, eds., Lecture Notes in Mathematics, Vol. 1432, Springer, Berlin, pp. 57–110.
[1991] The density of the low_2 n-r.e. degrees, *Arch. Math. Logic*, 30 (1), pp. 19–24.
[1992] A splitting theorem for the n-r.e. degrees, *Proc. Amer. Math. Soc.*, 115, pp. 461–471.
[1994] Rigidity and definability in the non-computable universe, in: *Proceedings of the 9th International Congress of Logic, Methodology and Philosophy of Science, Uppsala, Sweden, August, 1991*, D. Prawitz, B. Skyrms, D. Westerstahl, eds., North-Holland, Amsterdam, pp. 209–236.
[1996a] A characterisation of the jumps of minimal degrees below $\mathbf{0}'$, in: *Computability, Enumerability, Unsolvability: Directions in Recursion Theory*, S. B. Cooper, T. A. Slaman and S. S. Wainer, eds., London Mathematical Society Lecture Note Series 224, Cambridge University Press, Cambridge, pp. 81–92.
[1996b] Discontinuous phenomena and Turing definability, in: *Proceedings of the International Conference of Algebra and Analysis, Kazan, June 1994*, Walter de Gruyter, Berlin, pp. 41–55.
[1997] Beyond Gödel's Theorem: The failure to capture information content, in: *Complexity, Logic and Recursion Theory*, A. Sorbi, ed., Lecture Notes in Pure and Appl. Math., 187, Dekker, New York, pp. 93–122.
[ta1] On a conjecture of Kleene and Post, to appear in *Math. Logic Quarterly*.
[ta2] The Turing iniverse is not rigid, to appear.

S. B. COOPER AND C. S. COPESTAKE
[1988] Properly Σ_2 enumeration degrees, *Z. Math. Logik Grundlag. Math.*, 34, pp. 491–522.

S. B. COOPER AND R. L. EPSTEIN
[1987] Complementing below recursively enumerable degrees, *Ann. Pure Appl. Logic*, 34, pp. 15–32.

S. B. COOPER, L. HARRINGTON, A. H. LACHLAN, S. LEMPP AND R. I. SOARE
[1991] The D.R.E. degrees are not dense, *Ann. Pure Appl. Logic*, 55, pp. 125–151.

S. B. COOPER, S. LEMPP AND P. WATSON
[1989] Weak density and cupping in the d-r.e. degrees, *Israel J. Math.*, 67, pp. 137–152.

S. B. COOPER AND A. SORBI
[1996] Noncappable enumeration degrees below $\mathbf{0}'_e$, *J. Symbolic Logic*, 61, pp. 1347–1363.
[ta] Every incomplete Δ^0_2 *e*-degree is cappable, in preparation.

S. B. COOPER, A. SORBI AND X. YI
[1996] Cupping and noncupping in the enumeration degrees of Σ^0_2 sets, *Ann. Pure Appl. Logic*, 82, pp. 317–342.

S. B. COOPER AND X. YI
[ta] Isolated d-r.e. degrees, to appear.

C. S. COPESTAKE
[1988] 1-genericity in the enumeration degrees, *J. Symbolic Logic*, 53, pp. 878–887.
[1990] 1-generic enumeration degrees below $\mathbf{0}'_e$, in: *Mathematical Logic, Proceedings Heyting '88 Summer School and Conference on Math. Logic, September 1988, Chaika, Bulgaria*, P. P. Petkov, ed., Plenum Press, New York, pp. 257–265.

A. N. DEGTEV
[1973] tt- and *m*-degrees, *Algebra i Logika*, 12, pp. 143–161 (Russian); *Algebra and Logic*, 12, pp. 78–89 (English translation).

D. DING AND L. QIAN
[1996] Isolated d.r.e. degrees are dense in r.e. degree structure, *Archive for Math. Logic*, 36, pp. 1–10.
[ta1] An r.e. degree not isolating any d-r.e. degree, to appear.
[ta2] A splitting property of d-r.e. degrees, to appear.

R. G. DOWNEY
[1987] Δ_2^0 degrees and transfer theorems, *Illinois J. Math.*, 31, pp. 419–427.
[1989] D-r.e. degrees and the Nondiamond Theorem, *Bull. London Math. Soc.*, 21, pp. 43–50.

R. G. DOWNEY, S. LEMPP AND R. A. SHORE
[1996] Jumps of minimal degrees below $\mathbf{0}'$, *J. London Math. Soc. (2)*, 54, pp. 417–439.

R. L. EPSTEIN
[1975] *Minimal Degrees of Unsolvability and the Full Approximation Construction*, Memoirs Amer. Math. Soc. 3, no. 162, Amer. Math. Soc., Providence, RI.
[1979] *Degrees of Unsolvability: Structure and Theory*, Lecture Notes in Mathematics, Vol. 759, Springer, Berlin.
[1981] *Initial Segments of Degrees Below* $\mathbf{0}'$, Memoirs Amer. Math. Soc. 30, no. 241, Amer. Math. Soc., Providence, R. I.

R. L. EPSTEIN, R. HAAS AND R. KRAMER
[1981] Hierarchies of sets and degrees below $\mathbf{0}'$, in: *Logic Year 1979–80: University of Connecticut*, M. Lerman, J. H. Schmerl and R. I. Soare, eds., Lecture Notes in Mathematics, Vol. 859, Springer, Berlin, pp. 32–48.

Y. L. ERSHOV
[1968a] A hierarchy of sets, Part I, *Algebra i Logika*, 7, pp. 47–73 (Russian); *Algebra and Logic*, 7, pp. 24–43 (English translation).
[1968b] A hierarchy of sets, Part II, *Algebra i Logika*, 7, pp. 15–47 (Russian); *Algebra and Logic*, 7, pp. 212–232 (English translation).
[1970] A hierarchy of sets, Part III, *Algebra i Logika*, 9, pp. 34–51 (Russian); *Algebra and Logic*, 9, pp. 20–31 (English translation).
[1975] The upper semilattice of numerations of a finite set, *Algebra i Logika*, 14, pp. 258–284 (Russian); *Algebra and Logic*, 14, pp. 159–175 (English translation).

S. FEFERMAN
[1957] Degrees of unsolvability associated with classes of formalized theories, *J. Symbolic Logic*, 22, pp. 161–175.
[1965] Some applications of the notions of forcing and generic sets, *Fund. Math.*, 56, pp. 325–345.

P. FEJER
[1989] Embedding lattices with top preserved below non-GL_2 degrees, *Z. Math. Logik Grundlag. Math.*, 35, pp. 3–14.

R. M. FRIEDBERG AND H. ROGERS, JR.
[1959] Reducibility and completeness for sets of integers, *Z. Math. Logik Grundlag. Math.*, 5, pp. 117–125.

K. GÖDEL
[1931] Über formal unentscheidbare Sätze der Principia Mathematica und verwandter Systeme I, *Monatsh. Math. Phys.*, 38, pp. 173–198.
[1934] On undecidable propositions of formal mathematical systems, mimeographed notes, in: *The Undecidable. Basic Papers on Undecidable Propositions, Unsolvable Problems, and Computable Functions*, M. Davis, ed., Raven Press, New York, 1965, pp. 39–71.

E. M. GOLD
[1965] Limiting recursion, *J. Symbolic Logic*, 30, pp. 28–48.

L. GUTTERIDGE

[1971] *Some Results on Enumeration Reducibility*, Ph. D. Dissertation, Simon Fraser University.

W. HANF

[1965] Model theoretic methods in the study of elementary logic, in: *Symposium on the Theory of Models*, J. W. Addison, L. Henkin and A. Tarski, eds., North-Holland, Amsterdam, pp. 132–145.

L. HARRINGTON AND R. I. SOARE

[1991] Post's program and incomplete recursively enumerable sets, *Proc. Natl. Acad. Sci. USA*, 88, pp. 10242–10246.

C. A. HAUGHT

[1986] The degrees below a 1-generic degree and less than $\mathbf{0}'$, *J. Symbolic Logic*, 51, pp. 770–777.

P. G. HINMAN

[1969] Some applications of forcing to hierarchy problems in arithmetic, *Z. Math. Logik Grundlag. Math.*, 15, pp. 341–352.

D. F. HUGILL

[1969] Initial segments of Turing degrees, *Proc. London Math. Soc.*, 19, pp. 1–16.

SH. T. ISHMUKAMETOV

[1985] On differences of recursively enumerable sets, *Izv. Vyssh. Uchebn. Zaved. Mat.*, 279, pp. 3–12 (Russian).

Z. JIANG

[1993] Diamond lattice embedded into d.r.e. degrees, *Science in China (Series A)*, 36, pp. 803–811.

C. G. JOCKUSCH, JR.

[1969] The degrees of hyperhyperimmune sets, *J. Symbolic Logic*, 34, pp. 489–493.

[1973] Upward closure and cohesive degrees, *Israel J. Math.*, 15, pp. 332–335.

[1977] Simple proofs of some theorems on high degrees, *Canad. J. Math.*, 29, pp. 1072–1080.

[1980] Degrees of generic sets, in: *Recursion Theory: its Generalisations and Applications, Proceedings of Logic Colloquium '79, Leeds, August 1979*, F. R. Drake and S. S. Wainer, eds., London Mathematical Society Lecture Notes Series 45, Cambridge University Press, Cambridge, pp. 110–139.

C. G. JOCKUSCH, JR., M. LERMAN, R. I. SOARE AND R. M. SOLOVAY

[1989] Recursively enumerable sets modulo iterated jumps and extensions of Arslanov's completeness criterion, *J. Symbolic Logic*, 54, pp. 1288–1323.

C. G. JOCKUSCH, JR. AND D. POSNER

[1978] Double jumps of minimal degrees, *J. Symbolic Logic*, 43, pp. 715–724.

[1981] Automorphism bases for degrees of unsolvability, *Israel J. Math.*, 40, pp. 150–164.

C. G. JOCKUSCH, JR. AND R. A. SHORE

[1983] Pseudo jump operators I: The R.E. case, *Trans. Amer. Math. Soc.*, 275, pp. 599–609.

[1984] Pseudo jump operators II: Transfinite iterations, hierarchies, and minimal covers, *J. Symbolic Logic*, 49, pp. 1205–1236.

C. G. JOCKUSCH, JR. AND S. G. SIMPSON

[1980] Minimal degrees, hyperimmune degrees, and complete extensions of arithmetic, Preliminary report 781-E10, *Abstracts Amer. Math. Soc.*, 1, p. 546.

C. G. JOCKUSCH, JR. AND R. I. SOARE

[1972a] Degrees of members of Π^0_1 classes, *Pacific J. Math.*, 40, pp. 605–616.

[1972b] Π^0_1 classes and degrees of theories, *Trans. Amer. Math. Soc.*, 173, pp. 33–56.

C. G. JOCKUSCH, JR. AND F. STEPHAN

[1993] A cohesive set which is not high, *Math. Logic Quart.*, 39, pp. 515–530.

D. KADDAH

[1993] Infima in the d.r.e. degrees, *Ann. Pure Appl. Logic*, 62, pp. 207–263.

S. C. KLEENE

[1952] *Introduction to Metamathematics*, Van Nostrand, New York.

S. C. KLEENE AND E. L. POST

[1954] The upper semi-lattice of degrees of recursive unsolvability, *Ann. Math. (2)*, 59, pp. 379–407.

A. KUCERA

[1986] An alternative, priority-free, solution to Post's problem, in: *Proceedings MFCS '86*, Lecture Notes in Computer Science, Springer, Berlin, pp. 493–500.

[1989] On the use of diagonally nonrecursive functions, in: *Logic Colloquium '87*, H. D. Ebbinghaus et al., eds., North-Holland Amsterdam, pp. 219–239.

M. KUMABE

[1990] A 1-generic degree which bounds a minimal degree, *J. Symbolic Logic*, 55, pp. 733–743.

[1991] Relative recursive enumerability of generic degrees, *J. Symbolic Logic*, 58, pp. 1075–1084.

[1993] Every n-generic degree is a minimal cover of an n-generic degree, *J. Symbolic Logic*, 58, pp. 219–231.

[ta1] A 1-generic degree with a strong minimal cover, to appear.

[ta2] A fixed point free minimal degree, to appear.

A. H. LACHLAN

[1966] Lower bounds for pairs of recursively enumerable degrees, *Proc. London Math. Soc.*, 16, pp. 537–569.

[1972] Embedding nondistributive lattices in the recursively enumerable degrees, in: *Conference in Mathematical Logic, London, 1970*, W. Hodges, ed., Lecture Notes in Mathematics, Vol. 255, Springer, Berlin, pp. 149–177.

[1975] A recursively enumerable degree which will not split over all lesser ones, *Ann. Math. Logic*, 9, pp. 307–365.

A. H. LACHLAN AND R. LEBEUF

[1976] Countable initial segments of the degrees of unsolvability, *J. Symbolic Logic*, 41, pp. 289–300.

A. H. LACHLAN AND R. A. SHORE

[1992] The n-rea enumeration degrees are dense, *Arch. Math. Logic*, 31, pp. 277–285.

G. L. LAFORTE

[1995] *Phenomena in the n-R.E. and n-REA Degrees*, Ph. D. Dissertation, University of Michigan.

S. LEMPP, A. LERMAN AND R. A. SHORE

[ta] The existence of isomorphic cones in the r.e. wtt-degrees, to appear.

M. LERMAN

[1977] Automorphism bases for the semilattice of recursively enumerable degrees, A-251, Abstract #77T-E10, *Notices Amer. Math. Soc.*, 24.

[1983] *Degrees of Unsolvability*, Perspectives in Mathematical Logic, Omega Series, Springer, Berlin.

[1986] Degrees which do not bound minimal degrees, *Ann. Pure App. Logic*, 30, pp. 249–276.

M. LERMAN AND R. A. SHORE

[1988] Decidability and invariant classes for degree structures, *Trans. Amer. Math. Soc.*, 310, pp. 669–692.

A. LI

[ta] External center theorem of the recursively enumerable degrees, to appear.

S. S. MARCHENKOV

[1976] A class of incomplete sets, *Mat. Zametki*, 20, pp. 473–478 (Russian); *Math. Notes*, 20, pp. 823–825 (English translation).

D. A. MARTIN

[1967] *Measure, Category, and Degrees of Unsolvability*, unpublished manuscript.

JU. V. MATIJASEVIČ

[1970] Enumerable sets are diophantine, *Dokl. Akad. Nauk. SSSR*, 191, pp. 279–282 (Russian); *Sov. Math. Dokl.*, 11, pp. 354–357 (English translation).

K. MCEVOY

[1985] Jumps of quasi-minimal enumeration degrees, *J. Symbolic Logic*, 50, pp. 903–1001.

K. MCEVOY AND S. B. COOPER

[1985] On minimal pairs of enumeration degrees, *J. Symbolic Logic*, 50, pp. 839–848.

W. MILLER AND D. A. MARTIN

[1968] The degrees of hyperimmune sets, *Z. Math. Logik Grundlag. Math.*, 14, pp. 159–166.

J. MYHILL

[1955] Creative sets, *Z. Math. Logik Grundlag. Math.*, 1, pp. 97–108.

[1956] The lattice of recursively enumerable sets, *J. Symbolic Logic*, 21, pp. 215, 220 (abstract).

[1961] A note on degrees of partial functions, *Proc. Amer. Math. Soc.*, 12, pp. 519–521.

A. NERODE AND R. A. SHORE

[1980a] Second order logic and first order theories of reducibility orderings, in: *The Kleene Symposium*, J. Barwise et al., eds., North-Holland, Amsterdam, pp. 181–200.

[1980b] Reducibility orderings: theories, definability and automorphisms, *Ann. Math. Logic*, 18, pp. 61–89.

P. ODIFREDDI

[1989] *Classical Recursion Theory*, Studies in Logic and the Foundations of Mathematics, Vol. 125, North-Holland, Amsterdam.

[ta] *Classical Recursion Theory, II*, North-Holland, in preparation.

P. ODIFREDDI AND R. A. SHORE

[1989] Global properties of local structures of degrees, *Boll. Un. Mat. Ital.*

D. B. POSNER

[1977] *High Degrees*, Ph. D. Dissertation, University of California, Berkeley.

[1980] A survey of non-r.e. degrees $\leqslant \mathbf{0}'$, in: *Recursion Theory: its Generalisations and Applications, Proceedings of Logic Colloquium '79, Leeds, August 1979*, F. R. Drake and S. S. Wainer, eds., London Mathematical Society Lecture Notes Series 45, Cambridge University Press Cambridge, pp. 52–109.

[1981] The upper semilattice of degrees $\leqslant \mathbf{0}'$ is complemented, *J. Symbolic Logic*, 46, pp. 705–713.

D. B. POSNER AND R. W. ROBINSON

[1981] Degrees joining to $\mathbf{0}'$, *J. Symbolic Logic*, 46, pp. 714–722.

E. L. POST

[1944] Recursively enumerable sets of positive integers and their decision problems, *Bull. Amer. Math. Soc.*, 50, pp. 284–316.

[1948] Degrees of recursive unsolvability, preliminary report (abstract), *Bull. Amer. Math. Soc.*, 54, pp. 641–642.

H. PUTNAM

[1965] Trial and error predicates and the solution to a problem of Mostowski, *J. Symbolic Logic*, 30, pp. 49–57.

R. W. ROBINSON
[1968] A dichotomy of the recursively enumerable sets, *Z. Math. Logik Grundlag. Math.*, 14, pp. 339–356.

H. ROGERS, JR.
[1967] *Theory of Recursive Functions and Effective Computability*, McGraw-Hill, New York.

M. G. ROZINAS
[1978] Partial degrees of immune and hyperimmune sets, *Siberian Math. J.*, 19, pp. 613–616.

G. E. SACKS
[1961] A minimal degree less than $\mathbf{0}'$, *Bull. Amer. Math. Soc.*, 67, pp. 416–419.
[1963a] *Degrees of Unsolvability*, Ann. of Math. Stud. 55, Princeton University Press, Princeton, NJ, 1963.
[1963b] On the degrees less than $\mathbf{0}'$, *Ann. of Math. (2)*, 77, pp. 211–231.
[1963c] Recursive enumerability and the jump operator, *Trans. Amer. Math. Soc.*, 108, pp. 223–239.
[1964] The recursively enumerable degrees are dense, *Ann. of Math. (2)*, 80, pp. 300–312.
[1966] *Degrees of Unsolvability*, Ann. of Math. Stud. 55 (revised edition), Princeton University Press, Princeton, NJ, 1966.
[1967] On a theorem of Lachlan and Martin, *Proc. Amer. Math. Soc.*, 18, pp. 140–141.
[1971] Forcing with perfect closed sets, in: *Axiomatic Set Theory I, Proc. Symp. Pure Math., Los Angeles, 1967*, D. Scott, ed., Amer. Math. Soc., Providence, R.I., pp. 331–355.

L. P. SASSO
[1970] A cornucopia of minimal degrees, *J. Symbolic Logic*, 395, pp. 383–388.
[1974] A minimal degree not realising least possible jump, *J. Symbolic Logic*, 39, pp. 571–574.

D. SCOTT
[1962] Algebras of sets binumerable in complete extensions of arithmetic, in: *Proc. Symp. Pure Math., Vol. V, Recursive Function Theory*, Amer. Math. Soc., Providence, R.I., pp. 117–121.
[1975a] λ-calculus and recursion theory, in: *Third Scandinavian Logic Symposium*, Kanger, ed., North-Holland, Amsterdam, pp. 154–193.
[1975b] Data types as lattices, in: *Proc. Logic Conf., Kiel*, Lecture Notes in Mathematics, Vol. 499, Springer, Berlin, pp. 579–651.

D. SEETAPUN AND T. A. SLAMAN
[1992] *Minimal Complements*, unpublished manuscript.

V. L. SELIVANOV
[ta] *Hierarchies, Numerations and Index Sets*, to appear.

A. L. SELMAN
[1971] Arithmetical reducibilities I, *Z. Math. Logik Grundlag. Math.*, 17, pp. 335–350.

J. R. SHOENFIELD
[1959] On degrees of unsolvability, *Ann. of Math. (2)*, 69, pp. 644–653.
[1960] Degrees of models, *J. Symbolic Logic*, 25, pp. 233–237.
[1966] A theorem on minimal degrees, *J. Symbolic Logic*, 31, pp. 539–544.

R. A. SHORE
[1981] The theory of the degrees below $\mathbf{0}'$, *J. London Math. Soc.*, 24, pp. 1–14.
[1988] Defining jump classes in the degrees below $\mathbf{0}'$, *Proc. Amer. Math. Soc.*, 104, pp. 287–292.

R. A. SHORE AND T. A. SLAMAN
[1990] Working below a low_2 recursively enumerable degree, *Archive for Math. Logic*, 29, pp. 201–211.

S. G. SIMPSON

[1977] First-order theory of the degrees of recursive unsolvability, *Ann. of Math. (2)*, 105, pp. 121–139.

T. A. SLAMAN

[1991] Degree structures, in: *Proc. Int. Congress of Math., Kyoto, 1990*, Springer, Tokyo, pp. 303–316.

[1994] *Questions in Recursion Theory*, privately circulated manuscript.

T. A. SLAMAN AND J. R. STEEL

[1989] Complementation in the Turing degrees, *J. Symbolic Logic*, 54, pp. 160–176.

T. A. SLAMAN AND W. H. WOODIN

[1986] Definability in the Turing degrees, *Illinois J. Math.*, 30, pp. 320–334.

[1997] Definability in the enumeration degrees, *Archive for Mathematical Logic*, 36, pp. 255–267.

[ta] *Definability in Degree Structures*, in preparation.

R. I. SOARE

[1974] Automorphisms of the lattice of recursively enumerable sets, *Bull. Amer. Math. Soc.*, 80, pp. 53–58.

[1987] *Recursively Enumerable Sets and Degrees*, Perspectives in Mathematical Logic, Springer, Berlin.

[1994] *Redefining Recursion Theory*, privately circulated notes, June 1994.

R. I. SOARE AND M. STOB

[1982] Relative recursive enumerability, in: *Proceedings of the Herbrand Symposium Logic Colloquium '81*, J. Stern, ed., North-Holland, Amsterdam, pp. 299–324.

C. SPECTOR

[1956] On degrees of recursive unsolvability, *Ann. of Math. (2)*, 64, pp. 581–592.

A. M. TURING

[1936] On computable numbers, with an application to the Entscheidungsproblem, *Proc. London Math. Soc.*, 42, pp. 230–265.

L. V. WELCH

[1981] *A Hierarchy of Families of Recursively Enumerable Degrees and a Theorem on Bounding Minimal Pairs*, Ph.D. Dissertation, University of Illinois at Urbana-Champaign.

C. E. M. YATES

[1967] Recursively enumerable degrees and the degrees less than $\mathbf{0}'$, in: *Sets, Models, and Recursion Theory, Proceedings of the Summer School in Mathematical Logic and Logic Colloquium, Leicester, England, 1965*, J. N. Crossley, ed., North-Holland, Amsterdam, pp. 264–271.

[1970a] Initial segments of the degrees of unsolvability, Part I: A survey, in: *Mathematical Logic and Foundations of Set Theory, Proceedings of an International Colloquium, Jerusalem, November 11–14, 1968*, Y. Bar-Hillel, ed., North-Holland, Amsterdam, pp. 63–83.

[1970b] Initial segments of the degrees of unsolvability, Part II: Minimal degrees, *J. Symbolic Logic*, 35, pp. 243–266.

[1976] Banach–Mazur games, comeager sets, and degrees of unsolvability, *Math. Proc. Cambridge Philos. Soc.*, 79, pp. 195–220.

X. YI

[ta] Highness and the density property in the d.r.e. degrees, to appear.

CHAPTER 5

The Global Structure of the Turing Degrees

Theodore A. Slaman

Department of Mathematics, The University of California at Berkeley, Berkeley, CA 94720-3840, USA

Contents

1. Embedding theorems 157
   1.1. Partial order embeddings 157
   1.2. Initial segments 157
   1.3. Global questions 158
2. Coding and definability theorems 158
   2.1. A cone of minimal covers 159
   2.2. Coding in $\mathfrak{D}$ and undecidability 159
   2.3. Shore's program 160
   2.4. Cooper's definition of the Turing jump 161
3. Global definability and biinterpretability 162
   3.1. Biinterpretability with parameters 164
4. Automorphisms 165
   4.1. Cooper's automorphism 166

References 166

HANDBOOK OF COMPUTABILITY THEORY
Edited by E.R. Griffor

The Turing degrees $\mathfrak{D}$ were introduced by Kleene and Post [1954] to isolate and study those properties of the subsets of the natural numbers $\mathbb{N}$ which are expressed purely in terms of relative computability. Intuitively, we form $\mathfrak{D}$ by identifying subsets of $\mathbb{N}$ which are mutually computable and ordering the resulting equivalence classes by relative computability. The natural hierarchies of definability within arithmetic, analysis and higher fragments of set theory all have sharply focused images in $\mathfrak{D}$.

We will focus our attention on the properties of the Turing degrees at large, rather than on the properties of the Turing degrees of the definable sets. However, one of the most interesting features of $\mathfrak{D}$ is the way in which its global properties are reflected by those of its small ideals.

Formally, suppose that A and B are subsets of $\mathbb{N}$, henceforth called *reals*. We say that A is *Turing reducible* to B ($A \leqslant_T B$) if there is a computational procedure which takes an arbitrary input n from $\mathbb{N}$; over the course of its execution on input n, asks whether various numbers are in B; and, if it receives the correct responses to those questions, after finitely many computational steps returns the answer as to whether n is an element of A. In other words, if we were given B, then we would be able to compute A. We say that A and B are Turing equivalent ($A \equiv_T B$) if $A \leqslant_T B$ and $B \leqslant_T A$. The Turing degrees D are the $\equiv_T$-equivalence classes. The degree structure $\mathfrak{D}$ associated with Turing reducibility is the partial ordering $\langle D, \leqslant_T \rangle$, the Turing degrees with the ordering inherited from $\leqslant_T$.

In fact, $\mathfrak{D}$ has more structure than mere ordering. Any pair of degrees a and b has a least upper bound in $\mathfrak{D}$, which we denote by $a \vee b$. If a and b are represented by A and B, then $a \vee b$ is represented by $A \oplus B$, the effective disjoint union of A and B. So we may safely speak of $\mathfrak{D}$ as being equipped with the function $\vee$. Since $\vee$ is definable in $\mathfrak{D}$, adding it as a primitive operation does not change the class of definable relations in $\mathfrak{D}$. On occasion, we will pay attention to the complexity of definitions within $\mathfrak{D}$ and then we will restrict ourselves to the pure language of partially ordered sets.

In contrast, it is not clear whether notions such as the Turing jump, mapping x to x' the Turing degree of its associated complete Σ_1^0 set, or the ideal $\mathcal{A}$ of degrees of arithmetically definable sets should be definable in $\mathfrak{D}$, or even invariant under all automorphisms of $\mathfrak{D}$. These notions are based on hierarchies of definability rather than on the order theoretic properties of $\leqslant_T$.

In the first section, we will discuss the early results about the Turing degrees. For the most part, these were to the effect that $\mathfrak{D}$ has a rich existential theory in various languages, showing that $\mathfrak{D}$ plays various roles as a universal object. In the second section, we describe the calculation of the degree of the first-order theory of $\mathfrak{D}$, the development of REA-operators and the proof that the Turing jump is definable. In the final sections, we will discuss the automorphism group of $\mathfrak{D}$.

1. Embedding theorems

1.1. *Partial order embeddings*

In their 1954 paper, Kleene and Post used a Baire category argument to show that there are sets of incomparable Turing degree. In fact, Kleene and Post proved that every finite partial order can be embedded in $\mathfrak{D}$. Sacks [1961b] extended the Kleene and Post embedding theorem to show that every countable partial ordering, and even every one of size $\aleph_1$ which is locally countable, could be embedded in $\mathfrak{D}$.

Sacks [1963] conjectured:

CONJECTURE 1.1 (*Sacks* [1963]). A partially ordered set P is embeddable in the (Turing) degrees if and only if P has cardinality at most that of the continuum and each member of P has at most countably many predecessors.

The Kleene–Post and Sacks partial order embedding results indicated a universal quality of $\mathfrak{D}$. In Conjecture 1.1, Sacks expressed a belief in this quality by conjecturing that the strongest possible purely existential property would be satisfied by $\mathfrak{D}$. Surprisingly, this conjecture is still open.

1.2. *Initial segments*

Say that a subset $\mathcal{I}$ of $\mathfrak{D}$ is an *ideal* or an *initial segment* if $\mathcal{I}$ is closed downward in $\mathfrak{D}$ and is closed under join. $\mathcal{I}$ is a *principal ideal* if it is an ideal with a greatest element. Similarly, a subset $\mathcal{F}$ of $\mathfrak{D}$ is a *filter* if $\mathcal{F}$ is closed upward in $\mathfrak{D}$. A principal filter is also called a *cone*.

Spector [1956] answered a question of Kleene and Post [1954] by constructing a two element ideal, i.e., a minimal nontrivial degree. Spector's method could be extended to build many different examples of isomorphism types of initial segments in $\mathfrak{D}$. Prompted by Spector's results, Sacks [1966] made the following conjecture.

CONJECTURE 1.2 (*Sacks* [1966]). S is a finite, initial segment of the degrees if and only if S is order isomorphic to a finite, initial segment of some upper-semi-lattice with a least member.

Sacks's conjecture postulated another universal property for $\mathfrak{D}$: the initial segments of $\mathfrak{D}$ would realize all the finite possibilities. Extending work of Thomason and Lachlan, Lerman [1971] confirmed this conjecture. Lachlan and Lebeuf [1976] showed that every initial segment of a countable upper-semi-lattice with least element is isomorphic to an initial segment of $\mathfrak{D}$. Ultimately, Abraham and Shore [1986] showed that every initial segment of an upper-semi-lattice which is locally countable and of cardinality $\aleph_1$ is isomorphic to an initial segment of $\mathfrak{D}$. Groszek and Slaman [1983] had previously showed that the Abraham and Shore theorem is best possible; it is independent of $ZFC + 2^{\omega} > \aleph_1$ whether every initial segment of an upper-semi-lattice which is locally countable and of cardinality $\aleph_2$ can be embedded in $\mathfrak{D}$ as an upper-semi-lattice.

These successes prompted speculation that $\mathfrak{D}$ might have an algebraic characterization, or at least occupy a distinguished algebraic position among upper-semi-lattices.

1.3. *Global questions*

Early workers in the field faced several fundamental questions. Is there a global structure theory for $\mathfrak{D}$? Is the theory of $\mathfrak{D}$ specific to Turing reducibility or is it applicable to a general class of degree structures? Is the structure of $\mathfrak{D}$ tied to the continuum or is it reflected in the local substructures of $\mathfrak{D}$?

We recall some specific questions from those times.

Kleene and Post [1954]. Are the Turing jump and the relation *recursively enumerable in* absolutely definable in $\mathfrak{D}$?

Sacks [1966]. Is the theory of $\mathfrak{D}$ decidable?

Sacks [1966]. Are the Turing degrees and the Turing degrees of the arithmetic sets ($\mathcal{A}$) elementarily equivalent?

Rogers [1967]. For a degree a, let $\mathfrak{D}(\geqslant_T a)$ cone above a. Is it the case that for all a and b, $\mathfrak{D}(\geqslant_T a) \cong \mathfrak{D}(\geqslant_T b)$?

Rogers [1967]. Is there a nontrivial automorphism of $\mathfrak{D}$? (If not then we say that $\mathfrak{D}$ is *rigid*.)

We can give a partial positive answer to the second of these and discuss the others in the next section. Shore and Lerman independently showed that the $\exists\forall$-theory of $\mathfrak{D}$ is decidable; see Lerman [1983]. One reduces the problem by showing that it is enough to decide statements of the form *every embedding of the finite upper-semi-lattice L with a least element into $\mathfrak{D}$ can be extended to an embedding of the partial order Q*. Applying the Kleene and Post technology, such a statement is true whenever L is an initial segment of Q. Conversely, if L is not an initial segment of Q, then any embedding of L into $\mathfrak{D}$ as an initial segment cannot be extended to an embedding of Q. Jockusch and Slaman [1993] strengthened the Kleene and Post technology and then applied the same argument to conclude that the $\exists\forall$-theory of $\mathfrak{D}$ in the language with $\vee$ is also decidable.

2. Coding and definability theorems

The subsequent results of the 1970's to the mid 1980's ruled out any reasonable understanding of the Turing degrees in algebraic terms. Simultaneously, the exact properties of $\mathfrak{D}$ that make it algebraically intractable were used to settle almost all of the above questions. Their solutions illustrate the complexity of $\mathfrak{D}$: $\mathfrak{D}$ is not decidable; the theory of $\mathfrak{D}$ is not equal to the theory of the Turing degrees of the arithmetic sets; and there are a and b such that $\mathfrak{D}(\geqslant_T a)$ and $\mathfrak{D}(\geqslant_T b)$ are not isomorphic. Rogers's question whether the jump is definable was settled later (the jump is definable), but the techniques in its solution were steadily developed through this period. The only questions that remained open were whether $\mathfrak{D}$ is rigid and whether there is a characterization of the relations which are definable in $\mathfrak{D}$.

2.1. *A cone of minimal covers*

Of the questions in the previous section, one of the first to fall was whether $\mathfrak{D}$ is elementarily equivalent to $\mathcal{A}$, the ideal of degrees of the arithmetically definable sets.

DEFINITION 2.1. A degree m is a *minimal cover of* x if $x <_T m$ and (x, m) is empty. We say that m is a minimal cover if there is an x such that m is a minimal cover of x.

Jockusch and Soare [1970] showed that there is no cone of minimal covers in $\mathcal{A}$. In particular, $\{0^n\colon n \in \mathbb{N}\}$ is cofinal in $\mathcal{A}$ and and no 0^n is a minimal cover. One proves the latter fact by observing that the Sacks [1961b] Splitting Theorem implies if w is recursively enumerable in and above x then w cannot be a minimal cover of x, and then observing that for any n and any x, either $x \geqslant_T 0^n$ or there is a w such that $0^n \vee x \geqslant_T w >_T x$ and w is recursively enumerable in and above x. Consequently, $0^n \vee x$ cannot be a minimal cover of x.

In contrast, Jockusch [1973] showed that there is a cone of minimal covers. One proves this fact by relativizing Spector's construction to conclude that there are arbitrarily large degrees which are minimal covers and then applying Martin's [1975] theorem that any Borel subset of $\mathfrak{D}$ which is unbounded in $\mathfrak{D}$ must contain a cone.

The existence of a cone of minimal covers is an elementary property that separates $\mathcal{A}$ and $\mathfrak{D}$.

2.2. *Coding in $\mathfrak{D}$ and undecidability*

A primary ingredient in the work during this period was the proving and exploiting of *coding theorems*.

DEFINITION 2.2. Suppose that $\mathfrak{A}$ is a model of the finite language $\mathcal{L}$ and $\vec{p}$ is a finite sequence of parameters from $\mathfrak{D}$. $\mathfrak{A}$ is *coded by* $\vec{p}$ *in* $\mathfrak{D}$ if there is an isomorphic image of $\mathfrak{A}$ whose universe, relations, functions, constants and quantifiers are all first-order definable in the language of $\mathfrak{D}$ with additional symbols for the parameters $\vec{p}$.

A disparate sequence of coding schemes preceded the one due to Slaman and Woodin [1986] which we cite below. See Simpson [1977] or Nerode and Shore [1980]. Say that a relation R is countable if there is a countable subset of the degrees such that all of the elements of R come from that set.

LEMMA 2.3 (Slaman and Woodin [1986]. Coding lemma). *For any countable relation R on degrees there are parameters $\vec{p}$ such that R is definable in $\mathfrak{D}$ from $\vec{p}$.*

The coding lemma is uniform in the following sense. For each n, there is a fixed first-order formula φ such that for every countable n-ary relation R on $\mathfrak{D}$ there is a sequence $\vec{p}$ so that R is defined from $\vec{p}$ using the formula φ in $\mathfrak{D}$.

We can use the coding lemma to present the solution to Sacks's question whether $\mathfrak{D}$ is decidable. By the coding lemma, we can both code the standard model of arithmetic and also define the collection of codes of standard models. In addition, we can interpret second-order quantifiers over a coded countable model by quantifiers over sequences in $\mathfrak{D}$ which define unary relations on the universe of the coded model. Thus (Simpson [1977]), there is an interpretation of second-order arithmetic in the first-order theory of $\mathfrak{D}$, and in particular, the theory of $\mathfrak{D}$ is not recursive (Lachlan [1968]). The sharpest calculation is due to Schmerl, who showed that the $\exists\forall\exists$-theory of $\mathfrak{D}$ is not recursive; see Lerman [1983].

2.3. *Shore's program*

The next developments which we will discuss were initiated by Nerode and Shore [1980] pursued extensively by Shore and his collaborators.

Suppose that R is a countable relation on degrees. We note that the degrees which are produced during the proof of the coding lemma to define R in $\mathfrak{D}$ are recursion theoretically close to R. By this we mean the following. Let X be a real of degree x and let R be recursively presented relative to X. Then there is a sequence of degrees $\vec{p}$ which codes R in $\mathfrak{D}$ and which is arithmetic in x. As a special case, there is a sequence $\overrightarrow{P_X}$ of sets which is arithmetically definable from X and whose degrees code (in $\mathfrak{D}$) an isomorphic copy of the standard model of arithmetic with a unary predicate for X. Thus, the set X is coded in $\mathfrak{D}$ by parameters which are near its degree. If X is sufficiently complicated, say above $0'$, then a sequence of parameters whose degrees code X in $\mathfrak{D}$ can be found recursively in X, that is below x in $\mathfrak{D}$.

The next step is to find a notion of *neighborhood* which is first-order definable in $\mathfrak{D}$ and to link an arbitrary degree x to the reals coded in its neighborhood. A preliminary but very tangible success along this line was Shore [1979] refutation of Rogers's Homogeneity Conjecture: there are two principal filters in $\mathfrak{D}$ which are not isomorphic.

The Jockusch and Soare theorem and the Jockusch theorem pointed the direction in which to look for a precise implementation of the notion of neighborhood. There is a definable filter in $\mathfrak{D}$ that is nontrivial and disjoint from the degrees of the arithmetic sets, suggesting that there might be a related definable filter whose complement was an ideal of small degrees. Now, consider the property *there is an x such that $a \vee x$ is a minimal cover of x* as saying that x is not small. Let $\mathcal{I}$ be the ideal generated by the small degrees. By our previous remarks, no 0^n is large and so $\mathcal{I}$ contains the degrees of the arithmetic sets. On the other hand, there is a cone of large degrees and so $\mathcal{I}$ is not all of $\mathfrak{D}$.

So, we have the problem of determining whether for a given degree a there is an x such that $a \vee x$ is a minimal cover of x. From the proof of Spector's theorem and more importantly from Sacks's [1961a] proof that there is a minimal degree below $0'$, we have arithmetically definable functions M such that for any real X, $M(X)$ is a minimal cover of X. For any such M, we can apply determinacy and conclude that there is a cone in $\mathfrak{D}$ of degrees in the range of M.

Jockusch and Shore [1984] attacked the problem of determining for exactly which A is there an X such that $M(X) = A$, that is the problem of inverting the known minimal degree constructions. They succeeded in surprising generality and systematized the problem of inverting arithmetic operations with their development of the theory of iterated REA-operators (Recursively Enumerable in and Above). They showed that Sacks's minimal degree construction is an ω-REA operator, and in the process they proved that $\mathcal{I}$ is equal to $\mathcal{A}$.

THEOREM 2.4 (Jockusch and Shore [1984]). *The ideal of degrees of arithmetic sets is definable in* $\mathfrak{D}$.

2.4. *Cooper's definition of the Turing jump*

To show that the Turing jump is definable, we have to discard the ω-REA operator associated with the minimal degree construction and replace it with an appropriate 2-REA operator. Lachlan provided a natural class of 2-REA operators: for each d-RE set (Difference of Recursively Enumerable) $D = W_i - W_j$, the function $X \mapsto X \oplus W_i(X) - W_j(X)$ is equivalent to a 2-REA operator.

Now we find a new notion of size by finding a degree theoretic property which is realized by some d-RE set and which cannot be realized by $0'$.

On one hand, we have the strong form of the Sacks Splitting Theorem.

THEOREM 2.5 (Sacks [1961b]). *Suppose that A and B are recursive in $0'$ and $A \not\geqslant_T B$. There are W_1 and W_2 such that*

- $W_1 \geqslant_T A$ *and* $W_2 \geqslant_T A$,
- $W_1 \not\geqslant_T B$ *and* $W_2 \not\geqslant_T B$,
- $0' \equiv_T W_1 \oplus W_2$.

We say that $0'$ splits over A avoiding the cone above B. In contrast, Cooper produced an example of non-splitting in the d-RE degrees.

THEOREM 2.6 (Cooper [1990]). *There is a d-r.e. set D and sets A and B recursive in D such that the degree of D does not split over A avoiding the cone above B.*

For D and A as above, say *D is relatively unsplittable over A (r.u.o.)* and use the same phrase for the degrees of D and A. Theorem 2.6 gives a 2-REA operator J which maps each set A to a set $J(A)$ which is relatively unsplittable over it.

Now we have to show that all sufficiently large degrees are r.u.o. over some smaller degree. So, we need to prove that Cooper's d-RE construction can be inverted relative to any sufficiently large degree. Cooper [1990] extended arguments of Jockusch and Shore to prove the following, exactly as needed.

THEOREM 2.7 (Cooper [1990]). *Suppose that J is a 2-REA operator derived from a d-RE set. For all C and Z, if $C \geqslant_T Z + 0''$ and $0' \not\geqslant_T Z$ then*

$$(\exists X)[X \oplus Z \equiv_T C \equiv_T J(X)].$$

We can define $0'$ within $\mathfrak{D}$ using the following implications.

$$z \leqslant_T 0' \Rightarrow (\exists w \geqslant_T z)(\forall a)[w \vee a \text{ is not r.u.o. } a]. \tag{1}$$

Here (1) follows by letting w be $0'$ and using Theorem 2.5.

$$z \nleqslant_T 0' \Rightarrow (\forall w \geqslant_T z)(\exists a)[w \vee a \text{ is r.u.o. } a]. \tag{2}$$

And (2) follows from Theorem 2.7 and Theorem 2.6.

In fact, Cooper proved a stronger theorem to give a complete solution to the Kleene and Post question.

THEOREM 2.8 (Cooper [1990]). *The relation* x is recursively enumerable in and above y *is definable in* $\mathfrak{D}$.

We end this section by deducing Shore's Nonhomogeneity Theorem from Cooper's theorem and the fact that the Coding Lemma is local. Suppose that x is not arithmetic. Then there is a coding of a nonarithmetic set within the degrees between x and x'. So the structure $\mathfrak{D}(\geqslant_T x)$ satisfies the statement that there is a coding of a nonarithmetic set within the degrees between the least degree and it jump. Of course, $\mathfrak{D}$ does not satisfy this statement.

3. Global definability and biinterpretability

DEFINITION 3.1. $\mathfrak{D}$ has a *first-order assignment of representatives* if

(1) for every degree x there is a sequence $\vec{p}$ from $\mathfrak{D}$ which uniformly codes a representative of x (say using the formulas $\varphi_1, \ldots, \varphi_k$);

(2) the relation $\vec{p}$ *codes a representative of* x is a $\mathfrak{D}$-definable relation (say by the formula $\psi(\vec{p}, x)$).

We say that $\mathfrak{D}$ *has a first-order assignment of representatives in parameters* if the same conditions hold as above except that we allow $\varphi_1, \ldots, \varphi_k$ and ψ to mention fixed parameters from $\mathfrak{D}$.

If $\mathfrak{D}$ were to have a first-order assignment of representatives, then we would say that $\mathfrak{D}$ is *biinterpretable with second-order arithmetic*. We define biinterpretability with parameters similarly.

If $\mathfrak{D}$ were biinterpretable with second-order arithmetic, then all of the logical properties of $\mathfrak{D}$ would be reduced to those of the reals. For example, $\mathfrak{D}$ would be rigid.

We will describe the Slaman and Woodin proof that $\mathfrak{D}$ is biinterpretable with second-order arithmetic using a single parameter g_0 and draw the following conclusions.

- A relation is definable from finitely many parameters in $\mathfrak{D}$ if and only if it is induced by an $\equiv_T$-invariant relation on reals that is definable from finitely many parameters in second-order arithmetic, i.e., is induced by a *projective* relation.
- Any automorphism of $\mathfrak{D}$ is determined by its action on g_0. That is, g_0 is an *automorphism base*.

To begin, we state the coding lemma with its full effectivity. In particular, if x is above $0''$ then all the representatives of x are coded below x.

We continue with some countable algebra for $\mathfrak{D}$.

DEFINITION 3.2. Let $\mathcal{I}$ be a countable ideal in $\mathfrak{D}$ and let ρ be an automorphism of $\mathcal{I}$. We say that ρ is *persistent* if for every x in $\mathfrak{D}$ there exist an ideal $\mathcal{J}$ and an automorphism ρ^* of $\mathcal{J}$ such that $x \in \mathcal{J}$, $\mathcal{I} \subseteq \mathcal{J}$ and ρ^* agrees with ρ on $\mathcal{I}$.

Next we prove that every persistent countable automorphism extends to an automorphism of $\mathfrak{D}$. The proof employs a generalization of an insight due to Odifreddi and Shore [1989]: the coding lemma can be used to show that the restriction of an automorphism of $\mathfrak{D}$ to a countable ideal is recursion theoretically close to any uniform upper bound on that ideal. Specifically, we show that if ρ is a persistent automorphism of the ideal $\mathcal{I}$ then ρ is arithmetic in any upper bound of $\mathcal{I}$.

We now introduce some metamathematical methods. We show that there is a nontrivial automorphism of $\mathfrak{D}$ if and only if there is a nontrivial countable automorphism that is persistent. Further, any persistent automorphism of a countable ideal extends to a global automorphism. The latter condition is upwards absolute between well-founded models of *ZFC*.

The condition that there is a persistent countable automorphism ρ such that $\rho(x) = y$ is definable in $\mathfrak{D}$. Consequently, we have the Slaman and Woodin theorem that the relation *x and y are in the same orbit of the automorphism group of $\mathfrak{D}$* is definable.

Now we observe that applying results of Slaman and Woodin [1986], we obtain the following theorem.

THEOREM 3.3 (Slaman and Woodin [1986]). *$\mathfrak{D}$ is rigid if and only if $\mathfrak{D}$ is biinterpretable with second-order arithmetic.*

Continuing the metamathematical discussion, let V denote the universe of sets. Let $V[G]$ be a generic extension of V. We show that if π is an automorphism of $\mathfrak{D}^V$, the degrees in V, then π lifts to an automorphism π^* of $\mathfrak{D}^{V[G]}$, the degrees in the generic extension. By moving to a generic extension of V, we can use the definition of forcing to analyze π^*. In particular, if G is generic with respect to the partial order to add ω_1 Cohen reals to V then π^* is represented as a continuous function on the set of generic reals. In fact, we can use our proof that $\pi(x)$ is close to x to show the following. There is a recursive functional $\{e\}$ and an integer n such that if G is a Cohen generic real over V then the degree of $\{e\}(G \oplus 0'')$ is π^* of the degree of G.

Our next step is to extract a representation of π on a comeager set of reals in V from the representation of π^* on the comeager set of generic reals in $V[G]$. We prove that the same representation of π (using the functional $\{e\}$ relative to $0''$) holds on the set C of reals which are sufficiently generic relative to $0''$. We only require genericity for a bounded amount of arithmetic. Furthermore, if there is one G in C such that G and $\{e\}(G \oplus 0'')$ have the same degree then π is the identity on all the degrees represented in C.

Jockusch and Posner [1981] showed that the Turing degrees represented by any comeager set of reals generate the Turing degrees under the operations of meet and join. Consequently, C is an automorphism base.

Now, let g_0 be any degree of an element of C. If π maps g_0 to g_0 then π must be the identity on C and therefore π must be the identity on all of $\mathfrak{D}$. Thus, g_0 is a finite automorphism base for $\mathfrak{D}$. Second, from the arithmetic representation of π on C we can find an arithmetic function which represents π on all the reals. So we have the following.

THEOREM 3.4 (Slaman and Woodin [1986]).

- *There is an n such that if G is Cohen generic for all Σ^0_n sets, then every automorphism of $\mathfrak{D}$ is determined by its action on the degree of G.*
- *If π is an automorphism of $\mathfrak{D}$, then π is induced by an arithmetically definable function on reals. Consequently, the automorphism group of $\mathfrak{D}$ is countable.*

We see that absoluteness of the rigidity of $\mathfrak{D}$ may be explained by the fact that every automorphism of $\mathfrak{D}$ is arithmetically definable.

3.1. *Biinterpretability with parameters*

We use the fact that $\mathfrak{D}$ has a finite automorphism base in combination with the analysis of persistent automorphisms to show that $\mathfrak{D}$ is biinterpretable with second-order arithmetic in parameters.

Let G_0 denote the representative of an arithmetic and sufficiently generic degree g_0 as above. Suppose that ψ is a map from the reals onto $\mathfrak{D}$ which induces an automorphism on $\mathfrak{D}$, i.e., ψ is degree invariant, preserves $\leqslant_T$ and has distinct values on reals of distinct degree. If ψ maps G_0 to g_0, then ψ must induce the identity automorphism. In other words, ψ must be an assignment of representatives to $\mathfrak{D}$. With some finesse, we can use the coding lemma to express the following condition in D: *There is a persistent countable assignment of representatives sending G_0 to g_0 and X to x.* In our expression g_0 is a parameter, G_0 is replaced by its arithmetic definition, x is a free variable, and X is a free variable in the codes. By the characterization of persistent countable automorphisms as restrictions of global automorphisms, this statement is equivalent to one saying that there is a map from the reals to $\mathfrak{D}$ with the values as above that induces an automorphism. Of course, this automorphism is the identity. Thus, the statement above defines *X is a representative of x* as expressed in the codes for the real X. Consequently, $\mathfrak{D}$ is biinterpretable with second-order arithmetic using the parameter g_0.

THEOREM 3.5 (Slaman and Woodin [1986]). *The partial ordering of the Turing degrees $\mathfrak{D}$ is biinterpretable with second-order arithmetic in parameters.*

4. Automorphisms

From the previous section, the rigidity of $\mathfrak{D}$ is equivalent to its biinterpretability with second-order arithmetic. Thus, the remaining central problem is whether $\mathfrak{D}$ has a nontrivial automorphism.

We have already observed that any automorphism is given by an arithmetic function on representatives. The strongest result in this direction shows that for any automorphism π and any degree x, $\pi(x) \leqslant_T (x \vee 0'')$. Consequently, we have the following.

THEOREM 4.1 (Slaman and Woodin [1986]). *Any automorphism of the Turing degrees is the identity above $0''$.*

This theorem is one a sequence of ever stronger theorems. Without mentioning every step on the way, we note that Nerode and Shore [1980] first showed that every automorphism is eventually equal to the identity, Jockusch and Shore [1984] had the same conclusion as in Theorem 4.1 with 0^{ω} in place of $0''$, and Cooper [1990] had it with $0'''$ in place of $0''$.

In the above we sketched a metamathematical proof that any automorphism of $\mathfrak{D}$ is arithmetically definable. We can also give more traditional, purely recursion theoretic proof of this fact. Of course, this proof is a local version of its metamathematical progenitor. But with the sharper argument, we can replace the full structure of $\mathfrak{D}$ by any ideal in $\mathfrak{D}$ which has $0^{(7)}$ as element.

THEOREM 4.2. *Suppose the $\mathcal{I}$ is an ideal in $\mathfrak{D}$ and $0^{(7)} \in \mathcal{I}$.*

- *If π is an isomorphism from $\mathcal{I}$ to $\mathcal{I}$ then π is represented by an arithmetically definable function on reals.*
- *$\mathcal{I}$ is biinterpretable in parameters with the fragment of second-order arithmetic in which the second-order quantifiers range over the reals whose degrees lie in $\mathcal{I}$.*

As a corollary to the theorem, we can demonstrate a connection between the existence of a local automorphism and a global one. By the Kleene basis theorem, if an arithmetic function does not represent an automorphism of $\mathfrak{D}$ then there is a counter-example which is recursive in Kleene's $\mathcal{O}$. Thus, we can conclude that any automorphism of $\mathfrak{D}(\leqslant_T \mathcal{O})$ extends to an automorphism of $\mathfrak{D}$.

4.1. *Cooper's automorphism*

The previous section would seem to present solid evidence supporting the belief that $\mathfrak{D}$ is rigid. Slaman and Woodin were able to apply their techniques to conclude that the hyperdegrees are rigid, but were unable to make the same conclusion for $\mathfrak{D}$. Nevertheless, Slaman and Woodin conjectured that $\mathfrak{D}$ is rigid and hence biinterpretable with second-order arithmetic.

Cooper has announced a refutation of this conjecture and a construction of a nontrivial automorphism of $\mathfrak{D}$.

So we end in an interesting mathematical situation. We have an uncountable structure $\mathfrak{D}$, whose automorphism group is countable and in a very natural way described by a single real number, the codes for arithmetical formulas which define automorphisms of $\mathfrak{D}$.

What is the automorphism group of $\mathfrak{D}$? What are the definable relations in $\mathfrak{D}$? The best general characterization is given by the following.

THEOREM 4.3 (Slaman and Woodin [1986]). *A relation R on $\mathfrak{D}$ is definable within $\mathfrak{D}$ if and only if*

- *R is induced by a Turing degree invariant relation on subsets of $\mathbb{N}$ which is definable in second-order arithmetic,*
- *R is invariant under all automorphisms of $\mathfrak{D}$.*

The second clause can be dropped if R mentions only degrees above $0''$.

References

U. ABRAHAM AND R. A. SHORE
[1986] Initial segments of the degrees of size $\aleph_1$, *Israel J. Math.*, 53, pp. 1–51.

S. B. COOPER
[1990] The jump is definable in the structure of the degrees of unsolvability, *Bull. Amer. Math. Soc.*, 23, pp. 151–158.

M. J. GROSZEK AND T. A. SLAMAN

[1983] Independence results on the global structure of the Turing degrees, *Trans. Amer. Math. Soc.*, 277, pp. 579–587.

C. G. JOCKUSCH, JR.

[1973] An application of Σ_4^0 determinacy to the degrees of unsolvability, *J. Symbolic Logic*, 38, pp. 293–294.

C. G. JOCKUSCH, JR. AND D. POSNER

[1981] Automorphism bases for degrees of unsolvability, *Israel J. Math.*, 40, pp. 150–164.

C. G. JOCKUSCH, JR. AND R. A. SHORE

[1984] Pseudo-jump operators II: Transfinite iterations, hierarchies, and minimal covers, *J. Symbolic Logic*, 49, pp. 1205–1236.

C. G. JOCKUSCH, JR. AND T. A. SLAMAN

[1993] On the Σ_2-theory of the upper-semi-lattice of the Turing degrees, *J. Symbolic Logic*, 58, pp. 193–204.

C. G. JOCKUSCH, JR. AND R. I. SOARE

[1970] Minimal covers and arithmetical sets, *Proc. Amer. Math. Soc.*, 25, pp. 856–859.

S. C. KLEENE AND E. L. POST

[1954] The upper semi-lattice of degrees of recursive unsolvability, *Anal. of Math.*, 59, pp. 379–407.

A. H. LACHLAN

[1968] Degrees of recursively enumerable sets which have no maximal supersets, *J. Symbolic Logic*, 33, pp. 431–443.

A. H. LACHLAN AND R. LEBEUF

[1976] Countable initial segments of the degrees, *J. Symbolic Logic*, 41, pp. 289–300.

M. LERMAN

[1971] Initial segments of the degrees of unsolvability, *Anal. of Math.*, 93, pp. 311–389.

[1983] *Degrees of Unsolvability*, Perspectives in Mathematical Logic, Springer, Heidelberg.

D. A. MARTIN

[1975] Borel determinacy, *Ann. of Math. (2)*, 102, pp. 363–371.

A. NERODE AND R. A. SHORE

[1980] Reducibility orderings: theories, definability and automorphisms, *Ann. Math. Logic*, 18, pp. 61–89.

P. ODIFREDDI AND R. A. SHORE

[1989] *Global properties of local structures of degrees*, Preprint.

H. ROGERS, JR.

[1967] *Theory of Recursive Functions and Effective Computability*, McGraw-Hill, New York.

G. E. SACKS

[1961a] A minimal degree below $\mathbf{0}'$, *Bull. Amer. Math. Soc.*, 67, pp. 416–419.

[1961b] On suborderings of degrees of recursive unsolvability, *Z. Math. Logik Grundlag. Math.*, 17, pp. 46–56.

[1963] *Degrees of Unsolvability*, Annals of Mathematical Studies, Vol. 55, Princeton University Press.

[1966] *Degrees of Unsolvability*, 2nd ed., Annals of Mathematical Studies, Vol. 55, Princeton University Press.

R. A. SHORE

[1979] The homogeneity conjecture, *Proc. Natl. Ac. Sci. USA*, 76, pp. 4218–4219.

S. G. SIMPSON
[1977] First-order theory of the degrees of recursive unsolvability, *Anal. of Math.*, 105, pp. 121–139.

T. A. SLAMAN AND W. H. WOODIN
[1986] Definability in the Turing degrees, *Illinois J. Math.*, 30, pp. 320–334.

C. SPECTOR
[1956] On the degrees of recursive unsolvability, *Anal. of Math.*, 64, pp. 581–592.

CHAPTER 6

The Recursively Enumerable Degrees

Richard A. Shore*

Department of Mathematics, White Hall, Cornell University, Ithaca, NY 14853, USA

Contents

1. Introduction 170
2. Structure and decidability 171
3. Undecidability and beyond 180
4. Natural definability 188
References 193

*Partially supported by NSF Grants DMS-9204308, DMS-9503503, DMS-9802843 and the US ARO through ACSyAM at the Mathematical Sciences Institute of Cornell University Contract DAAL03-91-C-0027.

HANDBOOK OF COMPUTABILITY THEORY
Edited by E.R. Griffor

1. Introduction

Decision problems were the motivating force in the search for a formal definition of algorithm that constituted the beginnings of recursion (computability) theory. In the abstract, given a set A the decision problem for A consists of finding an algorithm which, given input n, decides whether or not n is in A. The classic decision problem for logic is whether a particular sentence is a theorem of a given theory T. Other examples arise in almost all branches of mathematics. In most settings one is almost immediately confronted by the notion of a *recursively* (*or computably*) *enumerable* (r.e.) *set* (the sets which can be listed (i.e. enumerated) by a computable (i.e. recursive) function): the theorems of a axiomatized theory, the solvable Diophantine equations, the true equations between words in a finitely presented group, etc. Typically, such decision problems amount to deciding if a particular r.e. set is computable (recursive). Indeed, the first examples of unsolvable decision problems provided examples of nonrecursive r.e. sets: the theorems of predicate logic, the word problem for groups, the halting problem. (For technical convenience, we code all expressions in formal languages, groups, etc. as natural numbers and so restrict our attention to sets of natural numbers.)

One can say that all these sets are simply noncomputable. Another view sees them as more complicated or harder to compute than the recursive sets. This is the view that leads to the notion of relative computability (reducibility) introduced by Turing [1936, 1939] and Post [1936, 1944]. The equivalence classes under this notion of relative computability were first called the degrees of recursive unsolvability. As Church's Thesis became widely accepted the word "recursive" was dropped and they became simply the degrees of unsolvability. As Turing's model of computation became the standard one, they became the Turing degrees. In view of the centrality of Turing's notion as the basic general definition of computability, the unqualified notion of degree eventually became that of Turing degrees. Other notions of relative computability whether stronger or weaker, from one-one to truth-table to arithmetic to constructibility, are refereed to by specifying the reducibility.

The starting point for the investigation of this fundamental notion of relative computability was the r.e. degrees (those equivalence classes containing r.e. sets). The classic results of logic (such as Gödel's incompleteness theorem, Church's proof of the undecidability of predicate logic and Turing's unsolvability of the Halting problem) each proved that there was a nonrecursive r.e. degree. All the natural examples, however, of nonrecursive r.e. sets supplied by standard theories which could be proven undecidable (e.g., Peano arithmetic) or from other natural definitions of noncomputable r.e. sets, turned out to have the same complexity. They were all complete, i.e. of the same degree as the halting problem $K = \{e \mid \phi_e(e)\downarrow\}$. The obvious question, first proposed by Post [1944], was whether there are any other classes of decision problems under the equivalence of relative computability, i.e. are there any r.e. degrees other than those of the recursive sets, $\mathbf{0}$, and $\mathbf{0}'$, the degree of K?

Post attacked this problem by trying to define set theoretic properties of r.e. sets such as simplicity or hypersimplicity which would guarantee incompleteness as well as nonrecursiveness. Post concentrated on thinness properties of the complement of the r.e. set. This particular approach was doomed to failure (Yates [1965]), but Post's work initiated the study of the structure of the r.e. sets under set inclusion and the connections between their set-theoretic structure and computational complexity (see Soare [1999]). The solution to Post's problem, however, came from another approach.

Friedberg [1957] and Muchnik [1956] independently solved the problem by constructing intermediate r.e. degrees. The construction technique they introduced is called the priority method. It has been extensively studied, expanded and developed over the years. The priority method has proven useful in many areas of recursion theory and in applications to other areas of logic as well. Indeed, this technique has been called the hallmark of recursion theory. The principle arena of both its development and application has been in the study of r.e. sets and in particular of the r.e. degrees. The great strides that have been made in the past forty years in the understanding of the structure of the r.e. degrees have gone hand in hand with the development of new types of priority arguments. In this chapter we try to present the important results contributing to our overall picture of the structure $\mathcal{R}$ of the r.e. degrees with just a word or two about the associated proof techniques followed by appropriate references.

2. Structure and decidability

$\mathcal{R}$, the structure of the r.e. degrees with the ordering $\leqslant$ induced by Turing reducibility, is clearly a partial ordering with least element 0 and greatest element 1 represented by the degrees $\mathbf{0}$ and $\mathbf{0}'$, respectively. It is also easily seen to be an uppersemilattice (*usl*) with join (least upper bound), $\vee$, induced by disjoint union on representatives: $\mathbf{a} \vee \mathbf{b} = \mathbf{c}$ where $\mathbf{c}$ is the degree of $A \oplus B = \{2x \mid x \in A\} \cup \{2x+1 \mid x \in B\}$ for any sets A, B of degree $\mathbf{a}$, $\mathbf{b}$, respectively. One way to describe such a structure (or to determine its complexity) is to decide which partial orderings or uppersemilattices (usls) can be embedded in it (preserving the appropriate relations).

Friedberg's and Muchnik's solutions to Post problem can be seen as the first such result: the p.o. with 0, 1 and two other incomparable elements can be embedded into $\mathcal{R}$. The technique they introduced is now known as the finite injury priority method. (In addition to the original papers, clear expositions of (different views of) the method can be found in Rogers [1967a, §10.2], Shoenfield [1971, §13–14], and Soare [1987, VI.1]. It has been used to prove many theorems about the r.e. degrees. In particular, it suffices to prove the following:

THEOREM 2.1 (Embedding Theorem) (Muchnik [1958], Sacks [1963a]). *Every countable partial ordering (and even usl) can be embedded into* $\mathcal{R}$.

In terms of the theory of $\mathcal{R}$, this result says that the existential ($\exists$) theory of $\mathcal{R}$ is decidable:

THEOREM 2.2. *There is an effective decision procedure for deciding the truth of sentences of the form* $\exists x_1 \cdots \exists x_n \phi$ *where* ϕ *is a quantifier free formula in the language of partial orderings (usls).*

PROOF. By the embedding theorem, any such sentence in the language of partial orderings ($\leqslant$) is true in $\mathcal{R}$ iff there is a partial ordering of size at most n (the number of distinct variables appearing in the sentence) in which it is true. It is clear that we can list all such orderings and then just check each one. If we consider the language of usls, then we must note that, in any usl $\mathcal{L}$, the substructure generated by a finite set $x_1, \ldots, x_n$ is finite and of size at most 2^n. Thus to decide if a given existential sentence in the language of usls ($\leqslant, \vee$) is true in $\mathcal{R}$ it suffices to see if it is true in one of the finitely many usls of size 2^n. □

Algebraically, after embedding problems, the next questions about the structure of a partial ordering are the extension of embedding problems such as density: Given two elements $a < b$ of the p.o., can we always find another c, such that $a < c < b$? This particular question was answered by Sacks [1964].

THEOREM 2.3 (Sacks Density Theorem) (Sacks [1964]). *For every pair of nonrecursive r.e. degrees* $\mathbf{a} < \mathbf{b}$ *there is one* $\mathbf{c}$ *such that* $\mathbf{a} < \mathbf{c} < \mathbf{b}$.

The proof of this theorem goes beyond the finite injury method by having both positive (put numbers into the set being constructed) and negative (keep numbers out of the set being constructed) infinitary requirements. It was the primary early example of what is now called the infinite injury method. (Interesting expositions of different approaches to this construction can be found in Shoenfield [1971, §16] and Soare [1987, VIII.4].) The construction requires simple coding (A into C), upward cone avoiding ($B \nleqslant C$), downward cone avoiding ($C \nleqslant A$) and some sort of control by the top set B to get $C \leqslant B$. The third of these four requirements also involves a coding procedure. The second was accomplished by an ingenious method (called Sacks preservation which, in its basic form, entails finite but unbounded action for both positive and negative requirements) introduced in an earlier important theorem:

THEOREM 2.4 (Sacks Splitting Theorem) ([Sacks 1963c]). *For every nonrecursive r.e. degree* $\mathbf{a}$ *there are r.e. degrees* $\mathbf{b}, \mathbf{c} < \mathbf{a}$ *such that* $\mathbf{b} \vee \mathbf{c} = \mathbf{a}$.

In 1963, these results lead Shoenfield [1965] to make the sweeping conjecture that the r.e. degrees, $\mathcal{R}$, are a "dense" (or more formally in the language of model theory, a countably saturated) usl with least and greatest elements:

CONJECTURE 2.5 (Shoenfield [1965]). *For every pair* $\mathcal{P} \hookrightarrow \mathcal{Q}$ *of finite usls with* 0, 1 *and every embedding* $f : \mathcal{P} \to \mathcal{R}$, *there is an extension* g *of* f *to an embedding of* $\mathcal{Q}$ *into* $\mathcal{R}$.

If true, this conjecture would have implied that the r.e. degrees had many of the familiar properties of structures like dense linear ordering or atomless Boolean algebras which satisfy the corresponding property for the appropriate family of structures (linear orderings and Boolean algebras). Such structures are countably categorical (i.e. there is a unique such countable structure up to isomorphism) and so (if axiomatizable) have decidable theories. They are countably homogeneous (every structure preserving map from one finite subset to another can be extended to an automorphism) and so there are continuum many automorphisms of the structure. A positive solution to Shoenfield's conjecture would thus have constituted an essentially complete characterization of the structure of the r.e. degrees.

In what may have seemed like an unfortunate development, however, the conjecture was refuted almost immediately by the construction of minimal pairs by Lachlan [1966a] and Yates [1966a]:

THEOREM 2.6 (Lachlan [1966a], Yates [1966a]). *There are nonrecursive r.e. degrees* $\mathbf{a}$ *and* $\mathbf{b}$ *such that* $\mathbf{a} \wedge \mathbf{b} = \mathbf{0}$, *i.e. any degree recursive in both* $\mathbf{a}$ *and* $\mathbf{b}$ *is recursive.*

(Shoenfield's Conjecture clearly implies that for any $\mathbf{a}, \mathbf{b} > \mathbf{0}$ there is a $\mathbf{c} > \mathbf{0}$ which is below both $\mathbf{a}$ and $\mathbf{b}$.)

This result was the beginning of a trend away from "simplicity" and towards "complexity". Its proof also introduced new techniques involving negative requirements that impose restraint that may be unbounded but that infinitely often drops back to a fixed finite value. After the original proofs, two important expository approaches to these constructions were discovered that would become the basis for further developments in both results and technology. The first (Lachlan [1973], Soare [1987, IX.1]) exploits an inductive definition of nested expansionary stages to make the restraint drop back simultaneously. The second involves the very important idea of priority trees and constructions introduced in Lachlan [1975a] for a more difficult theorem (Theorem 2.15). These latter techniques are exposited in Soare [1987, XIV].

First notice that a meet (greatest lower bound or infimum) operator $\wedge$ has been introduced. The existence of meets at first raises the hope that $\mathcal{R}$ might be a lattice. It is not.

THEOREM 2.7 (Lachlan [1966b], Yates [1966a]). *The r.e. degrees are not a lattice.*

The methods used to prove that there are minimal pairs also show that there is a strictly ascending sequence $\mathbf{c}_0 < \cdots < \mathbf{c}_n < \cdots$ with an *exact pair*, i.e. incomparable degrees $\mathbf{a}$ and $\mathbf{b}$ such that $\mathbf{c}_i \leqslant \mathbf{a}, \mathbf{b}$ for each i and every $\mathbf{c} \leqslant \mathbf{a}, \mathbf{b}$ is below some $\mathbf{c}_i$ (Yates [1966a]). No such $\mathbf{a}$ and $\mathbf{b}$ can have an infimum. (If $\mathbf{a} \wedge \mathbf{b} = \mathbf{c}$, then for some i, $\mathbf{c} \leqslant \mathbf{c}_i$. As $\mathbf{c}_i < \mathbf{c} < \mathbf{a}, \mathbf{b}$, we have contradicted the assumption that $\mathbf{c}$ is the infimum of $\mathbf{a}$ and $\mathbf{b}$.)

Lachlan's proof of this result was quite different. It was based on a relativization of the following:

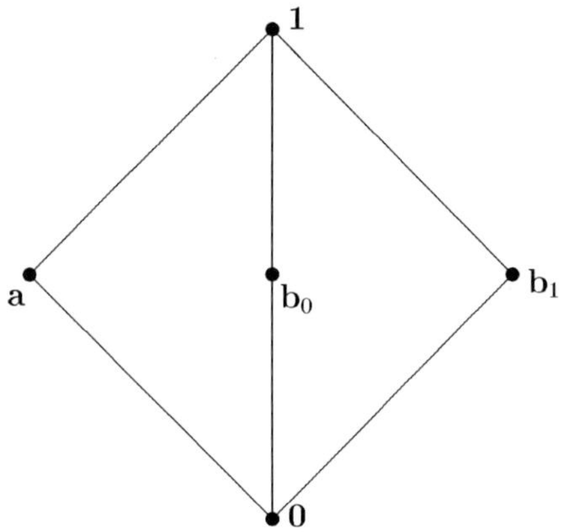

Fig. 2.1. The lattice M_5.

THEOREM 2.8 (Nondiamond Theorem) (Lachlan [1966b]). *There are no r.e. degrees* $\mathbf{a}$ *and* $\mathbf{b}$ *such that* $\mathbf{a} \mid \mathbf{b}$ (*i.e.* $\mathbf{a} \nleq \mathbf{b}$ *and* $\mathbf{b} \nleq \mathbf{a}$), $\mathbf{a} \vee \mathbf{b} = \mathbf{1}$, $\mathbf{a} \wedge \mathbf{b} = \mathbf{0}$.

This theorem is now known as the nondiamond theorem as it says that there is no embedding of the diamond shaped four element lattice (0, 1 plus two incomparable intermediate elements) into $\mathcal{R}$ preserving 0, 1 and the lattice structure.

These results opened up a new chapter in the embedding problem for the r.e. degrees which has yet to be finished: Which lattices can be embedded into $\mathcal{R}$ (preserving the lattice structure, of course, but perhaps also 0 and/or 1)? These initial results said that the diamond can be embedded preserving 0 but not preserving both 0 and 1. The techniques introduced to construct a minimal pair suffice to embed all countable distributive lattices:

THEOREM 2.9 (Lachlan, Lerman, Thomason; see Soare [1987, IX.2]). *Every countable distributive lattice can be embedded into* $\mathcal{R}$ *as a lattice preserving* 0.

The next step in the story was also taken by Lachlan who proved that the two basic nondistibutive lattices can be embedded in $\mathcal{R}$. M_5 (or the pentagon) and N_5 (or 1-3-1) pictured in Figs. 2.1 and 2.2 are nondistributive and every nondistributive lattice has one of them embedded in it.

THEOREM 2.10 (Lachlan [1972]). *Both* M_5 *and* N_5 *can be embedded* (*as lattices*) *in* $\mathcal{R}$ *preserving* 0.

Now, not every countable lattice can be embedded in any single countable structure. A sublattice $\mathcal{L}'$ of a lattice $\mathcal{L}$ is generated by a subset A of $\mathcal{L}$ if $\mathcal{L}'$ is the smallest sublattice of $\mathcal{L}$ containing A. It is obvious that any particular countable lattice $\mathcal{L}$ has

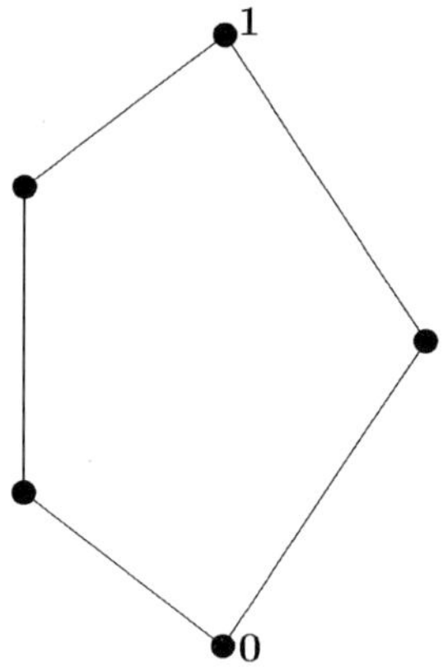

Fig. 2.2. The lattice N_5.

only countably many sublattices $\mathcal{L}'$ generated by finite subsets of $\mathcal{L}$. On the other hand, there is an uncountable set of lattices $\mathcal{L}$ such that each one is generated by four of its elements. (For an example, see Shore [1982].) So the obvious conjecture was simply that every finite lattice can be embedded in $\mathcal{R}$. The true state of affairs, however, proved not to be so simple.

THEOREM 2.11 (Lachlan and Soare [1980]). *The lattice S_8 (the diamond on top of* 1-3-1) *pictured in Fig.* 2.3 *below cannot be embedded in $\mathcal{R}$.*

The failure of the simplest form of a lattice embedding conjecture is the source of much of the difficulty in further progress in deciding the next level of the theory of $\mathcal{R}$. The $\exists$-theory of $\mathcal{R}$ (as a p.o. or usl) is decidable by Theorem 2.2. The next obvious goal is to decide the Σ_2 or $\exists\forall$-theory of $\mathcal{R}$, i.e. find an algorithm for deciding the truth of all sentences of the form $\exists x_1 \cdots \exists x_n \forall y_1 \cdots \forall y_m \phi$ with ϕ a quantifier free formula in the language of partial orderings (or, perhaps, even usls). It is routine to verify that, for every finite lattice $\mathcal{L}$, there is an $\exists\forall$ sentence ψ such that $\mathcal{L}$ is embeddable in $\mathcal{R}$ if and only if ψ is true in $\mathcal{R}$. For example, the embeddability of 1-3-1 is equivalent to the truth of the following sentence:

$$\begin{aligned}&\exists a_0, a_1 a_2, b, c\\ &\quad\{(c < a_0, a_1, a_2 < b) \;\&\; \forall x[(x \geqslant a_0, a_1 \;\vee\; x \geqslant a_0, a_2 \vee x \geqslant a_1, a_2)\\ &\quad \to x \geqslant b] \;\&\; \forall x[(x \leqslant a_0, a_1 \;\vee\; x \leqslant a_0, a_2 \vee x \leqslant a_1, a_2) \to x \leqslant c]\}.\end{aligned}$$

Thus the embedding problem for lattices is a crucial ingredient in any attempt at deciding the $\exists\forall$-theory of $\mathcal{R}$. Unfortunately, the current state of affairs is quite complicated. There are complex necessary conditions for embeddability as well sufficient ones (Ambos-Spies and Lerman [1986, 1989]). In an attempt to isolate what seemed

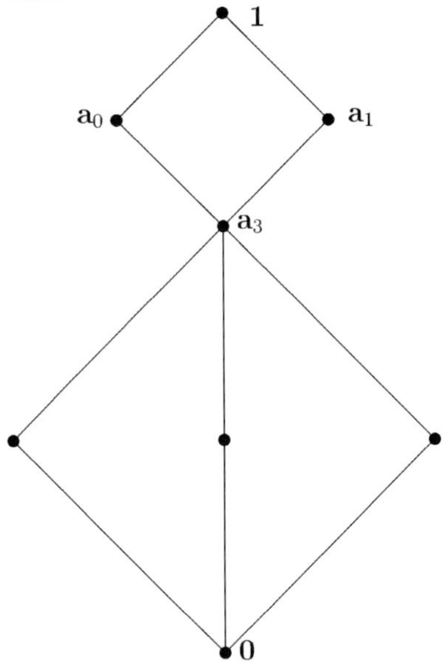

Fig. 2.3. The lattice S_8.

to be the crucial obstruction to embedding lattices in $\mathcal{R}$, Downey [1990a] (see also Weinstein [1988] for a similar notion) proposed the following definition:

DEFINITION 2.12. Three r.e. degrees $\mathbf{a}, \mathbf{b}, \mathbf{c}$ are called a *critical triple* if $\mathbf{a} \vee \mathbf{c} = \mathbf{a} \vee \mathbf{b}$ and $\mathbf{b} \wedge \mathbf{c} \leqslant \mathbf{a}$.

The nonembedding proofs show that the difficulties in carrying out the embedding construction arise if there are two degrees which "inf down" into a critical triple. This led to the following conjecture.

CONJECTURE 2.13 (Downey [1990a]). *A finite lattice is embeddable in $\mathcal{R}$ if and only if it does not contain a critical triple a, b, c and a pair p, q such that $b \leqslant p \wedge q \leqslant b \vee c$.*

Lempp and Lerman [1997] have shown this conjecture to be false by finding a twenty element lattice which has no critical triple but nonetheless cannot be embedded in $\mathcal{R}$. They suggested, however, that the way they found this counterexample offered some hope of finding a characterization of the finite lattices embeddable in $\mathcal{R}$. Lerman [1998, 1999] has now found a characterization of the finite lattices without critical triples that can be embedded in $\mathcal{R}$. The defining property, however, is not (at least on its face) recursive.

Now, every lattice that is known to be embeddable in $\mathcal{R}$ is known to be embeddable preserving 0. There has also been some work on lattice embedding problems preserving 1 (which is harder than preserving 0) and ones preserving both 0 and 1 (which is much harder and, as can be seen from the nondiamond theorem (Theorem 2.8), known to be much more restricted). For example, Lachlan [1980] and Shoenfield and Soare [1978] showed that the diamond is embeddable preserving 1. Ambos-Spies [1980] extended this result to all countable distributive lattices and Ambos-Spies, Lempp and Lerman [1994a] showed that both M_5 and N_5 can be embedded in $\mathcal{R}$ preserving 1. As for preserving both 0 and 1, Ambos-Spies, Lempp and Lerman [1994b] characterize the finite distributive lattices embeddable in $\mathcal{R}$ preserving both 0 and 1 as those containing a join irreducible noncappable element, i.e. an x which is not the join of elements below it and does not inf down to 0.

Let us reconsider the decision problem for the two quantifier theory of $\mathcal{R}$. Syntactic and algebraic manipulations show (Lerman [1983, pp. 156–158]) that deciding the truth of all $\forall\exists$ sentences (in the language of partial orderings) in an usl with 0 such as $\mathcal{R}$ or $\mathcal{D}$ is equivalent to determining for any given collection of extensions of finite partial orderings, $\mathcal{P} \hookrightarrow \mathcal{Q}_i$, $i \leqslant n$, whether every embedding $f:\mathcal{P} \to \mathcal{R}$ can be extended to an embedding $g:\mathcal{Q}_i \to \mathcal{R}$ for some $i \leqslant n$. The special case of this question for a single possible extension $\mathcal{Q}$ is called the *extension of embedding problem*. Slaman and Soare [1995, 1998] have recently solved this problem positively.

THEOREM 2.14 (Extension of Embedding Theorem) (Slaman and Soare [1995, 1998]). *There is an effective characterization of the pairs $\mathcal{P} \hookrightarrow \mathcal{Q}$ of partial orderings with* 0, 1 *such that, for every embedding $f:\mathcal{P} \to \mathcal{R}$, there is an extension g of f to an embedding of $\mathcal{Q}$ into $\mathcal{R}$.*

The idea of the decision procedure is that certain extensions are explicitly ruled out and all remaining ones are then proven extendible. The extension construction is an elaboration of the results and techniques embodied in the density theorem. The nonextension criteria have two parts. The first set of exclusions follows from the embedding of distributive lattices in $\mathcal{R}$. The second embodies various new results that rely on a level of priority arguments one higher than the infinite injury ones used for the density theorem and minimal pair arguments. These arguments are now called $\mathbf{0}'''$ constructions. The terminology refers to the difficulty of determining how the requirements are met by the construction. In this terminology, finite and infinite injury arguments are $\mathbf{0}'$ and $\mathbf{0}''$ ones, respectively. These techniques were first introduced by Lachlan to prove that the Sacks Splitting and Density Theorems cannot be combined.

THEOREM 2.15 (Lachlan Nonsplitting Theorem) (Lachlan [1975a]). *There are r.e. degrees $\mathbf{d} < \mathbf{a}$ for which there are no r.e. degrees $\mathbf{b}, \mathbf{c}$ such that $\mathbf{d} < \mathbf{b}, \mathbf{c} < \mathbf{a}$ and $\mathbf{b} \vee \mathbf{c} = \mathbf{a}$.*

Expositions of examples of these methods can be found, for example, in Soare [1987, XV.4], Shore and Slaman [1993] and Slaman [1991b] with other prototypes in Shore [1988] and Downey [1990b]. A typical instance of the facts needed to get Slaman and Soare's nonextension results is the following:

THEOREM 2.16 (Slaman and Soare [1998]). *There are incomparable r.e. degrees* $\mathbf{a}, \mathbf{b}$ *such that for every* $\mathbf{z} < \mathbf{a}$, *either* $\mathbf{z} < \mathbf{b}$ *or* $\mathbf{z} \vee \mathbf{b} = \mathbf{0}'$.

The gap between this solution to the extension of embedding problem and deciding the full two quantifier theory is precisely the problem of having more than one possible extension to consider. The reality of this possible problem is epitomized by the nondiamond theorem: Suppose $\mathcal{P}$ is the diamond p.o. with 0, 1 and incomparable elements a and b. Furthermore, suppose $\mathcal{Q}_0$ puts a nonzero element in below both a and b and $\mathcal{Q}_1$ puts one in which is below 1 but above both a and b. There are then embeddings of $\mathcal{P}$ which cannot be extended to ones of $\mathcal{Q}_0$ by the minimal pair theorem (Theorem 2.6) and there are ones which cannot be extended to $\mathcal{Q}_1$ by the Sacks Splitting Theorem (Theorem 2.4). On the other hand, the nondiamond theorem (Theorem 2.8) says that every embedding of $\mathcal{P}$ can be extended to either one of $\mathcal{Q}_0$ or one of $\mathcal{Q}_1$.

Now, in other recursion theoretic structures it has been possible to decide the full $\forall\exists$-theory by a combination of embedding results in stronger languages than just partial orderings and a series of extension of embedding results. (Examples include the weak truth-table degrees of the r.e. sets (Ambos-Spies, Fejer, Lempp and Lerman [1996]) where the essential ingredient is the characterization of the finite lattices which can be embedded preserving both 0 and 1; the Turing degrees as a whole (Lerman, Shore [1978]; see Lerman [1983, p. 157]) and those below $\mathbf{0}'$ (Lerman and Shore [1988]) where lattice embeddings as initial segments and partial order extensions of embeddings suffice; and the lattice of r.e. sets where more elaborate extensions of the language are necessary (Lachlan [1968]).) At times it was thought that it might be possible to similarly separate the lattice embedding and extension of embedding problems in $\mathcal{R}$, solve them individually and so decide the full two quantifier theory. In view of recent developments, this no longer seems a likely scenario. The two problems are more intimately connected than had previously been believed. The roots of this relationship are, on the one hand, the obstructions to extension of embeddings discovered by Slaman and Soare [1995, 1998] and, on the other, an extension of the nondiamond phenomena and the related notion of prompt simplicity introduced by Maass [1982] in the study of the set theoretic structure (and genericity properties) of the r.e. sets but phrased in terms of the dynamics of their enumerations.

DEFINITION 2.17. A coinfinite r.e. set A is *promptly simple* if there is there is a nondecreasing recursive function p and a recursive one-to-one function f enumerating A (i.e. $A = \operatorname{rg} f$) such that for every infinite r.e. W_e there is an s and an x such that x is enumerated in W at stage s (in some standard uniform enumeration of all the r.e. sets) and is also enumerated in A by stage $p(s)$, i.e. $x = n$ for some $n \leqslant p(s)$. An r.e. degree $\mathbf{a}$ is *promptly simple* if it contains a promptly simple r.e. set. We let $\mathbf{PS}$ denote the set of promptly simple r.e. degrees.

It turns out that the property that an r.e. degree is promptly simple can be characterized solely in terms of the ordering of r.e. degrees and has come to play an important role in the study of $\mathcal{R}$. For example, one purely order theoretic result provided by the following theorem is that no join of degrees which are halves of minimal pairs can be $\mathbf{0}'$. It is also connected with another *a priori* apparently extraneous consideration, the relationship between the jump of an r.e. degree and its place in the structure of $\mathcal{R}$. (The jump, $\mathbf{a}'$, of an r.e. degree $\mathbf{a}$ is the degree of $\{e \mid \phi_e^A(e)\}\downarrow$. It is the analog of $\mathbf{0}'$ for the sets and degrees r.e. in A. The role of this operator in the study of $\mathcal{R}$ will be considered at greater length in §4.)

THEOREM 2.18 (Ambos-Spies et al. [1984]). *Let* $\mathbf{M} = \{\mathbf{a} \mid \exists \mathbf{b}(\mathbf{a} \wedge \mathbf{b} = \mathbf{0})\}$ *be the set of* cappable *r.e. degrees, i.e. those which are halves of minimal pairs; let* $\mathbf{NC}$, *the* noncappable *r.e. degrees be its complement in* $\mathcal{R}$; *and let* $\mathbf{LC} = \{\mathbf{a} \mid \exists \mathbf{b}(\mathbf{a} \vee \mathbf{b} = \mathbf{0}' \,\&\, \mathbf{b}' = \mathbf{0}')\}$ *be the* low cuppable degrees, *i.e. those which can be cupped (joined) to* $\mathbf{0}'$ *by a low degree* $\mathbf{b}$, *i.e. one such that* $\mathbf{b}' = \mathbf{0}'$.

The three classes $\mathbf{PS}$, $\mathbf{NC}$ *and* $\mathbf{LC}$ *all coincide and together with their complement* $\mathbf{M}$ *partition* $\mathcal{R}$ *as follows*:

(i) $\mathbf{M}$ *is a proper ideal in* $\mathcal{R}$, *i.e. it is closed downward in* $\mathcal{R}$ *and if* $\mathbf{a}, \mathbf{b} \in \mathbf{M}$ *then* $\mathbf{a} \vee \mathbf{b} \in \mathbf{M}$ *and* $\mathbf{a} \vee \mathbf{b} < \mathbf{0}'$.

(ii) $\mathbf{NC}$ *is a strong filter in* $\mathcal{R}$, *i.e. it is closed upward in* $\mathcal{R}$ *and if* $\mathbf{a}, \mathbf{b} \in \mathbf{NC}$ *then there is a* $\mathbf{c} \in \mathbf{NC}$ *with* $\mathbf{c} \leqslant \mathbf{a}, \mathbf{b}$.

As Slaman has pointed out [personal communication], an essential interaction between the lattice embedding problem and the extension of embedding problem arises when this theorem is combined with Theorem 2.16 and the version of the Sacks Splitting Theorem (Theorem 2.4) (also in Sacks [1963c]) which says that any given r.e. degree $\mathbf{a}$ can be split into two low r.e. degrees. Thus if $\mathbf{a}$ and $\mathbf{b}$ are as in Theorem 2.16 we can split $\mathbf{a}$ into two low degrees $\mathbf{x}$ and $\mathbf{y}$ by the Sacks Splitting Theorem. As $\mathbf{a}$ and $\mathbf{b}$ are incomparable, not both $\mathbf{x}$ and $\mathbf{y}$ can be below $\mathbf{b}$. Thus, by the properties guaranteed by Theorem 2.16, one of them joins $\mathbf{b}$ up to $\mathbf{0}'$ and so by Theorem 2.18, $\mathbf{b}$ is not half of a minimal pair. Thus there are unexpected connections between the two problems that can already be expressed by an $\forall\exists$ sentence saying that if $\mathbf{b}$ is half of a minimal pair and $\mathbf{a}$ is incomparable with $\mathbf{b}$ then there is a $\mathbf{y}$ below $\mathbf{a}$ but not below $\mathbf{b}$ such that $\mathbf{b} \vee \mathbf{y} \neq \mathbf{0}'$. Such interactions make framing a specific plan to decide the $\forall\exists$ theory of $\mathcal{R}$ quite difficult. (For a more extensive discussion see Lerman [1996].)

On the other hand, we note that the usual techniques for proving such a theory (or fragment) undecidable cannot work for $\mathcal{R}$. Once one has proven the undecidability of any particular theory (e.g., predicate logic) by coding the computations of Turing machines or the like, one proves other theories undecidable by interpreting a theory that is known to be undecidable in the one of interest. As the general problem of the validity of $\forall\exists$ sentences in purely relational languages is decidable, no standard interpretation along these lines can show that the $\forall\exists$ theory of $\mathcal{R}$ is undecidable. Thus we are simply left with the open problem.

CONJECTURE 2.19. *The* $\forall\exists$*-theory of* $\mathcal{R}$ *is decidable.*

Should this conjecture prove correct, it will represent the limit of decidability results for $\mathcal{R}$ as Lempp, Nies and Slaman [1998] have shown that its $\forall\exists\forall$-theory is undecidable.

3. Undecidability and beyond

The minimal pair theorem (Theorem 2.6) refuted Shoenfield's Conjecture and, in retrospect at least, began a long series of developments in other directions. At the time, however, it was still hoped that the r.e. degrees would be "nice" in certain ways. In particular, even Sacks who had conjectured (Sacks [1963a]) that there were minimal pairs and that $\mathcal{R}$ is not a lattice, nonetheless continued (Sacks [1966]) to conjecture that the theory of $\mathcal{R}$ is decidable and that there is a strong sense of homogeneity for the notion of r.e. in the sense that "for each (not necessarily r.e.) degree **d**, the ordering of degrees r.e. in **d** and $\geqslant$ **d** is order isomorphic to the r.e. degrees". These conjectures all turned out to be false and some twenty years later a dramatically different view of the structure of the r.e. degrees (as well as of the degrees as a whole) became the prevailing paradigm.

Not only is $\mathcal{R}$ undecidable but its theory is as complicated as possible. It is recursively isomorphic to that of true arithmetic. That is, there are two recursive translations S and T such that S takes sentences ϕ in the language of arithmetic to sentences ϕ^S of partial orderings while T takes sentences ψ in the language of partial orderings to sentences ψ^T of arithmetic. Moreover, the translations preserve truth: $\mathcal{N} \models \phi \leftrightarrow \mathcal{R} \models \phi^S$ and $\mathcal{R} \models \psi \leftrightarrow \mathcal{N} \models \psi^T$. A similar result holds for the theory of all the Turing degrees $\mathcal{D}$. This latter theory is recursively isomorphic to that of true second order arithmetic. Moreover, almost all possible homogeneity conjectures are now known to fail both for relative recursive enumerability as well as relative computability.

At first, such results suggest that there is no hope of understanding or characterizing the structure of $\mathcal{R}$ or $\mathcal{D}$. Later, the opposite view took hold. The idea was that a strong enough proof of the complexity of a structure can characterize it as well as one of its simplicity. Instead of expecting the structure to be decidable and homogeneous, for all degrees to look the same and for there to be many automorphisms, one could look to prove that the theory is as complicated as possible, there are as many different types of degrees as possible (even that no two are alike but rather each is definable) and that the structure has no automorphisms.

In this section, we will describe how far we have been able to travel along this road and what the possible future developments might be like. (See Slaman [1999] for a discussion of these matters in the setting of the general Turing degrees.)

The first hints of the true complexity of $\mathcal{R}$ emerged with the lattice embedding results. These ideas where enough to show that there are many types of degrees in the model theoretic sense. (An n-type over a theory (or a structure) is a set of formulas in n free variables which is consistent with the theory (of the structure).

The type is realized in the structure if there is an n-tuple of elements of which every formula in the type is true.) Thus $\mathcal{R}$ is not countably categorical. (A structure M is *countably categorical* if all countable models of its theory are isomorphic.)

THEOREM 3.1 (Lerman, Shore and Soare [1984]). *There are countably many 3-types realized in $\mathcal{R}$. More precisely, there are infinitely many distinct finite partial lattices (i.e. infimum is not always defined) each of is generated (using $\vee$ and $\wedge$ where defined) by three elements that can all be embedded into $\mathcal{R}$ (preserving $\vee$ and $\wedge$ when it is defined).*

COROLLARY 3.2. *$\mathcal{R}$ is not countably categorical.*

PROOF. Each such finite three generated partial lattice embeddable in $\mathcal{R}$ shows that a distinct three type is realizable in $\mathcal{R}$. Thus, by the Ryll–Nardjewski Theorem, $\mathcal{R}$ is not countably categorical. □

The techniques introduced in this paper were then used to refute Sacks' Conjecture on the isomorphism of the r.e. degrees to those r.e. in and above any degree **d**.

THEOREM 3.3 (Shore [1982]). *For each degree **d** and each set A which is Π^0_2 in D, there is a partial lattice $\mathcal{P}_A$ generated by four elements which is embeddable in the degrees r.e. in and above **d**. Moreover, if $\mathcal{P}_A$ is embeddable in an usl $\mathcal{P}$, then A is computable in the jump of any presentation of $\mathcal{P}$ (even as a partial order). (So, in particular, there are continuum many four types in the theory of $\mathcal{R}$.)*

COROLLARY 3.4. *If **d** is sufficiently complicated (e.g., $\not\leq \mathbf{0}^{(5)}$), then the degrees r.e. in and above **d** are not isomorphic to $\mathcal{R}$.*

PROOF. The ordering of Turing reducibility on indices of r.e. sets is definable in arithmetic by a Σ_3 formula. Thus the standard presentation of $\mathcal{R}$ (even as an usl) is recursive in $0^{(4)}$. Consider then $\mathcal{P}_D$ for any D of degree **d**. $\mathcal{P}_D$ is embeddable in the degrees r.e. in and above **d**, but not in $\mathcal{R}$. □

(Improvements of this result follow from recent work of Nies, Shore and Slaman [1996, 1998] that will be discussed below.)

COROLLARY 3.5. *$\mathcal{R}$ is not recursively presentable, i.e. it is not isomorphic to any recursive partial ordering.*

PROOF. Let C be a Π^0_2 set of degree $\mathbf{0}''$. By the theorem, $\mathcal{P}_C$ is embeddable in $\mathcal{R}$. If $\mathcal{R}$ were recursively presentable, C would be recursive in $\mathbf{0}'$ for a contradiction. □

These results were all proven by lattice embedding and so $\mathbf{0}''$ methods. The real breakthrough came when Harrington began to exploit Lachlan's $\mathbf{0}'''$ methods. A particularly difficult application of these ideas was the first proof of the undecidability of $\mathcal{R}$ by Harrington and Shelah [1982].

THEOREM 3.6 (Harrington and Shelah [1982]). *The theory of $\mathcal{R}$ is undecidable.*

The original proof of this theorem was never published, in part because it was so complicated. Since then, simplifications in this proof were made by Harrington and Slaman and a couple of distinctly simpler proofs have been found (Slaman and Woodin [2000] (see Nies, Shore and Slaman [1996, 1998]); Ambos-Spies and Shore [1993]) which avoid the need for more than $\mathbf{0}''$ type priority arguments.

All of these proofs of undecidability have the same basic outline. One first constructs a tractable (at least pairwise incomparable) set of degrees which is definable from parameters. The next step is to combine the construction with a procedure for defining a coding of enough partial orderings to get undecidability. To be slightly more precise, each proof first describes a formula $\Phi(x, a, b, c)$ in the language of partial orderings. One then proves that for "enough" partial orderings $\mathcal{P}$, there are r.e. degrees $\mathbf{a}, \mathbf{b}, \mathbf{c}$ and $\mathbf{d}$ such that $\langle\{\mathbf{x} \vee \mathbf{d} \mid \Phi(\mathbf{x}, \mathbf{a})\}, \leqslant\rangle \cong \mathcal{P}$:

- Harrington and Shelah [1982]: $\Phi(\mathbf{x}, \mathbf{a}, \mathbf{b}, \mathbf{c}) \equiv \mathbf{x}$ is maximal among the degrees below $\mathbf{a}$ such that $\mathbf{c} \nleqslant \mathbf{b} \vee \mathbf{x}$; "enough" partial orderings $\equiv$ all Δ_2^0 partial orderings.
- Slaman and Woodin [2000]: $\Phi(\mathbf{x}, \mathbf{a}, \mathbf{b}, \mathbf{c}) \equiv \mathbf{x}$ is minimal among the degrees below $\mathbf{a}$ such that $\mathbf{c} \leqslant \mathbf{b} \vee \mathbf{x}$; "enough" partial orderings $\equiv$ all Δ_2^0 partial orderings.
- Ambos-Spies and Shore [1993]: $\Phi(\mathbf{x}, \mathbf{a}) \equiv \mathbf{x}$ is maximal among the degrees above $\mathbf{a}$ such that there is a $\mathbf{y}$ such that $\mathbf{x} \wedge \mathbf{y} = \mathbf{a}$; "enough" partial orderings $\equiv$ all finite partial orderings.

Each of these results suffices to prove the undecidability of $\mathcal{R}$ because each allows us to interpret enough of the theory of partial orderings in $\mathcal{R}$ to exploit the recursively inseparability of that theory. (Recursive inseparability means that there is no recursive set separating the set of theorems of the theory of partial orderings from the set of sentences of the language of partial orderings that are false in some finite partial ordering. In fact, the first two codings only require the undecidability of the theory of partial orderings since by the standard Henkin proof of the completeness theorem any sentence that is false in some partial ordering is false in some Δ_2^0 one.) While the first approach requires a difficult application of the $\mathbf{0}'''$ priority technique, the others are less complicated. The original proof of the second uses Boolean combinations of finitely many $0''$ type procedures for each requirement. The third one only requires something like the minimal pair method (branching degree construction) and another type of requirement which is finitary in nature (nonbranching degree). (A degree $\mathbf{a}$ is *branching* if there are degrees $\mathbf{b}, \mathbf{c}$ such that $\mathbf{b} \wedge \mathbf{c} = \mathbf{a}$. Otherwise, it is *nonbranching*.) Recently, Nies, Shore and Slaman [1998] have produced a construction for the second coding scheme that is essentially a minimal pair type argument (albeit a complex one).

Each of these codings of partial orderings automatically supplies us with continuum many different four types (two types for the third coding scheme) in the theory of $\mathcal{R}$. Another early manifestation of the complexity of the r.e. degrees was the proliferation of distinct (one) types of degrees. Along with his construction of a minimal pair, Lachlan [1966a] produced a nonzero branching degree. He also constructed degrees which are nonbranching. Another example that we have already seen is being half of a minimal pair. On the other hand, Yates [1966a] proved, along with the existence of minimal pairs, that there are r.e. degrees which are *noncappable*, i.e. not halves of minimal pairs. Other examples abound. There are *cuppable* degrees which join (cup) to $\mathbf{0}'$ (by the Sacks Splitting Theorem, Theorem 2.4) and *noncuppable* ones which do not (Lachlan [1966a]). There are ones which split over every lower degree (any low_2 degree by Shore and Slaman [1990]) and ones which do not (by the Lachlan Nonsplitting Theorem, Theorem 2.15). There are ones over which $\mathbf{0}'$ splits (any low degree by Robinson [1971]) and ones over which it does not (Harrington; see also Jockusch and Shore [1983]). There are ones below (or above) which we can embed 1-3-1 (Lachlan [1972]) and ones below (or above) which we cannot (Downey [1990a], Weinstein [1988], Cholak, Downey and Shore [1998]). The list seemed endless. The proof that it is in fact endless is given by the following theorems:

THEOREM 3.7 (Ambos-Spies and Soare [1989]). *There are infinitely many* 1*-types realized in the r.e. degrees.*

THEOREM 3.8 (Ambos-Spies and Shore [1993]). *There are continuum many* 1*-types over the theory of the r.e. degrees.*

The ultimate version of such results would be have been to show that every r.e. degree realizes a distinct type or even that each one is definable in $\mathcal{R}$. As we shall see, this possibility is also connected to view that the structure of $\mathcal{R}$ is as complicated as possible. The ultimate expression of this new paradigm was the following conjecture:

CONJECTURE 3.9 (Biinterpretability Conjecture for $\mathcal{R}$) (Harrington; Slaman and Woodin [2000], Slaman [1991a]). *There is a definable coding of a standard model of arithmetic in $\mathcal{R}$ for which the relation between degrees* $\mathbf{d}$ *and the codes of sets of degree* $\mathbf{d}$ *is definable.* (*For $\mathcal{R}$ this is equivalent to there being a definable map taking each r.e. degree* $\mathbf{d}$ *to the code in the standard model for* (*the index of*) *a set of degree* $\mathbf{d}$.)

This conjecture really expresses the strongest form of the view that the structure the r.e. degrees is as complicated as possible. More than simply saying that the r.e. degrees are complicated, it provides a strong characterization of the structure of $\mathcal{R}$. It would give complete information, for example, about definability in $\mathcal{R}$ (every degree in $\mathcal{R}$ would be definable as would every relation on $\mathcal{R}$ which is definable in arithmetic) and automorphisms for $\mathcal{R}$ (none other than the identity would exist). (To see that these results are corollaries of the conjecture, just use the definable mapping from $\mathcal{R}$ to the standard model of arithmetic. As there are no automorphisms of $\mathcal{N}$, there would then be none of $\mathcal{R}$. As every natural number is definable in $\mathcal{N}$, every degree would be definable in $\mathcal{R}$, etc.)

(There is also an appropriate second order version of the biinterpretability conjecture for $\mathcal{D}$ which would follow from the rigidity of $\mathcal{D}$ (see Slaman [1999]). Surprisingly, the conjecture for $\mathcal{R}$ implies the one for $\mathcal{D}$. Indeed, even more is true.

THEOREM 3.10 (Slaman and Woodin [2000]). *If $\mathcal{R}$ is rigid then so is $\mathcal{D}$. In fact, there are finitely many r.e. degrees $\mathbf{a}_1, \ldots, \mathbf{a}_n$ such that if Φ is an automorphism of $\mathcal{R}$ with $\Phi(\mathbf{a}_i) = \mathbf{a}_i$ for each $i \leqslant n$, then Φ is the identity map.*

Cooper [1996] has recently announced that there are automorphisms of both $\mathcal{R}$ and $\mathcal{D}$ and so also that the biinterpretability conjectures for both structures fail. Nonetheless, some progress has been made in the direction indicated by the conjecture which we will now describe. As we have mentioned, the theory of the structure $\mathcal{R}$ has already been shown to be as complicated as possible, in particular, it is possible to correctly interpret true arithmetic in the theory of $\mathcal{R}$:

THEOREM 3.11 (Harrington and Slaman; Slaman and Woodin [2000], Nies, Shore and Slaman [1998]). *There are recursive translations $S(T)$ taking sentences $\phi(\psi)$ of arithmetic (partial orderings) to sentences ϕ^S, ψ^T of partial orderings (arithmetic) such that $\mathcal{N} \models \phi \leftrightarrow \mathcal{R} \models \phi^S$ $(\mathcal{R} \models \psi \leftrightarrow \mathcal{N} \models \psi^T)$.*

Each of the various proofs of this theorem begins with one of the codings of partial orderings in $\mathcal{R}$ developed to prove its undecidability. They each provide a translation of the theory of partial orderings into $\mathcal{R}$. As the theory of partial orderings is rich enough to code all of predicate logic, we can view the codings as providing us with models of some finite axiomatization of arithmetic. The real problem, now, is to definably determine the (translations of) sentences true in the models which are standard models of arithmetic (i.e. isomorphic copies of $\mathcal{N}$). The most natural approach to this problem would seem to be to define a standard model or at least a class of models all of which are standard. One would then simply say that a sentence of arithmetic is true (in $\mathcal{N}$) iff the appropriate translation is true in (any of) the definable standard model(s). The first couple of proofs of this theorem (Harrington and Slaman; Slaman and Woodin [2000]) did not manage to define standard models and took much more indirect approaches to the theorem. We describe the most recent approach to this problem due to Nies, Shore and Slaman [1996, 1998] which not only defines a class of standard models in $\mathcal{R}$ but also (by an identification under a definable equivalence relation) a single definable standard model. It then goes on to establish a weak version of the biinterpretability conjecture that is, however, strong enough to prove many definability results for $\mathcal{R}$.

We begin with Slaman and Woodin's coding and a couple of technical additions:

THEOREM 3.12 (Slaman and Woodin [2000]). *Given any recursive partial ordering $\mathcal{P} = \langle \omega, \preceq \rangle$ there are r.e. degrees $\mathbf{p}, \mathbf{q}, \mathbf{r}, \mathbf{l}$ and $\mathbf{g}_i$ (for $i \in \omega$) such that:*

(1) *the $\mathbf{g}_i$ are the minimal degrees $\mathbf{x} \leqslant \mathbf{r}$ such that $\mathbf{q} \leqslant \mathbf{x} \vee \mathbf{p}$;*

(2) *for $i, j \in \omega$, $i \preceq j$ if and only if $\mathbf{g}_i \leqslant_T \mathbf{g}_j \oplus \mathbf{l}$;*

(3) *$\mathbf{r} \oplus \mathbf{p} \oplus \mathbf{q}$ is low, i.e., $(\mathbf{r} \oplus \mathbf{p} \oplus \mathbf{q})' = \mathbf{0}'$;*

(4) *if $\mathbf{a} > \mathbf{0}$ is any given r.e. degree, we can also make $\mathbf{r} < \mathbf{a}$.*

((3) and (4) are technical improvements that are needed later.)

We may as well think of the partial ordering $\mathcal{P}$ as coding a model of (a finitely axiomatized version of) arithmetic. The key to definably selecting a set of such models that are all standard is the ability to define comparison maps between (finite) initial segments of certain such models. The idea here is that the standard models are the models $\mathcal{M}$ such that each initial segment of $\mathcal{M}$ can be mapped into an initial segment of every model. The crucial technical lemma needed to define such maps is one that combines Slaman and Woodin coding with cone avoiding and permitting:

THEOREM 3.13 (Nies, Shore and Slaman [1998]). *Given any recursive partial ordering $\mathcal{P} = \langle \omega, \preceq \rangle$ and low r.e. degrees $\mathbf{q}_0, \ldots, \mathbf{q}_m$ there are r.e. degrees $\mathbf{p}, \mathbf{q}, \mathbf{r}, \mathbf{l}$ and $\mathbf{g}_i$ (for $i \in \omega$) as in Theorem 3.12 such that if $\mathbf{g}_{f(i)}$ is the degree corresponding to the natural number i in the model coded by $\mathcal{P}$, then $\mathbf{g}_{f(i)} \leqslant_T \mathbf{q}_i$ and $\mathbf{q}_i \nleqslant_T \mathbf{q}_j \Rightarrow \mathbf{g}_{f(i)} \nleqslant_T \mathbf{q}_j$.*

Given two coded low models $\mathcal{M}_1, \mathcal{M}_2$ one can use this theorem to interpolate a third model $\mathcal{M}$ so that one can define isomorphisms between the first n numbers of $\mathcal{M}_1$ and those of $\mathcal{M}$ and between the first n numbers of $\mathcal{M}_2$ and the second n numbers of $\mathcal{M}$. Together with the structure inherent in $\mathcal{M}$, these maps define the desired isomorphism between the first n elements of $\mathcal{M}_1$ and those of $\mathcal{M}_2$.

In this way, one can give a sufficient condition for a model to be standard and a definable scheme for maps between initial segments of such models. Thus Nies, Shore and Slaman are able to define a class of models which are all standard and such that there are definable isomorphisms between the natural numbers of any two models in the class:

A model $\mathcal{M}$ of arithmetic coded (as in Theorem 3.12) by parameters $\mathbf{p}, \mathbf{q}, \mathbf{r}, \mathbf{l}$ is *good via* $\mathbf{c}$ if $\mathbf{r} < \mathbf{c}$ and for every $\mathcal{M}'$ with its elements below $\mathbf{c}$ the comparison maps described above define maps for each initial segment of $\mathcal{M}$ to initial segments of $\mathcal{M}'$. $\mathcal{M}$ is *good* if it is good via some $\mathbf{c}$.

Now, every low standard model is good by Theorem 3.13. On the other hand, every good model is standard (as it can be mapped into some standard model). Moreover, given any two good models we can define an isomorphism between them. Thus we can define an equivalence relation on the (codes for) natural numbers in these models and interpretations of the language of arithmetic on these equivalence classes that make the structure so defined a standard model of arithmetic.

THEOREM 3.14 (Nies, Shore and Slaman [1998]). *There is a coding scheme interpreting arithmetic in $\mathcal{R}$ such that all the models so defined are standard. Moreover, there is a definable equivalence relation on the parameters coding these models and the degrees coding the natural numbers in these models such that the coding scheme defines a standard model $\mathcal{N}_0$ of arithmetic on the equivalence classes.*

We now have the definable copy $\mathcal{N}_0$ of $\mathcal{N}$ in $\mathcal{R}$ required by the biinterpretability conjecture. We next want to come as close as we can to associating each degree with an index for a set of that degree. The idea is to characterize, to the extent possible, a degree $\mathbf{a}$ by the isomorphism type of $\mathcal{R}(\leqslant \mathbf{a})$ (the ordering of r.e. degrees below $\mathbf{a}$) relative to certain other parameters.

The first ingredient is a coding scheme for a copy of $\mathcal{N}$ which is Σ_3 in the sense that the (codes for) the natural numbers can be enumerated recursively in $\mathbf{0}''$.

THEOREM 3.15 (Nies, Shore and Slaman [1998]). *Given any* $\mathbf{a} > \mathbf{0}$ *and any promptly simple* $\mathbf{u}$, *there are degrees* $\mathbf{b}, \mathbf{e}_0, \mathbf{e}_1, \mathbf{f}_0, \mathbf{f}_1, \mathbf{p}, \mathbf{q}, \mathbf{r}$ *and u.r.e.* $\mathbf{g}_i$ *(for* $i \in \omega$*) with* $\mathbf{p}, \mathbf{q} < \mathbf{u}$ *and all the other degrees below both* $\mathbf{a}$ *and* $\mathbf{u}$ *satisfying the following conditions*:

- *the minimal degrees* $\mathbf{x}$, $\mathbf{b} < \mathbf{x} < \mathbf{r}$ *such that* $\mathbf{q} \leq \mathbf{x} \vee \mathbf{p}$ *together with the partial ordering on them defined by* $\mathbf{x} \preceq \mathbf{y} \Leftrightarrow \mathbf{x} \leqslant \mathbf{y} \oplus \mathbf{l}$ *define a standard model of arithmetic as described above with the* $\mathbf{g}_i$ *as the elements* i;
- *for each* $i \in \omega$, $(\mathbf{g}_{2i} \vee \mathbf{e}_1) \wedge \mathbf{f}_1 = \mathbf{g}_{2i+1}$ *and* $(\mathbf{g}_{2i+1} \vee \mathbf{e}_0) \wedge \mathbf{f}_0 = \mathbf{g}_{2i+2}$.

Given the properties specified in this theorem, each $\mathbf{g}_i$ can be defined by an existential formula using the first eight degrees and $\mathbf{g}_0$ as parameters. For example, $\mathbf{g}_1 = (\mathbf{g}_0 \vee \mathbf{e}_1) \wedge \mathbf{f}_1$ and so $\mathbf{g}_1$ is the only degree $\mathbf{x}$ such that $\phi_1(\mathbf{x})$ holds where $\phi_1(\mathbf{x})$ says $\mathbf{x} \leqslant \mathbf{g}_0 \vee \mathbf{e}_1 \;\&\; \mathbf{x} \leqslant \mathbf{f}_1 \;\&\; \mathbf{q} \leqslant \mathbf{x} \vee \mathbf{p}$. Next, $\mathbf{g}_2$ is the only degree $\mathbf{y}$ such that $\exists \mathbf{x}(\phi_1(\mathbf{x}) \;\&\; \mathbf{y} \leqslant \mathbf{x} \vee \mathbf{e}_0 \;\&\; \mathbf{y} \leqslant \mathbf{f}_0 \;\&\; \mathbf{q} \leqslant \mathbf{y} \vee \mathbf{p})$. Similarly, we can define each $\mathbf{g}_i$ by such a formula. As the ordering of Turing reducibility relative to any set B is Σ_3^B and join is recursive on indices, we can make this generating procedure recursive in $\mathbf{0}''$ by choosing $\mathbf{u}$ to be low.

The next ingredient in the desired coding is a procedure that shows that every Σ_3^A set can be coded into such a set of degrees $\mathbf{g}_i$ in a positive way using $\leqslant$ and $\vee$ by degrees below $\mathbf{a}$. As the ordering on degrees below $\mathbf{a}$ is Σ_3^A (and join is recursive on indices) this would make the set coded Σ_3^A as well (and nothing better is possible).

THEOREM 3.16 (Nies, Shore and Slaman [1998]). *If* $\langle \mathbf{g}_i \rangle$ *is a uniformly r.e. antichain in* $\mathcal{R}$, $\oplus \mathbf{g}_i$ *is low and* $\mathbf{a} = \deg(A) \nleqslant_T \mathbf{g}_i$ *for each* i, *then, for each* Σ_3^A *set* S, *there are* $\mathbf{c}, \mathbf{d} \leqslant \mathbf{a}$ *such that* $S = \{i \mid \mathbf{c} \leqslant_T \mathbf{g}_i \vee \mathbf{d}\}$.

Together, these results show that precisely the Σ_3^A sets can be coded in this way. As this class of sets determines $\mathbf{a}''$, the isomorphism type of $\mathbf{a}$ in $\mathcal{R}$ determines $\mathbf{a}''$. As the coding scheme is amenable to the comparisons described above between the models of arithmetic, one can translate these codings into ones in the definable standard model and so convert this characterization of $\mathbf{a}''$ to a formula defining, from the degree $\mathbf{a}$, a (code for a) set of degree $\mathbf{a}''$ in the standard model of Theorem 3.14. The following results are then all among the corollaries of this work given in Nies, Shore and Slaman [1998].

THEOREM 3.17. *There is a definable map* $f: \mathcal{R} \to \mathcal{N}_0$ *such that, for every* $\mathbf{a}$, $f(\mathbf{a})$ *is (the code for) an index of an r.e. set* W *for which* $W'' \in \mathbf{a}''$.

COROLLARY 3.18. *The double jump is invariant in* $\mathcal{R}$, *i.e. if* φ *is an automorphism of* $\mathcal{R}$ *then* $\varphi(\mathbf{x})'' = \mathbf{x}''$ *for every* $\mathbf{x} \in \mathcal{R}$.

COROLLARY 3.19. *Any relation on $\mathcal{R}$ which is definable in arithmetic and invariant under the double jump is definable in $\mathcal{R}$.*

COROLLARY 3.20. *For each* **c** *r.e. in and above* $\mathbf{0}''$ *the set of r.e. degrees* **a** *with double jump* **c** *is definable in $\mathcal{R}$.*

COROLLARY 3.21. *The jump classes* $L_{n+1} = \{\mathbf{a} \mid \mathbf{a}^{(n+1)} = \mathbf{0}^{(n+1)}\}$ *and* $H_{n+1} = \{\mathbf{a} \mid \mathbf{a}^{(n)} = \mathbf{0}^{(n+1)}\}$ *are definable in $\mathcal{R}$ for* $n \geqslant 1$.

COROLLARY 3.22. *The jump class* $H_1 = \{\mathbf{a} \mid \mathbf{a}' = \mathbf{0}''\}$ *is definable in $\mathcal{R}$.*

PROOF. It follows from the Robinson Jump Interpolation Theorem (Robinson [1971] that $\mathbf{a}' = \mathbf{0}''$ if and only if for every **c** r.e. in and above $\mathbf{0}''$ there is a $\mathbf{b} < \mathbf{a}$ with $\mathbf{b}'' = \mathbf{c}$. Thus the definability of H_1 follows from Theorem 3.17. □

There are various *a priori* plausible schemes for defining even more in $\mathcal{R}$ and of proving results somewhat weaker than the full biinterpretability conjecture. Of particular interest is the question of biinterpretability with parameters. Here the definitions required in the conjecture are allowed to use finitely many individual r.e. degrees as parameters. Of course, this would give the definability results implied by the biinterpretability conjecture using parameters. It would also imply that there are at most countably many automorphisms of $\mathcal{R}$ as the finite set of parameters used in the definitions of the conjecture would constitute an *automorphism base* for $\mathcal{R}$, i.e. any automorphism would be completely determined by its action on this set. Of course, the discovery of any finite automorphism base for $\mathcal{R}$ would also imply that it has at most countably many automorphisms.

Now any set A which generates $\mathcal{R}$ under joins and meets is obviously an automorphism base for $\mathcal{R}$ and there are many known generating sets. Indeed, by the Sacks Splitting Theorem (Theorem 2.4), any dense set of degrees, such as the branching (i.e. the infima of pairs of incomparable degrees) (Slaman [1991b]) and nonbranching (Fejer [1983]) degrees, generates $\mathcal{R}$ and so is an automorphism base for it. More information on generating sets for $\mathcal{R}$ can be found in Ambos-Spies [1985]. There are however, many known automorphism bases for $\mathcal{R}$ that do not generate it. For example, each jump class H_n and L_n as defined above (in Corollary 3.21) is an automorphism base for $\mathcal{R}$ as is the class **PS** of promptly simple degrees (Ambos-Spies [1993]). Perhaps the most remarkable such result (Ambos-Spies [1993]) is that every downward cone in $\mathcal{R}$, i.e. the degrees below **a** for any $\mathbf{a} > \mathbf{0}$, is an automorphism base for $\mathcal{R}$. If this result could be extended to include any upper cone with a promptly simple base (the degrees above some promptly simple $\mathbf{a} < \mathbf{0}'$), then the following theorem would imply that there is a finite automorphism base for $\mathcal{R}$ and so there would be at most countably many automorphisms of $\mathcal{R}$.

THEOREM 3.23 (Shore and Slaman). *Given any promptly simple degree* **d**, *we can define (from parameters) a standard model of arithmetic and a one-one onto map from the integers of this model to the degrees above* **d**.

However, even before all of the above results on definability were proven, Cooper announced that he could construct an automorphism of $\mathcal{R}$ and indeed one that moves a low degree to a nonlow one so that the class of low degrees is not definable in $\mathcal{R}$ (Cooper [1996]). Clearly the existence of such automorphisms implies that the above definability results for the jump classes are the best possible. Given the construction of such an automorphism, one might be tempted to prove that no individual degree is definable in $\mathcal{R}$ by constructing the appropriate automorphisms. In any case, before moving on to the issue of "natural" definability (e.g., without using codings of arithmetic), we close this section with three conjectures (listed in order of increasing strength) which are all still weaker than the full biinterpretability conjecture and do not contradict the existence of a nontrivial automorphism of $\mathcal{R}$.

CONJECTURE 3.24. *There are only countably many automorphisms of $\mathcal{R}$.*

CONJECTURE 3.25. *There is a finite automorphism base for $\mathcal{R}$.*

CONJECTURE 3.26. *With parameters, $\mathcal{R}$ is biinterpretable with the standard model of arithmetic.*

The last conjecture would also imply that $\mathcal{R}$ is the prime model of its theory and that every relation on $\mathcal{R}$ is definable if and only if it is definable in arithmetic and invariant under automorphisms.

4. Natural definability

Whatever the ultimate achievements of such investigations in terms of definability in $\mathcal{R}$, the definitions produced will not be "natural" ones in terms of the structure of $\mathcal{R}$. Even if $\mathcal{R}$ were biinterpretable with arithmetic (perhaps with parameters) and every r.e. degree and arithmetic relation on the r.e. degrees is then definable (from finitely many parameters), the definitions produced in this way would all proceed by coding everything in models of arithmetic and then using arithmetic itself to produce the definitions. It would give no insight into the relations between the classes and their role in the structure of $\mathcal{R}$. This investigation is the provenance of another area of long term interest in the study of the r.e. degrees: the relationships between order theoretic properties of degrees in $\mathcal{R}$ and external properties of other sorts. Of interest have been set theoretic properties described in terms of the lattice of r.e. sets; dynamic properties of the enumerations of the r.e. sets; rates of growth of functions recursive in the various r.e. degrees; and relations with definability considerations in arithmetic as expressed by the jump operator. The last of these types of properties is often the one that connects all the others. For the r.e. degrees the relations with the jump operator are generally described in terms of the jump classes mentioned in the definability results of the last section:

DEFINITION 4.1. An r.e. degree $\mathbf{a}$ is $high_n$ iff $\mathbf{a}^{(n)} = \mathbf{0}^{(n+1)}$ (its n-th jump is as high as possible). The degree $\mathbf{a}$ is low_n if $\mathbf{a}^{(n)} = \mathbf{0}^{(n)}$ (its n-th jump is as low as possible). If $n = 1$, we usually omit the subscript.

There have been many important result connecting the jump classes with the lattice theoretic structure of the r.e. sets; with approximation procedures for functions recursive in different jumps of the given set; and with the growth rates of functions recursive in the sets themselves for several of these jump classes. For example, an r.e. degree $\mathbf{a}$ is high iff it contains a maximal set in $\mathcal{E}^*$ (the lattice of r.e. sets modulo the ideal of finite sets) iff there is a function f of degree $\mathbf{a}$ which dominates every recursive function iff every function h recursive in $\mathbf{0}''$ is approximable by one $g \leqslant \mathbf{a}$ in the sense that $h(x) = \lim_s g(x, s)$. Another class with a similar list of equivalent definitions consists of the low_2 r.e. degrees: $\mathbf{a}$ is low_2 iff every r.e. set of degree $\mathbf{a}$ has a maximal superset iff there is a function f recursive in $\mathbf{0}'$ which dominates every $g \leqslant \mathbf{a}$ iff every function $h \leqslant \mathbf{a}''$ is approximable by a recursive one g in the sense that $h(x) = \lim_s \lim_t g(x, s, t)$. (Proofs of these facts about high and low_2 r.e. sets and other similar ones can be found in Soare [1987, XI].) These interrelations have played an important role in the study of both the lattice of r.e. sets (Soare [1999]) and the degrees below $\mathbf{0}'$ (Cooper [1999]).

These classes and others have also played an important role in the study of $\mathcal{R}$. We want to mention some of the typical applications of a set being in each of several possible jump classes. The two earliest classes that were seen to have degree theoretic implications were the low degrees and the high ones. The intuition underlying most of these results is that the low degrees resemble the recursive one, $\mathbf{0}$, and the high degrees resemble the complete one, $\mathbf{0}'$. Thus the degrees above a low degree or below a high one should resemble the r.e. degrees as a whole.

The archetypic applications of lowness are in Robinson [1971]. One example was motivated by the attempt to combine the two most prominent early indications that the r.e. degrees are homogeneous—the Splitting and Density Theorems of Sacks (Theorems 2.3 and 2.4). Robinson proved that it is possible to combine these theorems if the bottom degree is low (and so like $\mathbf{0}$).

THEOREM 4.2 (Robinson's Splitting Theorem) (Robinson [1971]). *For every pair of r.e. degrees* $\mathbf{a} < \mathbf{b}$ *with* $\mathbf{a}$ *low, there are r.e. degrees* $\mathbf{b}_0$ *and* $\mathbf{b}_1$ *such that* $\mathbf{a} < \mathbf{b}_0, \mathbf{b}_1 < \mathbf{b}$ *and* $\mathbf{b}_0 \vee \mathbf{b}_1 = \mathbf{b}$.

Of course, we now know by Lachlan's Nonsplitting Theorem (Theorem 2.15) that it is not in general possible to combine these two results. Lowness is used to eliminate the potentially infinitary nature of the preservation requirements in the splitting theorem caused by changes in $\mathbf{a}$ destroying computations by recursively approximating answers to questions about $\mathbf{a}'$ ($\mathbf{a}$ is low iff every function h recursive in $\mathbf{a}'$ is approximable by a recursive g in the sense that $h(x) = \lim_s g(x, s)$ (see Soare [1987, III.3])). By similar arguments, Robinson [1971] also showed, for example, that every countable usl is embeddable in every interval $\mathbf{a} < \mathbf{b}$ of r.e. degrees where

a is low. Since then, many other results have shown that the low r.e. sets are much like the recursive ones. Some of these results have also been extended to the low_2 r.e. degrees. Indeed, Shore and Slaman [1990] even prove an extension of embedding theorem for the low_2 r.e. degrees that essentially says that any extension not ruled out by embeddings of finite distributive lattices can be carried out. Even the density and splitting theorems can be combined when the top degree is low_2 (Harrington and also Bickford and Mills; see Shore and Slaman [1990]) but, in contrast to the situation for low degrees, they cannot be combined just under the assumption that the bottom degree is low_2.

At the other end of the scale, there are results showing that various phenomena in the r.e. degrees occur below every high degree. Here the archetypic example is Cooper's theorem:

THEOREM 4.3 (Cooper [1974]). *If* **h** *is a high r.e. degree then there is a minimal pair* **a**, **b** *below* **h**.

A couple of other such results follow.

THEOREM 4.4 (Harrington; see Miller [1981]). *Every high degree* **h** *has the anticupping property, i.e., there is an r.e.* $\mathbf{b} < \mathbf{h}$ *such that for no r.e.* $\mathbf{c} < \mathbf{h}$ *does* $\mathbf{h} = \mathbf{b} \vee \mathbf{c}$.

THEOREM 4.5 (Shore and Slaman [1993]). *For every high r.e. degree* **h**, *there are r.e. degrees* $\mathbf{a}, \mathbf{b}, \mathbf{c} < \mathbf{h}$ *such that for every r.e.* $\mathbf{w}$, $\mathbf{0} < \mathbf{w} < \mathbf{a}$, $\mathbf{b} \vee \mathbf{w} \geqslant \mathbf{c}$.

The last result (together with the extension of embedding results below a low_2 mentioned above) definably separated the high r.e. degrees from the low_2 ones as the above results say that no such triple can exits if **h** is low_2. The early approach to exploiting highness as in Cooper [1974], was via the characterization by the existence of a function f recursive in the high degree **h** which dominates every recursive function. This function was then used to control a permitting procedure in which numbers are allowed into the set A being constructed only when the value of our approximation to f determined by its computation from an r.e. set $H \in \mathbf{h}$ changes. This guarantees that A is recursive in **h**. The domination properties of f were then used to show that the permitting was compatible with satisfying a positive requirement that wants to put all but finitely many of an infinite recursive set of numbers into A. Another view is presented in Shore and Slaman [1993]. There the idea is that **h** is high if and only if every function recursive in $\mathbf{0}''$ can be approximated recursively in **h**. This characterization is used to, recursively in **h**, approximate the outcome of infinitary (Π_2) requirements in a priority tree construction in such a way that **h** can recover enough of the construction (by enumerating when nodes will never be accessible again) to calculate the set being constructed.

Since this work separating high and low_2 (and before the definition of these classes given in Corollaries 3.22 and 3.21 by coding techniques) there were further developments of the techniques for working with low_2 r.e. degrees. In addition, new techniques have been introduced to exploit the hypothesis that an r.e. degree is nonlow_2.

The combination of these procedures has produced a natural order theoretic definition of the low$_2$ degrees in $\mathcal{R}_{tt}$, the structure of the r.e. truth table degrees (Downey and Shore [1995]): In $\mathcal{R}_{tt}$, the low$_2$ degrees are precisely those with minimal covers, i.e. the degrees $\mathbf{a}$ such that there is a $\mathbf{b} > \mathbf{a}$ with no degree $\mathbf{c}$ between $\mathbf{a}$ and $\mathbf{b}$. Another result whose proof introduces extensions of these techniques that we hope will eventually lead to a natural definition of this jump class in $\mathcal{R}$ is the following:

THEOREM 4.6 (Downey and Shore [1996]). *If* $\mathbf{a}$ *is r.e. and nonlow*$_2$*, then the lattice* 1-3-1 *can be embedded in the r.e. degrees below* $\mathbf{a}$.

This should be contrasted with the earlier result of Downey [1990a] and Weinstein [1988] that there are r.e. degrees $\mathbf{a}$ below which 1-3-1 cannot be embedded. Indeed, Cholak and Downey [1993] prove that for any r.e. $\mathbf{a} < \mathbf{b}$ there are r.e. $\mathbf{c}, \mathbf{d}$ such that $\mathbf{a} < \mathbf{c} < \mathbf{d} < \mathbf{b}$ such that 1-3-1 cannot be embedded in $[\mathbf{c}, \mathbf{d}]$. Although, a more recent result of Cholak, Downey and Shore [1998] shows that these ideas do not seem to lead to a definition of the low$_2$ degrees in $\mathcal{R}$ in terms of embeddings of 1-3-1, we hope that continued development of these techniques will lead to a positive solution of the following conjecture.

CONJECTURE 4.7. *The class of low*$_2$ *degrees is naturally definable in* $\mathcal{R}$.

The only externally defined classes of r.e. degrees that have actually been defined without the use of codings of arithmetic in $\mathcal{R}$ are those of the promptly simple and contiguous r.e. degrees. The former coincide with the noncappable (and low cuppable) ones (Theorem 2.18). Ambos-Spies et al. [1984] also showed that this class coincides with **ENC**, the effectively noncappable r.e. degrees (there is a recursive function on indices witnessing the noncappability) and two classes connected to the lattice of r.e. sets one of which, $\mathbf{SP\bar{H}}$, is definable in $\mathcal{E}^*$. Thus the class is definable in both $\mathcal{R}$ and $\mathcal{E}^*$. Not only is this collection of equivalences a remarkable instance of both uniformity and definability in the r.e. degrees, but it has also played a key role in other results including ones phrased solely in terms of order theoretical properties of $\mathcal{R}$. The first, of course, were the results mentioned in Theorem 2.18 that **NC** and its complement **M** (the class of r.e. degrees which are *halves of minimal pairs*) partition $\mathcal{R}$ into a pair of sets consisting of a complementary ideal and filter. As another example, we point out that it also played a crucial role in Slaman and Woodin's proof that the theory of $\mathcal{R}$ is recursively isomorphic to that of true arithmetic. The idea was that prompt simplicity allowed them to construct codes of models of arithmetic with particular properties (the codes for the natural numbers were taken to be a set of pairwise minimal degrees) that enabled them to guarantee the standardness of one of the models so constructed. The equivalence of **PS** and **NC** then made this a definable procedure in $\mathcal{R}$.

The contiguous degrees (those consisting of a single r.e. wtt-degree) have also played a role in many results about $\mathcal{R}$ including Theorem 3.7. Downey and Lempp [1997] have shown that they are the locally distributive degrees in $\mathcal{R}$, i.e. the r.e.

degrees **a** such that in $\mathcal{R}$

$$\forall \mathbf{a}_1, \mathbf{a}_2, \mathbf{b}(\mathbf{a}_1 \vee \mathbf{a}_2 = \mathbf{a} \;\&\; \mathbf{b} \leqslant \mathbf{a} \rightarrow \\ \exists \mathbf{b}_1, \mathbf{b}_2(\mathbf{b}_1 \vee \mathbf{b}_2 = \mathbf{b} \;\&\; \mathbf{b}_1 \leqslant \mathbf{a} \;\&\; \mathbf{b}_2 \leqslant \mathbf{a})).$$

Ambos-Spies and Fejer [1999] have strengthened this result by showing that the contiguous degrees are the ones which are not the tops of embeddings of N_5.

We now conclude where we began with Post's problem of an intermediate r.e. degree. The motivation for the question was that all naturally occurring r.e. sets were either recursive or complete. By now the problem has been solved over and over in the sense that we now are well aware of the richness of the structure of the r.e. degrees. Nonetheless, the question of whether there is a natural or definable (in some nice way) intermediate degree remains open. Of course, a negative solution requires a precise definition of "natural" or "nice". One suggestion (Steel [1982]) is that a natural degree should be definable and its definition should relativize to an arbitrary degree (and so, in particular, be defined on degrees independently of the choice of representative). Along these lines an old question of Sacks' [1963a] asks whether there is a degree invariant solution to Post's problem, i.e. a degree invariant computably enumerable operator W such that $A <_T W(A) <_T A'$. (A function $W: 2^\omega \to 2^\omega$ is a *computably enumerable operator* if there is an $e \in \omega$ such that, for each set A, $W(A) = W_e^A$, the e-th set computably enumerable in A. Any function $f: 2^\omega \to 2^\omega$ is *degree invariant* if, for every A and B, $A \equiv_T B$ implies that $f(A) \equiv_T f(B)$.) Such an operator would clearly be a candidate for a natural solution to Post's problem in this sense.

Lachlan [1975b] proved that if we require the degree invariance to be *uniform* in the sense that there is a function h that takes the (pairs of) indices of reductions between A and B to (pairs of) indices of reductions between $W(A)$ and $W(B)$ then there is no such operator. In fact, Lachlan proved that if **d** is the degree of the function h providing the uniformity then either $W(\mathbf{d}) = \mathbf{d}'$ or $W(W(\mathbf{d})) \leqslant W(\mathbf{d})$. In this way, Lachlan also characterized the jump operator as the only uniformly degree invariant computably enumerable operator which is always strictly greater than the identity *on a cone* (of degrees), i.e. on all $\mathbf{a} \geqslant_T \mathbf{c}$ for some degree **c**. Sacks' problem was greatly generalized by Martin (see Kechris and Moschovakis [1978, p. 279]) in the setting of the axiom of determinacy and Lachlan's results were generalized and extended to this context by Steel [1982], Slaman and Steel [1988] and Becker [1988]. A recent result by Downey and Shore [1997] refines this work in the context of the r.e. degrees in a way that puts severe restrictions on the types of degrees that could be defined in this way or, indeed, by any definition in the structure $\mathcal{R}$.

THEOREM 4.8 (Downey and Shore [1997]). *If W is a degree invariant computably enumerable operator such that $A \leqslant_T W(A)$ for all A Turing above some fixed degree (i.e. in a cone), then either $W(A)'' \equiv_T A''$ for all A in some cone or $W(A)'' \equiv_T A'''$ in some cone.*

THEOREM 4.9 (Downey and Shore [1997]). *For any formula* $\psi(x)$ *in the language of partial orderings which, for every* $\mathbf{z}$, *defines a degree* $\mathbf{x}$ (*other than* $\mathbf{z}$ *or* $\mathbf{z}'$) *in* $\mathcal{R}^{\mathbf{z}}$ (*the degrees r.e. in and above* $\mathbf{z}$) *must define a* low_2 *or* $high_2$ *degree* (*relative to* $\mathbf{z}$, *i.e.* $\mathbf{x}'' = \mathbf{z}''$ *or* $\mathbf{x}'' = \mathbf{z}'''$) *for all* $\mathbf{z}$ *above some fixed degree* $\mathbf{c}$.

Now this last result does not rule out definitions in $\mathcal{R}$ that make direct reference to specific arithmetic sets in some coded model of arithmetic and so would apply only to $\mathcal{R}$ or at best to $\mathcal{R}^{\mathbf{a}}$ for arithmetic $\mathbf{a}$. For example, the results in Corollary 3.20 on the definability of each class of r.e. degrees with double jump a particular $\mathbf{c}$ r.e. in and above $\mathbf{0}''$ are of this sort. On the other hand, the definitions for the jump classes L_{n+1} and H_n for $n \geqslant 1$ of Corollaries 3.21 and 3.22 can be stated so as to relativize properly to $\mathcal{R}^{\mathbf{a}}$ for any degree $\mathbf{a}$. However, the theorem perhaps does suggest that, whatever happens with biinterpretability and automorphisms, there may be no degree invariant solution to Post's problem and no "naturally" definable intermediate degree in $\mathcal{R}$.

References

K. AMBOS-SPIES

[1980] On the structure of the recursively enumerable degrees, Ph.D. Thesis, University of Munich.

[1985] Generators of the recursively enumerable degrees, in: *Recursion Theory Week, Proceedings, Oberwolfach 1984*, H.-D. Ebbinghaus, G. H. Müller and G. E. Sacks, eds., Lecture Notes in Mathematics, Vol. 1141, Springer, Berlin, pp. 1–28.

[1993] *Automorphism bases for the recursively enumerable degrees*, Preprint.

K. AMBOS-SPIES AND P. A. FEJER

[1999] Embeddings of N_5 and the contiguous degrees, in preparation.

K. AMBOS-SPIES, P. A. FEJER, S. LEMPP AND M. LERMAN

[1996] Decidability of the two-quantifier theory of the recursively enumerable weak truth-table degrees and other distributive upper semi-lattices, *J. Symbolic Logic*, 61, pp. 880–905.

K. AMBOS-SPIES, C. G. JOCKUSCH, JR., R. A. SHORE AND R. I. SOARE

[1984] *Trans. Amer. Math. Soc.*, 281, pp. 109–128.

K. AMBOS-SPIES, S. LEMPP AND M. LERMAN

[1994a] Lattice embeddings into the r.e. degrees preserving 1, in: *Logic and Philosophy of Science: Papers from the 9th ICLMPS*, D. Prawitz and D. Westerahl, eds., Kluwer Academic Publishers, Dordrecht, pp. 179–198.

[1994b] Lattice embeddings into the r.e. degrees preserving 0 and 1, *J. London Math. Soc. (2)*, 49, pp. 1–15.

K. AMBOS-SPIES AND M. LERMAN

[1986] Lattice embeddings into the recursively enumerable degrees, *J. Symbolic Logic*, 51, pp. 257–272.

[1989] Lattice embeddings into the recursively enumerable degrees II, *J. Symbolic Logic*, 54, pp. 735–760.

K. AMBOS-SPIES AND R. A. SHORE

[1993] Undecidability and 1-types in the r.e. degrees, *Ann. Pure Appl. Logic*, 63, pp. 3–37.

K. AMBOS-SPIES AND R. I. SOARE
[1989] The recursively enumerable degrees have infinitely many one-types, *Ann. Pure Appl. Logic*, 44, pp. 1–23.

H. A. BECKER
[1988] Characterization of jump operators, *J. Symbolic Logic*, 53, pp. 708–728.

P. CHOLAK AND R. G. DOWNEY
[1993] Lattice nonembeddings and intervals of the recursively enumerable degrees, *Ann. Pure Appl. Logic*, 61, pp. 195–221.

P. CHOLAK, R. G. DOWNEY AND R. A. SHORE
[1998] Intervals without critical triples, in: *Logic Colloquium '95*, J. A. Makowsky and E. V. Ravve, eds., Lecture Notes in Logic, vol. 11, Springer, Heidelberg, pp. 17–43.

S. B. COOPER
[1974] Minimal pairs and high recursively enumerable degrees, *J. Symbolic Logic*, 39, pp. 655–660.
[1990] The jump is definable in the structure of the degrees of unsolvability (research announcement), *Bull. Amer. Math. Soc.*, 23, pp. 151–158.
[1996] *Beyond Gödel's theorem: the failure to capture information content*, University of Leeds, Department of Pure Mathematics, Preprint Series, No. 4.
[1999] The degrees below $\mathbf{0}'$, this *Handbook*.

R. G. DOWNEY
[1990a] Lattice nonembeddings and initial segments of the recursively enumerable degrees, *Ann. Pure Appl. Logic*, 49, pp. 97–119.
[1990b] Notes on the $\mathbf{0}'''$ priority method with special attention to density results, in: *Recursion Theory Week*, K. Ambos-Spies, G. H. Müller, eds., Lecture Notes in Mathematics, Vol. 1432, Springer, Berlin, pp. 111–140.

R. G. DOWNEY AND S. LEMPP
[1997] Contiguity and distributivity in the enumerable Turing degrees, *J. Symbolic Logic*, 62, pp. 1215–1240.

R. G. DOWNEY AND R. A. SHORE
[1995] Degree theoretic definitions of the low_2 recursively enumerable sets, *J. Symbolic Logic*, 60, pp. 727–756.
[1996] Lattice embeddings below nonlow_2 recursively enumerable degrees, *Israel J. Math.*, 94, pp. 221–246.
[1997] There is no degree invariant half-jump, *Proc. Amer. Math. Soc.*, 125, pp. 3033–3037.

P. A. FEJER
[1983] The density of the nonbranching degrees, *Ann. Pure Appl. Logic*, 24, pp. 113–130.

R. M. FRIEDBERG
[1957] Two recursively enumerable sets of incomparable degrees of unsolvability, *Proc. Nat. Acad. Sci.*, 43, pp. 236–238.

L. HARRINGTON AND S. SHELAH
[1982] The undecidability of the recursively enumerable degrees (research announcement), *Bull. Amer. Math. Soc.*, (N.S.), 6, pp. 79–80.

L. HARRINGTON AND R. I. SOARE
[1991] Post's program and incomplete recursively enumerable sets, *Proc. Nat. Acad. Sci.*, 88, pp. 10242–10246.

C. G. JOCKUSCH, JR. AND R. A. SHORE

[1983] Pseudo-jump operators I: the r.e. case, *Trans. Amer. Math. Soc.*, 275, pp. 599–609.

[1984] Pseudo-jump operators II: Transfinite iterations, hierarchies and minimal covers, *J. Symbolic Logic*, 49, pp. 1205–1236.

S. C. KLEENE AND E. POST

[1954] The upper semi-lattice of degrees of recursive unsolvability, *Ann. Math. (2)*, 59, pp. 379–407.

A. S. KECHRIS AND Y. MOSCHOVAKIS (EDS.)

[1978] *Cabal Seminar 76–77*, Lecture Notes in Mathematics, Vol. 689, Springer, Berlin.

A. H. LACHLAN

[1966a] Lower bounds for pairs of recursively enumerable degrees, *Proc. London Math. Soc.*, 16, pp. 537–569.

[1966b] The impossibility of finding relative complements for recursively enumerable degrees, *J. Symbolic Logic*, 31, pp. 434–454.

[1968] The elementary theory of recursively enumerable sets, *Duke Math. J.*, 35, pp. 123–146.

[1972] Embedding nondistributive lattices in the recursively enumerable degrees, in: *Conference in Mathematical Logic, London, 1970*, W. Hodges, ed., Lecture Notes in Mathematics, Vol. 255, Springer, Berlin, pp. 149–172.

[1973] The priority method for the construction of recursively enumerable sets, in: *Cambridge Summer School in Mathematical Logic*, A. R. D. Mathias and H. Rogers Jr., eds., Lecture Notes in Mathematics, Vol. 337, Springer, Berlin, pp. 299–310.

[1975a] A recursively enumerable degree which will not split over all lesser ones, *Ann. Math. Logic*, 9, pp. 307–365.

[1975b] Uniform enumeration operators, *J. Symbolic Logic*, 40, pp. 401–409.

[1980] Decomposition of recursively enumerable degrees, *Proc. Amer. Math. Soc.*, 79, pp. 629–634.

A. H. LACHLAN AND R. I. SOARE

[1980] Not every finite lattice is embeddable in the recursively enumerable degrees, *Adv. Math.*, 37, pp. 74–82.

S. LEMPP AND M. LERMAN

[1997] A finite lattice without critical triples that cannot be embedded into the enumerable Turing degrees, *Ann. Pure Appl. Logic*, 87, pp. 167–185.

S. LEMPP, A. NIES AND T. SLAMAN

[1998] The Π_3 theory of the enumerable Turing degrees is undecidable, *Trans. Amer. Math. Soc.*, 350, pp. 2719–2736.

M. LERMAN

[1983] *Degrees of Unsolvability*, Springer, Berlin.

[1993] *General ∀∃-decision method for degree structures*, Unpublished manuscript.

[1996] Embeddings into the recursively enumerable degrees, in: *Computability, Enumerability, Unsolvability: Directions in Recursion Theory*, S. B. Cooper, T. A. Slaman and S. S. Wainer, eds., LMSLNS, Vol. 224, Cambridge University Press, Cambridge, UK, pp. 185–204.

[1998] A necessary and sufficient condition for embedding ranked finite lattices into the computably enumerable degrees, *Ann. Pure Appl. Logic*, 94, pp. 143–180.

[1999] A necessary and sufficient condition for embedding principally decomposable finite lattices into the computably enumerable degrees, to appear.

M. LERMAN AND R. A. SHORE

[1988] Decidability and invariant classes for degree structures, *Trans. Amer. Math. Soc.*, 310, pp. 669–692.

M. LERMAN, R. A. SHORE AND R. I. SOARE

[1984] The elementary theory of the recursively enumerable degrees is not $\aleph_0$-categorical, *Adv. Math.*, 53, pp. 301–320.

W. MAASS

[1982] Recursively enumerable generic sets, *J. Symbolic Logic*, 47, pp. 809–823.

D. MILLER

[1981] High recursively enumerable degrees and the anti-cupping property, in: *Logic Year 1979–1980: University of Connecticut*, M. Lerman, J. Schmerl and R. I. Soare, eds., Lecture Notes in Mathematics, Vol. 859, Springer, Berlin, pp. 230–245.

A. A. MUCHNIK

[1956] On the unsolvability of the problem of reducibility in the theory of algorithms, *Dokl. Akad. Nauk SSSR (N.S.)*, 108, pp. 29–32.

[1958] Solution of Post's reduction problem and of certain other problems in the theory of algorithms, *Trudy Moskov Mat. Obsc.*, 7, pp. 391–405.

A. NIES, R. A. SHORE AND T. SLAMAN

[1996] Definability in the recursively enumerable degrees, *Bul. Symbolic Logic*, 2, pp. 392–404.

[1998] Interpretability and definability in the recursively enumerable degrees, *Proc. London Math. Soc. (3)*, 77, pp. 241–249.

E. L. POST

[1936] Finite combinatory processes – formulation, *J. Symbolic Logic*, 1, pp. 103–105.

[1944] Recursively enumerable sets of positive integers and their decision problems, *Bull. Amer. Math. Soc.*, 50, pp. 284–316.

R. W. ROBINSON

[1971] Interpolation and embedding in the recursively enumerable degrees, *Ann. of Math. (2)*, 93, pp. 285–314.

H. ROGERS, JR.

[1967a] *Theory of Recursive Functions and Effective Computability*, McGraw-Hill, New York.

[1967b] Some problems of definability in recursive function theory, in: *Sets, Models, and Recursion Theory, Proceedings of the Summer School in Mathematical Logic and 10th Logic Colloquium, Leicester, August–Sept. 1965*, J. N. Crossley, ed., North-Holland, Amsterdam, pp. 183–201.

G. E. SACKS

[1963a] *Degrees of Unsolvability*, Ann. of Math. Stud., Vol. 55, Princeton University Press, Princeton, NJ.

[1963b] Recursive enumerability and the jump operator, *Trans. Amer. Math. Soc.*, 108, pp. 223–239.

[1963c] On the degrees less than $\mathbf{0}'$, *Ann. of Math. (2)*, 77, pp. 211–231.

[1964] The recursively enumerable degrees are dense, *Ann. of Math. (2)*, 80, pp. 300–312.

[1966] *Degrees of Unsolvability*, 2nd ed., Ann. of Math. Stud., Vol. 55, Princeton University Press, Princeton, NJ.

J. R. SHOENFIELD

[1959] On degrees of unsolvability, *Ann. Math. (2)*, 69, pp. 644–653.

[1961] Undecidable and creative theories, *Fund. Math.*, 49, pp. 171–179.

[1965] An application of model theory to degrees of unsolvability, in: *Symposium on the Theory of Models*, J. W. Addison, L. Henkin and A. Tarski, eds., North-Holland, Amsterdam, pp. 359–363.

[1971] *Degrees of Unsolvability*, Math. Stud., Vol. 2, North-Holland, Amsterdam.

J. R. SHOENFIELD AND R. I. SOARE

[1978] The generalized diamond theorem, *Recursive Function Theory Newsletter*, 19, abstract #219.

R. A. SHORE

[1978] On the ∀∃-sentences of α-recursion theory, in: *Generalized Recursion Theory II*, J. E. Fenstad, R. O. Gandy and G. E. Sacks, eds., Studies in Logic and the Foundations of Mathematics, Vol. 94, North-Holland, Amsterdam, pp. 331–354.

[1982] Finitely generated codings and the degrees r.e. in a degree **d**, *Proc. Amer. Math. Soc.*, 84, pp. 256–263.

[1988] A non-inversion theorem for the jump operator, *Ann. Pure Appl. Logic*, 40, pp. 277–303.

R. A. SHORE AND T. A. SLAMAN

[1990] Working below a low_2 recursively enumerable degree, *Archive for Math. Logic*, 29, pp. 201–211.

[1993] Working below a high recursively enumerable degree, *J. Symbolic Logic*, 58, pp. 824–859.

T. A. SLAMAN

[1991a] Degree structures, in: *Proc. Int. Cong. Math., Kyoto 1990*, Springer, Tokyo, pp. 303–316.

[1991b] The density of infima in the recursively enumerable degrees, *Ann. Pure Appl. Logic*, 52, pp. 155–179.

[1999] The global Turing degrees, in: *Handbook of Computability Theory*, E. R. Griffor, ed., Elsevier, Amsterdam, pp. 155–168.

T. A. SLAMAN AND R. I. SOARE

[1995] Algebraic aspects of the computably enumerable degrees, *Proc. Nat. Acad. Sci.*, 92, pp. 617–621.

[1998] Extension of embeddings in the recursively enumerable degrees, *Ann. Math.*, to appear.

T. SLAMAN AND J. R. STEEL

[1988] Definable functions on degrees, in: *Cabal Seminar 81–85*, A. S. Kechris, D. A. Martin and J. R. Steel, eds., Lecture Notes in Mathematics, Vol. 1333, Springer, Berlin, pp. 37–55.

T. A. SLAMAN AND W. H. WOODIN

[2000] *Definability in Degree Structures*, in preparation.

R. I. SOARE

[1987] *Recursively Enumerable Sets and Degrees*, Springer, Berlin.

[1999] The lattice of recursively enumerable sets, this *Handbook*.

J. R. STEEL

[1982] A classification of jump operators, *J. Symbolic Logic*, 47, pp. 347–358.

A. M. TURING

[1936] On computable numbers, with an application to the Entscheidungsproblem, *Proc. London Math. Soc. (2)*, 42, pp. 230–265.

[1939] Systems of logic based on ordinals, *Proc. London Math. Soc.*, 45, pp. 161–228.

B. J. WEINSTEIN

[1988] *On Embeddings of the 1-3-1 into the Recursively Enumerable Degrees*, PhD Thesis, University of California, Berkeley.

C. E. M. YATES

[1965] Three theorems on the degrees of recursively enumerable sets, *Duke Math. J.*, 32, pp. 461–468.

[1966a] A minimal pair of recursively enumerable degrees, *J. Symbolic Logic*, 31, pp. 159–168.

[1966b] On the degrees of index sets, *Trans. Amer. Math. Soc.*, 121, pp. 309–328.

CHAPTER 7

An Overview of the Computably Enumerable Sets

Robert I. Soare*

Department of Mathematics, University of Chicago, Chicago, IL 60637-1546, USA
E-mail: soare@math.uchicago.edu or soare@cs.uchicago.edu
Anonymous ftp: cs.uchicago.edu: ftp/pub/users/soare
World Wide Web: http://www.cs.uchicago.edu/~soare

Contents

1. A brief history of c.e. sets . . . 200
2. Definable properties of c.e. sets . . . 202
   2.1. Creative sets . . . 203
   2.2. Incomplete sets . . . 205
   2.3. Complete sets . . . 208
   2.4. Nonlow sets . . . 211
3. Automorphisms of $\mathcal{E}$ . . . 216
   3.1. Some results on automorphisms . . . 216
   3.2. A sketch of the Δ_3^0-automorphism method . . . 220
   3.3. The automorphism theorem . . . 234
   3.4. Using that A is noncomputable to obtain the set $\widehat{\mathcal{C}}_\alpha$ of coding states . . . 235
   3.5. Moving α-witnesses into B . . . 237
   3.6. Appointing α-witnesses and Step $\hat{7}$ . . . 238
   3.7. Coding states $\widehat{\mathcal{C}}_\alpha$ and Step $\hat{8}$ . . . 239
   3.8. The Coding Theorem . . . 240
4. Invariance . . . 241
5. Decidability and undecidability . . . 242
6. Further results . . . 244
References . . . 244

*The author was supported by National Science Foundation Grant DMS 94-00825.

HANDBOOK OF COMPUTABILITY THEORY
Edited by E.R. Griffor

The purpose of this article is to summarize some of the results on the algebraic structure of the computably enumerable (c.e.) sets since 1987 when the subject was covered in Soare [1987], particularly Chapters X, XI, and XV. We study the c.e. sets as a partial ordering under inclusion, $(\mathcal{E}, \subseteq)$. We do not study the partial ordering of the c.e. *degrees* under Turing reducibility, although a number of the results here relate the algebraic structure of a c.e. set A to its (Turing) degree in the sense of the information content of A.

We consider here various properties of $\mathcal{E}$: (1) definable properties; (2) automorphisms; (3) invariant properties; (4) decidability and undecidability results; (5) miscellaneous results. This is not intended to be a comprehensive survey of all results in the subject since 1987, but we give a number of references in the bibliography to other results.

1. A brief history of c.e. sets

Gödel [1934] introduced the definition of a (general) recursive function, and Church [1936a] proposed that this be taken as the definition of a computable function. Independently, Turing [1936–37] proposed that the computable functions be defined as those which could be computed by an *automatic machine* (now called a *Turing machine*). Gödel and most others accepted Turing's definition and analysis of computability as the most persuasive.[1]

In addition to defining computable functions, there was in interest in defining computable generated *sets*. Church [1936a] and Kleene [1936] defined a set of positive integers to be "*recursively enumerable*" if it is the range of a recursive function. Little more was done with these sets until Post [1943] proposed a formal system for *generating* sets rather than computing their characteristic functions. Post introduced his "production systems" and in a restricted form his "normal (production) systems" and defined a "normal set" to be one generated by a normal system. Post showed that the normal sets are exactly the recursively enumerable (r.e.) sets, providing the empty set is added as an r.e. set. Post, however, thought not so much in formal systems as in informal terms and described the corresponding *informal* concept of "effectively enumerable set" or "generated set." Post [1944] wrote,

> "Suffice it to say that each element of the set is at some time written down, and earmarked as belonging to the set, as a result of predetermined effective processes. It is understood that once an element is placed in the set, it stays there."

Post then [1944, p. 286] restated his thesis from [1943] that "*every generated set of positive integers is recursively enumerable.*" In his paper [1944] Post did not use the formalism of recursive functions. He used either his informal notion of generated set, or when formalism was required, his own formal definition of normal set. Never-

[1] We cite references in the usual convention by author and year, e.g., [Post, 1944] or simply Post [1944]. To save space we omit from our bibliography some references which appear in Soare [1987], and we cite them as there by year.

theless, he accepted the *terminology* of "recursively" enumerable which Kleene and Church had previously established.

In this paper we use the terminology "computably enumerable (c.e.)" to stress the intensional concepts and for the reasons explained in Soare [1996a]. A function is *Turing computable* if it is definable by a Turing machine, as defined by Turing [1936–37], see Soare [1987, p. 11]. A set A is *computably enumerable* (*c.e.*) if A is $\emptyset$ or is the range of a Turing computable function. By Turing's Thesis (T.T.) [1936–37] (which we accept) the (informal) class of *computable* functions is the same as the (formal) class of *Turing computable* functions. Hence, we shall use the term "*computable*" for either class of functions, and likewise we shall use the term "*computably enumerable*" for either class of corresponding sets.

With its informal style Post's paper [1944] gave new excitement to computably enumerable sets and indeed the whole subject of computability. Post stated his famous *Post's Problem*, "Does there exist a c.e. set A which is noncomputable but incomplete?" He initiated *Post's Program*, the study of the relationship between the structure of A as a *set* and its information content, usually measured by its degree under Turing reducibility, defined by Turing [1939]. Post proposed three properties he hoped might guarantee incompleteness, simple, h-simple, and hh-simple, but none succeeded. Post's Problem was solved by Friedberg [1957] and Muchnik [1956] with the introduction of the priority method. Meanwhile, Myhill [1956] had rediscovered the fact [Post, 1943] that the c.e. sets form a lattice $\mathcal{E}$ under inclusion, and he asked whether there is a maximal set (i.e. ua coatom of the quotient lattice $\mathcal{E}^*$ of $\mathcal{E}$ modulo finite sets) since this implies nondensity of $\mathcal{E}^*$. Let $\{W_e\}_{e\in\omega}$ be the standard enumeration of the c.e. sets. We let $\mathcal{E}$ denote this lattice

$$\mathcal{E} = \left(\{W_e\}_{e\in\omega}, \subset, \cup, \cap, \emptyset, \omega\right)$$

as a lattice under union, intersection, with least and greatest elements. Often we let $\mathcal{E}$ denote just the partial ordering $(\{W_e\}_{e\in\omega}, \subset)$ because the other four are definable from inclusion and because the lattice properties are rarely used in our paper.

The solution of Post's Problem stimulated more research into Post's Program. In the 1960's using ever stronger forms of the priority method Sacks, Yates, Martin, Lachlan, and others obtained deeper results about Post's Program, particularly about maximal sets. For example, Martin [1966] showed that the degrees of maximal sets are exactly the high c.e. degrees, and Lachlan [1968c] extended this to all hh-simple sets.

By the end of the 1960's the fascination with c.e. sets increased as Lachlan [1970] noted that all known constructions can be viewed as a game between two players each placing finitely many "balls" (integers) into "buckets" (c.e. sets) on each of ω many moves in the game. At the end the second player wins if the sets satisfy certain prearranged requirements stated as $\mathcal{E}$ conditions. Lachlan suggested that this reduced the emphasis on computability to merely finding winning strategies. We may think of the definability results here and the automorphism results [Harrington and Soare, 1996c] as a contest between two players the definability player (RED) and the automorphism player (BLUE) such that any particular theorem for one or the other is a Lachlan-type game.

Soare [1974] introduced machinery for generating automorphisms of $\mathcal{E}$ and showed that all maximal sets are automorphic. This machinery was then turned to the question of Rogers [1967] of whether all creative sets are automorphic, but here the automorphism method merely led to several false proofs in the 1970's that the creative sets did not form an orbit. In the mid 1980's Harrington, analyzing the failure of these attempted automorphisms viewed as games, distilled the obstacle and from it a winning strategy for the definability player. Harrington proved [Soare, 1987, p. 339] that there is an $\mathcal{E}$-definable property, $\mathrm{CRE}(A)$, such that A is creative if and only if $\mathcal{E} \models \mathrm{CRE}(A)$, a surprising fact considering how creativity appears to be so closely tied to a productive function, which does not seem to be preserved under automorphisms of $\mathcal{E}$.

This advance led eventually to the second definable property $Q(A)$ in Section 2.2. Echoing Post's Program, Sacks asked in his book on degrees [1963, p. 172, Q(3)] whether there is a property in the style of Post of a noncomputable c.e. set A which guarantees its incompleteness. Marchenkov [1976] proved that η-maximal semi-recursive c.e. sets are incomplete, and D. Miller [1981] later showed (see Soare [1987, p. 232]) they are all low$_2$. Soare [1987, p. 73] gave a property characterizing low c.e. sets, and Ambos-Spies and Nies [1992] gave a property characterizing c.e. sets whose degrees are *cappable* (i.e. halves of a minimal pair). However, these three properties are all non $\mathcal{E}$-definable by Harrington and Soare [1996b, Theorems 3.1 and 3.2], respectively.

The question for $\mathcal{E}$-definable properties remained open and several partial negative results were obtained using automorphisms (see Downey and Stob [1992]) in an unsuccessful attempt to show that every noncomputable c.e. set is automorphic to a complete set. Finally, by analyzing the failure of these automorphism attempts Harrington and Soare [1991] showed that the answer to the Post–Sacks question is actually yes; there exists an $\mathcal{E}$-definable property $Q(A)$ which guarantees that A is noncomputable and incomplete.

In the automorphism direction, if one interprets the Post–Sacks question as a property of $\overline{A}$ *alone*, namely a definable property of $\mathcal{L}(A) = (\{W\colon\ W \supseteq A\}, \subset)$, then the answer is no because Cholak [1995] has shown that for every noncomputable c.e. set A and every high c.e. degree $\mathbf{d}$ there is a $B \in \mathbf{d}$ such that $\mathcal{L}(A) \cong \mathcal{L}(B)$.

CONVENTION. From now on all sets will be c.e. sets unless specifically stated otherwise.

2. Definable properties of c.e. sets

DEFINITION 2.1. If $\{X_s\}_{s\in\omega}$ and $\{Y_s\}_{s\in\omega}$ are computable enumerations of c.e. sets X and Y define

$$X \backslash Y = \{z\colon\ (\exists s)[z \in X_s - Y_s]\},$$

the elements enumerated in X before Y, and

$$X \searrow Y = (X \backslash Y) \cap Y,$$

the elements enumerated first in X and later in Y.

All the definable properties could be written just in the language $L(\subset)$ over $\mathcal{E}$ because the functions $\cup$, $\cap$, and constants $\emptyset$ and ω are definable in $L(\subset)$ but for convenience we use the former also in order to improve readability. Likewise, we use the $\mathcal{E}$-definability of the properties "finite" and "computable" [Soare, 1987, p. 179].

2.1. *Creative sets*

Post [1944] defined a c.e. set C to be *creative* if there is a partial computable function ψ such that

$$(\forall e)\big[W_e \subset \overline{C} \Longrightarrow \psi(e)\downarrow \in \overline{C} - W_e\big].$$

It follows by Myhill's Theorem [Soare, 1987, p. 43] that C is creative iff C is m-complete, i.e. $W_e \leqslant_m C$ for every e, or equivalently $K \leqslant_m C$ for the complete set K. Although these properties of C at first appear to be very far from being $\mathcal{E}$-definable, Harrington [Soare, 1987, p. 339] exhibited the following $\mathcal{E}$-definable property CRE(A) which defines C being creative.

2.1.1. *The defining property for creative sets*

THEOREM 2.2 (Harrington). *A c.e. set A is creative iff*

$$\text{CRE}(A): \quad (\exists C \supset A)(\forall B \subseteq C)(\exists R)[R \textit{ is computable} \tag{1}$$
$$\& \ R \cap C \textit{ is noncomputable } \& \ R \cap A = R \cap B], \tag{2}$$

where all variables range over $\mathcal{E}$.

We may represent the property CRE(A) as a two person game in the sense of Lachlan [1970] between the $\exists$-player (called RED, the definability player) who plays the c.e. sets C, R (the red sets) and the $\forall$-player (called BLUE, the automorphism player) who plays the c.e. set B (the blue set).

THEOREM 2.3 (Blue). *If* CRE(A) *then $K \leqslant_m A$ so A is creative.*

PROOF. Suppose CRE(A). We may visualize R as dividing the universe ω into two halves and on the R half we visualize in the Venn diagram the following states (corresponding roughly to e-states) $\nu_1 = R \cap \overline{C}$, $\nu_2 = R \cap C - B$, $\nu_3 = R \cap B - A$,

$\nu_4 = R \cap A$. The static condition CRE(A) forces certain dynamic properties of the sets as follows. The condition that $R \cap C$ is noncomputable means that $R - C$ is not c.e. so there must be an infinite c.e. set of elements, say $\{x_n\}_{n\in\omega}$, which move from state ν_1 to state ν_2. Define $\psi(n) = x_n$. If n enters K, then enumerate $\psi(n)$ in B, from which the second conjunct of (2) eventually forces that $\psi(n) \in A$, so x_n passes from ν_1 to ν_2 to ν_3 to ν_4 in that order. If $n \in K$ then x_n remains in ν_2 forever. Hence, $K \leqslant_m A$ via ψ.

This strategy succeeds if BLUE knows an index $W_e = R$ while he defines B, but in general he must play against all possible sets W_i, $i \in \omega$, simultaneously, and define a possible function ψ_i for each $i \in \omega$. To prevent some $i < e$ from interfering with the minimal index e for R, we define a new value $\psi_{i,s+1}(x)$ only if

$$(\forall y < x)[\psi_{i,s}(y)\downarrow \ \& \ [y \in K_s \iff \psi_{i,s}(y) \in A_s]]. \tag{3}$$

Thus, except for finitely many elements taken for $i < e$ the strategy for e succeeds as before. The condition (3) is a Π^0_2 condition on i so the construction may be viewed as being done on a tree with e corresponding to the first infinite node and hence the node along the "true path." (For more details see Soare [1987, p. 339].) □

LEMMA 2.4 (Red). *If A is creative then* CRE(A) *holds.*

PROOF. Since all creative sets are computably isomorphic we can choose A to be a *specific* creative set. Define

$$A = \{\langle x, y\rangle \colon x \in K \ \& \ \langle x, y\rangle \in W_y\}.$$

Now $K \leqslant_1 A$ because if we choose $W_{y_0} = \omega$ then $x \in K$ iff $\langle x, y_0\rangle \in A$. Next define $C = \{\langle x, y\rangle \colon x \in K\}$. Given B c.e. and $B \subseteq C$ choose y_0 such that $W_{y_0} = B$. Define $R = \{\langle x, y_0\rangle \colon x \in \omega\}$. Now $R \cap C = \{\langle x, y_0\rangle \colon x \in K\}$ which is not computable, and

$$R \cap A = R \cap B = \{\langle x, y_0\rangle \colon x \in K \ \& \ \langle x, y_0\rangle \in W_{y_0} = B\},$$

because $B \subseteq C$. □

2.1.2. *Remarks on the method.* Notice that this property CRE(A), and all later $\mathcal{E}$-definable properties $P(A)$ presented here, begin with "$(\exists C \supset A)$." It is well known that in constructing a c.e. set A it is the *complement* $\overline{A}$ which matters during the construction because once an element x has entered A there is no further action with it. All the action in these constructions will be on $\overline{C}$ and particularly on $C - A$, where the two players will contest exactly which and how quickly elements will enter A. Thus, $C - A$ provides a convenient arena for this contest, and the properties will imply that $C - A$ is infinite.

Second, notice that all four results in this section show that a newly discovered $\mathcal{E}$-definable property $P(A)$ implies a well-known c.e. set condition $C(A)$ like creativeness, completeness, incompleteness, or non-lowness. In every case, after the definition of $P(A)$ we prove two theorems.

The first theorem by BLUE shows that $P(A)$ implies $C(A)$. We call this a BLUE theorem because if A satisfies the property $P(A)$ then BLUE can play certain blue sets to take control of certain red sets or A and can achieve $C(A)$. For example, in CRE(A), the condition (2) that $A \cap R = B \cap R$ means that on R the control of A has been turned over to B in the sense that on R: (i) $A \backslash B = \emptyset$, because any $x \in A \backslash B$ could be withheld by BLUE from B forever contradicting (2); and (ii) that A must "copy" B in the sense that any element in B must eventually enter A.

The second theorem by RED shows that there *exists* a set A with property $P(A)$ (and sometimes with additional properties) so that the first theorem is nontrivial. In the case of CRE(A) the second theorem showed that there is a *creative* set A satisfying $P(A)$, and hence that *every* creative set C satisfies $P(C)$.

2.2. *Incomplete sets*

After unsuccessful attempts to give a negative answer to the Post–Sacks question using automorphisms, Harrington and Soare produced the following $\mathcal{E}$-definable property $Q(A)$ which guarantees that A is incomplete but noncomputable.

DEFINITION 2.5. (i) A subset $A \subset C$ is a *major subset* of C (written $A \subset_{\mathrm{m}} C$) if $C - A$ is infinite and for all e,

$$\overline{C} \subseteq W_e \Longrightarrow \overline{A} \subseteq^* W_e.$$

(Note that if $A \subset_{\mathrm{m}} C$ then both A and C are noncomputable.)

(ii) $A \sqsubset B$ if there exists C such that $A \sqcup C = B$ (i.e. $A \cup C = B$ and $A \cap C = \emptyset$).

2.2.1. *The property $Q(A)$ guaranteeing incompleteness*

THEOREM 2.6 (Harrington and Soare [1991]). *There is a property $Q(A)$ which guarantees that $A <_{\mathrm{T}} K$ and which holds of some noncomputable set.*

Define the property $Q(A)$ by:

$$\begin{aligned} Q(A)\colon\ (\exists C)_{A \subset_{\mathrm{m}} C}\ (\forall B \subseteq C)(\exists D \subseteq C)(\forall S)_{S \sqsubset C}\big[\\ \big[B \cap (S - A) = D \cap (S - A)\big] \quad (4) \\ \Longrightarrow (\exists T)\big[\overline{C} \subset T \;\&\; A \cap (S \cap T) = B \cap (S \cap T)\big]\big]. \quad (5) \end{aligned}$$

THEOREM 2.7 (Blue). *If $Q(A)$ then A is incomplete (i.e. $A \not\leqslant_{\mathrm{T}} K$).*

PROOF. (Sketch only, see Harrington and Soare [1991] for details.) We may visualize the property $Q(A)$ as a two person game in the sense of Lachlan [1970] between the $\exists$-player (RED) who plays the c.e. sets A, C, D and T and the $\forall$-player (BLUE) who plays the c.e. sets B and S. For simplicity ignore all the sets but C, D, B and A, since the others are necessary only to give us a suitable domain on which to play the following strategy. Visualize $C \supseteq D \supseteq B \supseteq A$, and let $\nu_1, \nu_2, \ldots, \nu_5$ denote the differences of c.e. sets (called *d.c.e. sets*): $\omega - C$, $C - D$, $D - B$, $B - A$, A respectively, but viewed dynamically like e-states, so an element can pass from ν_i to ν_j, $i < j$. The oversimplified $Q(A)$ property now asserts that if BLUE plays: $(4)'$ $D = B$ on $\overline{A}$, then RED will play: $(5)'$ $B = A$. In particular, if both players are following their best strategies, then for an element x to enter A, it must pass through the ν-states in the order $\nu_1, \nu_2, \ldots, \nu_5$ as proved in [Harrington and Soare, 1991]. However, the set B acts like a wall of restraint, like the minimal pair restraint of Lachlan and Yates in [Soare, 1987, p. 153]. When presented with an $x \in D - B$, BLUE may hold x as long as he likes, but must eventually put x into B at which point RED is free to put x into A but not before. This implies that A is tardy (i.e. not of promptly simple degree, see Ambos-Spies, Jockusch, Shore, and Soare in [Soare, 1987, p. 284] or [Soare, 1987, Chap. XIII]), so A is incomplete. Furthermore, Harrington and Soare [1996d] have recently discovered that $Q(A)$ imposes a much stronger tardiness property on A (called 2-*tardy*) which helps us classify those sets which can be coded into any nontrivial orbit. □

THEOREM 2.8 (Red). *There exists a c.e. set A satisfying property $Q(A)$.*

PROOF. (Sketch only, see Harrington and Soare [1991] for details.) This proof is very similar to the standard proof (see Soare [1987, p. 194]) that every noncomputable c.e. set C has a small major subset A ($A \subset_{\mathrm{sm}} C$) to which the reader should now refer. (See Section 2.2.2 below for small sets.) Let C be any noncomputable c.e. set. (If we choose C simple (maximal) then A will be simple (r-maximal).) To make $A \subset_{\mathrm{m}} C$ it suffices to meet for every e the requirement,

$$P_e\colon \overline{C} \subseteq W_e \Longrightarrow \overline{A} \subseteq^* W_e.$$

Replace W_e by $V_e = \bigcup_{s \in \omega} V_{e,s}$ defined by

$$x \in V_{e,s} \Longrightarrow x \in V_{e,s-1} \vee \big[x \in (W_{e,s} - C_s) \mathbin{\&} (\forall y \leqslant x)\big[y \in W_{e,s} \cup C_s\big]\big]. \tag{6}$$

Note that $C \searrow V_i = \emptyset$, for every i. Define the e-state

$$\sigma(e, x, s) = \{i\colon i \leqslant e \mathbin{\&} x \in V_{i,s}\},$$

with the usual ordering of e-states. Let $C = f(\omega)$ for f a 1:1 computable function, and let $c_i = f(i)$. Let $C_s - A_s = \{d_0^s, d_1^s, \ldots\}$, in the ordering induced by $\{c_0, c_1, \ldots\}$. (Hence, if $x = d_i^s = d_j^t$ for $t > s$, then $j \leqslant i$.)

The strategy for P_e is as follows. If $i \geqslant e$, and $j > i$ is minimal such that $\sigma(e-1, d_i^s, s) = \sigma(e-1, d_j^s, s)$, $d_i^s \notin V_{e,s}$, and $d_j^s \in V_{e,s}$ then P_e *wants to enumerate* into A all the elements $\{d_k^s\colon i \leqslant k < j\}$ (but subject to the negative restraint by N_i, $i \leqslant e$, as described below).

Let $\{B_i\colon i \in \omega\}$ and $\{(S_j, \hat{S}_j)\colon j \in \omega\}$ be an effective listing of all c.e. sets and all pairs of c.e. sets respectively. Let RED play D_i against B_i, $D_i \subseteq C$, and also construct $T_{i,j}$ to meet (5) if BLUE satisfies (4). Let $\alpha = \langle i, j\rangle$, and let D_α, B_α, S_α, $\hat{S}_\alpha$, and T_α denote D_i, B_i, S_j, $\hat{S}_j$, and $T_{i,j}$ respectively. For each α the conjunction of the matrix of (4) for $(B_\alpha, D_\alpha, S_\alpha)$ with the conditions $B_\alpha \subseteq C$, and $S_\alpha \sqcup \hat{S}_\alpha = C$ is a Π_2^0 relation $F(\alpha)$. Let $\{Z_\alpha\}_{\alpha\in\omega}$ be an c.e. array of c.e. sets such that $F(\alpha)$ holds iff $|Z_\alpha| = \infty$.

Define T_α by

$$x \in T_{\alpha,s} \Longrightarrow x \in T_{\alpha,s-1} \vee \big[x \in \overline{C}_s \;\&\; x \leqslant |Z_{\alpha,s}|\big]. \tag{7}$$

The negative requirement N_α on A asserts that if (4) holds for $(B_\alpha, D_\alpha, S_\alpha)$ then (5) holds for $(B_\alpha, S_\alpha, T_\alpha)$. The strategy for N_α is this. If $x \in T_\alpha \searrow C$, then N_α restrains x from A until $x \in S_\alpha \sqcup \hat{S}_\alpha$. (If the latter never occurs then $F(\alpha)$ fails so Z_α and T_α are finite, and only finitely many such $x \in T_\alpha \cap C$ are permanently restrained by N_α.) If $x \in \hat{S}_\alpha$ then N_α imposes no further restraint on x.

If $x \in S_\alpha$, and while $x \in \overline{A}$, x is enumerated in $B_\alpha \backslash D_\alpha$ then N_α restrains x from both D_α and A forever (unless some P_e, $e < \alpha$, enumerates x in A). (If N_α successfully keeps $x \in \overline{D}_\alpha \cap \overline{A}$ then x violates (4) so $F(\alpha)$ fails and Z_α and T_α are finite.) Otherwise, suppose that $B \subseteq D \searrow B$ holds on $S_\alpha - A$.

If $x \in S_\alpha$ and some P_e, $e \geqslant \alpha$, wants to enumerate x in A then N_α first enumerates x in D_α and then restrains x from A until x is enumerated in B_α at which time N_α *releases* x (forever). (If x remains in $\overline{B}_\alpha$ then $x \in (D_\alpha - B_\alpha) \cap (S_\alpha - A)$ so x violates (4) and again T_α is finite.) Hence, in any of the three cases N_α permanently restrains at most finitely many elements.

We now combine the P_e and N_α-strategies to sketch the full construction of A. If $x \in C_s$, choose the least e (if any) such that P_e wants to enumerate x in A. First P_e waits until $x \in S_\alpha \sqcup \hat{S}_\alpha$ for all $\alpha \leqslant e$. Next let $\alpha_0, \alpha_1, \ldots, \alpha_n$ be a listing of all $\alpha \leqslant e$ such that $x \in S_\alpha$ and $D_\beta \neq D_\alpha$ for all $\beta < \alpha$, with $x \in S_\beta$. For each k, $0 \leqslant k \leqslant n$, we make x pass through the N_{α_k}-strategy above (also called the N_{α_k}*-gate*) in the order $N_{\alpha_n}, \ldots, N_{\alpha_1}, N_{\alpha_0}$. Hence, for example, when x is released by the N_{α_2}-gate by being enumerated in B_{α_2}, RED then enumerates $x \in D_{\alpha_1}$ and waits for x to be released by N_{α_1} by being enumerated in B_{α_1}. When x is released by the N_{α_0}-gate RED enumerates x in A.

Now it is easy to check that every requirement P_e and N_α is satisfied. (See Harrington and Soare [1991].) □

2.2.2. *Small sets and remarks on the $Q(A)$ property.* Lachlan [1968b] introduced small sets in his program to construct canonical examples of certain diagrams and then rule out possible extensions so as to give a decision procedure for the $\forall\exists$-theory of the lattice of c.e. sets. The following definition is clearly equivalent to the standard definition as in [Soare, 1987, Definition 4.10, p. 193].

DEFINITION 2.9. A subset $A \subset C$ is a *small subset* of C (written $A \subset_s C$) if $A \subset_\infty C$ and for all X and Y, if

(i) $X \cap (C - A) \subseteq Y$, then

(ii) $(X - C) \cup Y$ is c.e., namely

(ii)′ $(\exists Z)_{Z \subseteq X}[Z \supseteq (X - C) \,\&\, (Z \cap C) \subseteq Y]$, so $Z = (X - C) \cup Y$ and is c.e.

If A is both a small subset and major subset of C we say it is a *small major subset* and write $A \subset_{\mathrm{sm}} C$.

Harrington and Soare [1996a] showed that $A \subset_{\mathrm{sm}} C$ iff the following *dynamic* property small-tardy(A, C) holds, namely

$$(\forall f)(\exists T)\big[\overline{C} \subseteq T \,\&\, (\forall x)\big[x \in (T \cap C)_{\text{at } s} \Longrightarrow x \notin A_{f(s)}\big]\big].$$

It is easy to show [Soare, 1987, p. 194] that: (1) if $A \subset C$ and either A or C is computable then $A \subset_s C$; (2) if $A \subset_s C$ and C is noncomputable then $A \cup \overline{C}$ is noncomputable. Also if $A \subset_s C$ and C is noncomputable then $C - A$ is infinite. To see the latter, suppose $Y = C - A$ is finite; then (i) of Definition 2.9 is satisfied with $Y = C - A$ and $W = \omega$; choose c.e. Z satisfying (ii)′; then $Z - Y$ is c.e., but $Z - Y = \overline{C}$, contrary to C being noncomputable. Hence, the intuition is that $A \subset_s C$ guarantees among other things that the A boundary is far below the C boundary. Small sets will be used again in the next section.

Because of the similarity of the proof of Theorem 2.8 to the small major subset construction, it is natural to ask whether the Q-like property $\widehat{Q}(A)$: $(\exists C)[A \subset_{\mathrm{sm}} C]$ guarantees $A <_{\mathrm{T}} K$. This is false, but $\widehat{Q}(A)$ implies that A is not a promptly simple *set*. This is easy to see from the property small-tardy(A, C) because if small-tardy(A, C) then A is not promptly simple [Harrington and Soare, 1996a].

2.3. *Complete sets*

The property CRE(A) of Section 2.1 gives an orbit consisting of only complete sets, but all these are creative and therefore computably isomorphic, so there is no diversity including, for example, simple sets. In this section we present an $\mathcal{E}$-definable property $T(A)$ which implies that A is Turing complete, but which also allows A to be simple, even promptly simple, and to have other standard properties, such as being r-maximal or being a major subset of C.

This flexibility allows us to refute various automorphism conjectures and to limit the power of the automorphism building player (BLUE). For example, since the 1970's it has been known that one way to build an automorphism between certain promptly simple sets A and B is to first construct an isomorphism on their complements guaranteeing that $\mathcal{L}(A) \cong \mathcal{L}(B)$, and then to use prompt simplicity of A to ensure that A "covers" B and dually that B "co-covers" A in the sense of the automorphism machinery [Soare, 1987, p. 352]. Maass [1982] showed (see Soare [1987, p. 377]) that this works if A and B are low and promptly simple.

With this phenomenon in mind, Cholak raised the following question.

QUESTION 1. For all promptly simple high degrees $\mathbf{h}$ and for all promptly simple sets A is there a c.e. set $B \in \mathbf{h}$ such that $A \simeq_{\Delta_3^0} B$?

The present property $T(A)$ negatively answers this question, and indeed shows that it is false with such A for *every* degree $\mathbf{h} \neq \mathbf{0}'$.

THEOREM 2.10 (Harrington and Soare [1998]). *There is an $\mathcal{E}$-definable property T satisfied by a promptly simple set A such that for all W, $T(W)$ implies that $K \leqslant_{\mathrm{T}} W$.*

Define $T(A)$ by:

$$T(A)\colon\ (\exists C \supset A)(\forall B \subset C)(\exists \text{ a computable set } R)_{(R-C \text{ not c.e.})} \tag{8}$$

$$[B \subset_{\mathrm{s}} C \quad .\Longrightarrow. \quad A \supseteq (B \cap R)\ \&\ [(C - A) \cap R \text{ is not co-c.e.}]] \tag{9}$$

(We only sketch the proof which appears in Harrington and Soare [1998].) To understand the intuition behind $T(A)$ in an slightly oversimplified setting, assume that BLUE knows the true set R satisfying $T(A)$. On R consider the states $\overline{C}$, $C - (B \cup A)$, $B - A$, and A, denoted by ν_1, ν_2, ν_3, and ν_4, respectively. To show that $T(A)$ implies $K \leqslant_{\mathrm{T}} A$, BLUE defines a Turing reduction $\Psi^A = K$. When he defines $\Psi^A(n)$, he simultaneously defines the use function $\psi^A(n)$ to be an element x_n of $(C \cap R) - (B \cup A)$, namely in state ν_2. Since $A \subset C$ and $B \subset C$ we may assume $A \backslash C = \emptyset$ and $B \backslash C = \emptyset$. Hence, the clause "$R - C$ not c.e." guarantees that infinitely many elements pass (flow) from ν_1 to ν_2. Furthermore, $B \subset_{\mathrm{s}} C$ implies that ν_2 is infinite by the remarks in Section 2.2.2, so there will be a distinct position for each $\psi(n)$, and $\psi(n)$ eventually settles on some x_n in state ν_2 if $n \notin K$. If n later enters K then BLUE puts x_n into B but $A \supseteq B \cap R$ by the second clause of (9), so x_n eventually enters A allowing $\Psi(n)$ to be redefined. While $n \in \overline{K}$, if x_n prematurely enters A, then $\psi(n)$ becomes undefined and waits for a new element x'_n distinct from the other $\psi(m)$. The final clause of (9) guarantees that infinitely many such elements exist so for every $n \in \overline{K}$ we see that $\psi(n)$ comes to rest on some $x \in C - A$.

To prove the First Theorem for $T(A)$, that if $T(A)$ then $K \leqslant_{\mathrm{T}} A$, BLUE takes given sets A and C satisfying $T(A)$ and constructs the c.e. set B, Turing reductions Ψ_i, and c.e. sets Z_i, $i \in \omega$, to satisfy the following positive requirements, P_i, for showing $\Psi^A = K$ as above, and negative requirements N_i for showing $B \subset_{\mathrm{s}} C$, for $i \in \omega$.

P_i: If $T(A)$ and i is the minimal Δ_0-index of R
satisfying (9) then $\Psi_i(A) = K$,

N_i: $Y_i \supseteq X_i \cap (C - B) \Longrightarrow Z_i = (X_i - C) \cup Y_i$ (which is therefore c.e.).

To prove the second theorem for $T(A)$ that there is a promptly simple set A satisfying $T(A)$, the red player constructs C and A and the blue player a set B. Let $R_e = \omega^{[e]} = \{\langle x, y\rangle\colon\ y = e\}$ a computable decomposition of ω into disjoint, infinite, computable sets. Since RED does not know which set B BLUE is playing he considers all c.e. sets $\{W_e\}_{e\in\omega}$, and plays against the candidate $B = W_e$ on the computable set R_e. For convenience we now fix e and drop the subscript e. We let B_s denote $W_{e,s}$.

To achieve the clause $A \supseteq B \cap R$ of (9) strategy ρ commits to putting every element $x \in B \cap R$ into A. Since $A \subset C$ and $B \subset C$, we may assume that the enumerations satisfy $A\backslash C = \emptyset$ and $B\backslash C = \emptyset$. The main task of ρ is to arrange the final clause of (9) that $(C - A) \cap R$ is not co-c.e., namely we must meet for every i the requirement,

$$Q_i\colon\ (C - A) \cap R \neq \overline{W_i}.$$

To achieve this ρ will try to choose some element $x \in R_s - C_s$, wait for $x \in W_{i,s}$, then enumerate $x \in C$ and restrain x from A, so that either $x \in \overline{(C - A)} \cap \overline{W_i}$ or else $x \in (C - A) \cap W_i$, and either way requirement Q_i is satisfied. The problem is that once x enters C_s, then ρ can restrain x from A only if $x \notin B$ because ρ must play $A \supseteq (B \cap R)$. To force that $x \notin B$, ρ assumes that $B \subset_{\mathrm{s}} C$, namely that every requirement N_i, $i \in \omega$. Decompose R into an infinite disjoint union of infinite computable sets, $R = \sqcup_{j\in\omega} S_j$. Now the ρ-strategy is as follows.

In Phase 1 we begin to enumerate the red sets X_i and Y_i so that X_i includes an initial segment of $\overline{C_s}$, and Y_i includes an initial segment of $C_s - B_s$ until BLUE enumerates sufficiently many elements in one of his sets $Z_{i,j}$, $j \in \omega$, in reply, attempting to make $Z_{i,j} = (X_i - C) \cup Y_i$ for at least one j.

In Phase 2 we choose the least pair $\langle x, j\rangle$ (if it exists) such that

(1) $x \in W_{i,s}$,
(2) $x \in S_{j,s} - C_s$,
(3) $x \in Z_{i,j,s}$, and
(4) $\neg(\exists y)[y \in (Z_{i,j,s} \cap B_s) - Y_{i,s}]$.

Enumerate x in C_{s+1}; restrain x from Y_i; and temporarily cease all enumeration of X_i and Y_i, but if x later enters B, then return to Phase 1. (Necessarily, $x \notin (A_s \cup B_s \cup Y_{i,s})$ because $B\backslash C = A\backslash C = Y_i\backslash C = \emptyset$.) This completes the description of strategy ρ.

Now suppose $B \subset_{\mathrm{s}} C$. Then every N_i is met. Hence, for the sets X_i and Y_i above there must exist a $Z_{i,j}$ satisfying N_i. Suppose that $(C - A) \cap R = \overline{W_i}$. Then $W_i \supseteq (S_j - C) \cap R$, which will be infinite and $X_i \supseteq \overline{C}$. Hence, eventually an element $x \in S_j \cap Z_{i,j} \cap W_i$ is enumerated in C and permanently restrained from Y_i. Hence, $x \notin B$ because j cannot be bad for N_i. Hence, $x \notin A$ because nothing other than x entering B will cause RED to put x into A.

The effect of strategy ρ for requirement N_i is this. For every j we may imagine a movable marker $\Gamma_{i,j}$ resting on an element $\Gamma_{i,j,s} = x \in (S_{i,j,s} - C_s) \cap R_s$ until $x \in W_{i,s} \cap Z_{i,j,s}$, at which point x enters C. Now either x remains in $\overline{B}$ forever and

$\lim_s \Gamma_{i,j,s} = x$, or else x eventually enters B at which point $\Gamma_{i,j}$ is removed forever and never again attached to any y because j has been proved bad for N_i. Now it is easy to construct a promptly simple set A such that $T(A)$ in the usual fashion.

2.4. *Nonlow sets*

The previous property $T(A)$ guaranteed such a rapid flow of elements from $C - B$ into A that A was complete. The next property $NL(A)$ is more subtle but guarantees a sufficiently large flow into A so that A is nonlow, but A can still be low$_2$ and hence incomplete. It is also compatible with $NL(A)$ to make A promptly simple as for $T(A)$. (Recall that a c.e. set A is *low$_n$* if $A^{(n)} \equiv_T \emptyset^{(n)}$ and *high$_n$* if $A^{(n)} \equiv_T \emptyset^{(n+1)}$, and similarly for c.e. *degrees*.) Define $NL(A)$ by:

$$NL(A)\colon\ (\exists C \supseteq A)(\forall B_0, B_1)\big[\tag{10}$$

$$[B_0 \sqcup B_1 = C]\,\&\,[B_1 \subseteq A] \quad .\Longrightarrow (\exists \text{ computable } R)\big[\tag{11}$$

$$[A \supseteq B_0 \cap R]\,\&\,[R - B_0 \text{ is not c.e.}]\big]\big]. \tag{12}$$

THEOREM 2.11.
(i) $(\forall W)[NL(W) \Longrightarrow W$ *is not low*].
(ii) $(\exists A)[NL(A)\,\&\, A$ *is promptly simple and low*$_2$ $\&\ \overline{A}$ *is semi-low*$_{1.5}$].

This refutes the appealing conjecture based on Maass [1983] as discussed in Section 2.4.3. Note that in the statement of $NL(A)$ we can weaken the first clause of (11) to "$B_0 \sqcup B_1 =^* C$." Suppose $B_0 \cup B_1 =^* C$, and $F = C - (B_0 \cup B_1)$ is finite. Then $\widehat{B_0} = B_0 \cup F$ and B_1 satisfy (11) so R must exist satisfying (12) for $\widehat{B_0}$ and B_1, and therefore for B_0 and B_1.

PROOF. (Proof of Theorem 2.11, sketch only, see details in Harrington and Soare [1998].) As for the preceding properties we prove the theorem with two theorems. □

2.4.1. *The first theorem for NL(A).*

THEOREM 2.12 (Blue). $(\forall W)[NL(W) \Longrightarrow W$ *is not low*].

The informal intuition behind $NL(A)$ is this. BLUE will enumerate B_0 and B_1 to satisfy the hypotheses (10) and (11), and we let R be the reply by RED satisfying (12). Let all sets be restricted to R. Let ν_1 be $R - C$, ν_2 be $C - (B_0 \cup B_1)$, ν_3 be $B_0 - A$, ν_4 be $B_0 \cap A$, ν_5 be $B_1 - A$, and ν_6 be $B_1 \cap A$, again interpreted dynamically. The second clause of (12) guarantees that $R - C$ is not c.e., so there is a flow of infinitely many elements from state ν_1 to ν_2. When such an element x arrives in ν_2, BLUE can wait an arbitrarily long time but must eventually put x either into B_1 (providing $x \in A$ already because of the second clause of (11)) or into B_0 (state ν_3) from which RED must eventually move x into A (state ν_4) because of the first clause of (12). (Note that there is no flow from ν_2 to ν_5 only to ν_6.)

2.4.2. *The second theorem for NL(A).* The property $NL(A)$ can be used to show that certain automorphisms do not exist. To get the maximum power we wish to construct such an A which has as many other lowness properties as possible other than low$_1$, because these lowness properties tend to facilitate building automorphic copies of A. In addition to low$_2$ another lowness property which has been studied by Maass, Soare, Stob, and others in connection with automorphisms is the property defined in [Soare, 1987, p. 230] of $\overline{A}$ being semi-low$_{1.5}$, namely that $\{i\colon |W_i \cap \overline{A}| = \infty\} \leqslant_m \text{Inf}$, where $\text{Inf} = \{i\colon |W_i| = \infty\}$. Maass [1983] proved that a coinfinite set c.e. set A satisfies $\mathcal{L}^*(A) \cong_{\mathit{eff}} \mathcal{E}^*$ iff $\overline{A}$ is semi-low$_{1.5}$, where $\cong_{\mathit{eff}}$ denotes moreover that the isomorphism is *effective* [Soare, 1987, p. 244].

THEOREM 2.13 (Red). *There exists a c.e. set A such that*:
(i) $NL(A)$;
(ii) *A is promptly simple*;
(iii) *A is low*$_2$; *and*
(iv) $\overline{A}$ *is semi-low*$_{1.5}$.

We first sketch the strategy to achieve each of these four properties and then assemble them on a tree.

The strategy to guarantee NL(A). Player RED enumerates C and A. Let $\{(B_0^e, B_1^e)\}_{e\in\omega}$ be a computable enumeration of all disjoint pairs of c.e. sets (B_0, B_1) such that $B_0 \cup B_1 \subseteq C$. Next RED specifies the computable set $R_e = \omega^{[e]}$. To achieve $NL(A)$ as in RED must enumerate A to meet for every e the requirement,

$$\begin{aligned} Q_e\colon\ & [B_0^e \sqcup B_1^e = C] \,\&\, [B_1^e \subseteq A] \\ & . \Longrightarrow . \, [A \supseteq B_0^e \cap R_e] \,\&\, [R_e - B_0^e \text{ is not c.e.}] \end{aligned}$$

To achieve that $R_e - B_0^e$ is not c.e. it suffices to meet for all $i \in \omega$ the subrequirement,

$$M_{e,i}\colon \quad W_i \neq R_e - B_0^e.$$

Let $R_{e,i} = R_e^{[i]}$, $i \in \omega$, a computable decomposition of R_e into infinitely many infinite pieces. RED will guarantee that $C - R_{e,i}$ is infinite for all e and i. To meet $M_{e,i}$, RED waits until a sufficiently large initial segment of $R_{e,i,s} - C_s$ has been enumerated in $W_{i,s}$ and then selects some $x \in (W_{i,s} \cap R_{e,i,s}) - C_s$ and enumerates x into C_{s+1}, and waits for BLUE to enumerate x in B_0^e or B_1^e. If BLUE enumerates x into B_1^e before RED enumerates x into A, then RED restrains x from A forever so $B_1^e \not\subseteq A$ and the requirement Q_e is automatically met. If BLUE enumerates x into B_0^e then $B_0^e \cap W_i \neq \emptyset$ so requirement $M_{e,i}$ is satisfied forever. The remaining case is that BLUE never enumerates x into B_0^e or B_1^e in which case $B_0^e \cup B_1^e \neq C$ and again requirement Q_e is met.

The strategy to make A promptly simple. This is the standard strategy.

The strategy to make $\overline{A}$ semi-low$_{1.5}$. To ensure $\overline{A}$ semi-low$_{1.5}$ RED defines an array $\{Z_i\}_{i\in\omega}$ and defines h by $W_{h(i)} = Z_i$ so as to meet the requirement,

$$S_i\colon \quad |W_i \cap \overline{A}| = \infty \Longleftrightarrow |W_{h(i)}| = \infty.$$

Assume we have enumerated 0 through $k-1$ in $W_{h(i)}$. We wait until there are at least $g(i,k)$ many fresh members of $W_{i,s} - A_s$, where $g(i,k)$ is a predetermined computable function. Put these members in a set $V_{i,k}$, restrain them from A with priority $S_{i,k}$, and enumerate k in $W_{h(i),s+1}$. Positive requirements of higher priority may later enumerate some members of $V_{i,k}$ in A, but the size of $g(i,k)$ and the construction must be arranged so that once k enters $V_{i,k}$ at least one element of $V_{i,k}$ remains forever in $\overline{A}$. Hence, requirement S_i will be satisfied. The outcome of S_i is Π_2 if $W_{h(i)}$ is infinite and Σ_2 otherwise. These outcomes are denoted by 0 and 1, respectively.

The strategy to make A low$_2$. To simplify the proof define

$$\Psi^X_{e,s}(y) = \begin{cases} \Phi^X_{e,s}(y) & \text{if } (\forall z \leqslant y)[\Phi^X_{e,s}(z)\downarrow], \\ \text{undefined} & \text{otherwise.} \end{cases}$$

Define the use function, $\psi_{e,s}(y) = \varphi_{e,s}(y)$ if $\Psi_{e,s}(t)$ is defined, and let $\psi_{e,s}(y)$ be undefined otherwise. Note that Ψ^X_e is either total or has finite domain, and Ψ^X_e is total iff Φ^X_e is total. The first outcome is the Π_2-outcome for N_e and the second is the Σ_2-outcome. N_e can ensure its success against a single positive requirement P_i of lower priority as follows. P_i has two separate forms, P^1_i which guesses that N_e will have the Σ_2-outcome, and P^0_i which guesses that N_e will have the Π_2-outcome. Whenever a new value appears for $\psi_{e,s}$, the strategy P^1_i is completely reset. Since P^0_i believes that Ψ_e is total, it can afford to wait for arbitrarily many values $\psi_{e,s}(y)$ to appear before defining its next value x^e_n for Γ^e. Therefore, it can wait until $\psi_{e,s}(y)\downarrow$ for all $y \leqslant n$ before choosing x^e_n greater than all these values. Hence, a value $\psi_{e,s}(y)$ can be later destroyed by the entry of some smaller x^e_m into A only if $m < y$, and this can happen for a single m at most once. Thus, in the presence of only this single positive requirement P_i, N_e just notices that this strategy guarantees that

$$\Psi^A_e \text{ is total} \Longleftrightarrow (\forall y)(\forall k)(\exists s > k)\big[\Psi^{A_s}_{e,s}(y)\downarrow\big].$$

The right hand side is a Π_2 predicate so Tot^A is reduced to a Π_2 question and hence to $\emptyset''$. Now requirement N_e easily accommodates *all* lower priority requirements $\{P_i\}_{i>e}$ in the same fashion.

The tree for the construction. We convert these four strategies into a tree construction. Let the tree T be $\{0,1\}^{<\omega}$. Put the empty node λ on T.

(1) If $\alpha \in T$ and $|\alpha| = 4e$, then associate with α a version of the strategy N_e, to make A low$_2$, and put the nodes $\alpha\frown 0$ and $\alpha\frown 1$ on T which represent the Π_2 and Σ_2-outcomes, respectively, of the α strategy.

(2) If $\alpha \in T$ and $|\alpha| = 4e + 1$, then associate with α a version of the strategy for S_i to make $\overline{A}$ semi-low$_{1.5}$, and put the outcomes $\alpha\frown 0$ and $\alpha\frown 1$ on T representing the Π_2 and Σ_2-outcomes of the α strategy.

(3) If $\alpha \in T$ and $|\alpha| = 4e + 2$, then associate with α the strategy P_e for prompt simplicity of A, and put $\alpha\frown 1$ on T. (In cases 3 and 4 the outcomes are only Σ_2 so we put only one successor node on T.)

(4) If $\alpha \in T$ and $|\alpha| = 4k + 3$, and $k = \langle e, i\rangle$, then associate with α the strategy $M_{e,i}$, for ensuring that A has the NL property.

Nodes defined under the first two cases are *negative* nodes, and under the second two cases *positive* nodes. Each positive node β contributes at most one element to A. Hence, each negative node α will be injured at most finitely often by nodes $\beta \subset \alpha$ or $\beta <_{\mathrm{L}} \alpha$ and will restart its strategy after each such injury.

The construction is roughly as described above except in addition the positive nodes must respect negative restraint of higher priority nodes as follows. If β is a positive node, $|\beta| = 4j + 3$ or $4j + 4$, $\alpha\frown 0 \subseteq \beta$, or $\alpha <_{\mathrm{L}} \beta$, then in addition to the above conditions, when β chooses a witness x: if $|\alpha| = 4e$, then β must wait until $\psi_{\alpha,s}(k)\downarrow$ for all $k < j$, and must choose x to exceed all these values; and if $|\alpha| = 4e + 1$, then β must wait until α defines $V_{\alpha,k}$ for all $k \leqslant j$ and β must choose x not in these sets.

Thus, α will have the kth component of its strategy (i.e. $\psi_{\alpha,s}(k)$ or $V_{\alpha,k,s}$) disturbed by only finitely many nodes (those β with $|\beta| < k$), and at most once by each of these. On the other hand if $\beta \subset f$, the true path, then for every α such that $\alpha\frown 0 \subseteq \beta$ the action of the kth component of the α strategy will stabilize because the α strategy has the Π_2-outcome. Hence, there will be at most a finite obstacle of restraint for β to overcome, but an infinite reservoir of elements from which to choose a witness. Therefore, the β strategy will succeed. This completes the proof of Theorem 2.11.

2.4.3. *Applications of the property NL(A).* In a very influential paper Maass [1982] introduced the notion of a c.e. set being *generic* and proved that every generic c.e. set is low$_1$ and promptly simple. He also showed that if A and B are promptly simple and low$_1$ then A is automorphic to B ($A \simeq B$). Meanwhile, Soare [1982] proved that if A is a coinfinite c.e. set and $\overline{A}$ is low$_1$ or even semi-low$_1$, then $\mathcal{L}^*(A) \cong_{\mathit{eff}} \mathcal{E}^*$, where $\mathcal{L}^*(A)$ denotes the supersets of A under inclusion, and $\cong_{\mathit{eff}}$ denotes an effective isomorphism. Maass [1983] then sharpened the second result by weakening the hypothesis to semi-low$_{1.5}$ and proving that if A is a coinfinite c.e. set then $\mathcal{L}^*(A) \cong_{\mathit{eff}} \mathcal{E}^*$ iff $\overline{A}$ is semi-low$_{1.5}$. Combining these ideas of these results led one naturally to ask the following tempting question.

QUESTION 2. If A and B are both promptly simple and their complements are semi-low$_{1.5}$ then are A and B automorphic?

In a similar vein Cholak raised the following question. The outer splitting property was defined by Maass and follows from the semi-low$_{1.5}$ property.

QUESTION 3. Let A be a promptly simple set and $\mathbf{p}$ a promptly simple degree. If $\overline{A}$ is semi-low$_2$ and has the outer splitting property then is there a c.e. set B with $\deg(B) \leqslant \mathbf{p}$, such that $A \simeq_{\Delta_3^0} B$?

The point of these questions is that lowness properties on $\overline{A}$ guarantee that $\mathcal{L}(A)$ is well behaved, and prompt simplicity properties on A guarantee covering. Hence, one should be able to put the two halves together to produce an automorphism, at least a Δ_3^0 automorphism [Harrington and Soare, 1996c]. A plausibility argument for the assertion in Question 2 is the following. Since $\overline{A}$ and $\overline{B}$ are both semi-low$_{1.5}$ we know that

$$\mathcal{L}^*(A) \cong_{eff} \mathcal{L}^*(B),$$

and we begin to build a permutation h from $\overline{A}$ to $\overline{B}$ which induces this isomorphism. During this process elements will fall from $\overline{A}_s$ to A and from $\overline{B}_s$ to B but as in the maximal set automorphism [Soare, 1974], but using prompt simplicity of A and B we know that a stream of elements enters A promptly enough to cover the stream of elements entering B so as in the Extension Theorem [Soare, 1987, p. 352] we can cover the stream of elements entering B by a prompt stream entering A and conversely.

To refute both questions choose a c.e. set B which is low$_1$ and promptly simple and choose A to have the properties of Theorem 2.13, namely: $NL(A)$; A is promptly simple; A is low$_2$; and $\overline{A}$ is semi-low$_{1.5}$. Now B cannot satisfy property $NL(B)$ by Theorem 2.12 because B is low$_1$. But the property $NL(X)$ is $\mathcal{E}$-definable and therefore preserved under all automorphisms of $\mathcal{E}$. Hence, A and B cannot be automorphic even by a Δ_3^0 automorphism.

What goes wrong with the plausibility argument above? When we apply the plausibility argument to the case where A and B are as here we see that Maass's theorem guarantees that in the *limit* we have $\mathcal{L}^*(A) \cong \mathcal{L}^*(B)$ but we do not necessarily get that $\overline{B}$ co-covers $\overline{A}$ in real time *during* the construction. (Maass even explicitly remarks this, but his remark was apparently largely overlooked.) Therefore, on the $\overline{A}$ side a certain pathology develops which the $\overline{B}$ side cannot duplicate but need not duplicate because it will pass into A leaving the complements properly matched. However, this pathology already ruins the possibility of an automorphism and cannot be repaired after entering A by the mere hypothesis of prompt simplicity.

The deeper conclusion here is that to produce an automorphism from A to B we need much more than isomorphisms on the complements and promptly simple type properties to guarantee covering. We need to study in a much deeper the way the c.e. sets relate the complement $\overline{A}$ to A not merely how they behave in isolation. The negative answer to Question 1 by the property $T(A)$ in Section 2.4.2 reinforces this principle.

3. Automorphisms of $\mathcal{E}$

3.1. *Some results on automorphisms*

Automorphisms are useful for two reasons. First, if we are unable to exhibit an $\mathcal{E}$ definition of some property P (see $\overline{\mathbf{L}}_1$ in Section 4), then we may be able to produce an automorphism Φ of $\mathcal{E}$ mapping some A with property P to some B with $\neg P$, thereby proving P is undefinable in $\mathcal{E}$. The second use is in the spirit of Klein's *Erlangen Programm,* which is to classify some mathematical object such as a geometry in terms of the properties left invariant under its automorphisms. The first application of the automorphism method was to classify orbits of maximal sets following Klein's program.

To answer a question of Martin and Lachlan, Soare [1974] produced a new method for constructing automorphisms of $\mathcal{E}$ and used it to prove that any two maximal sets are automorphic. The method begins by choosing an appropriate skeleton for the c.e. sets (i.e. one member $U_{g(e)}$ from each class W_e^*) and then, by a fairly complicated construction, building an automorphism of $\mathcal{E}$ which is *effective* in the sense that there is a computable function $h(e)$ such that $\Phi(U_e) =^* W_{h(e)}$. An automorphism Φ is Δ_3^0 if there is a Δ_3^0 permutation h of ω such that $\Phi(W_e) = W_{h(e)}$ for all $e \in \omega$.

Recently, Harrington and Soare [1996c], and simultaneously Cholak [1995] building on some conversations with Harrington, combined the essence of the effective automorphism method with the tree method of Lachlan [1975] to produce a powerful new method for constructing Δ_3^0-automorphisms, and used it to prove, for example, the following.

THEOREM 3.1 (Harrington and Soare [1996c], Cholak [1995]). *For every noncomputable c.e. set A there is a c.e. set B which is high (i.e. deg*$(B') = \mathbf{0}''$) *such that A is Δ_3^0-automorphic to B.*

Theorem 3.1 asserts that every nontrivial orbit contains a high set. This has some interesting corollaries for noninvariant classes as we shall see in Section 4. Before considering this, let us consider the relation of Theorem 3.1 to Theorem 2.6. Theorem 3.1 implies that $Q(A)$ which prevents A from being complete cannot be extended to cause A to be low or even nonhigh. It says if we are willing to extend the target set from the complete degree to the high degrees, then the automorphism builder (BLUE) wins. On the other hand, if we insist on mapping A to a complete set, what stronger hypotheses must we place on A?

THEOREM 3.2 (Harrington and Soare [1996c]). *If A is any c.e. set which is prompt (i.e. of promptly simple degree) or even if A is* almost prompt *then A is automorphic to a complete set.*

Cholak, Downey, and Stob [1992] proved this result under the stronger hypothesis "A is a promptly simple *set*," and Harrington and Soare extended this with the much weaker hypothesis "of promptly simple degree". They then realized that the essence of the hypothesis is a much weaker promptness property still, which they named "almost promptly simple," and which is based on the following notion of n-c.e. (previously called n-r.e.) sets.

In related work Downey and Stob have proved that the class of sets known as HHM sets are all automorphic to complete sets. We discuss the relationship between HHM and almost prompt in [Harrington and Soare, 1996c, §12]. Also Harrington [ta] proved that there is no "fat" orbit, i.e. one containing a set in every nonzero degree.

DEFINITION 3.3. (i) A set $X \leqslant_T K$ is *n-c.e.* if $X = \lim_s X_s$ for some computable sequence $\{X_s\}_{s\in\omega}$ such that for all x, $X_0(x) = 0$ and

$$\operatorname{card}\big\{s\colon\ X_s(x) \neq X_{s+1}(x)\big\} \leqslant n.$$

For example, the only 0-c.e. set is $\emptyset$, the 1-c.e. sets are the usual c.e. sets, and the 2-c.e. sets are the d.c.e. sets (previously called d.r.e.).

(ii) Such a sequence $\{X_s\}_{s\in\omega}$ is called an *n-c.e. presentation* of X.

It is well-known and easy to show [Soare, 1987, Exercise III.3.8, p. 38] that for $n > 0$, X is n-c.e. iff

$$X = (W_{e_1} - W_{e_2}) \cup (W_{e_3} - W_{e_4}) \cup \cdots \cup W_{e_{2k+1}}, \text{ or} \tag{13}$$

$$X = (W_{e_1} - W_{e_2}) \cup (W_{e_3} - W_{e_4}) \cup \cdots \cup (W_{e_{2k+1}} - W_{e_{2k+2}}), \tag{14}$$

according as $n = 2k+1$ is odd or $n = 2k+2$ is even.

DEFINITION 3.4. For $n = 0$ let $X_0^0 = \emptyset$. For $n > 0$ and $e = \langle e_1, e_2, \ldots, e_n\rangle$ define

$$X_e^n = (W_{e_1} - W_{e_2}) \cup \cdots, \tag{15}$$

as in (13) or (14) according as n is odd or even. We say that $\langle n, e\rangle$ is an *n-c.e. index* for X_e^n. Let

$$X_{e,s}^n = (W_{e_1,s} - W_{e_2,s}) \cup \cdots. \tag{16}$$

DEFINITION 3.5. Let A be a c.e. set and let $\{A_s\}_{s\in\omega}$ be a computable enumeration of A. We say A is *almost prompt*, abbreviated *a.p.*, if there is a nondecreasing computable function $p(s)$ such that for all n and e,

$$X_e^n = \overline{A} \Longrightarrow (\exists x)(\exists s)[x \in X_{e,s}^n \ \&\ x \in A_{p(s)}]. \tag{17}$$

(ii) We say A is *very tardy* if A is *not* almost prompt, namely if for every nondecreasing computable function $p(s)$, the negation of (17) holds. In this case, if a fixed

n works uniformly for *all* such functions p then we say A is n-*tardy*. Note that for the case $n = 2$ this is equivalent to our definition of being 2-tardy [Harrington and Soare, 1996c, 1996a] which is a very important special case.

The results here on almost prompt sets and on 2-tardy sets stress the very important, but previously often hidden, connection between dynamic properties on one hand, and definable properties and automorphisms on the other.

Here is another unexpected connection. In Theorem 3.2 we put extra hypotheses on the set A so that it could be mapped to a complete set. Now we ask what hypotheses are necessary on a set D so that it can be computably coded into some set B in *every* nontrivial orbit. The answer involves the 2-tardy sets in an unexpected way. The *orbit* of A is the class $[A]$ of all sets B automorphic to A, written $A \simeq B$. By a *nontrivial* orbit we mean the orbit of a *noncomputable* c.e. set A.

DEFINITION 3.6. (i) We say X can be *coded into the orbit of* A, denoted by $X \leqslant_{\mathrm{T}} [A]$, if $X \leqslant_{\mathrm{T}} B$ for some $B \in [A]$.

(ii) We say X is *codable* if X can be coded in every nontrivial orbit, namely if $X \leqslant_{\mathrm{T}} [A]$ for every $A >_{\mathrm{T}} \emptyset$.

THEOREM 3.7 (Coding Theorem) (Harrington and Soare [1996d]). *If D is 2-tardy (say if $Q(D)$ holds) then D is codable.*

COROLLARY 3.8. *A set X is codable iff $X \leqslant_{\mathrm{T}} D$ for some D satisfying $Q(D)$ iff $X \leqslant_{\mathrm{T}} D$ for some 2-tardy D.*

PROOF. If $X \leqslant_{\mathrm{T}} D$ and D is 2-tardy, then

$$X \leqslant_{\mathrm{T}} D \leqslant_{\mathrm{T}} [A]$$

for every $A >_{\mathrm{T}} \emptyset$ by Theorem 3.7. If $X \leqslant_{\mathrm{T}} [A]$ for every $A >_{\mathrm{T}} \emptyset$ then $X \leqslant_{\mathrm{T}} C$ for some C and D such that $C \in [D]$ and $Q(D)$, hence D is 2-tardy. Hence, $Q(C)$ because Q is $\mathcal{E}$-definable. □

Note that $Q(D)$ implies that D is a major subset and hence D is high. Thus, Theorem 3.7 is a very strong generalization of Theorem 3.1 because by Theorem 3.7 if A is noncomputable and $Q(D)$ holds, then there exists $B \in [A]$ such that $D \leqslant_{\mathrm{T}} B$ so D and B are both high. Thus, $Q(D)$, and its associated property of D being 2-tardy, were originally introduced just to force the enumeration of elements into D to be sufficiently *slow* so that D would have to be incomplete. Now we see that it also forces a sufficiently slow enumeration so that the machinery building an automorphism $\Phi(A) = B$ has time to code into B the fact that x enters D. This connection of the *speed* of elements entering D to its *codability* into orbits is so interesting that we explore it further for a moment.

COROLLARY 3.9. *If S is a promptly simple set (or even of promptly simple degree) then S is not codable.*

PROOF. If S is promptly simple (or even of prompt, i.e. of promptly simple degree) and $S \leqslant_T C$ then C is also prompt [Soare, 1987, Corollary XIII.1.9, p. 287], thus not tardy, thus not 2-tardy, thus $\neg Q(C)$. Hence, by Corollary 3.8 S is not codable. □

Thus, codable sets can be high by Theorem 3.1, while noncodable sets can be low (choose a low promptly simple set in Corollary 3.9). Therefore, one of our main conclusions is that *the question of whether a set X can be coded* into an arbitrary orbit $[A]$ *depends more on the speed of enumeration of X (prompt or tardy) than on its information content (high or low).*

The fact that K is not codable has more to do with the fact that K is prompt (i.e. of promptly simple degree) than that K has complete information content. For example, to show that $K \leqslant_T B$ for some $B \in [A]$ we must do very rapid coding, but if $Q(A)$ holds then $Q(B)$ holds for every $B \simeq A$ (because Q is $\mathcal{E}$-definable although "2-tardy" is not). Thus, B is 2-tardy, hence tardy, and hence incomplete.

To code D into $[A]$, the orbit of a given noncomputable set A, we must construct $B \simeq A$ and a computable functional Ψ such that $D = \Psi(B)$. We must define the use function $\psi(n)$ to be a convenient element y not yet in B such that if n enters D, then we can gradually move y into B. The property of D being 2-tardy, as described later, will imply that there is a computable function g (played by BLUE) such that if n wants to enter D, it must first *declare* that intention at some stage s and then wait for some stage $t \geqslant p(s)$ before doing so. Since the automorphism machinery imposes considerable delay in putting $\psi(n)$ into B after first starting the process, BLUE arranges that when n declares its intention at stage s, BLUE starts $\psi(n)$ toward B immediately and makes $p(s)$ so large that $\psi(x)$ has arrived in B by stage $p(s)$ before n has arrived in D.

The entire automorphism construction is more complicated and is played on a tree T of nodes. Thus, our actual coding procedure is a bit more complicated as it is performed repeatedly for several nodes $\alpha \in T$. Perhaps, $\psi(n)$ begins in the region R_γ (defined in [Harrington and Soare, 1996c]) with witness $y_\gamma < \psi(n)$. Now in its journey toward D, element n passes through a series of "gates" G_α for $\alpha \in T$ with witnesses y_α, each time undergoing a delay as above with G_α in place of D. (In reality these sets G_α are simply different names for the set D at least for $\alpha \subset f$, where f is the true path through T.) After each successful entry y_α into B, the use function is redefined to some number above y_β where $\beta = \alpha^-$, the predecessor of α and $\psi(n)$ passes from region R_α to region R_β. Best of all, to complete the proof we do not need to know anything about the machinery for generating Δ^0_3-automorphisms of $\mathcal{E}$. Rather, in [Harrington and Soare, 1996c, §7] we develop the full automorphism machinery and a coding theorem, which can be applied here without further proof. Further details on coding can be found in [Harrington and Soare, 1996d]. We give a brief sketch here.

3.2. *A sketch of the Δ_3^0-automorphism method*

By [Soare, 1987, p. 343] building an automorphism of $\mathcal{E}$ is equivalent to building one of $\mathcal{E}^*$, the quotient lattice of $\mathcal{E}$ modulo the ideal $\mathcal{F}$ of finite sets. To do this we fix two copies of the natural numbers ω and $\hat{\omega}$. We let variables $x, y, \ldots$ $(\hat{x}, \hat{y}, \ldots)$ range over ω $(\hat{\omega})$. Normally, we shall specify the definitions and action for only one side (usually the ω-side) since those for the opposite side will be entirely dual.

We view the construction of the automorphism Φ as a game between two players in the sense of Lachlan [1970]. Player 1 (whom we call RED) produces two standard indexings $\{U_n\}_{n\in\omega}$ and $\{V_n\}_{n\in\omega}$ of the c.e. sets, where we view U_n as being on the ω-side and V_n on the $\hat{\omega}$-side. Player 2 (whom we call BLUE) responds by building c.e. sets $\{\widehat{U}_n\}_{n\in\omega}$ on the $\hat{\omega}$-side and $\{\widehat{V}_n\}_{n\in\omega}$ on the ω-side. The condition necessary to show that this correspondence $\Phi(U_n) = \widehat{U}_n$ and $\widehat{V}_n = \Phi^{-1}(V_n)$ is an automorphism is best stated in terms of the following notion of full e-state.

DEFINITION 3.10. Given two sequences of r.e. sets $\{X_n\}_{n\in\omega}$ and $\{Y_n\}_{n\in\omega}$. Define $\nu(e,x)$, the *full e-state* of x with respect to (w.r.t.) $\{X_n\}_{n\in\omega}$ and $\{Y_n\}_{n\in\omega}$ to be the triple $\langle e, \sigma(e,x), \tau(e,x)\rangle$, where

$$\sigma(e,x) = \{i\colon i \leqslant e \;\&\; x \in X_i\}, \quad \text{and} \quad \tau(e,x) = \{i\colon i \leqslant e \;\&\; x \in Y_i\}.$$

To see that Φ is an automorphism it suffices to satisfy the requirement,

$$\begin{aligned}&(\forall\nu)(\exists^\infty x \in \omega)\big[\nu(e,x) = \nu \text{ w.r.t. } \{U_n\}_{n\in\omega} \text{ and } \{\widehat{V}_n\}_{n\in\omega}\big]\\ &\iff (\exists^\infty \hat{y} \in \hat{w})\big[\nu(e,\hat{y}) = \nu \text{ w.r.t. } \{\widehat{U}_n\}_{n\in\omega} \text{ and } \{V_n\}_{n\in\omega}\big].\end{aligned} \tag{18}$$

DEFINITION 3.11. Given computable enumerations $\{X_s\}_{s\in\omega}$ and $\{Y_s\}_{s\in\omega}$ of c.e. sets X and Y define

(i) $X\backslash Y = \{z\colon (\exists s)[z \in X_s - Y_s\}$,

(ii) $X \searrow Y = (X\backslash Y) \cap Y$.

3.2.1. *Using a tree T to define the automorphism Φ.*

In the effective automorphism method $\{\widehat{U}_n\}_{n\in\omega}$ is a *computable* sequence of c.e. sets so that Φ has an effective presentation. For the Δ_3^0-automorphism method we combine the ideas of the effective automorphism method with the tree method of Lachlan [1970] as explained in [Soare, 1987, Chap. XIV]. We shall define in Section 3.2.8 a computable tree T with *true path* f. For each $n \in \omega$ there is some $m_n \in \omega$ such that for every $\alpha \in T$ of length m_n, $\widehat{U}_\alpha$ will be a potential candidate for $\widehat{U}_n$ and if $\alpha \subset f$ then $U_\alpha =^* U_n$ and $\widehat{U}_\alpha$ will be the correct candidate for $\widehat{U}_n$. Thus, f will specify the sequence $\{\widehat{U}_{f\restriction m_n}\}_{n\in\omega}$ which will be the desired sequence $\{\widehat{U}_n\}_{n\in\omega}$. In a tree construction f is not in general computable but only $\emptyset''$-computable so the sequence $\{\widehat{U}_n\}_{n\in\omega}$ will only have a $\emptyset''$-computable (i.e. Δ_3^0) presentation.

We use the usual notation for trees as in [Soare, 1987, p. 301]. By coding the intended nodes we may regard the tree T as a subset of $\omega^{<\omega}$. Let $[T]$ be the set of infinite paths through T, where h is a infinite path through T if $h \upharpoonright n \in T$ for all n. Let $\alpha, \beta, \gamma, \delta, \ldots$ range over T. Let $|\alpha|$ denote the length of α. Let $\alpha \subseteq \beta$ ($\alpha \subset \beta$) denote that string β extends (properly extends) α. Let λ denote the empty string, and α^- denote the predecessor of α if $\alpha \neq \lambda$. Let $\langle a \rangle$ denote the string consisting of element a alone. Let $\alpha^\frown\beta$ denote the concatenation of string α followed by string β.

DEFINITION 3.12. Let $\alpha, \beta \in T$.

(i) α is to the *left* of β ($\alpha <_{\mathrm{L}} \beta$) if

$$(\exists a, b \in \omega)(\exists \gamma \in T)\,[\gamma^\frown\langle a\rangle \subseteq \alpha \;\&\; \gamma^\frown\langle b\rangle \subseteq \beta \;\&\; a < b].$$

(ii) $\alpha \leqslant \beta$ if $\alpha <_{\mathrm{L}} \beta$ or $\alpha \subseteq \beta$.

(iii) $\alpha < \beta$ if $\alpha \leqslant \beta$ and $\alpha \neq \beta$.

(iv) If $h \in [T]$ we say $\alpha <_{\mathrm{L}} h$ ($h <_L \alpha$, $\alpha < h$, $h < \alpha$) if there exists $\beta \subset h$ such that $\alpha <_L \beta$ ($\beta <_L \alpha$, $\alpha < \beta$, $\beta < \alpha$, respectively).

Note that $\alpha \leqslant \beta$ is a kind of modified Kleene–Brouwer ordering. If $\alpha \subset \beta$ then α is a *predecessor* of β and β is a *successor* of α. (Thus, we view the tree T as growing downward with λ as the top node.)

3.2.2. *The α-section S_α, α-region R_α, and c.e. set Y_α.* We divide up the ω-side into disjoint α-*sections*, S_α, for $\alpha \in T$. We shall define during the construction a function $\alpha(x, s)$ with range T which indicates that x is in section $S_{\alpha(x,s)}$ at the end of stage s, and we shall guarantee that $\alpha(x) = \lim_s \alpha(s, x)$ exists. The α-*region* R_α consists of all S_γ such that $\alpha \subseteq \gamma$. For each stage s we define,

$$\begin{aligned} S_{\alpha,s} &= \{x\colon\ \alpha(x,s) = \alpha\},\\ R_{\alpha,s} &= \{x\colon\ \alpha(x,s) \supseteq \alpha\},\ \text{ and}\\ Y_{\alpha,s} &= \bigcup\{R_{\alpha,t}\colon\ t \leqslant s\}. \end{aligned}$$

Define $S_{\alpha,\infty} = \{x\colon\ \alpha(x) = \alpha\}$, and $R_{\alpha,\infty} = \{x\colon\ \alpha(x) \supseteq \alpha\}$. An element x will enter R_α at most once, but x may later leave R_α. Thus, $R_{\alpha,\infty}$ is a d.c.e. set, but the sets Y_α are c.e. with simultaneous computable enumeration $\{Y_{\alpha,s}\}_{\alpha \in T, s \in \omega}$ and Y_α consists of those x which enter R_α at some stage. If $\alpha \subset f$ then we shall ensure that $Y_\alpha =^* R_{\alpha,\infty}$ so $R_{\alpha,\infty}$ is c.e. It will follow that if $\alpha \neq \lambda$ then for all $x \in Y_\alpha$, $x > |\alpha|$.

We shall guarantee that for all $\alpha \in T$, $\alpha \neq \lambda$,

$$Y_\alpha \backslash Y_{\alpha^-} = \emptyset, \text{ and} \tag{19}$$

$$\alpha \subset f \Longrightarrow R_{\alpha,\infty} =^* Y_\alpha =^* \omega. \tag{20}$$

We shall ensure (19) by making x enter S_{α^-} before x enters R_α. Also x will enter R_α at most once (although x may later leave R_α). During the construction we shall define a computable sequence $\{f_s\}_{s \in \omega}$ such that $f = \liminf_s f_s$.

If $f_s <_L \alpha$ for some $s \geqslant x$ we say x is *α-ineligible* at all stages $t \geqslant s$, and we insist that $x \notin S_{\alpha,t}$. Hence, $R_{\alpha,\infty} = \emptyset$ for all α with $f <_L \alpha$. Secondly, Y_α will be finite for all $\alpha <_L f$. Finally, $S_{\alpha,\infty}$ will be finite for all α. These three facts imply (20).

3.2.3. *The α-states $\nu(\alpha, x, s)$, and lists $\mathcal{E}_\alpha$, $\mathcal{F}_\alpha$, $\mathcal{M}_\alpha$.* For conceptual simplicity we do as little action as possible at each node $\alpha \in T$. If $|\alpha| \equiv 1 \bmod 5$ ($|\alpha| \equiv 2 \bmod 5$) we consider one new U set (V set). If $|\alpha| \equiv 3 \bmod 5$ ($|\alpha| \equiv 4 \bmod 5$) we consider new α-states ν ($\hat{\nu}$) which may be non well-resided on Y_α ($\widehat{Y}_\alpha$). If $\alpha \equiv 0 \bmod 5$ we make no new commitments for the automorphism machinery but we may perform action for some additional requirement (such as coding information into B for Theorem 3.1). We shall arrange for all $n \in \omega$ that for $\alpha \subset f$,

$$|\alpha| = 5n + 1 \Longrightarrow U_\alpha =^* U_n, \text{ and} \tag{21}$$

$$|\alpha| = 5n + 2 \Longrightarrow V_\alpha =^* V_n. \tag{22}$$

We let U_α and $\widehat{U}_\alpha$ (V_α and $\widehat{V}_\alpha$) be undefined for α such that $|\alpha| \not\equiv 1 \bmod 5$ ($|\alpha| \not\equiv 2 \bmod 5$). We let e_α ($\hat{e}_\alpha$) correspond to n in (21) (respectively (22)). Namely, define $e_\lambda = \hat{e}_\lambda = -1$ and if $|\alpha| \equiv 1 \bmod 5$ then let $e_\alpha = e_{\alpha^-} + 1$, and otherwise let $e_\alpha = e_{\alpha^-}$. Define $\hat{e}_\alpha$ similarly with $|\alpha| \equiv 2 \bmod 5$ in place of $|\alpha| \equiv 1 \bmod 5$. Hence, $e_\alpha > e_{\alpha^-}$ ($\hat{e}_\alpha > \hat{e}_{\alpha^-}$) iff $|\alpha| \equiv 1 \bmod 5$ ($|\alpha| \equiv 2 \bmod 5$).

DEFINITION 3.13. An *α-state* is a triple $\langle \alpha, \sigma, \tau \rangle$ where $\sigma \subseteq \{0, \ldots, e_\alpha\}$ and $\tau \subseteq \{0, \ldots, \hat{e}_\alpha\}$. The only λ-state is $\nu_{-1} = \langle \lambda, \emptyset, \emptyset \rangle$.

The construction in Section 3.2.9 will produce a simultaneous computable enumeration $U_{\alpha,s}$, $V_{\alpha,s}$, $\widehat{U}_{\alpha,s}$, $\widehat{V}_{\alpha,s}$, for $\alpha \in T$ and $s \in \omega$, of these c.e. sets which we use in the following definition.

DEFINITION 3.14. (i) The *α-state of x at stage s*, $\nu(\alpha, x, s)$, is the triple

$$\langle \alpha, \sigma(\alpha, x, s), \tau(\alpha, x, s) \rangle$$

where

$$\sigma(\alpha, x, s) = \{e_\beta \colon \beta \subseteq \alpha \ \& \ e_\beta > e_{\beta^-} \ \& \ x \in U_{\beta,s}\},$$

$$\tau(\alpha, x, s) = \{\hat{e}_\beta \colon \beta \subseteq \alpha \ \& \ \hat{e}_\beta > \hat{e}_{\beta^-} \ \& \ x \in \widehat{V}_{\beta,s}\}.$$

(ii) The *final α-state of x* is $\nu(\alpha, x) = \langle \alpha, \sigma(\alpha, x), \tau(\alpha, x) \rangle$ where $\sigma(\alpha, x) = \lim_s \sigma(\alpha, x, s)$ and $\tau(\alpha, x) = \lim_s \tau(\alpha, x, s)$.

For each $\alpha \in T$ we define the following sets of α-states called *lists*,

$$\mathcal{E}_\alpha = \Big\{\nu \colon (\exists^\infty x)(\exists s)\Big[x \in S_{\alpha,s} - \bigcup\{S_{\alpha,t} \colon t < s\} \ \& \ \nu(\alpha, x, s) = \nu\Big]\Big\}, \text{ and}$$

$$\mathcal{F}_\alpha = \big\{\nu \colon (\exists^\infty x)(\exists s)\big[x \in R_{\alpha,s} \ \& \ \nu(\alpha, x, s) = \nu\big]\big\}.$$

Note that $\mathcal{E}_\alpha$ consists of states well visited by elements x when they first *enter* R_α and $\mathcal{F}_\alpha$ those states well-visited while they *remain* in Y_α so $\mathcal{E}_\alpha \subseteq \mathcal{F}_\alpha$. Each $\alpha \in T$ will have an associated list $\mathcal{M}_\alpha$ which is roughly α's "guess" at the true $\mathcal{F}_\alpha$ such that if $\alpha \subset f$ then $\mathcal{M}_\alpha = \mathcal{F}_\alpha$. For $\alpha \subset f$ we shall achieve $\mathcal{M}_\alpha = \mathcal{F}_\alpha$ by ensuring the following properties of $\mathcal{M}_\alpha$,

$$\mathcal{E}_\alpha \subseteq \mathcal{M}_\alpha, \tag{23}$$

$$\begin{aligned}(\text{a.e. } x) \quad & [\text{if } x \in Y_{\alpha,s},\ \nu_0 = \nu(\alpha, x, s) \in \mathcal{M}_\alpha, \\ & \text{and RED causes enumeration of } x \text{ so that} \\ & \nu_1 = \nu(\alpha, x, s+1) \text{ then } \nu_1 \in \mathcal{M}_\alpha],\end{aligned} \tag{24}$$

$$\begin{aligned}(\text{a.e. } x) \quad & [\text{if } x \in Y_{\alpha,s},\ \nu_0 = \nu(\alpha, x, s) \in \mathcal{M}_\alpha \\ & \text{and BLUE causes enumeration of } x \text{ so that} \\ & \nu_1 = \nu(\alpha, x, s+1) \text{ then } \nu_1 \in \mathcal{M}_\alpha].\end{aligned} \tag{25}$$

(Here (a.e. x) denotes "for almost every x".) Blue enumeration which satisfies (25) is called *α-legal*. Two main constraints on BLUE's moves will be (23) and (25). Clearly, (23), (24), and (25) guarantee

$$\mathcal{F}_\alpha \subseteq \mathcal{M}_\alpha. \tag{26}$$

During Step 1 of the construction in Section 3.2.9 we shall promptly pull elements $x \in Y_{\alpha^-,s}$ into $S_{\alpha,s+1}$ in order to ensure,

$$\mathcal{M}_\alpha \subseteq \mathcal{E}_\alpha. \tag{27}$$

Hence, by (26), (27), and $\mathcal{E}_\alpha \subseteq \mathcal{F}_\alpha$ we have,

$$\mathcal{M}_\alpha = \mathcal{F}_\alpha = \mathcal{E}_\alpha. \tag{28}$$

On the $\hat{\omega}$-side we have dual definitions for the above items by replacing $\omega, x, U_\alpha, \widehat{V}_\alpha$ by $\hat{\omega}, \hat{x}, \widehat{U}_\alpha, V_\alpha$ respectively. These dual items will be denoted by $\hat{\nu}(\alpha, \hat{x}, s)$, $\widehat{S}_\alpha, \widehat{R}_\alpha, \widehat{Y}_\alpha, \widehat{\mathcal{E}}_\alpha, \widehat{\mathcal{F}}_\alpha$, and $\widehat{\mathcal{M}}_\alpha$. We write hats over the α-states, e.g., $\hat{\nu}_1 = \nu(\alpha, \hat{x}, s)$, to indicate α-states for elements $\hat{x} \in \hat{\omega}$. We shall ensure

$$\widehat{\mathcal{M}}_\alpha = \{\hat{\nu}\colon\ \nu \in \mathcal{M}_\alpha\}, \tag{29}$$

which implies by (28) that the well visited α-states on both sides coincide.

DEFINITION 3.15. Given α-states $\nu_0 = \langle \alpha, \sigma_0, \tau_0 \rangle$ and $\nu_1 = \langle \alpha, \sigma_1, \tau_1 \rangle$.

(i) $\nu_0 \leqslant_R \nu_1$ if $\sigma_0 \subseteq \sigma_1$ and $\tau_0 = \tau_1$.
(ii) $\nu_0 \leqslant_B \nu_1$ if $\tau_0 \subseteq \tau_1$ and $\sigma_0 = \sigma_1$.
(iii) $\hat{\nu}_0 \leqslant_R \hat{\nu}_1$ if $\hat{\sigma}_0 = \hat{\sigma}_1$ and $\hat{\tau}_0 \subseteq \hat{\tau}_1$.
(iv) $\hat{\nu}_0 \leqslant_B \hat{\nu}_1$ if $\hat{\sigma}_0 \subseteq \hat{\sigma}_1$ and $\hat{\tau}_0 = \hat{\tau}_1$.

(v) $\nu_0 <_R \nu_1$ ($\nu_0 <_B \nu_1$) if $\nu_0 \leqslant_R \nu_1$ ($\nu_0 \leqslant_B \nu_1$) and $\nu_0 \neq \nu_1$, and similarly for $\hat{\nu}_0 <_R \hat{\nu}_1$ and $\hat{\nu}_0 <_B \hat{\nu}_1$.

The intuition is that if $\nu_0 = \nu(\alpha, x, s)$ and $\nu_0 <_R \nu_1$ ($\nu_0 <_B \nu_1$) then RED (BLUE) can enumerate x in the necessary U sets ($\widehat{V}$ sets) to cause $\nu_1 = \nu(\alpha, x, s+1)$. For $\hat{\nu}_0$ and $\hat{\nu}_1$ the role of σ and τ is reversed because on the $\hat{\omega}$-side BLUE (RED) plays the $\widehat{U}$ sets (V sets), and hence

$$[\nu_0 <_R \nu_1 \Longleftrightarrow \hat{\nu}_0 <_B \hat{\nu}_1] \,\&\, [\nu_0 <_B \nu_1 \Longleftrightarrow \hat{\nu}_0 <_R \hat{\nu}_1]. \tag{30}$$

DEFINITION 3.16. Given $\beta \subseteq \alpha \in T$ and an α-state $\nu_0 = \langle \alpha, \sigma_0, \tau_0 \rangle$ or a set $\mathcal{C}_\alpha$ of α-states,

(i) $\nu_0 \upharpoonright \beta = \langle \beta, \sigma_1, \tau_1 \rangle$ where we let $\sigma_1 = \sigma_0 \cap \{0, \ldots, e_\beta\}$ and we let $\tau_1 = \tau_0 \cap \{0, \ldots, \hat{e}_\beta\}$,

(ii) $\nu_1 \preceq \nu_0$ (read "ν_0 extends ν_1") if $\nu_0 \upharpoonright \beta = \nu_1$.

(iii) $\mathcal{C}_\alpha \upharpoonright \beta = \{\nu \upharpoonright \beta\colon \nu \in \mathcal{C}_\alpha\}$.

(iv) Given a finite set of α-states, $\{\nu(\alpha, \sigma_i, \tau_i)\colon i \in I\}$ let

$$\bigcup \bigl\{\nu(\alpha, \sigma_i, \tau_i)\colon i \in I\bigr\} =_{\mathrm{dfn}} \langle \alpha, \sigma, \tau \rangle,$$

where $\sigma = \bigcup\{\sigma_i\colon i \in I\}$, and $\tau = \bigcup\{\tau_i\colon i \in I\}$.

The combination of (23)–(28) and their duals together with (29) may cause additional upward closure of $\mathcal{M}_\alpha$ under $\leqslant_B$. For example, if $e_\alpha > e_{\alpha^-}$ (so α builds U_α and $\widehat{U}_\alpha$) suppose $\nu_0 \in \mathcal{M}_\alpha$ for some $\nu_0 = \langle \alpha, \sigma_0, \tau_0 \rangle$ with $e_\alpha \in \sigma_0$. Hence, $\hat{\nu}_0 \in \widehat{\mathcal{M}}_\alpha$ by (29). But if for infinitely many $\hat{y}$, $\nu(\alpha, \hat{y}, s) = \hat{\nu}_0$ and for some s, RED causes $\nu(\alpha, \hat{y}, s+1) = \hat{\nu}_1 >_{\mathrm{R}} \hat{\nu}_0$ (say by enumerating $\hat{y}$ in V_β for some $\beta \subset \alpha$) then $\hat{\nu}_1 \in \widehat{\mathcal{M}}_\alpha$ by the dual of (24) and hence $\nu_1 \in \mathcal{M}_\alpha$ by (29), and $\nu_0 <_B \nu_1$ by (30) because $\hat{\nu}_0 <_R \hat{\nu}_1$. We do not wait for RED to cause $\hat{\nu}_1 \in \widehat{\mathcal{M}}_\alpha$. Rather in the following definition we anticipate by *now* putting all such $\nu_1 \in \mathcal{M}_\alpha$ and many more as well. Indeed for each α which is *$\mathcal{M}$-consistent* in the following definition (which includes all $\alpha \subseteq f$) we put every ν_1 in $\mathcal{M}_\alpha$ if ν_1 is a blue move away from some $\nu_0 \in \mathcal{M}_\alpha$ (i.e. $\nu_0 <_B \nu_1$), so long as the blue move from ν_0 to ν_1 is β-legal, i.e. $\nu_1 \upharpoonright \beta \in \mathcal{M}_\beta$. But by (28) this means we must make all such ν_1 well-visited on R_α. Since there is no evidence that RED will actually make the proposed move, this extreme closure of $\mathcal{M}_\alpha$ seems unwarranted and outrageously bold. We shall prove that it is not.

DEFINITION 3.17. A node $\alpha \in T$ is *$\mathcal{M}$-inconsistent* if $e_\alpha > e_\beta$, where $\beta = \alpha^-$, and there are α-states $\nu_0 <_B \nu_1$ such that $\nu_0 \in \mathcal{M}_\alpha$ and $\nu_1 \upharpoonright \beta \in \mathcal{M}_\beta$ but $\nu_1 \notin \mathcal{M}_\alpha$. Otherwise α is *$\mathcal{M}$-consistent.*

We shall take action in Step 3 of the construction in Section 3.2.9 to ensure that α is $\mathcal{M}$-consistent if $\alpha \subset f$.

3.2.4. *Non well-resided α-states and the lists $\mathcal{R}_\alpha$ and $\mathcal{B}_\alpha$.* Define the set of non well-resided α-states,

$$\mathcal{K}_\alpha = \{\nu_1\colon \neg(\exists^\infty x)[x \in Y_\alpha \ \&\ \nu(\alpha, x) = \nu_1]\}. \tag{31}$$

Likewise define $\widehat{\mathcal{K}}_\alpha$ for the $\hat{\omega}$-side. To satisfy the automorphism requirement (18) we must show for $\alpha \subset f$ that

$$\widehat{\mathcal{K}}_\alpha = \{\hat{\nu}\colon \nu \in \mathcal{K}_\alpha\}. \tag{32}$$

To achieve (32) note that unlike $\mathcal{E}_\alpha$ and $\mathcal{F}_\alpha$ of Section 3.2.3 $\mathcal{K}_\alpha$ is Σ^0_3 not Π^0_2 so α cannot guess at $\mathcal{K}_\alpha$ directly but only at a certain Π^0_2 approximation $\mathcal{N}_\alpha \subseteq \mathcal{K}_\alpha$. We divide $\mathcal{N}_\alpha$ into the disjoint union of sets $\mathcal{R}_\alpha$ and $\mathcal{B}_\alpha$ which correspond to those $\nu \in \mathcal{N}_\alpha$ which α believes are being emptied by RED and BLUE respectively.

To define $\mathcal{R}_\alpha$ and $\mathcal{B}_\alpha$ fix $\alpha \in T$, let $\beta = \alpha^-$, and assume that $\mathcal{R}_\gamma$, $\mathcal{B}_\gamma$ and their duals $\widehat{\mathcal{R}}_\gamma$, $\widehat{\mathcal{B}}_\gamma$ have been defined for all $\gamma \subset \alpha$. We decompose $\mathcal{R}_\alpha$ into the disjoint union,

$$\mathcal{R}_\alpha = \mathcal{R}^\alpha_\alpha \sqcup \mathcal{R}^{<\alpha}_\alpha, \quad \text{where} \tag{33}$$

$$\mathcal{R}^{<\alpha}_\alpha =_{\text{dfn}} \{\nu\colon \nu \in \mathcal{M}_\alpha \ \&\ \nu \restriction \beta \in \mathcal{R}_\beta\}, \quad \text{and} \tag{34}$$

$$\mathcal{R}^\alpha_\alpha =_{\text{dfn}} \mathcal{R}_\alpha - \mathcal{R}^{<\alpha}_\alpha. \tag{35}$$

Note that $\mathcal{R}^{<\alpha}_\alpha$ is determined by $\mathcal{R}_\beta$, $\beta \subset \alpha$, but $\mathcal{R}^\alpha_\alpha$ may contain new elements and for $\alpha \subset f$ it has the meaning described below in (37). Likewise, let $\mathcal{B}_\alpha = \mathcal{B}^\alpha_\alpha \sqcup \mathcal{B}^{<\alpha}_\alpha$, where $\mathcal{B}^{<\alpha}_\alpha$ is defined as in (34) but with $\mathcal{B}_\beta$ in place of $\mathcal{R}_\beta$.

If $|\alpha| \not\equiv 3 \bmod 5$ define $\mathcal{R}^\alpha_\alpha = \widehat{\mathcal{B}}^\alpha_\alpha = \emptyset$. If $|\alpha| \equiv 3 \bmod 5$ we let $\mathcal{M}_\alpha = \mathcal{M}_\beta$ (since α-states are β-states because $e_\alpha = e_\beta$ and $\hat{e}_\alpha = \hat{e}_\beta$), we define the Π^0_2 predicate,

$$F(\beta, \nu) \equiv (\forall x)[[x > |\beta| \ \&\ x \in Y_\beta] \Longrightarrow \nu(\alpha, x) \neq \nu], \tag{36}$$

and we allow $\mathcal{R}^\alpha_\alpha \neq \emptyset$ with the intention that for $\alpha \subset f$,

$$\mathcal{R}^\alpha_\alpha = \{\nu\colon \nu \in \mathcal{M}_\alpha - (\mathcal{R}^{<\alpha}_\alpha \cup \mathcal{B}^{<\alpha}_\alpha) \ \&\ \mathcal{F}(\beta, \nu)\}. \tag{37}$$

We define

$$\widehat{\mathcal{B}}^\alpha_\alpha =_{\text{dfn}} \{\hat{\nu}\colon \nu \in \mathcal{R}^\alpha_\alpha\}. \tag{38}$$

If $|\alpha| \not\equiv 4 \bmod 5$ define $\widehat{\mathcal{R}}^\alpha_\alpha = \mathcal{B}^\alpha_\alpha = \emptyset$. If $\alpha \equiv 4 \bmod 5$ we allow $\widehat{\mathcal{R}}^\alpha_\alpha \neq \emptyset$ (using the duals of (33)–(37) where, e.g., in the dual of (36) we use $\widehat{Y_\beta}$ in place of Y_β), and we define

$$\mathcal{B}^\alpha_\alpha =_{\text{dfn}} \{\nu\colon \widehat{\nu} \in \widehat{\mathcal{R}}^\alpha_\alpha\}. \tag{39}$$

At most one of $\mathcal{R}_\alpha^\alpha$ and $\widehat{\mathcal{R}}_\alpha^\alpha$ is nonempty so by (39), (38) and (37),

$$\mathcal{R}_\alpha^\alpha \cap \mathcal{B}_\alpha^\alpha = \emptyset \ \& \ \big((\mathcal{R}_\alpha^\alpha \cup \mathcal{B}_\alpha^\alpha) \cap (\mathcal{R}_\alpha^{<\alpha} \cup \mathcal{B}_\alpha^{<\alpha}) = \emptyset\big), \tag{40}$$

and hence

$$\mathcal{R}_\alpha \cap \mathcal{B}_\alpha = \emptyset. \tag{41}$$

If $\alpha \subset f$ then $\nu \in \mathcal{R}_\alpha$ implies $F(\alpha^-, \nu)$ and hence

$$\begin{aligned}&(\forall \nu \in \mathcal{R}_\alpha)(\forall x \in Y_\alpha)(\forall s)\\ &\quad\big[\nu(\alpha, x, s) = \nu \Longrightarrow (\exists t > s)\big[\nu(\alpha, x, t) \neq \nu\big]\big].\end{aligned} \tag{42}$$

It will be BLUE's responsibility to change the α-state of x if $\nu(\alpha, x, s) \in \mathcal{B}_\alpha$, and $x \in R_\alpha$. However, $\mathcal{B}_\alpha \cap \mathcal{R}_\alpha = \emptyset$ so if $\nu(\alpha, x, s) = \nu \in \mathcal{R}_\alpha$ then BLUE can wait for RED to change the α-state of x to meet (42), namely

$$\begin{aligned}&(\forall \nu \in \mathcal{R}_\alpha)(\forall x \in R_\alpha)(\forall s)\\ &\quad[\text{if } \nu(\alpha, x, s) = \nu \text{ then}\\ &\quad\text{it is an } \alpha\text{-admissible move for BLUE to restrain}\\ &\quad x \text{ from further BLUE enumeration until}\\ &\quad(\exists t > s)[\nu(\alpha, x, s) <_R \nu(\alpha, x, t)].\end{aligned} \tag{43}$$

DEFINITION 3.18. A node $\alpha \in T$ is $\mathcal{R}$-*consistent* if

$$(\forall \nu_0 \in \mathcal{R}_\alpha)(\exists \nu_1)[\nu_0 <_R \nu_1 \ \& \ \nu_1 \in \mathcal{M}_\alpha], \tag{44}$$

and $\mathcal{R}$-*inconsistent* otherwise.

By applying (43) BLUE will ensure that α is $\mathcal{R}$-consistent for $\alpha \subset f$. Now (44) (38), and (30) imply for $\alpha \subset f$ that

$$(\forall \hat{\nu}_0 \in \widehat{\mathcal{B}}_\alpha)(\exists \hat{\nu}_1)\big[\hat{\nu}_0 <_B \hat{\nu}_1 \ \& \ \hat{\nu}_1 \in \widehat{\mathcal{M}}_\alpha\big]. \tag{45}$$

By repeatedly applying (45) BLUE can achieve $\hat{\nu}_1 \in \widehat{\mathcal{M}}_\alpha - \widehat{\mathcal{B}}_\alpha$, namely

$$(\exists \text{ function } \hat{h}_\alpha)\big[\hat{h}_\alpha\colon \widehat{\mathcal{B}}_\alpha \to \big(\widehat{\mathcal{M}}_\alpha - \widehat{\mathcal{B}}_\alpha\big) \ \& \ \big(\forall \hat{\nu} \in \widehat{\mathcal{B}}_\alpha\big)\big[\hat{\nu} <_B \hat{h}_\alpha(\hat{\nu})\big]\big]. \tag{46}$$

It is BLUE's responsibility to move any $\hat{x} \in \widehat{R}_\alpha$ for which $\nu(\alpha, \hat{x}, s) = \hat{\nu}_0 \in \widehat{\mathcal{B}}_\alpha$ to the *target state* $\hat{\nu}_1 = \hat{h}_\alpha(\hat{\nu}_0)$ (and where $\hat{h}$ is called the *target function*) so that BLUE can achieve,

$$(\forall \hat{x} \in \hat{R}_\alpha)(\forall s)$$
$$\big[\nu(\alpha, \hat{x}, s) \in \widehat{\mathcal{B}}_\alpha \Longrightarrow (\exists t > s)\big[\nu(\alpha, \hat{x}, t) \in \widehat{\mathcal{M}}_\alpha - \widehat{\mathcal{B}}_\alpha\big]\big], \tag{47}$$

and hence BLUE will cause every state $\hat{\nu}_0 \in \widehat{\mathcal{B}}_\alpha$ to be emptied. To achieve (47) on $\widehat{R}_\alpha$ it suffices to achieve the following on $\widehat{S}_\gamma$ for each $\gamma \supseteq \alpha$,

$$(\forall \hat{x} \in \hat{S}_\gamma)(\forall s)$$
$$\big[\nu(\gamma, \hat{x}, s) \in \widehat{\mathcal{B}}_\gamma \Longrightarrow (\exists t > s)\big[\nu(\gamma, \hat{x}, t) \in \widehat{\mathcal{M}}_\gamma - \widehat{\mathcal{B}}_\gamma\big]\big]. \tag{48}$$

(For BLUE to achieve (48) from the hypothesis of (47) there is a subtle but crucial point. Suppose $\nu_0 \in \mathcal{R}_\alpha$ so $\hat{\nu}_0 \in \widehat{\mathcal{B}}_\alpha$. Hence, $\hat{\nu}_0' \in \widehat{\mathcal{B}}_\gamma$ for all $\gamma \supset \alpha$ such that $\hat{\nu}_0' \restriction \alpha = \hat{\nu}_0$. Now by (47) BLUE is required for every $\hat{x}$ in region $\widehat{R}_\alpha$ such that $\nu(\alpha, \hat{x}, s) = \hat{\nu}_0 \in \widehat{\mathcal{B}}_\alpha$ to enumerate $\hat{x}$ in blue sets to achieve $\nu(\alpha, \hat{x}, t) = \hat{\nu}_1 >_B \hat{\nu}_0$ for some $t > s$. However, if $\hat{x} \in \hat{S}_{\gamma,s}$ for some $\gamma \supset \alpha$ then BLUE can only make γ*-legal* moves, namely BLUE must ensure that $\nu(\gamma, \hat{x}, s) \in \widehat{\mathcal{M}}_\gamma$. Hence, on the γ-level if $\hat{\nu}_0' = \nu(\gamma, \hat{x}, s)$ and $\hat{\nu}_0' \restriction \alpha = \hat{\nu}_0 \in \hat{\mathcal{B}}_\alpha$ then $\nu_0 \in \mathcal{R}_\alpha$ so $\nu_0' \in \mathcal{R}_\gamma$ and BLUE needs a γ*-target* $\hat{\nu}_1' >_B \hat{\nu}_0'$ for $\hat{x}$ not merely an α-target $\hat{\nu}_1 >_B \hat{\nu}_0$. To obtain this γ-target ν_1', BLUE can hold some $y \in S_\gamma$ in γ-state ν_0' until, by $(43)_\gamma$, RED is forced to cause $\nu(\alpha, y, t) = \nu_1 >_R \nu_0$, for some $t > s$, and hence $\nu(\gamma, y, t) = \nu_1' >_R \nu_0'$, thus ensuring that γ is $\mathcal{R}$-consistent and giving a target γ-state $\hat{\nu}_1'$ for $\hat{x}$. This action may have to be repeated for each of the infinitely many $\gamma \supseteq \alpha$ even for those $\gamma <_L f$. Hence, (47) constitutes a very strong BLUE constraint on the entire downward cone $\widehat{R}_\alpha$. This procedure for producing an appropriate target j-state ν_1' for $j > e$ when an e-state ν_0 is emptied is taken from the effective automorphism machinery in [Soare, 1987, Chap. XV; 1974], where it also plays a central role.)

We often refer to the dual of (46) which asserts

$$(\exists \text{ function } h_\alpha)\big[h_\alpha\colon \mathcal{B}_\alpha \to (\mathcal{M}_\alpha - \mathcal{B}_\alpha)\ \&\ (\forall \nu \in \mathcal{B}_\alpha)\big[\nu <_B h_\alpha(\nu)\big]\big], \tag{49}$$

and which enables us to achieve the dual of (48), namely

$$(\forall x \in S_\gamma)(\forall s)$$
$$\big[\nu(\gamma, x, s) \in \mathcal{B}_\gamma \Longrightarrow (\exists t > s)\big[\nu(\gamma, x, t) \in \mathcal{M}_\gamma - \mathcal{B}_\gamma\big]\big]. \tag{50}$$

Finally, we have ensured

$$(\forall \gamma \subset f)(\forall \nu_0 \in \mathcal{M}_\gamma)(\exists^{<\infty} x)\big[x \in Y_\gamma\ \&\ \nu(\gamma, x) = \nu_0\big]$$
$$\Longrightarrow (\exists \alpha)_{\gamma \subset \alpha \subset f}\big[\{\nu_1 \in \mathcal{M}_\alpha\colon \nu_1 \restriction \gamma = \nu_0\} \subseteq \mathcal{R}_\alpha \cup \mathcal{B}_\alpha\big]\big]. \tag{51}$$

To check (51) fix $\gamma \subset f$ and $\nu_0 \in \mathcal{M}_\gamma$. By (20) $Y_\gamma =^* \omega$ so if the hypothesis of (51) holds then we can choose b such that

$$(\forall x \in \omega)[x > b \Longrightarrow \nu(\gamma, x) \neq \nu_0].$$

Choose $\alpha \subset f$ such that $\alpha \supset \gamma$, $|\alpha| > b$ and $|\alpha| \equiv 3 \bmod 5$. Consider any $\nu_1 \in \mathcal{M}_\alpha$ such that $\nu_1 \upharpoonright \gamma = \nu_0$. If $\nu_1 \notin \mathcal{R}_\alpha^{<\alpha} \cup \mathcal{B}_\alpha^{<\alpha}$ then $F(\alpha^-, \nu_1)$ holds so $\nu_1 \in \mathcal{R}_\alpha^\alpha$ by (37), and hence $\nu_1 \in \mathcal{R}_\alpha$ by (33).

Equations (38), (42), (47), (51) and their duals guarantee (32).

3.2.5. *Verifying the automorphism requirement* (1). We shall arrange that

$$\lim_{\alpha \subset f} e_\alpha = \infty.$$

By (21) and (22) the sets $\{U_\alpha\}_{\alpha \subset f}$ and $\{V_\alpha\}_{\alpha \subset f}$ constitute skeletons for $\{W_n\}_{n \in \omega}$. By (28), its dual, and (29) we know that the well-visited α-states on the ω-side and $\hat{\omega}$-side coincide. By (32) the non well-resided α-states also coincide so (18) is satisfied. The construction in Section 3.2.9 and verification demonstrate that the equations of Section 3.2.2, Section 3.2.3, and Section 3.2.4 are satisfied. First we need a few more definitions in Section 3.2.7 and Section 3.2.8.

3.2.6. *Splitting S_α into S_α^0 and S_α^1.* We divide the α-section S_α into two subsections S_α^0 and S_α^1. For $k \in \{0, 1\}$ let $S_{\alpha,s}^k$ denote the set of elements $x \in S_{\alpha,s}$ which lie in S_α^k at the end of stage s. The elements $x \in S_\alpha^0$ may be appointed as α-witnesses (e.g., the position of an α-coding marker), and may require special enumeration into or restraint from certain blue sets to meet certain additional requirements (such as making B high) beyond the automorphism requirements. The other elements of S_α, namely $x \in S_\alpha^1$, are available to be moved into S_γ for any $\gamma \supset \alpha$. For $k \in \{0, 1\}$ define,

$$\mathcal{E}_\alpha^k = \Big\{\nu\colon (\exists^\infty x)(\exists s)\Big[x \in S_{\alpha,s}^k - \bigcup\{S_{\alpha,t}^k\colon t < s\} \,\&\, \nu(\alpha, x, s) = \nu\Big]\Big\}. \tag{52}$$

We shall arrange that the stream of elements entering S_α is split into two equivalent streams entering S_α^0 and S_α^1 so that $\mathcal{E}_\alpha = \mathcal{E}_\alpha^0 = \mathcal{E}_\alpha^1$. Similarly, we define

$$R_{\alpha,s}^1 = \big\{x\colon x \in S_{\alpha,s}^1 \text{ or } (\exists \gamma \supset \alpha)[x \in S_{\gamma,s}]\big\}, \quad \text{and} \tag{53}$$

$$Y_{\alpha,s}^1 = \bigcup\big\{R_{\alpha,t}^1\colon t \leqslant s\big\}. \tag{54}$$

We shall arrange that $S_{\alpha,\infty}^0$, the set of permanent residents of S_α^0, is finite. Thus, it will suffice to use $R_{\alpha,s}^1$ and $Y_{\alpha,s}^1$ (rather than the slightly larger $R_{\alpha,s}$ and $Y_{\alpha,s}$) in Section 3.2.7 and Steps 1 and 2 of the construction in Section 3.2.9, since the former are the elements truly available to those $\gamma \supset \alpha$.

3.2.7. *The set $\mathcal{F}^+_\beta$ and the definition of $\mathcal{M}_\alpha$.* In Section 3.2.3 we said that every $\alpha \in T$ would have an associated set $\mathcal{M}_\alpha$ such that $\mathcal{M}_\alpha = \mathcal{F}_\alpha$ if $\alpha \subset f$. However, although this is the *property* we want $\mathcal{M}_\alpha$ to have we cannot simply *define* $\mathcal{M}_\alpha$ to be α's guess at $\mathcal{F}_\alpha$ because that definition would be circular. Rather we must define here a certain set $\mathcal{F}^+_\beta$ which depends only on β, and then let $\mathcal{M}_\alpha$ be α's guess at $\mathcal{F}^+_\beta$ so that $\mathcal{M}_\alpha = \mathcal{F}^+_\beta$ $(= \mathcal{F}_\alpha)$ for $\alpha \subset f$.

Fix $\alpha \in T$ such that $e_\alpha > e_\beta$ for $\beta = \alpha^-$. Define the c.e. set $Z_{e_\alpha} = \bigcup_s Z_{e_\alpha,s}$ where

$$Z_{e_\alpha,s+1} =_{\text{dfn}} \{x\colon\ x \in U_{e_\alpha,s+1}\ \&\ x \in Y^1_{\beta,s}\}. \tag{55}$$

Define the α-state function $\nu^+(\alpha, x, s)$ exactly as for $\nu(\alpha, x, s)$ in Definition 3.14 but with $Z_{e_\alpha,s}$ in place of $U_{\alpha,s}$.

Define

$$\mathcal{F}^+_\beta = \{\nu\colon\ (\exists^\infty x)(\exists s)[x \in Y^1_{\beta,s}\ \&\ \nu^+(\alpha, x, s) = \nu]\}, \tag{56}$$

$$\begin{aligned} k^+_\beta \;=\; \min\{y\colon\ & (\forall x > y)(\forall s) \\ & [[x \in Y^1_{\beta,s}\ \&\ \nu^+(\alpha, x, s) = \nu_1] \Longrightarrow \nu_1 \in \mathcal{F}^+_\beta]\}. \end{aligned} \tag{57}$$

If $e_\alpha > e_\beta$ we also define $\widehat{\mathcal{F}}^+_\beta = \{\hat{\nu}\colon\ \nu \in \mathcal{F}^+_\beta\}$. (Note that Z_{e_α} and hence $\mathcal{F}^+_\beta$ and k^+_β depend only upon β not α and thus α can make guesses $\mathcal{M}_\alpha$ and k_α for $\mathcal{F}^+_\beta$ and k^+_β.)

If $\hat{e}_\alpha > \hat{e}_\beta$ we first define $\widehat{\mathcal{F}}^+_\beta$ and k^+_β using the duals of (56) and (57) (with $\hat{Y}_{\beta,s}$, $V_{\hat{e}_\alpha}$, $\hat{Z}_{\hat{e}_\alpha}$, and $\nu^+(\alpha, \hat{x}, s)$ in place of $Y_{\beta,s}$, U_{e_α}, Z_{e_α}, and $\nu^+(\alpha, x, s)$, respectively), and then we define $\mathcal{F}^+_\beta = \{\nu\colon\ \hat{\nu} \in \mathcal{F}^+_\beta\}$. (Note that there is no $\hat{k}^+_\beta$ only k^+_β.)

Every $\alpha \in T$ will have associated items $\mathcal{M}_\alpha$ and k_α such that $\mathcal{M}_\alpha = \mathcal{F}^+_\beta$ and $k_\alpha = k^+_\beta$ for $\alpha \subset f$. We allow x to enter Y_α only if $x > k_\alpha$. If $e_\alpha = e_\beta$ and $\hat{e}_\alpha = \hat{e}_\beta$ we define $\mathcal{F}^+_\beta = \mathcal{F}_\beta$, $\widehat{\mathcal{F}}^+_\beta = \widehat{\mathcal{F}}_\beta$, and $k^+_\beta = k_\beta$. If

$$(\exists x)(\exists s)[x \in Y_{\alpha,s}\ \&\ \nu(\alpha, x, s) \notin \mathcal{M}_\alpha] \tag{58}$$

then we say that α is *provably incorrect* at all stages $t \geqslant s$ and we ensure that $\alpha \not\subset f$.

3.2.8. *The definition of the tree T.*

DEFINITION 3.19. For $\alpha \in T$ to say α is *consistent* means α is $\mathcal{M}$-consistent (Definition 3.17), $\mathcal{R}$-consistent (Definition 3.18), and also $\mathcal{C}$-consistent (Definition 3.27).

Note that by clause (i) in the following Definition 3.20 of T,

$$\beta \in T \Longrightarrow [\beta \text{ inconsistent} \Longleftrightarrow \beta \text{ is a terminal node on } T]. \tag{59}$$

We shall show that if $\alpha \subset f$ then α is consistent, and hence $\lim_{\alpha \subset f} e_\alpha = \infty$, so the argument of Section 3.2.5 applies.

DEFINITION 3.20. Put $\lambda \in T$ and define $\mathcal{M}_\lambda = \mathcal{R}_\lambda = \mathcal{B}_\lambda = \emptyset$, and $k_\lambda = e_\lambda = \hat{e}_\lambda = -1$. If $\beta \in T$ we put $\alpha = \beta^\frown \langle \mathcal{M}_\alpha, \mathcal{R}_\alpha, \mathcal{B}_\alpha, k_\alpha \rangle$ in T providing the following conditions hold:

(i) β is consistent (as defined in Definition 3.19),

(ii) $\mathcal{M}_\alpha$ is a set of α-states, $\mathcal{R}_\alpha \subseteq \mathcal{M}_\alpha$, $\mathcal{B}_\alpha \subseteq \mathcal{M}_\alpha$, and $\mathcal{R}_\alpha \cap \mathcal{B}_\alpha = \emptyset$,

(iii) $\mathcal{M}_\alpha \upharpoonright \beta \subseteq \mathcal{M}_\beta$,

(iv) $[e_\alpha = e_\beta \ \&\ \hat{e}_\alpha = \hat{e}_\beta] \Longrightarrow \mathcal{M}_\alpha = \mathcal{M}_\beta$,

(v) $\mathcal{R}_\alpha^{<\alpha} =_{\text{dfn}} \{\nu \in \mathcal{M}_\alpha \colon \nu \upharpoonright \beta \in \mathcal{R}_\beta\} \subseteq \mathcal{R}_\alpha$,

(vi) $\mathcal{B}_\alpha^{<\alpha} =_{\text{dfn}} \{\nu \in \mathcal{M}_\alpha \colon \nu \upharpoonright \beta \in \mathcal{B}_\beta\} \subseteq \mathcal{B}_\alpha$,

(vii) $\mathcal{R}_\alpha^{\alpha} =_{\text{dfn}} \mathcal{R}_\alpha - \mathcal{R}_\alpha^{<\alpha} \neq \emptyset \Longrightarrow |\alpha| \equiv 3 \bmod 5$,

(viii) $\mathcal{B}_\alpha^{\alpha} =_{\text{dfn}} \mathcal{B}_\alpha - \mathcal{B}_\alpha^{<\alpha} \neq \emptyset \Longrightarrow |\alpha| \equiv 4 \bmod 5$.

In addition, each $\alpha \in T$ has associated dual sets $\widehat{\mathcal{M}}_\alpha$, $\widehat{\mathcal{R}}_\alpha$, and $\widehat{\mathcal{B}}_\alpha$ which are determined from $\mathcal{M}_\alpha$, $\mathcal{B}_\alpha$ and $\mathcal{R}_\alpha$ by (29), (39), and (38), respectively. Also α has associated integers e_α and $\hat{e}_\alpha$ (depending only on $|\alpha|$) as defined at the beginning of Section 3.2.3. (We identify the finite object $\langle \mathcal{M}_\alpha, \mathcal{R}_\alpha, \mathcal{B}_\alpha, k_\alpha \rangle$ with an integer under some effective coding so we may regard $T \subseteq \omega^{<\omega}$.)

DEFINITION 3.21. The *true path* $f \in [T]$ is defined by induction on n. Let $\beta = f \upharpoonright n$ be consistent. Then $f \upharpoonright (n+1)$ is the $<_L$-least $\alpha \in T$, $\alpha \supset \beta$, of length $m = n+1$ such that:

(i) $m \equiv 1 \bmod 5 \Longrightarrow \mathcal{M}_\alpha = \mathcal{F}_\beta^+ \ \&\ k_\alpha = k_\beta^+$,

(ii) $m \equiv 2 \bmod 5 \Longrightarrow \widehat{\mathcal{M}}_\alpha = \widehat{\mathcal{F}}_\beta^+ \ \&\ k_\alpha = k_\beta^+$,

(iii) $m \equiv 3 \bmod 5 \Longrightarrow \mathcal{R}_\alpha^\alpha = \{\nu \colon \nu \in \mathcal{M}_\alpha - (\mathcal{R}_\alpha^{<\alpha} \cap \mathcal{B}_\alpha^{<\alpha}) \ \&\ F(\beta, \nu)\}$,

(iv) $m \equiv 4 \bmod 5 \Longrightarrow [\widehat{\mathcal{R}}_\alpha^\alpha = \{\hat{\nu} \colon \hat{\nu} \in \widehat{\mathcal{M}}_\alpha - (\widehat{\mathcal{R}}_\alpha^{<\alpha} \cup \widehat{\mathcal{B}}_\alpha^{<\alpha}) \ \&\ \widehat{F}(\beta, \nu)\}$ and $\mathcal{B}_\alpha^\alpha = \{\nu \colon \hat{\nu} \in \widehat{\mathcal{R}}_\alpha^\alpha\}]$,

(v) unless otherwise specified in (i)–(iv), $\mathcal{M}_\alpha$, $\mathcal{R}_\alpha$, $\mathcal{B}_\alpha$, and k_α take the values $\mathcal{M}_\beta$, $\mathcal{R}_\beta$, $\mathcal{B}_\beta$, and k_β, respectively.

For a consistent $\beta = f \upharpoonright n$, note that $\mathcal{F}_\beta^+$ is just a finite set of states and k_β^+ is an integer, so clearly α exists. We shall prove that if $\alpha \subset f$ then α is consistent, so the true path f exists and is infinite. Note that each of the conditions in Definition 3.21 is Π_2^0. Hence, there is a computable collection of c.e. sets $\{C_\alpha\}_{\alpha \in T}$ such that $\alpha \subset f$ iff $|C_\alpha| = \infty$. Fix a simultaneous computable enumeration $\{C_{\alpha,s}\}_{\alpha \in T, s \in \omega}$ which will be used in Section 3.2.9 to define a computable sequence $\{f_s\}_{s \in \omega}$ such that $f = \liminf_s f_s$.

REMARK 3.22. It does not hurt the present construction if we expand the tree T to include other components for action which will not interfere with the automorphism construction. For example, in [Harrington and Soare, 1996d] we modify the tree T by putting $\alpha = \beta^\frown\langle \mathcal{M}_\alpha, \mathcal{R}_\alpha, \mathcal{B}_\alpha, k_\alpha, n_\alpha \rangle$ in T providing $\beta \in T$, $n_\alpha \in \omega$, and conditions (i)–(viii) of Definition 3.20 hold as before. The Definition 3.21 is the same but with a new clause (vi) which asserts that n_α must have a certain property depending on β.

To ensure that $\mathcal{M}_\alpha \subseteq \mathcal{E}_\alpha$ for (27) we have a list $\mathcal{L}$ to be defined in Section 3.2.9. Very roughly when $\alpha \subset f_s$ we add to the bottom of $\mathcal{L}$ an (unmarked) α-entry of the form $\langle \alpha, \nu_1 \rangle$ for each $\nu_1 \in \mathcal{M}_\alpha$. At some later stage $t + 1 > s$ if we see some $x \in Y_{\beta,t} - Y_{\alpha,t}$ such that $\nu^+(\alpha, x, t) = \nu_1$ and $\nu(\alpha, x, t) \restriction \beta = \nu_1 \restriction \beta$, then (under Step 1 of Section 3.2.9) we move x to S_α, enumerate x in $U_{\alpha,t+1}$ if necessary so that $\nu(\alpha, x, t + 1) = \nu_1$, and we *mark* the α-entry $\langle \alpha, \nu_1 \rangle$ on $\mathcal{L}$. When each α-entry $\langle \alpha, \nu_1 \rangle$ on $\mathcal{L}$ has been marked we say that $\mathcal{L}$ has been *α-marked*, and we repeat the process by adding new (unmarked) entries $\langle \alpha, \nu_1 \rangle$ to $\mathcal{L}$ when next $\alpha \subset f_v$. We define $m(\alpha, s)$ to be the number of times $\mathcal{L}$ has been α-marked at stages $\leqslant s$, and we prove that $\lim_s m(\alpha, s) = \infty$ for $\alpha \subset f$. Let $\mathcal{L}_s$ denote that portion of $\mathcal{L}$ defined by the end of stage s.

3.2.9. *The construction.* To *initialize* node α means: to remove every $x \in S_{\alpha,s}$ ($\hat{x} \in \hat{S}_{\alpha,s}$), and put x in S^1_β ($\hat{x}$ in $\hat{S}^1_\beta$) for $\beta = \alpha \cap f_{s+1}$ (where $\alpha \cap \delta$ denotes the longest γ such that $\gamma \subseteq \alpha$ and $\gamma \subseteq \delta$); and if x ($\hat{x}$) is an *α-witness* as explained later, then cancel it as an α-witness.

We present in this section Steps 1–5 for the construction and a final Step 11 at which we define f_{s+1}. (Steps $\hat{1}$–$\hat{5}$ are the obvious duals to Steps 1–5, and will not be stated. There is no dual of Step 11.) These properties will produce the automorphism. In later sections we may add additional Steps $n(\hat{n})$, $5 < n < 11$, to achieve additional properties.

Stage $s = 0$. For all $\alpha \in T$ define $U_{\alpha,0} = V_{\alpha,0} = \widehat{U}_{\alpha,0} = \widehat{V}_{\alpha,0} = \emptyset$, and define $m(\alpha, 0) = 0$. Define $Y_{\lambda,0} = \widehat{Y}_{\lambda,0} = \emptyset$, and $f_0 = \lambda$.

Stage $s + 1$. Find the least $n < 11$ such that Step n applies to some $x \in Y_{\alpha,s}$, and perform the indicated action. If there is no such n then likewise find the least $n < 11$ such that Step $\hat{n}$ applies to some $\hat{x} \in \widehat{Y}_{\alpha,s}$, and perform the indicated action. If none of these steps applies then apply Step 11, and go to stage $s + 1$. (It is important that these steps be performed in the indicated order.)

In the following Steps 1–5 (Steps $\hat{1}$–$\hat{5}$) we let $\alpha \in T$, $\alpha \neq \lambda$, be arbitrary, let $\beta = \alpha^-$, and let $x \in Y_{\lambda,s}$ ($\hat{x} \in \widehat{Y}_{\lambda,s}$) be arbitrary.

Step 1. (Prompt pulling of x from R^1_β to S_α to ensure $\mathcal{M}_\alpha \subseteq \mathcal{E}_\alpha$.) Suppose $\langle \alpha, \nu_1 \rangle$ is the first unmarked entry on the list $\mathcal{L}_s$ such that the following conditions hold for some x, where $\nu_1 = \langle \alpha, \sigma_1, \tau_1 \rangle$,

(1.1) $x \in R^1_{\beta,s} - Y_{\alpha,s}$,
(1.2) $x > k_\alpha$ and $x > |\alpha|$,
(1.3) x is α-eligible (i.e. $\neg(\exists t)[x \leqslant t \leqslant s \;\&\; f_t < \alpha]$),
(1.4) $\neg[\alpha(x, s) <_L \alpha]$,
(1.5) $x > m(\alpha, s)$,
(1.6) $\nu(\beta, x, s) = \nu_1 \restriction \beta$,
(1.7) $e_\alpha > e_\beta \Longrightarrow \nu^+(\alpha, x, s) = \nu_1$.

Action. Choose the least x corresponding to $\langle\alpha, \nu_1\rangle$, and do the following:

(1.8) mark the α-entry $\langle\alpha, \nu_1\rangle$ on $\mathcal{L}_s$, and suppose this is the k-th occurrence of $\langle\alpha, \nu_1\rangle$ on $\mathcal{L}_s$,

(1.9) move x to S^i_α, where $k \equiv i \bmod 2$,

(1.10) if $e_\alpha > e_\beta$ and $e_\alpha \in \sigma_1$ then enumerate x in $U_{\alpha,s+1}$, and

(1.11) if $\hat{e}_\alpha > \hat{e}_\beta$ and $\hat{e}_\alpha \in \tau_1$ then enumerate x in $\widehat{V}_{\alpha,s+1}$. (Hence, $\nu(\alpha, x, s+1) = \nu_1$. Also $\nu_1 \in \mathcal{M}_\alpha$ because $\langle\alpha, \nu_1\rangle \in \mathcal{L}$ implies $\nu_1 \in \mathcal{M}_\alpha$.)

(1.12) if $\alpha <_L \alpha(x, s)$ then for every γ such that $\alpha <_L \gamma$, cancel all γ-witnesses if any exist, where the latter are defined in Section 3.5.

Step 2. (Move x from S^1_β to S^1_α so $Y_\alpha =^* \omega$.) Suppose there is an x such that,

(2.1) $x \in S^1_{\beta,s}$,

(2.2) $x > |\alpha|$ and $x > k_\alpha$,

(2.3) x is α-eligible,

(2.4) $x < m(\alpha, s)$,

(2.5) α is the $<_L$-least $\gamma \in T$ with $\gamma^- = \beta$ satisfying (2.1)–(2.4).

Action. Choose the least pair $\langle\alpha, x\rangle$ and

(2.6) move x from S^1_β to S^1_α.

(In Step 2 we need (2.4) so Y_α will not grow while α is waiting for another prompt pulling under Step 1.)

Step 3. (For α $\mathcal{M}$-inconsistent to ensure $\alpha \not\subset f$.) Suppose for $\alpha \in T$ there exists x such that,

(3.1) $e_\alpha > e_\beta$,

(3.2) $x \in S_{\alpha,s}$,

(3.3) $\nu(\alpha, x, s) = \nu_0 \in \mathcal{M}_\alpha$,

(3.4) $(\exists \nu_1)[\nu_0 <_{\mathrm{B}} \nu_1 \ \& \ \nu_1 \restriction \beta \in \mathcal{M}_\beta \ \& \ \nu_1 \notin \mathcal{M}_\alpha]$.

Action. Choose the least such pair $\langle\alpha, x\rangle$ and,

(3.5) enumerate x in $\widehat{V}_{\delta,s+1}$ for all $\delta \subset \alpha$ such that $e_\delta \in \tau_1$. (This action causes $\nu(\alpha, x, s+1) = \nu_1$. Hence, α is provably incorrect at all stages $t \geqslant s+1$ so $\alpha \not\subset f$.)

Step 4. (Delayed RED enumeration into U_α.) Suppose $x \in R_{\alpha,s}$ and

(4.1) $e_\alpha > e_\beta$,

(4.2) $x \notin U_{\alpha,s}$,

(4.3) $x \in Z_{e_\alpha,s} =_{\mathrm{dfn}} U_{e_\alpha,s} \cap Y_{\beta,s-1}$.

Action. Choose the least such pair $\langle\alpha, x\rangle$ and,

(4.4) enumerate x in $U_{\alpha,s+1}$.

Step 5. (BLUE emptying of state $\nu \in \mathcal{B}_\alpha$.) Suppose for $\alpha \in T$ there exists x such that either Case 1 or Case 2 holds.

Case 1. Suppose

(5.1) $\nu(\alpha, x, s) = \nu_0 \in \mathcal{B}_\alpha$, say $\nu_0 = \langle\alpha, \sigma_0, \tau_0\rangle$,

(5.2) $x \in S_{\alpha,s}$,

(5.3) α is $\mathcal{M}$-consistent and $\mathcal{R}$-consistent.

Action. Choose the least such pair $\langle\alpha, x\rangle$. Let $\nu_1 = h_\alpha(\nu_0) >_B \nu_0$, where h_α is a target function satisfying (49). In Section 3.4 and thereafter we shall assume that h_α also satisfies (61). In Section 3.5 and thereafter we shall assume that $\hat{h}_\alpha$ also satisfies (69). Let $\nu_1 = \langle\alpha, \sigma_1, \tau_1\rangle$.

(5.4) Enumerate $x \in \widehat{V}_\delta$ for all $\delta \subseteq \alpha$ such that $\hat{e}_\delta > \hat{e}_{\delta^-}$ and $e_\delta \in \tau_1 - \tau_0$. (Hence, $\nu(\alpha, x, s+1) = \nu_1$.)

Case 2. Suppose that (5.1) holds and

(5.5) $x \in S_{\gamma,s}$ where $\gamma^- = \alpha$, and

(5.6) γ is either $\mathcal{M}$-inconsistent or $\mathcal{R}$-inconsistent.

Action. Perform the same action as in Case 1 to achieve $\nu(\alpha, x, s+1) = \nu_1$.

(In (5.6) note that by (59) $\gamma \in T$ implies (5.3) for $\alpha = \gamma^-$, so h_α exists in Case 2. Note in Step 5 Case 2 that the enumeration may not be γ-legal, i.e. perhaps $\nu(\gamma, x, s+1) \notin \mathcal{M}_\gamma$, but this will not matter because we shall prove that $\gamma \not\subset f$ if γ is inconsistent. Hence, it only matters that the enumeration is α-legal, i.e. $\nu(\alpha, x, s) \in \mathcal{M}_\alpha$.)

Step 11. (Defining f_{s+1}, $m(\alpha, s+1)$, $\mathcal{L}_{s+1}$ and $Y_{\lambda,s+1}$.)

Substep 11A. (Defining f_{s+1}.) First we define δ_t by induction on t for $t \leqslant s+1$. Let $\delta_0 = \lambda$. Given δ_t let $v \leqslant s$ be maximal such that $\delta_t \subseteq f_v$ if v exists and let $v = 0$ otherwise. (Let $\{C_{\gamma,v}\}_{\gamma \in T, v \in \omega}$ be the simultaneous computable enumeration specified at the end of Section 3.2.8.) Choose the $\leqslant_L$-least $\alpha \in T$ such that $\alpha^- = \delta_t$ and $C_{\alpha,s} \neq C_{\alpha,v}$ if α exists and define $\delta_{t+1} = \alpha$. If α does not exist define $\delta_{t+1} = \delta_t$. Finally, define $f_{s+1} = \delta_{s+1}$.

Substep 11B. (Defining $m(\alpha, s+1)$, $\mathcal{L}_{s+1}$, and their duals.) For every $\alpha \subseteq f_{s+1}$ if every α-entry $\langle \alpha, \nu \rangle$ on $\mathcal{L}_s$ and every α-entry $\langle \alpha, \hat{\nu} \rangle$ on $\widehat{\mathcal{L}}_s$ is marked we say that the lists are *α-marked* and we

(11.1) define $m(\alpha, s+1) = m(\alpha, s) + 1$, and

(11.2) add to the bottom of list $\mathcal{L}_s$ ($\widehat{\mathcal{L}}_s$) a new (unmarked) α-entry $\langle \alpha, \nu \rangle$ ($\langle \alpha, \hat{\nu} \rangle$) for every such α and every $\nu \in \mathcal{M}_\alpha$. Let the resulting list be $\mathcal{L}_{s+1}(\widehat{\mathcal{L}}_{s+1})$.

If the lists are not both α-marked then let $m(\alpha, s+1) = m(\alpha, s)$, $\mathcal{L}_{s+1} = \mathcal{L}_s$ and $\widehat{\mathcal{L}}_{s+1} = \widehat{\mathcal{L}}_s$.

Substep 11C. (Emptying R_α to the right of f_{s+1}.) For every α such that $f_{s+1} <_L \alpha$, *initialize* α.

Substep 11D. (Moving from S^0_α to S^1_α.) If currently x is in S^0_α but x is not an α-witness then move x to S^1_α.

(Steps $n(\hat{n})$, $5 < n < 11$, to be defined in later sections, will determine when $x \in S^0_\alpha$ starts and stops being an α-witness. Up through the present section there are no α-witnesses so every $x \in S^0_\alpha$ is eventually moved to S^1_α under Substep 11D, unless x is first removed from S_α by some other step such as Step 11C or Step 1_β for $\beta <_L \alpha$.)

Substep 11E. (Filling Y_λ and $\hat{Y}_\lambda$.) Choose the least $x \notin Y_{\lambda,s}$ ($\hat{x} \notin \hat{Y}_{\lambda,s}$) and $x < s$. Put x in S_λ ($\hat{x}$ in $\hat{S}_\lambda$).

For each $x \in Y_{\lambda,s+1}$ ($\hat{x} \in \hat{Y}_{\lambda,s+1}$) let $\alpha(x, s+1)$ ($\alpha(\hat{x}, s+1)$) denote the unique γ such that $x \in S_{\gamma,s+1}$. This completes stage $s+1$ and the construction.

Note that after each application of Step 11, the other Steps 1–5 and Steps $\hat{1} - \hat{5}$ can apply only finitely often until the next application of Step 11 as is easily verified.

3.3. *The automorphism theorem*

From now on we assume that $A = U_0$ is a nonrecursive r.e. set. In Section 3.4 we introduce Step 6 to exploit this hypothesis. Step 6 together with Steps 1–5, Steps $\hat{1}$–$\hat{5}$, and Step 11 of Section 3.2.9 constitute the *basic construction* designed to ensure that we achieve an automorphism. We may also want to add in later sections of this paper (and in subsequent papers) certain additional Steps $n(\hat{n})$, $6 < n < 11$, to ensure special properties about $B = \widehat{U}_0$, such as B is high or $D \leqslant_{\mathrm{T}} B$, for a given set D. We now wish to isolate certain minimal conditions which these additional steps must satisfy so that the resulting construction will still produce an automorphism.

CONVENTION 3.23. From now on Step n $(\hat{n})$ denotes one of these new steps for $6 \leqslant n < 11$. In addition we assume that given finitely many elements in $Y_{\lambda,s}$, Step n can apply for at most finitely many stages until another element is put in Y_λ, and similarly for Step $\hat{n}$.

THEOREM 3.24 (Automorphism theorem). *Assume that $A = U_0$ is a noncomputable c.e. set. Suppose c.e. sets $\{U_\alpha\}_{\alpha\in T}$, $\{V_\alpha\}_{\alpha\in T}$, $\{\widehat{U}_\alpha\}_{\alpha\in T}$, and $\{\widehat{V}_\alpha\}_{\alpha\in T}$ are enumerated by the construction in Section* 3.2.9 *using Steps* 1–5, *Steps* $\hat{1}$–$\hat{5}$, *and Step* 11 *of Section* 3.2.9, *Step* 6 *of Section* 3.4, *and possibly also some additional Steps n $(\hat{n})$, $6 < n < 11$, such that for all n, $6 \leqslant n < 11$, Steps $n(\hat{n})$ satisfy the following conditions* P1–P4 (*and their duals* $\widehat{\mathrm{P1}}$–$\widehat{\mathrm{P4}}$ *for* $\widehat{S}_\alpha$). *Then the correspondence $U_\alpha \leftrightarrow \widehat{U}_\alpha$ and $\widehat{V}_\alpha \leftrightarrow V_\alpha$, $\alpha \subset f$, defines an automorphism of $\mathcal{E}$.*

(P1) *If α is $\mathcal{R}$-inconsistent or $\mathcal{M}$-inconsistent then Step n does not apply to α. If α is $\mathcal{C}$-inconsistent then Step n applies to α only if $n = 6$. (Step* 6 *and $\mathcal{C}$-inconsistent are defined in Section* 3.4.)

(P2) *Step n cannot enumerate x in any red set U_α. If Step n at stage $s+1$ enumerates x in a blue set $\widehat{V}_\alpha$, then $x \in R_{\alpha,s}$, and this enumeration must be α-legal, i.e. must satisfy* (25), *so that $\nu(\alpha, x, s+1) \in \mathcal{M}_\alpha$.*

(P3) *Step n cannot move x from from S_α to S_γ for $\alpha \neq \gamma$, or from S^1_α to S^0_α, but can only appoint some x already in S^0_α as an α-witness, and can later cancel x as an α-witness and simultaneously move x from S^0_α to S^1_α.*

(P4) *For all α, $S^0_{\alpha,\infty} =^* \emptyset$.*

The importance of the Automorphism Theorem 3.24 is that from now on we need only verify that the new Steps $n(\hat{n})$, $6 \leqslant n < 11$, satisfy conditions (P1)–(P4) (and their duals) and we need not mention anything about automorphisms explicitly. For our purposes in this paper conditions (P1)–(P3) for some new Step n will be immediately verifiable by inspection, and (P4) will verified later. On the other hand the new Steps $n(\hat{n})$ have great latitude to enumerate and restrain elements, subject primarily to (P2), Step 5, (P4), and their duals. Namely, suppose that Step n operates on S_α, where $\alpha \subset f$, and that after some stage v_α, α is not initialized, and no $\beta <_L \alpha$ acts.

First, Step n may cause certain elements $x \in S_\alpha$ (not just $x \in S^0_\alpha$) to be enumerated in various blue sets, so long as this enumeration is α-legal by (P2). Second, Step n

may cause certain elements $x \in S^0_\alpha$ to become α-witnesses, i.e. the positions of α-markers, whereupon by holding x as an α-witness Step n may restrain x from leaving S^0_α, and hence restrain x from being enumerated in $\widehat{V}_\gamma$ for any $\gamma \supset \alpha$, and may also restrain x from entering any further blue sets $\widehat{V}_\gamma$, $\gamma \subseteq \alpha$, subject only to Step 5. Note that Steps 1 and 2 cannot apply to $x \in S^0_\alpha$ after stage v_α, and Step 3 only applies to an α which is $\mathcal{M}$-inconsistent but such $\alpha \not\subset f$. Hence, only Steps 4 and 5 from the basic Steps 1–5, and 11, can apply to $x \in S^0_\alpha$ after stage v_α. The latter will still hold after we add to the basic construction Step 6 in Section 3.4, because Step 6 only applies to an α which is $\mathcal{C}$-inconsistent and such $\alpha \not\subset f$ by Lemma 3.28.

An element x enters S^0_α at most once (when it is first pulled to S_α by Step 1), x becomes an α-witness at most once, and if x ceases to be an α-witness then x moves from S^0_α to S^1_α. Finally, the new steps must satisfy (P4), that $S^0_{\alpha,\infty} =^* \emptyset$, so that at most finitely many elements are permanently restrained in S_α and thus almost every $x \in S_\alpha$ is available to be passed to S_γ for $\gamma \supset \alpha$. Hence, the new steps will not interfere with the basic construction which produces an automorphism.

The details of the proof of the Automorphism Theorem may be found in [Harrington and Soare, 1996c]. The point is that this is a very general theorem which can be used to prove additional specific theorems by adding appropriate rules.

3.4. *Using that A is noncomputable to obtain the set $\widehat{\mathcal{C}}_\alpha$ of coding states*

For the rest of this section we assume that RED specifies a noncomputable c.e. set A and BLUE replies by constructing an r.e. set B automorphic to A such that B also codes certain additional information (such as B is high) as in the conclusion of Theorem 3.1. We let $U_0 = A$ and $B = \widehat{U}_\rho$, where $\rho = f \restriction 1$. Define $B_s = \widehat{U}_{\rho,s}$. (From now on we consider only nodes $\alpha \in T$ such that $\rho \subset \alpha$.) To code this information into B BLUE will choose an α-state $\hat{\nu}_1$ with certain properties, choose a witness $\hat{y} \in \widehat{S^0_\alpha}$ in α-state $\hat{\nu}_1$, begin by holding $\hat{y}$ in $\overline{B}$ and in α-state $\hat{\nu}_1$, and perhaps later attempt to move $\hat{y}$ into B. To see that $\widehat{\mathcal{C}}_\alpha \neq \emptyset$ where $\widehat{\mathcal{C}}_\alpha$ is the set of α-states $\hat{\nu}_1$ (called *coding* states) with the necessary properties we now use the noncomputability of A to verify that the dual set $\mathcal{C}_\alpha \neq \emptyset$, where $\widehat{\mathcal{C}}_\alpha = \{\hat{\nu}\colon \nu \in \mathcal{C}_\alpha\}$.

DEFINITION 3.25.

(i) Let $\mathcal{W}_\alpha$ be that subset of $\mathcal{M}_\alpha$ which is generated by the following three clauses:

(1) $[\nu_1 = \langle \alpha, \sigma_1, \tau_1\rangle \;\&\; 0 \in \sigma_1] \Longrightarrow \nu_1 \in \mathcal{W}_\alpha$,

(2) $(\exists \nu_2)[\nu_1 <_R \nu_2 \;\&\; \nu_2 \in \mathcal{W}_\alpha] \Longrightarrow \nu_1 \in \mathcal{W}_\alpha$,

(3) $[\nu_1 \in \mathcal{B}_\alpha \;\&\; (\forall \nu_2 \in \mathcal{M}_\alpha)[\nu_1 <_B \nu_2 \Longrightarrow \nu_2 \in \mathcal{W}_\alpha]] \Longrightarrow \nu_1 \in \mathcal{W}_\alpha$.

(ii) Define $\mathcal{W}^\#_\alpha = \{\nu_1\colon \nu_1 = \langle \alpha, \sigma_1, \tau_1\rangle \in \mathcal{W}_\alpha \;\&\; 0 \notin \sigma_1\}$.

(iii) Define $\mathcal{V}_\alpha = \mathcal{M}_\alpha - \mathcal{W}_\alpha$.

Note that $\mathcal{W}_\alpha$ consists of the α-states $\nu_1 \in \mathcal{M}_\alpha$ for which RED has a *winning strategy* F_α to force any x in α-state ν_1 into A. Namely, if $\nu(\alpha, x, s) = \nu_1$ and (1) holds then $x \in U_0 = A$ already; if (2) then RED can change the α-state of x from ν_1 to $F_\alpha(\nu_1) = \nu_2 >_R \nu_1$; and if (3) then by (22) and (28) RED can wait for BLUE to change x from α-state ν_1 to some $\nu_2 >_B \nu_1$ and then RED can apply F_α to ν_2. Hence, this winning strategy can be identified with a function,

$$F_\alpha : (\mathcal{W}_\alpha^\# - \mathcal{B}_\alpha) \to \mathcal{W}_\alpha \;\&\; (\forall \nu_1 \in (\mathcal{W}_\alpha^\# - \mathcal{B}_\alpha))[\nu_1 <_R F_\alpha(\nu_1)]. \tag{60}$$

Similarly, if $\nu(\alpha, x, s) = \nu_1 \in \mathcal{V}_\alpha$ then BLUE has a winning strategy G_α to keep x out of A. Namely, BLUE keeps x in α-state ν_1 unless $\nu_1 \in \mathcal{B}_\alpha$ in which case by the negation of (3), BLUE can change x to some α-state $G_\alpha(\nu_1) = \nu_2 >_B \nu_1$ such that $\nu_2 \in \mathcal{V}_\alpha$. Meanwhile if RED causes $\nu(\alpha, x, t) = \nu_3 >_R \nu_1$ at some $t > s$, then by the negation of (2), $\nu_3 \in \mathcal{V}_\alpha$ so BLUE continues to play as for ν_1. By repeatedly applying G_α if necessary we may assume that $\text{range}(G_\alpha) \cap \mathcal{B}_\alpha = \emptyset$. Hence, from now on we may assume that BLUE's target function h_α of (49) agrees with the function G_α on their common domain, namely h_α also satisfies

$$(\forall \nu \in \mathcal{V}_\alpha \cap \mathcal{B}_\alpha)[\nu_1 <_B h_\alpha(\nu_1) = G_\alpha(\nu_1) \in \mathcal{V}_\alpha - \mathcal{B}_\alpha]. \tag{61}$$

(Thus, by using this h_α any BLUE enumeration under Step 5 of Section 3.2.9 is automatically following BLUE's winning strategy G_α for all $\nu_1 \in \mathcal{V}_\alpha$.)

DEFINITION 3.26. If $\alpha \neq \lambda$, let $\mathcal{C}_\alpha$ be the set of $\nu_1 \in \mathcal{M}_\alpha$ such that,

(i) $\nu_1 \in \mathcal{W}_\alpha^\#$,

(ii) $\neg(\exists \nu_2 \in \mathcal{M}_\alpha)[\nu_1 <_B \nu_2]$,

(iii) $\nu_1 \notin \mathcal{N}_\alpha =_{\text{dfn}} \mathcal{R}_\alpha \cup \mathcal{B}_\alpha$.

Property (ii) asserts that ν_1 is maximal with respect to α-legal enumeration in blue sets, and the import of (iii) is that $\nu_1 \notin \mathcal{R}_\alpha$. Hence, if x is in state ν_1 then RED can hold x forever in that state (and hence in $\overline{A}$), or by (i) RED can later force x to eventually enter A.

DEFINITION 3.27. (i) A node $\alpha \in T$ is *$\mathcal{C}$-consistent* if $\alpha = \lambda$ or $\mathcal{C}_\alpha \neq \emptyset$ and *$\mathcal{C}$-inconsistent* otherwise.

(ii) A node $\alpha \in T$ is *consistent* if α is $\mathcal{M}$-consistent (Definition 3.17), $\mathcal{R}$-consistent, (Definition 3.18), and also $\mathcal{C}$-consistent.

Any inconsistent α is terminal by (59). From now on we assume that the following Step 6 has been added to the construction in Section 3.2.9. (Step 6 will ensure that α is $\mathcal{C}$-consistent for $\alpha \subset f$. There is no dual Step $\hat{6}$.)

Step 6. Suppose $\alpha \in T$, α is $\mathcal{C}$-inconsistent, but $\mathcal{M}$-consistent and $\mathcal{R}$-consistent, $x \in S_{\alpha,s}$, $\nu(\alpha, x, s) = \nu_1$, and

$$(\exists \nu_2 \in \mathcal{M}_\alpha)[\nu_1 <_B \nu_2].$$

Action. Choose the least such pair $\langle\alpha, x\rangle$, and the first such $\nu_2 \in \mathcal{M}_\alpha$. Let $\nu_2 = \langle\alpha, \sigma_2, \tau_2\rangle$. Enumerate x in $\widehat{V}_{\beta,s+1}$ for all $\beta \subseteq \alpha$ such that $\hat{e}_\beta > \hat{e}_{\beta^-}$ and $\hat{e}_\beta \in \tau_2$. (Hence, $\nu(\alpha, x, s+1) = \nu_2$.)

Clearly, Step 6 satisfies conditions (P1)–(P4) of the Automorphism Theorem 3.24, because $\nu_2 \in \mathcal{M}_\alpha$.

LEMMA 3.28. *If $\alpha \subset f$ then α is $\mathcal{C}$-consistent.*

PROOF. Assume for a contradiction that $\alpha \subset f$ and α is $\mathcal{C}$-inconsistent. Now α is terminal on T, $S_\alpha = R_\alpha$, $S_{\alpha,\infty} =^* \omega$, and no $x \in S_{\alpha,s}$, $s > \upsilon_\alpha$, later leaves S_α. Thus, neither Step 1 nor Step 2 can apply to any $x \in S_{\alpha,s}$ after stage υ_α, and neither Step 3 nor Step 5 Case 2 can ever apply because α is $\mathcal{M}$-consistent and $\mathcal{R}$-consistent. For each $\nu_1 \in \mathcal{M}_\alpha$ define the c.e. set

$$D_{\nu_1} = \{x\colon (\exists s > \upsilon_\alpha)[x \in S_{\alpha,s} \ \&\ \nu(\alpha, x, s) = \nu_1]\}.$$

Now $D_{\nu_1} \subseteq \overline{A}$ for every $\nu_1 \in \mathcal{V}_\alpha$ because by (P1) the only red (blue) enumeration of x after $x \in D_{\nu_1,s}$ comes from Step 4 (Step 5), but in Step 5 the target function h_α now satisfies (61) so $\nu(\alpha, x, t) \in \mathcal{V}_\alpha$ for all $t \geqslant s$.

Let $\mathcal{K}_\alpha$ be as in (31). For each $\nu_1 \in \mathcal{M}_\alpha - \mathcal{K}_\alpha$ such that $0 \notin \sigma_1$ (i.e. each ν_1 well-resided on $\overline{A}$) let $E_{\nu_1} = \{x\colon \nu(\alpha, x) = \nu_1\}$. If $\nu_1 \in \mathcal{V}_\alpha$ then $E_{\nu_1} \subseteq D_{\nu_1} \subseteq \overline{A}$. Since A is noncomputable there must exist $\nu_1 \in (\mathcal{M}_\alpha - \mathcal{K}_\alpha) \cap \mathcal{W}^{\#}_\alpha$. Hence,

$$\nu_1 \notin \mathcal{N}_\alpha = \mathcal{R}_\alpha \cup \mathcal{B}_\alpha$$

because $\mathcal{N}_\alpha \subseteq \mathcal{K}_\alpha$. By Step 6, every such ν_1 must satisfy Definition 3.26(ii), and hence $\nu_1 \in \mathcal{C}_\alpha$. Thus, α is $\mathcal{C}$-consistent. □

3.5. *Moving α-witnesses into B*

Let A and B be as in the beginning of Section 3.4. Let the set of coding states $\widehat{\mathcal{C}_\alpha}$ be the dual of $\mathcal{C}_\alpha$ of Section 3.4, namely $\widehat{\mathcal{C}_\alpha} = \{\hat{\nu}\colon\ \nu \in \mathcal{C}_\alpha\}$. To code information into B we define Step $\hat{7}$ in Section 3.6, which determines when an element $\hat{x} \in \hat{S}^0_\alpha$ in some state $\hat{\nu}_1 \in \widehat{\mathcal{C}_\alpha}$ becomes an α-witness; various versions of Steps $\hat{n}$, $9 \leqslant n < 11$, defined in later sections (to prove one of several different theorems about B) will determine when an α-witness $\hat{x}$ later becomes *activated* indicating that $\hat{x}$ wants to enter B; Step $\hat{8}$ defined in Section 3.7 processes an activated witness until it enters B; and finally the Coding Theorem 3.33 in Section 3.8 proves that this coding procedure succeeds. (There are no dual Steps 7, or 8.) Let $\widehat{L_\alpha}$ ($\widehat{J_\alpha}$) denote the d.c.e. set of α-witnesses (activated α-witnesses) and $\widehat{L_{\alpha,s}}$ ($\widehat{J_{\alpha,s}}$) the set of elements in $\widehat{L_\alpha}$ (respectively $\widehat{J_\alpha}$) at the end of stage s. We shall assume from now on that any additional Steps $\hat{n}$, $9 \leqslant n < 11$, cannot add elements to or remove elements from $\widehat{L_\alpha}$.

3.6. *Appointing α-witnesses and Step $\hat{7}$*

As input to Step $\hat{7}$ we have a computable function $g(\alpha, s)$. The choice of g will depend on the theorem being proved. For example, in Theorem 3.1, $g(\alpha, s) = 1$ for all α and s. Furthermore, g may even be defined *during* the construction providing that for all α and s, $g(\alpha, s)$ is defined by the end of stage s.

We order the set $\widehat{L}_\alpha$ of α-witnesses so that for all α and every i, $1 \leqslant i \leqslant g(\alpha, s)$, we attempt to define a *primary* witness $\hat{y}_{\alpha,i,s}$ and a *backup* witness $\hat{y}'_{\alpha,i,s}$. First we divide the witness set $\widehat{L}_\alpha$ into the disjoint union of subsets $\widehat{L}'_{\alpha,i}$, $1 \leqslant i$, such that $\widehat{L}'_{\alpha,i,s}$ is the set of elements in $\widehat{L}_{\alpha,i}$ at the end of stage s, and $\widehat{L}'_{\alpha,i,s} = \{\hat{y}_{\alpha,i,s}, \hat{y}'_{\alpha,i,s}\}$ if these are defined. We define $\widehat{L}'_{\alpha,i,s}$, $\hat{y}_{\alpha,i,s}$, and $\hat{y}'_{\alpha,i,s}$, by induction on s as follows.

DEFINITION 3.29.

(i) If $\hat{x} \in \widehat{L}_{\alpha,s+1} - \widehat{L}_{\alpha,s}$ (necessarily because Step $\hat{7}$ Case 1 applies at stage $s+1$ so there will be at most one such $\hat{x}$) then let i be the least $j \geqslant 1$ such that $|\widehat{L}'_{\alpha,j,s}| < 2$. Put $\hat{x}$ in $\widehat{L}'_{\alpha,i,s+1}$.

(ii) $\hat{x}$ remains in $\widehat{L}'_{\alpha,i}$ until if ever $\hat{x}$ is removed from $\widehat{L}_\alpha$ at which time $\hat{x}$ is also removed from $\widehat{L}'_{\alpha,i}$.

(iii) Define $\hat{y}_{\alpha,i,s}$ and $\hat{y}'_{\alpha,i,s}$ by

$$\begin{aligned} \hat{y}_{\alpha,i,s} &= (\mu\hat{x})\big[\hat{x} \in \widehat{L}'_{\alpha,i,s}\big], \quad \text{and} \\ \hat{y}'_{\alpha,i,s} &= (\mu\hat{x})\big[\hat{x} > \hat{y}_{\alpha,i,s} \ \&\ \hat{x} \in \widehat{L}'_{\alpha,i,s}\big], \end{aligned} \tag{62}$$

if these elements exist.

Step $\hat{7}$. (Putting $\hat{x}$ into $\widehat{L}_\alpha$.) Assume α is consistent as defined in Definition 3.27.

Case 1. If $1 \leqslant i \leqslant g(\alpha, s)$, $|\widehat{L}'_{\alpha,i,s}| < 2$, and there exists $\hat{x} \in \widehat{S}^0_{\alpha,s}$, such that,

(7.1) $\nu(\alpha, \hat{x}, s) \in \widehat{C}_\alpha$, and

(7.2) $\hat{x} > \max(\bigcup_{t \leqslant s} \widehat{L}_{\alpha,t})$,

then put the least such $\hat{x}$ into $\widehat{L}'_{\alpha,i,s+1}$.

Case 2. For all $i > g(\alpha, s)$, remove $\hat{y}_{\alpha,i,s}$ and $\hat{y}'_{\alpha,i,s}$ from $\widehat{L}_\alpha$ and from $\widehat{S}^0_\alpha$, and put them in $\widehat{S}^1_\alpha$.

LEMMA 3.30. *Assume that the construction of Section* 3.2.9 *is performed but also with Step* 6 *and Step* $\hat{7}$ *and perhaps with additional Steps* n $(\hat{n})$, $8 \leqslant n < 11$. *Let* $g(\alpha, s)$ *be the function for Step* $\hat{7}$. *Suppose* $(\forall \gamma \subset f)[\liminf_s g(\gamma, s) < \infty]$. *Then for all* $\alpha \in T$, $S^0_{\alpha,\infty} = \emptyset$, *and* $\widehat{S}^0_{\alpha,\infty} =^* \emptyset$, *so conditions* (P4) *and* ($\widehat{\text{P}}$4) *of the Automorphism Theorem* 3.24 *are satisfied.*

PROOF. There is no Step 7 so $L_{\alpha,s} = \emptyset$ for all s, and every element $x \in S^0_{\alpha,s}$ is eventually removed from S^0_α by Step 11D, so $S^0_{\alpha,\infty} = \emptyset$. Hence, condition (P4) is satisfied. If $\alpha \not\subset f$ then $\widehat{S}_{\alpha,\infty} =^* \emptyset$. Now consider $\alpha \subset f$. Let $g(\alpha) = \liminf_s g(\alpha, s)$. Step $\hat{7}$ infinitely often has an opportunity to act. By Step $\hat{7}$ Case 2, $|\widehat{L}_{\alpha,\infty}| \leqslant 2g(\alpha)$, and hence by Step 11D, $|\widehat{S}^0_{\alpha,\infty}| \leqslant 2g(\alpha)$. Thus, $\widehat{S}^0_{\alpha,\infty} =^* \emptyset$, and condition ($\widehat{\text{P}}4$) is satisfied. □

Note that if $g(\alpha, s) = m$, for all α and s (for example, $m = 1$ in Theorem 3.1, then $|\widehat{L}_{\alpha,s}| \leqslant 2m$ for all s and hence Case 2 of Step $\hat{7}$ will never apply.

3.7. *Coding states $\widehat{\mathcal{C}}_\alpha$ and Step $\hat{8}$*

In this section we use the coding states $\widehat{\mathcal{C}}_\alpha$ to produce a strategy formalized in Step $\hat{8}$ below for moving $\hat{x} \in \widehat{J}_\alpha$ into B. Assume $\alpha \subset f$. Since $\mathcal{C}_\alpha \neq \emptyset$ by Lemma 3.28 we have by (29), (38), and (39) that $\widehat{\mathcal{C}}_\alpha \neq \emptyset$ where $\widehat{\mathcal{C}}_\alpha = \{\hat{\nu}\colon\ \nu \in \mathcal{C}_\alpha\}$. Choose any $\hat{\nu}_1 \in \widehat{\mathcal{C}}_\alpha$. By the dual of Definition 3.26 and (30) we have,

$$\hat{\nu}_1 \in \widehat{\mathcal{W}}^\#_\alpha, \tag{63}$$

$$\neg(\exists \hat{\nu}_2 \in \widehat{\mathcal{M}}_\alpha)[\hat{\nu}_1 <_{\text{R}} \hat{\nu}_2], \tag{64}$$

$$\hat{\nu}_1 \notin \widehat{\mathcal{N}}_\alpha =_{\text{dfn}} \widehat{\mathcal{R}}_\alpha \cup \widehat{\mathcal{B}}_\alpha. \tag{65}$$

If $\nu(\alpha, \hat{x}, s) = \hat{\nu}_1 \in \widehat{\mathcal{C}}_\alpha$ then by (64) $\hat{\nu}_1$ is maximal with respect to α-legal red moves so RED cannot change the α-state of $\hat{x}$, and by (65), $\hat{\nu}_1 \notin \widehat{\mathcal{B}}_\alpha$ so BLUE does not have to change the state; and hence BLUE can hold $\hat{x}$ in α-state $\hat{\nu}_1$ forever if he chooses. However, BLUE can later force $\hat{x}$ into B as follows. By the dual of Definition 3.25 and the remarks following it, if $\hat{\nu}_2 = \langle \alpha, \sigma_2, \tau_2 \rangle \in \widehat{\mathcal{W}}^\#_\alpha$ then

$$0 \notin \sigma_2 \quad \text{(so if } \nu(\alpha, \hat{x}, s) = \hat{\nu}_2 \text{ then } \hat{x} \notin B_s\text{), and} \tag{66}$$

$$\begin{aligned}\text{BLUE} \quad &\text{has a winning strategy, } \widehat{F}_\alpha\text{, to force}\\ &\text{any element } \hat{x} \text{ in } \alpha\text{-state } \hat{\nu}_2 \text{ into } B.\end{aligned} \tag{67}$$

Namely, by the dual of (60), we have

$$\big(\forall \hat{\nu}_2 \in \big(\widehat{\mathcal{W}}^\#_\alpha - \widehat{\mathcal{R}}_\alpha\big)\big)\big[\hat{\nu}_2 <_{\text{B}} \widehat{F}_\alpha(\hat{\nu}_2) \in \widehat{\mathcal{W}}_\alpha\big]. \tag{68}$$

By repeatedly applying $\widehat{F}_\alpha$ if necessary we may assume in (68) that $\widehat{F}_\alpha(\hat{\nu}_3) \notin \widehat{\mathcal{B}}_\alpha$. Hence, from now on we may assume that the target function $\hat{h}_\alpha$ for (46) used in Step $\hat{5}_\alpha$ agrees with $\widehat{F}_\alpha$ on their common domain, namely,

$$\big(\forall \hat{\nu}_2 \in \big(\widehat{\mathcal{W}}^\#_\alpha \cap \widehat{\mathcal{B}}_\alpha\big)\big)\big[\hat{\nu}_2 <_{\text{B}} \hat{h}_\alpha(\hat{\nu}_2) = \widehat{F}_\alpha(\hat{\nu}_2) \in \widehat{\mathcal{W}}_\alpha - \widehat{\mathcal{B}}_\alpha\big], \tag{69}$$

so that while $\hat{x} \in \widehat{S}_\alpha$ any BLUE enumeration under Step $\hat{5}_\alpha$ Case 1 automatically follows strategy $\widehat{F}_\alpha$.

Step $\hat{8}$. (To move $\hat{x} \in \widehat{J}_\alpha$ toward B.) Suppose

(8.1) $\hat{x} \in \widehat{J}_{\alpha,s} - B_s$, and

(8.2) $\nu(\alpha, \hat{x}, s) = \hat{\nu}_1 \in \widehat{\mathcal{W}}^{\#}_{\alpha} - \widehat{\mathcal{R}}_{\alpha}$.

Action. Choose the least such pair $\langle \alpha, s \rangle$. Let $\widehat{F}_{\alpha}(\hat{\nu}_1) = \hat{\nu}_2 = \langle \alpha, \widehat{\sigma}_2, \widehat{\tau}_2 \rangle$. (Necessarily $\widehat{F}_{\alpha}(\hat{\nu}_1)\downarrow$ because $\hat{\nu}_1 \in \widehat{\mathcal{W}}^{\#}_{\alpha}$.)

(8.3) Enumerate $\hat{x}$ in $\widehat{U}_{\delta,s+1}$ for all $\delta \subseteq \alpha$ such that $\hat{e}_{\delta} \in \widehat{\sigma}_2$.

(8.4) If $\hat{x} \in B_{s+1} - B_s$ then move $\hat{x}$ from $\widehat{S}^0_{\alpha}$ to $\widehat{S}^1_{\alpha}$, and let $\hat{x}$ be cancelled as an α-witness (i.e. remove $\hat{x}$ from $\widehat{L}_{\alpha}$, and hence from $\widehat{J}_{\alpha}$).

Clearly, Step $\hat{8}$ satisfies $(\widehat{\text{P}}1)$–$(\widehat{\text{P}}4)$ of the Automorphism Theorem 3.24 because $\widehat{\mathcal{W}}_{\alpha} \subseteq \widehat{\mathcal{M}}_{\alpha}$ so $(\widehat{\text{P}}2)$ is satisfied; the others are obvious.

3.8. *The Coding Theorem*

DEFINITION 3.31. Let the *basic coding construction* denote the construction in Section 3.2.9 (consisting of Steps 1–5, $\hat{1}$–$\hat{5}$, and 11) but also with Step 6, Step $\hat{7}$, and Step $\hat{8}$ (as defined in Section 3.4, Section 3.6, and Section 3.7, respectively).

DEFINITION 3.32. For the following theorem define $t(\alpha, i)$ by

$$t(\alpha, i) = \begin{cases} (\mu t)(\forall s \geqslant t)[i \leqslant g(\alpha, s)] & \text{if } t \text{ exists,} \\ \infty & \text{otherwise.} \end{cases}$$

THEOREM 3.33 (Coding Theorem). *Let $A = U_0$ be a given noncomputable c.e. set, and let $B = \widehat{U}_{\rho}$ where $\rho = f \restriction 1$. Let $g(\alpha, s)$ be a computable function (to be used in Step $\hat{7}$). Perform the basic coding construction consisting of Steps* 1–6, 11, $\hat{1}$–$\hat{5}$, $\hat{7}$, $\hat{8}$, *and possibly with additional Steps $\hat{n}$, $9 \leqslant n < 11$, defined later which satisfy conditions* $(\widehat{\text{P}}1)$–$(\widehat{\text{P}}3)$ *from the Automorphism Theorem* 3.24.

(i) $(\forall \gamma \subseteq f)[\liminf_s g(\gamma, s) < \infty] \Longrightarrow$ *A is Δ^0_3-automorphic to B.*

In addition, if the Steps $\hat{n}$, $9 \leqslant n < 11$, satisfy the following conditions $(\widehat{\text{Q}}1)$–$(\widehat{\text{Q}}4)$ *then conclusions* (ii)–(viii) *hold.*

$(\widehat{\text{Q}}1)$ *Step $\hat{n}$ may not put any element $\hat{x}$ into the witness set $\widehat{L}_{\alpha}$.*

$(\widehat{\text{Q}}2)$ *Step $\hat{n}$ may not remove any element $\hat{x}$ from the witness set $\widehat{L}_{\alpha}$.*

$(\widehat{\text{Q}}3)$ *If Step $\hat{n}$ puts $\hat{x}$ into $\widehat{J}_{\alpha,s+1} - \widehat{J}_{\alpha,s}$ then $\hat{x} \in \widehat{L}_{\alpha,s}$, and Step $\hat{n}$ may not remove any element $\hat{x}$ from $\widehat{J}_{\alpha}$.*

$(\widehat{\text{Q}}4)$ *If $\hat{x} \in \widehat{S}^0_{\alpha,s}$ then Step $\hat{n}$ may not enumerate $\hat{x} \in \widehat{U}_{\alpha,s+1} - \widehat{U}_{\alpha,s}$ for any blue set $\widehat{U}_{\alpha}$.*

Assume $\alpha \subset f$, $\alpha \neq \lambda$. Choose υ_{α} such that for all $s \geqslant \upsilon_{\alpha}$, α is not initialized and no $\beta <_L \alpha$ acts at stage s. Then for all $\hat{x}$ and s and all $i \geqslant 1$,

(ii) $\widehat{J}_{\alpha,s} \subseteq \widehat{L}_{\alpha,s} \subseteq \widehat{S}^0_{\alpha,s}$, *and $\widehat{L}_{\alpha}$ and $\widehat{J}_{\alpha}$ are d.c.e. sets*;

(iii) $[\hat{x} \in (\widehat{S}^0_{\alpha,s} - \widehat{J}_{\alpha,s})\ \&\ \nu(\alpha, \hat{x}, s) = \hat{\nu}_1 \in \widehat{\mathcal{C}}_{\alpha}] \Longrightarrow \nu(\alpha, \hat{x}, s+1) = \hat{\nu}_1$;

(iv) $\hat{x} \in \widehat{L}_{\alpha,s} - \widehat{J}_{\alpha,s} \Longrightarrow [\nu(\alpha, \hat{x}, s) \in \widehat{\mathcal{C}}_{\alpha}\ \&\ \hat{x} \in \overline{B}_s]$;

(v) $\hat{x} \in \widehat{J}_{\alpha,s} - B_s \Longrightarrow \nu(\alpha, \hat{x}, s) \in \widehat{\mathcal{W}}^{\#}_{\alpha}$;

(vi) $[s \geqslant \max\{\upsilon_{\alpha}, t(\alpha, i)\}\ \&\ \hat{y}_{\alpha,i,s} \in \widehat{J}_{\alpha,s}] \Longrightarrow \hat{y}_{\alpha,i,s} \in B$;

(vii) $i \leqslant \liminf_s g(\alpha, s) \Longrightarrow (\exists^{\infty} s)[\hat{y}_{\alpha,i,s}\downarrow\ \&\ \nu\ (\alpha, \hat{y}_{\alpha,i,s}, s) \in \widehat{\mathcal{C}}_{\alpha}]$;

(viii) $[i \leqslant \liminf_s g(\alpha, s)$ & $(\exists^{<\infty} s)[\hat{y}_{\alpha,i,s} \in B_{s+1} - B_s]] \Longrightarrow [\lim_s \hat{y}_{\alpha,i,s} < \infty]$.
In addition if Steps $\hat{n}$, $9 \leqslant n < 11$, *satisfy the following condition* ($\widehat{Q}5$) *then conclusion* (ix) *holds for* α *and* i *as above.*

($\widehat{Q}5$) *Step* $\hat{n}$ *may not put* $\hat{y}'_{\alpha,i,s}$ *into* $\widehat{J}_{\alpha,s+1} - \widehat{J}_{\alpha,s}$, *and may put* $\hat{y}_{\alpha,i,s}$ *into* $\widehat{J}_{\alpha,s+1} - \widehat{J}_{\alpha,s}$ *only if* $\hat{y}'_{\alpha,i,s}$ *is defined.*

(ix) $i \leqslant \liminf_s g(\alpha, s) \Longrightarrow [(\text{a.e. } s)[\hat{y}_{\alpha,i,s}\downarrow]$ & $(\exists^{\infty} s)[\hat{y}_{\alpha,i,s}\downarrow$ & $\hat{y}'_{\alpha,i,s}\downarrow]]$.

The proof of the Coding Theorem can be found in [Harrington and Soare, 1996c]. Like the Automorphism Theorem, the Coding Theorem gives a general criterion for mapping a set A to a set B while simultaneously coding information into B and adding additional steps so long as these steps satisfy very minimal properties.

4. Invariance

A property of c.e. sets or class $\mathcal{C} \subseteq \mathcal{E}$ is *invariant* if it is invariant under $\text{Aut}(\mathcal{E})$, and $\mathcal{E}$*-definable* if there is a first order property in the language $L(\subset)$ which defines it over $\mathcal{E}$. A class $\mathbf{C}$ of c.e. degrees is *invariant* if it is the class of degrees of sets in some class $\mathcal{C} \subseteq \mathcal{E}$ which is invariant (e.g., if $\mathcal{C}$ is $\mathcal{E}$-definable). For $\mathbf{R}$ the c.e. degrees and $\mathbf{C} \subset \mathbf{R}$ define

$$\mathbf{H}_n = \{\mathbf{a} \in \mathbf{R}\colon \mathbf{a}^{(n)} = \mathbf{0}^{(n+1)}\},$$
$$\mathbf{L}_n = \{\mathbf{a} \in \mathbf{R}\colon \mathbf{a}^{(n)} = \mathbf{0}^{(n)}\},$$
$$\mathbf{L}_0 = \{\mathbf{0}\}, \quad \mathbf{H}_0 = \{\mathbf{0}'\} \quad \text{and} \quad \overline{\mathbf{C}} = \mathbf{R} - \mathbf{C}.$$

The degrees in $\mathbf{H}_n$ ($\mathbf{L}_n$) are called *high*$_n$ (*low*$_n$) and the high$_1$ (low$_1$) degrees are called *high* (*low*).

A set $A \in \mathcal{E}$ is *maximal* if A is maximal in the inclusion ordering, i.e. $\neg(\exists W)[A \subset^* W \subset^* \omega]$, and a coinfinite A is *atomless* if A has no maximal superset. Martin [1966] showed that the degrees of maximal sets are exactly $\mathbf{H}_1$, see [Soare, 1987, p. 217]. Lachlan [1968a] and Shoenfield [1976] showed that the degrees of atomless sets are exactly the nonlow$_2$ c.e. degrees $\overline{\mathbf{L}}_2$, see [Soare, 1987, p. 231]. Thus, $\mathbf{H}_1$ and $\overline{\mathbf{L}}_2$ are invariant. For the trivial jump classes corresponding to $n = 0$, $\mathbf{L}_0$ $\overline{\mathbf{L}}_0$, and $\mathbf{H}_0$ are invariant, while $\overline{\mathbf{H}}_0$ is noninvariant by Theorem 3.2. These three results and his work at the time on projective determinacy led Martin to make the following conjecture.

CONJECTURE 4.1 (Martin's invariance conjecture). *Among the jump classes* $\mathbf{H}_n$ *and* $\mathbf{L}_n$ *for* $n > 0$, *and their complements* $\overline{\mathbf{H}}_n$ *and* $\overline{\mathbf{L}}_n$, *the invariant classes are exactly* $\mathbf{H}_{2n-1}$ *and* $\overline{\mathbf{L}}_{2n}$.

As first stated, the conjecture stated that these were the only invariant classes among *all* nontrivial classes of degrees, but this was soon refuted by Lerman and Soare [1980] who showed that the d-simple sets form an $\mathcal{E}$-definable class which splits $\mathbf{L}_1$. Therefore, the conjecture was modified to be restricted to just the jump classes and their complements. It was also modified to exclude the case of $n = 0$ because these classes tend to be pathological. The alternation of every odd $\mathbf{H}_n$ and every even $\overline{\mathbf{L}}_n$ was inspired by the behavior of projective determinacy. The following immediate corollary of Theorem 3.1 confirms the Invariance Conjecture prediction for the *downward* closed jump classes for $n > 0$.

COROLLARY 4.2. *For all $n > 0$ the* downward *closed jump classes of c.e. degrees $\mathbf{L}_n$ and $\overline{\mathbf{H}}_n$ are noninvariant.*

For the *upward* closed classes $\mathbf{H}_n$ and $\overline{\mathbf{L}}_n$, $n > 0$, after the discovery of invariance of $\mathbf{H}_1$ and $\overline{\mathbf{L}}_2$, attention has been focused on $\overline{\mathbf{L}}_1$ because of the important role played by the low c.e. sets. Researchers had tried unsuccessfully to find a property for the definability player (i.e. the RED player) which would define the class of degrees $\overline{\mathbf{L}}_1$ analogously as the property "atomless" defines $\overline{\mathbf{L}}_2$. The property $NL(A)$ in Section 2.4 almost succeeded, but not quite. After years of unsuccessful efforts by several researchers, Harrington and Soare discovered that $\overline{\mathbf{L}}_1$ is *non*invariant.

THEOREM 4.3 (Harrington and Soare [ta]). *There is a c.e. set $D \in low_2 – low_1$ such that every c.e. set $A \leqslant_T D$ is automorphic to a low set B.*

COROLLARY 4.4. *$\overline{\mathbf{L}}_1$ is not an invariant class of c.e. degrees.*

5. Decidability and undecidability

Lachlan [1968b, Theorem 1] showed that the decision problem for Th($\mathcal{E}$) is reducible to that for Th($\mathcal{E}^*$) and conversely. (Indeed he proved the first reduction for any sublattice $\mathcal{L}$ of $\mathcal{N}$ closed under symmetric difference with finite sets where $\mathcal{L}^* = \mathcal{L}/\mathcal{F}$. The first reduction relies on elimination of quantifiers while the second is an easy consequence of the fact that the property of being finite is definable in $\mathcal{E}$.)

The first progress about the decision problem for Th($\mathcal{E}$) was made by Lachlan [1968b]. Recall that $\mathcal{B}$ denotes the Boolean algebra generated by $\mathcal{E}$. Lachlan added to the language a unary predicate $E(x)$ which is to be interpreted over $(\mathcal{E}^*, \mathcal{B}^*)$ as "$x \in \mathcal{E}^*$". An $\overrightarrow{\forall}\overrightarrow{\exists}$-sentence in this language is one of the form $(\forall x_1)\cdots(\forall x_n)(\exists y_1)\cdots(\exists y_m)\, P(\overrightarrow{x}, \overrightarrow{y})$, where P is quantifier free.

THEOREM 5.1 (Lachlan [1968b]). *There is a decision procedure for the $\overrightarrow{\forall}\overrightarrow{\exists}$-sentences true in $\mathcal{B}^*$ where quantifiers range over $\mathcal{E}^*$.*

Lachlan began by proving that all consistent existential sentences are true in $\mathcal{E}^*$. A lattice $\mathcal{L}$ is *separated* if for any pair x, y of elements of $\mathcal{L}$ there exists a disjoint pair x_1, y_1 of elements of $\mathcal{L}$ such that $x_1 \leqslant x$, $y_1 \leqslant y$, and $x_1 \cup y_1 = x \cup y$. Clearly, $\mathcal{E}^*$ is separated. Lachlan reduced the $\overrightarrow{\forall}\overrightarrow{\exists}$-decision problem for $\mathcal{E}^*$ to the following.

Given finite separated lattices $L, L_1, L_2, \dots, L_k$, such that each L_i is a refinement of L, when is it true that for all sublattices $\mathcal{L}$ of $\mathcal{E}^*$ such that $\mathcal{L} \cong L$, there exists a sublattice $\mathcal{L}'$ of $\mathcal{E}^*$ such that $\mathcal{L}' \supset \mathcal{L}$ represents the extension of embeddings solution isomorphic to the pair $L_i \supset L$.

Lerman and Soare [1980] extended Lachlan's method to give a decision procedure for the $\vec{\forall}\,\vec{\exists}$-sentences of an extended language which are true in $\mathcal{E}^*$. The extended language contains new predicates such as $\mathrm{Max}(x)$ and $\mathrm{Hhs}(x)$ which are to be interpreted as "x is maximal" and "x is hh-simple". The idea is to construct canonical realizations and to prove a refinement theorem as in Lachlan's proof but each step is more complicated now, and requires new structural information about $\mathcal{E}$, because the $\vec{\forall}\,\vec{\exists}$-sentences of the extended language include such statements as "there exists an atomless hh-simple set with an r-maximal subset," or "there exists an r-maximal set".

The next major development was the proof of undecidability of $\mathrm{Th}(\mathcal{E})$ given by Herrmann and independently by Harrington. (Harrington's proof appeared slightly later and exists only in the form of unpublished notes.)

THEOREM 5.2 (Herrmann [1984], Harrington [1983]). *The first order theory of scr E is undecidable.*

Both proofs depend upon the undecidability of the theory of a Boolean algebra with a distinguished subalgebra (see M. Rubin [1976] and Burris and McKenzie [1981]). A *Boolean pair* is a pair $(\mathcal{A}_1, \mathcal{A}_2)$ where $\mathcal{A}_1$ is a Boolean algebra and $\mathcal{A}_2$ is a subalgebra defined over $\mathcal{A}_1$ by some unary relation $R(x)$. A Boolean pair is *computable* if the usual Boolean operations and the relation $R(x)$ are all computable. Burris and McKenzie [1981] constructed a class of Boolean pairs (isomorphic to computable Boolean pairs) such that every larger class has an undecidable theory. Thus, if $\mathcal{C}$ is a class of Boolean pairs including all isomorphism types of computable Boolean pairs then $\mathcal{C}$ has an undecidable theory. Herrmann proved Theorem 2.2 by finding a class $\mathcal{C}$ of Boolean pairs such that: $\mathcal{C}$ contains isomorphism types of all computable Boolean pairs; and $\mathcal{C}$ is elementarily definable with parameters (e.d.p.) in $\mathcal{E}$. Herrmann achieved this after a series of preliminary results about definability in $\mathcal{E}$ contained in various papers such as Herrmann [1981, 1983, 1984]. For example, he first proved that (Σ_4, Σ_3) is e.d.p. in $\mathcal{E}$ where Σ_4 is the lattice of Σ_4 sets together with a unary predicate defining exactly the Σ_3 sets. The proof uses ideas from Lachlan's method of Theorem X.7.2 for embedding Σ_3 relations in $\mathcal{E}$. Harrington proved Theorem 2.2 by showing that any Boolean pair $(\mathcal{A}_1, \mathcal{A}_2)$ of Δ^0_2 Boolean algebras is e.d.p. in $\mathcal{E}$. Given $\mathcal{A}_1$ and $\mathcal{A}_2$ Harrington used a $0'''$-construction to construct an hh-simple set A and an c.e. set $B \subseteq A$ defining $(\mathcal{A}_1, \mathcal{A}_2)$ as follows. Let the ideal determined by A and B be

$$\mathcal{I} = \{X \in \mathcal{E}\colon (\exists \text{ rec. } C \subseteq A)\,(\exists D \subset_m A)[X \subseteq C \cup B \cup (A - D)]\}.$$

For $X \in \mathcal{E}(\overline{A})$ choose any computable set Y_x such that $Y_x - A = X$. He showed that if $\hat{Y}_x$ is another such set then $Y_x \triangle \hat{Y}_x \in \mathcal{I}$. Hence, the map $f(X) = Y_x$ produces a lattice homomorphism from $\mathcal{E}(\overline{A})$ to $\mathcal{E}(A)/\mathcal{I}$. He constructed A and B such that $(\mathcal{A}_1, \mathcal{A}_2) \cong (\mathcal{E}(A)/\mathcal{I}, f(\mathcal{E}(\overline{A})))$. Recently Harrington solved a major open question concerning the degree of this theory.

THEOREM 5.3 (Harrington). *The degree of $Th(\mathcal{E})$ is equal to $\mathbf{0}^{(\omega)}$.*

The proof has not been published, but a different version of the proof appears in another paper by Harrington and Nies [ta]. In this paper Harrington and Nies develop methods for coding with first-order formulas into the lattice $\mathcal{E}$ of c.e. sets under inclusion. First they use these methods to reprove and generalize the result of Harrington that the elementary theory of $\mathcal{E}$ has the same computational complexity as the theory of the natural number (namely $\mathbf{0}^{(\omega)}$). Relativized versions of the coding methods show that the partial ordering of Σ^0_p and Σ^0_q sets are not elementarily equivalent for natural numbers $p \neq q$. As a further application, definability of the class of quasimaximal sets in $\mathcal{E}$ is obtained. On the other side, they prove theorems limiting coding and definability in $\mathcal{E}$, thereby establishing a sharp contrast between $\mathcal{E}$ and other structures occurring in computability theory.

6. Further results

The automorphisms constructed here are all Δ^0_3. The question was raised of whether there exist c.e. sets A and B which are automorphic by a Δ^0_4 automorphism but not by any Δ^0_3 automorphism. Cholak and Downey [ta] claimed to give a positive answer to this question, but their proof had an error. Cholak and Harrington conjecture that the claim is still true. Furthermore, the methods of Cholak and Downey were extended by Cholak, Downey, and Harrington to answer a conjecture of Slaman and Woodin that the set $\{\langle e, i\rangle\colon W_e \simeq W_i\}$ is Σ^1_1-complete. Slaman and Woodin had showed the analogous result for $\mathcal{L}(W_e) \cong \mathcal{L}(W_i)$ in place of $W_e \simeq W_i$.

Harrington (unpublished) proved that there is no fat orbit, namely an orbit which hits every nonzero c.e. degree. Extensions of this result and related results are derived in Downey and Harrington [ta].

Nies [ta] has shown that each closed interval of the lattice $\mathcal{E}$ of enumerable sets is either a Boolean algebra or has an undecidable theory. This answers a question from an article by Maass and Stob. Nies also gives an example of a subclass of $\mathcal{E}^*$ which is not first-order definable, but arithmetical and closed under automorphic images.

References

K. AMBOS-SPIES AND A. NIES
[1992] Cappable recursively enumerable degrees and Post's program, *Arch. Math. Logic*, 32, pp. 51–56.

S. BURRIS AND R. MCKENZIE
[1981] Decidability and Boolean representations, *Memoirs Amer. Math. Soc.*, 32, No. 246.

P. CHOLAK
[1994] The translation theorem, *Arch. Math. Logic*, 33, pp. 87–108.

P. A. CHOLAK
[1995] Automorphisms of the lattice of recursively enumerable sets, *Mem. Amer. Math. Soc.*, No. 113.
[ta] The dense simple sets are orbit complete, in: *Proceedings of the Oberwolfach Conference on Computability Theory in 1996*; *Journal of Pure and Applied Logic*, to appear.

P. CHOLAK AND R. DOWNEY
[ta] *A pair of automorphic computably enumerable sets which are not Δ_3-automorphic*, Preprint.

P. A. CHOLAK, R. DOWNEY AND L. A. HARRINGTON
[ta] Automorphisms of the computably enumerable sets: the Slaman–Woodin Conjecture, in preparation.

P. CHOLAK, R. DOWNEY AND M. STOB
[1992] Automorphisms of the lattice of recursively enumerable sets: Promptly simple sets, *Trans. Amer. Math. Soc.*, 332, pp. 555–570.

P. CHOLAK AND A. NIES
[prep] r-maximal sets, in preparation.

A. CHURCH
[1936a] An unsolvable problem of elementary number theory, *Amer. J. Math.*, 58, pp. 345–363; reprinted in Davis [1965], pp. 88–107.
[1936b] A note on the Entscheidungsproblem, *J. Symbolic Logic*, 1, pp. 40–41 and 101–102; reprinted in Davis [1965], pp. 108–115.

M. DAVIS (ED.)
[1965] *The Undecidable*. Basic Papers on Undecidable Propositions, Unsolvable Problems, and Computable Functions, Raven Press, Hewlitt, New York.

R. DOWNEY AND L. HARRINGTON
[ta] There is no fat orbit, *Ann. Pure Appl. Logic*, to appear.

R. DOWNEY AND M. STOB
[1992] Automorphisms of the lattice of recursively enumerable sets: Orbits, *Adv. in Math.*, 92, pp. 237–265.

R. FRIEDBERG
[1957] Two recursively enumerable sets of incomparable degrees of unsolvability, *Proc. Natl. Acad. Sci. USA*, 43, pp. 236–238.

K. GÖDEL
[1934] On undecidable propositions of formal mathematical systems, Notes by S.C. Kleene and Barkley Rosser on lectures at the Institute for Advanced Study, Princeton, NJ; reprinted in Davis [1965], pp. 39–71.

L. HARRINGTON
[1983] *The Undecidability of the Lattice of Recursively Enumerable Sets*, Handwritten notes.

L. HARRINGTON AND A. NIES
[ta] Coding in the lattice of enumerable sets, *Advances in Math.*, to appear.

L. HARRINGTON AND R. I. SOARE

[1991] Post's Program and incomplete recursively enumerable sets, *Proc. Natl. Acad. Sci. USA*, 88, pp. 10242–10246.

[1996a] Dynamic properties of computably enumerable sets, in: *Computability, Enumerability, Unsolvability: Directions in Recursion Theory, Proceedings of the Recursion Theory Conference, University of Leeds, July 1994*, S. B. Cooper, T. A. Slaman and S. S. Wainer, eds., London Math. Soc. Lecture Notes Series, Cambridge University Press.

[1996b] Definability, automorphisms, and dynamic properties of computably enumerable sets, *Bull. Symbolic Logic*, 2, pp. 199–213.

[1996c] The Δ_3^0 automorphism method and noninvariant classes of degrees, *J. Amer. Math. Soc.*, to appear in 1996.

[1996d] Codable sets and orbits of computably enumerable sets, *J. Symbolic Logic*, to appear.

[1998] Definable properties of the computably enumerable sets, in: *Proceedings of the Oberwolfach Conference on Computability Theory, 1996*; *Journal of Pure and Applied Logic*, 94, pp. 97–125.

[ta] Martin's Invariance Conjecture and low sets, in preparation.

E. HERRMANN

[1981] *Die Verbandseigenschaften der rekursiv aufzählbaren Mengen*, Seminarbericht Nr. 36, Humboldt-Universität Sektion Mathematik, Berlin, 1981.

[1983] *Major Subsets of Hypersimple Sets and Ideal Families*, Dissertation B, Humboldt-Universität Sektion Mathematik, Berlin.

[1984] The undecidability of the elementary theory of the lattice of recursively enumerable sets (abstract), in: *Frege Conference 1984, Proceedings of the International Conference at Schwerin* (*GDR*), Akademie-Verlag, Berlin, pp. 66–72.

S. C. KLEENE

[1936] General recursive functions of natural numbers, *Math. Ann.*, 112 (5), pp. 727–742.

A. H. LACHLAN

[1968a] Degrees of recursively enumerable sets which have no maximal superset, *J. Symbolic Logic*, 33, pp. 431–443.

[1968b] The elementary theory of recursively enumerable sets, *Duke Math. J.*, 35, pp. 123–146.

[1968c] On the lattice of recursively enumerable sets, *Trans. Amer. Math. Soc.*, 130, pp. 1–37.

[1970] On some games which are relevant to the theory of recursively enumerable sets, *Ann. of Math.* (2), 91, pp. 291–310.

[1975] A recursively enumerable degree which will not split over all lesser ones, *Ann. Math. Logic*, 9, pp. 307–365.

M. LERMAN AND R. I. SOARE

[1980] d-simple sets, small sets, and degree classes, *Pacific J. Math.*, 87, pp. 135–155.

W. MAASS

[1982] Recursively enumerable generic sets, *J. Symbolic Logic*, 47, pp. 809–823.

[1983] Characterization of recursively enumerable sets with supersets effectively isomorphic to all recursively enumerable sets, *Trans. Amer. Math. Soc.*, 279, pp. 311–336.

[1984] On the orbits of hyperhypersimple sets, *J. Symbolic Logic*, 49, pp. 51–62.

[1985] Variations on promptly simple sets, *J. Symbolic Logic*, 50, pp. 138–148.

W. MAASS, R. A. SHORE AND M. STOB

[1981] Splitting properties and jump classes, *Israel J. Math.*, 39, pp. 210–224.

W. MAASS AND M. STOB

[1983] The interval of the lattice of r.e. sets determined by major subsets, *Ann. Pure Appl. Logic*, 24, pp. 189–212.

S. S. MARCHENKOV

[1976] A class of incomplete sets, *Mat. Zametki*, 20, pp. 473–478 (Russian); *Math. Notes*, 20, pp. 823–825 (English translation).

D. A. MARTIN

[1966] Classes of recursively enumerable sets and degrees of unsolvability, *Z. Math. Logik Grundlag. Math.*, 12, pp. 295–310.

D. MILLER

[1981] The relationship between the structure and degrees of recursively enumerable sets, PhD Dissertation, University of Chicago.

A. A. MUCHNIK

[1956] On the unsolvability of the problem of reducibility in the theory of algorithms, *Dokl. Akad. Nauk SSR*, 108, pp. 194–197 (Russian).

J. MYHILL

[1956] The lattice of recursively enumerable sets, *J. Symbolic Logic*, 21, pp. 215, 220 (abstract).

A. NIES

[ta] Intervals of the lattice of computably enumerable sets and effective boolean algebras, submitted for publication.

E. L. POST

[1936] Finite combinatory processes – formulation I, *J. Symbolic Logic*, 1, pp. 103–105; reprinted in Davis [1965], pp. 288–291.

[1943] Formal reductions of the general combinatorial decision problem, *Amer. J. Math.*, 65, pp. 197–215.

[1944] Recursively enumerable sets of positive integers and their decision problems, *Bull. Amer. Math. Soc.*, 50, pp. 284–316; reprinted in Davis [1965], pp. 304–337.

H. ROGERS, JR.

[1967] *Theory of Recursive Functions and Effective Computability*, McGraw-Hill, New York, 482 pp.

M. RUBIN

[1976] The theory of Boolean algebras with a distinguished subalgebra is undecidable, *Ann. Sci. Univ. Clermont-Ferrand II Math.*, 13, pp. 129–134.

G. E. SACKS

[1963] *Degrees of Unsolvability*, Ann. of Math. Stud., Vol. 55, Princeton University Press, Princeton, NJ (see revised edition, 1966).

J. R. SHOENFIELD

[1976] Degrees of classes of r.e. sets, *J. Symbolic Logic*, 41, pp. 695–696.

R. I. SOARE

[1974] Automorphisms of the recursively enumerable sets, Part I: Maximal sets, *Ann. of Math. (2)*, 100, pp. 80–120.

[1982] Automorphisms of the lattice of recursively enumerable sets, Part II: Low sets, *Ann. Math. Logic*, 22, pp. 69–107.

[1987] *Recursively Enumerable Sets and Degrees: A Study of Computable Functions and Computably Generated Sets,* Springer, Heidelberg.

[1996a] Computability and recursion, *Bull. Symbolic Logic*, 2, pp. 284–321.

[1996b] Computability and enumerability, in: *Proceedings of the 10th International Congress for Logic, Methodology and the Philosophy of Science, Section 3: Recursion Theory and Constructivism, Florence, August 19–25, 1995*, to appear.

A. M. TURING

[1936–37] On computable numbers, with an application to the Entscheidungsproblem, *Proc. London Math. Soc.*, 42 (1936), pp. 230–265; A correction, *ibid.,* 43 (1937), pp. 544–546; reprinted in Davis [1965], pp. 115–154.

[1937] Computability and λ-definability, *J. Symbolic Logic*, 2, pp. 153–163.

[1939] Systems of logic based on ordinals, *Proc. London Math. Soc.*, 45, pp. 161–228; reprinted in Davis [1965], pp. 154–222.

Part 3
Generalized Computability Theory

CHAPTER 8

The Continuous Functionals

Dag Normann

Department of Mathematics, University of Oslo, P.O. Box 1053 Blindern, N-0316 Oslo, Norway

Contents

1. Some elements from the history 252
2. The functionals 253
3. Complexity 256
4. Computations 258
5. The fan functional 262
6. Bar recursion 265
7. Associates 266
8. The finitary aspects 268
9. Functional interpretation of analysis 271
References 274

HANDBOOK OF COMPUTABILITY THEORY
Edited by E.R. Griffor

1. Some elements from the history

Recursion theory as defined, e.g., by Turing or Post is concerned with operations on the strings or words in some finite alphabet. Since natural numbers can be represented in various ways as strings in an alphabet, the general definition of a computable function also defines a set of computable or recursive functions on the natural numbers. This set is the same as the set of μ-recursive functions of Gödel, where the algorithms were based directly on the structure of the natural numbers.

Kleene defined the recursive functions in an analogue way. He used a finite set of *schemes*, i.e. general rules for how to construct new recursive functions from old ones. He also systematically assigned a number, or an *index*, to each algorithm, defining the relation

$$\{e\}(x_1, \ldots, x_n) = y$$

meaning that the recursive function with index e and input $x_1, \ldots, x_n$ has y as the output. Among the important results are the $S_{n,m}$-theorem, the recursion theorem and the enumeration theorem which states that there is one index e_0 such that for all $x_1, \ldots, x_n$ we have

$$\{e_0\}(e, x_1, \ldots, x_n) \approx \{e\}(x_1, \ldots, x_n)$$

($\approx$ means that either both sides are undefined or both sides are defined and take the same value.) This last result is the analogue of the existence of a universal Turing Machine.

In order to make the problem about "how unsolvable is the halting problem" precise, one had to relativize the notion of an algorithm to sets and functions. In Kleene's setting we define algorithms relativized to a function $f : \mathbb{N} \to \mathbb{N}$ by accepting $\{e\}(f, x) = f(x)$ as a new initial computation.

Kleene [1959a] suggested an extension of the notion of computation to all functionals of pure finite type. This work was continued in [Kleene, 1963].

We define

$$Tp(0) = \mathbb{N},$$
$$Tp(k+1) = Tp(k) \to \mathbb{N} = \text{the set of total functions from } Tp(k) \text{ to } \mathbb{N}.$$

This notion of computation, which we will define precisely in Section 4, enabled Kleene to discuss the computational power of, e.g., higher type quantification. The choice of algorithms is however not a canonical one, and Kleene [1978, 1980, 1982, 1985, 1991] has later suggested an alternative definition, at least for the first few levels.

The main features of Kleene's original definition are:

(1) If Φ is of type $k+2$ then $\Phi(\phi)$ is only defined when ϕ is a total object $\phi : Tp(k) \to \mathbb{N}$.

(2) Oracle calls $\Phi(\phi)$ for higher type functionals are permitted uniformly in any algorithm for ϕ.

(3) The existence of a uniform algorithm is made explicit as a part of the definition (Scheme S9).

Although one important aspect of Kleene's definition is that it enables us to define algorithms relative to higher type objects, it will also define a class of computable objects of higher types. Kleene [1959b] isolated a subclass of functionals called *the countable functionals*, and he proved that all functionals computable in countable functionals will themselves be countable.

In the same year Kreisel [1959] defined a typed hierarchy of what he called *continuous functionals*. The continuous functionals were defined via a set of formal neighbourhoods, and technically as equivalence classes of certain sets of such formal neighbourhoods. With the canonical interpretation of application, he actually defined a typed structure of functions. The main difference between the two approaches is that Kleene's countable functionals are total objects in the full type-hierarchy, i.e. from type 3 and above they will also be defined on objects that are not countable as well, while Kreisel's continuous functionals of type $k+1$ will be defined exactly on the continuous functionals of type k. Apart from this difference it was clear from the time of publication that the two structures are essentially the same, Kreisel's functionals will be the transitive, extensional, collapse of Kleene's functionals.

Kreisel's motivation for defining the continuous functionals was to develop a tool for discussing the constructive content of statements in analysis or 2nd order number theory. In this paper we will concentrate on the recursion theoretic aspects of the continuous functionals, but Kreisel's application is still relevant, and with the growth of various typed formal theories with applications in computer science, a semantical analysis of constructivity will be of importance.

None of the results in this chapter are new. We give some basic definitions in a "modern" setting and we prove a few theorems laying out some basic tools useful in a further investigation of the continuous functionals. The list of references is not inclusive, but then so is not the collection of aspects that we concentrate on.

2. The functionals

There are several ways of defining the continuous functionals. Kleene [1959b] used a set of associates, elements of $\mathbb{N} \to \mathbb{N}$ that code the behaviour of the functionals. If we use a set A of functions to code the elements of a set X, the canonical topology on A will induce a topology on X. If $\mathbb{N}$ is the topological space of natural numbers with discrete topology, any continuous function from X to $\mathbb{N}$ will correspond to a continuous function F from A to $\mathbb{N}$. Such functions can again be determined by a set of pairs σ, n where σ is a finite sequence of natural numbers, and $F(f) = n$ for any f extending σ. The associates will code sets of such pairs that are sufficiently rich to give a full description of F.

Then Kleene [1959b] defines

$Ct(0) = \mathbb{N}$.

$Ct(1) = \mathbb{N} \to \mathbb{N}$ with any f as its own associate.

$Ct(k+1)$ is the set of continuous maps from $Ct(k)$ to $\mathbb{N}$ with associates as discussed above.

We will not give the details of Kleene's definition. Kreisel [1959] used a set of formal neighbourhoods with a consistency relation defined for each type, and defined the continuous functionals as equivalence classes of consistent sets of neighbourhoods.

Ershov [1972, 1973, 1974] characterized the continuous functionals as the hereditarily total functionals in a hierarchy of partial continuous functionals. The Ershov approach essentially use what we now call Scott–Ershov domains.

Later Hyland [1975, 1979] gave various characterisations of the continuous functionals as limit-spaces and filter spaces. Originally due to Hyland but only published in Normann [1980], there is also a characterisation using a standard topology on the set of continuous functions from one topological space to another. A survey of these results, together with a characterisation via a hyperfinite type structure in the sense of nonstandard analysis, was given in [Normann, 1983]. Finally Berger [1993] defined the continuous functionals completely in the setting of domain theory, with the main purpose of extending the Kreisel–Lacombe–Shoenfield theorem to the setting of the continuous functionals.

We will give the precise definition in the setting of domains. We will assume some familiarity with complete partial orderings, compact elements, the definition of domains and the most fundamental ingredients of the theory of domains. Stoltenberg-Hansen, Lindström and Griffor [1994] will provide a sufficient background for what we need from domain theory.

DEFINITION 1. Let $(D_0, \sqsubseteq_0)$ be the domain representing the natural numbers, with a bottom element $\perp$ and with the numbers n as pairwise incomparable objects above $\perp$.

Let $D_{k+1} = D_k \to D_0$.

The domains D_k will form a hierarchy of functionals. If we let $\perp$ signify the value "undefined" we see that we have constructed a hierarchy of hereditarily partial functionals.

We see that a functional $\Phi \in D_{k+2}$ will be monotone with respect to the ordering on D_{k+1} defined by

$$\phi \sqsubseteq_{k+1} \psi \Leftrightarrow \phi(F) \sqsubseteq_0 \psi(F) \text{ for all } F \in D_k,$$

and moreover, Φ will be continuous in the sense that

$$\Phi(\sup X) = \sup\{\Phi(\phi) \mid \phi \in X\}$$

for any bounded set $X \subseteq D_{k+1}$.

All this is standard domain theory.

We will primarily be interested in the total objects, and we define the hierarchy T of hereditarily total elements in D:

DEFINITION 2. T_0 is the set of natural numbers $\mathbb{N}$ in D_0. $T_{k+1} = \{F \in D_{k+1} \mid F(G) \in \mathbb{N}$ for all $G \in T_k\}$.

We will call the elements of T_k *total*. Two total elements of type $k+1$ are to be considered as equivalent if they give the same values to all total objects of type k. We will define this relation and prove that all total elements of type $k+2$ will respect the relation. This property, called extensionality, is traditionally proved by proving that T_k is dense in D_k, essentially [Kleene, 1959b; Kreisel, 1959]. Longo and Moggi [1984] observed that we can prove the extensionality of the total objects without proving the density theorem. We will give the proof from [Longo and Moggi, 1984].

DEFINITION 3. For $n, m \in T_0$ we let $n \sim m$ if $n = m$. For $F, G \in T_{k+1}$ we let $F \sim G$ if for all $x \in T_k$ we have that $F(x) = G(x)$.

The relation $\sim$ is clearly an equivalence relation for each T_k.

LEMMA 1. (a) *Let F and G be in T_k. Then $F \sim G$ if and only if $F \sqcap G \in T_k$.* (b) *If F is in T_{k+1} and x and y are in T_k with $x \sim y$, then $F(x) = F(y)$.*

PROOF. We use induction on k to prove (a). For $k = 0$ the lemma is trivial since $n \sqcap n = n$ and $n \sqcap m = \bot$ when $n \neq m$.

If $F, G \in T_{k+1}$ and $x \in T_k$, we have that $(F \sqcap G)(x) = F(x) \sqcap G(x)$, and the induction step is trivial.

In order to prove (b) we see that if $x \sim y$ then by (a), $x \sqcap y \in T_k$ so $F(x \sqcap y) \in \mathbb{N}$. Then $F(x) = F(x \sqcap y) = F(y)$. □

DEFINITION 4. The hierarchy of $\sim$-equivalence classes with the inherited application operator will be isomorphic to a hierarchy of functionals. We let $\{Ct(k)\}_{k\in\mathbb{N}}$ be this isomorphic hierarchy.

Whenever convenient we will however consider the elements of $Ct(k)$ as equivalence classes of elements of T_k.

One important property is the *density theorem*. Essentially equivalent versions of this theorem can be found in the two initial papers [Kleene, 1959b; Kreisel, 1959].

The next lemma is a standard result from domain theory, and is left without proof:

LEMMA 2. *Let $C \in D_{k+1}$ be compact. Then there are compacts $p_1, \ldots, p_n \in D_k$ and numbers $k_1, \ldots, k_n$ such that:*

(i) *If p_i and p_j are consistent, then $k_i = k_j$.*

(ii) *$C(F) = \sqcup\{k_i \mid p_i \sqsubseteq F\}$ for all $F \in D_k$.*

THEOREM 1 (Density). *Let $C \in D_k$ be compact. Then there is an $F \in T_k$ with $C \sqsubseteq F$.*

PROOF. If $k = 0$ or $k = 1$ the theorem is trivial, so let $k > 1$ and assume that the theorem holds for $k - 2$.

Let $(p_1, k_1), \ldots, (p_n, k_n)$ be as in Lemma 2 for C. We use Lemma 2 for each p_i as well, and let p_i be the compact element in D_{k-1} determined from $(q_{i,1}, m_{i,1}), \ldots, (q_{i,n_i}, m_{i,n_i})$ where each $q_{i,j}$ is compact in D_{k-2}.

Let $R = \{((i,i'),(j,j')) \mid q_{i,i'}$ and $q_{j,j'}$ are consistent$\}$. For each $r = ((i,i'),(j,j')) \in R$ let $x_r \in T_{k-2}$ with $q_{i,i'} \sqsubseteq x_r$ and $q_{j,j'} \sqsubseteq x_r$. Now let $y \in D_{k-1}$ be such that $y(x_r) \in \mathbb{N}$ for all $r \in R$. We say that p_i is *semiconsistent* with y if for all j and $r = ((i,i'),(j,j')) \in R$ we have that $y(x_r) = p_i(x_r)$.

If p_i and p_j are semiconsistent with the same y, then p_i and p_j will be consistent, because otherwise there will be i' and j' such that $q_{i,i'}$ and $q_{j,j'}$ are consistent, but $m_{i,i'} \neq m_{j,j'}$, and then they cannot both agree with y. Moreover, if $p_i \sqsubseteq y$ and $y(x_r) \in \mathbb{N}$ for all $r \in R$, then p_i is semiconsistent with y.

We then define $F(y) = k_i$ if $p_i \sqsubseteq y$ or y is semiconsistent with p_i for some i, and $F(y) = 0$ if $y(x_r) \in \mathbb{N}$ for all $r \in R$, but y is not semiconsistent with p_i for any i.

Then F will be in T_k and extend C. □

The density theorem can be made effective, we may uniformly compute a total extension of a compact set using a primitive recursive function. In Berger [1993] there is a proof of the density theorem based on a conceptual analysis of totality. Berger also gives details of the effective version of the density theorem in the setting of domains.

3. Complexity

It is well known that a normal form for Π^1_1-sets $A \subseteq \mathbb{N} \to \mathbb{N}$ is

$$f \in A \Leftrightarrow \forall g \exists n R(f, g, n)$$

where R is primitive recursive. We will use the existence of an effective countable dense subset of $Ct(k)$ to prove a similar normal form theorem for Π^1_k-sets. We have not yet defined the recursion theory of these functionals, but the constructions in the proof of the next theorem are so simple that it is clear that the statement is correct as soon as it is made precise. The results of this section are essentially due to Kreisel [1959]. First we need a lemma:

LEMMA 3. *For $k < n$ there are effective operators*

$$\phi_{k,n} : Ct(k) \to Ct(n) \quad \text{and} \quad \psi_{k,n} : Ct(n) \to Ct(k)$$

such that for all $F \in Ct(k)$

$$\psi_{k,n}\big(\phi_{k,n}(F)\big) = F.$$

PROOF. We define $\phi_{k,k+1}$ and $\psi_{k,k+1}$ by recursion on k. These are one-step embeddings and projections, and the general embeddings and projections are obtained by composition of the one-step ones.

$$\phi_{0,1}(n) = c_n \text{ (the constant } n \text{ function).}$$
$$\psi_{0,1}(f) = f(0).$$

If F and G are of type $k+1$, we let

$$\phi_{k+1,k+2}(F)(G) = F\big(\psi_{k,k+1}(G)\big).$$

For Ψ of type $k+2$ and x of type k, let

$$\psi_{k+1,k+2}(\Psi)(x) = \Psi\big(\phi_{k,k+1}(x)\big).$$

The lemma now follows by induction on k, where the induction start is trivial and the induction step follows by a straightforward calculation based on the definitions. □

THEOREM 2. *Let $A \subseteq \mathbb{N} \to \mathbb{N}$ be Π^1_k for $k > 0$. Then there is a primitive recursive set R such that for all f:*

$$f \in A \Leftrightarrow \forall G \in Ct(k) \exists n R(f, G, n).$$

PROOF. We use induction on k, and the induction start is the representation theorem for Π^1_1. Assume that the theorem holds for k. First let B be Σ^1_k. By the induction hypothesis there is a primitive recursive set S such that

$$f \notin B \Leftrightarrow \forall G \in Ct(k) \exists n S(f, G, n).$$

By considering

$$F(G) = \mu n S(f, G, n)$$

we see that

$$\forall G \in Ct(k) \exists n S(f, G, n) \Leftrightarrow \exists F \in Ct(k+1) \forall G \in Ct(k) S\big(f, G, F(G)\big).$$

Now $\{G \in Ct(k) \mid S(f, G, F(G))\}$ will be primitive recursive uniformly in f and F, and thus closed and open in $Ct(k)$. By the effective version of the density-theorem, there is a primitive recursive enumeration $\{G_n\}_{n\in\mathbb{N}}$ of a dense subset of $Ct(k)$. Thus

$$f \notin B \Leftrightarrow \exists F \in Ct(k+1)\forall n S\big(f, G_n, F(G_n)\big).$$

We obtain the theorem for Σ^1_k-sets as follows:

Let

$$R'(f, F, n) \Leftrightarrow \neg S\big(f, G_n, F(G_n)\big).$$

Then

$$f \in B \Leftrightarrow \forall F \in Ct(k+1)\exists n R'(f, F, n).$$

Using the embedding $\phi_{1,k+1}$ from Lemma 3 of $Ct(1)$ into $Ct(k+1)$, and the pairing

$$\langle F_1, F_2\rangle(G) = \langle F_1(G), F_2(G)\rangle$$

we may reduce a pair of universal quantifiers over $Ct(1)$ and $Ct(k+1)$ to one over $Ct(k+1)$. It is then trivial to complete the proof of the theorem. □

COROLLARY 1. *We have the following normal form for Σ^1_k-sets: There is a primitive recursive set R' such that*

$$f \in B \Leftrightarrow \forall F \in Ct(k+1)\exists n R'(f, F, n)$$

where we have a uniform algorithm that produces a counterexample $F \in Ct(k+1)$ such that $\forall n \neg R'(f, F, n)$ whenever $f \notin B$.

PROOF. In the proof of this case of the theorem we used

$$F(G) = \mu n S(f, G, n)$$

as a possible counterexample, and this function is clearly computable. □

4. Computations

We have defined two hierarchies, the hierarchy of domains D_k and the hierarchy $\{Ct(k)\}_{k\in\mathbb{N}}$ of total, continuous functionals. We will now give the definition of computations in higher types from Kleene [1959a]:

DEFINITION 5. By a transfinite induction we define the set of *computations*

$$\{e\}(\phi_1, \ldots, \phi_n)$$

as follows, where $\phi_1, \ldots, \phi_n$ are functionals, σ code the *signature*, i.e. the sequence of types for $\phi_1, \ldots, \phi_n$ and x and q are natural numbers.

S1 If $e = \langle 1, \sigma \rangle$ then $\{e\}(x, \phi_1, \ldots, \phi_n) = x + 1$.

S2 If $e = \langle 2, \sigma, q \rangle$ then $\{e\}(\phi_1, \ldots, \phi_n) = q$.

S3 If $e = \langle 3, \sigma \rangle$ then $\{e\}(x, \phi_1, \ldots, \phi_n) = x$.

S4 If $e = \langle 4, e_1, e_2, \sigma \rangle$ then

$$\{e\}(\phi_1, \ldots, \phi_n) = \{e_1\}\big(\{e_2\}(\phi_1, \ldots, \phi_n), \phi_1, \ldots, \phi_n\big).$$

S5 Primitive recursion. This can be omitted in the presence of S9, and we will do so.

S6 If $e = \langle 6, e_1, t, \sigma \rangle$ and $t \in \mathbb{N}$ codes a permutation τ of n elements, then

$$\{e\}(\phi_1, \ldots, \phi_n) = \{e_1\}(\phi_{\tau(1)}, \ldots, \phi_{\tau(n)}).$$

S7 If $e = \langle 7, \sigma \rangle$ and $f : \mathbb{N} \to \mathbb{N}$, then $\{e\}(x, f, \phi_1, \ldots, \phi_n) = f(x)$.

S8 If $e = \langle 8, e_1, \sigma \rangle$ then $\{e\}(\phi_1, \ldots, \phi_n) = \phi_1(\lambda\psi(\{e_1\}(\psi, \phi_1, \ldots, \phi_n)))$.

S9 If $e = \langle 9, m, \sigma \rangle$ and $m \leqslant n$, then $\{e\}(e_1, \phi_1, \ldots, \phi_n) = \{e_1\}(\phi_1, \ldots, \phi_m)$.

If for some x, $\{e\}(\phi_1, \ldots, \phi_k) = x$ we say that $\{e\}(\phi_1, \ldots, \phi_k)$ *terminates*.

REMARK. This definition makes sense for both the hierarchy of domains and the hierarchy of total, continuous functionals. In S8, we will assume that the types of the variables are such that the type of ϕ_1 is two above the type of ψ.

For the D-hierarchy Kleene's schemes are nothing more than a set of positive equations that will have a least fix-point $\{e\}$ for each e, where the value in each case will be $\perp$ or a natural number. In the hierarchy of total continuous functionals we will assume that the input is total before we consider an instance of S8 as giving a defined value. When this requirement is satisfied, we will however not break out of the continuous functionals.

LEMMA 4.

(a) *If $\phi_1, \ldots, \phi_n, \psi_1, \ldots, \psi_n$ are functionals in the D-hierarchy with ϕ_i consistent with ψ_i for each i, and if e, k_1, k_2 are numbers such that $\{e\}(\phi_1, \ldots, \phi_n) = k_1$ and $\{e\}(\psi_1, \ldots, \psi_n) = k_2$, then $k_1 = k_2$.*

(b) *Let $\phi_1, \ldots, \phi_n$ be total continuous functionals, $\psi_1, \ldots, \psi_n$ corresponding elements in the T-hierarchy. For any e and k, if $\{e\}(\phi_1, \ldots, \phi_n) = k$ in the sense of the Ct-hierarchy, then $\{e\}(\psi_1, \ldots, \psi_n) = k$ in the sense of the D-hierarchy*

PROOF. In order to prove (a), it is sufficient to prove it for compact elements in the D-hierarchy. For this we may use induction on the complexity of the compact objects.

In order to prove (b), we see that the computations in the Ct-hierarchy are defined via a transfinite, positive inductive definition. We may then use induction on the rank of the computations, and (b) will be trivial. □

LEMMA 5. *Let e be an index for a Kleene-computation, let $\phi_1, \ldots, \phi_n$ be total, continuous functionals, and assume that for all $\psi \in Ct(k)$ we have that*

$$\{e\}(\psi, \phi_1, \ldots, \phi_n) \text{ takes a value.}$$

Then $\lambda\psi \in Ct(k)\{e\}(\psi, \phi_1, \ldots, \phi_n) \in Ct(k+1)$.

PROOF. As remarked, by standard domain theory we have

$$\Phi = \lambda\psi \in D(k)\{e\}(\psi, \phi_1, \ldots, \phi_n) \in D(k+1).$$

But Φ will be in T_k, and by Lemma 4 its equivalence class in $Ct(k)$ will be exactly the functional

$$\lambda\psi \in Ct(k)\{e\}(\psi, \phi_1, \ldots, \phi_n) \in Ct(k+1).$$

□

We will not give a full account of the topology of $Ct(k)$. We will, however, find it useful to give a criterion for when a sequence of functionals is convergent. Our criterion is actually a characterisation, but we will not give the full proof of that. Details for a further discussion can be found in [Hyland, 1975, 1979; Normann, 1980].

THEOREM 3. *Let $k > 0$, ϕ and $\{\phi_n\}_{n\in\mathbb{N}}$ be functionals in $Ct(k)$.*

Then the following are equivalent:

(i) *For all $\Psi \in Ct(k+1)$ we have $\Psi(\phi) = \lim_{n\to\infty} \Psi(\phi_n)$.*

(ii) *ϕ is the pointwise limit of $\{\phi_n\}_{n\in\mathbb{N}}$ and there is a* modulus of convergency *$\psi \in Ct(k)$ such that for all $x \in Ct(k-1)$ and all m*

$$m \geqslant \psi(x) \Rightarrow \phi_m(x) = \phi(x).$$

PROOF. Assume (i). For $x \in Ct(k-1)$, let $\Psi(\psi) = \psi(x)$. $\Psi \in Ct(k+1)$ so by the assumption

$$\phi(x) = \Psi(\phi) = \lim_{n\to\infty} \Psi(\phi_n) = \lim_{n\to\infty} \phi(x).$$

This shows pointwise convergency.

Let $\psi(x) = \mu n(\forall m \geqslant n)(\phi_m(x) = \phi(x))$. Assume that ψ is not continuous. Then for some $x \in Ct(k-1)$ and for all compacts $C \sqsubseteq x$ there is a total x_C with $C \sqsubseteq x_C$ and $\psi(x_C) \neq \psi(x)$.

CLAIM. *$\{\psi(x_C) \mid C \sqsubseteq x\}$ is unbounded.*

PROOF. Let m be such that $\phi_n(x) = \phi(x)$ for all $n \geqslant m$. We will prove that $\{\psi(x_C) \mid C \sqsubseteq x\}$ is not bounded by m. There is a compact $C \sqsubseteq x$ such that $\phi, \phi_0, \dots, \phi_m$ are all constant on $\{y \mid C \sqsubseteq y\}$. Then $\psi(x_C) \neq \psi(x)$. If $\psi(x_C) \leqslant m$ we see that

$$\begin{aligned}\psi(x_C) &= \mu n \forall k(n \leqslant k \leqslant m)\big(\phi_k(x_C) = \phi(x_C)\big)\\ &= \mu n \forall k(n \leqslant k \leqslant m)\big(\phi_k(x) = \phi(x)\big) = \psi(x).\end{aligned}$$

This contradicts the choice of x_C, so $\psi(x_C) > m$. □

We may then pick an increasing sequence $\{C_i\}_{i \in \mathbb{N}}$ of compacts, and total extensions $\{x_i\}_{i \in \mathbb{N}}$ such that $x = \sqcup\{C_i \mid i \in \mathbb{N}\}$ and $\psi(x_i) \geqslant i$ for all i.

By construction, x is the topological limit of x_i in D_{k-1}, and we may use the induction hypothesis on x and the sequence $\{x_i\}_{i \in \mathbb{N}}$. Thus

$$\Psi(\xi) = \mu n(\forall m \geqslant n)\big(\xi(x_m) = \xi(x)\big)$$

is continuous. By the construction, however, $\Psi(\phi_m) \geqslant m$ when $m \geqslant \max\{\Psi(\phi), \psi(x)\}$, and this contradicts the assumption. This proves (ii) from (i).

Now assume that (ii) holds, but that we have a counterexample Ψ to (i). We will obtain a contradiction by computing the discontinuous functional

$$^2E(f) = \begin{cases} 0 & \text{if } f \text{ is constant } 0,\\ 1 & \text{if } f \text{ is not constant } 0\end{cases}$$

from $\Psi, \psi, \phi, \{\phi_i\}_{i \in \mathbb{N}}$.

Let f be given. We define $\phi_f \in Ct(k)$ by the following algorithm:

Let $x \in Ct(k-1)$ and let $m = \psi(x)$. If for some $i \leqslant m$ we have that $f(i) \neq 0$, pick the least one, and let $n \geqslant i$ be minimal such that $\Psi(\phi) \neq \Psi(\phi_n)$. This is possible since (i) fails. Then let $\phi_f(x) = \phi_n(x)$. If there is no such i, let $\phi_f(x) = \phi(x)$. If f is constant 0, clearly $\phi_f = \phi$. Assume that f is not constant 0 and let i be minimal with $f(i) \neq 0$. Let $n \geqslant i$ be minimal with $\Psi(\phi_n) \neq \Psi(\phi)$. Let $x \in Ct(k-1)$ be given. There are two cases:

$i \leqslant \psi(x)$: Then by the algorithm, $\phi_f(x) = \phi_n(x)$.

$\psi(x) < i$: Then $\phi_f(x) = \phi(x) = \phi_n(x)$ since $\psi(x) < n$.

This shows that

$$f \text{ is constant } 0 \Leftrightarrow \Psi(\phi_f) = \Psi(\phi)$$

and this can be used to compute $^2E(f)$. This ends the proof. □

This method of proof is originally due to Grilliot [1971]. It has been developed further by Bergstra [1976], Wainer [1978] and Normann [1979]. We will use the method in Section 7 in order to get a primitive recursive approximation to computations.

On one hand there is a clear connection between the computations in the two typed hierarchies, any computation that works for inputs from $Ct(k)$ will also work for corresponding inputs from D_k. On the other hand the nature of the computations are quite different. For the total, continuous functionals we have a clear concept of a computation tree and of a subcomputation. Computations given by S1–S3 or S7 are called *initial* computations. They have no subcomputations, and the computation tree consists of the initial computation as the single node.

In the case of S4, $\{e\}(\phi_1, \ldots, \phi_n) = \{e_1\}(\{e_2\}(\phi_1, \ldots, \phi_n), \phi_1, \ldots, \phi_n)$, there are two immediate subcomputations, $\{e_2\}(\phi_1, \ldots, \phi_n)$, and if the value of that computation is m, $\{e_1\}(m, \phi_1, \ldots, \phi_n)$. The computation tree will then have the given computation as the top-node, with a branching to the computation trees of the immediate subcomputations.

In the case of S6, the picture is obvious.

In the case of S8, $\{e\}(\phi_1, \ldots, \phi_n) = \phi_1(\lambda\psi\{e_1\}(\psi, \phi_1, \ldots, \phi_n))$, we have a branching to the computation trees of all $\{e_1\}(\psi, \phi_1, \ldots, \phi_n)$ where ψ varies over all total objects of the appropriate type. We will say that Φ defined by

$$\Phi(\psi) = \{e_1\}(\psi, \phi_1, \ldots, \phi_n) \quad \text{for all total } \psi$$

is *used* in the computation. If any functional is used in a subcomputation we will also say that it is used in the computation.

LEMMA 6. *If Φ is used in the computation $\{e\}(\phi_1, \ldots, \phi_n)$ then the type of Φ is at most one less than the maximal type of the $\phi_1, \ldots, \phi_n$.*

The proof is by a trivial induction on the computation tree.

In contrast with the situation for the Ct-hierarchy, there is no clear notion of a subcomputation for computations in the D-hierarchy. The functionals will accept partial inputs, and then it is not clear how much information about the input that is needed in order to get a value. We will illustrate this by defining a total functional, a simplified fan functional, that will not be computable over the Ct-hierarchy, but computable over the D-hierarchy. Tait (unpublished) showed that the fan functional is recursive, i.e. it has a recursive associate in the sense of Kleene, but it is not S1–S9 computable.

5. The fan functional

In this section we will use the topology on $\mathbb{N}^{\mathbb{N}}$. We let $C = \{0, 1\}^{\mathbb{N}}$ be the compact set of functions bounded by 1. Any functional $F \in Ct(2)$ will be uniformly continuous on C. Our version of the fan functional will be the functional giving us the modulus for uniform continuity on C.

DEFINITION 6. We define the *fan functional* Φ as follows: If F is a total functional in $Ct(2)$ we let $\Phi(F)$ be the least number n such that for all f and g in C, if $f(m) = g(m)$ for all $m < n$, then $F(f) = F(g)$.

LEMMA 7. *The fan functional is in* $Ct(3)$.

PROOF. We will have to produce a corresponding element in T_3. Let $F \in T_2$. For each $f \in C$ there is a finite $\sigma \subset f$ such that $F(\sigma)$ is defined. The set of g extending σ will be an open set, so by the compactness of C, there is a finite list $\sigma_1, \dots, \sigma_n$ such that $F(\sigma_i)$ is defined for all i, and all f in C will extend some σ_i. Let $G = F$ restricted to $\{f \mid f \text{ extends some } \sigma_i\}$. Then G is compact in D_2, $G \sqsubseteq F$ and $\Phi(G) = \Phi(F)$ is well defined. This proves the lemma. □

REMARK. We see from the argument above that the set of compact elements in D_2 for which Φ is defined will be recursive in any natural enumeration of all the compact elements of D_2. This corresponds to the result that Φ has a recursive associate.

THEOREM 4. *The functional* Φ *is not* S1–S9 *computable in the Ct-hierarchy.*

PROOF. We will assume that Φ is computable via index e, and derive a contradiction. Let O be the constant zero functional of type two. Then $\Phi(O) = 0$, so by the assumption, $\{e\}(O) = 0$. Since O is recursive, we will only use recursive functions in any subcomputations of $\{e\}(O)$. Thus for any F in D_2, if $F(f) = 0$ for all recursive f, then $\{e\}(F) = 0$. We may use the same computation tree to verify that the computation really is defined. Choose one such F that is undefined at some nonrecursive point in C. Since we are working in D_2 there will be a compact $G \sqsubseteq F$ such that $\{e\}(G) = 0$. G will be defined on a closed-open set that does not cover all of C. It is then easy to extend G to a total G' that is not constant, and then $\Phi(G') \neq 0$, contradicting that $\{e\}(G') = 0$. This proves the theorem. □

Berger (unpublished) observed that this result does not hold when we work in the D-hierarchy:

THEOREM 5. *The fan functional seen as an element of* D_3 *is* S1–S9 *computable.*

PROOF. We will use the following notation:

If F is in D_2 we let $F_0(f) = F(0 * f)$ and $F_1(f) = F(1 * f)$, where $*$ just ads the number in front of the infinite sequence f.

We let c_i be the constant i function of type 1.

Using the recursion theorem, we find two indices d and e satisfying

$$\{d\}(F) = \begin{cases} 0 & \text{if } F(c_0) = F(\lambda x\{e\}(F, x)), \\ 1 + \max\{\{d\}(F_0), \{d\}(F_1)\} & \text{if } F(c_0) \neq F(\lambda x\{e\}(F, x)) \end{cases}$$

and

$$\{e\}(F,0) = \begin{cases} 0 & \text{if } \{d\}(F_0) \neq 0 \text{ or if } \{d\}(F_0) = \{d\}(F_1) = 0 \text{ and} \\ & F(c_0) = F(c_1), \\ 1 & \text{if } (\{d\}(F_0) = 0 \text{ and } \{d\}(F_1) \neq 0) \text{ or} \\ & (\{d\}(F_0) = \{d\}(F_1) = 0 \text{ and } F(c_0) \neq F(c_1)), \end{cases}$$

$$\{e\}(F, x+1) = \{e\}(F_0, x) \quad \text{if } \{e\}(F,0) = 0,$$

$$\{e\}(F, x+1) = 0 \quad \text{if } \{e\}(F,0) = 1 \text{ and } F(c_0) \neq F(1 * c_0),$$

$$\{e\}(F, x+1) = \{e\}(F_1, x) \quad \text{if } \{e\}(F,0) = 1 \text{ and } F(c_0) = F(1 * c_0).$$

Now let $F \in T_2$ be given. Let k be such that for all 0–1 sequences σ of length $\geqslant k$ we have that $F(\sigma)$ is defined.

By induction on k we show that $\{d\}(F) = \Phi(F)$ and that if $\Phi(F) = 0$ then $\lambda x < k\{e\}(F,x)$ is constant zero, while if $\Phi(F) \neq 0$ then $\lambda x < k\{e\}(F,x)$ is the least σ in the lexicographical ordering for which $F(c_0) \neq F(\sigma)$. If $k = 0$ we have that F is defined on the empty function. Thus we do not need to compute $\{e\}(F,x)$ for any x in order to see that $F(c_0) = F(\lambda x\{e\}(F,x))$, and we give out $\{d\}(F) = \Phi(F) = 0$.

In this case we do not have to prove anything about e. If $k > 0$ we have to consider several cases.

Case 1 F_0 and F_1 are constants with the same values. Then $\{e\}(F,0) = 0$ using the instruction for e and the first part of the induction hypothesis. By the second part of the induction hypothesis $\{e\}(F, x+1) = \{e\}(F_0, x) = 0$ for all $x < k-1$. This proves the second part of the claim in this case. The first part follows by inspecting the algorithm given by d.

Case 2 F_0 is constant but F_1 is not constant, or constant with a different value. By the first part of the induction hypothesis we get that $\{e\}(F,0) = 1$ and if F_1 is constant we have $\{e\}(F, x+1) = 0$. If F_1 is not constant, there are two subcases, depending on whether $F_1(c_0)$ is the constant value of F_0 or not. For both cases we see that we get the right value for $\{e\}(F, x+1)$. We will use the second part of the definition of $\{d\}(F)$, and by the first part of the induction hypothesis, this will be $\Phi(F)$.

Case 3 F_0 is not constant.Then by the first part of the induction hypothesis, $\{e\}(F,0) = 0$, and by the second part of the induction hypothesis, $\{e\}(F, x+1)$ will take the appropriate value for $x < k-1$. This proves the second part of the claim. For the first part we may argue as in case 2.

This shows that $\Phi(F) = \{d\}(F)$ for all F, and thus that the fan functional is computable. □

6. Bar recursion

In this section we will give a very brief introduction to the concept of *bar recursion*. Bar recursion is a special recursion scheme for type two functionals and represents a generalisation of primitive recursion to well founded trees.

Before defining the precise recursion scheme for bar recursion, let us consider tree recursion. By a *tree* we will mean a nonempty set of finite sequences of natural numbers closed under subsequences. A *branch* will be a function f such that all finite segments of f is in the tree. The tree is *wellfounded* if there are no branches.

Let W be the set of wellfounded trees. If X is a set, $x_0 \in X$ and $F : (\mathbb{N} \to X) \to X$, we may define $\Phi : W \to X$ by recursion as follows:

$$\begin{aligned}\Phi(T) &= x_0 \text{ if } T \text{ is the tree consisting only of the empty sequence.}\\ \Phi(T) &= F\big(\lambda n.\Phi(T_n)\big) \text{ otherwise.}\end{aligned}$$

It can be shown that if X and W have reasonable topologies, and if F is continuous, then Φ will be continuous uniformly in F and x_0.

If σ is a finite sequence of natural numbers, we let B_σ be the set of functions extending σ. Using this notation, we can associate a wellfounded tree to any total continuous functional F of type 2 as follows: σ is in T_F if F is not total on B_σ.

Bar recursion is essentially recursion over this well founded tree. This tree is, however, not even continuously dependent on F, so we must show some care in formulating the recursion scheme:

DEFINITION 7. Let $G : \mathbb{N} \to \mathbb{N}$, $H : \mathbb{N}^{\mathbb{N}} \to \mathbb{N}^{\mathbb{N}}$ be continuous such that for all natural numbers m we have

$$H\big(\lambda n.G(m)\big) = G(m).$$

Then we define $\Phi : Ct(z) \to \mathbb{N}$ by

$$\begin{aligned}\Phi(F) &= G(m) \quad \text{if } F \text{ is constant } m,\\ \Phi(F) &= H(\lambda n \Phi(F_n)) \quad \text{otherwise,}\end{aligned}$$

where $F_n(f) = F(n * f)$, $(n * f)(0) = n$ and $(n * f)(k+1) = f(k)$.

THEOREM 6. *Φ as defined above is uniformly continuous in G and H.*

INDICATION OF PROOF. We may give an alternative definition of Φ on D_2, replacing

$$F \text{ is constant } m$$

with

$$F \text{ is the compact determined by } (\bot, m).$$

We may then use the assumption on G and H to show that for total F, the alternative definition gives the same output. Thus there is an element of D_3 that represents Φ. This ends our indication of proof. □

One example of a functional defined by Bar-recursion is the Γ-functional defined by Gandy:

$$\Gamma(F) = F_0\big(\lambda n \Gamma(F_{n+1})\big)$$

with the understanding that $\Gamma(F) = m$ when F is constant m.

Hyland [1975] proved that this functional is not S1–S9-computable in the fan functional and any $f : \mathbb{N} \to \mathbb{N}$ in the hierarchy of hereditarily total continuous functionals. On the other hand it is easily seen that Γ is S1–S9-computable in the D-hierarchy.

The author recently proved that every Bar-recursive functional of type 3 is S1–S9-computable in the D-hierarchy.

7. Associates

Kleene [1959b] defined the countable functionals via a system of *associates*. The associates are functions $f : \mathbb{N} \to \mathbb{N}$ that contains the information about how the functional behave on various inputs. On the basis of our approach via domains, we will give an alternative definition of the associates.

Throughout this section we will let $\{C_{n,k}\}_{k\in\mathbb{N}}$ be an enumeration of the compact elements in D_k such that the following relations are primitive recursive uniformly in k:

$$C_{n,k} \sqsubseteq C_{m,k}, \qquad C_{n,k} = C_{m_1,k} \sqcup C_{m_2,k}.$$

For $\Phi \in D_k$ we let $X_\Phi = \{i \mid C_{i,k} \sqsubseteq \Phi\}$.

DEFINITION 8. An *associate* for $\Phi \in D_k$ will be an enumeration of X_Φ. If $\phi \in Ct(k)$, an associate for ϕ will be an associate for any element $\Phi \in T_k$ that is in the equivalence class representing Φ.

We let $As(k)$ be the set of associates for elements in $Ct(k)$, and we let ρ_k be the function that maps an element in $As(k)$ to its corresponding functional.

LEMMA 8. (a) *If $\Phi \in Ct(k)$, there is a continuous map $\hat{\Phi} : As(k-1) \to \mathbb{N}$ such that whenever f is an associate for $\phi \in Ct(k-1)$ then $\hat{\Phi} = \Phi(\phi)$.*

(b) *If there is a continuous* $\hat{\Phi} : As(k-1) \to \mathbb{N}$ *such that the function* Φ *defined by*

$$\Phi(\phi) = \hat{\Phi}(f) \text{ whenever } f \text{ is an associate for } \phi$$

is well defined and total, then $\Phi \in Ct(k)$.

PROOF. (a) Let $\Phi \in Ct(k)$ and let $\Phi_1 \in T_k$ be in the equivalence class of Φ.

Let $f \in As(k-1)$ be an associate for $\phi \in T_{k-1}$. Then $\Phi_1(\phi) \in \mathbb{N}$ and there is a compact $C_{n,k-1} \sqsubseteq \phi$ with $\Phi_1(C_{n,k-1}) \in \mathbb{N}$. Then there is a compact $C_{m,k} \sqsubseteq \Phi_1$ with $C_{m,k}(C_{n,k-1}) \in \mathbb{N}$. Let $\hat{\Phi}(f) = n$ if and only if

$$\exists m \exists i \big(C_{m,k} \sqsubseteq \Phi_1 \wedge C_{m,k}(C_{f(i),k-1}) = n\big).$$

Then $\hat{\Phi}$ is continuous, and clearly $\hat{\Phi}(f) = \Phi(\phi)$.

(b) Now let $\hat{\Phi}$ be given, and Φ well defined. We will have to produce an element $\Phi_1 \in T_k$. Let $\Phi_1(C_{m,k-1}) = n$ if and only if there exist a finite $\sigma : \{0, \dots, l-1\} \to \mathbb{N}$ with $\hat{\Phi}(\sigma) = n$ and $C_{\sigma(i),k-1} \sqsubseteq C_{m,k-1}$ for all $i < l$.

CLAIM. *If* $\Phi_1(C_{m_1,k-1}) = n_1$ *and* $\Phi_1(C_{m_2,k-1}) = n_2$ *and* $\{C_{m_1,k-1}, C_{m_2,k-1}\}$ *is bounded, then* $n_1 = n_2$.

PROOF. Let σ and τ be the two finite enumerations used to produce n_1 and n_2. By the density theorem there is a $\phi \in T_{k-1}$ such that $C_{m_1,k-1} \sqsubseteq \phi$ and $C_{m_2,k-1} \sqsubseteq \phi$ and we may find enumerations f and g of X_ϕ extending σ and τ respectively. Then $\hat{\Phi}(f) = n_1$ and $\hat{\Phi}(g) = n_2$, so $n_1 = \Phi(\phi) = n_2$ since Φ is well defined. This proves the claim. □

Φ_1 is obviously monotone and if $\phi \in T_{k-1}$ and f is an enumeration of X_ϕ, then $\hat{\Phi}(f)$ is defined, so for some n, $\hat{\Phi}(\bar{f}(n))$ is defined.

$C_{f(0),k-1} \sqcup \cdots \sqcup C_{f(n-1),k-1}$ will be a compact $C \sqsubseteq \phi$ and $\Phi_1(C) \in \mathbb{N}$. Thus $\Phi(\phi) \in \mathbb{N}$. This ends the proof of the lemma. □

DEFINITION 9. A functional $\phi \in Ct(k)$ is *recursive* if and only if ϕ has a recursive associate.

As a consequence of Theorem 2 and Lemma 8 we get

COROLLARY 2. *The set* $As(k+1)$ *of associates for elements in* $Ct(k+1)$ *will be complete* Π^1_k.

It is easy to see from the proof of Lemma 5 that if ϕ is S1–S9 computable over $\{Ct(k)\}_{k \in \mathbb{N}}$ then ϕ is recursive. The converse is not true, the fan functional is a counterexample.

The argument from Lemma 5 also shows that if $\phi \in T_n$ and ϕ is S1–S9 computable over $\{D_k\}_{k\in\mathbb{N}}$, then ϕ has a recursive associate. The converse has recently been proved by the author:

> If $\Phi \in Ct(n)$ is recursive, then there is a $\Phi_1 \in T_n$ in the equivalence class of Φ that is S1–S9 computable over $\{D_k\}_{k\in\mathbb{N}}$.

8. The finitary aspects

One important aspect of standard recursion theory is that any successful computation is finite. This is used in two almost undistinguishable ways, we talk about finite computation trees, and we talk about computations in n steps. When we work with computations where the input is a list of continuous functionals, the computation trees are infinite, and it makes no immediate sense to define a total approximation to a partial computable function Φ, e.g. by saying:

> "Compute $\Phi(\phi)$ in k steps.
> If we get a value m, use that, otherwise give output 0."

The finitary aspects of standard recursion theory will have corresponding aspects for the continuous functionals, but here the concept of a finite computation tree and the concept of a k-step computation diverge.

We have shown that if $\{e\}(\phi_1, \dots, \phi_n)$ takes a value, then there is a well founded computation tree for the computation, and, by induction on the rank of this computation tree, we showed that there are compact sets $C_1, \dots, C_n$ in $\phi_1, \dots, \phi_n$ such that $\{e\}(C_1, \dots, C_n)$ is defined and with the same value. We can use this proof to show that there is a finite computation tree only operating on compacts, that can be embedded in the full computation tree. In this sense the computation is finite. The problem with this approach is that the existence of a successful finite approximation to an alleged computation does not guarantee that the alleged computation terminates, since the full computation tree does not need to be well-founded. We just know that the computation will terminate in the hierarchy of the partial continuous functionals, and we have narrowed down the possible values to just one number.

Let us now turn to the higher type analogue of k-step computations. The basic idea is that given a proposed computation $\{e\}(\phi_1, \dots, \phi_n)$, we may develop the computation tree top-down (the root is at the top in these trees), where we cut off each branch after k steps, either with the correct value of a basic computation, or with the value 0. We pretend that it takes 1 step to carry out a basic computation.

DEFINITION 10. Let $\{e\}_0(\Phi_1, \dots, \Phi_n) = 0$. If $\{e\}(\Phi_1, \dots, \Phi_n)$ is an initial computation, we let

$$\{e\}_{k+1}(\Phi_1, \dots, \Phi_n) = \{e\}(\Phi_1, \dots, \Phi_n).$$

If

$$\{e\}(\Phi_1, \ldots, \Phi_n) = \{e_1\}\big(\{e_2\}(\Phi_1, \ldots, \Phi_n), \Phi_1, \ldots, \Phi_n\big),$$

we let

$$\{e\}_{k+1}(\Phi_1, \ldots, \Phi_n) = \{e_1\}_k\big(\{e_2\}_k(\Phi_1, \ldots, \Phi_n), \Phi_1, \ldots, \Phi_n\big).$$

For the other non-initial cases we also define $\{e\}_{k+1}$ by using $\{e_1\}_k$ in the subcomputations. If e does not fit into one of the schemes or is incompatible with the signature of $(\Phi_1, \ldots, \Phi_n)$, we let $\{e\}_{k+1}(\Phi_1, \ldots, \Phi_n) = 0$.

We observe that $\{e\}_k$ is defining a total function for all e and all signatures.

THEOREM 7. (a) *If* $\{e\}(\Phi_1, \ldots, \Phi_n) = m$ *then* $\lim_{k\to\infty}\{e\}_k(\Phi_1, \ldots, \Phi_n) = m$.
(b) *The modulus of convergency*

$$M(e, \Phi_1, \ldots, \Phi_n) = \mu m(\forall k \geqslant m)\big(\{e\}_k(\Phi_1, \ldots, \Phi_n) = \{e\}(\Phi_1, \ldots, \Phi_n)\big)$$

is uniformly computable and takes a value whenever $\{e\}(\Phi_1, \ldots, \Phi_n)$ *takes a value.*

PROOF. We use induction on the ordinal rank of the computation tree.

We will use the recursion theorem to produce an index for the modulus function M.

For initial computations, the theorem is trivial, and it is clear how to produce an index for M in this case.

All other cases except case 8 are trivial, and are left for the reader. So assume that

$$\{e\}(\Phi_1, \ldots, \Phi_n) = \Phi_1\big(\lambda\phi\{e_1\}(\phi, \Phi_1, \ldots, \Phi_n)\big).$$

Let $\Psi(\phi) = \{e_1\}(\phi, \Phi_1, \ldots, \Phi_n)$. Let $\Psi_k(\phi) = \{e_1\}_k(\phi, \Phi_1, \ldots, \Phi_n)$. By Theorem 3 and the induction hypothesis (b): $\Phi_1(\Psi) = \lim_{k\to\infty} \Phi_1(\Psi_k)$. This proves (a) in this case.

We now prove (b): Given m we can give an algorithm for the function Ψ^m that will be Ψ if $\Phi_1(\Psi) = \Phi_1(\Psi_k)$ for all $k \geqslant m$, and Ψ_k for the least k that is a counterexample, otherwise. We use the same methods as in the proof of Theorem 3.

Then the modulus will be the least m for which $\Phi_1(\Psi^m) = \Phi_1(\Psi)$. This gives us an algorithm for computing the modulus in this case, and this ends the proof of the theorem. □

REMARK. Almost all aspects of this method are used in Normann [1979] where it is shown for the full type hierarchy that if the set of functions recursive in a functional Φ is closed under jump, then ${}^2E$ is computable in Ψ, where ${}^2E$ is as in the proof of Theorem 3.

DEFINITION 11. Let Φ be a functional of type k. By *the* 1-*section* of Φ we mean

$$^1sc(\Phi) = \{f \in \mathbb{N} \to \mathbb{N} \mid f \text{ is computable in } \Phi\}.$$

COROLLARY 3. *Let Φ be a continuous functional of type k. Then there is an $f \in {}^1sc(\Phi)$ such that for all $g \in {}^1sc(\Phi)$ there is an r.e. (f) set $A \in {}^1sc(\Phi)$ such that g is recursive in A.*

PROOF. Let $f(e,k,x) = \{e\}_k(x, \Phi)$. If g is recursive in Φ via index e, let

$$A = \{(x,n) \mid \exists m > n\big(f(e,m,x) \neq f(e,n,x)\big)\}.$$

By the claim of the proof of the theorem, we see that A is computable in Φ, and A is clearly r.e. (f). In order to compute $g(x)$ from A and f, take the least m such that $(x,m) \notin A$ and then $g(x) = f(e,m,x)$. □

We have used the existence of a continuous modulus of convergency as a way of saying that a sequence of functionals converges towards one functional. Following Hyland, this is a characterisation: A sequence $\{\Phi_k\}_{k\in\mathbb{N}}$ will have Φ as a limit if the sequence converges pointwise to Φ and there is a continuous modulus for the pointwise convergency. We will not prove this here, see [Normann, 1980] for a proof. We will however use this concept to see that there is a dense set of very simple total functionals:

DEFINITION 12. We define the restriction $\Phi \downarrow n$ of Φ to n as follows:

$$m \downarrow n = m \text{ if } m \leqslant n,$$
$$m \downarrow n = n \text{ if } m > n,$$
$$(\Phi \downarrow n)(\phi) = (\Phi(\phi \downarrow n)) \downarrow n \text{ for higher type functionals } \Phi.$$

THEOREM 8. *Let $\Phi \in Ct(k+1)$. Then Φ is the pointwise limit of $\Phi \downarrow n$, and uniformly in Φ we can compute a modulus function for the pointwise convergency.*

PROOF. We use induction on $k+1$. The induction start is trivial and the induction step is as in the proof of Theorem 6.

We define the hyperfinitary functionals as follows:

- All numbers are hyperfinitary.
- $f \in Ct(1)$ is hyperfinitary if f is constant except on a finite set.
- $\Psi \in Ct(k+2)$ is hyperfinitary if there are hyperfinitary $\xi_1, \ldots, \xi_n \in Ct(k)$ and a function $f : \mathbb{N}^n \to \mathbb{N}$ that is constant except on a finite set, such that for all $\psi \in Ct(k+1)$

$$\Psi(\phi) = f\big(\phi(\xi_1), \ldots, \phi(\xi_n)\big).$$

We see that the functions $\Phi \downarrow n$ are hyperfinitary, and thus every functional Φ will be the limit of a sequence of hyperfinitary functionals that can be found effectively in Φ. □

A closer look at the proof of the density theorem, also shows that the functionals constructed there are hyperfinitary.

9. Functional interpretation of analysis

Kreisel's [1959] key motivation for introducing the continuous functionals was to use them to give a theory-independent constructive interpretation of statements of analysis or second order number theory. Kreisel's interpretation is based on an analysis of intuitionistic arithmetic due to Gödel [1932]. In order to describe the interpretation, we need to extend the hierarchy of continuous functionals to mixed types:

DEFINITION 13.
(i) 0 is a type.
(ii) If σ and τ are types, then $\sigma \to \tau$ and $\sigma \times \tau$ are types.

DEFINITION 14. If σ is a type we define
(a) $\mathrm{Tp}(\sigma)$ as the full classical interpretation

$$\mathrm{Tp}(0) = \mathbb{N},$$
$$\mathrm{Tp}(\sigma \to \tau) = \{F \mid F : \mathrm{Tp}(\sigma) \to \mathrm{Tp}(\tau)\},$$
$\mathrm{Tp}(\sigma \times \tau)$ as the cartesian product of $\mathrm{Tp}(\sigma)$ and $\mathrm{Tp}(\tau)$.

(b) $\mathrm{Ct}(\sigma)$ as the canonical set of hereditarily total continuous objects of type σ.

We omit the details of the construction of $\mathrm{Ct}(\sigma)$, it is based on first constructing the underlying domain and then isolate the hereditarily total elements and the canonical equivalence relation on the total elements.

We define the associates in complete analogy with $\mathrm{Ct}(k)$ and we call $\phi \in \mathrm{Ct}(\sigma)$ *recursive* if ϕ has a recursive associate.

We will now define the language for analysis that we will use. We will use variables f, g etc. for functions in $\mathbb{N}^{\mathbb{N}}$, and n, m etc. for numbers. We will let $\vec{x}$ denote an arbitrary list of variables.

The atomic formulas will be $A(\vec{x})$, where A is a primitive recursive predicate of the appropriate signature.

We will use the connectives $\wedge$, $\vee$, $\neg$ and $\to$ and the quantifiers $\exists$ and $\forall$ in the usual way. Since we will give the constructive interpretation we will distinguish between statements that are classically equivalent. Thus we cannot eliminate any of the connectives or quantifiers.

DEFINITION 15. For each type σ we extend the language with variables ϕ^σ, ψ^σ etc. for objects of type σ. These will be terms of type σ. We define other terms by "breaking down":

If t is a term of type $\sigma \times \tau$ we let $(t)_0$ be a term of type σ and $(t)_1$ a term of type τ.

If t is a term of type $\sigma \to \tau$ and s is a term of type σ, we let $t(s)$ be a term of type τ.

The atomic expressions will now be expressions of the form $A(t_1, \ldots, t_n)$, where each t_i is a term of type 0 or $0 \to 0$.

Whenever we interpret each functional variable in an expression as a functional in one of our type hierarchies, we get a true or false statement. Thus we may extend our language with quantification over each type, and when the range of each quantifier is fixed, each closed statement will get a truth value in the standard way.

We will now transform any statement $\Phi(\vec{x})$ in analysis to a statement of the form

$$\Phi^c = \exists\phi^\sigma \forall\psi^\tau A(\phi, \psi, \vec{x})$$

following Kreisel [1959], where the types σ and τ will depend on Ψ.

For the sake of notational simplicity we may use sequences of quantifiers instead of contracting them to one via the $\times$-constructions.

In the constructive interpretation of Ψ we let ϕ^σ range over the recursive objects of type σ while ψ^τ will range over the continuous objects of type τ.

DEFINITION 16.

(i) If $\Psi(\vec{x})$ is atomic, we let Ψ^c be Ψ (with some dummy quantifiers).

(ii) If $\Psi(\vec{x}) = \Psi_1(\vec{x}) \wedge \Psi_2(\vec{x})$, $\Psi_1^c = \exists\phi^{\sigma_1}\forall\psi^{\tau_1} A_1$ and $\Psi_2^c = \exists\phi^{\sigma_2}\forall\psi^{\tau_2} A_2$ we let

$$\Psi^c(\vec{x}) = \exists\phi^{\sigma_1}\exists\phi^{\sigma_2}\forall\psi^{\tau_1}\forall\psi^{\tau_2}\big(A_1(\phi^{\sigma_1}, \psi^{\tau_1}, \vec{x}) \wedge A_2(\phi^{\sigma_2}, \psi^{\tau_2}, \vec{x})\big).$$

(iii) If $\Psi(\vec{x}) = \Psi_1(\vec{x}) \vee \Psi_2(\vec{x})$, $\Psi_1^c = \exists\phi^{\sigma_1}\forall\psi^{\tau_1} A_1$ and $\Psi_2^c = \exists\phi^{\sigma_2}\forall\psi^{\tau_2} A_2$ we let

$$\begin{aligned}\Psi^c(\vec{x}) &= \exists n \exists\phi^{\sigma_1\times\sigma_2}\forall\psi^{\tau_1\times\tau_2}((n = 0 \wedge A_1((\phi)_0, (\psi)_0, \vec{x})) \vee \\ &\quad (n = 1 \wedge A_2((\phi)_1, (\psi)_1, \vec{x}))).\end{aligned}$$

(iv) If $\Psi(\vec{x}) = \Psi_1(\vec{x}) \to \Psi_2(\vec{x})$, $\Psi_1^c = \exists\phi^{\sigma_1}\forall\psi^{\tau_1} A_1$ and $\Psi_2^c = \exists\phi^{\sigma_2}\forall\psi^{\tau_2} A_2$ we let

$$\begin{aligned}\Psi^c(\vec{x}) &= \exists\Phi_1^{\sigma_1\to\sigma_2}\exists\Phi_2^{(\sigma_1\times\tau_2)\to\tau_1}\forall\phi^{\sigma_1}\forall\psi^{\tau_2} \\ &\quad \big(A_1(\phi, \Phi_2(\phi, \psi)) \to A_2(\Phi_1(\phi), \psi)\big).\end{aligned}$$

(v) If $\Psi(\vec{x}) = \neg\Psi_1(\vec{x})$ where $\Psi^c(\vec{x}) = \exists\phi^\sigma\forall\psi^\tau A(\phi, \psi, \vec{x})$, we let

$$\Psi^c(\vec{x}) = \exists\Phi^{\sigma\to\tau}\forall\phi^\sigma \neg A(\phi, \Psi(\phi), \vec{x}).$$

(vi) If $\Psi(\vec{x}) = \forall f \Psi_1(f, \vec{x})$ where $\Psi_1^c(\vec{x}) = \exists \phi^\sigma \forall \psi^\tau A(\phi, \psi, f, \vec{x})$, we let

$$\Psi^c(\vec{x}) = \exists F^{(0\to 0)\to\sigma} \forall \psi^\tau \forall f A(F(f), \psi, f, \vec{x}).$$

(vii) Universal quantification over $\mathbb{N}$ is treated likewise.

(viii) Existential quantifiers in front of a statement are just preserved, i.e. $(\exists f \Psi)^c = \exists f \Psi^c$.

The justification for this definition of the constructive interpretation is discussed at length in Kreisel [1959], and we will not get involved in this here. Kreisel points out that the case of $\to$ is not fully justifiable from a constructivists point of view, so special care has to be taken in his justification for that case.

If we interpret Ψ^c over the full typestructure $\mathrm{Tp}(\sigma)$, we get that Ψ^c is true if and only if Ψ is classically true. This is proved by a simple induction using the axiom of choice.

Kreisel [1959] shows, using the density of the set of total objects, that if Ψ does not contain disjunctions or existential quantifiers, then Ψ is constructively true if and only if Ψ is classically true. Since every statement of second order number theory is classically equivalent to a statement of the above form, and since there are such statements that are not absolute with respect to transitive models for set theory, the constructive interpretation of an analytic statement is in general dependent of the underlying set theory.

On the other hand there will be statements that are constructively true (proved classically) that are not theorems of ZFC. We will offer one example, a statement stating $(\neg \mathrm{V} = \mathrm{L})$.

Let WO be the set of codes for countable well orderings, and let $|g|$ be the ordinal corresponding to $g \in \mathrm{WO}$.

Consider the statement

$$\forall f \exists g (g \in \mathrm{WO} \wedge f \in L_{|g|}).$$

The matrix here is a Π_1^1-statement $\forall h \exists n A(f, g, h, n)$. Thus we consider the statement

$$\Psi = \forall f \exists g \forall h \exists n A(f, g, h, n)$$

where we cannot find a continuous function G that selects an appropriate g from each f. This fact is provable in ZFC.

We then have

$$\Psi^c = \exists G \exists \Phi \forall f \forall h A\big(f, G(f), h, \Phi(f)(h)\big).$$

For each F and Φ we can effectively find f and h negating the matrix. Thus the constructive interpretation of $\neg\Psi$ will be true, and provable in ZFC.

Similar examples can be constructed for statements where the statement is provably false in ZFC while the constructive interpretation is provably true, still in ZFC. We leave this construction as an exercise for the reader, bearing in mind that some knowledge of higher recursion theory may be useful in constructing an example.

References

U. BERGER

[1993] Total sets and objects in domain theory, *Ann. Pure Appl. Logic*, 60, pp. 91–117.

J. BERGSTRA

[1976] *Computability and Continuity in Finite Types*, Thesis, University of Utrecht.

YU. L. ERSHOV

[1972] Computable functionals of finite type, *Algebra and Logic*, 11, pp. 203–277.

[1973] *The Theory of Numerations*, Vol. 2 (in Russian), Novosibirsk.

[1974] Maximal and everywhere defined functionals, *Algebra and Logic*, 13, pp. 210–225.

K. GÖDEL

[1932] Zur intuitionistischen Arithmetik und Zahlentheorie, *Ergebnisse eines mathematischen Kollokuiums.*

T. GRILLIOT

[1971] On effectively discontinuous type-2 objects, *J. Symbolic Logic*, 36, pp. 245–248.

J. M. E. HYLAND

[1975] *Recursion on the Countable Functionals*, Dissertation, Oxford.

[1979] Filter spaces and continuous functionals, *Ann. Math. Logic*, 16, pp. 101–143.

S. C. KLEENE

[1959a] Recursive functionals and quantifiers of finite types I, *Trans. Amer. Math. Soc.*, 91, pp. 1–52.

[1959b] Countable functionals, in: *Constructivity in Mathematics*, A. Heyting, ed., North-Holland, Amsterdam, pp. 81–100

[1963] Recursive functionals and quantifiers of finite types II, *Trans. Amer. Math. Soc.*, 108, pp. 106–142.

[1978] Recursive functionals and quantifiers of finite types revisited I, in: *Generalized Recursion Theory II*, J. E. Fenstad, R. O. Gandy and G. E. Sacks, eds., North-Holland, Amsterdam, pp. 185–222.

[1980] Recursive functionals and quantifiers of finite types revisited II, in: *The Kleene Symposium*, J. Barwise, H. J. Keisler and K. Kunen, eds., North-Holland, Amsterdam, pp. 1–29.

[1982] Recursive functionals and quantifiers of finite types revisited III, in: *Patras Logic Symposium*, G. Metakides, ed., North-Holland, Amsterdam, pp. 1–40.

[1985] Unimonotone functions of finite types (Recursive functionals and quantifiers of finite types revisited IV), in: *Recursion Theory*, A. Nerode and R. A. Shore, eds., AMS Proceedings of Symposia in Pure Mathematics, Vol. 42, pp. 119–138.

[1991] Recursive functionals and quantifiers of finite types revisited V, *Trans. Amer. Math. Soc.*, 325, pp. 593–630.

G. KREISEL

[1959] Interpretation of analysis by means of functionals of finite type, in: *Constructivity in Mathematics*, A. Heyting, ed., North-Holland, Amsterdam, pp. 101–128.

G. LONGO AND E. MOGGI

[1984] The hereditary partial effective functionals and recursion theory in higher types, *J. Symbolic Logic*, 49, pp. 1319–1332.

D. NORMANN

[1979] A classification of higher type functionals, in: *Proceedings from 5th Scandinavian Logic Symposium*, F. V. Jensen, B. H. Mayoh and K. K. Møller, eds., Aalborg University Press, pp. 301–308.

[1980] *Recursion on the Continuous Functionals*, Lecture Notes in Mathematics, Vol. 811, Springer, Berlin.

[1983] Characterising the continuous functionals, *J. Symbolic Logic*, 48, pp. 965–969.

V. STOLTENBERG-HANSEN, I. LINDSTRÖM AND E. GRIFFOR

[1994] *Mathematical Theory of Domains*, Cambridge Tracts in Theor. Comp. Sci., 22.

S. S. WAINER

[1978] The 1-section of a non-normal type-2 object, in: *Generalized Recursion Theory II*, J. E. Fenstad, R. O. Gandy and G. E. Sacks, eds., North-Holland, Amsterdam, pp. 407–417.

CHAPTER 9

Ordinal Recursion Theory

C. T. Chong

National University of Singapore, 10 Kent Ridge Crescent, Singapore 119260

S. D. Friedman*

Massachusetts Institute of Technology

Contents

1. α-recursion theory 278
   1.1. Regularity 280
   1.2. Definability 280
   1.3. The α-finite injury priority method 282
   1.4. The Density Theorem 284
   1.5. Non-existence of maximal sets 286
   1.6. Post's problem above $\emptyset'$ and set-theoretic methods 288
   1.7. Applications to fragments of Peano arithmetic 289
2. β-recursion theory 290
3. The admissibility spectrum 293
References 297

*Preparation of this paper was supported by NSF Grant #9205530.

HANDBOOK OF COMPUTABILITY THEORY
Edited by E.R. Griffor

Introduction

In a fundamental paper, Kreisel and Sacks [1965] initiated the study of "metarecursion theory", an analog of classical recursion theory where ω is replaced by Church–Kleene ω_1, the least non-recursive ordinal. Subsequently, Sacks and his school developed recursion theory on arbitrary Σ_1-admissible ordinals, now known as "α-recursion theory".

In Section 1 of the present article, we present the basic concepts and techniques of this theory, putting particular emphasis on the main new ideas that have been introduced to study recursion-theoretic problems assuming only Σ_1-admissibility on a domain greater than ω. As Σ_1-admissibility is easily lost under relativization, we turn to "β-recursion theory" (Section 2) which attempts to develop recursion theory on arbitrary limit ordinals. In Section 3, the final part of this article, we take up the topic of "admissibility spectra", where instead of studying the definability of subsets of a fixed Σ_1-admissible ordinal, we ask: given a set X, which are the ordinals Σ_1-admissible relative to X?

The reader will notice that Jensen's work on the fine structure theory of Gödel's L features prominently throughout. Indeed a major development of ordinal recursion theory is the infusion of set-theoretic ideas in studying recursion-theoretic problems. The unmistakeable presence of a strong set-theoretic flavor in the subject of admissibility spectra is especially pronounced. We thus view the appearance of Jensen's paper (Jensen [1972], preliminary copies of which had been circulated earlier), at a time when ordinal recursion theory was being developed, to be a fortuitous happening.

Some of the techniques and ideas which were invented in ordinal recursion theory have recently found applications in "recursion theory on fragments of Peano arithmetic". This is an unexpected turn of events which signal a basic unity among various fields in recursion theory and fine structure theory. We touch briefly on this work at the end of Section 1.

1. α-recursion theory

We begin with some basic notions. Recall Gödel's constructible universe L, defined as $\bigcup\{L_\alpha \mid \alpha \text{ an ordinal}\}$. A limit ordinal α is *Σ_n-admissible* if L_α satisfies the replacement axiom for Σ_n formulas (with parameters in L_α) in ZF set theory. If α is Σ_n-admissible for some $n \geqslant 1$, there is a $\Sigma_1(L_\alpha)$ bijection between α and L_α, allowing one to identify these two objects if and when necessary. Σ_1-admissible ordinals are sometimes referred to simply as admissible ordinals. Unless otherwise specified, we fix α to be an admissible ordinal henceforth.

A set $K \subset \alpha$ is *α-finite* if $K \in L_\alpha$. A function is partial *α-recursive* if its graph is $\Sigma_1(L_\alpha)$. A set is *α-recursively enumerable* (α-RE) if it is the domain of a partial α-recursive function. $A \subset \alpha$ is α-recursive if both A and $\alpha \setminus A$ are α-RE. In terms of definability, a set is α-recursive if and only if it is $\Delta_1(L_\alpha)$. It is α-finite if and only if it is α-recursive and bounded in α.

All the basic results in classical recursion theory, for example those covered in the first seven chapters of Rogers [1967], hold for all Σ_1 admissible ordinals. Thus, a set $K \subset \alpha$ is RE if and only if it is the range of a total α-recursive function; there is an effective (i.e. $\Sigma_1(L_\alpha)$ definable) enumeration of all α-finite sets and all α-RE sets; Kleene's Recursion Theorem is true for each α. We denote by W_e the eth α-RE set and by K_e the e-th α-finite set under the respective effective enumerations.

Reducibility. The notion of reducibility provides a means of comparing the relative complexity of subsets of α. Given $A \subset \alpha$, define by the collection of neighborhood conditions of A the set

$$N(A) = \{(c,d) \mid K_c \subset A \ \& \ K_d \subset \alpha \setminus A\}.$$

We say that A is α-RE in $B \subset \alpha$ if there is an e such that for all $x < \alpha$,

$$x \in A \leftrightarrow (\exists c)(\exists d)\big[(x,c,d) \in W_e \ \& \ (c,d) \in N(B)\big].$$

A is *weakly α-recursive in* B, written $A \leqslant_{w\alpha} B$, if A and $\bar{A}$ are α-RE in B. Define by A^* the set $\{u \mid K_u \subset A\}$. Then A is *α-recursive in* B, written $A \leqslant_\alpha B$, if A^* and $\bar{A}^*$ are α-RE in B.

Thus A is α-recursive in B provided there is an algorithm such that for any given α-finite set K, it is possible to use the algorithm, with B as an oracle, to conclude within α-finite time whether K is a subset of A or disjoint from A. It is not difficult to verify that $\leqslant_\alpha$ is reflexive and transitive. A and B are said to have the same α-degree, written $A \equiv_\alpha B$, if $A \leqslant_\alpha B$ and $B \leqslant_\alpha A$. Although $\leqslant_{w\omega}$ is equivalent to $\leqslant_\omega$ and therefore transitive, $\leqslant_{w\alpha}$ is not transitive in general.

The least complicated α-degree, which we denote by $\mathbf{0}$, is the α-recursive degree, which consists of α-recursive sets. A degree is an α-RE degree if it contains an α-RE set. There is a *greatest* α-RE degree $\mathbf{0}'$ which contains the α-RE set $\emptyset' = \{(x,e) \mid x \in W_e\}$, in which every α-RE set is α-recursive.

There is an analog of Church's thesis for α-recursion theory which we shall appeal to in this article. This thesis allows a more informal presentation of the topics to be covered, emphasizing intuition over formalism.

The key motivation of ordinal recursion lies in the search for a "generalized" recursion theory. It is evident that the notion of effective computation applies to a wider class of mathematical structures, as exemplified in Kleene's work on ordinal notations (Kleene [1938]). Kreisel and Sacks [1965] initiated the study of recursion theory on Church–Kleene ω_1, and this led to the subsequent introduction of the theory of Σ_1 admissible ordinals by Sacks and his school.

A closer examination reveals that if the full replacement axiom is assumed in L_α, then many difficult proofs in classical recursion theory go through almost routinely, without major modifications of the classical construction (there are exceptions: cf. the section on maximal sets). From the point of view of effective computation, where "Σ_1"-ness is identified with "effectively enumerable", it should be sufficient to assume only Σ_1-replacement axiom to arrive at a satisfactory recursion theory (though all is not lost even when this crucial assumption is removed in β-recursion theory, see Section 2). This view is supported by the successful solution of Post's problem for all admissible ordinals (Theorem 2 below).

We give here some examples of admissible ordinals:

(a) $\alpha = \omega$, the classical case. Then α is Σ_n-admissible for all $n < \omega$. The same conclusion holds for any regular constructible cardinal;

(b) $\alpha = \omega_1^{CK}$, Church–Kleene ω_1. Here α is Σ_1 but not Σ_2-admissible. There is also a $\Sigma_1(L_\alpha)$ map from α into ω (ω is called the Σ_1-projectum of α). A subset of ω is α-RE if and only if it is Π_1^1 definable;

(c) $\alpha = \delta_2^1$, the least ordinal which is not the order type of a Δ_2^1 set of natural numbers. In this case a set of natural numbers is α-RE if and only if it is Σ_2^1 definable;

(d) $\alpha = \aleph_\omega^L$, the ωth constructible cardinal of L. There is a $\Sigma_2(L_\alpha)$ cofinal map from ω into α. Hence α is not Σ_2-admissible. On the other hand, every infinite cardinal in a well-founded model of ZF is Σ_1-admissible.

1.1. *Regularity*

A set $A \subset \alpha$ is *regular* (in α) if its restriction to every $\gamma < \alpha$ is α-finite. It follows that every set of natural numbers is regular. On the other hand, in ω_1^{CK}, Kleene's $\mathcal{O}$, a complete Π_1^1-set of natural numbers, is not regular (in ω_1^{CK}), even though it is bounded and ω_1^{CK}-RE. Non-regularity is a major feature which distinguishes ordinal recursion theory from the classical theory. Non-regular α-RE sets are sets with bounded parts which cannot be enumerated in α-finite time. Their existence renders some of the standard techniques ineffective. Nevertheless, at least for α-RE sets and for the study of α-degrees, this difficulty can be circumvented:

THEOREM 1 (Sacks [1966]). *Let α be admissible. Then every α-RE degree contains a regular α-RE set.*

Maass [1978a] showed that there is a parameter free Σ_1 function f such that for any α and $e < \alpha$, W_e and $W_{f(e)}$ have the same α-degree and $W_{f(e)}$ is regular.

1.2. *Definability*

Jensen's work on the fine structure of L (see Jensen [1972]) turns out to be a key component in the development of ordinal recursion theory, a development which arguably exemplifies the successful integration of set-theoretic and recursion-theoretic ideas. In retrospect, the secret to the solutions of such basic problems as Post's problem, Sacks Splitting Theorem, and the Density Theorem for all admissible α, rests on the insight that the complexities of the classical constructions, with the intervention of fine structure theory, may be refined to achieve the goals, provided that one chooses the appropriate definable objects within α to carry out the necessary priority arguments. On the other hand, in certain cases where such an approach fails, it is shown that the problems being considered have negative solutions. Problems such as the existence of maximal sets, and ordering of α-degrees above $\mathbf{0}'$, are examples.

We list here several important objects in fine structure theory that play pivotal roles in ordinal recursion. Let $B \subset \alpha$.

The Σ_n-cofinality of (L_α, B) is defined to be the least γ for which there is a $\Sigma_n(L_\alpha, B)$ function from γ cofinally into α. We denote this ordinal by $\kappa_n(B)$, or simply write it as Σ_n-cofinality (α, B). Clearly $\kappa_n(\emptyset) = \alpha$ if and only if α is Σ_n-admissible.

The Σ_n^B-projectum of α, denoted $\alpha_n^*(B)$ (or sometimes Σ_n-projectum (α, B)), is the least ordinal $\gamma \leqslant \alpha$ for which there is a $\Sigma_n(L_\alpha, B)$ map from α into γ. If $B = \emptyset$, Jensen's theory provides several characterizations of this ordinal: (a) it is the largest limit ordinal less than or equal to α in which every bounded $\Sigma_n(L_\alpha)$ set is α-finite; (b) it is the least ordinal γ for which there is a $\Sigma_n(L_\alpha)$ map from a subset of γ onto α. When B is α-RE and regular, the above characterization continues to hold with $\Sigma_n(L_\alpha)$ replaced by $\Sigma_n(L_\alpha, B)$, given and used in Shore's proof of the Density Theorem (Shore [1976a]).

We use the notations α_n^* and κ_n when B is empty. When $n = 1$, we omit the subscript 1 and simply write α^* and $\alpha^*(B)$ instead. In α-recursion theory, it is important to present the set of requirements with as short a list as possible. The ordinal α^* or $\alpha^*(B)$ are often used for this purpose.

We say that $\lambda < \alpha$ is an α-cardinal if there is no α-finite injection of λ into a smaller ordinal. If $\alpha_n^* < \alpha$ (or if $\kappa_n(B) < \alpha$), it is not difficult to prove that it is an α-cardinal. And $\alpha^* < \alpha$ implies that it is the greatest α-cardinal.

A set B is *hyperregular* if $\kappa_1(B) = \alpha$. In other words, B is hyperregular if (L_α, B) is a Σ_1-admissible structure. It is not difficult to verify that every α-recursive set is hyperregular. However, there are α-RE sets which do not satisfy hyperregularity. As an example, consider the set B of non-cardinals in $\alpha = \aleph_\omega^L$. This is an α-RE set whose complement is of order type ω. Then $f(n) = n$th member of $\alpha \setminus B$ is a function weakly α-recursive in B, mapping ω unboundedly into α. It turns out that for $\alpha = \aleph_\omega^L$, the only nonhyperregular α-RE set is of complete degree $\mathbf{0}'$, and every set of α-degree above this is nonhyperregular, while any set which does not compute B defined above is hyperregular (under the axiom of constructibility). In particular, every incomplete α-RE set is hyperregular.

Hyperregularity is a strong condition which ensures that computations carried out on α-finite sets using oracles are completed in α-finite steps. Its recursion-theoretic consequences are significant: for example, for the α considered above, there is no incomplete α-RE set whose "jump" (an analog of the classical notion) is strictly above $\emptyset'$ (hence no incomplete "high set") (Shore [1976b]). On the other hand, many tools, such as priority arguments, are not relativizable to nonhyperregular sets. This introduces additional complications to the study of α-recursion theory. Different techniques are needed in many cases and, in the most extreme case, nonhyperregularity leads to radically different degree-theoretic results (see Section 1.5 below).

1.3. *The α-finite injury priority method*

The method of finite injury priority argument introduced by Friedberg and Muchnik to solve Post's problem marked the advent of modern recursion theory. This method has since been joined by a variety of highly complex and ingenious techniques invented to handle problems about RE sets and their degrees, of which the Friedberg–Muchnik proof is now seen to be the simplest. Solving Post's problem may indeed be considered the first important test for any reasonable ordinal recursion theory.

THEOREM 2 (Friedberg–Muchnik Theorem). *Let α be an admissible ordinal. There exist α-RE sets A and B with incomparable α-degrees.*

COROLLARY (Solution of Post's Problem). *There exists an incomplete, α-RE and non-α-recursive degree.*

We sketch the solution by Sacks and Simpson [1972]. This is a proof which has a strong model-theoretic and set-theoretic flavor, in contrast to that of Lerman [1972] which has a stronger recursion-theoretic tilt (it is worth noting that Lerman's approach may be refined to provide a parameter free construction of the sets A and B, yielding a uniform solution, for all admissible ordinals, to Post's problem (Lerman, unpublished)).

PROOF OF THEOREM 1. Consider requirements of the type R_e: $\{e\}^A \neq B$ and those with the roles of A and B reversed. The basic strategy is to diagonalize against equalities whenever possible while preserving computations, respecting requirements of higher priority if and when necessary. The strategy succeeds in the classical theory because (a) for any e_0, there is a stage after which no requirement of higher priority than e_0 gets injured, and (b) it can be established that each requirement R_e gets injured at most finitely many (indeed 2^e) times. Closer inspection shows that (a) is essentially a Σ_2 condition, and when satisfied, is sufficient to derive (b). What the construction demands then is for a Σ_1-admissible ordinal α to perform a Σ_2 task. (We will see later that with the Density Theorem, the required task is even more onerous—at almost the Σ_3 level.)

Since α is in general not Σ_2-admissible, a straightforward adaptation of the original approach will clearly fail. Instead, the following two lemmas provide an insight into how the difficulties may be overcome:

LEMMA (Sacks–Simpson Lemma). *Let κ be a regular α-cardinal. Suppose that K is an α-finite set of α-cardinality less than κ such that $\{I_d \mid d \in K\}$ is a simultaneous*

α-RE sequence of α-finite sets each of which has α-cardinality less than κ. Then $\bigcup_{d \in K} I_d$ is α-finite and of α-cardinality less than κ.

An ordinal $\sigma < \alpha$ is said to be α-stable if L_σ is a Σ_1-elementary substructure of L_α.

LEMMA (α-Stability Lemma). *If $\omega < \alpha = \alpha^*$, then it is a limit of α-stable ordinals.*

The first step to the solution is to provide a short indexing of requirements with its associated list of priorities. There are several cases to consider. First suppose that α has a greatest α-cardinal κ.

(a) $\kappa = \alpha^* < \alpha$. In this case we use the Σ_1-projectum α^* of α to provide a list of requirements. Let p be a one-one α-recursive map from α into α^*. Requirement R_d is said to have higher priority than requirement R_e if $p(d) < p(e)$. This shorter list of indices ensures that every α-RE set bounded in α^* is α-finite, an essential feature that is needed during the inductive stage to verify that every requirement is satisfied.

Now commence with the construction using the revised indexing of requirements. For each $e < \alpha$, let I_e denote the injury set (defined in the usual sense) associated with R_e. The main observation here is that if there is a regular α-cardinal $\rho \leqslant \alpha^*$ such that $p(e) < \rho$ and $p(d) < p(e)$ implies that I_d has α-cardinality less than ρ, then the Sacks–Simpson Lemma ensures that I_e has α-cardinality less than ρ as well. This is sufficient to show that the requirement with the highest priority after that R_e is injured less than ρ times. Induction hypothesis then allows one to conclude that every requirement is eventually satisfied.

By the α-Stability Lemma 2.4, let β be the order type of α-stable ordinals above κ.

(b) $\alpha^* = \alpha$. If $\beta = \alpha$ then we use the identity function for priority listing, and modify the classical construction slightly. Lemma 2.4 and the construction provides the necessary tool to argue that every requirement settles down before the next α-stable ordinal. If $\beta < \alpha$, then there is a $\Sigma_2(L_\alpha)$ bijection p between α and $\kappa \cdot \beta$ (since the property of "being α-stable" is $\Pi_1(L_\alpha)$). We use $\kappa \cdot \beta$ to index the requirements and say that R_d has higher priority than R_e if $p(d) < p(e)$. The positions of the priorities are given by an α-recursive approximation p' of p. This gives meaning to "the priority of R_e at stage σ is ν". The "final priority" of R_e is then $p(e)$, which is the limit of $p'(\sigma, e)$ as σ tends to α.

Construction proceeds as before, using p' to guide the priority ordering at each stage. The rules governing the injury of requirements in order of priority at each stage are observed. Exploiting the property of Σ_1-stability, coupled with Lemma 2.4, ensures that all requirements of priority at least $\kappa \cdot \nu$ settle down by the $(\nu + 1)$th α-stable ordinal.

Finally, if α is a limit of α-cardinals (analogous to ω), one uses an indexing provided by the identity function on α. The argument then proceeds as in Case (a).

1.4. *The Density Theorem*

THEOREM 3 (Shore [1976a]). *Let* $\mathbf{b} < \mathbf{c}$ *be* α*-RE degrees. Then there is an* α*-RE degree* $\mathbf{a}$ *such that* $\mathbf{b} < \mathbf{a} < \mathbf{c}$.

This theorem is one of the first successful liftings of infinite injury priority argument to ordinal recursion theory. We sketch the key ideas here. Fix $B <_\alpha C$ to be regular α-RE sets (Theorem 1). An α-RE set A of intermediate α-degree is to be constructed.

LEMMA (Shore Incompleteness Lemma). *Suppose* B *is an incomplete* α*-RE set. Then* $\kappa_1(B) \geqslant \alpha^*(B)$. *Furthermore there is a* $\Sigma_1(L_\alpha, B)$ *map from* $\kappa_1^*(B)$ *onto* α.

PROOF. We sketch the proof of the first half of the lemma. Assume $\kappa_1^*(B) < \alpha^*(B)$. Let D be a regular α-RE set. We show that $D \leqslant_\alpha B$. Fix $g : \kappa_1(B) \to \alpha$ to be cofinal. Consider

$$K = \{(\gamma, \delta) \mid D \cap g(\gamma) \subset D^{g(\delta)}\}.$$

Now K is a $\Pi_1(B)$ set bounded below $\alpha^*(B)$, and so is α-finite. Using it as a parameter set, we see that $D \leqslant_\alpha B$. Hence B is complete. □

The lemma says essentially that if B is incomplete, then (L_α, B) is a *weakly admissible* structure. Weak admissibility allows many $\Sigma_2(B)$ constructions, with suitable modifications, to go through (for example, Post's problem in β-recursion theory, cf. Section 2).

There are essentially three key ingredients used in the proof of Theorem 3: the use of $\alpha^*(B)$ for a sufficiently short listing of the set of requirements; the exploitation of the blocking technique, in which requirements are grouped into $\kappa_2(B)$ many blocks of the same priority; and the use of $\kappa_1(B)$ and its associated cofinal function to measure lengths of agreements between computations in the course of the construction. We elaborate the points below.

(a) Let $p \leqslant_{w\alpha} B$ be an injection from α into $\alpha^*(B)$. There is a simultaneous α-recursive approximation $\{p_\sigma\}$ of p such that for all x, $p_\sigma(x) = p(x)$ for all sufficiently large σ. Requirements are given a short list of length $\alpha^*(B)$ using p. The principal feature of this ordinal exploited in the proof of the Density Theorem is that every set α-RE in B and bounded below $\alpha^*(B)$ is α-finite.

(b) In the construction there are altogether $\kappa_2(B)$-many blocks of requirements. Requirements in the same block are accorded the same priority. This reduces at once the number of injury sets to a manageable level. During verification step, one does induction on $z < \kappa_2(B)$, and argues first of all that the set of permanent injuries inflicted on the computations is bounded within each block, and secondly that such a bound may be found in a $\Sigma_2(L_\alpha, B)$ manner as a function of z. The fact that $z < \kappa_2(B)$ then ensures that a uniform bound exists for all blocks $z' \leqslant z$.

(c) $\kappa_1(B)$ is also known as the *recursive cofinality of* B. Let $k:\kappa_1(B) \to \alpha$ be a cofinal map weakly α-recursive in B. There is a simultaneous α-recursive sequence of α-recursive functions $\{k_\sigma\}$ such that for each $y < \kappa_1(B)$, $k_\sigma|y = k|y$ for all sufficiently large σ. By the Shore Incompleteness Lemma we may choose k to be a surjective map. The calculations of lengths of agreement between two computations will be based on $k|y$, for $y < \kappa_1(B)$. Furthermore, the surjectivity of k allows the construction to pick up every α-finite set contained in C. During the construction, such sets are coded into A (which in turn causes complications). Each of these strategies is designed to ensure that should B be able to compute A, or A compute C, then in fact $C \leqslant_\alpha B$, a contradiction.

There are two types of requirements. The positive requirements $\{e\}^B \neq A$ for each e, which attempts to ensure that the set A to be constructed is not α-recursive in B, and negative requirements $\{e\}^A \neq C$ for each e, which arranges that C is not α-recursive in A. These requirements are grouped into blocks indexed by $\kappa_2(B)$ with the aid of the following lemma. Denote $B^{<\sigma}$ to be the set of ordinals enumerated in B before stage σ. Assume $\kappa_1(B) > \omega$. Then $B^{<\sigma} = B \cap \sigma$ (i.e. σ is B-correct) for unboundedly many σ.

LEMMA (Blocking Lemma). *There is a function* $g:\kappa_2(B) \to \alpha^*(B)$ *which is* $\Sigma_2(L_\alpha, B)$, *together with a simultaneous* α*-recursive sequence* $\{g_\sigma\}$ *of* g *such that*

(a) $g_\sigma(z) \geqslant g(z)$ *for all sufficiently large* σ;

(b) *for all* $z < \kappa_2(B)$, *and for all sufficiently large* B*-correct* σ, $g_\sigma|z = g|z$.

We say that $\{g_\sigma\}$ is a *tame approximation* of g in view of the Blocking Lemma. We shall only consider $\kappa_1(B) > \omega$ here. The case when $\kappa_1(B) = \omega$ is considerably simpler. We say that a requirement with index e is in block z if $p(e) < g(z)$. With the blocking lemma, it makes sense using $\{g_\sigma\}$ to say that a requirement is "in block z at stage σ". Indeed by the Blocking Lemma, if a requirement is in block z, then it is in block z for all sufficiently large B-correct σ. Furthermore, by tameness property, this occurs uniformly in z,

As in the classical construction, we code the set B into the even part of A, and think of the odd part of A as consisting of triples (z, x, σ). The three ordinals are related via a length of agreement function: suppose at stage σ there is an e in block z, with $A^{<\sigma}|k_\sigma(x)$ agreeing with $\{e\}_\sigma^{B^{<\sigma}}|k_\sigma(x)$. Then ordinal $k_\sigma(x)$ is said to be the length of agreement of computation for e at stage σ.

This length of agreement is destroyed (i.e. computation restarts) at a later stage if new elements below σ enter either A or B, since such occurrences are likely to invalidate any computations reached so far. Those agreements which are never destroyed are called permanent. They turn out to be α-recursively identifiable by B. Precaution is taken so that if (z, x, σ) enters A, then $K_{k_\sigma(x)}$, the $k_\sigma(x)$th α-finite set, is contained in C. Since C is regular, this can be verified at some stage. And since k is a surjective map, all the relevant α-finite sets will be considered at some stage. The objective here is to code enough of C into A so as to obtain the following lemma:

LEMMA.
(i) *For each e in block z,* $\{e\}^B \neq A$.
(ii) *Within each block, the permanent lengths of agreement are bounded below* α.

PROOF. The idea is that if (i) fails, then using B to identify permanent lengths of agreement, one is able to compute C (which are coded in A) from B, a contradiction.

A special feature of the blocking technique is that requirements within the same block work together to achieve collectively a longer length of agreement. To prove (ii), one uses the fact that the set K of e's in block z for which $\{e\}^B$ is total on $k(x)$ and not equal to $A|k(x)$ after an agreement had been reached earlier, is a set Σ_1 in B and bounded below $\alpha^*(B)$, hence α-finite. It is then sufficient to consider only $e \in z \setminus K$. Repeating an argument similar to that for (i) above on the set $z \setminus K$, but this time collectively on all the computations that provide permanent lengths of agreement, shows that if (ii) is false, then again $C \leqslant_\alpha B$. □

Consider requirements e in block z. A negative requirement is intended to preserve computations of the form $\{e\}^{A^{<\sigma}}|k_\sigma(x)$ to make it different from $C|k_\sigma(x)$. At stage σ, each requirement e is assigned a marker which is placed at the least ordinal $\nu_{e,\sigma}$ greater than the negative facts used about $A^{<\sigma}$ in the computation above. The idea is that for as long as markers stay, then no new ordinal below their positions is allowed to enter A. On the other hand, should a new element below σ enter B at a later stage, then all markers assigned at stage σ are removed, clearing the way for ordinals below $\nu_{e,\sigma}$ to be added to A if and when necessary. These markers may reappear subsequently (say at $\zeta > \sigma$) occupying different positions provided that, roughly speaking, there is a *collective* length of agreement between $C^{<\zeta}$ and $\{\{e\}_\zeta^{A^{<\zeta}}\}$, e in block z, longer than those achieved before.

The construction of the set A involves the coding of B into the even part of A (to ensure $B \leqslant_\alpha A$), and the manipulation of positive and negative requirements. A negative requirement (marker) is permanent if it is never removed. The set of permanent negative requirements within a block has to be bounded else one argues that C is α-recursive in B. Furthermore, it can be arranged that within a block, the limit inferior of the positions of the markers that stay behind at the end of each stage of construction is bounded below α, and may be computed from C. This allows unboundedly many opportunities for ordinals above certain level to enter A, and is crucial to the success of the construction. With this it is also possible to show that C is not α-recursive in A.

The final thread is to establish $A \leqslant_\alpha C$. This is achieved through arranging the construction so that the set of permanent negative requirements is α-recursive in C. We omit the details.

1.5. *Non-existence of maximal sets*

In this and the next section, we give two examples of problems which have negative solutions in ordinal recursion. The first, due to Lerman [1974], states roughly that there is a lattice-theoretic property of α-RE sets which is inherently definably countable. More precisely,

THEOREM 4. *There is a maximal α-RE set if and only if there is a function f which is S_3-definable mapping α onto ω.*

The notion of maximality is derived from the classical one: M is maximal if and only if its complement $\overline{M}$ is unbounded, and there is no α-RE set which splits $\overline{M}$ into two non-α-finite parts. We say that f is S_3-definable if there is an α-recursive function f' such that for all $x < \alpha$,

$$\lim_{\tau} \lim_{\sigma} f'(\tau, \sigma, x) = f(x).$$

Thus in our terminology, we may say that there is a maximal α-RE set if and only if the S_3-projectum of α is ω. This complete characterization of the existence of maximal sets raises a very interesting but apparently quite difficult question: is there a classification of recursion-theoretic problems which are inherently linked to the cardinality of the universe?

The following weak form of Theorem 4 shows how the size of α has a bearing on the existence of maximal sets:

THEOREM 5. *If there is a maximal α-RE set, then α is countable.*

To prove this theorem, we consider κ_2 which is Σ_2-cofinality (α), and α_2^*, the Σ_2-projectum of α.

PROOF OF THEOREM 5. Let M be a maximal set. We first claim that $\kappa_2 \geqslant \alpha_2^*$. To do this, build a simultaneous α-recursive sequence of pairwise disjoint α-finite sets $\{H_\nu\}_{\nu<\kappa_2}$ such that $\overline{M} \cap (\bigcup_{\nu<\kappa_2} H_\nu)$ is not α-finite, and each H_ν contains at most one member of $\overline{M}$. Now the set

$$K = \{\nu \mid H_\nu \cap \overline{M} \neq \emptyset\}$$

is a Σ_2 definable subset of κ_2. If $\kappa_2 < \alpha_2^*$, then K is α-finite, in which case it is possible to split K into two non-empty parts K_1 and K_2 so that $\bigcup_{\nu\in K_1} H_\nu$ and $\bigcup_{\nu\in K_2} H_\nu$ each contains a non-α-finite unbounded subset of $\overline{M}$, contradicting maximality of M. Thus $\kappa_2 \geqslant \alpha_2^*$.

Next we argue that κ_2 is in fact countable. To do this, let β be the order type of $\overline{M}$. Partition α into an α-recursive sequence of pairwise disjoint α-RE sets $\{A_\nu\}_{\nu<\kappa_2}$. Define

$$B_\nu = \{\gamma \mid \exists\sigma[\text{order type of } \gamma \setminus M^\sigma] \in A_\nu\}.$$

It can be shown that for each ν, unboundedly many members of $\overline{M}$ belongs to B_ν. By maximality, $\overline{M} \setminus B_\nu$ is α-finite for all $\nu < \kappa_2$. Let $h(\nu)$ be the supremum of this α-finite set. We claim:

For each $y \in \overline{M}$, there are only finitely many ν's such that $h(\nu) < y$.

Fix a $y \in \overline{M}$. Suppose there are infinitely many ν's such that $h(\nu) < y$. This means that $y \in B_\nu$ for each of these ν's. Since the A_ν's are pairwise disjoint, y must have entered the B_ν's at different stages σ exhibiting infinitely many different order types for $y \setminus M^\sigma$. But this contradicts the well-ordering of ordinals. This proves the claim.

It follows from the claim that κ_2 and hence α_2^* is countable. We conclude that α is countable. □

1.6. *Post's problem above $\emptyset'$ and set-theoretic methods*

The second example in the negative direction concerns α-degrees above $\mathbf{0}'$. We discuss how Silver's work on singular cardinals of uncountable cofinality when merged with Jensen's theory is exploited to derive a strong structural difference in degree theory for a class of admissible ordinals. Further applications are discussed in Section 2.

The problem to consider is simple: does Post's problem hold above any α-degree? In other words, for any set A, do there exist sets B and C RE in A such that $A <_\alpha B$ and $A <_\alpha C$, and B, C have incomparable α-degrees? A related, and more general, question asks if there exist incomparable α-degrees above any given degree. A basic theorem of Kleene–Post states that this holds when $\alpha = \omega$. For $\alpha = \aleph_{\omega_1}$, the answer turns out to be negative in a very strong way:

THEOREM 6 (Friedman [1981a]). *Assume $V = L$. If $\alpha = \aleph_{\omega_1}$, then the α-degrees above $\mathbf{0}'$ are well-ordered, with successor provided by the jump operator.*

PROOF. A complete proof requires a heavy dose of Jensen's fine structure theory. We give a sketch here of the proof of the easy half. Given $A, B \geqslant_\alpha \emptyset'$, define the growth function g_A of A so that $g_A(\delta)$ is the least ordinal u such that $A \cap \aleph_\delta \in L_u$. Define g_B similarly. Then either $g_A(\delta) \geqslant g_B(\delta)$ for stationarily many δ, or $g_A(\delta) < g_B(\delta)$ for closed and unboundedly many δ. Silver's analysis [1974] of growth functions shows that in the former case $A \leqslant_\alpha B$, while in the latter case $A >_\alpha B$. As a consequence, if $A <_\alpha B$, then $g_A(\delta) < g_B(\delta)$ for a closed and unbounded set of δ's. Using this, the well-ordering property follows from the observation that a countable intersection of closed and unbounded sets is closed and unbounded. Hence a countable descending chain of α-degrees above $\emptyset'$ has a least element.

In Friedman [1981a] it is shown that the well-ordering of α-degrees above $\mathbf{0}'$ is actually achieved through the jump operator, and these α-degrees are represented by "master codes" in Jensen's sense.

The situation for countable cofinality turns out to be radically different. Harrington and Solovay have independently shown that incomparable α-degrees exist above $\mathbf{0}'$ for $\alpha = \aleph_\omega^L$. The following result (Chong and Mourad [in preparation]) solves Post's problem above $\mathbf{0}'$:

THEOREM 7. *Let $\alpha = \aleph_\omega^L$. Then there exist sets A and B, α-RE in and above $\emptyset'$, which are of incomparable α-degree.*

Since such sets A and B are necessarily nonhyperregular, the classical approach of finite injury argument no longer applies. Instead, a refinement of the method first used in establishing the Friedberg–Muchnik Theorem for $B\Sigma_1$-models of arithmetic (Chong and Mourad [1992]), called unions of intervals, is exploited to ensure that all requirements are met within ω-steps. Since A and B lie above $\emptyset'$ and are therefore able to "climb up" α in ω-many steps, the construction succeeds.

1.7. *Applications to fragments of Peano arithmetic*

One of the most interesting applications of techniques of α-recursion theory in recent years has been in the area of *reverse recursion theory*. Starting with the basic axioms of Peano arithmetic without the induction scheme, one asks:

> What is the proof-theoretic strength of a given theorem in recursion theory? In particular, how much of the induction scheme is required to prove the theorem?

Kirby and Paris [1978] have provided a hierarchy of theories of increasing proof-theoretic strength, and this hierarchy forms the basis for the study of subrecursive recursion theory. Let P^- be axioms of Peano arithmetic with exponentiation but without the induction scheme. Let $I\Sigma_n$ denote the induction scheme for all Σ_n formulas, and $B\Sigma_n$ to be replacement (collection) axiom for Σ_n formulas: every Σ_n-function maps a "finite set" (in the sense of the given model) onto a "finite set". Then with P^- as the underlying theory, one has ($n \geqslant 0$) $B\Sigma_{n+1}$ to be strictly stronger than $I\Sigma_n$, which is in turn strictly stronger than $B\Sigma_n$.

Slaman and Woodin [1989] initiated the study of recursion theory on fragments of Peano arithmetic. We illustrate here how techniques of ordinal recursion theory are adapted to investigate problems in this area.

THEOREM 8 (Chong and Mourad [1992]). *$P^- + B\Sigma_1$ proves the Friedberg–Muchnik theorem.*

PROOF. Simpson (unpublished) observed that $I\Sigma_1$ was sufficient to verify that the standard construction works. Thus let $\mathcal{M}$ be a model of $P^- + B\Sigma_1$ in which Σ_1-induction fails. There is then a cofinal $\Sigma_1(\mathcal{M})$ map f defined on a $\Sigma_1(\mathcal{M})$-definable "cut" X. This map f on $\mathcal{M}$ acts very much like a Σ_2-cofinal function of $\aleph_\omega^L$ (with domain ω), or indeed a Σ_1-cofinal map on a rudimentarily closed

β which is not admissible (β-recursion theory in Section 2). The idea now is to treat $\mathcal{M}$ as having "cofinality X" (so that $\mathcal{M} = \bigcup_{t \in X} M_t$, and $M_t \subset M_{t+1}$), and construct a "Friedberg–Muchnik pair" by satisfying the requirements successively within each M_x. □

The following example shows how the methods of Shore [1976b] is applied.

THEOREM 9 (Mytilinaios and Slaman [1988]). *$P^- + B\Sigma_2$ does not prove the existence of an incomplete high RE set.*

PROOF. There is a model $\mathcal{M}$ of $P^- + B\Sigma_2$ in which Σ_2-induction fails (with ω as the domain of a $\Sigma_2(\mathcal{M})$-cofinal function f), and in which every real is "coded" (meaning it is the initial segment of a "finite" set). The function f is recursive in $\emptyset'$. If A is an incomplete RE set in $\mathcal{M}$, then for each $n \in \omega$, there is a least $g(n)$ such that $e \in A' | f(n)$ if and only if $\{e\}^{A^{g(n)}}_{g(n)}(e)\downarrow$, else A will be complete. Now $n \mapsto g(n)$ is coded, and so one may use it to compute A' from $\emptyset'$. □

Chong and Yang ([ta] and [1997]) have recently shown that the existence of a maximal set, as well as that of an incomplete high set, is equivalent to $P^- + I\Sigma_2$. In general, just as for α-recursion theory, infinite injury priority method is less well understood. Groszek, Mytilinaios and Slaman [1996] have recently shown that $P^- + B\Sigma_2$ proves the Density Theorem. The proof-theoretic classification of this theorem is not known.

We refer the reader to Chong [1984] and Sacks [1990] for more complete treatments on α-recursion theory.

2. β-recursion theory

Studying the global structure of the α-degrees clearly exposes the need to deal with failures of admissibility: even though an ordinal is admissible it may fail to be relative to a set whose degree we wish to analyze. Indeed, the main thrust of the work in α-recursion theory has been to demonstrate that recursion-theoretic constructions from classical recursion theory which seem to require a large amount of admissibility, say Σ_2 or even Σ_3, can actually be refined so as to succeed with only the assumption of Σ_1-admissibility. In view of this it is natural to ask: can the assumption of Σ_1-admissibility be eliminated?

However on hindsight it is fair to say that a stronger motivation for the development of β-recursion theory was to find new applications of the beautiful work of Jensen [1972] on the fine structure of L, to ordinal recursion theory. Jensen's work ignores admissibility distinctions but concentrates only on iterations of the jump operator ("master codes"); β-recursion theory extends his idea to degree theory in general.

The basic notions in β-recursion theory are defined using Jensen's hierarchy for L, the J_α-hierarchy, which enjoys the following properties:

(a) $J_0 = \emptyset$, $J_{\alpha+1} \cap P(J_\alpha) =$ Definable subsets of J_α (with parameters), $J_\lambda = \bigcup\{J_\alpha \mid \alpha < \lambda\}$ for limit λ.

(b) J_α obeys $\Sigma_0(J_\alpha)$-comprehension and is closed under pairing.

Of course the improvement over the L_α-hierarchy is closure under pairing. Unfortunately $J_\alpha \cap \mathrm{ORD}$ is $\omega\alpha$ and not α. So we define, for limit β: $S_\beta = J_\alpha$ where $\beta = \omega\alpha$. β-recursion theory takes place on the set S_β.

The notions Σ_n-cofinality and Σ_n-projection apply to β as they do in the admissible case: Σ_n-cofinality $(\beta) =$ least γ such that there is a cofinal $f : \gamma \to \beta$ which is $\Sigma_n(S_\beta)$; Σ_n-projectum $(\beta) =$ least γ such that there is a one-one $f : \beta \to \gamma$ which is $\Sigma_n(S_\beta)$. These are either equal to β or are β-cardinals (cardinals in the sense of S_β). An important result of Jensen [1972] states that Σ_n projectum (β) is also the least γ such that some $\Sigma_n(S_\beta)$ subset of γ is not an element of S_β.

We are ready to define the basic notions of β-recursion theory. As in α-recursion theory, $A \subseteq S_\beta$ is β-recursively enumerable, β-recursive, β-finite if and only if A is $\Sigma_1(S_\beta)$, $\Delta_1(S_\beta)$, an element of S_β, respectively. However when β is inadmissible (i.e. Σ_1 cofinality $(\beta) < \beta$), a new and stronger notion of β-RE (β-recursively enumerable) arises: A is *tamely* β-RE if $A^* = \{x \in S_\beta \mid x \subseteq A\}$ is β-RE. This is equivalent to saying that A is the union of a β-recursive sequence $\langle A^\sigma \mid \sigma < \beta\rangle$ with the property that if $x \subseteq A$ and β-finite then $x \subseteq A^\sigma$ for some $\sigma < \beta$.

The weak and strong reducibilities $\leqslant_{w\beta}$, $\leqslant_\beta$ are defined as they are in α-recursion theory: one way of achieving these definitions is through the use of "neighborhood conditions": define $N(A) = \{\langle x, y\rangle \mid x, y$ are β-finite, $x \subseteq A$, $y \subseteq \bar{A}\}$. B is β-RE in A if for some β-RE W, $x \in B$ if and only if $\exists z \in N(A)\,[(x, z) \in W]$. Then $B \leqslant_{w\beta} A$ if and only if B, $\overline{B}$ are both β-RE in A, and $B \leqslant_\beta A$ if and only if B^* and $\overline{B}^*$ are both β-RE in A (if and only if B, $\overline{B}$ are both "tamely" β-RE in A). The strong reducibility $\leqslant_\beta$ is transitive.

Now some genuinely new phenomena arise in the inadmissible case, with regard to β-reducibility. These are summarized in the following result.

THEOREM 1 (Friedman [1979]). *Assume that β is inadmissible. Then there is a β-recursive set A such that*:

(i) $\emptyset <_\beta A <_\beta C$ *where C is a complete β-RE set.*

(ii) *Any tamely β-RE set and any β-recursive set is β-reducible to A.*

(iii) $C \leqslant_{w\beta} A$.

Thus β-recursiveness does not imply β-reducibility to $\emptyset$, and the complete β-RE set is weakly β-reducible to a β-recursive set!

It is easy to define A (in fact A can be taken to be a Δ_1 master code in the sense of Jensen [1972]). Let $f : \Sigma_1$-cofinality$(\beta) \to \beta$ be $\Sigma_1(S_\beta)$ and cofinal, and take $A = \{(e, x, \gamma) \mid x \in W_e$ by stage $f(\gamma)$, $\gamma < \Sigma_1$-cofinality $(\beta)\}$, where W_e is the eth β-RE set. Then A is β-recursive and since $x \notin W_e$ if and only if $\{e\} \times \{x\} \times \gamma \subseteq \bar{A}$ we get $C \leqslant_{w\beta} A$. The other properties are not difficult to verify.

The β-degree of A is referred to as $\mathbf{0}^{1/2}$ and serves as a new type of jump operator in β-recursion theory. Of course $\mathbf{0}^{1/2}$ provides an easy solution to a version of Post's

Problem in the inadmissible case; however it does not answer the following question, which has come to be adopted as the official version of Post's Problem in β-recursion theory.

POST'S PROBLEM. *Do there exist β-RE sets A, B such that $A \not\leqslant_{w\beta} B$, $B \not\leqslant_{w\beta} A$?*

As in α-recursion theory, Post's Problem has served as a driving force behind much of the work in β-recursion theory.

Early on it became apparent that with regard to questions such as Post's Problem the inadmissible ordinals divide into two very different classes. (This distinction occurred earlier in Jensen's proof of Σ_2 uniformization.) β is *weakly admissible* if Σ_1-cofinality $(\beta) \geqslant \Sigma_1$-projectum (β). Otherwise β is *strongly inadmissible*. In the former case many arguments from α-recursion theory can be adapted, for the following reason: if β is weakly admissible (but inadmissible) then there is a β-recursive bijection between S_β and Σ_1-cofinality (β). Moreover there is a β-recursive $A \subseteq \Sigma_1$-cofinality $(\beta) = \kappa$ which is a Δ_1 master code for S_β in Jensen's sense: $B \subseteq K$ is β-RE iff B is $\Sigma_1(L_\kappa, A)$. Thus β-recursion theory is closely related to κ-recursion theory, relativized to A and the structure (L_κ, A) is admissible. This is sufficient to reduce the solution to Post's Problem for β to the previously known (positive) solution for (L_κ, A). Exactly how much can be reduced from β to (L_κ, A) is analyzed in Maass [1978b].

The greater challenges in β-recursion theory arise in the strongly inadmissible case. Techniques from admissibility theory no longer apply; instead methods from combinatorial set theory are needed. The first attack on Post's Problem in the strongly inadmissible case appears in Friedman [1980].

THEOREM 2 (Friedman [1980]). *Suppose β has regular projectum: Σ_1-projectum (β) is regular with respect to β-recursive functions. Then Post's Problem has a positive solution.*

The proof uses an effective analog of Jensen's $\diamondsuit$-principle. We provide here a sketch of the proof, in the special case where Σ_1-projectum $(\beta) = \aleph_1^L$. We may assume that Σ_1-cofinality $(\beta) = \omega$ (else β is weakly admissible) but actually the proof makes no use of this.

We build β-RE $A, B \subseteq \aleph_1^L$ so as to meet the requirements R_e^A: $\overline{B} \neq W_e^A$ and R_e^B: $\bar{A} \neq W_e^B$, where of course W_e^A is the eth set β-RE in A. To achieve R_e^B we want an $x \notin A$ and a neighborhood condition $y \subseteq B$, $z \subseteq \overline{B}$ so that $(x, (y, t)) \in W_e$. One difference from the admissible case is that we may in fact have to actively guarantee $y \subseteq B$ as otherwise there may be no stage $\sigma < \beta$ where $y \subseteq B^\sigma$, due to the lack of tameness. It is possible however to arrange a weak form of tameness (through use of additional requirements) to insure that in fact $y - B^\sigma$ is countable at some stage, so we need only act to put a countable set into A or B for the sake of each requirement.

The second and most striking difference from the admissible case is that we act on each requirement *at most once*. What enables us to make this restriction is the following. Requirements can be listed in a sequence $\langle R_\delta \mid \delta < \aleph_1^L \rangle$ and as we are only putting countable sets into A or B to satisfy requirements there will be a closed unbounded set of requirements R_δ such that all action taken by R'_δ, $\delta' < \delta$ takes place below δ. Moreover, R_δ will only seek to protect ordinals $\geqslant \delta$ from entering A or B so will never be injured. If we arrange that each requirement appears as R_δ for a stationary set of δ's then each requirement will have the opportunity to act without injury. (So in fact this is not really an injury argument at all.)

Finally notice however that we have prohibited requirement R_δ from taking any action below δ; this requires that R_δ has a way of "guessing" at $A \cap \delta$, $B \cap \delta$. The necessary guesses are provided by Jensen's $\diamondsuit$-principle. We end this sketch with no more than a statement of $\diamondsuit$.

$\diamondsuit$-PRINCIPLE. *Suppose $E \subseteq \aleph_1^L$ is stationary. Then there exists $\langle G_\alpha \mid \alpha \in E \rangle$ such that:*

(a) *$G_\alpha \subseteq \alpha$ for $\alpha \in E$.*

(b) *If $A \subseteq \aleph_1^L$ then $\{\alpha \in E \mid A \cap \alpha = G_\alpha\}$ is stationary.*

(*In the general case of Theorem 2 we must weaken this somewhat but the general idea is the same.*)

The final case, where β is strongly inadmissible with *singular* projectum is entirely different. In fact Post's Problem may have a negative solution! We illustrate the result with a typical example: $\beta = \alpha \cdot \omega$, where $\alpha = \aleph_{w_1}^L$.

THEOREM 3 (Friedman [1978]). *Let C be the complete β-RE set. If A is β-RE then either $A \leqslant_\beta \emptyset$ or $C \leqslant_{w\beta} A$.*

The proof makes use of the work in Silver [1974] on the singular cardinal problem in set theory (as was for Theorem 5.1 in Section 1). We confine ourselves here to only a very rough sketch of the proof. The main idea is to look at *growth rates* for subsets of α. Specifically, suppose $A \subseteq \alpha$ is constructible and define $f_A(\alpha) =$ least δ such that $A \cap \aleph_\gamma^L$ belongs to L_δ. Then it can be shown that if $f_A(\gamma) \leqslant f_B(\gamma)$ for unboundedly many γ then in fact A is weakly β-reducible to B. This can be extended to $\beta = \alpha \cdot \omega$ to show that in fact any two subsets of β are $\leqslant_{w\beta}$-comparable. If C is the complete β-RE set then associated to C is a growth rate f which is the limit of β-finite growth rates f_n, $n \in \omega$. Thus either f_A is dominated by some f_n and is hence β-finite or f_A dominates f in which case $C \leqslant_{w\beta} A$. The uncountable cofinality of α is used both to apply Silver's work and to simultaneously bound the f_n's in this last argument.

3. The admissibility spectrum

Until now we have fixed an ordinal α (admissible or not) and studied definability for subsets of α. In this section we invert the process: fix a subset x of some cardinal κ, a theory T and consider the *T-spectrum* of $x = \Lambda_T(x) = \{\alpha \mid L_\alpha[x] \models T\}$.

Thus natural classes of ordinals can be defined from sets x and we can ask for a characterization of which classes arise in this way.

Most of the work in this area has concentrated on the case $\kappa = \omega$, $T = KP =$ Admissible Set Theory. However there is a good understanding of $\alpha_T(x) = \min \Lambda_T(x)$ for arbitrary κ and other theories such as $KP_n = \Sigma_n$-Admissibility, ZF. We will mention some of the latter work as well.

The first result in this area is due to Sacks.

THEOREM 1 (Sacks [1976]). *If $\alpha > \omega$ is admissible and countable then $\alpha = \omega_1^R = \alpha_{KP}(R)$ for some real R.*

There are many proofs of Theorem 1, but the most adaptable (see Friedman [1986]) is via the method of almost disjoint forcing. As a first step we can add $A_0 \subseteq \alpha$ so that α is A_0-admissible and $L_\alpha[A_0] \models$ every set is countable. This is done by (Levy) forcing with finite conditions $\mathcal{P}$ from $\alpha \times \omega$ into α such that $\mathcal{P}(\beta, n) < \beta$. Second, we can add $A_1 \subseteq \alpha$ so that α is (A_0, A_1)-admissible and $\beta < \alpha$ implies β is not $(A_0 \cap \beta, A_1 \cap \beta)$-admissible. This is done with conditions $\mathcal{P} : \beta \to 2$ such that $\beta' \leqslant \beta$ implies β' is not $(A_0 \cap \beta', \mathcal{P} \cap \beta')$-admissible.

Now we can canonically assign a real R_β to each $\beta < \alpha$ so that if $\beta_1 \neq \beta_2$ then $R_{\beta_1} \cap R_{\beta_2}$ is finite. By "canonical" we mean that R_β is defined in $L_{\beta+1}[A_0 \cap \beta, A_1 \cap \beta]$, uniformly. Then we code $A = A_0 \vee A_1$ by a real R using conditions $(r, \bar{r})$ where r is a finite subset of $\omega, \bar{r}$ a finite subset of $\{R_\beta \mid \beta \in A\}$ and $(r_0, \bar{r}_0) \leqslant (r_1, \bar{r}_1)$ if $r_0 \supseteq r_1$, $\bar{r}_0 \supseteq \bar{r}_1$ and $n \in r_0 - r_1$ implies $n \notin R_\beta$ for each $R_\beta \in \bar{r}_1$. The result is that $\beta \in A$ if and only if $R \cap R_\beta$ is finite and thus $A \cap B$ is Δ_1 over $L_\beta[R]$ for each $\beta < \alpha$. So β is not R-admissible for $\beta < \alpha$. Preserving the admissibility of α requires a bit of care, but is based on the simple fact that almost disjoint forcing satisfies the countable chain condition.

Jensen extended Sacks' result to countable sequences of countable admissibles. For a proof of the following result see Friedman [1986].

THEOREM 2 (Jensen [1972]). *Suppose X is a countable set of countable admissibles greater than ω and $\alpha \in X \to \alpha$ is $X \cap \alpha$-admissible. Then for some real R, X is an initial segment of $\Lambda_{KP}(R)$.*

The proof strategy for Theorem 2 is similar to that used in Theorem 1: first add $A \subseteq \alpha$ preserving admissibility so that $\beta < \alpha$ is $A \cap \beta$-admissible if and only if $\beta \in X$, and then code A by a real using almost disjoint forcing. But as we must preserve the admissibility of ordinals in X (while destroying admissibility for ordinals not in X) the argument is more delicate and has the interesting feature that extendibility of conditions for the desired forcing is established using forcing.

There are severe limitations on how much more can be done concerning admissibility spectra in ZFC alone. This is illustrated by the next result. A class $X \subseteq$ ORD is Σ_1-complete if Y is $\Delta_1([X], X)$ whenever $Y \subseteq$ ORD is $\Sigma_1(L)$.

THEOREM 3. *Let $\Lambda(R)$ abbreviate $\Lambda_{KP}(R) = \{\alpha \mid \alpha$ is R-admissible$\}$.*

(a) $R \in L \to \Lambda(R) \supseteq \Lambda(0) - \beta$ *for some* $\beta < \aleph_1^L$.

(b) *If* $R \in L[G]$, G *is* $\mathcal{P}$*-generic over* L, *and* $\mathcal{P} \in L$, *then* $\Lambda(R) \supseteq \Lambda(0) - \beta$ *for some* β.

(c) *Suppose that* $R \in L[G]$ *and* $G \subseteq \mathcal{P}$. *If* G *is* $\mathcal{P}$*-generic over the amenable structure* $(L, \mathcal{P})$, *then* $\Lambda(R)$ *is not* Σ_1*-complete.*

PROOF. (a) Let β be large enough so that $R \in L_\beta$. (b) Let β be large enough so that $\mathcal{P} \in L_\beta$. (c) If $\Lambda(R)$ is Σ_1-complete then L-Card $= \{\kappa \mid L \models \kappa$ is a cardinal$\}$ is $\Delta_1(L[R])$ and hence by reflection, $(\kappa^+)^L < \kappa^+$ for large enough cardinals κ. By Jensen's Covering Theorem, $0^\# \in L[R]$. But $0^\#$ does not satisfy the hypothesis of (c) (see Beller, Jensen and Welch [1982]). □

By (a), (b) of this result we see that class-forcing is required to get a nontrivial admissibility spectrum (without assuming $0^\#$) and we should not expect such a spectrum to be Σ_1-complete.

Using a variant of Jensen coding, R. David and S. Friedman independently obtained a class-generic real R such that $\Lambda_{KP}(R) \subseteq$ Admissible Limits of Admissibles. This is a special case of the following result which appeared in David [1989].

THEOREM 4 (David and Friedman [1985]). *Suppose* $\varphi(\alpha)$ *is* Σ_1 *and* $\alpha \in L$*-Card* $\to L \models \varphi(\alpha)$. *Then there is a real* R *class-generic over* L *such that* $\Lambda_{KP}(R) \subseteq \{\alpha \mid L \models \varphi(\alpha)\}$.

This result is optimal in the sense that if $\varphi(\alpha)$ is the Π_1 formula "α is a cardinal" then the conclusion must fail by Theorem 3(c).

We give some idea of the proof of Theorem 4. The desired forcing is made up of certain "building blocks" that are not difficult to describe. Jensen coding methods are used to put these building blocks together.

We wish to arrange that if α is R-admissible then α is a limit of admissibles. Suppose that we have $D \subseteq \aleph_1^L$ so that if α is D-admissible then α is a limit of admissibles. Then we could hope to choose R so as to code D and satisfy the desired property.

The problem is that if we code D by R in the usual way (with almost disjoint forcing) we only get: for all α, $D \cap (\aleph_1)^{L_\alpha}$ is $\Delta_1(L_\alpha[R])$. So in fact what we need about D is: $L_\alpha[D \cap \xi] \models KP + \xi = \aleph_1$ implies α is a limit of admissibles. For then we need only recover $D \cap (\aleph_1)^{L_\alpha}$ inside $L_\alpha[R]$ to guarantee that α is a limit of admissibles.

How do we obtain D? The natural thing is to force with conditions d which are initial segments of a potential D. Now we come to the main points in the proof.

(1) Extendibility is easy for this forcing because given d and $\gamma < \aleph_1^L$ we are free to extend d to length γ by killing the admissibility of all ordinals between $\sup(d)$ and γ. It is crucial for this argument that we are only concerned with killing admissibility, not with preserving it.

(2) Cardinal-preservation for this forcing is easy to prove assuming there is $D_2 \subseteq (\aleph_2)^L$ such that: $L_\alpha[D \cap \xi] \models KP + \xi = \aleph_2 \to \alpha$ a limit of admissibles.

Thus we are faced with the original problem, one cardinal higher! The solution (due to Jensen in the proof of his Coding Theorem) is to build $R, D_1, D_2, \ldots$ simultaneously.

Finally we introduce the requirement of admissibility preservation into the above. Note that in the conclusion of Theorem 4 we have $\subseteq$ and not equality; indeed the freedom to kill admissibility is crucial to the extendibility argument in (1) above.

Nonetheless we can ask for a real R for which we can control its (nontrivial) admissibility spectrum. This requires the method of strong coding.

THEOREM 5 (Friedman [1987]). *There is a real R, class-generic over L, such that $\Lambda_{KP}(R) =$ Admissible limits of admissibles.*

To prove Theorem 5 we can approach the problem much as in the proof of Theorem 4, however extendibility of conditions is much more difficult. The desired extension of d to length γ must be made generically, so as to preserve the admissibility of admissible limits of admissibles. (Note that this idea was foreshadowed by Jensen's proof of Theorem 2.) Thus conditions must be constructed out of generic sets for "local" versions of the very same forcing. So in fact we construct a strong coding $\mathcal{P}^\beta \subseteq L_\beta$ at each admissible β and then inductively build $\mathcal{P}^\beta$ out of generic sets for various $\mathcal{P}^{\beta'}$, $\beta' < \beta$.

A complete characterization of admissibility spectra is not known. A related question, which may indeed be a prerequisite for such a characterization, is the following: which $A \subseteq \mathrm{ORD}$ can be Δ_1-definable in a real class-generic over L? On this latter problem there has been some significant progress. The following is proved in Friedman and Velickovic [1997].

THEOREM 6 (Friedman). *Suppose $V = L$ and that $A \subseteq \mathrm{ORD}$ obeys the Condensation Condition. Then A is Δ_1 in a real class-generic over L, preserving cardinals.*

We refer the reader to Friedman and Velickovic [1997] for a definition of the Condensation Condition and a proof of Theorem 6.

Other work. Much is known about $\alpha_T(x) = \min \Lambda_T(x)$, $x \subseteq \kappa$, for $T = KP_n$, ZF and arbitrary infinite cardinals κ, assuming $V = L$. We confine ourselves here to only a brief account.

First we consider the (remaining) cases when $\kappa = \omega$.

THEOREM 7A (Sacks [1976]). (a) *$\alpha_{KP_n}(R)$, $R \subseteq \omega$, can be any countable Σ_n-admissible ordinal greater than ω.*

THEOREM 7B (David [1982], Beller, Jensen and Welch [1982]). *$\alpha_{ZF}(R)$, $R \subseteq \omega$, can be any countable α such that $L_\alpha \models ZF$.*

Theorem 7(a) can be proved much like Theorem 1. For Theorem 7(b) note that it suffices to first find R_0 such that β an L-cardinal implies that $L_\beta[R_0] \not\models ZF$ and then apply Theorem 4 (relativized to R_0). The former is not hard to arrange using only the statement of Jensen's Coding Theorem. (Of course historically Theorem 7(b) was proved directly as Theorem 4 was not available.)

Next suppose that κ is regular and uncountable.

THEOREM 8 (Friedman [1982]). $\alpha = \alpha_{KP_n}(x)$ *for some* $x \subseteq \kappa$ *if and only if* $\alpha < \kappa^+$, α *is* Σ_n*-admissible, cofinality* $(\alpha) = \kappa$ *and* L_α *is closed under the function* $\beta \mapsto \beta^{<\kappa}$.

The difficult part of Theorem 8 is the necessity of the stated condition, which draws heavily on Jensen's fine structure theory. The sufficiency is based on an almost disjoint forcing argument, not unlike Theorem 7(a).

THEOREM 9 (David and Friedman [1985]). $\alpha = \alpha_{ZF}(x)$ *for some* $x \subseteq \kappa$ *if and only if* $\kappa < \alpha < \kappa^+$, $L_\alpha \models ZF$ *and there are* $\beta < \alpha$, $\langle X_n \mid n \in \omega \rangle$ *such that*

(i) $\forall \gamma < \kappa \forall f : \gamma \to \beta$ $(f$ *bounded* $\to f \in L_\alpha)$,

(ii) $X_n \in L_\alpha$, L_α*-Card* (X_n) *is less than* β *for all* n, *and* $L_\alpha = \bigcup\{X_n \mid n \in \omega\}$, *and*

(iii) β *is a regular cardinal in* L_α.

The proof of Theorem 9 makes use of almost disjoint forcing, the Covering Theorem (relativized to some $L_\alpha[x]$) and Jensen's fine structure theory.

When κ is singular of cofinality ω then methods from infinitary model theory come into play.

THEOREM 10 (Friedman [1981b]). $\alpha = \alpha_{KP}(x)$ *for some* $x \subseteq \kappa$ *if and only if*

(i) $\kappa < \alpha < \kappa^+$,

(ii) *if there is a largest* L_α*-cardinal* γ *then cofinality* $(\gamma) = \omega$, *and*

(iii) *there is a* 1–1 *function* $f : L_\alpha \to \kappa$ *such that* $f^{-1}[\delta] \in L_\alpha$ *for each* $\delta < \kappa$.

Under the conditions stated in Theorem 10, a version of the Barwise Compactness Theorem is established, which can then be used to obtain the desired x. A related result appears in Magidor, Shelah and Stavi [1984].

For $n > 1$ a surprising thing occurs: for any $x \subseteq \aleph_\omega$, $x \in L_{\alpha_{KP_2}}$! And an even stronger fact holds for $x \subseteq \aleph_{\omega_1}$, namely $x \in L_{\alpha_{KP}}(x)$. Both of these facts follow from an effective version of Jensen's Covering Theorem. This puts severe restrictions on the possible values for $\alpha_{KP_n}(x)$, $x \subseteq \kappa$ for $n > 1$, κ singular of cofinality ω and for $n \geqslant 1$, κ singular of uncountable cofinality (as well as for $\alpha_{ZF}(x)$). The reader is referred to Friedman [1981a], David and Friedman [1985] for complete characterizations.

References

A. BELLER, R. B. JENSEN AND P. WELCH

[1982] *Coding the Universe*, London Mathematical Society Lecture Notes, Vol. 47.

C. T. CHONG
[1984] *Techniques of Admissible Recursion Theory*, Lecture Notes in Mathematics, Vol. 1106, Springer.

C. T. CHONG AND K. J. MOURAD
[1992] Σ_n-definability without Σ_n-induction, *Trans. Amer. Math. Soc.*, 334, pp. 349–363.
[ta] Post's problem and singularity.

C. T. CHONG AND Y. YANG
[1997] Σ_2-induction and infinite injury priority arguments, Part II: Tame Σ_2 sets and the jump operator, *Ann. Pure Appl. Logic*, 87, pp. 163–116.
[ta1] Maximal sets, high RE sets, and Σ_2-induction, *J. Symbolic Logic*, to appear.

R. DAVID
[1982] Some applications of Jensen's Coding Theorem, *Ann. Math. Logic*, 22, pp. 177–196.

R. DAVID AND S. D. FRIEDMAN
[1985] Uncountable *ZF*-ordinals, in: *Proc. Symposia in Pure Math.*, Vol. 42, Amer. Math. Soc., Providence, RI, pp. 217–222.

R. DAVID
[1989] A functional Π^1_2-singleton, *Adv. in Math.*, 74, pp. 258–268.

S. D. FRIEDMAN
[1978] Negative solutions to Post's Problem, I, in: *Generalized Recursion Theory II*, Fenstad, Gandy and Sacks, eds., North-Holland, Amsterdam.
[1979] β-recursion theory, *Trans. Amer. Math. Soc.*, 255, pp. 173–200.
[1980] Post's Problem without admissibility, *Adv. in Math.*, 35, pp. 30–49.
[1981a] Negative solutions to Post's problem, II, *Ann. Math.*, 113, pp. 25–43.
[1981b] Uncountable admissibles, II: Compactness, *Israel J. Math.*, 40, pp. 129–149.
[1982] Uncountable admissibles, I: Forcing, *Trans. Amer. Math. Soc.*, 270, pp. 61–73.
[1983] Some recent developments in higher recursion theory, *J. Symbolic Logic*, 48, pp. 629–642.
[1986] An introduction to the admissibility spectrum, in: *Logic, Methodology and Philosophy of Science VII*, Marcus, Dorn and Weingartner, eds., North-Holland, Amsterdam, pp. 129–139.
[1987] Strong coding, *Ann. Pure Appl. Logic*, pp. 1–98.

S. D. FRIEDMAN AND VELICKOVÍC
[1997] Δ_1-definability, *Ann. Pure Appl. Logic*, 89, pp. 93–99.

M. GROSZEK, M. MYTILINAIOS AND T. A. SLAMAN
[1996] The Sacks density theorem and Σ_2 bounding, *J. Symbolic Logic*, 61, pp. 450–467.

R. B. JENSEN
[1972] The fine structure of the constructible universe, *Ann. Math. Logic*, 4, pp. 229–308.

S. C. KLEENE
[1938] On notation for ordinal numbers, *J. Symbolic Logic*, 3, pp. 150–155.

G. KREISEL AND G. E. SACKS
[1965] Metarecursive sets, *J. Symbolic Logic*, 10, pp. 318–336.

M. LERMAN
[1972] On suborderings of the α-recursively enumerable degrees, *Ann. Math. Logic*, 4, pp. 369–392.
[1974] Maximal α-RE sets, *Trans. Amer. Math. Soc.*, 188, pp. 341–386.

W. MAASS
[1978a] The uniform regulary set theorem in α-recursion theory, *J. Symbolic Logic*, 43, pp. 270–279.
[1978b] Inadmissibility, tame RE sets and the admissible collapse, *Ann. Math. Logic*, 13, pp. 149–170.

M. MAGIDOR, S. SHELAH AND J. STAVI
[1984] Countably decomposable admissible sets, *Ann. Math. Logic*, 26, pp. 287–361.

M. MYTILINAIOS AND T. A. SLAMAN
[1988] Σ_2-collection and the infinite injury priority method, *J. Symbolic Logic*, 53, pp. 212–221.

J. B. PARIS AND L. A. KIRBY
[1978] Σ_n collection schemas in models of arithmetic, in: *Logic Colloquium '77*, North-Holland, Amsterdam.

H. ROGERS, JR.
[1967] *Theory of Recursive Functions and Effective Computability*, McGraw-Hill, New York.

G. E. SACKS
[1966] Post's problem, admissible ordinals, and regularity, *Trans. Amer. Math. Soc.*, 124, pp. 1–23.
[1976] Countable admissible ordinals and hyperdegrees, *Adv. Math.*, 20, pp. 231–262.
[1990] *Higher Recursion Theory*, Springer, Berlin.

G. E. SACKS AND S. G. SIMPSON
[1972] The α-finite injury method, *Ann. Math. Logic*, 4, pp. 323–367.

R. A. SHORE
[1976a] The recursively enumerable α-degrees are dense, *Ann. Math. Logic*, 9, pp. 123–155.
[1976b] On the jump of an α-recursively enumerable set, *Trans. Amer. Math. Soc.*, 217, pp. 351–363

J. H. SILVER
[1974] On the singular cardinals problem, in: *Proc. International Congress of Mathematicians 1974*, pp. 265–268.

T. A. SLAMAN AND W. H. WOODIN
[1989] Σ_1-collection and the finite injury priority method, in: *Mathematical Logic and Its Applications*, Lecture Notes in Mathematics, Vol. 1388, Springer, Berlin.

CHAPTER 10

E-recursion

Gerald E. Sacks*

Harvard University, Department of Mathematics, Cambridge, MA 02138, USA

*The author is grateful to B. Dreben, A. Kanamori and T. Slaman for their longtime encouragement.

HANDBOOK OF COMPUTABILITY THEORY
Edited by E.R. Griffor

"The longer the computation the better."
G. K.

Time to tell you about E-recursion in fair detail[1]. The subject began with Kleene's [1959] notion of recursion in objects of finite type in the so-called normal case. The objects of type 0 are the natural numbers. An object of type $n+1$ is a set of objects of type n. Kleene defines $\{e\}(x)$ for x an object of finite type by means of schemes. Normality allows each computation to invoke the appropriate equality predicate. The computation of $\{e\}(x)$, for x of type n, draws on ${}^nE$, the equality predicate for objects of type $n-1$ $(n>0)$. For an x of type 1, normality is unavoidable since ${}^1E$ is recursive a priori. ${}^nE$ increases in power with n.

THEOREM 1 (Kleene [1959]). *Let $x \subseteq \omega$. Then x is recursive in ${}^2E$ iff x is hyperarithmetic.*

The proof of Theorem 1 has two parts. First show that x', the Turing jump of x, is recursive in ${}^2E$ if x is. Second show by induction on the length of computations that every countable, convergent computation tree derived from ${}^2E$ can be encoded by a hyperarithmetic real. Thus the hyperarithmetic hierarchy of reals corresponds in a natural way to the set of countable, convergent computations from ${}^2E$.

Kleene was not the sort to declare equality recursive by fiat. He preferred to find out what was computable from equality. In the world of E-recursion, where $\{e\}(x)$ has a meaning for every set x, it is necessary to regard equality as recursive. Nonetheless the E-recursion theorist discovers, as did Kleene, what can be computed from equality. The definition of $\{e\}(x)$ is in terms of schemes introduced by Normann [1978], and subsequently and independently by Moschovakis. The first three schemes are projection, difference and pairing. The fifth is composition, and the sixth is enumeration:

$$\{e\}(c, x_1, \ldots, x_n, y_1, \ldots, y_m) \simeq \{c\}(x_1, \ldots, x_n)$$

if $e = \langle 6, n, m \rangle$. To some, enumeration is a theorem, not a scheme. Casting it as a scheme makes it possible to omit the least number operator and primitive recursion, two schemes well abandoned when there is no underlying effective wellordering of the sets. Bounding with union, the fourth scheme, is the sole source of infinitely long computations:

$$\{e\}(x_1, \ldots, x_n) \simeq \bigcup \{\{c\}(y, x_2, \ldots, x_n) \mid y \varepsilon x_1\}$$

if $e = \langle 4, n, c \rangle$. If x_1 is infinite, then $\{c\}$ has to be applied infinitely often. A partial function from V, the class of all sets, into V is *partial E-recursive* if it belongs to the least class of partial functions closed under the Normann schemes. The graph of

[1] For more-than-fair detail consult Sacks [1990].

such a function is Σ_1^{ZF}, but the converse is false. A worthy example is $O(x)$, Gödel's order of constructibility function

$$O(x) \simeq \mu\gamma[x \varepsilon L(\gamma+1) - L(\gamma)].$$

$O(x)$ is Σ_1^{ZF} but not partial E-recursive. If $x \varepsilon L$, then $O(x)$ is found by an unbounded search devoid of effective content. Theorem 30 (van de Wiele [1982]) below explains the gap between Σ_1^{ZF} and partial E-recursive.

Let $f : \omega \to \omega$ be a partial function. f is partial recursive in the sense of ordinary recursion theory iff f is Σ_1^{ZF} over *HF*, the set of hereditarily finite sets, iff f is partial E-recursive.

Let A be a transitive set. A is said to be *E-closed* if

$$\vec{x} \varepsilon A \wedge f(\vec{x})\downarrow \longrightarrow f(\vec{x}) \varepsilon A \quad (\vec{x} \text{ is } x_1, \ldots, x_n)$$

for every partial E-recursive f. ($f(x)\downarrow$ means $f(x)$ converges.) Every Σ_1 admissible set is E-closed but not conversely. As will be seen below, E-closed structures possess sufficient expressive power to support forcing and priority arguments. $E(x)$, the *E-closure* of x, is the least E-closed set that contains the transitive closure of $\{x\}$. Thus $x \varepsilon E(x)$ and $E(x)$ is transitive. $E(\emptyset)$ is *HF*. $E(\omega)$ is $L(\omega_1^{CK)})$. $E(2^\omega)$ is not Σ_1 admissible.

A *computation instruction* is an $(n+1)$-tuple of the form $\langle e, x_1, \ldots, x_n\rangle$, or more simply, $\langle e, x\rangle$. $T_{\langle e,x\rangle}$ is the *computation tree* associated with $\langle e, x\rangle$. Each node of $T_{\langle e.x\rangle}$ is a computation instruction. The top node is $\langle e, x\rangle$. Each Normann scheme give rise to finite branching except for the fourth. If e is $\langle 4, n, c\rangle$, then the nodes immediately below $\langle e, \vec{x}\rangle$ are $\langle c, y, x_2, \ldots, x_n\rangle$ for all $y \varepsilon x_1$. If e is not the index of a scheme, then $\langle e, x\rangle$ is immediately below $\langle e, x\rangle$.

PROPOSITION 2. *$\{e\}(x)\downarrow \leftrightarrow T_{\langle e,x\rangle}$ is wellfounded.*

LEMMA 3. *There exists a partial E-recursive function g such that for all $d < \omega$ and all x:*

(i) $\{d\}(x)\downarrow \leftrightarrow g(d, x)\downarrow$;
(ii) $\{d\}(x)\downarrow \leftrightarrow g(d, x) = T_{\langle d,x\rangle}$.

The proof of Lemma 3 is an effective transfinite recursion. Suppose A is E-closed. Lemma 3 implies: if $x \varepsilon A$ and $\{e\}(x)\downarrow$, then $T_{\langle e,x\rangle} \varepsilon A$. A well-founded $T_{\langle e,x\rangle}$ has an ordinal height, denoted by $|T_{\langle e,x\rangle}|$ or by $|\{e\}(x)|$, and called the length of the computation of $\{e\}(x)$. If $\{e\}(x)\uparrow$ (diverges), then $|\{e\}(x)| = \infty$.

LEMMA 4. *The predicates, $|\{e\}(x)| < \gamma$ and $|\{e\}(x)| = \gamma$, are E-recursive.*

PROPOSITION 5. *If $\{e\}(x)\downarrow$ and $|\{e\}(x)| \geqslant \omega$, then $\{e\}(x)$ and $T_{\langle e,x\rangle}$ are first order definable over $L(|\{e\}(x)|, TC(\{x\}))$.*

y is said to be *E-recursive in* $x_1, \ldots, x_n$ (in symbols $y \leqslant_E x_1, \ldots, x_n$) if $y = \{e\}(x_1, \ldots, x_n)$ for some e. If f is a total E-recursive function, then $f(x) \leqslant_E x$ for all x uniformly. If $x, y \subseteq \omega$, then y is hyperarithmetic in x iff $y \leqslant_E x, \omega$. An ordinal γ is *E-recursive in* x if $\gamma \leqslant_E x$. The Church–Kleene recursive ordinals are the ordinals E-recursive in ω. Define

$$\kappa_0^x = \sup\{\gamma \mid \gamma \leqslant_E x\}.$$

$\kappa_0^{\omega_1}$ is not a Σ_1 admissible ordinal. A, a class, is *E-recursively enumerable in* x if for some e,

$$A = \{y \mid \{e\}(y, x)\downarrow\}.$$

Define

$$\kappa^x = \sup\big\{\gamma \mid \gamma \leqslant_E x,\ a_1, \ldots, a_n \text{ for some } a_1, \ldots, a_n \varepsilon TC(x)\big\}.$$

PROPOSITION 6. $E(x) = L(\kappa^x,\ TC(\{x\}))$.

An ordinal δ is said to be *x-reflecting* if

$$L\big(\delta, TC(\{x\})\big) \models \mathcal{F} \rightarrow L\big(\kappa_0^x, TC(\{x\})\big) \models \mathcal{F}$$

for every Σ_1^{ZF} sentence $\mathcal{F}$ whose only parameter is x. Define

$$\kappa_r^x = \text{the greatest } x\text{-reflecting ordinal.}$$

If a Σ_1 fact about x is true in $L(\kappa_r^x,\ TC(\{x\}))$, then it is true in $L(\gamma,\ TC(\{x\}))$ for some $\gamma \leqslant_E x$.

PROPOSITION 7. (i) $\kappa_0^x \leqslant \kappa_r^x$. (ii) $\kappa_r^{x,a} \leqslant \kappa^x$ *for all* $a \varepsilon TC(x)$.

The proof of (ii) uses: there exists a Π_3^{ZF} sentence $\mathcal{F}$ such that for every transitive set M,

$$M \text{ is E-closed} \leftrightarrow M \models \mathcal{F}.$$

THEOREM 8 (Gandy selection). *There exists a partial E-recursive function* $\phi(e, x)$ *such that for all* $e < \omega$ *and all* x:

(i) $\exists n_{n<\omega}[\{e\}(n, x)\downarrow] \leftrightarrow \phi(e, x)\downarrow$;
(ii) $\phi(e, x)\downarrow \rightarrow \{e\}(\phi(e, x), x)\downarrow$.

Gandy [1967] proved Theorem 8 for normal type 2 objects, and Moschovakis extended it to higher types. The nature of selection is simple. A member of a nonempty re set is selected by enumerating the set and selecting the first element to appear. Sometimes the best one can do is to select a "small" nonempty recursive subset of the re set.

COROLLARY 9. *Suppose* $\kappa_r^{x,a} = \kappa^x$ *for all* $a \varepsilon TC(x)$. *Then* $E(x)$ *is* Σ_1 *admissible.*

Suppose $\{e\}(x)\uparrow$. A *Moschovakis witness* to the divergence of $\{e\}(x)$ is an infinite descending path in $T_{\langle e,x\rangle}$. Such a witness has the form $\lambda n \mid t(n)$, where $t(0) = \langle e, x\rangle$ and for each n, $t(n+1)$ is an immediate subcomputation instruction of $t(n)$. The predicate,

$$t \text{ witnesses } \{e\}(x)\uparrow,$$

is E-recursively enumerable.

THEOREM 10. *Assume some wellordering of* $TC(x)$ *is E-recursive in* x. *If* $\{e\}(x)\uparrow$, *then some Moschovakis witness to its divergence is first order definable over* $L(\kappa_r^x, TC(x))$.

Theorem 10 is the most important structural fact in E-recursion theory. It was inspired by Harrington [1973] in the setting of Kleene recursion in finite types. Harrington showed: if $a \varepsilon 2^\omega$ and $\{e\}(a, 2^\omega)\uparrow$, then some Moschovakis witness to $\{e\}(a, 2^\omega)\uparrow$ is enumerated via a computation of height at most $\kappa_r^{2^\omega, a}$. His argument makes use of a basis theorem due to Kechris that reappears in the proof of Theorem 10.

With the aid of Theorem 10 all E-closed $L(\kappa)$'s split into two disjoint classes.

Class I: $L(\kappa)$ *admits Moschovakis witnesses*, i.e. if $x \varepsilon L(\kappa)$ and $\{e\}(x)\uparrow$, then a witness to $\{e\}(x)\uparrow$ belongs to $L(\kappa)$.

Class II: $L(\kappa)$ is Σ_1 admissible; and for all $A \subseteq L(\kappa)$, A is Σ_1 definable over $L(\kappa)$ iff A is *E-recursively enumerable on* $L(\kappa)$, i.e. there is an e and a $p \varepsilon L(\kappa)$ such that for all $x \varepsilon L(\kappa)$,

$$x \varepsilon A \leftrightarrow \{e\}(p, x)\downarrow.$$

In short either $L(\kappa)$ admits divergence witnesses, or $\Sigma_1 = E - re$ on $L(\kappa)$, but not both. Thus the pursuit of re degree theory on an E-closed $L(\kappa)$ reduces to the well understood admissible case when $L(\kappa)$ does not admit divergence witnesses.

Time now for forcing over an E-closed, but not Σ_1 admissible, $L(\kappa)$. Let $gc(\kappa)$ denote the greatest cardinal in the sense of $L(\kappa)$. For set forcing the generic object G can be regarded as a subset of $gc(\kappa)$. $L(\kappa, G)$ need not be E-closed when G is set generic. Let $L(\kappa) = E(\omega_1)$ (a non-Σ_1 admissible set) and G collapse ω_1. If $L(\kappa, G)$ is E-closed, then $L(\kappa, G) = E(b)$ for some $b \subseteq \omega$, and so $L(\kappa, G)$, hence $L(\kappa)$, is

Σ_1 admissible. Recall that set forcing over a Σ_1 admissible set always preserves Σ_1 admissibility.

All members of $L(\kappa, G)$ are of the form $\{e\}(a, G)$, where $a \varepsilon L(\kappa)$ and

$$\big|\{e\}(a, G)\big| < \kappa.$$

The principal terms of the forcing language, $\mathcal{L}(\kappa, \mathcal{G})$, are of the form $\{e\}(\underline{a}, \mathcal{G})$. A typical ranked formula is $|\{e\}(\underline{a}, \mathcal{G})| = \underline{\sigma}$ for $\sigma < \kappa$. Let $\mathcal{P} = \langle P, \leqslant \rangle \varepsilon L(\kappa)$ be a forcing notion. The forcing conditions are $p, q, r, \ldots$; $p \geqslant q$ (q extends p) means q says at least what p says about $\mathcal{G}$. $\mathcal{P}$ determines a forcing relation $\Vdash$ defined by effective transfinite recursion on $\sigma < \kappa$. (a), (b) and (c) are defined simultaneously.

(a) $p \Vdash |\{e\}(\underline{a}, \mathcal{G})| = \underline{\sigma}$.

(b) $\mathcal{F}(p, e, a, \mathcal{G}, \sigma)$.

(c) $q \Vdash s \varepsilon \{e\}(\underline{a}, \mathcal{G})$.

(b) is a set of terms adequate for naming the elements of $\{e\}(\underline{a}, \mathcal{G})$. In (c) s is a member of (b) and q extends the p of (a). The clauses of the definition of $\Vdash$ parallel the Normann schemes. $\Vdash$ extends to ranked formulas as usual.

PROPOSITION 11. (a), (b) *and* (c) *above are E-recursive in* σ, $\mathcal{P}$ *uniformly.*

$\mathcal{P}$ is said to satisfy *effective bounding* if

$$p \Vdash^* \exists \sigma \big[|\{e\}(t)| = \sigma \big]$$

implies $p \Vdash^* |\{e\}(t)| \leqslant \underline{\gamma}$ for some $\gamma \leqslant_E p, t, \mathcal{P}$ (uniformly). $\mathcal{G}$ is $\mathcal{P}$*-generic* if for every sentence F,

$$\exists p_{p \varepsilon G} [p \Vdash \mathcal{F} \text{ or } p \Vdash \neg \mathcal{F}].$$

LEMMA 12. *If* $\mathcal{P}$ *satisfies effective bounding and* $\mathcal{G}$ *is* $\mathcal{P}$*-generic, then* $L(\kappa, G)$ *is E-closed.*

$\mathcal{P}$ is *countably closed* if for every $(\lambda n \mid p_n) \varepsilon L(\kappa)$,

$$\forall n [p_n \geqslant p_{n+1}] \rightarrow \exists q \forall n [p_n \geqslant q].$$

Countably closed forcing over an E-closed structure occurs in Sacks [1980]. Slaman was the first to explain the role of the countably closed condition in the latter paper.

THEOREM 13. *Suppose* $\mathcal{P} \varepsilon L(\kappa)$ *is countably closed and* $\mathcal{G}$ *is* $\mathcal{P}$*-generic. Then* $L(\kappa, \mathcal{G})$ *is E-closed.*

The proof of Theorem 13 is based on $>_V$, the *tree of possibilities*. A node on $>_V$ is of the form $\langle p, e, t\rangle$, where p is a forcing condition and t is a term of $\mathcal{L}(\kappa, \mathcal{G})$. $>_V$ below $\langle p, e, t\rangle$ contains all the possibilities for $T_{\langle e,t\rangle}$ for all generic $\mathcal{G}\varepsilon p$.

LEMMA 14. *If* $>_V$ *is well-founded below* $\langle p, e, t\rangle$ *whenever*

$$p \Vdash^* \exists\sigma\big[|\{e\}(t)| = \sigma\big],$$

then $\mathcal{P}$ *satisfies effective bounding.*

To complete the proof of Theorem 13 via Lemmas 12 and 14, one need only show the hypothesis of Lemma 14 holds for countably closed $\mathcal{P}$'s. Suppose not. Then an infinite descending path below $\langle p, e, t\rangle$ in $>_V$ can be transformed into a $q \leqslant p$ that forces the existence of a Moschovakis witness to the divergence of $\{e\}(t)$ contrary to the hypothesis on p in Lemma 14. The n-th leg of the witness is forced by p_n, and $\forall n[p \geqslant p_n \geqslant p_{n+1} \geqslant q]$. The next result is an application of Theorem 13.

THEOREM 15 (Sacks [1986]). *Assume* $L(\kappa)$ *is not* Σ_1 *admissible and*

$$L(\kappa) \models [gc(\kappa) \textit{ is regular}].$$

Then $2^{gc(\kappa)} \cap L(\kappa)$ *is not E-recursively enumerable in any* $b\varepsilon L(\kappa)$.

One interpretation of Theorem 15 is that certain naturally enumerable sets are not E-recursively enumerable. A case for regarding $E(\omega_1)$ as naturally enumerable can be made as follows. Begin with ω_1 and iterate first order definability. At limit stage λ, look back and see if there is an x already enumerated and an e such that $|\{e\}(x)| = \lambda$. If yes, then collect everything enumerated before stage λ and continue. If no, stop. The procedure at λ is predicative because $|\{e\}(x)| = \lambda$ only if all immediate subcomputations of $\{e\}(x)$ converge and were enumerated prior to λ. This way of laying out $E(\omega_1)$ resembles the usual enumeration of all finite computations in classical recursion theory. The next theorem explains why $E(\omega)$, unlike $E(\omega_1)$, is E-recursively enumerable.

THEOREM 16. *If* z *is a set of ordinals and* $E(z)$ *is* Σ_1 *admissible, then* $E(z)$ *is E-recursively enumerable in some* $b\varepsilon E(z)$.

The next two results on set forcing are general in nature.

THEOREM 17 (Sacks and Slaman). *Assume* $L(\kappa)$ *is E-closed but not* Σ_1 *admissible. Let* $\mathcal{P}\varepsilon L(\kappa)$. *Then* (i) $\equiv$ (ii).

(i) $\mathcal{P}$ *satisfies effective bounding.*

(ii) *The relation,* $p \Vdash^* \exists\sigma[|\{e\}(\underline{a}, \mathcal{G})| = \sigma]$, *is E-recursively enumerable on* $L(\kappa)$.

Assume $L(\kappa)$ is E-closed but not Σ_1 admissible. Let $\mathcal{P}\varepsilon L(\kappa)$. If

$$p \Vdash^* |\{e\}(\underline{a}, \mathcal{G})| \leqslant \kappa_r^{p,a,\mathcal{P}},$$

then

$$p \Vdash^* |\{e\}(\underline{a}, \mathcal{G})| \leqslant \underline{\gamma}$$

for some $\gamma \leqslant_E p, a, \mathcal{P}$.

If $p, q\varepsilon P$ and $\neg\exists r[p \geqslant r \wedge q \geqslant r]$, then p and q are said to be *incompatible.* An *antichain* is a set of mutually incompatible conditions. $\mathcal{P}$ satisfies the *countable chain condition* (c.c.c.) if every antichain in $L(\kappa)$ is countable in $L(\kappa)$. For c.c.c. forcing: E-closure is preserved by Theorem 18, but $>_V$, the tree of possibilities, does not always possess the wellfoundedness property expressed by the hypothesis of Lemma 14. To see the difference between countably closed, and c.c.c., forcing, look at the conclusion of Theorem 18.

$\mathcal{P}$ satisfies the *ρ-chain condition* if every antichain in $L(\kappa)$ has cardinality less than ρ in $L(\kappa)$. $L(\kappa)$ obeys *less-than-γ selection* if there exists a partial E-recursive (in γ and possibly other parameters from $L(\kappa)$) function f such that for all $e < \omega$, $\delta < \gamma$ and $p\varepsilon L(\kappa)$:

$$\exists x_{x<\delta}\big[\{e\}(p, x)\downarrow\big] \rightarrow \big[f(e, \delta)\downarrow \wedge \{e\}(p, f(e, \delta)\downarrow\big].$$

Thanks to Gandy selection (Theorem 8), $L(\kappa)$ obeys $< \omega_1$ selection. Normann showed that $E(\gamma)$ obeys $< \gamma$ selection if γ is a regular cardinal inside $E(\gamma)$.

THEOREM 18 (Sacks [1986]). *Let $L(\kappa)$ be E-closed. Assume $\mathcal{P}\varepsilon L(\kappa)$ satisfies the ρ-chain condition, and $L(\kappa)$ obeys $< \rho$ selection. For each p, if there exist r and δ such that*

$$p \geqslant r \Vdash |\{e\}(t)| = \underline{\delta},$$

then there exist such r and δ E-recursive in $p, t, \mathcal{P}, \rho$ (and some background parameters).

COROLLARY 19 (Slaman). *Let $L(\kappa)$ be E-closed and $\mathcal{P}\varepsilon L(\kappa)$. Assume $L(\kappa) = E(gc(\kappa)), \mathcal{P}$ satisfies the ρ-chain condition, and $L(\kappa)$ obeys $< \rho$ selection. Then*

$$\kappa_r^{G,a} \leqslant \kappa_r^a$$

for all $\mathcal{P}$-generic G and $a\varepsilon L(\kappa)$. (Parameters ρ, $gc(\kappa)$ and $\mathcal{P}$ are suppressed.)

Note that the conclusion of Corollary 19 is much stronger than "every $\mathcal{P}$-generic extension of $L(\kappa)$ is E-closed".

Slaman [1981] defined the *closed κ_r spectrum.*

$$\kappa_{r,0} = \kappa_r,$$
$$\kappa_{r,\delta+1} = \kappa_r^{\kappa_{r,\delta}},$$
$$\kappa_{r,\lambda} = \sup \kappa_{r,\delta}.$$

He observed $\forall a_{a\varepsilon L(\kappa)} \exists \delta_{\delta \varepsilon L(\kappa)}[\kappa_r^a = \kappa_{r,\delta}]$. According to Corollary 19 every c.c.c. set notion of forcing preserves the κ_r spectrum of $L(\kappa)$.

Slaman [1985a] applied class forcing over an E-closed structure to settle many questions. Let $L(\kappa)$ be E-closed and countable. Then there exists an $x \subseteq 2^\omega$ such that:

(i) $L(\kappa, x) = E(x)$;
(ii) $L(\kappa, x)$ does not admit Moschovakis witnesses;
(iii) if $\kappa_r^a < \kappa$ for all $a \varepsilon L(\kappa)$, then $\kappa_r^a < \kappa$ for all $a \varepsilon L(\kappa, x)$;
(iv) there is no Σ_1 formula $\mathcal{F}$ of ZF such that for all $a \varepsilon L(\kappa, x)$,

$$L(\kappa_r^a, x) \vDash \neg \mathcal{F}(\underline{a}) \quad \text{and} \quad L(\kappa_r^a + 1, x) \vDash \mathcal{F}(a).$$

The next result is an application of class forcing with E-pointed, perfect trees.

THEOREM 20 (Sacks and Slaman [1987]). *Let $L(\kappa)$ be countable, E-closed and not Σ_1 admissible. If*

$$L(\kappa) \vDash [\mathit{cofinality}(gc(\kappa)) > \omega],$$

then $L(\kappa, G) = E(G)$ for some $G \subseteq \omega_1^{L(\kappa)}$.

The evolution of selection in the setting of E-recursion owes much to Grilliot [1969]. Let $f, g, \ldots$ be functions from x into $\omega \times (2^x \times \{2^x\})$. For simplicity let x be transitive, closed under pairing and assume $\omega \subseteq x$. Define $g < f$ by

$$\forall z_{z\varepsilon x}[g(z) \text{ is an immediate subcomp. instruc. of } f(z)].$$

Let $\min f$ be $\min\{|f(z)| \mid z \varepsilon x\}$. Assume $\min f < \infty$. Grilliot saw that $\min f$ was computable from $f, 2^x$. His approach was by recursion on the ordinals less than $\min f$. His recursion equation was

$$\min f = \max_{g<f}{}^{+}(\min g).$$

An immediate obstacle is the fact that

$$\{g \mid g < f\} \text{ is not in general E-recursive in } f, 2^x.$$

The solution is to approximate $g < f$ by $g <^{\beta} f$ defined by: for all $z \varepsilon x$, $g(z)$ is seen to be a subcomputation instruction of $f(z)$ by a computation from $f, 2^x$ of length at most β. $\{g \mid g <^{\beta} f\}$ is E-recursive in $f, 2^x, \beta$ uniformly. If β is too small, then the use of $<^{\beta}$ instead of $<$ in Grilliot's recursion equation can produce a false minimum less than the true minimum. Harrington and MacQueen added an iteration trick to the development of selection. $<^{\beta}$ generates a positive contribution to min f. A sequence of length less than $(card\, x)^+$ of such contributions adds up to min f. Note that $(card\, x)^+$, the least ordinal onto which x cannot be mapped, is E-recursive in 2^x uniformly. The ordinal computed at each stage of the iteration is the β for the next stage.

THEOREM 21 (Harrington and MacQueen [1976]).

$$\min f \leqslant_E f, 2^x \textit{ uniformly.} \qquad \textit{(Grilliot selection)}$$

Recall $R(0) = \emptyset$, $R(\delta+1) = 2^{R(\delta)}$, and $R(\lambda) = \bigcup\{R(\delta) \mid \delta < \lambda\}$. An *$\omega$-sequence through* λ is a function k with domain ω and range unbounded in λ.

THEOREM 22 (Moschovakis). *Suppose* $E(R(\lambda)) \models [\textit{cofinality}(\lambda) = \omega]$. *Let* $k \varepsilon E(R(\lambda))$ *be an* ω*-sequence through* λ. *Assume*

$$f : R(\lambda) \to \omega \times R(\lambda) \times \{R(\lambda)\}.$$

If $\min_{x \varepsilon R(\lambda)} |f(x)| < \infty$, *then*

$$\min_{x \varepsilon R(\lambda)} \left| f(x) \right| \leqslant_E f, R(\lambda), k \textit{ uniformly.}$$

COROLLARY 23 (Moschovakis). *If* $E(R(\lambda)) \models [\textit{cofinality}(\lambda) = \omega]$, *then* $E(R(\lambda))$ *is* Σ_1 *admissible.*

As in Kleene [1959] define $tp(0) = \omega$ and $tp(n+1) = 2^{tp(n)}$. The *objects of type* n are the elements of $tp(n)$. Suppose $0 < k < n+2$. The *k-section* of a type $n+2$ F is:

$$k - sc(F) = \left\{ z \mid z \varepsilon tp(k) \wedge z \leqslant_E tp(n);\, F \right\}.$$

The semi-colon (;) between $tp(n)$ and F means F is present as an additional class; thus z is $\{e\}^F(tp(n))$ for some $e < \omega$. Kleene's ${}^{n+2}E$, a type $(n+2)$ object, is defined by:

$$^{n+2}E(x) = \begin{cases} 0 & \text{if } x = \emptyset, \\ 1 & \text{if } x \neq \emptyset \end{cases}$$

for all x of type $n+1$. Kleene calls a type $n+2$ F *normal* if ${}^{n+2}E$ is recursive in F. E-recursion theory has normality built in.

THEOREM 24 (Sacks [1974, 1977]). *Let F be an object of type $n+2$. Assume $0 < k < n+2$. Then there exists a G of type $k+1$ such that*

$$k - sc(F) = k - sc(G). \quad (\textit{Plus-One})$$

The proof of Theorem 24 splits into two cases: $k = 1$ and $k > 1$. The first is class forcing over a countable Σ_1 admissible set. The second is forcing with a reflection trick derived from Grilliot selection. The next result follows from the proof of the first case.

COROLLARY 25. *Let A be a countable Σ_1 admissible set. Then there exists a type 2 F such that $1 - sc(F) = 2^\omega \cap A$.*

It follows from Corollary 25 that the set of all Δ^1_2 subsets of ω is the 1-section of some type 2 object.

Assume $0 < k < n+2$ and F is a type $n+2$ object. The *k-envelope of F*, denoted by $k - en(F)$, was defined by Moschovakis to be:

$$\{z \mid z \varepsilon tp(k) \wedge z \text{ is E-r.e. in } tp(k-1);\ F\}.$$

The *k-envelope* of F is equivalent to the complete E-r.e. subset of $tp(k-1)$ (E-r.e. in $tp(k-1)$; F). Long ago Moschovakis showed that the k-envelope of a type $k+2$ object is not the k-envelope of any type $k+1$ object.

THEOREM 26 (Harrington [1973]). *Assume $0 < k \leqslant n$ and F is a type $n+2$ object. Then there exists a G of type $k+2$ such that*

$$k - en(F) = k - en(G). \quad (\textit{Plus-Two})$$

The proof of Harrington's theorem is a delicate r.e. enumeration argument based on a reflection principle of his own devising. A typical instance of the principle is: let A be the complete E-r.e. (in 2^ω) subset of ω; suppose $P(X)$ is a Σ_1 formula of class-set theory whose only class variable is the free variable X, and whose only parameter is 2^ω; then

$$L(\kappa_0^{2^\omega}, 2^\omega) \models P(A) \longrightarrow L(\gamma, 2^\omega) \models P(A_{<\gamma})$$

for some $\gamma \leqslant_E 2^\omega$ (A_γ is that part of A enumerated via computations of length less than γ).

THEOREM 27 (Sacks and Slaman [1987]). *If x is a set of ordinals and*

$$E(x) \models [\textit{cofinality (greatest cardinal)} = \omega],$$

then $E(x)$ is Σ_1 admissible.

The above was inspired by Corollary 23 and was proved by the so-called elemental approach to selection. For more of the same see Griffor and Normann [1982].

The theory of E-r.e. degrees has only one surprise, but it is a big one. The use of Moschovakis divergence witnesses reduces the complexity of priority arguments. Infinite injury becomes finite injury, and finite injury becomes wait-and-see. One last time assume $L(\kappa)$ is E-closed. Suppose $A \subseteq L(\kappa)$. A is said to be *E-r.e.* on $L(\kappa)$ if

$$A = \{x \mid x \varepsilon L(\kappa) \wedge \{e\}(x, p)\downarrow\}$$

for some $e < \omega$ and $p \varepsilon L(\kappa)$. Recall that if $x, p \varepsilon L(\kappa)$ and $\{e\}(x, p)\downarrow$, then the associated wellfounded computation belongs to $L(\kappa)$. *E-recursive on* $L(\kappa)$ means both the set and its complement are E-r.e. on $L(\kappa)$.

The next two definitions are needed for Post's problem. The *greater E-r.e. projectum* is

$$\rho^\kappa = \mu\gamma_{\gamma \leqslant \kappa} \exists f\big[f : \gamma \to L(\kappa) \text{ is partial E-recursive and onto}\big].$$

The *lesser E-r.e. projectum is*

$$\eta^\kappa = \mu\gamma_{\gamma \leqslant \kappa} \exists A\big[A \varepsilon 2^\gamma - L(\kappa) \text{ and } A \text{ is E-r.e. on } L(\kappa)\big].$$

LEMMA 28. $\rho^\kappa = \eta^\kappa$.

LEMMA 29. *Assume* $L(\kappa)$ *admits Moschovakis witnesses. If* $p \varepsilon \kappa$ *and* $\gamma < \rho^\kappa$, *then*

$$\sup\{\kappa_r^{p,\delta} \mid \delta < \gamma\} < \kappa.$$

The last two lemmas are applied in Sacks [1985] to obtain a positive solution for Post's problem for every E-closed $L(\kappa)$. The argument assigns priorities to requirements using ordinals less than ρ^κ, but there are no injuries. Deeper structural results are unearthed by Slaman [1985b] for his proofs of splitting and density; both are finite injury arguments.

The most troubling open question about E-r.e. degrees is: does there exist an infinite injury argument despite the presence of Moschovakis divergence witnesses? I am confident the answer is (editor's note: the next phrase in the author's original manuscript, recovered from his plane abandoned in the Congo, was illegible).

Let $f : V \to V$ be a total function. f is *uniformly* Σ_1 *definable on every* Σ_1 *admissible set* if there is a lightface Σ_1^{ZF} formula $\mathcal{F}(x, y)$ such that for every Σ_1 admissible A:

$$f[A] \subseteq A;$$

$$f \restriction A = \{\langle a, b\rangle \mid \langle A, \varepsilon\rangle \models \mathcal{F}(\underline{a}, \underline{b})\}.$$

THEOREM 30 (van de Wiele [1982]). *Let $f: V \to V$ be total. Then* (i) $\Leftrightarrow$ (ii).

(i) *f is E-recursive.*

(ii) *f is uniformly Σ_1 definable on every Σ_1 admissible set.*

Theorem 30 clarifies the nature of "effective unbounded search".

References

J. E. FENSTAD

[1980] *General Recursion Theory*, Springer, Heidelberg.

R. O. GANDY

[1967] General recursive functionals of finite type and hierarchies of functionals, *Ann. Fac. Sci. Univ. Clermont-Ferrand*, 35, pp. 202–242.

E. R. GRIFFOR

[1980] *E-recursively Enumerable Degrees*, Ph.D. Thesis, Massachusetts Institute of Technology.

E. R. GRIFFOR AND D. NORMANN

[1982] *Effective cofinalities and admissibility in E-recursion*, Preprint, University of Oslo.

T. GRILLIOT

[1969] Selection functions for recursive functionals, *Notre Dame J. Formal Logic*, pp. 225–234.

L. HARRINGTON

[1973] *Contributions to Recursion Theory in Higher Types*, Ph.D. Thesis, Massachusetts Institute of Technology.

L. HARRINGTON AND D. MACQUEEN

[1976] Selection in abstract recursion theory, *J. Symbolic Logic*, 41, pp. 153–158.

S. C. KLEENE

[1959] Recursive functionals and quantifiers of finite types I, *Trans. Amer. Math. Soc.*, 91, pp. 1–52.

[1963] Recursive functionals and quantifiers of finite types II, *Trans. Amer. Math. Soc.*, 108, pp. 106–142.

D. MACQUEEN

[1972] *Recursion in Finite Types*, Ph.D. Thesis, Massachusetts Institute of Technology.

Y. N. MOSCHOVAKIS

[1967] Hyperanalytic predicates, *Trans. Amer. Math. Soc.*, 138, pp. 249–282.

[1980] *Descriptive Set Theory*, North-Holland, Amsterdam.

[1981] On the Grilliot–Harrington–MacQueen theorem, in: *Logic Year 79–80*, Lecture Notes in Mathematics, Vol. 859, Springer, Berlin, pp. 246–267.

D. NORMANN

[1975] *Degrees of functionals*, Preprint Series in Math. 22, University of Oslo.

[1978] Set recursion, in: *Generalized Recursion Theory II*, North-Holland, Amsterdam, pp. 303–320.

G. E. SACKS

[1970] Recursion in objects of finite type, *Proc. Internat. Cong. Math.*, 80, pp. 193–205.

[1974] The 1-section of a type n object, in: *Generalized Recursion Theory*, North-Holland, Amsterdam, pp. 81–96.

[1977] The k-section of a type n object, *Amer. J. Math.*, 99, pp. 901–917.

[1980] Post's problem, absoluteness and recursion in finite types, in: *The Kleene Symposium*, North-Holland, Amsterdam, pp. 201–222.

[1985] Post's problem in E-recursion, in: *Proc. Symposia Pure Math. Amer. Math. Soc.*, Vol. 42, pp. 177–193.

[1986] On the limits of E-recursive enumerability, *Ann. Pure Appl. Logic*, 31, pp. 87–120.

[1990] *Higher Recursion Theory*, Springer, Heidelberg.

[1996] Effective versus Proper Forcing, *Ann. Pure Appl. Logic*, 81, pp. 171–185.

G. E. SACKS AND T. A. SLAMAN

[1987] Inadmissible forcing, *Adv. in Math.*, 66, pp. 1–30.

T. A. SLAMAN

[1981] *Aspects of E-recursion*, Ph.D. Thesis, Harvard University.

[1985a] Reflection and forcing in E-recursion theory, *Ann. Pure Appl. Logic*, 29, pp. 79–106.

[1985b] The E-recursively enumerable degrees are dense, in: *Proc. Symposia Pure Math. Amer. Math. Soc.*, Vol. 42, pp. 195–213.

J. VAN DE WIELE

[1982] Recursive dilators and generalized recursion, in: *Proc. Herbrand Symposium*, North-Holland, Amsterdam, pp. 325–332.

CHAPTER 11

Recursion on Abstract Structures

Peter G. Hinman

Mathematics Department, University of Michigan, Ann Arbor, MI 48109, USA

Contents

1. Introduction 316
2. Structures and functionals 318
3. Register machines over first-order structures 323
4. Recursion 326
5. The main examples 331
6. Structures with arithmetic 339
7. Sections and envelopes 344
8. Appendix 354
References 357

HANDBOOK OF COMPUTABILITY THEORY
Edited by E.R. Griffor

We develop and compare two models for computability over an abstract structure. The first, characterized in term of generalized register machines, provides a good theory for a large class of first-order structures. The second, defined in terms of minimal solutions for functional equations, is more versatile and handles many common second-order examples.

1. Introduction

From its origins as a theory of computable functions of natural numbers, Recursion Theory has been extended and generalized to include many other sorts of 'computable' functions. The first such extensions were directed towards establishing theories of computability over particular domains, such as initial segments of the ordinal numbers or the finite-type structure over the natural numbers. Later work developed theories of computability over abstract structures, initially first-order structures, then special sorts of second-order structures. Along the way appeared several still more abstract axiomatic computation theories. Our goal here is to survey some of these developments with an emphasis on the role of inductive definability as the basic principle of general computability. We make no claim to completeness.

One of the striking features of ordinary (or 'classical') Recursion Theory is the diversity of equivalent formal characterizations. Most of these fall into one of three general classes by characterizing a (partial) recursive function(al) as one which

(1) is computable by an abstract machine, such as a Turing or register machine;

(2) is definable in existential form in an interpreted formal language such as first-order arithmetic or an equational calculus;

(3a) belongs to the smallest class of functions which contains certain very simple functions and is closed under a small number of generating principles;

(3b) is represented by objects belonging to some other inductively defined set.

Attempts to develop generalized recursive notions of computability have made use of extended versions of each of these. A full review of these efforts would merit a (long!) article to itself, and we mention here only some of the highlights.

Characterizations of computability based on abstract machines have a natural appeal, although they have not been uniformly successful. For example, a version of Turing machines for recursion in finite types was worked out carefully in [Kleene, 1962a, 1962b], but this approach was largely neglected by subsequent research in favor of the inductive definability (3b) approach of [Kleene, 1959, 1963]. On the other hand, Friedman's notion [1971] of a finite algorithmic procedure (FAP) over any first-order structure was elaborated in the articles [Moldestad, Stoltenberg-Hansen and Tucker, 1980a, 1980b] and [Tucker, 1980] (or see the excellent summary in [Fenstad, 1980, Chapter 0]) and more recently the article [Blum, Shub and Smale, 1989] characterizes a natural notion of computability over the real numbers – and more generally over any ring – in terms of a kind of register machine. We shall develop this approach in several sections below.

In the book [1988] and the series of papers [1992a, 1992b, 1992c, 1995] Tucker and Zucker explore several other notions of computability over abstract algebras.

Among other topics these works consider versions of μ-recursion, while-programs, semi-computability, flow charts, infinitary quantifier-free definability, and computable functionals. Some of these build on earlier work of Engeler [1968, 1975] and Herman and Isard [1970]. Most of these notions coincide extensionally with one of those described below.

Definability has been a very successful framework for certain parts of Generalized Recursion Theory. It was early recognized that for an admissible ordinal α, existential definability over the segment L_α of the hierarchy of constructible sets is a natural and useful characterization of the partial α-recursive functions. The point is that just as the computation process for obtaining the value of an ordinary recursive function can be coded as a natural number, so the corresponding process over an admissible ordinal α can be coded as a member of L_α. This idea was extended in the Companion Theorem [Moschovakis, 1974, 9E1] which shows that for any Spector class Γ of relations (an analogue of the semi-recursive relations) over a transitive infinite set A (in particular an infinite ordinal), there exists an admissible set $\mathcal{M} \supseteq A$ such that (among other properties) the members of Γ are exactly the relations Σ_1 definable over $\mathcal{M}$ relative to a single fixed relation. This more complex method is necessary because most structures are not rich enough to code the relevant computations. This approach will not play a big role in this article.

It has turned out that approach (3) has the widest range of application. Not only does it work in more contexts, but it seems more intrinsic in avoiding reference to other constructs such as machines or formulas. The ordinary partial recursive functions over ω form the smallest class which contains some simple functions (for example, the primitive recursive functions) and is closed under composition and search. In developing the theory of recursion for finite types, Kleene ([1959, 1963]) relied on method (3b) in defining inductively a class $\mathcal{C}$ of tuples of the form (a, x, n), where $a, n \in \omega$ and x is a finite sequence of finite-type objects, and calling a function F partial recursive iff for some (index) a, $F = \{a\}$, where for all x and n,

$$\{a\}(x) \simeq n \iff (a, x, n) \in \mathcal{C}.$$

The closure conditions for $\mathcal{C}$ correspond to simple initial functions, composition and substitution, and the key principle of data-as-program which introduces the function

$$H(a, x) \simeq \{a\}(x).$$

This approach is convenient for many purposes because it introduces indexing (Gödel numbering) at the beginning of the development and makes available the Second Recursion Theorem as a basic tool for complex computations. The details for finite types and ordinals are worked out in [Hinman, 1978]. The drawback to this approach is that it relies on the richness of the underlying structure to code the 'programs' as indices belonging to the universe of computation. At the very least this requires a copy of the natural numbers and some kind of coding for finite sequences.

It was the insight of Platek [1966] that many of these problems are avoided if we take as our basic generating principle the *First* Recursion Theorem instead of the Second. This idea was further refined by Moldestad [1977], much improved by Kechris and Moschovakis in [1977], and further developed in the papers [Moschovakis, 1983, 1989]. The key property of this approach is that it is inherently second-order; computable functions are generated as least fixed points of computable functionals. This means that computation relative to fixed functions and even functionals is included in the theory from the beginning. This approach will be developed in some detail below.

The plan for the rest of this article is as follows. In Section 2 we set out our basic framework and discuss the simplest functions over a structure, the explicit ones. In Sections 3 and 4 we introduce the two main models of computation based on register machines and recursion. Section 5 contains some examples. In Section 6 we show that much of basic recursion theory generalizes to computation over a structure which can be considered an extension of ordinary arithmetic. Section 7 explores some of the basic properties of semi-recursive relations in an abstract setting, and Section 8 gives two technical proofs.

2. Structures and functionals

Much of our notation will be relatively but not entirely standard, and we review it here. A *first-order structure* is a quadruple $\mathfrak{A} = (A, \mathsf{Rel}_{\mathfrak{A}}, \mathsf{Fn}_{\mathfrak{A}}, \mathsf{Dis}_{\mathfrak{A}})$, where A is a non-empty set, $\mathsf{Rel}_{\mathfrak{A}}$ is a finite (possibly empty) set of finitary *relations* on A – that is, subsets of the Cartesian power A^k, for some natural number $k \in \omega$ – $\mathsf{Fn}_{\mathfrak{A}}$ is a finite (possibly empty) set of finitary *functions* on A with values in A, and $\mathsf{Dis}_{\mathfrak{A}}$ is a finite subset of A, the set of *distinguished elements* of $\mathfrak{A}$. The equality relation $=_A$ will be included explicitly rather than implicitly as is common in first-order logic. Associated with such a structure is a standard *first-order language* with non-logical relation, function and constant symbols corresponding to each of relations, functions, and distinguished elements, respectively. The *terms* and *formulas* of this language are defined as usual. We shall always assume that A has two distinct elements denoted as $0_{\mathfrak{A}}$ and $1_{\mathfrak{A}}$, which are either distinguished elements or values of closed terms. Particular structures will be described in the form $(A, R, \ldots, F, \ldots, a \ldots)$.

Let $\mathsf{Pf}^l_{\mathfrak{A}}$ denote the set of all l-ary partial functions on A – that is, partial maps $f : A^l \to A$. Then a *finitary functional on* A is a partial mapping

$$F : A^k \times \mathsf{Pf}^{l_0}_{\mathfrak{A}} \times \cdots \times \mathsf{Pf}^{l_{n-1}}_{\mathfrak{A}} \to A,$$

for some *type* $(k; \boldsymbol{l}) = (k; l_0, \ldots, l_{n-1})$. In particular, a k-ary partial function is also a functional of type $(k;)$. To specify the value of such a functional we often write $F(\boldsymbol{a}, \boldsymbol{f}) \simeq e$. Thus $\boldsymbol{a}$ represents $a_0, \ldots, a_{k-1}$ and $\boldsymbol{f}$ represents $f_0, \ldots, f_{n-1}$ (for unspecified k and n, which are either determined by context or are arbitrary). The relation $\simeq$ is read 'is defined and equal to'.

We shall be concerned exclusively with *monotone* functionals, which have the property

$$F(\boldsymbol{a}, \boldsymbol{f}) \simeq e \text{ and } \bigwedge_{i<n} [f_i \subseteq g_i] \Longrightarrow F(\boldsymbol{a}, \boldsymbol{g}) \simeq e.$$

Here, of course, $f \subseteq g$ means inclusion as a set of ordered pairs or equivalently that g is a function which extends f. The restriction to monotonicity is dictated by our interest in *deterministically computable* functionals, which we view as given by algorithms equipped with 'black box' subroutines or oracles for providing values of the function arguments as needed. Thus if the oracles for $\boldsymbol{f}$ provide all the requested values in the computation of $F(\boldsymbol{a}, \boldsymbol{f})$, then the oracles for $\boldsymbol{g}$ will provide exactly the same values and hence yield the same computation.

A *functional structure* is a pair $\mathfrak{A} = (A, \mathsf{Fnl}_{\mathfrak{A}})$, where $\mathsf{Fnl}_{\mathfrak{A}}$ is a finite set of monotone finitary functionals on A. We consider every first-order structure $\mathfrak{A} = (A, \mathsf{Rel}_{\mathfrak{A}}, \mathsf{Fn}_{\mathfrak{A}}, \mathsf{Dis}_{\mathfrak{A}})$ to be also a functional structure by identifying it with the functional structure whose functionals are: the characteristic function χ_R, a functional of type $(k;)$, for each $R \in \mathsf{Rel}_{\mathfrak{A}}$; the members of $\mathsf{Fn}_{\mathfrak{A}}$, which are already functionals of some type $(k;)$; and a constant functional of type $(0;)$ for each $a \in \mathsf{Dis}_{\mathfrak{A}}$.

As examples and for future reference, we consider several more or less familiar functionals and functional structures.

(1) Probably the most basic example is the *application* or *evaluation functional* of type $(k; k)$ which exercises the call to the oracle for f:

$$\mathsf{Ev}^{k,k}(\boldsymbol{a}, f) \simeq f(\boldsymbol{a}).$$

(2) Over the set ω of natural numbers we have the *search functional* of type $(0; 1)$:

$$\begin{aligned} \mathsf{Se}_{\omega}(g) &\simeq \text{'least' } n[g(n) \simeq 0] \\ &\simeq n \Longleftrightarrow g(n) \simeq 0 \text{ and } (\forall p < n) g(p)\!\downarrow \neq 0. \end{aligned}$$

Here $g(p)\!\downarrow \neq 0$ is read '$g(p)$ is defined and different from 0' and means that for some $q \neq 0$, $g(p) \simeq q$.

(3) If H is a functional of type $(k + l; \ldots)$, then for each $\boldsymbol{a}$ and $\boldsymbol{g}$, $\lambda \boldsymbol{b}.H(\boldsymbol{a}, \boldsymbol{b}, \boldsymbol{g})$ denotes the function $\boldsymbol{b} \mapsto H(\boldsymbol{a}, \boldsymbol{b}, \boldsymbol{g})$ of type $(l;)$. Thus if G is of type $(k; l, \ldots)$, then

$$F(\boldsymbol{a}, \boldsymbol{g}) \simeq G\big(\boldsymbol{a}, \lambda \boldsymbol{b}.H(\boldsymbol{a}, \boldsymbol{b}, \boldsymbol{g}), \boldsymbol{g}\big)$$

is of type $(k; \ldots)$. A class $\mathcal{F}$ of functionals has the *substitution property* or is *closed under substitution* iff $G, H \in \mathcal{F}$ implies also $F \in \mathcal{F}$ (together with more general instances as described below). Substitution is a powerful property of a class $\mathcal{F}$ since it implies other closure properties. For example, if $\mathcal{F}$ is a class of functionals over ω with the substitution property and $\mathsf{Se}_\omega \in \mathcal{F}$, then for any $G \in \mathcal{F}$, the functional

$$F(\boldsymbol{m}, \boldsymbol{f}) \simeq \text{`least'}\, n\big[G(\boldsymbol{m}, n, \boldsymbol{f}) \simeq 0\big] \simeq \mathsf{Se}_\omega\big(\lambda n.G(\boldsymbol{m}, n, \boldsymbol{f})\big)$$

is also in $\mathcal{F}$. In other words, $\mathcal{F}$ is closed under search over ω.

(4) Another nice example of a particular functional whose main importance is to join with substitution to guarantee a closure property is the *primitive recursion* functional over ω of type (2; 1):

$$\mathsf{Prim}_\omega(p, n, h) \simeq \begin{cases} n & \text{if } p = 0; \\ h(\mathsf{Prim}_\omega(p-1, n, h)) & \text{otherwise.} \end{cases}$$

Thus if $\mathsf{Prim}_\omega, G, H \in \mathcal{F}$, $\mathcal{F}$ is closed under substitution, and

$$F(p, \boldsymbol{m}, \boldsymbol{f}) \simeq \begin{cases} G(\boldsymbol{m}, \boldsymbol{f}) & \text{if } p = 0; \\ H(F(p-1, \boldsymbol{m}, \boldsymbol{f}), p, \boldsymbol{m}, \boldsymbol{f}) & \text{otherwise,} \end{cases}$$

then $F(p, \boldsymbol{m}, \boldsymbol{f}) \simeq \mathsf{Prim}_\omega(p, G(\boldsymbol{m}, \boldsymbol{f}), \lambda q.H(q, p, \boldsymbol{m}, \boldsymbol{f}))$, so also $F \in \mathcal{F}$.

(5) Let κ be a limit ordinal. The *supremum functional* over κ of type (1; 1) is

$$\begin{aligned} \mathsf{Sup}_\kappa(\rho, f) &\simeq \mathrm{Sup}^+_{\pi<\rho} f(\pi) \\ &\simeq \nu < \kappa \iff (\forall \pi < \rho) f(\pi)\!\downarrow\, < \nu \text{ and} \\ &\qquad (\forall \mu < \nu)(\exists \pi < \rho) f(\pi) \geqslant \mu. \end{aligned}$$

Note that even if f is a function on κ defined for all $\pi < \rho$, $\mathsf{Sup}_\kappa(\rho, f)$ is undefined when $\{f(\pi)\colon \pi < \rho\}$ is cofinal in κ. The characteristic property of recursion over recursively regular or admissible ordinals κ is that if f is partial κ-recursive and defined for all $\pi < \rho$, then $\mathsf{Sup}_\kappa(\rho, f)$ is defined. We shall see below that the usual notion of partial κ-recursion can be characterized as recursion over the functional structure $(\kappa, \mathsf{Sc}, 0, <, \mathsf{Sup}_\kappa, \mathsf{Se}_\kappa)$, where Sc is the usual successor function $\mu \mapsto \mu + 1$ and

$$\begin{aligned} \mathsf{Se}_\kappa(g) &\simeq \text{`least'}\, \nu[g(\nu) \simeq 0] \\ &\simeq \nu \iff g(\nu) \simeq 0 \text{ and } (\forall \pi < \nu) g(\pi)\!\downarrow\, \neq 0. \end{aligned}$$

(6) Let A be a transitive set of sets. The *replacement functional* over A of type (1; 1) is

$$\begin{aligned} \mathsf{Repl}_A(a, f) &\simeq \{f(b)\colon b \in a\}_A \\ &\simeq c \in A \iff (\forall b \in a) f(b)\!\downarrow\, \in c \text{ and} \\ &\qquad (\forall d \in c)(\exists b \in a) f(b) \simeq d. \end{aligned}$$

The notion of E-recursion over A will be seen below to be characterized as recursion over the structure $\mathfrak{A} = (A, \emptyset, \subseteq, \mathsf{pair}, \cup, \mathsf{Repl}_A)$ and A is E-closed iff for any $a \in A$ and $\mathfrak{A}$-recursive function f, if $f(b)$ is defined for all $b \in a$, then also $\mathsf{Repl}_A(a, f)$ is defined.

(7) Over any set A there is the functional $\mathsf{E}^\sharp_A$ of type (0;1):

$$\mathsf{E}^\sharp_A(f) \simeq \begin{cases} 1_\mathfrak{A} & \text{if } (\exists a \in A)\, f(a) \simeq 1_\mathfrak{A}; \\ 0_\mathfrak{A} & \text{if } (\forall a \in A)\, f(a) \simeq 0_\mathfrak{A}. \end{cases}$$

We shall see in Section 5 that if $\mathfrak{A}$ is a first-order structure, then recursion over $(\mathfrak{A}, =_A, \mathsf{E}^\sharp_A)$ corresponds to first-order inductive definability over $\mathfrak{A}$.

The first class of functionals that we associate with a functional structure consists of those that are explicitly defined from the basic ones in $\mathsf{Fnl}_\mathfrak{A}$.

DEFINITION 2.1. The class $\mathsf{Expl}_\mathfrak{A}$ of $\mathfrak{A}$-*explicit functionals* is the smallest class of functionals over $\mathfrak{A}$ such that

(i) $\mathsf{Expl}_\mathfrak{A}$ contains all of the *initial functionals*:

(a) $\mathsf{Pr}^{k,\boldsymbol{l}}_i(\boldsymbol{a}, \boldsymbol{f}) = a_i$ $(i < k)$;

(b) $\mathsf{Ev}^{k,\boldsymbol{l}}_j(\boldsymbol{a}, \boldsymbol{f}) \simeq f_j(\boldsymbol{a})$ $(j < n,\ l_j = k)$;

(c) all members of $\mathsf{Fnl}_\mathfrak{A}$.

(ii) $\mathsf{Expl}_\mathfrak{A}$ is closed under *composition*, *substitution*, and *definition by cases*:

(a) for any $G, H_0, \ldots, H_{m-1} \in \mathsf{Expl}_\mathfrak{A}$, if

$$F(\boldsymbol{a}, \boldsymbol{f}) \simeq G\big(H_0(\boldsymbol{a}, \boldsymbol{f}), \ldots, H_{m-1}(\boldsymbol{a}, \boldsymbol{f}), \boldsymbol{f}\big),$$

then also $F \in \mathsf{Expl}_\mathfrak{A}$;

(b) for any $G, H_0, \ldots, H_{m-1} \in \mathsf{Expl}_\mathfrak{A}$, if

$$F(\boldsymbol{a}, \boldsymbol{f}) \simeq G\big(\boldsymbol{a}, \lambda \boldsymbol{b}^0.H_0\big(\boldsymbol{a}^0, \boldsymbol{b}^0, \boldsymbol{f}\big), \ldots, \lambda \boldsymbol{b}^{m-1}.H_{m-1}\big(\boldsymbol{a}^{m-1}, \boldsymbol{b}^{m-1}, \boldsymbol{f}\big)\big),$$

where $\boldsymbol{a}^0, \ldots, \boldsymbol{a}^{m-1}$ are (possibly empty) lists selected from $\boldsymbol{a}$, then also $F \in \mathsf{Expl}_\mathfrak{A}$;

(c) for any G, H, and $I \in \mathsf{Expl}_\mathfrak{A}$, if

$$F(c, \boldsymbol{a}, \boldsymbol{f}) \simeq \begin{cases} G(\boldsymbol{a}, \boldsymbol{f}) & \text{if } c = 1_\mathfrak{A}; \\ H(\boldsymbol{a}, \boldsymbol{f}) & \text{if } c = 0_\mathfrak{A}; \\ I(\boldsymbol{a}, \boldsymbol{f}) & \text{otherwise}; \end{cases}$$

then also $F \in \mathsf{Expl}_\mathfrak{A}$.

We note that easily

PROPOSITION 2.2. $\mathsf{Expl}_\mathfrak{A}$ *is closed under argument rearrangement and expansion – that is, for any* $G \in \mathsf{Expl}_\mathfrak{A}$, *if*

$$F(a_0, \ldots, a_{k-1}, f_0, \ldots, f_{l-1}) \simeq G(a_{i_0}, \ldots, a_{i_{m-1}}, f_{j_0}, \ldots, f_{j_{n-1}}),$$

where $i_0, \ldots, i_{m-1} < k$ *and* $j_0, \ldots, j_{n-1} < l$, *then also* $F \in \mathsf{Expl}_{\mathfrak{A}}$.

PROOF. We illustrate the proof with an example: if $G \in \mathsf{Expl}_{\mathfrak{A}}$ and $F(a, b, c, f, g) \simeq G(c, b, g)$, with f and g of type $(k;)$ and $(l;)$, respectively, then set

$$\begin{aligned} H(c, b, \boldsymbol{d}, f, g) &\simeq \mathsf{Ev}_1^{k,k,l}\big(\mathsf{Pr}_2^{k+2,k,l}(c, b, \boldsymbol{d}, f, g), \ldots, \\ &\qquad\qquad \mathsf{Pr}_{k+1}^{k+2,k,l}(c, b, \boldsymbol{d}, f, g), f, g\big) \\ &\simeq g(\boldsymbol{d}), \\ G_0(c, b, f, g) \simeq G(c, b, g) &\simeq G\big(c, b, \lambda \boldsymbol{d}.H(c, b, \boldsymbol{d}, f, g)\big). \end{aligned}$$

Then $H, G_0 \in \mathsf{Expl}_{\mathfrak{A}}$ and since

$$F(a, b, c, f, g) \simeq G_0\big(\mathsf{Pr}_2^{3,k,l}(a, b, c, f, g), \mathsf{Pr}_1^{3,k,l}(a, b, c, f, g), f, g\big),$$

also $F \in \mathsf{Expl}_{\mathfrak{A}}$. □

REMARK 2.3. It is easy to derive that the functionals

$$\mathsf{Cmp}^{k,m}(\boldsymbol{a}, g, h_0, \ldots, h_{m-1}) \simeq g\big(h_0(\boldsymbol{a}), \ldots, h_{m-1}(\boldsymbol{a})\big)$$

and

$$\mathsf{Cases}^k(c, \boldsymbol{a}, g, h, i) \simeq \begin{cases} g(\boldsymbol{a}) & \text{if } c = 1_{\mathfrak{A}}; \\ h(\boldsymbol{a}) & \text{if } c = 0_{\mathfrak{A}}; \\ i(\boldsymbol{a}) & \text{otherwise}; \end{cases}$$

are $\mathfrak{A}$-explicit. Conversely, any class of functionals which includes these functionals and is closed under substitution (in the form given above) is also closed under composition and definition by cases, since, for example,

$$G\big(H(\boldsymbol{a}, \boldsymbol{f}), \boldsymbol{f}\big) \simeq \mathsf{Cmp}^{k,1}\big(\boldsymbol{a}, \lambda b.G(b, \boldsymbol{f}), \lambda \boldsymbol{c}.H(\boldsymbol{c}, \boldsymbol{f})\big).$$

Although the placement of variables in the schemas for composition and substitution is precise and somewhat arbitrary, by argument rearrangement and expansion (2.2), all reasonable variations are derivable. For example, if G, H, and I are $\mathfrak{A}$-explicit and

$$F(\boldsymbol{a}, \boldsymbol{b}, \boldsymbol{f}, \boldsymbol{g}) \simeq G\big(\boldsymbol{a}, H(\boldsymbol{a}, \boldsymbol{g}), \lambda \boldsymbol{c}.I(\boldsymbol{b}, \boldsymbol{c}, \boldsymbol{f}), \boldsymbol{g}, \boldsymbol{f}\big),$$

then also F is $\mathfrak{A}$-explicit. Similarly, we have closure under much more general schemas of definition by cases – for example, if G, H, I, J, and K are $\mathfrak{A}$-explicit (or partial $\mathfrak{A}$-recursive), J and K are total functions, and

$$F(\boldsymbol{a}, \boldsymbol{f}) \simeq \begin{cases} G(\boldsymbol{a}, \boldsymbol{f}) & \text{if } J(\boldsymbol{a}) = 1_{\mathfrak{A}}; \\ H(\boldsymbol{a}, \boldsymbol{f}) & \text{if } J(\boldsymbol{a}) = 0_{\mathfrak{A}} \text{ and } K(\boldsymbol{a}) = 1_{\mathfrak{A}}; \\ I(\boldsymbol{a}, \boldsymbol{f}) & \text{otherwise}, \end{cases}$$

then also F is $\mathfrak{A}$-explicit. We shall use these extensions of the basic schemas without mention. In particular, we will need the following functions which are $\mathfrak{A}$-explicit for every $\mathfrak{A}$:

$$\mathsf{Not}(a) \simeq \begin{cases} 0_{\mathfrak{A}} & \text{if } a = 1_{\mathfrak{A}}; \\ 1_{\mathfrak{A}} & \text{if } a = 0_{\mathfrak{A}}; \end{cases}$$

$$\mathsf{And}(a,b) \simeq \begin{cases} 1_{\mathfrak{A}} & \text{if } a = 1_{\mathfrak{A}} = b; \\ 0_{\mathfrak{A}} & \text{if } a = 0_{\mathfrak{A}} \text{ or } b = 0_{\mathfrak{A}}; \end{cases}$$

$$\mathsf{Or}(a,b) \simeq \begin{cases} 1_{\mathfrak{A}} & \text{if } a = 1_{\mathfrak{A}} \text{ or } b = 1_{\mathfrak{A}}; \\ 0_{\mathfrak{A}} & \text{if } a = 0_{\mathfrak{A}} = b; \end{cases}$$

$$\mathsf{IsZero}_{\mathfrak{A}}(a) \simeq \begin{cases} 1_{\mathfrak{A}} & \text{if } a = 0_{\mathfrak{A}}; \\ 0_{\mathfrak{A}} & \text{otherwise}; \end{cases}$$

$$\mathsf{IsOne}_{\mathfrak{A}}(a) \simeq \begin{cases} 1_{\mathfrak{A}} & \text{if } a = 1_{\mathfrak{A}}; \\ 0_{\mathfrak{A}} & \text{otherwise}. \end{cases}$$

Note that composing these functions with total functions gives the expected result, but while

$$\mathsf{And}\big(F(x), G(x)\big)\!\downarrow \iff F(x)\!\downarrow \text{ and } G(x)\!\downarrow,$$

the analogous equivalence for Or is not in general true. Indeed, we shall see in Section 7 that although the class of $\mathfrak{A}$-semi-recursive relations is always closed under intersection, an additional hypothesis is needed to guarantee closure under union.

PROPOSITION 2.4. *If* $\mathfrak{A}' = (A, \mathsf{Fnl}_{\mathfrak{A}'})$, *and* $\mathsf{Fnl}_{\mathfrak{A}} \subseteq \mathsf{Fnl}_{\mathfrak{A}'} \subseteq \mathsf{Fnl}_{\mathfrak{A}} \cup \mathsf{Expl}_{\mathfrak{A}}$, *then* $\mathsf{Expl}_{\mathfrak{A}} = \mathsf{Expl}_{\mathfrak{A}'}$.

PROOF. Immediate. □

We introduce one other notion here for later reference:

DEFINITION 2.5. A functional F is *continuous* iff whenever $F(\boldsymbol{a}, \boldsymbol{f}) \simeq e$, there exist finite $f_0' \subseteq f_0, \ldots, f_{n-1}' \subseteq f_{n-1}$ such that also $F(\boldsymbol{a}, \boldsymbol{f}') \simeq e$.

PROPOSITION 2.6. *For any functional structure* $\mathfrak{A}$, *if every* $F \in \mathsf{Fnl}_{\mathfrak{A}}$ *is continuous, then also every* $F \in \mathsf{Expl}_{\mathfrak{A}}$ *is continuous.*

PROOF. A straightforward induction. □

3. Register machines over first-order structures

For all of this section we restrict $\mathfrak{A}$ to be a first-order structure. We recall the notion of an *unlimited register machine* (URM) due to Shepherdson and Sturgis [1963];

we follow essentially the excellent presentation of Cutland [1980]. The machine M has an infinite sequence $R_0, R_1, \ldots$ of *registers* (memory cells) each of which contains at each discrete time step a natural number $r_0, r_1, \ldots$. At the beginning of a computation all $r_p = 0$. A *program* $\pi = (I_0, \ldots, I_m)$ consists of a finite sequence of *instructions*, each of one of the following four types:

$$r_p := r_p + 1,$$
$$r_p := 0,$$
$$r_p := r_q,$$
$$\text{if } r_p = r_q, \text{ go to } I_s, \text{ else go to } I_t.$$

A computation $\pi(m_0, \ldots, m_{k-1})$ proceeds by setting $r_p := m_p$ for $p < k$ and $r_p := 0$ for $p \geqslant k$ and executing the instructions starting with I_0 in numerical order except as dictated by a conditional instruction. The computation terminates if and when the machine would attempt to execute a non-existent instruction, either as the result of a conditional instruction or by executing the last instruction of the program, which is not a conditional instruction. If it terminates, the computation has output r_0. Thus every program π computes, for each k, a k-ary partial function (also denoted) π defined by: $\pi(m_0, \ldots, m_{k-1}) \simeq n \Longleftrightarrow \pi(m_0, \ldots, m_{k-1})$ terminates with $r_0 = n$. The functions so computed are exactly the partial recursive functions.

Note that a computation $\pi(m_0, \ldots, m_{k-1})$ addresses only the finitely many registers mentioned in the program π so can be considered to take place on a machine with only finitely many registers. It is easy to extend this formalism to compute also the partial recursive functionals by introducing new instructions of the form

$$r_p := f_j(r_{q_0}, \ldots, r_{q_{l_j-1}}) \quad (j < n).$$

Note that although a program may be considered to act on a sequence of numerical arguments of any length, the type $\boldsymbol{l} = (l_0, \ldots, l_{n-1})$ of its function arguments is determined by the program.

For any first-order structure $\mathfrak{A} = (A, \mathsf{Fn}_{\mathfrak{A}}, \mathsf{Rel}_{\mathfrak{A}}, \mathsf{Dis}_{\mathfrak{A}})$, we define an *unlimited register machine over* $\mathfrak{A}$ (URM$_{\mathfrak{A}}$) of type $\boldsymbol{l}$ as follows. Each register now contains at each time step an element of A. A program $\pi = (I_0, \ldots, I_m)$ consists of a finite sequence of instructions, each of one of the following five types:

$$r_p := f(r_{q_0}, \ldots, r_{q_{k-1}}) \quad \text{for a } k\text{-ary } f \in \mathsf{Fn}_{\mathfrak{A}},$$
$$r_p := a \quad \text{for } a \in \mathsf{Dis}_{\mathfrak{A}},$$
$$r_p := f_j(r_{q_0}, \ldots, r_{q_{l_j-1}}) \quad (j < n),$$
$$r_p := r_q,$$
$$\text{if } R(r_{q_0}, \ldots, r_{q_{k-1}}), \text{ go to } I_s, \text{ else go to } I_t \quad \text{for a } k\text{-ary } R \in \mathsf{Rel}_{\mathfrak{A}}.$$

A program π for a URM$_{\mathfrak{A}}$ of type $\boldsymbol{l}$ computes in the same way for each k a partial functional $\pi : A^k \times \mathsf{Pf}_{\mathfrak{A}}^{l_0} \times \cdots \times \mathsf{Pf}_{\mathfrak{A}}^{l_{n-1}} \to A$, of type $(k; \boldsymbol{l})$. We call the resulting partial functionals $\mathfrak{A}$-*register computable* and denote the set of them by $\mathsf{Reg}_{\mathfrak{A}}$.

COROLLARY 3.1. *If* $\Omega = (\omega, \mathsf{Sc}, =_\omega, 0)$, *then the* Ω*-register computable functionals are exactly the partial recursive functionals.*

Although the elementary parts of a theory of computability can be established for the register computable functions over any $\mathfrak{A}$, without further assumptions this is a rather weak notion of computability. First we have

PROPOSITION 3.2. *For any first-order structure* $\mathfrak{A}$, $\mathsf{Reg}_{\mathfrak{A}}$ *is closed under composition, substitution, and definition by cases. Hence all* $\mathfrak{A}$*-explicit functions are* $\mathfrak{A}$*-register computable.*

PROOF. These are all simple exercises in programming. Closure of the class of register computable functions under definition by cases follows easily using the conditional instruction, and closure under composition can be proved exactly as in [Cutland, 1980, 2.3.1]. For substitution, consider the special case where $F(\boldsymbol{a}) \simeq G(\boldsymbol{a}, \lambda \boldsymbol{b}.H(\boldsymbol{a}, \boldsymbol{b}))$. The program for G contains instructions of the form $r_p := h(r_{q_0}, \ldots, r_{q_{l-1}})$. We construct a program for F by replacing each such instruction by the program for computing H modified to use the input of G together with $r_{q_0}, \ldots, r_{q_{l-1}}$ for its input, otherwise to use registers with indices larger than any used by G, and to store its output value in r_p. □

Over the natural numbers it is easy to show [Cutland, 1980, 2.5.2] that the register computable functionals are also closed under search and hence include all of the partial recursive functions. The analogous result here will be that the class of register computable functions is closed under least-fixed-point recursion, as defined in the next section. We shall be able to prove this (in Section 6 below) only for structures which contain a copy of the natural numbers. Partly in preparation for this, we sketch here how to assign indices to the register computable functionals over an arbitrary first-order structure.

First, it is a standard exercise to assign Gödel numbers (indices) to programs. Fix a primitive recursive coding $\langle m_0, \ldots, m_{k-1} \rangle$ of finite sequences of natural numbers with the usual properties. Then assign numbers to URM$_{\mathfrak{A}}$ instructions as follows:

(i) To "$r_p := f(r_{q_0}, \ldots, r_{q_{k-1}})$" assign the number $\langle 0, i, p, q_0, \ldots, q_{k-1} \rangle$, where f is the i-th member of $\mathsf{Fn}_{\mathfrak{A}}$;

(ii) to "$r_p := a$" assign $\langle 1, i, p \rangle$, where a is the i-th member of $\mathsf{Dis}_{\mathfrak{A}}$;

(iii) to "$r_p := f_j(r_{q_0}, \ldots, r_{q_{l_j-1}})$" assign $\langle 2, j, p, q_0, \ldots, q_{l_j-1} \rangle$;

(iv) to "$r_p := r_q$" assign $\langle 3, p, q \rangle$;

(v) to "if $R(r_{q_0}, \ldots, r_{q_{k-1}})$, go to I_s, else go to I_t" assign $\langle 4, i, q_0, \ldots, q_{k-1}, s, t \rangle$, where R is the ith member of $\mathsf{Rel}_{\mathfrak{A}}$.

Then each program π is assigned a number $\#(\pi)$ which is the code for the finite sequence of the numbers of its instructions. For any natural number b it is primitively recursively computable whether or not b is the number of a program for a URM$_{\mathfrak{A}}$, and if so the minimum possible type $\boldsymbol{l}$. For each k, let $\{b\}^{\mathfrak{A}}_{\mathsf{Reg}}$ denote the empty function if b is not the number of a URM$_{\mathfrak{A}}$ program, otherwise the functional of type $(k; \boldsymbol{l})$ computed by the program. Clearly $\langle \{b\}^{\mathfrak{A}}_{\mathsf{Reg}}$: b is the number of a program$\rangle$ is an enumeration of $\mathsf{Reg}_{\mathfrak{A}}$. The effectiveness of the proof of the preceding proposition yields immediately

PROPOSITION 3.3. *The closure of* $\mathsf{Reg}_{\mathfrak{A}}$ *under composition, substitution, and definition by cases is effective in terms of these indices – that is, for each closure condition there is a primitive recursive function which computes an index for the resulting functional from indices of the component functionals.*

4. Recursion

In this section we develop the basic properties of the class $\mathsf{Rec}_{\mathfrak{A}}$ of functionals recursive over an arbitrary functional structure $\mathfrak{A}$. The fundamental generating principle is a version of the First Recursion Theorem, which asserts the existence of computable fixed points for computable functionals.

Consider first a monotone functional I of type $(k; k)$ over a set A. We call a k-ary partial function i a *fixed point* of I iff for all $\boldsymbol{a} \in A$,

$$I(\boldsymbol{a}, i) \simeq i(\boldsymbol{a}).$$

The terminology derives from regarding I as a mapping on functions: $I[i] = \lambda \boldsymbol{a}.I(\boldsymbol{a}, i)$. The monotonicity of I guarantees that I has, in fact, a *least fixed point*

$$I^{\infty} = \bigcap \{i\colon\ I[i] \subseteq i\}.$$

Clearly if I^{∞} is a fixed point at all, then it is least (under inclusion). To see that I^{∞} *is* a fixed point, note that for any i such that $I[i] \subseteq i$, we have by definition $I^{\infty} \subseteq i$, so by monotonicity, $I[I^{\infty}] \subseteq I[i] \subseteq i$. Hence, since I^{∞} is the intersection of all such i, also $I[I^{\infty}] \subseteq I^{\infty}$. Then, again by monotonicity, $I[I[I^{\infty}]] \subseteq I[I^{\infty}]$, so that $I[I^{\infty}]$ is itself such an i and hence $I^{\infty} \subseteq I[I^{\infty}]$. Thus $I^{\infty} = I[I^{\infty}]$.

The function I^{∞} may also be described ‘from below’ by *iterating* I starting with the empty function. For each ordinal σ, set

$$I^{<\sigma} = \bigcup_{\tau<\sigma} I^{\tau} \quad \text{and} \quad I^{\sigma} = I[I^{\sigma}].$$

Thus $I^{<0} = \emptyset$ and an easy induction shows that for all σ, $I^{<\sigma} \subseteq I^{\sigma}$ and that if $I^{<\sigma} = I^{\sigma}$, then also $I^{\sigma} = I^{\rho}$ for all $\rho > \sigma$. Since the domain of each I^{σ} is a subset of A^k, so of power $\mathsf{Card}(A)$, it follows easily that $I^{<\bar{\sigma}} = I^{\bar{\sigma}}$ for some $\bar{\sigma} < \mathsf{Card}(A)^+$, whence $I^{\bar{\sigma}}$ is a fixed point of I. Another straightforward induction shows that $I^{\sigma} \subseteq I^{\infty}$ for all σ, so since I^{∞} is the *least* fixed point of I, necessarily $I^{\bar{\sigma}} = I^{\infty}$. The least such $\bar{\sigma}$ is called the *closure ordinal* of I. As an important special case we have

PROPOSITION 4.1. *If I is continuous, then the closure ordinal of I is finite or ω.*

PROOF. Continuity easily implies that $I[I^{<\omega}] = I^{<\omega}$. □

The general notion of fixed point goes beyond this only in the inclusion of parameters. We call a functional I *iterable* iff it is of some type $(k+l; l_0, \dots, l_{n-1}, k)$ and a partial function i is a *fixed point of I with respect to* (parameters) $x = (\boldsymbol{b}, \boldsymbol{f})$ iff

$$I(\boldsymbol{a}, x, i) \simeq i(\boldsymbol{a}).$$

For any functional H and sequences of parameters x and y we define H_x by $H_x(y) := H(x, y)$ and call F a *fixed point of I* iff for each x, F_x is a fixed point of I with respect to x (or equivalently a fixed point of I_x) – in symbols,

$$I(\boldsymbol{a}, x, F_x) \simeq I_x(\boldsymbol{a}, F_x) \simeq F_x(\boldsymbol{a}) \simeq F(\boldsymbol{a}, x).$$

The *least fixed point* I^{∞} of I is simply the unique functional such that for each x, I_x^{∞} is the least fixed point of I_x. The corresponding iterative description of I^{∞} is then

$$I^{<\sigma} = \bigcup_{\tau<\sigma} I^{\tau} \quad \text{and} \quad I^{\sigma}(\boldsymbol{a}, x) \simeq I\big(\boldsymbol{a}, x, I_x^{<\sigma}\big).$$

As above, for each x there exists an ordinal $\sigma_x < \mathsf{Card}(A)^+$ such that $I_x^{<\sigma_x} = I_x^{\sigma_x} = I_x^{\infty}$, and hence an ordinal $\bar{\sigma} \leqslant \mathsf{Card}(A)^+$ such that $I^{<\bar{\sigma}} = I^{\bar{\sigma}} = I^{\infty}$.

Finally, we introduce the notion of simultaneous fixed points. Call a sequence $(I_0, \dots, I_n)$ *iterable* iff each I_p is of a type of the form $(k_p + l; l_0, \dots, l_{m-1}, k_0, \dots, k_n)$. A sequence of functionals $(F_0, \dots, F_n)$ is a *simultaneous fixed point* of $(I_0, \dots, I_n)$ iff for all x and $p = 0, \dots, n$

$$I_p(a_0, \dots, a_{k_p-1}, x, F_{0,x}, \dots, F_{n,x}) \simeq F_p(a_0, \dots, a_{k_p-1}, x).$$

An easy extension of the arguments above shows that there is a unique sequence $(I_0^{\infty}, \dots, I_n^{\infty})$ which is a *simultaneous least fixed point* of $(I_0, \dots, I_n)$ in the sense that for any other simultaneous fixed point $(F_0, \dots, F_n)$, we have $I_p^{\infty} \subseteq F_p$ for all $p \leqslant n$. The functionals I_p^{∞} are approximated from below by functionals I_p^s which satisfy the equations

$$I_p^s(a_0, \dots, a_{k_p-1}, x) \simeq I_p\big(a_0, \dots, a_{k_p-1}, x, I_{0,x}^{<\sigma}, \dots, I_{n,x}^{<\sigma}\big).$$

We call I_0^∞ the *principal least fixed point* of $I_0, \dots, I_n$ and write

$$I_0^\infty = \mathsf{Rec}(I_0, \dots, I_n).$$

DEFINITION 4.2. The class $\mathsf{Rec}_{\mathfrak{A}}$ of *partial $\mathfrak{A}$-recursive functionals* is the smallest class of functionals over $\mathfrak{A}$ such that

(i) $\mathsf{Expl}_{\mathfrak{A}} \subseteq \mathsf{Rec}_{\mathfrak{A}}$;

(ii) $\mathsf{Rec}_{\mathfrak{A}}$ is closed under substitution;

(iii) $\mathsf{Rec}_{\mathfrak{A}}$ is closed under *recursion* – that is, if $I_0, \dots, I_n \in \mathsf{Rec}_{\mathfrak{A}}$ and $(I_0, \dots, I_n)$ is iterable, then $I_0^\infty = \mathsf{Rec}(I_0, \dots, I_n) \in \mathsf{Rec}_{\mathfrak{A}}$.

COROLLARY 4.3.

(i) $\mathsf{Rec}_{\mathfrak{A}}$ *is closed under composition, definition by cases, and argument rearrangement and expansion*;

(ii) *if* $\mathfrak{A}' = (A, \mathsf{Fnl}_{\mathfrak{A}'})$, *and* $\mathsf{Fnl}_{\mathfrak{A}} \subseteq \mathsf{Fnl}_{\mathfrak{A}'} \subseteq \mathsf{Fnl}_{\mathfrak{A}} \cup \mathsf{Rec}_{\mathfrak{A}}$, *then* $\mathsf{Rec}_{\mathfrak{A}} = \mathsf{Rec}_{\mathfrak{A}'}$.

PROOF. The first two parts of (i) follow by Remark 2.3; the remainder is proved exactly as for $\mathsf{Expl}_{\mathfrak{A}}$. □

In classical Recursion Theory a vital role is played by various normal forms which provide a relatively simple and uniform representation for all partial recursive functions. For example, for every partial recursive function f, there is a primitive recursive function g such that

$$f(\boldsymbol{m}) \simeq \text{‘least’}\, n\big[g(\boldsymbol{m}, n) = 0\big].$$

The key point here is that since g is primitive recursive its definition does not involve the search operator, so that the computation of $f(\boldsymbol{m})$ can be arranged to use only one search. The corresponding normal form here demonstrates that every partial $\mathfrak{A}$-recursive functional can be obtained by a single application of the operator Rec to a finite sequence of $\mathfrak{A}$-explicit functionals.

DEFINITION 4.4. A functional F over A is *simply partial $\mathfrak{A}$-recursive* iff for some iterable sequence $(I_0, \dots, I_n)$ of $\mathfrak{A}$-explicit functionals, $F = I_0^\infty = \mathsf{Rec}(I_0, \dots, I_n)$.

The normal form may then be stated:

THEOREM 4.5. *Every partial $\mathfrak{A}$-recursive functional is simply partial $\mathfrak{A}$-recursive.*

PROOF. We need to establish three facts:

(i) every $\mathfrak{A}$-explicit functional is simply partial $\mathfrak{A}$-recursive;

(ii) the class of simply partial $\mathfrak{A}$-recursive functionals is closed under substitution;

(iii) the class of simply partial $\mathfrak{A}$-recursive functionals is closed under recursion.

The first of these is immediate: if I is $\mathfrak{A}$-explicit of type $(k; \boldsymbol{l})$, let ${}^*I(x, i) \simeq I(x)$, with ${}^*I$ of type $(k; \boldsymbol{l}, 0)$. Clearly ${}^*I^\infty = I$ and is simply partial $\mathfrak{A}$-recursive.

The proofs of parts (ii) and (iii) are somewhat tedious and are sketched in the Appendix. □

We have also the following alternative normal form.

THEOREM 4.6. *A functional F is partial $\mathfrak{A}$-recursive iff for some $\mathfrak{A}$-explicit functional I and some finite sequence $\boldsymbol{e}$ of the distinguished elements $0_\mathfrak{A}$ and $1_\mathfrak{A}$,*

$$F(\boldsymbol{a}, \boldsymbol{f}) \simeq I^\infty(\boldsymbol{a}, \boldsymbol{e}, \boldsymbol{f}).$$

PROOF. Since the (constant functions with values) $0_\mathfrak{A}$ and $1_\mathfrak{A}$ are $\mathfrak{A}$-explicit, any such functional is partial $\mathfrak{A}$-recursive. For the converse, assume that F is partial $\mathfrak{A}$-recursive; by Theorem 4.5, there exists an iterable sequence $(I_0, \ldots, I_n)$ of $\mathfrak{A}$-explicit functionals such that $F = I_0^\infty = \mathsf{Rec}(I_0, \ldots, I_n)$. Each I_p $(p \leqslant n)$ has type of the form $(k_p + l; \boldsymbol{l}, k_0, \ldots, k_n)$. Choose $k > \max\{k_p\colon\ p \leqslant n\}$ and distinct sequences $\boldsymbol{e}^0, \ldots, \boldsymbol{e}^n$ of the elements $0_\mathfrak{A}$ and $1_\mathfrak{A}$ such that each $\boldsymbol{e}_p$ has length $k - k_p$. Let $\boldsymbol{a}^p$ denote a sequence of length k_p. Then there is an $\mathfrak{A}$-explicit functional J such that for $p \leqslant n$,

$$J(\boldsymbol{a}^p, \boldsymbol{e}^p, x, j) \simeq I_p\big(\boldsymbol{a}^p, x, j^0, \ldots, j^n\big),$$

where

$$j^p = \lambda \boldsymbol{a}^p . j\big(\boldsymbol{a}^p, \boldsymbol{e}^p\big).$$

Now it is straightforward to prove by induction that

$$J^\sigma\big(\boldsymbol{a}^p, \boldsymbol{e}^p, x\big) \simeq I_p^\sigma\big(\boldsymbol{a}^p, x\big)$$

so in particular

$$F\big(\boldsymbol{a}^0, x\big) \simeq I_0^\infty\big(\boldsymbol{a}^0, x\big) \simeq J^\infty\big(\boldsymbol{a}^0, \boldsymbol{e}^0, x\big)$$

as desired. □

We conclude this section by indicating two ways to assign natural numbers as indices to the members of $\mathsf{Rec}_\mathfrak{A}$. First we need to assign indices to the $\mathfrak{A}$-explicit functionals. As usual in such situations, we actually do the reverse by assigning functionals to all natural numbers regarded as codes for programs.

DEFINITION 4.7. For and functional structure $\mathfrak{A}$ and each $b \in \omega$, we define recursively a functional $[b]^\mathfrak{A}$ as follows, where $\boldsymbol{l} = l_0, \ldots, l_{m-1}$:

(i) $[\langle 0, \langle k, \boldsymbol{l}\rangle, i\rangle]^{\mathfrak{A}} = \mathsf{Pr}_i^{k,\boldsymbol{l}}$ $(i < k)$;
(ii) $[\langle 1, \langle k, \boldsymbol{l}\rangle, j\rangle]^{\mathfrak{A}} = \mathsf{Ev}_j^{k,\boldsymbol{l}}$ $(j < m, l_j = k)$;
(iii) $[\langle 2, \langle k, \boldsymbol{l}\rangle, i\rangle]^{\mathfrak{A}} =$ the i-th member of $\mathsf{Fnl}_{\mathfrak{A}}$;
(iv) $[\langle 3, \langle k, \boldsymbol{l}\rangle, c, d_0, \ldots, d_{m-1}\rangle]^{\mathfrak{A}} =$ the composition of $[c]^{\mathfrak{A}}$ with $[d_0]^{\mathfrak{A}}, \ldots, [d_{m-1}]^{\mathfrak{A}}$;
(v) $[\langle 4, \langle k, \boldsymbol{l}\rangle, c, d_0, \ldots, d_{m-1}, e\rangle]^{\mathfrak{A}} =$ the substitution of $[c]^{\mathfrak{A}}$ with $[d_0]^{\mathfrak{A}}, \ldots, [d_{m-1}]^{\mathfrak{A}}$;
(vi) $[\langle 5, \langle k+1, \boldsymbol{l}\rangle, c, d, e\rangle]^{\mathfrak{A}} =$ the functional defined by cases from $[c]^{\mathfrak{A}}$, $[d]^{\mathfrak{A}}$, and $[e]^{\mathfrak{A}}$.

In (v), the number e codes information to describe the parameter lists for the substituted functionals. In each case the assigned functional is completely undefined (the empty functional) if the ranks of the component parts do not match. For all other $b \in \omega$, $[b]^{\mathfrak{A}}$ is the empty functional.

PROPOSITION 4.8. *The class of functionals $[b]^{\mathfrak{A}}$ for $b \in \omega$ coincides with the class of $\mathfrak{A}$-explicit functionals.*

Now we can use either one of the normal forms to assign indices to all $\mathfrak{A}$-partial recursive functionals:

DEFINITION 4.9. For all natural number values of the parameters,

$$\{\langle 6, b_0, \ldots, b_n\rangle\}_{\mathsf{Rec}}^{\mathfrak{A}} = [b_0]^{\mathfrak{A},\infty} = \mathsf{Rec}\big([b_0]^{\mathfrak{A}}, \ldots, [b_n]^{\mathfrak{A}}\big),$$

and

$$\{\langle 7, b, i_0, \ldots, i_{n-1}\rangle\}_{\mathsf{Rec}}^{\mathfrak{A}}(\boldsymbol{a}, \boldsymbol{f}) \simeq [b]^{\mathfrak{A},\infty}(\boldsymbol{a}, e_0, \ldots, e_{n-1}, \boldsymbol{f}),$$

where for $j < n$, either $i_j = 0$ and $e_j = 0_{\mathfrak{A}}$ or $i_j = 1$ and $e_j = 1_{\mathfrak{A}}$. As above, in all other cases, $\{b\}_{\mathsf{Rec}}^{\mathfrak{A}}$ is the empty functional.

It then follows immediately that

PROPOSITION 4.10. *Each of the classes of functionals $\{b\}_{\mathsf{Rec}}^{\mathfrak{A}}$ with $(b)_0 = 6$ or with $(b)_0 = 7$ coincides with the class of partial $\mathfrak{A}$-recursive functionals.*

Note that each of these indexings has the property that the type of the functional can be effectively recovered from the index.

We conclude this section with

THEOREM 4.11. *For any first-order structure $\mathfrak{A}$, $\mathsf{Reg}_{\mathfrak{A}} \subseteq \mathsf{Rec}_{\mathfrak{A}}$ – that is, every $\mathfrak{A}$-register computable functional is $\mathfrak{A}$-partial recursive*

PROOF. Let $\pi = (I_0, \ldots, I_m)$ be a fixed URM$_{\mathfrak{A}}$-program which computes a functional F. Let $R_0, \ldots, R_k$ include all registers mentioned in π. Each non-conditional instruction I_p can be regarded as an $\mathfrak{A}$-explicit function which maps the sequence $r_0, \ldots, r_k$ of contents of these registers to another such sequence, which we denote by $I_p(r_0, \ldots, r_k)$. Let $J_0, \ldots, J_n$ be the following iterable sequence of functionals: if I_p is non-conditional, then

$$J_p(r_0, \ldots, r_k, \boldsymbol{f}, j_0, \ldots, j_k) \simeq \begin{cases} j_{p+1}(I_p(r_0, \ldots, r_k)) & \text{if } p < m; \\ r_0 & \text{otherwise;} \end{cases}$$

and if I_p is the instruction "if $R(r_{q_0}, \ldots, r_{q_{k-1}})$, go to I_s, else go to I_t", then

$$J_p(r_0, \ldots, r_k, \boldsymbol{f}, j_0, \ldots, j_k) \simeq \begin{cases} j_s(r_0, \ldots, r_k) & \text{if } R(r_{q_0}, \ldots, r_{q_{k-1}}) \\ & \quad \text{and } s \leqslant m; \\ j_t(r_0, \ldots, r_k) & \text{if not } R(r_{q_0}, \ldots, r_{q_{k-1}}) \\ & \quad \text{and } t \leqslant m; \\ r_0 & \text{otherwise.} \end{cases}$$

It follows easily that for each $p \leqslant m$, $J_p^\infty(r_0, \ldots, r_k, \boldsymbol{f})$ is the halting value of r_0 (if any) obtained by starting at instruction I_p with register contents $r_0, \ldots, r_k$ and function arguments $\boldsymbol{f}$. Hence

$$F(\boldsymbol{a}, \boldsymbol{f}) \simeq J_0^\infty(\boldsymbol{a}, \boldsymbol{e}),$$

where $\boldsymbol{e}$ is the sequence of $0_{\mathfrak{A}}$'s of the appropriate length. It follows that F is partial $\mathfrak{A}$-recursive. □

5. The main examples

Of course the first example is classical recursion over the natural numbers. We saw in Corollary 3.1 that the Ω-register computable functionals coincide with the ordinary partial recursive functionals. We have also

THEOREM 5.1. *The class* Rec_Ω *of partial* Ω*-recursive functionals coincides with the usual class of partial recursive functionals over* ω.

PROOF. We take as our characterization of the partial recursive functionals the one of [Hinman, 1978, II.3.3]: the smallest class which contains the primitive recursive functionals and is closed under composition and unbounded search. The functionals there take only total functions as arguments, but the characterization extends easily to the more general case considered here. This class surely contains the Ω-explicit functionals and is closed under substitution (see, for example, [Hinman, 1978, II.3.9] and the discussion preceding it). Furthermore, by an easy extension of Kleene's First

Recursion Theorem (see [Hinman, 1978, II.4.31]) this class is closed under recursion in the sense of Definition 4.2(iii). Hence every partial Ω-recursive functional is partial recursive.

For the opposite inclusion, it will suffice to show that the class of Ω-recursive functionals is closed under primitive recursion and search. By Remark 2.3, this is equivalent to showing that the two particular functionals Se_ω and Prim_ω are partial Ω-recursive. For Se_ω, let

$$I(p, g, i) \simeq \begin{cases} 0 & \text{if } g(p) \simeq 0; \\ \mathsf{Sc}(i(\mathsf{Sc}(p))) & \text{if } g(p)\downarrow \neq 0. \end{cases}$$

It is straightforward to prove by induction on m that

$$I^m(p, g) \simeq n \iff n \leqslant m \text{ and } g(p+n) \simeq 0 \text{ and } (\forall q < n) g(p+q)\downarrow \neq 0.$$

Hence,

$$I^\infty(p, g) \simeq n \iff g(p+n) \simeq 0 \text{ and } (\forall q < n) g(p+q)\downarrow \neq 0,$$

and $\mathsf{Se}_\omega(g) \simeq I^\infty(0, g)$.

Towards Prim_ω, we note first that the predecessor function on ω is Ω-recursive:

$$\mathsf{Pred}(m) \simeq \begin{cases} 0 & \text{if } m = 0; \\ \text{`least' } n[\mathsf{Sc}(n) = m] & \text{otherwise.} \end{cases}$$

Now if we set

$$I(p, n, h, i) \simeq \begin{cases} n & \text{if } p = 0; \\ h(i(\mathsf{Pred}(p), n)) & \text{otherwise;} \end{cases}$$

then it is straightforward to prove that $\mathsf{Prim}_\omega = I^\infty$. □

We turn next to the class of partial κ-recursive functionals for an admissible ordinal κ. There is no full treatment of computable functionals on ordinals in print, so we shall need to be somewhat more detailed here. We start with the characterization of the partial κ-recursive functions of [Hinman, 1978, VIII.1.3]. There we define a set $\Omega_{\kappa,\kappa}$ of finite sequences of (natural numbers and) ordinals and set

$$\{a\}_\kappa(\boldsymbol{\mu}) \simeq \nu \iff (a, \boldsymbol{\mu}, \nu) \in \Omega_{\kappa,\kappa}.$$

The set $\Omega_{\kappa,\kappa}$ is inductively defined as the smallest set containing certain tuples $(a, \boldsymbol{\mu}, \nu)$ which represent some simple initial functions and is closed under clauses corresponding to composition, indexing, and the functionals Sup_κ and Se_κ. The latter, for example, takes the form

$$\nu < \kappa \text{ and } (b, \nu, \boldsymbol{\mu}, 0) \in \Omega_{\kappa,\kappa} \text{ and } (\forall \pi < \nu)(\exists \xi > 0)\big[(b, \nu, \boldsymbol{\mu}, \xi) \in \Omega_{\kappa,\kappa}\big] \\ \Longrightarrow \big(\langle 4, k, b\rangle, \boldsymbol{\mu}, \nu\big) \in \Omega_{\kappa,\kappa}.$$

It is straightforward to extend this definition to partial κ-recursive functionals. All that is required is to extend the coding of the indices to record the types of functionals and the addition of a clause for evaluating the function arguments, say

$$\big(\langle 0, k, \boldsymbol{l}, 3, j\rangle, \boldsymbol{\mu}, \boldsymbol{f}, f_j(\boldsymbol{\mu})\big) \in \Omega_{\kappa,\kappa}.$$

Now the development of [Hinman, 1978, VIII.1.3–13] can be repeated essentially verbatim for the partial κ-recursive functionals.

The new results which must be established are closure of the class of partial κ-recursive functionals under substitution and recursion (in the sense of 4.2(iii)). Closure under substitution can be established by the method of [Hinman, 1978, VI.2] – one shows that there is a primitive recursive function ι such that

$$\begin{aligned}&\{\iota(a, \boldsymbol{c})\}_\kappa(\boldsymbol{\mu}, \boldsymbol{f})\\ &\quad\simeq \{b\}\big(\boldsymbol{\mu}, \lambda\boldsymbol{\nu}^0.\{c_0\}\big(\boldsymbol{\mu}, \boldsymbol{\nu}^0, \boldsymbol{f}\big), \ldots, \lambda\boldsymbol{\nu}^{m-1}.\{c_{m-1}\}\big(\boldsymbol{\mu}, \boldsymbol{\nu}^{m-1}, \boldsymbol{f}\big)\big).\end{aligned}$$

Note that because function arguments here are partial, none of the problems of totality which plague [Hinman, 1978, VI.2] arise here.

Towards closure under recursion, let I be an iterable partial κ-recursive functional and let G be the partial κ-recursive functional defined by:

$$\begin{aligned}G(e, 0, \sigma, \boldsymbol{\mu}, x) &\simeq \{e\}_\kappa(1, \tau^*, \boldsymbol{\mu}, x),\\ G(e, 1, \sigma, \boldsymbol{\mu}, x) &\simeq I\big(\boldsymbol{\mu}, x, \lambda\boldsymbol{\mu}'.\{e\}_\kappa(0, \sigma, \boldsymbol{\mu}', x)\big),\end{aligned}$$

where using the selection function Sel_κ of [Hinman, 1978, VIII.2.9] $\tau^* \simeq \tau^*(e, \sigma, \boldsymbol{\mu}, x)$ is a κ-partial recursive functional such that

$$(\exists\tau < \sigma)\{e\}_\kappa(1, \tau, \boldsymbol{\mu}, x)\!\downarrow\; \Longrightarrow \tau^* < \sigma \text{ and } \{e\}_\kappa(1, \tau^*, \boldsymbol{\mu}, x)\!\downarrow.$$

By the Second Recursion Theorem, choose an index $\bar{e}$ such that

$$\{\bar{e}\}_\kappa(i, \sigma, \boldsymbol{\mu}, x) \simeq G(\bar{e}, i, \sigma, \boldsymbol{\mu}, x).$$

Then it is straightforward to prove by induction on σ that

$$\begin{aligned}\{\bar{e}\}_\kappa(0, \sigma, \boldsymbol{\mu}, x) &\simeq I^{<\sigma}(\boldsymbol{\mu}, x),\\ \{\bar{e}\}_\kappa(1, \sigma, \boldsymbol{\mu}, x) &\simeq I^{s}(\boldsymbol{\mu}, x),\end{aligned}$$

so that if, again using Sel_κ, $\sigma^* \simeq \sigma^*(\boldsymbol{\mu}, x)$ is κ-partial recursive such that $\{\bar{e}\}_\kappa(1, \sigma^*, \boldsymbol{\mu}, x)\!\downarrow$ whenever any such $\sigma < \kappa$ exists, then

$$I^{<\kappa}(\boldsymbol{\mu}, x) \simeq \{e\}_\kappa(1, \sigma^*, \boldsymbol{\mu}, x)$$

is partial κ-recursive. Now, if there are no partial function arguments and κ is recursively regular, it follows as in [Hinman, 1978, VIII.2.5] that the closure ordinal of I is at most κ, so that $I^\infty = I^{<\kappa}$ and thus also I^∞ is partial κ-recursive. The extension of this argument to an iterable sequence $(I_0, \ldots, I_n)$ of partial κ-recursive functionals is straightforward.

Let $\mathfrak{A}_\kappa$ denote $(\kappa, \mathsf{Sc}, 0, \mathsf{Sup}_\kappa, \mathsf{Se}_\kappa)$.

THEOREM 5.2. *For any admissible ordinal κ, the partial $\mathfrak{A}_\kappa$-recursive functions coincide with the usual partial κ-recursive functions.*

PROOF. We have essentially shown in the preceding paragraphs that every partial $\mathfrak{A}_\kappa$-recursive function is partial κ-recursive. For the opposite inclusion, we can verify that the universal partial κ-recursive function

$$U\big(a, \langle \boldsymbol{\mu} \rangle, i\big) \simeq \{a\}_\kappa(\boldsymbol{\mu})$$

is partial $\mathfrak{A}_\kappa$-recursive. In fact, U is the least fixed point of a single functional I, with $I(a, \langle \boldsymbol{\mu} \rangle, i)$ defined by cases on a. If a is an index corresponding to an initial function, $I(a, \langle \boldsymbol{\mu} \rangle, i)$ is defined outright. If $a = \langle 3, k+1, b \rangle$ is an index for an application of the search operator Se_κ, then we take

$$I\big(a, \langle \boldsymbol{\mu} \rangle\big) \simeq \mathsf{Se}_\kappa\big(\lambda\pi . i\big(b, \langle \pi, \boldsymbol{\mu} \rangle\big)\big).$$

The other cases are handled similarly and it is straightforward to verify that $U = I^\infty$ as required. □

We consider next positive elementary (first-order) inductive definition over a first-order structure $\mathfrak{A}$ as developed in [Moschovakis, 1974]. In accord with the conventions of that book, we assume that $\mathsf{Dis}_\mathfrak{A} = A$ – that is, every element of A is distinguished. Note that this conflicts with our earlier restriction to finite languages, but the only place we have made use of this is in assigning indices. Let $\mathfrak{A}^+$ be the functional structure $(\mathfrak{A}, =_A, \mathsf{E}^\sharp_A)$. Roughly, the correspondence is

$$\begin{aligned} \mathfrak{A}\text{-positive elementary} &= \mathfrak{A}^+\text{-explicit},\\ \mathfrak{A}\text{-inductive} &= \mathfrak{A}^+\text{-semi-recursive},\\ \mathfrak{A}\text{-hyperelementary} &= \mathfrak{A}^+\text{-recursive}, \end{aligned}$$

but a little preparation is needed to make this precise. Let L be the usual first-order language associated with $\mathfrak{A}$ and L^+ the language obtained from L by adjoining a countable sequence of relation variables. We use letters p, q for the relation symbols of L and L^+, t (with scripts) for terms, u, v, w for individual variables, and U, V, W for relation variables. The terms of L^+ are just those of L and the formulas of each language are as usual with the stipulation that the relation variables occur positively.

Thus the L^+-formulas comprise the smallest class containing all atomic and negated atomic formulas of L, formulas $V(t_0, \ldots, t_n)$, and closed under $\wedge$, $\vee$, $\exists v$ and $\forall v$.

For any l-ary partial function f over A, let Gr_f denote the usual $(l+1)$-ary relation called the *graph* of f. If F is a $(k; \boldsymbol{l})$-ary functional over A, Gr_F is the second-order relation such that

$$F(\boldsymbol{a}, \boldsymbol{f}) \simeq e \Longleftrightarrow \mathsf{Gr}_F(\boldsymbol{a}, e, \mathsf{Gr}_{\boldsymbol{f}}).$$

Here, $\mathsf{Gr}_{\boldsymbol{f}} = \mathsf{Gr}_{f_0,\ldots,f_{m-1}} = (\mathsf{Gr}_{f_0}, \ldots, \mathsf{Gr}_{f_{m-1}})$. If P is a relation on A, the *semi-characteristic function* of P is

$$\chi_P^+(\boldsymbol{a}) \simeq \begin{cases} 1_{\mathfrak{A}} & \text{if } P(\boldsymbol{a}); \\ \uparrow & \text{otherwise.} \end{cases}$$

We write $\chi_{\boldsymbol{P}}^+$ for $\chi_{P_0}^+, \ldots, \chi_{P_{l-1}}^+$.

PROPOSITION 5.3.

(i) *For any $\mathfrak{A}^+$-explicit functional F, there exists an L^+-formula ϕ_F which defines* Gr_F *over $\mathfrak{A}$ – that is,*

$$F(\boldsymbol{a}, \boldsymbol{f}) \simeq e \Longleftrightarrow \underset{\mathfrak{A}}{\models} \phi_F[\boldsymbol{a}, e, \mathsf{Gr}_{\boldsymbol{f}}];$$

(ii) *for any L^+-formula ϕ, there exists an $\mathfrak{A}^+$-explicit functional F_ϕ such that*

$$\underset{\mathfrak{A}}{\models} \phi[\boldsymbol{a}, \boldsymbol{P}] \Longleftrightarrow F_\phi(\boldsymbol{a}, \chi_{\boldsymbol{P}}^+) \simeq 1_{\mathfrak{A}};$$

and in case ϕ has no relation variables, also

$$\underset{\mathfrak{A}}{\not\models} \phi[\boldsymbol{a}] \Longleftrightarrow F_\phi(\boldsymbol{a}) \simeq 0_{\mathfrak{A}}.$$

PROOF. The proof of (i) is by induction over the $\mathfrak{A}^+$-explicit functionals and is relatively straightforward. For example, suppose

$$F(\boldsymbol{a}, f) \simeq G\big(H(\boldsymbol{a}, f), f\big).$$

Then from ϕ_G and ϕ_H, we define $\phi_F(\boldsymbol{u}, v, W)$ as

$$\exists w\big[\phi_H(\boldsymbol{u}, w, W) \wedge \phi_G(w, v, W)\big].$$

$\phi_{\mathsf{E}_A^\sharp}(w, V)$ is the formula

$$\big[w \doteq 1_{\mathfrak{A}} \wedge \exists v V(v, 1_{\mathfrak{A}})\big] \vee \big[w \doteq 0_{\mathfrak{A}} \wedge \forall v V(v, 0_{\mathfrak{A}})\big].$$

To see that the class of F for which ϕ_F exists is closed under substitution, suppose

$$F(\boldsymbol{a}, \boldsymbol{f}) \simeq G\big(\boldsymbol{a}, \lambda \boldsymbol{b}.H(\boldsymbol{a}, \boldsymbol{b}, \boldsymbol{f})\big) \quad \text{and} \quad \phi_G \text{ and } \phi_H \text{ exist.}$$

Then we may take for ϕ_F the result of replacing each subformula of the form $W(\boldsymbol{s}, \boldsymbol{t}, u)$ of ϕ_G by $\phi_H(\boldsymbol{s}, \boldsymbol{t}, u)$ (changing bound variables if necessary to avoid collisions).

For (ii), we proceed by induction on L^+-formulas. First note that for any L-term t, there is an $\mathfrak{A}$-explicit function F_t such that

$$\models_{\mathfrak{A}} t[\boldsymbol{a}] \doteq e \iff F_t(\boldsymbol{a}) = e.$$

If ϕ is an atomic formula of L, say $p(\boldsymbol{t})$, then set

$$F_\phi(\boldsymbol{a}) \simeq \chi_{p^{\mathfrak{A}}}\big(F_{\boldsymbol{t}}(\boldsymbol{a})\big).$$

(Technically, we account for differing variables in the terms $t_0, \ldots, t_{k-1}$.) Because of the explicit inclusion of the equality on A in the structure $\mathfrak{A}^+$, this works also for atomic equations. If ϕ is $V(\boldsymbol{t})$, then

$$F_\phi(\boldsymbol{a}, f) \simeq \begin{cases} 1_{\mathfrak{A}} & \text{if } f(F_{\boldsymbol{t}}(\boldsymbol{a})) \simeq 1_{\mathfrak{A}}; \\ \uparrow & \text{otherwise.} \end{cases}$$

Now we set recursively,

$$\begin{aligned} F_{\phi\wedge\psi}(x) &\simeq \mathsf{And}\big(F_\phi(x), F_\psi(x)\big); \\ F_{\phi\vee\psi}(x) &\simeq \mathsf{E}_A^\sharp\big(\lambda b.G(b, x)\big); \end{aligned}$$

where

$$\begin{aligned} G(b, x) &\simeq \begin{cases} F_\phi(x) & \text{if } b = 0_{\mathfrak{A}}, \\ F_\psi(x) & \text{otherwise;} \end{cases} \\ F_{\exists v\phi}(x) &\simeq \mathsf{E}_A^\sharp\big(\lambda b.F_\phi(b, x)\big); \\ F_{\forall v\phi}(x) &\simeq \mathsf{Not}\Big(\mathsf{E}_A^\sharp\big(\lambda b.\mathsf{Not}\big(F_\phi(b, x)\big)\big)\Big). \end{aligned}$$

□

To complete the picture, we need to recall from [Moschovakis, 1974, 1983] the definitions of the $\mathfrak{A}$-inductive and $\mathfrak{A}$-hyperelementary relations. With any L^+-formula $\phi(v_0, \ldots, v_{k-1}, V)$, with V a k-ary relation variable, we associate a mapping $\Gamma_\phi : \wp(A^k) \to \wp(A^k)$ defined by

$$\Gamma_\phi(R) = \Big\{\boldsymbol{a} \in A^k : \models_{\mathfrak{A}} \phi[\boldsymbol{a}, R]\Big\}.$$

Because V occurs positively in ϕ, Γ_ϕ is monotone in R and hence by arguments very similar to those of Section 2 has a least fixed point Γ_ϕ^∞ – that is, $\Gamma_\phi(\Gamma_\phi^\infty) = \Gamma_\phi^\infty$ and for any R such that $\Gamma_\phi(R) = R$, $\Gamma_\phi^\infty \subseteq R$. Similarly, for a finite sequence of formulas $\phi_i(v_0, \dots, v_{k_i-1}, V_0, \dots, V_n)$ $(i \leqslant n)$, we have mappings

$$\Gamma_{\phi_i}(R_0, \dots, R_n) = \left\{ \boldsymbol{a} \in A^{k_i} \colon \models_{\mathfrak{A}} \phi_i[\boldsymbol{a}, R_0, \dots, R_n] \right\}$$

which have a simultaneous least fixed point $(\Gamma_{\phi_0}^\infty, \dots, \Gamma_{\phi_n}^\infty)$. A relation $R \subseteq A^k$ is $\mathfrak{A}$*-inductive* $\mathfrak{A}$ iff for some such sequence, $R = \Gamma_{\phi_0}^\infty := \mathsf{Ind}(\phi_0, \dots, \phi_n)$, and $\mathfrak{A}$*-hyperelementary* iff both R and $A^k \sim R$ are $\mathfrak{A}$-inductive. These definitions can also be relativized to individual and relational parameters, but to simplify the presentation here we leave this generalization to the reader.

These definitions are slightly different from those of [Moschovakis, 1974, 1D]. There, R is called *inductive on* $\mathfrak{A}$ iff for some formula $\phi(\boldsymbol{v}, \boldsymbol{w}, V)$ and some $\boldsymbol{b} \in A^l$,

$$R(\boldsymbol{a}) \iff (\boldsymbol{a}, \boldsymbol{b}) \in \Gamma_\phi^\infty.$$

Using the techniques of the proof of Theorem 4.6 it is easy to show that these two notions are the same, at least for infinite structures. The key point is that here every $a \in A$ is a distinguished element and the equality relation is definable. In Theorem 4.6 we used the special distinguished elements $0_\mathfrak{A}$ and $1_\mathfrak{A}$ for which the predicates $\cdot = 0_\mathfrak{A}$ and $\cdot = 1_\mathfrak{A}$ are by definition $\mathfrak{A}$-recursive (because they are $\mathfrak{A}$-explicit) even if equality on A is not in general $\mathfrak{A}$-recursive.

PROPOSITION 5.4.
(i) *For any partial* $\mathfrak{A}^+$*-recursive function* f, Gr_f *is* $\mathfrak{A}$*-inductive*; *hence every* $\mathfrak{A}^+$*-semi-recursive relation is* $\mathfrak{A}$*-inductive*;
(ii) *every* $\mathfrak{A}$*-inductive relation is* $\mathfrak{A}^+$*-semi-recursive.*

PROOF. Let f be partial $\mathfrak{A}^+$-recursive; then by Theorem 4.5 there exists an iterable sequence $(I_0, \dots, I_n)$ of $\mathfrak{A}^+$-explicit functionals such that $f = I_0^\infty = \mathsf{Rec}(I_0, \dots, I_n)$. For each $p \leqslant n$, let $\phi_p = \phi_{I_p}$ be an L^+-formula as in the preceding proposition such that

$$\begin{aligned} &I_p(a_0, \dots, a_{k_p-1}, i_0, \dots, i_n) \simeq e \\ &\quad\iff \models_{\mathfrak{A}} \phi_p[a_0, \dots, a_{k_p-1}, e, \mathsf{Gr}_{i_0}, \dots, \mathsf{Gr}_{i_n}]. \end{aligned}$$

Then with notation as above, by induction on σ, $\mathsf{Gr}_{I_p^\sigma} = \Gamma_{\phi_p}^\sigma$ so that $\mathsf{Gr}_{I_p^\infty} = \Gamma_{\phi_p}^\infty$ and thus in particular, $\mathsf{Gr}_f = \mathsf{Gr}_{I_0^\infty}$ is $\mathfrak{A}$-inductive. Now if R is any $\mathfrak{A}^+$-semi-recursive relation, then for some $(I_0, \dots, I_n)$ as above,

$$R = \mathsf{Dom}\, I_0^\infty = \{\boldsymbol{a} \colon \exists e[(\boldsymbol{a}, e) \in \Gamma_{\phi_0}^\infty]\},$$

which by [Moschovakis, 1974, 1D1] is also $\mathfrak{A}$-inductive.

For (ii), we use similarly the second half of the preceding proposition. For any iterable sequence $(\phi_0, \ldots, \phi_n)$ of L^+-formulas with no parameters let $I_p = F_{\phi_p}$ be $\mathfrak{A}^+$-explicit functionals such that

$$\models_{\mathfrak{A}} \phi_p[\boldsymbol{a}^p, R_0, \ldots, R_n] \Longleftrightarrow I_p(\boldsymbol{a}^p, \chi^+_{R_0}, \ldots, \chi^+_{R_n}) \simeq 1_{\mathfrak{A}}.$$

Then by induction on σ, we have $I_p^\sigma = \chi^+_{\Gamma^\sigma_{\phi_p}}$ so that $I_p^\infty = \chi^+_{\Gamma^\infty_{\phi_p}}$. In particular, $\Gamma^\infty_{\phi_0} = \mathrm{Dom}\, I_0^\infty$ and is therefore $\mathfrak{A}^+$-semi-recursive. □

Another very important example, which motivated much of the development of this approach to abstract recursion, is recursion on the finite type structure over a functional structure. This example is fully developed in [Kechris and Moschovakis, 1977], to which we refer the interested reader.

Our final example is the computation theory over ordered commutative rings with unit introduced in [Blum, Shub and Smale, 1989]. It would take us too far afield to lay out here the entire formalism of this theory, which is quite different from ours, so we shall restrict ourselves to a general description. Computation of a BSS-*machine* is carried out on a *state space* $\overline{S}$ and defines a partial map from an *input space* $\overline{I}$ to an *output space* $\overline{O}$. Each of these spaces is some finite Cartesian power of the domain R of the underlying ring. (The theory of [Blum, Shub and Smale, 1989] embraces also infinite dimensional input, output, and state spaces, but we shall not discuss this aspect.) The machine M acts by generating a path through a finite directed graph. Each node of the graph is either a (unique) input node, an output node, a computation node, or a branch node. With each time step s is associated a *state* $\boldsymbol{r}^s = (r_0^s, \ldots, r_n^s) \in \overline{S}$; $\boldsymbol{r}^0$ is the input. With each computation node is associated a sequence $f_0, \ldots, f_n$ of polynomials with coefficients from R, and the effect of the node is to alter the state as follows:

$$(r_0, \ldots, r_n) \mapsto \big(f_0(r_0, \ldots, r_n), \ldots, f_n(r_0, \ldots r_n)\big),$$

and to point to a unique next node. With a branch node is associated a polynomial g and two potential next nodes. A branch node does not alter the state but chooses the next node depending as $g(r_0, \ldots, r_n) < 0$ or not.

A BSS-machine can be converted in a straightforward way to a register machine. A state $(r_0, \ldots, r_n)$ corresponds to the contents of registers $R_0, \ldots, R_n$. The action of a computation node corresponds to a finite sequence of instructions $r_i := f_i(r_0, \ldots, r_n)$. This, in turn, can be expanded to a longer finite sequence of instructions involving only the addition and multiplication operators of the ring and the constants of the polynomials f_i. Similarly, the action of a branch node corresponds to a conditional instruction "if $r_q < 0$, go to I_s, else go to I_t" preceded by a sequence of instructions with the effect $r_q := g(r_0, \ldots, r_n)$. The resulting program governs a register machine over the structure

$$\mathcal{R}_{a_0,\ldots,a_{n-1}} = (R, +, \times, <, =_R, 0, 1, a_0, \ldots, a_{n-1}),$$

where $a_0, \ldots, a_{n-1}$ are all of the constant coefficients of any of the polynomials of the BSS-machine. Conversely, any $\mathcal{R}_{a_0,\ldots,a_{n-1}}$-register machine is (essentially) a BSS-machine. Thus we have

THEOREM 5.5. *For any ordered ring* $(R, +, \times, <, =_R, 0, 1)$, *a function* $f : R^k \to R$ *is BSS-computable iff for some* $a_0, \ldots, a_{n-1} \in R$, f *is* $\mathcal{R}_{a_0,\ldots,a_{n-1}}$*-register computable.*

6. Structures with arithmetic

A basic feature of most intuitive notions of computability is that at some level the algorithms are specified in a finite way. This is reflected by the existence of an indexing of the computable partial functionals by natural numbers. In ordinary recursion theory over ω, such an indexing results in names for functions which are elements of the underlying domain and hence in a large group of results such as the existence of universal computable functions, the Second Recursion Theorem, etc. To extend this part of the theory to an abstract structure $\mathfrak{A}$, we will need to require that $\mathfrak{A}$ contain a copy of the natural numbers.

DEFINITION 6.1. A functional [first-order] structure is a *structure with arithmetic* iff $\mathfrak{A}$ includes a unary relation (set) $\omega_{\mathfrak{A}} \subseteq A$, a unary (successor) function $\mathsf{Sc}_{\mathfrak{A}}$, and a binary relation $=_{\omega_{\mathfrak{A}}}$ such that the structure $(\omega_{\mathfrak{A}}, \mathsf{Sc}_{\mathfrak{A}}, =_{\omega_{\mathfrak{A}}}, 0_{\mathfrak{A}})$ is isomorphic to the standard structure $(\omega, \mathsf{Sc}, =, 0)$.

For simplicity of notation, we shall assume that the two structures $(\omega_{\mathfrak{A}}, \mathsf{Sc}_{\mathfrak{A}}, =_{\omega_{\mathfrak{A}}}, 0_{\mathfrak{A}})$ and $(\omega, \mathsf{Sc}, =, 0)$ are actually identical; this can always be arranged by replacing $\mathfrak{A}$ by an isomorphic copy with domain including ω.

Many of the examples of the preceding section satisfy this condition, including the structures $\mathfrak{A}_\kappa$ and $\mathcal{R}_{a_0,\ldots,a_{n-1}}$.

PROPOSITION 6.2. *For any functional [first-order] structure* $\mathfrak{A}$ *with arithmetic,* $\mathsf{Rec}_{\mathfrak{A}}$ [$\mathsf{Reg}_{\mathfrak{A}}$] *includes all of the ordinary partial recursive functionals.*

PROOF. For $\mathsf{Rec}_{\mathfrak{A}}$ this follows from Theorem 5.1 and for $\mathsf{Reg}_{\mathfrak{A}}$ from Corollary 3.1. □

For structures with arithmetic it makes sense to ask if the *universal* functionals

$$U^{\mathfrak{A},k,l}_{\mathsf{Reg}}(b, \boldsymbol{a}, \boldsymbol{f}) \simeq \{b\}^{\mathfrak{A}}_{\mathsf{Reg}}(\boldsymbol{a}, \boldsymbol{f}) \quad \text{and}$$
$$U^{\mathfrak{A},k,l}_{\mathsf{Rec}}(b, \boldsymbol{a}, \boldsymbol{f}) \simeq \{b\}^{\mathfrak{A}}_{\mathsf{Rec}}(\boldsymbol{a}, \boldsymbol{f})$$

are computable in the relevant sense. For $\mathsf{Reg}_{\mathfrak{A}}$ (over first-order structures) this turns out to be true for functions (without function arguments) under mild additional hypotheses. For $\mathsf{Rec}_{\mathfrak{A}}$, there are two conditions under which universal functionals are partial $\mathfrak{A}$-recursive. The first concerns functional structures $\mathfrak{A}$ for which there is an $\mathfrak{A}$-recursive coding of finite sequences of elements of A; this case is treated thoroughly in [Kechris and Moschovakis, 1977, Section 8]. We consider here first-order structures $\mathfrak{A}$, for which the functionals $U^{\mathfrak{A},k,l}_{\mathsf{Rec}}$ are partial $\mathfrak{A}$-recursive without restriction. The proofs are less straightforward than one might expect; in both cases the difficulty lies in verifying that certain aspects of computations are bounded.

Consider first register computations. A program π for a universal function must effectively simulate all programs π_b. But π must work within some fixed number of registers, while the programs that it is simulating will generally use an unbounded number of registers. In the case of ordinary register machines on ω, this difficulty is overcome by the use of sequence coding; the information in any finite number of registers can be coded into a single number and stored in a single register. In particular, the whole configuration of a URM after execution of s steps of program π_b with input $\boldsymbol{m}$ can be coded as a single number. Thus a simulating machine needs only to calculate the sequence of these configuration codes as s increases and read off the answer when a halting state is reached. In slightly more detail, following [Cutland, 1980, Section 5.1], let

$$\sigma_k(b, \boldsymbol{m}, s) = \langle\langle r^s_0, \ldots, r^s_{n_b}\rangle, p^s\rangle,$$

where r^s_i denotes the contents of the i-th register after s steps of the computation of π_b with input $\boldsymbol{m}$, and p^s denotes the number of the instruction to be executed at the $s+1$-st step. It is easy to verify that the functions σ_k are primitive recursive and that

$$U^k_{\mathsf{Reg}}(b, \boldsymbol{m}) \simeq r^s_0$$
$$\text{for the least } s \text{ such that } p^s \text{ is not a valid instruction number}$$

is partial recursive and hence register computable.

For a $\mathrm{URM}_{\mathfrak{A}}$ this approach is not directly available, since we cannot in general code the contents of finitely many registers containing elements of A as a single element of A. The solution to this problem has been discovered at least twice independently in [Moldestad, Stoltenberg-Hansen and Tucker, 1980b, Section 3] and [Michaux, 1989] and lies in the observation that for any computation without function arguments the contents of the registers of a $\mathrm{URM}_{\mathfrak{A}}$ can be described in a finitary way as the interpretations of terms of the first-order language of $\mathfrak{A}$ evaluated at the input points. We may therefore describe the configuration of a $\mathrm{URM}_{\mathfrak{A}}$ by a number coding the sequence of Gödel numbers of these terms and a simulating machine may manipulate these configuration codes as before. For example, suppose that at step s of the computation of a program π acting on input $\boldsymbol{a}$, the instruction to be executed is $r_p := f(r_{q_0}, \ldots, r_{q_{k-1}})$ and the configuration description contains the information that the contents of each r_{q_i} is the value $t_i^{\mathfrak{A}}[\boldsymbol{a}]$ of the term t_i. Then the configuration description at step $s+1$ will be changed to show the contents of r_p to be the value $f(t_0 \ldots t_{k-1})^{\mathfrak{A}}[\boldsymbol{a}]$ of this compound term. We set

$$\sigma_k^{\mathfrak{A}}(b, \boldsymbol{a}, s) = \langle\langle \mathsf{gn}(t_0^s), \ldots, \mathsf{gn}(t_{n_b}^s)\rangle, p^s\rangle,$$

where t_i^s denotes the term such that after s steps of the computation of π_b with input $\boldsymbol{a}$, R_i contains $t_i^{s,\mathfrak{A}}[\boldsymbol{a}]$ and gn denotes any standard Gödel numbering.

This scheme encounters one difficulty in verifying that the functions $\sigma_k^{\mathfrak{A}}$ are register computable: at the execution of a conditional instruction we need the actual values of the register contents in order to decide which branch to follow. Thus, we need to be able to compute $t^{\mathfrak{A}}[\boldsymbol{a}]$ from the Gödel number $\mathsf{gn}(t)$ of the term t. The same sort of computation suffices at the end of the simulation to produce the value of the computed function.

DEFINITION 6.3. A first-order structure $\mathfrak{A}$ *admits* URM$_{\mathfrak{A}}$ *term evaluation* iff for each $k \in \omega$ there exists $F_k \in \mathsf{Reg}_{\mathfrak{A}}$ such that for all terms t containing at most the variables $v_0, \ldots, v_{k-1}$ and all $\boldsymbol{a} \in A^k$, $F_k(\mathsf{gn}(t), \boldsymbol{a}) = t^{\mathfrak{A}}[\boldsymbol{a}]$.

THEOREM 6.4. *For any first-order structure $\mathfrak{A}$, if $\mathfrak{A}$ admits* URM$_{\mathfrak{A}}$ *term evaluation, then for each $k \in \omega$ there exists a universal $\mathsf{Reg}_{\mathfrak{A}}$-computable function*

$$U_{\mathsf{Reg}}^{\mathfrak{A},k}(b, \boldsymbol{a}) \simeq \{b\}_{\mathsf{Reg}}^{\mathfrak{A}}(\boldsymbol{a}).$$

Consider next the existence of functionals $U_{\mathsf{Rec}}^{\mathfrak{A},k,\boldsymbol{l}}$ universal for the functionals in $\mathsf{Rec}_{\mathfrak{A}}$. The basic idea for constructing these functionals is relatively simple, although the details are messy. Consider first universal functionals $U_{\mathsf{Expl}}^{\mathfrak{A},k,\boldsymbol{l}}$ for $\mathsf{Expl}_{\mathfrak{A}}$. The natural approach is to define this as a fixed point I^∞ for an $\mathfrak{A}$-explicit functional I which mirrors the inductive definition of $\mathsf{Expl}_{\mathfrak{A}}$ – for example, corresponding to composition would be a clause

$$I(\langle 3, \langle k, \boldsymbol{l}\rangle, c, d_0, \ldots, d_{m-1}\rangle, \boldsymbol{a}, \boldsymbol{f}, i) \simeq i(c, i(d_0, \boldsymbol{a}), \ldots, i(d_{m-1}, \boldsymbol{a})),$$

but we see immediately that this is not a legal application of recursion, since i is applied to argument lists of different lengths $k+1$ and $m+1$.

If we assume that $\mathfrak{A}$ admits $\mathfrak{A}$-recursive coding of finite sequences, this problem can be overcome by converting all functions to unary ones using this coding; this is the approach of [Kechris and Moschovakis, 1977] and works for functional as well as first-order structures. Without this assumption we can still deal with first-order structures as follows. With each $F \in \mathsf{Expl}_{\mathfrak{A}}$ we can associate a finite *derivation*

$F_0, \ldots, F_{n-1}, F$ such that each term of this sequence is either an initial functional or arises from some of the preceding terms via one of the closure principles. Suppose for a moment that every functional of type $(k; \boldsymbol{l})$ has such a derivation with each F_i of some type $(m; \boldsymbol{l})$ with $m \leqslant k$. Then the clause above could be written

$$I(\langle 3, \langle k, \boldsymbol{l}\rangle, c, d_0, \ldots, d_{m-1}\rangle, \boldsymbol{a}, \boldsymbol{f}, i) \\ \simeq i(c, i(d_0, \boldsymbol{a}), \ldots, i(d_{m-1}, \boldsymbol{a}), 0_{\mathfrak{A}}, \ldots, 0_{\mathfrak{A}}),$$

with $(k-m)$-many $0_{\mathfrak{A}}$'s. In other words, we could exhibit a functional universal for functionals of all types $(m; \boldsymbol{l})$ with $m \leqslant k$. Of course, this condition will not generally be satisfied for two reasons: some of the functionals in the signature of $\mathfrak{A}$ may have types other than these and the composition and substitution schemas do not preserve types. For first-order structures, we can get around these difficulties as follows.

DEFINITION 6.5. For any structure $\mathfrak{A}$, a functional $F \in \mathsf{Expl}_{\mathfrak{A}}$ of type $(k; \boldsymbol{l})$ is *conservative* iff it has a derivation $F_0, \ldots, F_{n-1}, F$ such that each F_i is of type $(m; \boldsymbol{l})$ or $(m;)$ with $m \leqslant k$, and the only uses of the substitution schema are trivial ones of the form $F(\boldsymbol{a}, \boldsymbol{f}) \simeq g(\boldsymbol{a})$, for g a given function of $\mathfrak{A}$.

PROPOSITION 6.6. *For any first-order structure $\mathfrak{A}$, any $\boldsymbol{l}$, and all sufficiently large k, every $F \in \mathsf{Expl}_{\mathfrak{A}}$ of type $(k; \boldsymbol{l})$ is conservative.*

PROOF (*Sketch*). Define a functional F to be *strongly conservative* iff F is conservative and every functional which results from composing or substituting F with conservative functionals is also conservative. Then it is straightforward (if messy) to prove by induction that for k larger than all of the l_j and all of the ranks of the functions and relations of $\mathfrak{A}$, every $F \in \mathsf{Expl}_{\mathfrak{A}}$ of type $(k; \boldsymbol{l})$ is strongly conservative. □

Now for sufficiently large k it is easy to define along the lines suggested above an $\mathfrak{A}$-explicit functional I such that I^{∞} is universal for $\mathfrak{A}$-explicit functionals of types $(m; \boldsymbol{l})$ for all $m \leqslant k$; shorter lists are "padded" with $0_{\mathfrak{A}}$'s, and we have

THEOREM 6.7. *For any first-order structure $\mathfrak{A}$ with arithmetic, for each k and $\boldsymbol{l}$, the functional $U^{\mathfrak{A},k,\boldsymbol{l}}_{\mathsf{Rec}}$ universal for partial $\mathfrak{A}$-recursive functionals of type $(k; \boldsymbol{l})$ is partial $\mathfrak{A}$-recursive.*

PROOF. Since for $m \leqslant k$, any m-ary function can be regarded also as a k-ary function, it suffices to prove the result for large k. Let I be as in the preceding paragraph and set

$$J(b, \boldsymbol{a}, \boldsymbol{f}, j) \simeq \begin{cases} I^{\infty}(b, \boldsymbol{a}, \boldsymbol{f}, j) & \text{if } (b)_1 = \langle k + l; \boldsymbol{l}, k\rangle; \\ 0_{\mathfrak{A}} & \text{otherwise.} \end{cases}$$

Then by an easy induction, for all ordinals σ and iterable $[b]$,

$$J^{\sigma}(b, \boldsymbol{a}, \boldsymbol{f}) \simeq [b]^{\sigma}(\boldsymbol{a}, \boldsymbol{f}),$$

whence J^{∞} is universal for fixed points of $\mathfrak{A}$-explicit functionals. It follows immediately that $U^{\mathfrak{A},k,l}_{\mathsf{Rec}}$ is partial $\mathfrak{A}$-recursive. □

We conclude this section by showing that for all first-order structures with arithmetic which admit $\mathrm{URM}_{\mathfrak{A}}$ term evaluation, a function is $\mathfrak{A}$-register computable iff it is $\mathfrak{A}$-partial recursive. By Proposition 3.2 and Theorem 4.11, it suffices to show that $\mathsf{Reg}_{\mathfrak{A}}$ is closed under recursion. The outline of the proof is quite similar to a standard proof of the First Recursion Theorem in ordinary recursion theory, for example, [Cutland, 1980, 10.3]. First we have

PROPOSITION 6.8. *For any first-order structure $\mathfrak{A}$ with arithmetic and any $\mathfrak{A}$-register computable function F, there exists a $\mathfrak{A}$-register computable relation R such that*

$$F(\boldsymbol{a})\downarrow \iff (\exists s \in \omega) R(s, \boldsymbol{a}).$$

PROOF. We define $R(s, \boldsymbol{a})$ to hold just in case $F(\boldsymbol{a})$ is defined by a computation which executes at most s many instructions. A $\mathrm{URM}_{\mathfrak{A}}$ program for the characteristic function of R is easily constructed from one for F by inserting after each instruction a routine which increments a counter, compares it with the stored value of s, and either continues with the computation of F if the counter value is less than s or halts with the value $0_{\mathfrak{A}}$. If the computation terminates because $F(\boldsymbol{a})$ is defined, the value 1 is produced as output. □

NUMBER SELECTION THEOREM 6.9. *For any first-order structure $\mathfrak{A}$ with arithmetic and any $\mathfrak{A}$-register computable function F, there exists a $\mathfrak{A}$-register computable function Sel_F such that*

$$(\exists n \in \omega) F(n, \boldsymbol{a})\downarrow \Longrightarrow \mathsf{Sel}_F(\boldsymbol{a})\downarrow \textit{ and } F\big(\mathsf{Sel}_F(\boldsymbol{a}), a\big)\downarrow.$$

PROOF. With R as in the proposition, we take simply

$$\mathsf{Sel}_F(\boldsymbol{a}) \simeq \big(\text{least}\, p \,.\, R\big((p)_0, (p)_1, \boldsymbol{a}\big)\big)_1.$$

□

PROPOSITION 6.10. *For any first-order structure $\mathfrak{A}$ with arithmetic which admits $\mathrm{URM}_{\mathfrak{A}}$ term evaluation, and any iterable $\mathfrak{A}$-register computable functional I, also I^{∞} is $\mathfrak{A}$-register computable.*

PROOF. Let b be a register index for I. By Proposition 3.3, there exists a primitive recursive function h such that for any index c,

$$\{b\}^{\mathfrak{A}}_{\mathsf{Reg}}(\boldsymbol{a}, [c]_{\mathfrak{A}}) \simeq \{h(b,c)\}^{\mathfrak{A}}_{\mathsf{Reg}}(\boldsymbol{a}).$$

Hence if $f(0)$ is an index for a totally undefined function and $f(n+1) = h(b, f(n))$, then $f(n)$ is an index for I^n, and by Theorem 6.4 the function

$$F(n, \boldsymbol{a}) \simeq \{f(n)\}^{\mathfrak{A}}_{\mathsf{Reg}}(\boldsymbol{a}) \simeq I^n(\boldsymbol{a})$$

is $\mathfrak{A}$-register computable. By Propositions 2.6 and 4.1. $I^\infty = I^{<\omega}$, so

$$I^\infty(\boldsymbol{a})\downarrow \implies (\exists n \in \omega) F(n, \boldsymbol{a})\downarrow \implies \mathsf{Sel}_F(\boldsymbol{a})\downarrow \text{ and } I^{\mathsf{Sel}_F(\boldsymbol{a})}(\boldsymbol{a})\downarrow.$$

Thus $I^\infty(\boldsymbol{a}) \simeq F(\mathsf{Sel}_F(\boldsymbol{a}), \boldsymbol{a})$ and is thus $\mathfrak{A}$-register computable. □

COROLLARY 6.11. *For any first-order structure $\mathfrak{A}$ with arithmetic which admits* URM$_{\mathfrak{A}}$ *term evaluation, a function is $\mathfrak{A}$-register computable iff it is partial $\mathfrak{A}$-recursive.*

PROOF. By the preceding proposition and Theorem 4.11. □

The ordered rings of [Blum, Shub and Smale, 1989] discussed at the end of Section 5 are examples of structures which satisfy the hypothesis of this corollary. It is easy to see that term evaluation – which is simply polynomial evaluation – is register computable over any ring. Roughly speaking, each monomial may be computed by accumulating partial products in a single register and the whole polynomial similarly computed by accumulating partial sums. The number of registers needed for scratchwork does not depend on the complexity of the polynomial. For details, see [Blum, Shub and Smale, 1989, Section 8]. For an example of a structure $\mathfrak{A}$ which does not admit URM$_{\mathfrak{A}}$-term evaluation, see [Moldestad, Stoltenberg-Hansen and Tucker, 1980b, Theorem 4.3], and for numerous algebraic examples which satisfy the hypotheses of the corollary, see [Tucker, 1980].

7. Sections and envelopes

Every theory of computability uses functions to characterize two classes of relations.

DEFINITION 7.1. For any relation $R \subseteq A^k$,

(i) R is *$\mathfrak{A}$-register decidable* (*$\mathfrak{A}$-recursive*) iff its characteristic function χ_R is an $\mathfrak{A}$-register computable ($\mathfrak{A}$-recursive) function. The classes of these relations are called *sections* and denoted

$$\mathsf{Sec}_{\mathsf{Reg}}(\mathfrak{A}) = \{R\colon\ R \text{ is an } \mathfrak{A}\text{-register decidable relation}\},$$
$$\mathsf{Sec}_{\mathsf{Rec}}(\mathfrak{A}) = \{R\colon\ R \text{ is an } \mathfrak{A}\text{-recursive relation}\}.$$

(ii) R is $\mathfrak{A}$-*semi-register decidable* ($\mathfrak{A}$-*semi-recursive*) iff for some partial $\mathfrak{A}$-register computable (partial $\mathfrak{A}$-recursive) function f, $R = \mathsf{Dom}\, f$, the domain of f. The classes of these relations are called *envelopes* and denoted

$$\mathsf{Env}_{\mathsf{Reg}}(\mathfrak{A}) = \{R\colon\ R \text{ is an } \mathfrak{A}\text{-semi-register decidable relation}\},$$
$$\mathsf{Env}_{\mathsf{Rec}}(\mathfrak{A}) = \{R\colon\ R \text{ is an } \mathfrak{A}\text{-semi-recursive relation}\}.$$

In the following, when either subscript Rec or Reg applies, the subscript will be omitted.

In classical recursion theory over $\Omega = (\omega, \mathsf{Sc}, 0)$, the first simple observations about the classes of recursive and semi-recursive relations are:

(1) $\mathsf{Sec}(\Omega)$ is a Boolean algebra;
(2) $\mathsf{Sec}(\Omega) \subseteq \mathsf{Env}(\Omega)$;
(3) $\mathsf{Env}(\Omega)$ is closed under the positive Boolean operations;
(4) $R \in \mathsf{Sec}(\Omega)$ iff both R and its complement $\omega^k \sim R$ belong to $\mathsf{Env}(\Omega)$;
(5) $\mathsf{Env}(\Omega) \not\subseteq \mathsf{Sec}(\Omega)$;
(6) $\mathsf{Env}(\Omega)$ is not closed under complementation.

In the general case, some of these results remain trivially true, while others are true only with additional hypotheses and more complicated proofs.

First the easy parts:

PROPOSITION 7.2. *For any functional structure* $\mathfrak{A}$,
(i) $\mathsf{Sec}(\mathfrak{A})$ *is a Boolean algebra*;
(ii) $\mathsf{Sec}(\mathfrak{A}) \subseteq \mathsf{Env}(\mathfrak{A})$;
(iii) $\mathsf{Env}(\mathfrak{A})$ *is closed under intersection.*

PROOF. For (i) it suffices to show that $\mathsf{Sec}(\mathfrak{A})$ is closed under intersection and complement. These follow from the equations

$$\chi_{\sim R}(\boldsymbol{a}) = \begin{cases} 1_{\mathfrak{A}} & \text{if } \chi_R(\boldsymbol{a}) = 0_{\mathfrak{A}}; \\ 0_{\mathfrak{A}} & \text{otherwise}; \end{cases}$$
$$\chi_{R\cap S}(\boldsymbol{a}) = \begin{cases} 1_{\mathfrak{A}} & \text{if } \chi_R(\boldsymbol{a}) = 1_{\mathfrak{A}} \text{ and } \chi_S(\boldsymbol{a}) = 1_{\mathfrak{A}}; \\ 0_{\mathfrak{A}} & \text{otherwise.} \end{cases}$$

For (ii), if $R \in \mathsf{Sec}(\mathfrak{A})$ and

$$F(\boldsymbol{a}) \simeq \begin{cases} 0_{\mathfrak{A}} & \text{if } R(\boldsymbol{a}); \\ \uparrow & \text{otherwise}; \end{cases}$$

then $R = \mathsf{Dom}\, F \in \mathsf{Env}(\mathfrak{A})$. For (iii), note that if $F(\boldsymbol{a}) \simeq \mathsf{Pr}_0^2(G(\boldsymbol{a}), H(\boldsymbol{a}))$, then $\mathsf{Dom}\, F = \mathsf{Dom}\, G \cap \mathsf{Dom}\, H$. □

PROPOSITION 7.3. *For any of the cases discussed above for which there exist universal partial computable functions,*

(i) $\mathsf{Env}(\mathfrak{A}) \not\subseteq \mathsf{Sec}(\mathfrak{A})$,

(ii) $\mathsf{Env}(\mathfrak{A})$ *is not closed under complementation.*

PROOF. By the standard diagonal argument. □

The easiest example that shows that (4) does not hold in general is the following. Let $A \subseteq \omega$ be a non-recursive recursively enumerable set and let f and g be the semi-characteristic functions of A, $\omega \sim A$, respectively:

$$f(m) \simeq \begin{cases} 1 & \text{if } m \in A; \\ \uparrow & \text{otherwise;} \end{cases} \quad \text{and} \quad g(m) = \begin{cases} 1 & \text{if } m \notin A; \\ \uparrow & \text{otherwise;} \end{cases}$$

and $\Omega_{f,g} = (\omega, \mathsf{Sc}, 0, f, g)$. Clearly both A and $\omega \sim A$ belong to $\mathsf{Env}(\Omega_{f,g})$, but we claim that $\mathsf{Sec}(\Omega_{f,g}) = \mathsf{Sec}(\Omega)$, so that $A \notin \mathsf{Sec}(\Omega_{f,g})$. To see this, note that for any monotone functional K over ω,

$$K(x, f, g) \simeq n \Longrightarrow K(\lambda m.1, \lambda m.1) \simeq n.$$

Hence, if $\lambda x.K(x, f, g)$ is total, then it coincides with $\lambda x.K(x, \lambda m.1, \lambda m.1)$. It is easy to prove by induction that

$$F \text{ is partial } \Omega_{f,g}\text{-recursive} \Longrightarrow F(x) \simeq K(x, f, g)$$
$$\text{for some partial } \Omega\text{-recursive } K.$$

Hence if F is (total) $\Omega_{f,g}$-recursive, then F is also Ω-recursive. In particular, this holds when F is the characteristic function of $R \in \mathsf{Sec}(\Omega_{f,g})$.

The key to establishing (3) (closure under union) and (4) is some kind of selection theorem. From the results of the preceding section we have

THEOREM 7.4. *For any first-order structure $\mathfrak{A}$ with arithmetic,*

(i) $\mathsf{Env}_{\mathsf{Reg}}(\mathfrak{A})$ *is closed under union;*

(ii) $R \in \mathsf{Sec}_{\mathsf{Reg}}(\mathfrak{A})$ *iff both R and its complement $A^k \sim R$ belong to* $\mathsf{Env}_{\mathsf{Reg}}(\mathfrak{A})$.

PROOF. Let $R = \mathsf{Dom}\, f$ and $S = \mathsf{Dom}\, g$ be $\mathfrak{A}$-semi-register-decidable and define

$$F(n, \boldsymbol{a}) \simeq \begin{cases} f(\boldsymbol{a}) & \text{if } n = 0; \\ g(\boldsymbol{a}) & \text{if } n > 0. \end{cases}$$

Then (using Theorem 6.9) $(R \cup S)(\boldsymbol{a}) \iff (\exists n \in \omega)F(n, \boldsymbol{a})\downarrow \iff \mathsf{Sel}_F(\boldsymbol{a})\downarrow$, so $R \cup S$ is also $\mathfrak{A}$-semi-register-decidable. Similarly, for (ii), if $R = \mathsf{Dom}\, f$ and $\sim R = \mathsf{Dom}\, g$ and we set

$$G(n, \boldsymbol{a}) \simeq \begin{cases} f(\boldsymbol{a}) & \text{if } n = 1; \\ g(\boldsymbol{a}) & \text{if } n = 0, \end{cases}$$

then Sel_G is $\mathfrak{A}$-register computable and is the characteristic function of R, so R is $\mathfrak{A}$-register decidable. □

One of the main results of [Blum, Shub and Smale, 1989] (Section 4, Proposition 2), together with [Michaux, 1990], is a characterization of the BSS-semi-computable relations over the real numbers as the countable unions of semi-algebraic sets whose definitions require only finitely many non-rational parameters. One direction of this characterization is a general fact about register computability, which we sketch briefly. Fix a first-order structure $\mathfrak{A}$ and a program $\pi = (I_0, \ldots, I_n)$ for an $\mathfrak{A}$-register machine. For each finite sequence $\boldsymbol{i} = (i_0, \ldots, i_s)$, let

$$V_{\boldsymbol{i}}^k = \left\{\boldsymbol{a} \in A^k \colon \ \pi(\boldsymbol{a})\downarrow \text{ via the computation path } I_{i_0}, \ldots, I_{i_s}\right\}.$$

By an analysis similar to that of Section 6, with each p and $\boldsymbol{i}$ we may associate a fixed term $t_{p,\boldsymbol{i}}$ such that a computation with input $\boldsymbol{a}$ which follows the computation path $I_{i_0}, \ldots, I_{i_s}$ leaves register R_p containing the value $t_{p,\boldsymbol{i}}^{\mathfrak{A}}[\boldsymbol{a}]$ of this term evaluated at $\boldsymbol{a}$. If I_{i_j} is a conditional instruction with hypothesis $R(r_{q_0}, \ldots, r_{q_{k-1}})$, then whether or not a computation which has reached I_{i_j} continues to $I_{i_{j+1}}$ depends on whether or not

$$R^{\mathfrak{A}}\big(t_{q_0, i_0, \ldots, i_j}^{\mathfrak{A}}[\boldsymbol{a}], \ldots, t_{q_{k-1}, i_0, \ldots, i_j}^{\mathfrak{A}}[\boldsymbol{a}]\big)$$

holds – in other words, whether the atomic formula $Rt_{q_0,i_0,\ldots,i_j} \cdots t_{q_{k-1},i_0,\ldots,i_j}$ is satisfied by $\boldsymbol{a}$. It follows that for each k and $\boldsymbol{i}$ there exists a formula $\phi_{\boldsymbol{i}}^k$ which is a conjunction of atomic and negated atomic formulas such that

$$V_{\boldsymbol{i}}^k = \left\{\boldsymbol{a} \in A^k \colon \ \models_{\mathfrak{A}} \phi_{\boldsymbol{i}}^k[\boldsymbol{a}]\right\}.$$

Then, since the domain of π is the union of such sets over all possible paths, we have

THEOREM 7.5. *For any first-order structure $\mathfrak{A}$, every $\mathfrak{A}$-semi-register-decidable relation is a countable union of relations quantifier-free definable over $\mathfrak{A}$.*

In the case of ordered rings, where the only relations are $<$ and $=$, this reduces to [Blum, Shub and Smale, 1989, Section 4, Proposition 2] that every semi-computable relation is a countable union of basic semi-algebraic sets – that is, sets defined by finitely many polynomial inequalities of the types $g(\boldsymbol{a}) < 0$ and $h(\boldsymbol{a}) \leqslant 0$. This fact is used in [Blum, Shub and Smale, 1989, Section 1] to derive that various natural sets of real numbers, such as the complements of Cantor sets, are semi-computable but not computable over the ring of real numbers. For this special case there is also a converse to the theorem due to C. Michaux:

THEOREM 7.6 ([Michaux, 1990]). *A relation P is BSS-semi-computable over the ring of real numbers if and only if for some reals $a_0, \ldots, a_j$, P is a countable union of relations quantifier-free definable over* $(\mathbb{R}, +, \times, <, =_{\mathbb{R}}, 0, 1, a_0, \ldots, a_{n-1})$.

The proof is based on the fact that the countable amount of information required to specify such a union can be coded as a single real number.

For the $\mathfrak{A}$-recursive case, the key idea, introduced in [Kechris and Moschovakis, 1977] is normality.

DEFINITION 7.7. A functional F is $\mathfrak{A}$-*normal* iff there exists a partial $\mathfrak{A}$-recursive *normalizing functional* Δ_F as follows. For any partial function δ, set

$$Z_\delta = \{\delta\colon (\boldsymbol{b}) \simeq 0_{\mathfrak{A}}\}.$$

Then (when F has only one partial function argument),

(i) $F(\boldsymbol{a}, f \restriction Z_\delta)\downarrow \Longrightarrow \Delta_F(\boldsymbol{a}, f, \delta) \simeq 1_{\mathfrak{A}}$;

(ii) δ is total, $Z_\delta \subseteq \mathsf{Dom}\, f$, and $F(\boldsymbol{a}, f \restriction Z_\delta)\uparrow \Longrightarrow \Delta_F(\boldsymbol{a}, f, \delta) \simeq 0_{\mathfrak{A}}$.

The intuition behind this definition is that in case (i), if $f \restriction Z_\delta$ is enough of f for $F(\boldsymbol{a}, f \restriction Z_\delta)$ to be defined, this can be verified by delaying the exercise of an evaluation of a value $f(\boldsymbol{b})$ during the computation until we verify that $\delta(\boldsymbol{b}) \simeq 0$. In case (ii), we cannot expect in general to be able to verify that $f \restriction Z_\delta$ is *not* enough of f for $F(\boldsymbol{a}, f \restriction Z_\delta)$ to be defined, since δ itself my contain too little information; the extra hypotheses rule this out.

The extension of the definition to functionals with several partial function arguments is straightforward, for example, if F is of type $(k; l, m)$, then Δ_F is of type $(k; l, m, l, m)$.

EXAMPLES 7.8. (i) If F has no partial function arguments, then Δ_F is just the characteristic function of $\mathsf{Dom}\, F$, so F is $\mathfrak{A}$-normal iff $\mathsf{Dom}\, F$ is $\mathfrak{A}$-recursive. In particular, if F is total, then F is $\mathfrak{A}$-normal.

(ii) The evaluation functionals are $\mathfrak{A}$-normal – for example, let

$$\Delta_{\mathsf{Ev}_j^{k,l}}(\boldsymbol{a}, f, \delta) \simeq \begin{cases} 1_{\mathfrak{A}} & \text{if } \delta(\boldsymbol{a}) \simeq 0_{\mathfrak{A}}; \\ 0_{\mathfrak{A}} & \text{if } \delta(\boldsymbol{a})\downarrow \neq 0_{\mathfrak{A}}. \end{cases}$$

It is easily verified that this is a normalizing functional.

(iii) The composition functionals are normal – for example, for any γ and δ, we have

$$\begin{aligned}&\mathsf{Cmp}^{k;1}(\boldsymbol{a}, g \restriction Z_\gamma, h \restriction Z_\delta) \simeq (g \restriction Z_\gamma)\big((h \restriction Z_\delta)(\boldsymbol{a})\big)\downarrow \\ &\quad\iff \delta(\boldsymbol{a}) \simeq 0_{\mathfrak{A}} \text{ and } \gamma\big(h(\boldsymbol{a})\big) \simeq 0_{\mathfrak{A}} \text{ and } g\big(h(\boldsymbol{a})\big)\downarrow.\end{aligned}$$

Set

$$\Delta_{\mathsf{Cmp}^{k;1}}(\boldsymbol{a}, g, h, \gamma, \delta) \simeq \begin{cases} \mathsf{IsZero}_{\mathfrak{A}}(\gamma(h(\boldsymbol{a}))) & \text{if } \delta(\boldsymbol{a}) \simeq 0_{\mathfrak{A}}; \\ 0_{\mathfrak{A}} & \text{if } \delta(\boldsymbol{a})\downarrow \neq 0_{\mathfrak{A}}. \end{cases}$$

If $(g \restriction Z_\gamma)((h \restriction Z_\delta)(\boldsymbol{a}))\downarrow$, then clearly $\Delta_{\mathsf{Cmp}^{k;1}}(\boldsymbol{a}, g, h, \gamma, \delta) \simeq 1$ by virtue of the first clause. Suppose now that $(g \restriction Z_\gamma)((h \restriction Z_\delta)(\boldsymbol{a}))\uparrow$, and the other conditions of (ii) of the definition are satisfied. If $\delta(\boldsymbol{a}) \not\simeq 0_{\mathfrak{A}}$, then since δ is total,

$$\Delta_{\mathsf{Cmp}^{k;1}}(\boldsymbol{a}, g, h, \gamma, \delta) \simeq 0_{\mathfrak{A}}$$

by virtue of the second clause. Otherwise, $\delta(\boldsymbol{a}) \simeq 0_{\mathfrak{A}}$, so $\boldsymbol{a} \in Z_\delta \subseteq \mathsf{Dom}\, h$ and since γ is total, $\gamma(h(\boldsymbol{a}))\downarrow$. If $\gamma(h(\boldsymbol{a})) \simeq 0_{\mathfrak{A}}$, then $h(\boldsymbol{a}) \in Z_\gamma \subseteq \mathsf{Dom}\, g$, so $(g \restriction Z_\gamma)((h \restriction Z_\delta)(\boldsymbol{a})) \simeq g(h(\boldsymbol{a}))\downarrow$, contrary to assumption. Hence $\gamma(h(\boldsymbol{a}))\downarrow \neq 0_{\mathfrak{A}}$ and $\Delta_{\mathsf{Cmp}^{k;1}}(\boldsymbol{a}, g, h, \gamma, \delta) \simeq 0_{\mathfrak{A}}$ by the first clause.

DEFINITION 7.9. A functional structure $\mathfrak{A}$ is *normal* iff every $F \in \mathsf{Fnl}_{\mathfrak{A}}$ is $\mathfrak{A}$-normal.

COROLLARY 7.10. *Every first-order structure is normal.*

PROPOSITION 7.11. *If $\mathfrak{A}$ is normal, then all $\mathfrak{A}$-explicit functionals are $\mathfrak{A}$-normal.*

PROOF. From the examples above and by Remark 2.3, it remains only to show that each Cases^k is $\mathfrak{A}$-normal and that the class of $\mathfrak{A}$-normal functionals is closed under substitution. The first of these is immediate:

$$\Delta_{\mathsf{Cases}^k}(c, \boldsymbol{a}, f, g, h, \gamma, \delta, \varepsilon) \simeq \begin{cases} \mathsf{IsZero}_{\mathfrak{A}}(\gamma(\boldsymbol{a})) & \text{if } c = 1_{\mathfrak{A}}; \\ \mathsf{IsZero}_{\mathfrak{A}}(\delta(\boldsymbol{a})) & \text{if } c = 0_{\mathfrak{A}}; \\ \mathsf{IsZero}_{\mathfrak{A}}(\varepsilon(\boldsymbol{a})) & \text{otherwise} \end{cases}$$

is easily seem to be a normalizing functional.

For substitution we prove a special case. Suppose G and H are $\mathfrak{A}$-normal and

$$F(\boldsymbol{a}, f) \simeq G(\boldsymbol{a}, H_{\boldsymbol{a},f}),$$

where, as usual, $H_{\boldsymbol{a},f} = \lambda \boldsymbol{b}.H(\boldsymbol{a}, \boldsymbol{b}, f)$. Set (abusing notation)

$$\delta'(\boldsymbol{b}) \simeq \mathsf{IsZero}_{\mathfrak{A}}\big(\Delta_H(\boldsymbol{a}, \boldsymbol{b}, f, \delta)\big) \quad \text{and} \quad \Delta_F(\boldsymbol{a}, f, \delta) \simeq \Delta_G(\boldsymbol{a}, H_{\boldsymbol{a},f}, \delta').$$

To see that Δ_F is a normalizing functional for F, suppose first that $F(\boldsymbol{a}, f \upharpoonright Z_\delta)\downarrow$ and set

$$\gamma(\boldsymbol{b}) \simeq 0 \Longleftrightarrow H(\boldsymbol{a}, \boldsymbol{b}, f \upharpoonright Z_\delta)\downarrow.$$

Then $G(\boldsymbol{a}, H_{\boldsymbol{a},f} \upharpoonright Z_\gamma)\downarrow$ and $Z_\gamma \subseteq Z_{\delta'}$, so by monotonicity, also $G(\boldsymbol{a}, H_{\boldsymbol{a},f} \upharpoonright Z_{\delta'})\downarrow$, from which follows that $\Delta_F(\boldsymbol{a}, f, \delta) \simeq 1$ as desired.

If, on the other hand, δ is total, $Z_\delta \subseteq \mathsf{Dom}\, f$ and $F(\boldsymbol{a}, f \upharpoonright Z_\delta)\uparrow$, then also $G(\boldsymbol{a}, H_{\boldsymbol{a},f} \upharpoonright Z_\gamma)\uparrow$. In this case, $Z_\gamma = Z_{\delta'}$, so also $G(\boldsymbol{a}, H_{\boldsymbol{a},f} \upharpoonright Z_{\delta'})\uparrow$. By the hypotheses, $Z_{\delta'} \subseteq \mathsf{Dom}\, H_{\boldsymbol{a},f}$, so we have $\Delta_F(\boldsymbol{a}, f, \delta) \simeq 0$ as desired. □

For any iterable sequence $\mathcal{I} = (I_0, \ldots, I_m)$, the *$\mathcal{I}$-norm* is the function $|\cdot|_{\mathcal{I}}$ defined by

$$|x|_{\mathcal{I}} = \begin{cases} \text{least}\, \sigma\, [I_0^\sigma(x)\downarrow] & \text{if } I_0^\infty(x)\downarrow; \\ \infty & \text{otherwise;} \end{cases}$$

where ∞ denotes any fixed large ordinal, say $\mathsf{Card}(A)^+$. Thus $|x|_{\mathcal{I}}$ measures "how long" it takes to compute $I_0^\infty(x)$. For two iterable sequences $\mathcal{I}$ and $\mathcal{J}$, set

$$x \leqslant_{\mathcal{I},\mathcal{J}} y \Longleftrightarrow I_0^\infty(x)\downarrow \text{ and } |x|_{\mathcal{I}} \leqslant |y|_{\mathcal{J}},$$
$$y <_{\mathcal{I},\mathcal{J}} x \Longleftrightarrow J_0^\infty(y)\downarrow \text{ and } |y|_{\mathcal{J}} < |x|_{\mathcal{I}}.$$

Note that the condition '$J_0^\infty(y)\downarrow$' is redundant. Let

$$\mathsf{C}_{\mathcal{I},\mathcal{J}}(x, y) \simeq \begin{cases} 0_{\mathfrak{A}} & \text{if } x \leqslant_{\mathcal{I},\mathcal{J}} y, \\ 1_{\mathfrak{A}} & \text{if } y <_{\mathcal{I},\mathcal{J}} x. \end{cases}$$

$\mathsf{C}_{\mathcal{I},\mathcal{J}}(x, y)$ is called the *stage comparison functional* for $\mathcal{I}$ and $\mathcal{J}$.

STAGE COMPARISON THEOREM 7.12 ([Kechris and Moschovakis, 1977]). *If $\mathfrak{A}$ is normal, then for any iterable sequences $\mathcal{I}$ and $\mathcal{J}$ of $\mathfrak{A}$-explicit functionals, the stage comparison functional $\mathsf{C}_{\mathcal{I},\mathcal{J}}$ is partial $\mathfrak{A}$-recursive.*

PROOF. We prove the special, but typical, case $\mathcal{I} = (I)$ and $\mathcal{J} = (J)$. Let Δ_I and Δ_J be normalizing functionals for I and J. Let

$$G(\boldsymbol{a}, \boldsymbol{b}, g, h) \simeq \Delta_I\big(\boldsymbol{a}, I^\infty, \lambda \boldsymbol{c}.h(\boldsymbol{c}, \boldsymbol{b})\big);$$
$$H(\boldsymbol{a}, \boldsymbol{b}, g, h) \simeq \Delta_J\big(\boldsymbol{a}, J^\infty, \lambda \boldsymbol{d}.g(\boldsymbol{a}, \boldsymbol{d})\big).$$

This is an iterable pair of partial $\mathfrak{A}$-recursive functionals; we prove by induction on σ that

$$\boldsymbol{a} \leqslant_{IJ} \boldsymbol{b} \wedge |\boldsymbol{a}|_I = \sigma \Longrightarrow G^\infty(\boldsymbol{a}, \boldsymbol{b}) \simeq 1_{\mathfrak{A}}, \tag{1}$$

$$b <_{IJ} a \wedge |b|_J = \sigma \Longrightarrow G^\infty(a, b) \simeq 0_{\mathfrak{A}}, \tag{2}$$

$$b \leqslant_{JI} a \wedge |b|_J = \sigma \Longrightarrow H^\infty(a, b) \simeq 1_{\mathfrak{A}}, \tag{3}$$

$$a <_{JI} b \wedge |a|_I = \sigma \Longrightarrow H^\infty(a, b) \simeq 0_{\mathfrak{A}}. \tag{4}$$

Assume as induction hypothesis that (1)–(4) hold for all $\tau < \sigma$ and assume first the hypothesis of (1) for a fixed $\boldsymbol{a}$ and $\boldsymbol{b}$. Then $I(\boldsymbol{a}, I^{<\sigma}) \simeq I^s(\boldsymbol{a})\downarrow$, and by (4) of the induction hypothesis

$$|c|_I < \sigma \Longrightarrow |c|_I < |b|_J \Longrightarrow H^\infty(a, b) \simeq 0_{\mathfrak{A}}.$$

In other words,

$$I^{<\sigma} \subseteq I^\infty \upharpoonright Z_{\lambda c. H^\infty(c,b)},$$

whence also

$$I\big(a, I^\infty \upharpoonright Z_{\lambda c. H^\infty(c,b)}\big)\downarrow,$$

so by the definition of Δ_I,

$$G^\infty(a, b) \simeq G\big(a, b, G^\infty, H^\infty\big) \simeq \Delta_I\big(a, I^\infty, \lambda c. H^\infty(c, b)\big) \simeq 1_{\mathfrak{A}},$$

as desired.

We may prove (3) in a parallel fashion. Now assume the hypotheses of (2). From these, (4) of the induction hypothesis, and the just established (3), it follows that $\lambda \boldsymbol{c}. H^\infty(\boldsymbol{c}, \boldsymbol{b})$ is a total function. Furthermore,

$$H^\infty(c, b) \simeq 0_{\mathfrak{A}} \Longrightarrow |c|_I < \sigma \Longrightarrow c \in \mathsf{Dom}\, I^{<\sigma} \subseteq \mathsf{Dom}\, I^\infty,$$

so $I^\infty \upharpoonright Z_{\lambda c. H^\infty(c,b)} \subseteq I^{<\sigma}$. The hypotheses of (2) mean that $I(\boldsymbol{a}, I^{<\sigma}) \simeq I^s(\boldsymbol{a})\uparrow$ and we have all of the conditions satisfied to guarantee

$$G^\infty(a, b) \simeq G\big(a, b, G^\infty, H^\infty\big) \simeq \Delta_I\big(a, I^\infty, \lambda c. H^\infty(c, b)\big) \simeq 0_{\mathfrak{A}},$$

as desired. The proof of (4) is parallel.

Now we have

$$C_{I,J}(a, b) \simeq \begin{cases} 1_{\mathfrak{A}}(I^\infty(a)) & \text{if } G^\infty(a, b) \simeq 1_{\mathfrak{A}}; \\ 0_{\mathfrak{A}}(J^\infty(a)) & \text{if } G^\infty(a, b) \simeq 0_{\mathfrak{A}}, \end{cases}$$

where $1_{\mathfrak{A}}$ and $0_{\mathfrak{A}}$ denote constant unary functions. □

COROLLARY 7.13. *If $\mathfrak{A}$ is normal, then*

(i) $\mathsf{Env}_{\mathsf{Rec}}(\mathfrak{A})$ *is closed under finite union;*

(ii) $R \in \mathsf{Sec}_{\mathsf{Rec}}(\mathfrak{A})$ *iff both* R *and* $A^k \sim R$ *belong to* $\mathsf{Env}_{\mathsf{Rec}}(\mathfrak{A})$.

PROOF. For (i), let $R = \mathsf{Dom\,Rec}(\mathcal{I})$ and $S = \mathsf{Dom\,Rec}(\mathcal{J})$ be two $\mathfrak{A}$-semi-recursive relations on A. Then easily $R \cup S = \mathsf{Dom}\,\lambda \boldsymbol{a}.C_{\mathcal{I},\mathcal{J}}(\boldsymbol{a},\boldsymbol{a})$ and is thus $\mathfrak{A}$-semi-recursive. For (ii), suppose that $R = \mathsf{Dom\,Rec}(\mathcal{I})$ and $A^k \sim R = \mathsf{Dom\,Rec}(\mathcal{J})$. Then easily, $C_{\mathcal{I},\mathcal{J}}(\boldsymbol{a},\boldsymbol{a})$ is the characteristic function of R, which is thus $\mathfrak{A}$-recursive. □

For first-order structures with arithmetic (for which we have universal functions), this result may also be proved by establishing a number selection theorem. We first introduce a modified indexing of the partial $\mathfrak{A}$-recursive functions:

$$\big\{\{\langle c, s\rangle\}\big\}^{\mathfrak{A}}_{\mathsf{Rec}}(x) \simeq \{c\}^{\mathfrak{A}}_{\mathsf{Rec}}(s, x),$$

for which we can prove the Second Recursion Theorem. We follow here the presentation of [Kechris and Moschovakis, 1977].

PROPOSITION 7.14. *For any first-order structure* $\mathfrak{A}$ *with arithmetic, there exist primitive recursive functions* $S^{k,l}_n$ *such that for all* $b, m_0, \ldots, m_{n-1} \in \omega$,

$$\big\{\{S^{k,l}_n(b, \boldsymbol{m})\}\big\}^{\mathfrak{A}}_{\mathsf{Rec}}(x) \simeq \big\{\{b\}\big\}^{\mathfrak{A}}_{\mathsf{Rec}}(\boldsymbol{m}, x).$$

PROOF. There exists an index $\bar{d}$ such that

$$\{\bar{d}\}^{\mathfrak{A}}_{\mathsf{Rec}}\big(\langle b, \boldsymbol{m}\rangle, x\big) \simeq \big\{\{b\}\big\}^{\mathfrak{A}}_{\mathsf{Rec}}(\boldsymbol{m}, x).$$

Now if $S^{k,l}_n(b, \boldsymbol{m}) := \langle \bar{d}, \langle b, \boldsymbol{m}\rangle\rangle$, we have

$$\big\{\{S^{k,l}_n(b, \boldsymbol{m})\}\big\}^{\mathfrak{A}}_{\mathsf{Rec}}(x) \simeq \{\bar{d}\}^{\mathfrak{A}}_{\mathsf{Rec}}\big(\langle b, \boldsymbol{m}\rangle, x\big) \simeq \big\{\{b\}\big\}^{\mathfrak{A}}_{\mathsf{Rec}}(\boldsymbol{m}, x),$$

as desired. □

SECOND RECURSION THEOREM 7.15. *For any first-order structure* $\mathfrak{A}$ *with arithmetic, for any partial* $\mathfrak{A}$*-recursive functional* F*, there exists an index* $\bar{b}$ *such that for all* x,

$$\big\{\{\bar{b}\}\big\}^{\mathfrak{A}}_{\mathsf{Rec}}(x) \simeq F\big(\bar{b}, x\big).$$

PROOF. As usual, set $\bar{b} = S^{k,l}_1(\bar{c}, \bar{c})$, where $\{\{\bar{c}\}\}^{\mathfrak{A}}_{\mathsf{Rec}}(b, x) \simeq F\big(S^{k,l}_1(b, b), x\big)$. □

NUMBER SELECTION THEOREM 7.16. *For any first-order structure with arithmetic, for each partial* $\mathfrak{A}$*-recursive functional* F *there exists a partial* $\mathfrak{A}$*-recursive functional* Sel_F *such that*

$$(\exists n \in \omega) F(n, \boldsymbol{a})\!\downarrow \Longrightarrow \mathsf{Sel}_F(\boldsymbol{a})\!\downarrow \text{ and } F(\mathsf{Sel}_F(\boldsymbol{a}), a)\!\downarrow.$$

PROOF. Since $\mathfrak{A}$ is first-order, it is normal and the stage-comparison functional is partial $\mathfrak{A}$-recursive. Let $\mathcal{I}$ be an iterable sequence of $\mathsf{Expl}_{\mathfrak{A}}$ functionals such that $I_0^\infty(b, a, x) \simeq \{\{b\}\}^{\mathfrak{A}}_{\mathsf{Rec}}(a, x)$ is the appropriate universal functional and fix an index $\bar{b}$ for F. Set

$$G(c, a, x) \simeq \begin{cases} 0 & \text{if } a \in \omega \text{ and } \mathcal{C}_{\mathcal{I},\mathcal{I}}(\bar{b}, a, x, c, a+1, x) \simeq 0; \\ \{\{c\}\}^{\mathfrak{A}}_{\mathsf{Rec}}(a+1, x) + 1 & \text{if } a \in \omega \text{ and } \mathcal{C}_{\mathcal{I},\mathcal{I}}(\bar{b}, a, x, c, a+1, x) \simeq 1; \\ 0 & \text{if } a \notin \omega. \end{cases}$$

By the Second Recursion Theorem, choose $\bar{c}$ such that

$$I_0^\infty(\bar{c}, a, x) \simeq \{\{\bar{c}\}\}^{\mathfrak{A}}_{\mathsf{Rec}}(a, x) \simeq G(\bar{c}, a, x).$$

Suppose that $F(n, x)\downarrow$ so $I_0^\infty(\bar{b}, n, x)\downarrow$ and hence $\mathcal{C}_{\mathcal{I},\mathcal{I}}(\bar{b}, n, x, c, n+1, x) \simeq 0$ or 1. In the first case $G(\bar{c}, n, x) \simeq 0$ and in the second, since

$$|\bar{c}, n+1, x|_{\mathcal{I}} < |\bar{b}, n, x|_{\mathcal{I}}, G(\bar{c}, n+1, x) \simeq \{\{\bar{c}\}\}^{\mathfrak{A}}_{\mathsf{Rec}}(n+1, x)\downarrow.$$

Furthermore, it follows easily that if $G(\bar{c}, m, x)\downarrow$, then also $G(\bar{c}, m', x)\downarrow$ for all $m' < m$. Hence in any case $G(\bar{c}, 0, x)\downarrow$, say $G(\bar{c}, 0, x) \simeq p$. Then clearly

$$G(\bar{c}, 1, x) \simeq p - 1, \ldots, G(\bar{c}, p, x) \simeq 0,$$

and this must be due to the first clause – that is $\mathcal{C}_{\mathcal{I},\mathcal{I}}(\bar{b}, p, x, \bar{x}, p+1, x) \simeq 0$. Hence $F(p, x) \simeq I_0^\infty(\bar{b}, p, x)\downarrow$ and thus $\mathsf{Sel}_F(x) :\simeq G(\bar{c}, 0, x)$ has the desired property. □

As a final example we show

THEOREM 7.17. *For any first-order structure,* $\mathfrak{A}^+ = (\mathfrak{A}, =_A, \mathsf{E}^\sharp_A)$ *is normal.*

PROOF. Let G^0 and G^1 be the functionals defined by

$$G^i(a, f, \delta) \simeq \begin{cases} i_{\mathfrak{A}} & \text{if } \delta(a)\downarrow \neq 0_{\mathfrak{A}}; \\ \mathsf{IsOne}_{\mathfrak{A}}(f(a)) & \text{if } \delta(a) \simeq 0_{\mathfrak{A}}. \end{cases}$$

Then we shall show that

$$\Delta_{\mathsf{E}^\sharp_A}(f, \delta) \simeq \begin{cases} 1_{\mathfrak{A}} & \text{if } \mathsf{E}^\sharp_A(G^0_{f,\delta}) \simeq 1_{\mathfrak{A}}; \\ \mathsf{IsZero}_{\mathfrak{A}}(\mathsf{E}^\sharp_A(G^1_{f,\delta})) & \text{if } \mathsf{E}^\sharp_A(G^0_{f,\delta}) \simeq 0_{\mathfrak{A}}; \end{cases}$$

is a normalizing function for $\mathsf{E}^\sharp_A$ which is $\mathfrak{A}^+$-explicit. (Recall that $G^i_{f,\delta} = \lambda a.\, G^i(a, f, \delta)$.)

Suppose first that $\mathsf{E}^\sharp_A(f \restriction Z_\delta) \simeq \mathsf{1}_{\mathfrak{A}}$. Then

$$\exists a\big[f(a) \simeq \mathsf{1}_{\mathfrak{A}} \wedge \delta(a) \simeq \mathsf{0}_{\mathfrak{A}}\big] \quad \text{so} \quad \exists a\big[G^0(a, f, \delta) \simeq \mathsf{1}_{\mathfrak{A}}\big],$$

whence $\mathsf{E}^\sharp_A(G^0_{f,\delta}) \simeq \mathsf{1}_{\mathfrak{A}}$ and thus $\Delta_{\mathsf{E}^\sharp_A}(f, \delta) \simeq \mathsf{1}_{\mathfrak{A}}$.

Next, if $\mathsf{E}^\sharp_A(f \restriction Z_\delta) \simeq \mathsf{0}_{\mathfrak{A}}$, then

$$\forall a\big[f(a) \simeq \mathsf{0}_{\mathfrak{A}} \wedge \delta(a) \simeq \mathsf{0}_{\mathfrak{A}}\big], \quad \text{so} \quad \text{for } i = 0, 1,\ \forall a\big[G^i(a, f, \delta) \simeq \mathsf{0}_{\mathfrak{A}}\big],$$

whence for $i = 0, 1$, $\mathsf{E}^\sharp_A(G^i_{f,\delta}) \simeq \mathsf{0}_{\mathfrak{A}}$ and thus $\Delta_{\mathsf{E}^\sharp_A}(f, \delta) \simeq \mathsf{1}_{\mathfrak{A}}$.

Now suppose that δ is total, $Z_\delta \subseteq \mathsf{Dom}\, F$ and $\mathsf{E}^\sharp_A(f \restriction Z_\delta)\uparrow$. Since

$$\mathsf{E}^\sharp_A(f \restriction Z_\delta) \not\simeq \mathsf{1}_{\mathfrak{A}},$$

we have

$$\forall a\big[\delta(a) \simeq \mathsf{0}_{\mathfrak{A}} \Longrightarrow f(a) \not\simeq \mathsf{1}_{\mathfrak{A}}\big],$$
$$\text{so since } \delta \text{ is total, } \forall a\big[G^0(a, f, \delta) \simeq \mathsf{0}_{\mathfrak{A}}\big],$$

whence $\mathsf{E}^\sharp_A(G^0_{f,\delta}) \simeq \mathsf{0}_{\mathfrak{A}}$. Since $\mathsf{E}^\sharp_A(f \restriction Z_\delta) \not\simeq \mathsf{0}_{\mathfrak{A}}$,

$$\neg\forall a\big[\delta(a) \simeq \mathsf{0}_{\mathfrak{A}} \wedge f(a)\downarrow \neq \mathsf{1}_{\mathfrak{A}}\big]$$
$$\text{so either } \exists a\big[\delta(a)\downarrow \neq \mathsf{0}_{\mathfrak{A}}\big] \text{ or } \exists a\big[\delta(a) \simeq \mathsf{0}_{\mathfrak{A}} \wedge f(a) \simeq \mathsf{1}_{\mathfrak{A}}\big].$$

In either case

$$\exists a\big[G^1(a, f, \delta) \simeq \mathsf{1}_{\mathfrak{A}}\big],$$

whence $\mathsf{E}^\sharp_A(G^1_{f,\delta}) \simeq \mathsf{1}_{\mathfrak{A}}$, and thus $\Delta_{\mathsf{E}^\sharp_A}(f, \delta) \simeq \mathsf{0}_{\mathfrak{A}}$. □

8. Appendix

For completeness, we include here sketches of proofs of two technical results used in the proof of Theorem 4.5.

PROPOSITION 8.1. *The class of simply partial $\mathfrak{A}$-recursive functionals is closed under substitution.*

PROOF. To keep the notation under control, we shall prove explicitly a special case and hope that the extension to the general case is clear. Let G and H be simply partial $\mathfrak{A}$-recursive, say

$$G = I^{\infty} = \mathsf{Rec}(I, J) \quad \text{and} \quad H = K^{\infty} = \mathsf{Rec}(K),$$

with I, J, and K, all $\mathfrak{A}$-explicit, and

$$F(a, x) \simeq G(a, x, H_x) \simeq G\big(a, x, \lambda c.H(c, x)\big).$$

Let $({}^*I, {}^*J, {}^*K)$ be the following iterable sequence of $\mathfrak{A}$-explicit functionals:

$$\begin{aligned} {}^*I(a, x, i, j, k) &\simeq I(a, x, k, i, j), \\ {}^*J(b, x, i, j, k) &\simeq J(b, x, k, i, j), \\ {}^*K(c, x, i, j, k) &\simeq K(c, x, k). \end{aligned}$$

We claim that $F = {}^*I^{\infty} = \mathsf{Rec}({}^*I, {}^*J, {}^*K)$ and is thus partial $\mathfrak{A}$-recursive – in other words, that for all x,

$${}^*I_x^{\infty} = I_{x,H_x}^{\infty}. \tag{1}$$

It will suffice to show that for all σ,

$${}^*K^{\sigma} = K^{\sigma} \quad \text{whence} \quad {}^*K^{\sigma} \subseteq H \tag{2}$$

and for all σ and x,

$${}^*I_x^{\sigma} \subseteq I_{x,H_x}^{\sigma} \subseteq {}^*I_x^{\infty} \quad \text{and} \quad {}^*J_x^{\sigma} \subseteq J_{x,H_x}^{\sigma} \subseteq {}^*J_x^{\infty}. \tag{3}$$

The intuition behind (3) (for I) is that ${}^*I^{\sigma}(a, x)$ is obtained by applying

$$I\big(a, x, {}^*K^{<\sigma}, \ldots\big)$$

to earlier stages. By (2) this is 'weaker' than applying $I(a, x, H_x, \ldots)$, which is used to compute I_{x,H_x}^{σ}, but in the limit the result is the same.

The proof of (3) is also by induction; assume as induction hypothesis all of these inclusions for $\tau < \sigma$. Thus we have

$${}^*I_x^{<\sigma} \subseteq I_{x,H_x}^{<\sigma} \subseteq {}^*I_x^{\infty} \quad \text{and} \quad {}^*J_x^{<\sigma} \subseteq J_{x,H_x}^{<\sigma} \subseteq {}^*J_x^{\infty}. \tag{4}$$

Now if for some e,

$$\begin{aligned} {}^*I^{\sigma}(a, x) &\simeq {}^*I\big(a, x, {}^*I_x^{<\sigma}, {}^*J_x^{<\sigma}, {}^*K_x^{<\sigma}\big) \\ &\simeq I\big(a, x, {}^*K_x^{<\sigma}, {}^*I_x^{<\sigma}, {}^*J_x^{<\sigma}\big) \simeq e, \end{aligned}$$

then by monotonicity, (2), and the first and third inclusions of (4) we have

$$I^\sigma(a, x, H_x) \simeq I\big(a, x, H_x, I^{<\sigma}_{x,H_x}, J^{<\sigma}_{x,H_x}\big) \simeq e,$$

which gives the first inclusion of (3). Now, by (2) and the second and fourth inclusions of (4), if $I^\sigma(a, x, H_x) \simeq e$, we have also

$$^*I^\infty(a, x) \simeq {}^*I\big(a, x, {}^*I^\infty_x, {}^*J^\infty_x, {}^*K^\infty_x\big) \simeq I\big(a, x, {}^*K^\infty_x, {}^*I^\infty_x, {}^*J^\infty_x\big) \simeq e,$$

and hence the second inclusion of (3). The corresponding calculations for J are similar. □

PROPOSITION 8.2. *The class of simply partial $\mathfrak{A}$-recursive functionals is closed under recursion.*

PROOF. We again prove a special case. Let I be simply partial $\mathfrak{A}$-recursive, say $I = J^\infty = \mathsf{Rec}(J, K)$, with J and K both $\mathfrak{A}$-explicit, and $F = I^\infty = \mathsf{Rec}(I)$. Let $({}^*I, {}^*J, {}^*K)$ be the following iterable sequence of $\mathfrak{A}$-explicit functionals:

$$\begin{aligned} {}^*I(a, x, i, j, k) &\simeq j(a), \\ {}^*J(a, x, i, j, k) &\simeq J(a, x, i, j, k), \\ {}^*K(b, x, i, j, k) &\simeq K(b, x, i, j, k). \end{aligned}$$

We claim that $F = {}^*I^\infty = \mathsf{Rec}({}^*I, {}^*J, {}^*K)$ and is thus simply partial $\mathfrak{A}$-recursive. To establish this it will suffice to show that for all σ,

$$^*I^\sigma \subseteq I^\sigma \subseteq {}^*I^\infty. \tag{1}$$

We need first to establish that for all σ and x,

$$J^\sigma_{x,{}^*I^\infty_x} \subseteq {}^*J^\infty_x \quad \text{and} \quad K^\sigma_{x,{}^*I^\infty_x} \subseteq {}^*K^\infty_x.$$

Assuming this for $\tau < \sigma$, if $J^\sigma(a, x, {}^*I^\infty_x) \simeq e$, then

$$\begin{aligned} {}^*J\big(a, x, {}^*I^\infty_x, J^{<\sigma}_{x,{}^*I^\infty_x}, K^{<\sigma}_{x,{}^*I^\infty_x}\big) &\simeq J\big(a, x, {}^*I^\infty_x, J^{<\sigma}_{x,{}^*I^\infty_x}, K^{<\sigma}_{x,{}^*I^\infty_x}\big) \\ &\simeq J^\sigma\big(a, x, {}^*I^\infty_x\big) \simeq e, \end{aligned}$$

so by the induction hypothesis and monotonicity, also

$$^*J^\infty(a, x) \simeq {}^*J\big(a, x, {}^*I^\infty_x, {}^*J^\infty_{x,{}^*I^\infty_x}, {}^*K^\infty_{x,{}^*I^\infty_x}\big) \simeq e.$$

The calculation for K is analogous. Thus we have

$$J^\infty_{x,{}^*I^\infty_x} \subseteq {}^*J^\infty_x \quad \text{and} \quad K^\infty_{x,{}^*I^\infty_x} \subseteq {}^*K^\infty_x. \tag{2}$$

Now we prove (1) simultaneously by induction with

$$^*J^\sigma_x \subseteq J^\sigma_{x,I^\sigma_x} \quad \text{and} \quad ^*K^\sigma_x \subseteq K^\sigma_{x,I^\sigma_x}. \tag{3}$$

Assume as induction hypothesis that these hold for $\tau < \sigma$, so we have

$$^*I^{<\sigma} \subseteq I^{<\sigma} \subseteq {}^*I^\infty, \tag{4}$$

and

$$^*J^{<\sigma}_x \subseteq J^{<\sigma}_{x,I^{<\sigma}_x} \quad \text{and} \quad ^*K^{<\sigma}_x \subseteq K^{<\sigma}_{x,I^{<\sigma}_x}. \tag{5}$$

Toward the first inclusion of (1), suppose that $^*I^\sigma(a,x) \simeq e$. Then

$$^*J^{<\sigma}(a,x) \simeq {}^*I\big(a,x,{}^*I^{<\sigma}_x,{}^*J^{<\sigma}_x,{}^*K^{<\sigma}_x\big) \simeq {}^*I^\sigma(a,x) \simeq e,$$

so by (5), also $J^{<\sigma}(a,x,I^{<\sigma}_x) \simeq e$, whence

$$I^s(a,x) \simeq I\big(a,x,I^{<\sigma}_x\big) \simeq J^\infty\big(a,x,I^{<\sigma}_x\big) \simeq e,$$

as desired. Next, for the second inclusion of (1), suppose that

$$I^\sigma(a,x) \simeq e \quad \text{so also} \quad J^\infty\big(a,x,I^{<\sigma}_x\big) \simeq I\big(a,x,I^{<\sigma}_x\big) \simeq e.$$

Then by (4), $J^\infty(a,x,{}^*I^\infty_x) \simeq e$, so by (2), $^*J^\infty(a,x) \simeq e$. Hence

$$^*I^\infty(a,x) \simeq {}^*I\big(a,x,{}^*I^\infty_x,{}^*J^\infty_x,{}^*K^\infty_x\big) \simeq {}^*J^\infty(a,x) \simeq e.$$

Finally, for (3), suppose that

$$^*J^\sigma(a,x) \simeq {}^*J\big(a,x,{}^*I^{<\sigma}_x,{}^*J^{<\sigma}_x,{}^*K^{<\sigma}_x\big) \simeq e.$$

Then by (4) and (5), also

$$J^\sigma\big(a,x,I^{<\sigma}_x\big) \simeq J\big(a,x,I^{<\sigma}_x,J^{<\sigma}_{x,I^{<\sigma}_x},K^{<\sigma}_{x,I^{<\sigma}_x}\big) \simeq e,$$

whence also (by monotonicity), $J^\sigma(a,x,I^\sigma_x) \simeq e$ as desired. □

References

P. ACZEL, H. SIMMONS AND S. S. WAINER (EDS.)
[1992] *Proof Theory: Papers from the International Summer School, Leeds, 1990*, Cambridge Univ. Press, Cambridge.

J. BARWISE (ED.)
[1977] *Handbook of Mathematical Logic*, North-Holland, Amsterdam.

E. BOERGER, W. OBERSCHELP, M. M. RICHTER, B. SCHINZEL AND W. THOMAS (EDS.)
[1983] *Computation and Proof Theory: Proceedings of the Logic Colloquium, Vol. 2*, Lecture Notes in Mathematics, Vol. 1104, Springer, Berlin.

L. BLUM, M. SHUB AND S. SMALE
[1989] On a theory of computation and complexity over the real numbers: NP-completeness, recursive functions, and universal machines, *Bull. Amer. Math. Soc.*, 21, pp. 1–46.

N. J. CUTLAND
[1980] *Computability: An Introduction to Recursive Function Theory*, Cambridge Univ. Press, Cambridge.

F. R. DRAKE AND S. S. WAINER
[1980] *Recursion Theory: its Generalizations and Applications*, London Mathematical Society Lecture Notes, Vol. 45, Cambridge Univ. Press, Cambridge.

E. ENGELER
[1968] *Formal Languages: Automata and Structures*, Markham Publ. Co., Chicago.

E. ENGELER
[1975] On the solvability of algorithmic problems, in: Rose and Shepherdson [1975], pp. 231–251.

J.-E. FENSTAD
[1980] *General Recursion Theory: An Axiomatic Approach*, Springer, Berlin.

H. M. FRIEDMAN
[1971] Algorithmic procedures, generalized Turing algorithms, and elementary recursion theory, in: Gandy and Yates [1971], pp. 113–137.

R. O. GANDY AND C. M. E. YATES (EDS.)
[1971] *Logic Colloquium '69*, North-Holland, Amsterdam.

G. T. HERMAN AND S. D. ISARD
[1970] Computability over arbitrary fields, *J. London Math. Soc., Ser. 2*, 2, pp. 73–79.

P. G. HINMAN
[1978] *Recursion-Theoretic Hierarchies*, Springer, Berlin.

A. S. KECHRIS AND Y. N. MOSCHOVAKIS
[1977] Recursion in higher types, in: Barwise [1977], pp. 681–737.

S. C. KLEENE
[1959] Recursive functionals and quantifiers of finite types I, *Trans. Amer. Math. Soc.*, 91, pp. 1–52.
[1962a] Turing-machine computable functionals of finite types I, in: Nagel, Suppes and Tarski [1962], pp. 38–45.
[1962b] Turing-machine computable functionals of finite types II, *Proc. London Math. Soc., Ser. 3*, 12, pp. 245–258.
[1963] Recursive functionals and quantifiers of finite types II, *Trans. Amer. Math. Soc.*, 108, pp. 106–142.

C. MICHAUX
[1989] Une remarque à propos des machines sur **R** introduites par Blum, Shub, et Smale, *C. R. Acad. Sci. Paris, Série I*, 309, pp. 435–437.
[1990] Machines sur les réels et problèmes NP-complets, in: *Séminaire de Structures Algébriques Ordonnées*, Prépubl. de l'Equipe de Logique Math. de Paris 7.

Y. N. MOSCHOVAKIS

[1974] *Elementary Induction on Abstract Structures*, North-Holland, Amsterdam.

[1983] Abstract recursion as a foundation for the theory of algorithms, in: Boerger et al. [1983], pp. 289–364.

[1989] The formal language of recursion, *J. Symbolic Logic*, 54, pp. 1216–1252.

J. MOLDESTAD

[1977] *Computations in Higher Types*, Lecture Notes in Mathematics, Vol. 574, Springer, Berlin.

J. MOLDESTAD, V. STOLTENBERG-HANSEN AND J. V. TUCKER

[1980a] Finite algorithmic procedures and inductive definability, *Math. Scand.*, 46, pp. 62–76.

[1980b] Finite algorithmic procedures and computation theories, *Math. Scand.*, 46, pp. 77–94.

J. P. MYERS, JR. AND M. J. O'DONNELL (EDS.)

[1992] *Constructivity in Computer Science: Proceedings of the Summer Symposium, San Antonio, 1991*, Lecture Notes in Comput. Sci., Vol. 613, Springer, Berlin.

E. NAGEL, P. SUPPES AND A. TARSKI (EDS.)

[1962] *Proceedings of 1st International Congress for Logic, Methodology, and the Philosophy of Science*, Stanford Univ. Press.

R. A. PLATEK

[1966] *Foundations of Recursion Theory*, PhD Dissertation, Stanford University.

H. E. ROSE AND J. C. SHEPHERDSON

[1975] *Logic Colloquium '73*, North-Holland, Amsterdam.

H. SCHWICHTENBERG (ED.)

[1995] *Proof and Computation: Proceedings of the NATO Advanced Study Institute and Summer School, Marktoberdorf, 1993*, Springer, Berlin.

J. C. SHEPHERDSON AND H. E. STURGIS

[1963] Computability of recursive functions, *J. Assoc. Comput. Mach.*, 10, pp. 217–255.

J. V. TUCKER

[1980] Computing in algebraic systems, in: Drake and Wainer [1980], pp. 215–235.

J. V. TUCKER AND J. I. ZUCKER

[1988] *Program Correctness over Abstract Data Types*, North-Holland, Amsterdam.

[1992a] Examples of semicomputable sets of real and complex numbers, in: Myers and O'Donnell [1992], pp. 179–198.

[1992b] Deterministic and nondeterministic computation and Horn programs on abstract data types, *J. Logic Programming*, 13, pp. 23–55.

[1992c] Provable computable selection functions on abstract structures, in: Aczel, Simmons and Wainer [1992], pp. 277–306.

[1995] Computable functions on stream algebras, in: Schwichtenberg [1995], pp. 397–437.

Part 4
Mathematics and Computability Theory

CHAPTER 12

Computable Rings and Fields

V. Stoltenberg-Hansen

Department of Mathematics, Uppsala University, Box 480, S-751 06 Uppsala, Sweden

J. V. Tucker

Department of Computer Science, University of Wales Swansea, Singleton Park, Swansea, SA2 8PP, Wales, UK

Contents

1. Introduction 365
   1.1. Computable algebraic structures 366
   1.2. Historical notes on computable rings and fields 369
   1.3. Structure and prerequisites 371
2. Computable commutative rings 372
   2.1. Primary definitions 372
   2.2. Invariance 378
   2.3. Subrings and ideals 384
   2.4. Direct sums and limits 387
3. Computable fields 391
   3.1. Field extensions 391
   3.2. Splitting algorithms 399
   3.3. Algebraically closed fields 403
   3.4. Some undecidable problems 405
4. Computable Noetherian rings 408
   4.1. Noetherian rings and modules 409
   4.2. Computable coherence 410
   4.3. The ideal membership relation 413
   4.4. Polynomial rings 417
5. Further reading 421
   5.1. Computable groups 421
   5.2. Linear algebra 424
   5.3. Other computable structures 426
   5.4. Computable universal algebras 426
   5.5. Algebraic theory of data in computer science 427
   5.6. Theory of numberings 428

HANDBOOK OF COMPUTABILITY THEORY
Edited by E.R. Griffor

5.7. Computable numbers . . . 428
5.8. General algebraic framework for computations on algebras . . . 429
5.9. Primitive recursive algebra . . . 430
5.10. Polynomial-time algebras . . . 431
5.11. Generalised computability theories and exact computation in uncountable algebras . . . 432
5.12. Computable approximations of topological algebras . . . 433
5.13. Constructive algebra . . . 434
5.14. Other matters . . . 435
6. Concluding remarks . . . 435
References . . . 435

This is a survey of the theory of computable rings and fields. First, we introduce the concept of a computable ring and explain the basic ideas and results about computable homomorphisms, computable subrings, computable ideals and quotient rings, and the computability of polynomial rings and direct limits. We also discuss the invariance of computability across different choices of numbering. Secondly, we survey the theory of finite and infinite field extensions. Special attention is paid to the role of splitting algorithms for computable fields, algebraic closures and undecidability theorems. Thirdly, we survey the theory of computable Noetherian rings, focusing on the concept of computable coherence. Finally we survey 13 related topics ranging from computable groups to computable approximations to topological algebras.

1. Introduction

Effective Algebra aims at establishing the scope and limits of finite computation by means of algorithms on *any* set of data. A set of data, together with some basic functions, forms an algebraic structure A. In Effective Algebra, the approach to analysing computation in the algebra A is to apply the theory of computable functions on $\mathbb{N}$, using a surjective map

$$\alpha : \Omega_\alpha \to A$$

called a *numbering*, where $\Omega_\alpha \subseteq \mathbb{N}$. Algorithmic properties of A are measured by the algorithmic properties of the number-theoretic representation of A given by α. In particular, the concept of a *computable algebraic structure* can be defined. Thus, Effective Algebra studies what data can be represented algorithmically, and what sets and functions can be defined by algorithms, using the same concepts as those that underpin the Church–Turing Thesis for algorithms on $\mathbb{N}$. It also studies algebraic structures that can be algorithmically approximated.

The theory of the computable functions on $\mathbb{N}$ is a general theory of digital computation and it can be used to characterise the functions implementable on digital computers. Effective Algebra can be used to establish when an algebra A is implementable on a computer. Equivalently, precise answers can be given to the general question:

QUESTION. What sets of data and functions on that data can be implemented on a computer, in principle?

Algebraic structures can be found throughout mathematics and computer science, and their applications. Effective Algebra encompasses a wide range of subjects, some of which are well developed mathematical theories, while others are awaiting systematic investigation.

In this chapter we have chosen to focus on the *theory of computable rings and fields* because it is an excellent mathematical subject that is of wide interest and use. Furthermore, the history of the subject is important for the history of mathematics. Specifically, the chapter has the following objectives:

(1) To explore the basic ideas about computable rings and fields and their numberings.

(2) To survey the theory of computable fields.

(3) To survey the theory of computable Noetherian rings and modules.

(4) To introduce the larger field of Effective Algebra through suggestions for further reading.

We do not discuss the important but specialised topic of lattices of r.e. substructures of these structures: see the survey Nerode and Remmel [1985]; but for the related case of vector spaces we provide some information in Section 5 on Further Reading.

The subject is intimately linked with the theories of computable universal algebras and computably approximable topological universal algebras; these we have surveyed in Stoltenberg-Hansen and Tucker [1995].

The chapter is related to chapters on the Theory of *Numberings*, Recursive Model Theory, and Computable Analysis, and draws on several other chapters of this *Handbook*.

The prerequisites for this chapter are a knowledge of the computable functions on $\mathbb{N}$ and of basic ring, field and module theory.

In this Introduction we will discuss the general approach taken in Effective Algebra and its relation with abstract algebra. We will define the concept of a computable universal algebra and outline the history of the theory of computable rings and fields.

1.1. *Computable algebraic structures*

The theory of the computable functions establishes the scope and limits of finite computation by means of algorithms on the set $\mathbb{N} = \{0, 1, 2, \ldots\}$ of natural numbers. It is based on the idea that this set of data, equipped with some very simple operations, forms an algebraic structure that is computable in a fundamental sense:

Finiteness: The natural numbers can be represented by explicit finite objects, and the operations can be carried out on these data representations in an explicit way in finitely many steps.

Stability: There are several methods for explicitly representing natural numbers, and they are equivalent in the sense that data representations in one method can be transformed into corresponding data representations in another method, in an explicit way in finitely many steps.

An objective of the theory is to analyse and classify the functions and sets that can be defined, using a variety of models of computation and formal systems, over some simple algebras of natural numbers (regardless of the data representations). According to the Church–Turing Thesis, the class of recursive functions of the form $f : \mathbb{N}^n \to \mathbb{N}$, for $n > 0$, is precisely the class of functions definable by means of algorithms on the natural numbers.

We wish to explore the scope and limits of computation on other sets of data. First, there are numerical data: the integer, rational, real and complex numbers. With these numbers are associated matrices, polynomials and power series; geometric objects such as vectors, algebraic curves and Lie groups; and analytic objects such as continuous, differentiable and analytic functions. Secondly, there is syntactic data: finite and infinite strings, words, terms, well-formed formulae, programs and parse trees. We wish to answer the question:

QUESTION. Which algebras of data are computable? Which functions and sets can be computed on those algebras?

The study of data in mathematics over many centuries has led to abstract algebra in which the many different data representations are formulated, related and classified, and their essential features abstracted. Two tools of abstract algebra are

(a) axiomatic theories that define classes of algebras of data; and

(b) homomorphisms and isomorphisms that compare and classify the algebras.

The achievement of abstract algebra is to allow the study of properties of calculations that are independent of the *nature* of the data and its *representation*. This is well illustrated by Dedekind's analysis of the natural numbers (see Dedekind [1888]).

Notice that to tighten up the meaning of the above finiteness and stability conditions on the natural numbers we need these tools from abstract algebra. More generally, to explore computation on data we must have a theory of computation that can be applied to classes of algebras in a uniform way, especially to the classes of rings and fields, modules and vector spaces, semigroups and groups, lattices and Boolean algebras, and so on.

The approach in Effective Algebra is to apply the theory of computable functions on $\mathbb{N}$ to analyse computation on an algebraic structure A using the idea of a numbering. The key concept is that of a *computable algebra* which is an algebra that can be faithfully represented in a recursive way using the natural numbers. Here is the exact definition:

1.1.1. DEFINITION. An algebra $A = (A; a_1, \ldots, a_p, \sigma_1, \ldots, \sigma_q)$ is *computable* if

(i) the elements of A can be *computably enumerated*: there exists a recursive subset $\Omega_\alpha \subseteq \mathbb{N}$ and a surjection $\alpha : \Omega_\alpha \to A$ that lists or enumerates, possibly with repetitions, all the elements of A;

(ii) the operations of A are *computable in the enumeration*: for each operation $\sigma_i : A^{n(i)} \to A$ of A there exists a recursive function $f_i : \Omega_\alpha^{n(i)} \to \Omega_\alpha$ that *tracks* the σ_i in the set Ω_α of numbers in the sense that for all $x_1, \ldots, x_{n(i)} \in \Omega_\alpha$,

$$\sigma_i\bigl(\alpha(x_1), \ldots, \alpha(x_{n(i)})\bigr) = \alpha\bigl(f_i(x_1, \ldots, x_{n(i)})\bigr);$$

equivalently, Diagram 1 commutes.

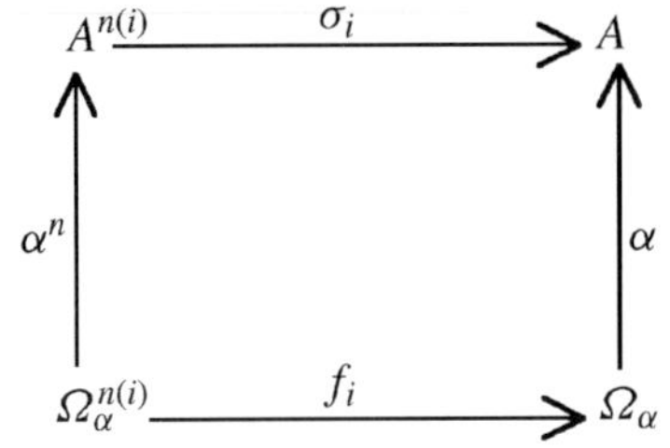

Diagram 1.

(iii) *the equality relation of* A *is decidable*: equivalence of numerical representations of data in A defined by

$$x_1 \equiv_\alpha x_2 \Leftrightarrow \alpha(x_1) = \alpha(x_2)$$

is recursive.

The numbers in the set Ω_α code the elements of A, and the tracking functions simulate the operations of the algebra A in the set Ω_α of numerical codes. This idea of simulation is familiar in algebra. The functions and sets based on A that can be shown to be computable, or non-computable, are precisely those whose representations, based on the set Ω_α, can be shown to be computable, or non-computable, using all the means available in recursion theory. This definition is that given in Mal'cev [1961]. It can be applied easily to rings, fields, groups, Boolean algebras and so on.

Two examples of computing constructs that are available and crucially important for the theory are the following:

Finite Sequences: Via a numbering, we can computably process finite sequences of elements of A using pairing on $\mathbb{N}$.

Global Search: Via a numbering, we can computably search the set A using the effective least number operator on $\mathbb{N}$.

We can also express algorithmic uniformities using the enumeration of the partial recursive functions, and establish subtle undecidability results using various sophisticated methods. It is easy to define the notion of computable homomorphism between computable algebras and, in fact, investigate the computability of many aspects of abstract algebra.

Two points stand out about this general approach to the definition of a computable algebra that correspond with the conditions of finiteness and stability for the natural numbers. First, the computability is defined by the *existence* of a computable representation. Second, that there may be several such representations that may (or may not!) be effectively equivalent. This latter eventuality, concerning invariance, is a special subject of study. Thus the general approach of Effective Algebra is based on the same line of thought that underlies computability theory on $\mathbb{N}$.

1.2. *Historical notes on computable rings and fields*

The systematic study of computable rings and fields, based on recursive function theory, originates in Fröhlich and Shepherdson [1956]. This paper is an important study of rings and fields that introduces many of the basic notions and results on computable field extensions. For example, Fröhlich and Shepherdson analysed the computability of algebraic closures, using the Steinitz construction (Steinitz [1910]) and introduced the fundamental notion of *canonical computable field extension*, which *allowed them to prove that a computable field with a splitting algorithm has a computable algebraic closure which is unique up to computable isomorphism.* A splitting algorithm for a field F is an algorithm that decides whether or not a polynomial over F is irreducible. It plays a fundamental role in polynomial rings and fields. Fröhlich and Shepherdson constructed a computable field that did not possess a splitting algorithm. We will explain their work in Section 3.

The definition of a computable ring that they use rests on the idea of representing structures by symbols and computing with Turing machines. They use the term *explicit* ring and field for a computable ring and field, which reflects the idea of an explicit representation, and the literature of the time.

The move to the numbering or enumeration of arbitrary sets using natural numbers was made in Rabin [1960a, 1960b] and Mal'cev [1961]. The idea is an obvious generalisation of the Gödel numbering of logical syntax. M.O. Rabin defined indexings of the form $\iota : A \to \mathbb{N}$, and proved several interesting results about computable groups, rings and fields. In particular, Rabin established that *the algebraic closure of any computable field is computable*, using Artin's construction of the algebraic closure.

A.I. Mal'cev studied computable and other effective universal algebras using numberings of the form described in Section 1.1. Most of the notions that we use are adaptations of those of Mal'cev, who began the theory of numberings, numbered sets and structures in a series of papers (see his selected works Mal'cev [1971a]). Thorough mathematical accounts of computable sets and universal algebras and their numberings have been developed by Yu. Ershov and others (see Ershov [1973, 1975, 1977a, 1977b] and this *Handbook*). We will discuss computable universal algebra in Section 5 on Further Reading.

The theory of computable rings, modules and fields is the work of many hands, and it enjoys connections with constructive algebra, computable analysis and other parts of logic. We will try to survey a great deal of it in this chapter.

Fröhlich and Shepherdson's paper brings to an end a long period in which some vagueness, confusion and error had crept into writings on algorithmic aspects of fields and polynomials (for example, about the topics of splitting algorithms and primary ideal decompositions).

The concern for algorithms and explicit computations is fundamental in algebra. Roughly speaking, until the 1880's, to do algebra was to compute. The need for increasing rigour in mathematics was recognised as urgent early in the nineteenth century, and many mathematicians desired more precise definitions of concepts and more careful reasoning in proofs. The process of clarification and improvement for mathematical reasoning led to the extensive use of abstract and infinitistic concepts and proofs in mathematics, including algebra. This is because these methods were so *practically useful* in the mathematical process of defining ideas and building arguments that are clearly expressed.

However, problems on the nature of algebra arose early on, in connection with the manipulation of data such as complex numbers. Discussion of the fundamental role of symbols and rules were advanced in works such as George Peacock's *Treatise on Algebra* (Peacock [1830]). Charles Babbage had investigated the foundations of algebra earlier and formulated a rather formalist approach to algebra. A symbolic and abstract view is also present in Ada Lovelace's notes on programming of 1845 (see Babbage [1989]).

The theory of equations represented by the works of Lagrange [1772], Ruffini [1779], Abel [1824] and Galois [1846] are all concrete and algorithmic. The work of Gauss and Kummer in the first half of the nineteenth century concerning ideals in special number fields is algorithmic. The algorithmic nature of algebra is well illustrated by the dramatic influence of the work of Charles–François Sturm on the number of roots of equations (Sturm [1835]).

But by the 1880's the situation had changed. Important changes occurred as infinitistic and non-explicit methods were used in algebra through the contributions of Richard Dedekind and others. There is also Hilbert's celebrated solution to Gordan's Problem in 1880 (which developed into the *Basis Theorem*) and which is said to have established that algebra need not be limited to explicit constructions (see Reid [1970]). On the other hand, L. Kronecker was consciously interested in explicit constructions (see Novy [1973, pp. 131–150]) and Kronecker [1882] contains a computational study of polynomial ideals. He had shown that every finite dimensional field was explicit and had a splitting algorithm. He had been engaged in the development of algebraic systems since the 1850's.

Dedekind thought of fields as fields of complex numbers; abstract fields were introduced in Weber [1893]. The p-adic fields of K. Hensel appeared in 1904 (see the book Hensel [1908]). E. Steinitz proved the existence and uniqueness of the algebraic closures of abstract fields in Steinitz [1910], and E. Artin and O. Schreier developed the theory of real closed fields and the Galois theory of infinite field extensions in Artin and Schreier [1927]. In this period, explicit or constructive work was developed and published: König [1903] on elimination theory, Henzelt [1922] and Hermann [1926] on polynomial ideals.

The lack of a precise understanding of the concept of an explicit field, and of experience with constructive reasoning, hampered progress. In G. Hermann's basic paper (Hermann [1926]) there is a standing but erroneous assumption that every explicit (i.e. computable) field has a splitting algorithm. This was used in a non-trivial way in her work on primary decomposition. Shortly thereafter, van der Waerden [1930b] provided an *intuitionistic* counter-example to Hermann's assumption. In 1930 B.L. van der Waerden published his text-book *Moderne Algebra* which is a convenient landmark for the shape of contemporary algebra. In the early editions, there is a section on effective processes in field theory. There are papers, such as Vandiver [1936], on constructive fields, but a theory did not emerge. Work on polynomials is taken up in Krull [1953a, 1953b, 1954]; which brings us to Fröhlich and Shepherdson [1955, 1956].

The theory of fields and polynomial rings leads to the theory of Noetherian rings and modules. The constructive theory of Noetherian rings was taken up and further developed by J. Tennenbaum and F. Richman, and in a series of papers by A. Seidenberg (see Seidenberg [1985] for a summary). Further information is available in Bridges and Richman [1987] and Mines, Richman and Ruitenburg [1988].

The history of computable algebra is much broader than that of the theories of rings and fields. It involves many algorithmic questions: the word problem for groups in algebraic topology described by Max Dehn in 1911; combinatorial systems of Axel Thue in 1914; and some of Hilbert's Problems of 1900. It also involves problems in the foundations of mathematics including the conception and development of intuitionistic thinking by L.E.J. Brouwer; and the development of computability theory from 1936 and the emergence of undecidable problems (see Kleene [1981] and Gandy [1988]). The study of constructive and recursive analysis is also important to the development of effective algebra (see Goodstein [1961] and Pour-El and Richards [1989]).

At a deeper level of historical analysis, the origins of computable algebra cannot be separated from the origins of abstract algebra. The history is in need of a great deal of difficult research.

1.3. *Structure and prerequisites*

In Section 2 we consider rings. We define various types of numberings for rings and, of course, the notion of a computable ring. Numberings and their equivalence are discussed, and the extent to which computability properties are invariant across numberings. The computability of various ring concepts and constructions are described.

In Section 3 we describe the computability of finite and infinite field extensions, and that of algebraic closures. The key role of a splitting algorithm is emphasised. Some undecidable problems for fields are described.

In Section 4 we consider Noetherian rings and modules. The theory is focused by the notions of computable coherence and decidable ideal membership problem.

Section 5 we devote to further reading, attempting to cover 13 topics; and in Section 6 we offer some concluding remarks.

After this introduction, the algebraic level rises steadily and the level of detail varies accordingly. In Section 2 basic algebraic concepts are discussed rather carefully to show how the effective and computable numberings are used. In Section 3 the theory of computable field extensions is introduced efficiently with many arguments outlined. Section 4 is a new account of theorems on Noetherian rings and modules that we think should be better known and more easily accessible; our discussion is therefore more detailed.

First, we assume the reader is familiar with the theory of the recursive functions on the natural numbers. Introductions are available in handbooks (e.g., Enderton [1977], Phillips [1992]) and textbooks (e.g., Odifreddi [1989], Soare [1987], Cutland [1980], Machtey and Young [1978], Mal'cev [1970], and Rogers [1967]).

Secondly, we assume the reader is familiar with the necessary algebra of rings, fields and modules. For field theory, see Stewart [1973] and, especially, Edwards [1984] which is an excellent companion for Section 3. For Noetherian rings and modules see Atiyah and MacDonald [1969]. For a general introduction to algebra see Lang [1965] and Cohn [1977].

We thank M. Djordjevic, E. Palmgren, I.A. Stewart and J.I. Zucker for useful comments and information.

2. Computable commutative rings

This section will examine the basic ideas of a computable commutative ring, and its computable numberings, in preparation for the theories of computable fields (in Section 3), and computable Noetherian rings and modules (in Section 4). The primary definitions will be given in Section 2.1. An important topic in Section 2.2 is the invariance of computable sets and functions under transformations of numberings. In Section 2.3 we discuss computable aspects of *subrings, ideals, factor rings*, and *homomorphisms*. We describe some computable properties of *direct products* and *direct limits* of rings in Section 2.4. It will be easy to adapt the ideas about rings to the corresponding concepts and results for fields.

Throughout we consider only *commutative rings*, usually with identity, which, for brevity, we call rings.

2.1. *Primary definitions*

First, we will define numberings, effective numberings and computable, semicomputable and cosemicomputable numberings for a ring.

2.1.1. NUMBERINGS. A *numbering* of a ring R consists of a set Ω_α of natural numbers and a surjection $\alpha : \Omega_\alpha \to R$ such that:

(i) For the operations of addition, additive inverse, and multiplication in the ring there exist total *tracking functions* $f_+ : \Omega_\alpha \times \Omega_\alpha \to \Omega_\alpha$, $f_- : \Omega_\alpha \to \Omega_\alpha$ and $f_\times : \Omega_\alpha \times \Omega_\alpha \to \Omega_\alpha$, respectively, such that for all $x_1, x_2, x \in \Omega_\alpha$,

$$\begin{aligned} \alpha(x_1) + \alpha(x_2) &= \alpha\big(f_+(x_1, x_2)\big); \\ -\alpha(x) &= \alpha\big(f_-(x)\big); \\ \alpha(x_1) \times \alpha(x_2) &= \alpha\big(f_\times(x_1, x_2)\big). \end{aligned}$$

(ii) For the additive identity 0 and multiplicative identity 1 in R there are distinguished numbers $c_0, c_1 \in \Omega_\alpha$, such that

$$\alpha(c_0) = 0 \quad \text{and} \quad \alpha(c_1) = 1.$$

A numbering is usually denoted by $\alpha : \Omega_\alpha \to R$ or simply α. We write (R, α) for a ring with its numbering and call it a *numbered ring*.

A numbering α provides a representation of the ring R in $\mathbb{N}$. We combine the set Ω_α with the tracking functions for the ring operations, and the numbers that label the identity elements, to make an algebraic structure

$$\Omega_\alpha = (\Omega_\alpha;\ f_+,\ f_-,\ f_\times,\ c_0, c_1)$$

with the same signature Σ as a ring. However, this algebra Ω_α of natural numbers is *not* necessarily a ring. A numbering is simply a Σ-epimorphism $\alpha : \Omega_\alpha \to R$: condition (i) says that α preserves the operations of Ω_α and (ii) says that α preserves the constants. The fact that numberings are epimorphisms of ring-like Σ-structures is invaluable in constructing numberings; for example, the following is obvious:

2.1.2. LEMMA. *Let R and S be rings and let $\phi : R \to S$ be a ring epimorphism. If $\alpha : \Omega_\alpha \to R$ is a numbering of R then $\phi\alpha : \Omega_\alpha \to S$ is a numbering of S.*

Consider the relation $\equiv_\alpha$ on the numerical Σ-algebra Ω_α, defined for $x, y \in \Omega_\alpha$ by

$$x \equiv_\alpha y \quad \text{if, and only if,} \quad \alpha(x) = \alpha(y) \text{ in } R.$$

Since α is a Σ-homomorphism, the relation is a Σ-congruence on Ω_α; the relation is called the kernel of α. By the Homomorphism Theorem (for universal algebras), $R \cong \Omega_\alpha / \equiv_\alpha$ and, thus, we have constructed an isomorphic copy of the numbered ring using sets of natural numbers.

Having represented a countable ring R using $\mathbb{N}$, we may now consider its effectiveness.

2.1.3. EFFECTIVE NUMBERINGS. An *effective numbering* α of R consists of a recursive set Ω_α of natural numbers and, for each operation $+$, $-$ and $\times$ of R, recursive tracking functions f_+, f_- and $f_\times$, respectively.

Let us note that by a recursive function $f : \Omega_1 \to \Omega_2$ between arbitrary subsets Ω_1 and Ω_2 of $\mathbb{N}$, we mean a partial recursive function $f : \mathbb{N} \to \mathbb{N}$ such that $\Omega_1 \subseteq dom(f)$ and $f(\Omega_1) \subseteq \Omega_2$. When Ω_1 is recursive we can take f to be a total recursive function.

2.1.4. LEMMA. *Let R and S be rings and let $\phi : R \to S$ be a ring epimorphism. If $\alpha : \Omega_\alpha \to R$ is an effective numbering of R then $\phi\alpha : \Omega_\alpha \to S$ is an effective numbering of S.*

Later (2.1.12) we will see that *any* countable ring R can be given an effective numbering. Thus, the complexity of a ring can be measured by the complexity of its equality relation, through the congruence of an effective numbering.

2.1.5. COMPUTABLE NUMBERINGS. Let (R, α) be an effectively numbered ring. The numbering α is called a *computable numbering* if, and only if, the relation $\equiv_\alpha$ is recursive; in this case the ring R is said to be *computable under* α.

Furthermore, the numbering α is called a *semicomputable*, or *cosemicomputable, numbering* if, and only if, the relation $\equiv_\alpha$ is recursively enumerable, or co-recursively enumerable, respectively; in this case the ring R is said to be *semicomputable*, or *cosemicomputable*, *under* α, respectively.

From the computability of numberings we define the computability of rings in an obvious way.

2.1.6. COMPUTABLE RINGS. A ring is *computable*, *semicomputable* or *cosemicomputable* if there exists a computable, semicomputable or cosemicomputable numbering for the ring, respectively.

Thus, a computable ring is simply an image of a recursive Σ-algebra of natural numbers under a Σ-homomorphism $\alpha : \Omega_\alpha \to R$ with a recursive kernel $\equiv_\alpha$. Lemma 2.1.4, and an examination of kernels, shows the following important fact:

2.1.7. LEMMA. *Let R and S be rings and let $\phi : R \to S$ be a ring isomorphism. If $\alpha : \Omega_\alpha \to R$ is a computable numbering of R then $\phi\alpha : \Omega_\alpha \to S$ is a computable numbering of S.*

In fact the same is true for semicomputable and cosemicomputable numberings and so our computability notions are isomorphism invariants:

Let R and S be rings and suppose $R \cong S$. If R is computable, semicomputable or cosemicomputable then S is computable, semicomputable or cosemicomputable, respectively.

We note that since every finite ring is computable, the concepts of computable, semicomputable and cosemicomputable algebras are *finiteness conditions* in algebra, i.e. isomorphism invariants possessed of all finite structures.

2.1.8. SIMPLE EXAMPLES.

(i) The ring $\mathbb{Z}$ of integers is computable. The polynomial rings $\mathbb{Z}[X_1, \ldots, X_n]$ in finitely many indeterminates $X_1, \ldots, X_n$ are computable. The polynomial ring

$\mathbb{Z}[X_1, X_2, \ldots]$ in countably infinitely many indeterminates $X_1, X_2, \ldots$ is computable.

(ii) The field $\mathbb{Q}$ of rationals is computable. The polynomial rings $\mathbb{Q}[X_1, \ldots, X_n]$ and $\mathbb{Q}[X_1, X_2, \ldots]$, and the fields $\mathbb{Q}(X_1, \ldots, X_n)$ and $\mathbb{Q}(X_1, X_2, \ldots)$ of rational functions with rational number coefficients are computable.

(iii) Any finite extension field $\mathbb{Q}(a_1, \ldots, a_n)$ of the field $\mathbb{Q}$ is computable. Any countably infinite purely transcendental extension field $\mathbb{Q}(t_1, t_n, \ldots)$ is computable. Some infinite algebraic extension subfields within the fields of real and complex numbers are computable, for example: the field $\mathbb{Q}(\sqrt{p} : p \text{ prime})$; the field $\mathbb{A}$ of algebraic numbers; the field $\mathbb{A}_{\mathbb{R}}$ of real algebraic numbers; the field of numbers constructible by ruler and compass.

(iv) The fields of recursive real and complex numbers are not computable (or semicomputable or cosemicomputable).

(v) Every semicomputable field is computable.

(vi) Many of the above examples we will see as special cases of general theorems in later sections. For example, *any finitely generated commutative ring is computable* (in Section 2.3); the *polynomial rings with countably many indeterminates over a computable ring are computable* (in Section 2.2); and *any finite extension field of any computable field is computable* (in Section 3.1).

(vii) For each computable ring R there exists some numerical machinery of the form

$$(\Omega_\alpha;\ f_+,\ f_-,\ f_\times,\ c_0,\ c_1, \equiv_\alpha)$$

for computing in R. There are at most countably many such tuples of recursive sets, recursive functions and recursive relations. Since at least one appropriate set of algorithms must be allocated to every ring isomorphism type, the class of all isomorphism types of computable rings is countably infinite. However, the collection of distinct isomorphism classes of countable rings and fields is uncountable. Hence most isomorphism types of countable rings and fields are not computable (or even semicomputable or cosemicomputable).

(viii) There are associated algebras that are computable, such as the non-commutative matrix rings $M(n, \mathbb{Q})$ and the classical groups $GL(n, \mathbb{Q})$, $SL(n, \mathbb{Q})$, and $O(n, \mathbb{Q})$ of matrices over the computable field $\mathbb{Q}$ of rationals. Some associated rings such as the rings $\mathbb{Z}[[X_1, \ldots, X_n]]$ and $\mathbb{Z}[[X_1, X_2, \ldots]]$ of integer power series are uncountable but can be computably approximated. These kinds of algebras are outside the scope of the chapter, but are taken up in Section 5.12 of further reading.

A word on terminology: other terms for computable ring, etc., are *recursive ring, recursively presentable ring, constructive ring* and *explicit ring* (the latter is also known in the literature on constructive algebra). A little care is needed by the beginner: in the literature, a recursively presentable ring is usually a computable ring but a recursively presented group is always a semicomputable group.

Let us define the computable subsets and functions for an effectively numbered ring.

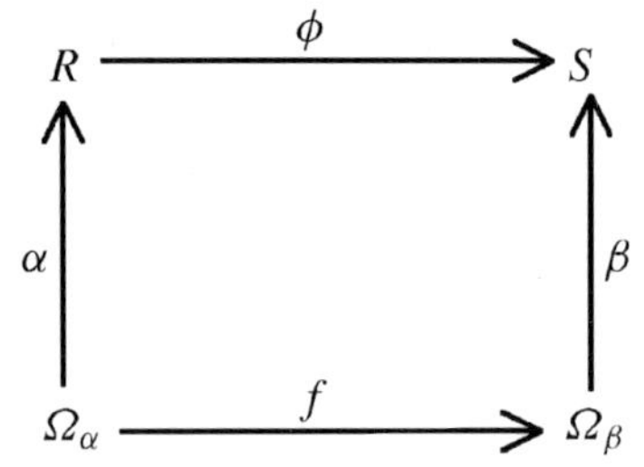

Diagram 2.

2.1.9. SUBSETS. Let (R, α) be an effectively numbered ring. Then a set $S \subseteq R^k$ is *α-decidable*, *α-semidecidable*, or *α-cosemidecidable* if its set

$$\alpha^{-1}(S) = \{(x_1, \ldots, x_k) \in \Omega_\alpha^k \colon (\alpha(x_1), \ldots, \alpha(x_k)) \in S\}$$

of numbers is recursive, r.e., or co-r.e., respectively.

Let F be a computable subfield of the complex numbers under α. The set $U(F)$ of all roots of unity in F is α-semidecidable but need not be α-decidable (see Theorem 3.4.1).

2.1.10. MAPPINGS. Let (R, α) and (S, β) be effectively numbered rings, and let $\phi: R \to S$ be any total function. Then ϕ is *(α, β)-computable* if there exists a recursive function $f: \Omega_\alpha \to \Omega_\beta$ such that for all $x \in \Omega_\alpha$,

$$\phi(\alpha(x)) = \beta(f(x));$$

or, equivalently, f commutes Diagram 2.

Let $Comp_{\alpha,\beta}(R, S)$ be the set of all (α, β)-computable maps from R to S. In the case that $R = S$ and $\alpha = \beta$ we let $Comp_\alpha(R)$ be the set of all α-computable maps on R. The composition of computable maps is again computable so $Comp_\alpha(R)$ is a subsemigroup of the semigroup of all functions on R.

An (α, β)*-computable ring homomorphism* between R and S is a homomorphism that is an (α, β)-computable map.

An automorphism $\phi: R \to R$ is an *α-computable automorphism* if both ϕ and ϕ^{-1} are α-computable. The set $CAut_\alpha(R)$ of all computable automorphisms is a subgroup of $Aut(R)$.

Now, suppose that R is semicomputable under α. Then the inverse of a computable bijection is computable and

$$CAut_\alpha(A) = Aut(A) \cap Comp_\alpha(A).$$

2.1.11. POLYNOMIAL RINGS. Consider the construction of the polynomial rings $R[X_1, \ldots, X_n]$ and $R[X_1, X_2, \ldots]$ over a ring R. If R is computable then it is easy to prove that $R[X_1, \ldots, X_n]$ and $R[X_1, X_2, \ldots]$ are computable. More specifically, from any computable numbering $\alpha : \Omega_\alpha \to R$ we can make, via constructions of the rings, a computable numbering $\alpha_* : \Omega_{\alpha_*} \to R[X_1, \ldots, X_n]$ or $\alpha_* : \Omega_{\alpha_*} \to R[X_1, X_2, \ldots]$ that has the following properties:

(i) given $x \in \Omega_{\alpha_*}$ we can compute a list of α_*-codes for the non-zero terms of the polynomial $\alpha_*(x)$;

(ii) we may decompose the non-zero terms of $\alpha_*(x)$ and list the α-codes for their coefficients and the powers and indices of the indeterminates in the monomials;

(iii) the canonical embeddings $R \to R[X_1, \ldots, X_n]$ and $R \to R[X_1, X_2, \ldots]$ of the ring R are (α, α_*)-computable homomorphisms.

A computable numbering of the polynomial ring satisfying the properties (i)–(iii) we call a *standard numbering* with respect to the numbering of the coefficient ring. A similar concept exists for rational function field constructions. The uniqueness of the choice of standard numberings will be discussed in the next section.

From any numbering of the ring of integers we can obtain a standard numbering of the polynomial rings $\mathbb{Z}[X_1, \ldots, X_n]$ and $\mathbb{Z}[X_1, X_2, \ldots]$. The integer polynomial rings are the free rings for the class of commutative rings and play a special role. To illustrate their use in providing numberings, we prove that any countable ring has an effective numbering.

Let R be any countable ring. Let $a = \{a_i\colon i \in I\}$ be any set of generators for the ring R (for example, an enumeration of all the elements of R). Let $X = \{X_i\colon i \in I\}$ be a corresponding list of indeterminates. Because $\mathbb{Z}[X]$ is free on X for the class of all rings, we have the map $v(a) : \mathbb{Z}[X] \to R$ which is the unique homomorphic extension of the assignment map $a(X_i) = a_i$ that substitutes a_i for X_i in the polynomials and evaluates the polynomials. Because $a = \{a_i\colon i \in I\}$ generates R, the map $v(a)$ is an epimorphism.

Let α be any computable numbering of $\mathbb{Z}[X]$. We define the composition

$$\alpha_a = v(a) \circ \alpha.$$

Since α is also an effective numbering, $\alpha_a : \Omega_\alpha \to A$ is an effective numbering of R, by Lemma 2.1.4:

2.1.12. PROPOSITION. *Each countable ring R has an effective numbering.*

The notion of numbering and effective numbering can be refined usefully by various logical and computational methods. The numerical sets, functions and relations belonging to a numbering of an algebra can be classified using their definability in the arithmetical, hyperarithmetical or analytical hierarchies, or by using many-one or Turing degrees of unsolvability. For example, numberings in which the code sets are Π_2, the congruences are Π_1 relative to the code sets, and only the operations are recursive, are used to study the field of recursive reals. Refinements are also possible in the direction of subrecursive computations: for instance, we can define *primitive recursive* and *polynomial-time* rings. In each case the corresponding classification of numbered algebras is preserved by isomorphisms. We take up these topics in Section 5 on Further Reading.

2.2. *Invariance*

A ring R is computable if *there exists* some computable numbering α for R. Let $C(R)$ be the set of all computable numberings of the ring R. The *choice* of a numbering $\alpha \in C(R)$ suggests that the effectiveness of a subset or function on R may vary over the set $C(R)$ of possible computable numberings. To illustrate, for a subset $S \subseteq R$ we may ask the following questions: *Is S decidable in all computable numberings of R? Is S decidable in some computable numberings of R, but undecidable in others? Is S undecidable in all computable numberings of R?* To understand invariance it is necessary to study the theory of numberings in detail (see the chapter in this *Handbook* by Ershov, and Section 5). We will consider just two equivalence relations on effective numberings, those of *recursive equivalence* and *recursive autoequivalence*, and we will apply them to computable numberings. There are several equivalence relations that yield interesting classification theories for numberings, indeed the theory of numberings is astonishingly rich (see Ershov [1977b] and Section 5.6).

2.2.1. RECURSIVE EQUIVALENCE OF EFFECTIVE NUMBERINGS. Let R be a ring with effective numberings $\alpha : \Omega_\alpha \to R$ and $\beta : \Omega_\beta \to R$. Then α *recursively reduces* to β (written $\alpha \leqslant \beta$) if there exists a recursive function $f : \Omega_\alpha \to \Omega_\beta$ such that for all $x \in \Omega_\alpha$,

$$\alpha(x) = \beta f(x).$$

Furthermore, α is *recursively equivalent* to β (written $\alpha \sim \beta$) if $\alpha \leqslant \beta$ and $\beta \leqslant \alpha$.

The idea of the reduction $\alpha \leqslant \beta$ is that for any α-code of any element of R we can compute some β-code for that element, using the *reduction map* f. The idea of the equivalence $\alpha \sim \beta$ is that the α-codes for any element of R can be computably interchanged with the β-codes for that element: there exist recursive reduction functions f, g which commute Diagrams 3.

Recall that if $X \subseteq \mathbb{N}^p$ and $Y \subseteq \mathbb{N}^q$ we say that X is *many-one reducible* to Y if there exists a total recursive function $f : \mathbb{N}^p \to \mathbb{N}^q$ such that for all $x \in \mathbb{N}^p$,

$$x \in X \Leftrightarrow f(x) \in Y.$$

2.2.2. LEMMA. *Let R be a ring with effective numberings α and β. Let $S \subseteq R^k$ be any subset. If $\alpha \leqslant \beta$ then $\alpha^{-1}(S)$ is many-one reducible to $\beta^{-1}(S)$ and if $\alpha \sim \beta$ then $\alpha^{-1}(S)$ is many-one equivalent to $\beta^{-1}(S)$.*

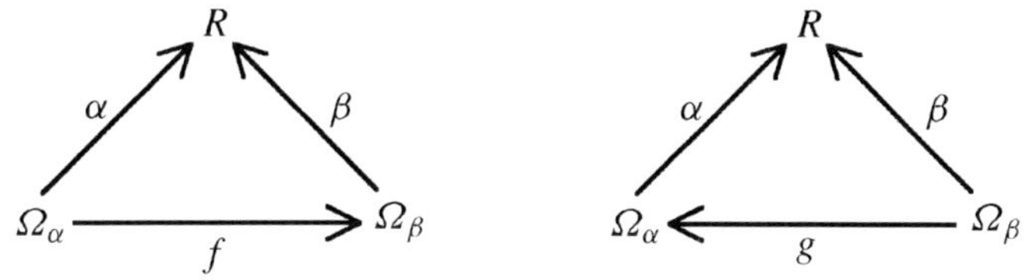

Diagrams 3.

This fact has important consequences for the invariance of decidability properties. If $\alpha \leqslant \beta$ and S is β-decidable then S is α-decidable. In particular, let S be equality on R. Then the congruence $\equiv_\alpha$ of α is many-one reducible to the congruence $\equiv_\beta$ of β. So if β is a computable numbering of R then α is also a computable numbering of R.

If $\alpha \sim \beta$ then S is β-decidable if, and only if, S is α-decidable. Again, let S be equality on R. Then β is a computable numbering of R if, and only if, α is a computable numbering of R. Similar consequences for semidecidable and cosemidecidable sets and relations are obvious.

We are mainly interested in properties of *computable* numberings. The faithful representation of the computable rings by isomorphic copies of rings of natural numbers is accomplished by transforming numberings.

2.2.3. LEMMA (Representation Lemma). *Let (R, α) be an infinite computable ring. Then there exists a bijective computable numbering β with code set $\mathbb{N}$ such that $\beta \sim \alpha$. In this case, the numbering β is an isomorphism between a recursive ring on the set $\mathbb{N}$ of natural numbers and the computable ring R.*

Obviously, no such isomorphic representation is possible for the semicomputable rings for otherwise they would be computable.

2.2.4. DEFINITION. A ring R is said to be *computably stable* if any two computable numberings of R are recursively equivalent. A ring R is said to be *semicomputably stable* if any two semicomputable numberings of R are recursively equivalent.

Semicomputable stability implies computable stability and both are finiteness conditions.

2.2.5. EXAMPLES.

(i) The ring $\mathbb{Z}$ of integers and the polynomial rings $\mathbb{Z}[X_1, \ldots, X_n]$ in finitely many indeterminates $X_1, \ldots, X_n$ are computably stable.

(ii) The field $\mathbb{Q}$ of rational numbers, the polynomial rings $\mathbb{Q}[X_1, \ldots, X_n]$ and the fields $\mathbb{Q}(X_1, \ldots, X_n)$ of rational functions, with rational number coefficients are computably stable.

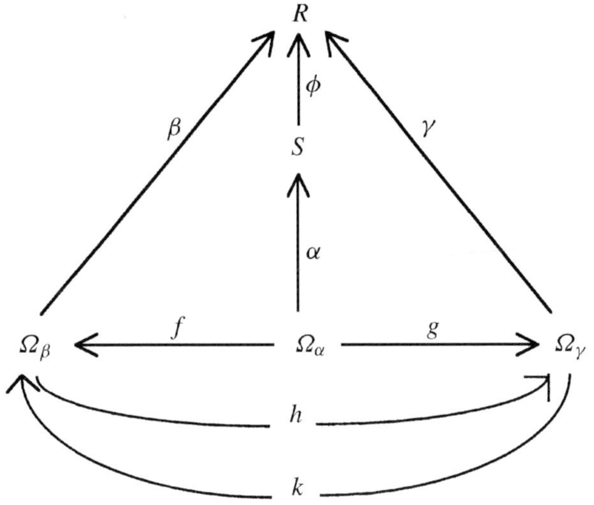

Diagram 4.

(iii) Any finite extension field $\mathbb{Q}(a_1, \ldots, a_n)$ of the field $\mathbb{Q}$ of rationals is computably stable.

(iv) Any finitely generated computable ring is computably stable. This is a special case of a result from Mal'cev [1961]: *any finitely generated semicomputable universal algebra is semicomputably stable*. This general result provides the examples above and many more.

The notion of computable stability needs to be generalised to apply to a range of algebraic constructions.

2.2.6. Definition. Let S and R be rings and $\phi: S \to R$ a homomorphism that is an embedding. Suppose (S, α) is a computable ring. Then R said to be *computably stable* over (S, α) and with respect to ϕ if

(i) there is a computable numbering β of R that makes R a computable extension of (S, α) via ϕ, i.e. ϕ is an (α, β)-computable map;

(ii) any computable numberings β and γ of R that make R a computable extension of (S, α) via ϕ are recursively equivalent.

Condition (ii) is displayed in Diagram 4, wherein f and g track $\phi: S \to R$ with respect to the numberings β and γ of R, and h and k are the recursive reduction maps for $\beta \leqslant \gamma$ and $\gamma \leqslant \beta$, respectively. Note that not all of Diagram 4 commutes.

If R is computably stable over (S, α), and S is a semidecidable subset of R in all computable numberings of R, and S is computably stable, then R is computably stable.

A simple construction is that of polynomial rings. Let $R[X]$ be the polynomial ring over the set X of indeterminates. In Section 2.1 we defined the idea of a computable numbering of $R[X]$ being standard with respect to a computable numbering of R. Given a computable numbering α of R, any two computable numberings of $R[X]$ that are standard with respect to α are recursively equivalent. If X is finite then if $(R[X], \beta)$ is a computable extension of (R, α) then β is recursively reducible to a computable numbering of $R[X]$ that is standard with respect to α. It follows that *if X is finite then $R[X]$ is computably stable over R for any computable numbering of R.*

To produce examples of rings and fields that are not computably stable, we use automorphisms. Let (R, α) be a computable ring. As shown in Lemma 2.1.7, for any $\phi \in Aut(R)$, $\phi\alpha$ is a computable numbering. Now

$$\alpha \sim \phi\alpha \quad \text{if, and only if,} \quad \phi \in CAut_\alpha(A).$$

If R is computably stable then

$$Aut(R) = CAut_\alpha(R).$$

(The converse is not true.) If there is a computable numbering α such that

$$CAut_\alpha(R) \neq Aut(R)$$

then R is *not* computably stable. *If R is a computable ring with uncountably many automorphisms then $CAut_\alpha(R) \neq Aut(R)$ and R is not computably stable.*

A simple example of a computable ring with uncountably many automorphisms is the polynomial ring $\mathbb{Z}[X_1, X_2, \ldots]$. This is because any bijective map $\phi_0 : \{X_1, X_2, \ldots\} \to \{X_1, X_2, \ldots\}$ extends uniquely to an automorphism $\phi : \mathbb{Z}[X_1, X_2, \ldots] \to \mathbb{Z}[X_1, X_2, \ldots]$.

The second important method of classifying effective and computable numberings is based on transformations of an algebra by automorphisms.

A set $S \subseteq R^k$ is an *automorphism invariant* subset of R if for all $(a_1, \ldots, a_k) \in R^k$ and any $\phi \in Aut(R)$,

$$(a_1, \ldots, a_k) \in S \quad \text{if, and only if,} \quad \big(\phi(a_1), \ldots, \phi(a_k)\big) \in S.$$

For example, any subset of R defined by a first order formula is an automorphism invariant.

2.2.7. Recursive Autoequivalence of Effective Numberings. Let R be a ring with effective numberings α and β. Then α is *recursively autoequivalent* to β if there exists some automorphism $\phi \in Aut(R)$ such that the effective numbering $\phi\alpha$ is recursively equivalent to β. We write $\alpha \approx \beta$.

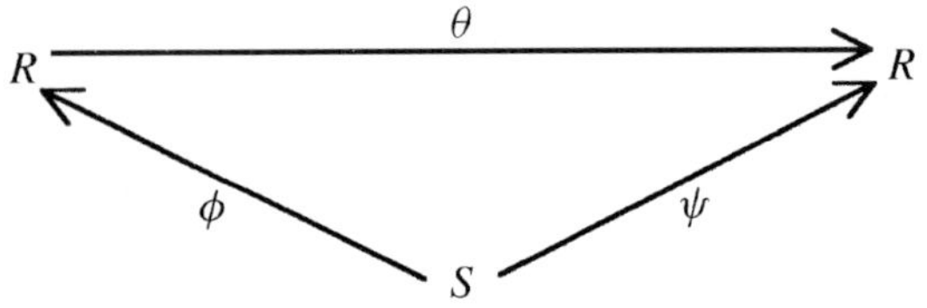

Diagram 5.

2.2.8. LEMMA. *Let R be a ring with effective numberings α and β. Let $S \subseteq R^k$ be an automorphism invariant subset. If $\alpha \approx \beta$ then $\alpha^{-1}(S)$ is many-one equivalent to $\beta^{-1}(S)$.*

Consequently, S is β-decidable if, and only if, S is α-decidable; and, in particular, when S is equality on R, β is a computable numbering if, and only if, α is a computable numbering.

2.2.9. DEFINITION. A computable ring R is said to be *computably autostable* if any two computable numberings are recursively autoequivalent.

Another characterisation is: A computable ring R *is computably autostable if, and only if, for any ring S, if S is isomorphic with R then S is computably isomorphic with R*. Computable stability implies computable autostability. Computable autostability is a finiteness condition. The field $\mathbb{Q}(\sqrt{p}$: p prime) is computably autostable but not computably stable.

This property can also be relativised to apply to constructions. We consider a general form of relativised autostability for pairs of embeddings satisfying the following extension property.

Let R and S be rings and $\phi : S \to R$ and $\psi : S \to R$ homomorphisms. Then ϕ and ψ are said to satisfy the *extension property* if there is $\theta \in Aut(R)$ such that $\theta\phi = \psi$, i.e. Diagram 5 commutes.

2.2.10. DEFINITION. Let (S, α) be a computable ring and let R be a computable extension via $\phi : S \to R$. Then R is *computably autostable* over (S, α) with respect to ϕ if whenever

(a) $\psi : S \to R$ is a computable homomorphism,

(b) ϕ and ψ satisfy the extension property, and

(c) β and γ are computable numberings of R such that ϕ is (α, β)-computable and ψ is (α, γ)-computable,

then there exists $\theta \in Aut(R)$ that is (β, γ)-computable such that $\theta\phi = \psi$, i.e. Diagram 6 commutes.

Another characterisation is: *R is computably autostable over S with respect to ϕ if, and only if, whenever T is S isomorphic to R then T is computably S-isomorphic*

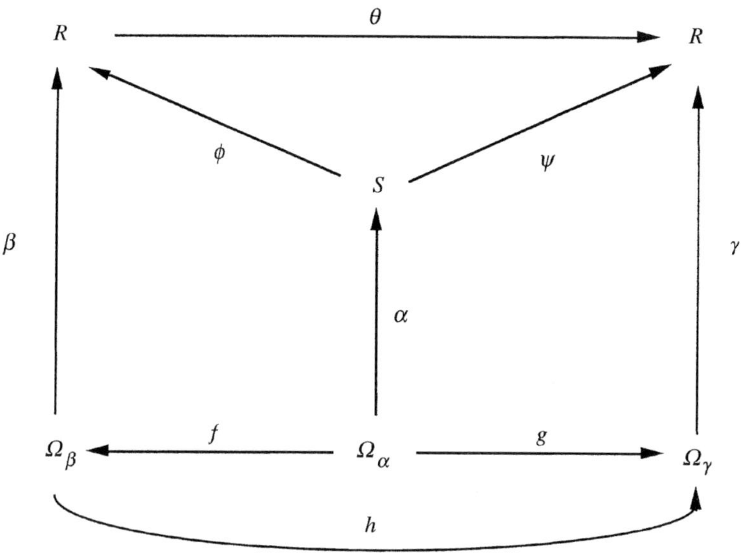

Diagram 6.

to R with respect to any computable numberings of R and T making the homomorphisms computable over S.

To return to the classification of the numberings under recursive equivalence. Let $E(R)$ be the set of all effective numberings of R. Then $\leqslant$ is a partial ordering on $E(R)$ and induces an equivalence relation $\sim$ on $E(R)$. We call the factor structure $E(R)/\sim$ the *effective spectrum* of R; this set is ordered by $\leqslant$. The structure of $E(R)$ can be complicated (as the structure of the set of Gödel numberings of the partial recursive functions might suggest).

The subsets $C(R)$ and $SC(R)$ of $E(R)$, containing the computable and semicomputable numberings, are similarly factored and ordered to form $C(R)/\sim$ and $SC(R)/\sim$, the *computable spectrum* and *semicomputable spectrum*, respectively. Their classification is much easier than that of effective numberings in general. It turns out that the computable spectrum and semicomputable spectrum are discrete in their induced orderings. This is because of the following simple fact:

2.2.11. LEMMA. *Let R be a ring with effective numberings α and β. Suppose that β is semicomputable. If α is recursively reducible to β then β is recursively reducible to α and, hence, α and β are recursively equivalent and α is semicomputable. Moreover, if β is computable then α is computable.*

The cardinality $|C(R)/\sim|$ of the computable spectrum of R is called the *computable dimension* of R. It is easy to construct rings whose computable dimensions are 0, 1, or infinite. It is a remarkable and surprising fact that for each $n \in \mathbb{N}$ there is a ring R whose computable dimension is n. The construction of such rings builds on the following deep theorem proved by a complicated priority argument.

2.2.12. THEOREM (Goncharov [1980a, 1980b]). *For each $n \in \mathbb{N}$ there is a family $\mathcal{S}$ of r.e. sets of natural numbers with exactly n computable numberings up to recursive equivalence.*

To be precise, $\alpha : \Omega_\alpha \to \mathcal{S}$ is a computable numbering if the relation $x \in \alpha(n)$ is r.e. and $\equiv_\alpha$ is recursive. One should think of α as a 1–1 enumeration of $\mathcal{S}$ in terms of r.e.-indices of the sets in $\mathcal{S}$.

By a clever construction of rings encoding Goncharov's families of sets we obtain the following theorem, due independently to Khusainov and Kudinov [1996].

2.2.13. THEOREM. *For each $n \in \mathbb{N}$ there is a computable integral domain R whose computable dimension equals n.*

2.3. *Subrings and ideals*

What are the computable subrings of a computable ring?

2.3.1. DEFINITION. Let (R, α) be a computable ring. Let S be a subring of R. Then S is an *α-computable subring* of R if S is an α-semidecidable subset of R. Furthermore, S is an *α-decidable subring* of R if S is an α-decidable subset of R.

The term α-computable subring is appropriate because *if S is an α-computable subring then S has a computable numbering β for which the inclusion map $i : S \to R$ is (β, α)-computable.*

Let R be a ring and $H \subseteq R$. The subring generated by H is denoted $\langle H \rangle$ and is the image of the map that evaluates all polynomials under substitution by elements of H.

2.3.2. LEMMA. *Let (R, α) be a computable ring. The following are equivalent:*

(i) *S is an α-computable subring of R;*

(ii) *S is the subring $\langle H \rangle$ of R generated by an α-semidecidable subset H of R.*

In consequence, every finitely generated subring of a computable ring is a computable subring.

The fact that (i) implies (ii) is obvious since we may take $H = S$. To see that (ii) implies (i) we take $\Omega = \alpha^{-1}(H)$ in the following useful lemma.

2.3.3. LEMMA. *Let R be a ring computable under $\alpha : \Omega_\alpha \to R$. Let Ω be an r.e. subset of Ω_α. Then the subring $\langle \alpha(\Omega) \rangle$ of R generated by $\alpha(\Omega)$ is an α-computable subring of R.*

The argument is easy. Membership in a subring is definable by a projection over the polynomial ring and a finite sequence of elements from the set $\alpha(\Omega)$. Now let us consider ideals.

2.3.4. DEFINITION. Let (R, α) be a computable ring. Let I be an ideal of R. Then I is an *α-semicomputable ideal* of R if I is an α-semidecidable subset of R. Furthermore, I is an *α-decidable ideal* of R if I is an α-decidable subset of R.

Let R be a ring and $H \subseteq R$. The ideal generated by H is denoted (H) and

$$(H) = \{r_1a_1 + \cdots + r_ka_k\colon\ k \geqslant 1 \text{ and } r_i \in R,\ a_i \in H\}.$$

2.3.5. LEMMA. *Let (R, α) be a computable ring. The following are equivalent*:

(i) *I is an α-semidecidable ideal of R*;

(ii) *I is the ideal (H) of R generated by an α-semidecidable subset H of R.*

In consequence, every finitely generated ideal of a computable ring is a semicomputable ideal.

2.3.6. LEMMA. *Let R be a ring computable under $\alpha : \Omega_\alpha \to R$. Let Ω be an r.e. subset of Ω_α. Then the ideal $(\alpha(\Omega))$ of R generated by $\alpha(\Omega)$ is an α-semidecidable ideal of R.*

Which subrings and ideals of a computable ring are decidable? More specifically, the following are important questions:

2.3.7. IDEAL MEMBERSHIP PROBLEMS. *Let (R, α) be a computable ring.*

(1) *Given any $a_1, \ldots, a_n \in R$, is the finitely generated ideal $(a_1, \ldots, a_n)$ of R an α-decidable ideal?*

(2) *Given any $a_1, \ldots, a_n \in R$, is the ideal $(a_1, \ldots, a_n)$ α-decidable uniformly in the $a_1, \ldots, a_n \in R$?*

Every ideal of the ring $\mathbb{Z}$ is finitely generated being of the form $(n) = n\mathbb{Z}$ and, thus, is α-decidable in every computable numbering α of $\mathbb{Z}$ in a uniform way. Kronecker's constructive methods can be employed to show that *every finitely generated ideal of $\mathbb{Z}[X_1, \ldots, X_k]$ is decidable in every computable numbering of the polynomial ring*. Many algorithms for this problem have been studied (see, for example, Ayoub [1983]) and have been efficiently implemented using Gröbner bases (see, for example, Adams and Loustaunau [1994] and Becker and Weispfenning [1993]).

Let F be a field and $F[X_1, \ldots, X_n]$ be the ring of polynomials in indeterminates $X_1, \ldots, X_n$ with coefficients in F.

THEOREM (Hilbert Basis Theorem). *Every ideal of $F[X_1, \ldots, X_n]$ is finitely generated.*

In connection with this is the following important result in Hermann [1926].

2.3.8. THEOREM (Hermann's Theorem). *There is a primitive recursive function ϕ such that for any $n, k, f, g_1, \ldots, g_k \in F[X_1, \ldots, X_n]$, $f \in (g_1, \ldots, g_k)$ if, and only if, there exist $r_1, \ldots, r_k \in F[X_1, \ldots, X_n]$ such that*

(i) $f = r_1 g_1 + \ldots + r_k g_k$ *and*

(ii) $degree(r_i) \leqslant \phi(n, \deg(f), \max(\deg(g_1), \ldots, \deg(g_k)))$ *for* $i = 1, \ldots, k$.

Using Hermann's Theorem and a little linear algebra, we can show:

2.3.9. COROLLARY. *If (F, α) is computable then the ideal membership problem for $F[X_1, \ldots, X_n]$ is decidable uniformly in the generators, in any standard computable numbering α_* of $F[X_1, \ldots, X_n]$.*

The Hilbert Basis Theorem is a motivation for the development of the theory of Noetherian rings, where it becomes the following:

THEOREM (Generalised Hilbert Basis Theorem). *If R is a Noetherian ring then $R[X_1, \ldots, X_k]$ is Noetherian.*

Section 4 is devoted to Noetherian rings; there we will study generalisations of Corollary 2.3.9:

Each ideal in a computable Noetherian ring is decidable, but not necessarily decidable uniformly in the generators.

To complement the Generalised Hilbert Basis Theorem we need an effective version of the Noetherian condition; this is the condition of *computable coherence* (defined in Section 4.2).

Let R be a Noetherian ring which is computably coherent. If R has a uniform ideal membership algorithm then $R[X_1, \ldots, X_k]$ has a uniform ideal membership algorithm.

Now we turn to homomorphisms and ideals to establish a computable version of the Homomorphism Theorem.

2.3.10. LEMMA. *Let (R, α) and (S, β) be computable rings. Let the homomorphism $\phi: R \to S$ be (α, β)-computable. Then the image $im(\phi)$ of R under ϕ is a β-computable subring of S.*

If ϕ is tracked by the recursive function f then $f(\Omega_\alpha)$ is a recursively enumerable subset of Ω_β. Clearly, $\langle \beta(f(\Omega_\alpha)) \rangle = im(\phi)$. By Lemma 2.3.3, $im(\phi)$ is a β-computable subring of R.

Let I be an ideal of R and consider the factor ring R/I. The factor ring consists of the equivalence classes of elements of R under the equivalence relation

$$a \equiv b \Leftrightarrow a - b \in I.$$

Let $v: R \to R/I$ be the natural homomorphism defined by $v(a) = [a]$ where $[a]$ is the equivalence class of $a \in R$; note $I = ker(v)$. Let (R, α) be a computable ring.

Then composing α with the natural homomorphism υ we obtain an effective numbering

$$\alpha^* = \upsilon\alpha : \Omega_\alpha \to R/I$$

of R/I by Lemma 2.1.4. The natural homomorphism is computable with respect to α and α^* with the identity as tracking function.

Consider the congruence of α^*. For $x, y \in \Omega_{\alpha*}$,

$$\begin{aligned} x \equiv_{\alpha*} y &\Leftrightarrow \upsilon\alpha(x) = \upsilon\alpha(y) \text{ in } R/I \\ &\Leftrightarrow [\alpha(x)] = [\alpha(y)] \\ &\Leftrightarrow \alpha(x) - \alpha(y) \in I. \end{aligned}$$

In summary:

2.3.11. LEMMA. *Let R be a computable ring and I an ideal of R. If I is α-decidable, α-semidecidable, or α-cosemidecidable then the factor ring R/I is α^*-computable, α^*-semicomputable, or α^*-cosemicomputable, respectively.*

2.3.12. THEOREM (Computable Homomorphism Theorem). *Let R and S be rings computable under α and β, respectively. Let the homomorphism $\phi : R \to S$ be (α, β)-computable. Then the image $im(\phi)$ of R under ϕ is a β-computable subring of S that is computably isomorphic with $R/ker(\phi)$, with respect to β and α^* under the natural isomorphism.*

We will apply the above observations on decidable ideals by proving the following.

2.3.13. THEOREM. *Any finitely generated ring is computable.*

PROOF. Let R be a ring finitely generated by $a = \{a_1, \ldots, a_n\}$. Because $\mathbb{Z}[X_1, \ldots, X_n]$ is free on $X_1, \ldots, X_n$ for the class of all rings, we have the unique homomorphic extension $\upsilon(a) : \mathbb{Z}[X_1, \ldots, X_n] \to R$ of the assignment map that substitutes a_i for X_i in the polynomials. Because a generates R, the map $\upsilon(a)$ is an epimorphism and we have

$$R \cong \mathbb{Z}[X_1, \ldots, X_n]/ker(\upsilon(a))$$

by the Homomorphism Theorem. Each ideal of $\mathbb{Z}[X_1, \ldots, X_n]$ is a finitely generated ideal and is decidable. Hence the ideal $ker(\upsilon(a))$ is decidable and R is computable. □

2.4. *Direct sums and limits*

A number of ring-theoretic results depend upon direct sums and limits of families of rings. In general, these constructions are defined up to isomorphism, being characterised as the unique solutions ("universal objects") to certain equations expressed by diagrams. However, to discuss their computability, it is necessary to take an interest in specific concrete representations of these rings. We have to choose a construction and build it a numbering. We begin by looking at the simple fact that the direct sum of finitely many computable rings is computable.

The direct sum $R \oplus S$ of rings R and S consists of the Cartesian product set $R \times S$ equipped with pointwise operations, e.g., addition is $(r_1, s_1) + (r_2, s_2) = (r_1 + r_2, s_1 + s_2)$ for $r_1, r_2 \in R$ and $s_1, s_2 \in S$. The rings R and S embed in $R \oplus S$ using the standard embeddings, e.g., $e_R : R \to R \oplus S$ defined by $e_R(r) = (r, 0)$ for $r \in R$.

Suppose (R, α) and (S, β) are computable rings. Then we construct a numbering for $R \oplus S$ as follows. Define a map

$$\alpha \oplus \beta : \Omega_\alpha \times \Omega_\beta \to R \oplus S$$

by $(\alpha \oplus \beta)(x, y) = (\alpha(x), \beta(y))$ for $x \in \Omega_\alpha$ and $y \in \Omega_\beta$. Strictly speaking, to have a computable numbering we must recode the product $\Omega_\alpha \times \Omega_\beta$ into a recursive subset Δ of $\mathbb{N}$, using pairing, and the map $\alpha \oplus \beta$ into $\delta : \Delta \to R \oplus S$. The pointwise operations of $R \oplus S$ are tracked by applying their tracking functions pointwise. It is easy see that

$$(x_1, y_1) \equiv_{\alpha\oplus\beta} (x_2, y_2) \Leftrightarrow x_1 \equiv_\alpha x_2 \text{ and } y_1 \equiv_\beta y_2$$

and that the congruence is recursive. For finite sums we have the following:

2.4.1. LEMMA. *Let $R_1, \ldots, R_n$ be computable rings under numberings $\alpha_1, \ldots, \alpha_n$. Then the direct sum $R_1 \oplus \cdots \oplus R_n$ is computable under a numbering $\alpha_1 \oplus \cdots \oplus \alpha_n$. The standard embeddings $e_i : R_i \to R_1 \oplus \cdots \oplus R_n$ are computable with respect to α_i and $\alpha_1 \oplus \cdots \oplus \alpha_n$, for $i = 1, \ldots, n$.*

A natural way of generalising the definition of the finite direct sum is to extend the pointwise operations of the rings to the infinite Cartesian product $\prod_{i\in I} R_i$ where I is an infinite index set; for simplicity, we take $I = \mathbb{N}$. This then results in the infinite *direct product* ring $\bigotimes_{i\in\mathbb{N}} R_i$ which is uncountable.

The infinite *direct sum* $\bigoplus_{i\in\mathbb{N}} R_i$ is constructed as a subring of the infinite *direct product* ring $\bigotimes_{i\in\mathbb{N}} R_i$ by restricting the operations to the countable subset

$$\begin{aligned} F &= \left\{ r = (r_i) \in \prod_{i\in\mathbb{N}} R_i \colon r_i = 0 \text{ for all but finitely many } i \in \mathbb{N} \right\} \\ &= \left\{ r = (r_i) \in \prod_{i\in\mathbb{N}} R_i \colon \exists k \in \mathbb{N}\ r_i = 0 \text{ for } i \geqslant k \right\}. \end{aligned}$$

If the infinite family of rings is computable then the infinite direct sum may also be computable.

2.4.2. DEFINITION. Let $\langle (R_i, \alpha_i) \mid i \in \mathbb{N} \rangle$ be an infinite family of computable rings with their numberings. The family is said to be a *computable family* if the code sets, tracking operations and congruences are recursive, uniformly in the indexing. This means that there exist recursive functions to compute the following:

(i) a uniform membership relation $M(i, x) \Leftrightarrow x \in \Omega_{\alpha_i}$;
(ii) a uniform congruence relation $E(i, x, y) \Leftrightarrow x \equiv_{\alpha_i} y$;
(iii) a uniform addition operation $F_+(i, x, y) = f^{\alpha_i+}(x, y)$ if $x, y \in \Omega_{\alpha_i}$;
(iv) a uniform additive inverse operation $F_-(i, x) = f^{\alpha_i-}(x)$ if $x, y \in \Omega_{\alpha_i}$;
(v) a uniform multiplication operation $F_\times(i, x, y) = f^{\alpha_i\times}(x, y)$ if $x, y \in \Omega_{\alpha_i}$;
(iv) the uniform identity elements $F_0(i) = c^{\alpha_i 0}$ and $F_1(i) = c^{\alpha_i 1}$.

2.4.3. LEMMA. *Let $\langle (R_i, \alpha_i) \mid i \in \mathbb{N} \rangle$ be a computable family of computable rings. Then the direct sum $\bigoplus_{i\in\mathbb{N}} R_i$ is computable under a numbering $\delta = \bigoplus_{i\in\mathbb{N}} \alpha_i$. The standard embeddings $e_i : R_i \to \bigoplus_{i\in\mathbb{N}} R_i$ are computable with respect to α_i and δ, for each $i \in \mathbb{N}$.*

Let $(D, \leqslant)$ be an ordered set that is directed. Let $\langle (R_i, \phi_{ij}) \mid i, j \in D \rangle$ be an infinite family of rings and homomorphisms $\phi_{ij} : R_i \to R_j$ for $i \leqslant j$ such that $\phi_{jk} \circ \phi_{ij} = \phi_{ik}$, and ϕ_{ii} is the identity, for $i \leqslant j \leqslant k \in D$. The direct limit construction produces a ring $\mathrm{Lim}_{i\in D}(R_i, \phi_{ij})$ which is the smallest ring into which all the rings R_i of the family may be mapped by

$$\phi_{i\infty} : R_i \to \operatorname*{Lim}_{i\in D}(R_i, \phi_{ij})$$

in a way that is compatible with their homomorphisms, viz., for $i \leqslant j \in D$, $\phi_{j\infty} \circ \phi_{ij} = \phi_{i\infty}$. For simplicity, we look at the case when D is $\mathbb{N}$ with its usual ordering.

The direct limit $\mathrm{Lim}_{i\in\mathbb{N}}(R_i, \phi_{ij})$ and its homomorphisms

$$\phi_{i\infty} : R_i \to \operatorname*{Lim}_{i\in\mathbb{N}}(R_i, \phi_{ij})$$

can be constructed as a quotient ring of a subring of the direct product ring.

Let

$$C = \left\{ r = (r_i) \in \prod_{i\in\mathbb{N}} R_i \colon \exists k \in \mathbb{N}\, \forall j \geqslant i \geqslant k \in \mathbb{N}\ \phi_{ij}(r_i) = r_j \right\}$$

of the eventually consistent sequences in the direct product. The set C forms a subring of the direct product ring $\bigotimes_{i\in\mathbb{N}} R_i$.

Recall the set

$$F = \left\{ r = (r_i) \in \prod_{i\in\mathbb{N}} R_i \colon \exists k \in \mathbb{N}\, \forall i \geqslant k \in \mathbb{N}\ r_i = 0 \right\}$$

of finite elements. F is an ideal of the ring C. The factor ring C/F is a copy of the direct limit. The congruence relation of the ring is

$$\begin{aligned}(r_i) \equiv (s_i) &\Leftrightarrow (r_i) - (s_i) \in F \\ &\Leftrightarrow (r_i - s_i) \in F \\ &\Leftrightarrow \exists k \in \mathbb{N}\, \forall i \geqslant k\ r_i = s_i.\end{aligned}$$

Note we may use sequences of the form $(r_1, \ldots, r_k, \ldots, \phi_{k,k+i}(r_k), \ldots)$ as coset representatives, where the $(k+i)$th coordinate is the embedding $\phi_{k,k+i}(r_k)$ of the kth coordinate r_k, and define the morphisms $\phi_{i\infty}(r) = [(r_j)]$ where (r_j) is any sequence such that $r_j = \phi_{ij}(r)$ for all $j \geqslant i$.

2.4.4. Definition. Let $\langle (R_i, \alpha_i)(\phi_{ij}, f_{ij}) \mid i, j \in \mathbb{N}\rangle$ be an infinite family of computable rings with their numberings, and computable homomorphisms with their recursive tracking functions. The directed family is said to be a *computable directed family* if $\langle (R_i, \alpha_i) \mid i \in \mathbb{N}\rangle$ is a computable family of rings (in the sense of Definition 2.4.2) and, in addition, there exists a recursive uniform tracking function $F(i, j, x) = f_{ij}(x)$ if $x \in \Omega_{\alpha_i}$ for the homomorphisms ϕ_{ij}.

2.4.5. Lemma. *Let $\langle (R_i, \alpha_i)(\phi_{ij}, f_{ij}) \mid i, j \in \mathbb{N}\rangle$ be a computable directed family of computable rings and homomorphism with their numberings and tracking functions. Then the direct limit $\mathrm{Lim}_{i\in\mathbb{N}}(R_i, \phi_{ij})$ is computable under a numbering γ. The standard homomorphisms $\phi_{i\infty} : R_i \to \mathrm{Lim}_{i\in\mathbb{N}}(R_i, \phi_{ij})$ are computable with respect to α_i and γ, for all $i \in \mathbb{N}$.*

Proof. Let the limit be represented by C/F. The recursive set of finite sequences

$$\Delta = \{(x_1, \ldots, x_k) \in \mathbb{N}^*\colon\ k \geqslant 1 \text{ and } x_i \in \Omega_{\alpha_i} \text{ for } i = 1, \ldots, k\}$$

can be used to code the factor ring C/F. The set is recursive because the membership relation M is recursive. Recalling the coset representative for C/F, we use the obvious map $\gamma : \Delta \to C/F$ defined by

$$\gamma((x_1, \ldots, x_k)) = [(\alpha_1(x_1), \ldots, \alpha_k(x_k), \ldots, \alpha_{k+i}(f_{k,k+i}(x_k)), \ldots)].$$

Consider the congruence of γ:

$$\begin{aligned}&(x_1, \ldots, x_n) \equiv_\gamma (y_1, \ldots, y_m) \\ &\quad\Leftrightarrow \text{if } n = \min(n, m),\ y_{n+i} \equiv_{\alpha_{n+i}} f_{n,n+i}(x_n) \text{ all } i \leqslant m - n \text{ or} \\ &\quad\ \text{if } m = \min(n, m),\ x_{m+i} \equiv_{\alpha_{m+i}} f_{m,m+i}(y_n) \text{ all } i \leqslant n - m\end{aligned}$$

is a recursive predicate because of the recursive functions $E(i, x, y)$ and $F(i, j, x)$ tracking the equality and homomorphisms in a uniform way. Again appropriate tracking functions can be made for γ. □

A full version of the lemma, based on any computable directed set $(D, \leqslant)$, is easy to obtain. Direct limits have many applications. Any ring is the direct limit of its local system of finitely generated subrings (which is a directed set under subring inclusion). Since every finitely generated ring is computable we have that *a ring is computable if, and only if, its local system of finitely generated subrings forms a computable directed family of computable rings.*

3. Computable fields

This section will survey the theory of field extensions from an effective point of view. A central theme is the computable content of Steinitz's construction of the algebraic closure of a field. An analysis of the computability of the Steinitz construction leads naturally to the notion of *canonical computable field extension*, due to Fröhlich and Shepherdson [1956]. This is developed in Section 3.1. A property essential for the computability of the Steinitz construction is the existence of a *splitting algorithm*, which allows us to factor polynomials into irreducible factors. This will be studied in Section 3.2. In Section 3.3 we consider algebraic closures and present a full account of the computable content of the Steinitz construction. Finally in Section 3.4 we briefly consider some undecidable problems for fields.

As mentioned in the introduction there are many contributors to the effective theory of fields. Let us again emphasise the importance of Fröhlich and Shepherdson [1956] which introduced many of the basic notions and results. Further essential contributions are Rabin [1960a], Ershov [1977a] and Metakides and Nerode [1979].

3.1. *Field extensions*

Let (F, α) and (K, β) be computable fields and suppose K is an extension field of F, that is there is a field embedding $\iota : F \to K$. As usual, we often identify F with its image $\iota(F)$ and assume for simplicity that K contains F. Then (K, β) is said to be a *computable extension field* of (F, α) if the embedding ι is (α, β)-computable. We say that K is a *computable extension field* of (F, α) if there is a computable numbering β such that (K, β) is a computable extension field of (F, α).

Suppose F is a subfield of a field K and suppose $X \subseteq K$. Then $F(X)$ denotes the smallest subfield of K containing F and X. In case X is $\{a_1, \ldots, a_n\}$ or $\{a_1, a_2, \ldots\}$ we write $F(a_1, \ldots, a_n)$ or $F(a_1, a_2, \ldots)$, respectively, for $F(X)$. $F(a_1, \ldots, a_n)$ is said to be a *finite* extension of F and $F(a)$, for $a \in K$, is a *simple* extension.

First we consider the computability of finite extensions. Let (F, α) be a computable field and let K be some extension field of F. Let $a \in K$ and consider the field extension $F(a)$.

If a is transcendental over F then $F(a) \cong F(X)$, the field of rational functions in the indeterminate X. It follows, using a standard numbering of $F(X)$ derived from the numbering α of F, that $F(a)$ is a computable extension field of (F, α) and that F is a decidable subfield of $F(a)$.

On the other hand, if a is algebraic over F then

$$F(a) \cong F[X]/(f)$$

where $F[X]$ is the polynomial ring in the indeterminate X and $f \in F[X]$ is the irreducible polynomial of a over F. Providing the polynomial ring with a standard numbering α it follows by the division algorithm for polynomials that the ideal (f) is α-decidable. Hence $F(a)$ is a computable field extension of (F, α) and F is a decidable subfield of $F(a)$. Applying this argument finitely many times we have the following:

3.1.1. PROPOSITION. *Let (F, α) be a computable field and let K be an extension field of F. Let $a_1, \ldots, a_n \in K$. Then $F(a_1, \ldots, a_n)$ is a computable extension field of F and F is a decidable subfield of $F(a_1, \ldots, a_n)$.*

In order to construct a computable numbering for a simple extension field $F(a)$ it suffices to know whether or not a is algebraic, and, in case a is algebraic, the irreducible polynomial of a over F. When K is an extension field of F we use the notation $[K : F]$ for the dimension of K as a vector space over F. Recall that $[F(a) : F]$ is finite just in case a is algebraic over F and then $[F(a) : F]$ equals the degree of the irreducible polynomial of a over F.

In order to build an infinite computable extension field $F(a_1, a_2, \ldots)$ of (F, α) in stages, by adjoining a_n at each stage n to the previously constructed extension $F(a_1, \ldots, a_{n-1})$ using Proposition 3.1.1, we are led to the following important concept.

3.1.2. DEFINITION (*Fröhlich and Shepherdson* [1956]). Let (F, α) be a computable field and let K be an extension field of F. Then (K, β) is a *canonical computable extension field* of (F, α) if

(i) (K, β) is a computable extension field of (F, α);

(ii) there is a β-computable sequence $a_1, a_2, \ldots$ of elements in K such that $K = F(a_1, a_2, \ldots)$; and

(iii) the function $f(n) = [F(a_1, \ldots, a_n) : F(a_1, \ldots, a_{n-1})]$ is recursive.

The β-computable sequence $a_1, a_2, \ldots$ asserted to exist in (ii) is called a *canonical basis* for (K, β) over (F, α).

Every finitely generated extension field of a computable field (F, α) is a canonical computable extension. Trivially, every countable purely transcendental extension of a computable field (F, α) is also canonical for a standard numbering. Such a field is isomorphic to the field of rational functions in countably many indeterminates over F.

A straight-forward sufficient condition for obtaining a canonical computable field extension is the following.

3.1.3. PROPOSITION. *Let (F, α) be a computable field and let $K = F(a_1, a_2, \ldots)$ be an extension field of F. Then there is β such that (K, β) is a canonical computable extension field of (F, α) if*

(i) *the set $A = \{n \in \omega\colon a_n$ is algebraic over $F(a_1, \ldots, a_{n-1})\}$ is recursive, and*

(ii) *the function $\phi : A \to F[X_1, X_2, \ldots]$ is computable, where, for $n \in A$, $\phi(n) = g_n \in F[X_1, \ldots, X_n]$ is such that $h_n(X_n) = g_n(a_1, \ldots, a_{n-1}, X_n)$ is the irreducible polynomial in the indeterminate X_n of a_n over $F(a_1, \ldots, a_{n-1})$.*

The computability of ϕ refers to the standard numbering α_* of the polynomial ring $F[X_1, X_2, \ldots]$ obtained from α. The proof is straight-forward. The conditions provide us with the uniformities needed to build a computable directed system of fields consisting of the $F(a_1, \ldots, a_n)$, and hence the direct limit $F(a_1, a_2, \ldots)$ is computable by Lemma 2.4.5.

The problem of providing a nice characterisation of when a computable extension field (K, β) is a canonical extension of (F, α) is more subtle. In particular we need to consider the computability of algebraic dependence.

Let K be an extension field of F. Recall that an element $a \in K$ is said to be *algebraically dependent on $b_1, \ldots, b_n \in K$ over F* if there is a non-zero polynomial $f \in F[X_0, X_1, \ldots, X_n]$ such that $f(a, b_1, \ldots, b_n) = 0$. The definition extends to arbitrary $A \subseteq K$ by saying that a is *algebraically dependent on A* if there are $b_1, \ldots, b_n \in A$ such that a is algebraically dependent on $b_1, \ldots, b_n$. A set $A \subseteq K$ is *algebraically independent over F* if for each $a \in A$, a is not algebraically dependent on $A - \{a\}$ over F.

It is a basic fact that there is a set $A \subseteq K$ algebraically independent over F such that K is an algebraic extension of $F(A)$. Such a set A is called a *transcendence basis* for K over F. Any two transcendence bases for K over F have the same cardinality. This unique cardinal is called the *transcendence degree* of K over F.

3.1.4. DEFINITION. Let (K, β) be a computable extension field of (F, α). An *algebraic dependence algorithm for (K, β) over (F, α)* is an algorithm which given β-indices for $a, b_1, \ldots, b_n \in K$ decides, uniformly in n, whether or not a is algebraically dependent on $b_1, \ldots, b_n$ over F.

Dependence algorithms have been studied by many people in various settings, for example, by Fröhlich and Shepherdson [1956] for fields, Metakides and Nerode [1977, 1979] for vector spaces and fields, and by Metakides and Nerode [1980] for an abstract dependence relation.

It is clear that, for (K, β) a computable extension field of (F, α), the set

$$AD(K/F) = \{(a, b_1, \ldots, b_n)\colon a \text{ is algebraically dependent on } b_1, \ldots, b_n \text{ over } F, n \geqslant 0\}$$

is β-semidecidable. It can be shown that $AD(K/F)$ can take any r.e. degree of unsolvability.

The effective connection between algebraic dependence and transcendence bases is given by the following theorem.

3.1.5. THEOREM (Fröhlich and Shepherdson [1956]). *Let (K, β) be a computable extension field of (F, α). Then the following are equivalent.*

(i) *There is an algebraic dependence algorithm for (K, β) over (F, α).*

(ii) *There is a β-semidecidable transcendence basis for K over F.*

(iii) *There is a β-decidable transcendence basis for K over F.*

By producing a semidecidable transcendence basis it is easily seen that a canonical computable extension has an algebraic dependence relation. Using this and some intricate concerns involving separability of fields we obtain the following characterisation theorem, due to Fröhlich and Shepherdson [1956].

3.1.6. THEOREM. *Let (K, β) be a computable extension field of (F, α). Then (K, β) is a canonical computable extension of (F, α) if, and only if, the function*

$$f(a, D) = \big[F(D)(a) : F(D)\big]$$

is β-computable, for $a \in K$ and D a finite subset of K.

An immediate consequence is that every finite extension field $F(D)$ is a β-decidable subfield of K, uniformly in the finite set $D \subseteq K$, simply because $a \in F(D) \Leftrightarrow f(a, D) = 1$.

As always it is important to consider the extent a computable property is dependent on a particular numbering. We say that K is a *canonical computable extension* of (F, α) if there is a numbering β such that (K, β) is a canonical computable extension of (F, α). Recall that an extension field K of F is a *normal* extension if K is algebraic over F and whenever an irreducible polynomial $f \in F[X]$ has a root in K then f has all its roots in K, i.e. f *splits* in K.

3.1.7. PROPOSITION. *Let K be a normal extension of F and suppose K is a canonical computable extension of (F, α). If (K, γ) is a computable extension of (F, α) then (K, γ) is a canonical extension of (F, α).*

There are algebraic canonical computable extensions which are not normal and there are normal computable extensions which are not canonical.

Next we turn to questions of invariance for field extensions. Recall the notions of computable stability and autostability from Section 2.2.

3.1.8. THEOREM.

(i) *Every prime field is computably stable.*

(ii) *Every finite extension of a computable field (F, α) is computably stable over (F, α).*

The proof of (ii) is easy and builds on the fact that each element in $F(a_1, \dots, a_n)$ can be written as $p(a_1, \dots, a_n)/q(a_1, \dots, a_n)$, where $p, q \in F[X_1, \dots, X_n]$ and $q(a_1, \dots, a_n) \neq 0$. It follows that every field which is a finite extension of its prime subfield is computably stable.

Consider the computable field $K = \mathbb{Q}(\sqrt{p_n}\colon n \in \omega)$, where p_n is an increasing enumeration of the prime integers. For each $B \subseteq \omega$ define the field automorphism $\phi_B \in Aut(K)$ by

$$\phi_B(\sqrt{p_n}) = \begin{cases} -\sqrt{p_n} & \text{if } n \in B, \\ \sqrt{p_n} & \text{if } n \notin B. \end{cases}$$

Thus $Aut(K)$ is uncountable and hence, by the remarks following 2.2.6, K *is not computably stable*. Let α and β be computable numberings of K and define recursive functions $f : \omega \to \Omega_\alpha$ and $g : \omega \to \Omega_\beta$ by

$$f(n) = \mu m[\alpha(m)^2 = p_n] \quad \text{and} \quad g(n) = \mu m[\beta(m)^2 = p_n].$$

Then the automorphism ϕ defined by $\phi(\alpha f(n)) = \beta g(n)$ is (α, β)-computable, showing that K *is computably autostable*. This is just a special case of the following important fact about relative autostability (recall 2.2.10).

3.1.9. THEOREM. *Let (F, α) be a computable field. If K is a normal canonical computable field extension of (F, α) then K is computably autostable over (F, α).*

PROOF. Suppose we are given numberings β and γ of K and (α, β)-computable and (α, γ)-computable embeddings of F into K. Let $\{a_i\colon i = 1, 2, \dots\}$ be a canonical basis for (K, β) over (F, α). For each $n \geqslant 0$ consider the subfield $F_n = F(a_1, \dots, a_n)$ of K which is β-computable uniformly in n. Then inductively define γ-computable subfields L_n of K and (β, γ)-computable isomorphisms $\theta_n : F_n \to L_n$ such that θ_n extends θ_m when $m \leqslant n$. The extension θ_{n+1} is obtained effectively from θ_n and the irreducible polynomial of a_{n+1} over F_n. This is possible by the normality of K over F. Then $\theta = \bigcup \theta_n : K \to K$ is an (α, β)-computable F-automorphism. □

None of the conditions can be dropped. We give an example, from Ershov [1977a], of an algebraic canonical computable extension of $\mathbb{Q}$ which is not computably autostable.

Let A and B be recursively inseparable r.e. sets and let $f : \omega \to A \cup B$ be a recursive function enumerating $A \cup B$ without repetitions such that the set $\{n\colon f(n) \in A\}$ is recursive. Let $F = \mathbb{Q}(\sqrt{p_n}\colon n \in \omega)$, where p_n is an increasing enumeration of the prime integers. Then F is a canonical computable extension of $\mathbb{Q}$. Let $K = F(a_0, a_1, \dots)$ where $a_n^2 = \sqrt{p_{f(n)}}$ and let $L = F(b_0, b_1, \dots)$ where

$$b_n^2 = \begin{cases} \sqrt{p_{f(n)}} & \text{if } f(n) \in A, \\ -\sqrt{p_{f(n)}} & \text{if } f(n) \in B. \end{cases}$$

Then K and L are canonical computable extensions of F, by Proposition 3.1.3, and hence of $\mathbb{Q}$. Note that the only roots of the polynomial $X^4 - p_{f(n)}$ in K are $\pm a_n$, and in L they are $\pm b_n$.

To show that $K \cong L$ we first define $\phi \in Aut(F)$ by

$$\phi(\sqrt{p_n}) = \begin{cases} \sqrt{p_n} & \text{if } n \in A, \\ -\sqrt{p_n} & \text{if } n \in B, \\ \sqrt{p_n} & \text{if } n \notin A \cup B. \end{cases}$$

Then we extend ϕ to an isomorphism $\psi : K \to L$ by setting $\psi(a_n) = b_n$ for each n.

To show that K is not computably autostable suppose there were a computable isomorphism $\theta : K \to L$. Then clearly $\theta(\sqrt{p_n}) = \pm\sqrt{p_n}$ since θ must be the identity on $\mathbb{Q}$. Similarly, $\theta(a_n) = \pm b_n$. Let $n \in A \cup B$ and let $m = f^{-1}(n)$. Then

$$\theta(\sqrt{p_n}) = \theta(a_m^2) = \theta(a_m)^2 = b_m^2.$$

Let $C = \{n \in \omega \colon \theta(\sqrt{p_n}) = \sqrt{p_n}\}$. Then C is a recursive set by the assumption that θ is computable. If $n \in A$ then $b_m^2 = \sqrt{p_n}$ and hence $\theta(\sqrt{p_n}) = \sqrt{p_n}$, i.e. $A \subseteq C$. If $n \in B$ then $b_m^2 = -\sqrt{p_n}$ and hence $\theta(\sqrt{p_n}) = -\sqrt{p_n}$. Thus $B \cap C = \emptyset$ and C recursively separates A and B, which is impossible.

In the remaining part of this section we consider the following general and fundamental question.

QUESTION. Let (F, α) be a computable field and let K be an extension field of F. When is K a computable extension field of (F, α)?

Ershov has given the following beautiful and powerful characterisation for algebraic extensions.

3.1.10. THEOREM (Ershov [1977a]). *Let (F, α) be a computable field and let K be an algebraic extension of F. Then K is a computable extension field of (F, α) if, and only if, the set $\{f \in F[X] \colon (\exists a \in K)(f(a) = 0)\}$ is α_*-semidecidable.*

3.1.11. COROLLARY (Rabin [1960a]). *The algebraic closure $\overline{F}$ of a computable field (F, α) is a computable extension field of (F, α).*

The proof of Ershov's theorem uses model theoretic methods. Rabin's original proof of the corollary is an effectivisation of Artin's construction of the algebraic closure. We will here give a main ingredient of Rabin's proof.

Let (F, α) be a computable field and let $f \in F[X]$ be a fixed polynomial over F. We aim to show that the splitting field K_f of f over F, i.e. the smallest field containing F in which f splits into linear factors, is a computable extension of (F, α) with a numbering obtained uniformly from f and α.

3.1.12. LEMMA. *Let (F, α) be a computable field. Then the function $\nu : F[X] \to \omega$, which gives the number $\nu(f)$ of distinct roots of f in the algebraic closure $\overline{F}$ of F, is computable.*

PROOF. Assume F has characteristic 0. The characteristic p case is only slightly more involved. Given a non-zero polynomial f compute its formal derivative f' and then compute $g = (f, f')$, the greatest common divisor of f and f'. If $g = 1$ then f has only simple roots so $\nu(f) = \deg(f)$, and the latter is computed from a standard code for f. If $g \neq 1$, then g is a proper factor of f, and we effectively find $h \in F[X]$ such that $f = gh$. The roots common to g and h are precisely the roots of $r = (g, h)$, that is,

$$\nu(f) = \nu(g) + \nu(h) - \nu(r).$$

Thus $\nu(f)$ is computed by first obtaining $r = (g, h)$ and then inductively calculating $\nu(g)$, $\nu(h)$ and $\nu(r)$. □

The above lemma also has a simple model-theoretic proof using the decidability of the theory of algebraically closed fields.

By the above lemma we calculate $\nu(f) = k$, the number of distinct roots of f in K_f. Then we construct the computable polynomial ring $F[X_0, X_1, \ldots, X_k]$ and give it a standard numbering α_*. The basic idea is to construct an α_*-decidable maximal ideal M in $F[X_0, X_1, \ldots, X_k]$ such that the polynomial f splits over the field

$$K = F[X_0, X_1, \ldots, X_k]/M.$$

Of course, K is *computable* under the natural numbering β obtained from α_* and the natural homomorphism, since M is decidable and the embedding $\iota : F \to K$ is (α, β)-computable. The splitting field K_f is then obtained as the computable subfield of K generated by F and the k distinct roots of f.

Recall the common notation for cosets: for $g \in F[X_0, X_1, \ldots, X_k]$ the coset $g + M$ is denoted by $\bar{g}$. In case $a \in F$ we identify $\bar{a}$ with a.

The cosets $\overline{X}_1, \ldots, \overline{X}_k$ are to play the role of distinct roots of f in K. This is arranged by simply placing the polynomials

$$f(X_1), \ldots, f(X_k)$$

in M. For suppose we evaluate f for $\overline{X}_i$ in the field K, then

$$f(\overline{X}_i) = \overline{f(X_i)} = \bar{0}.$$

To ensure that the roots $\overline{X}_1, \ldots, \overline{X}_k$ are distinct in K, in the case $k \geqslant 2$, we also include in M the separating polynomial

$$s(f) = X_0 \cdot \prod_{1 \leqslant i < j \leqslant k} (X_i - X_j) + 1.$$

If $\overline{X}_i = \overline{X}_j$ in K for some $i \neq j$ then $\overline{s(f)} = 1$. But since $s(f) \in M$ we have $\overline{s(f)} = 0$.

Here is the construction of M. Let $g_0, g_1, \ldots$ be an α_*-computable enumeration of the elements in $F[X_0, X_1, \ldots, X_k]$. For $A \subseteq F[X_0, X_1, \ldots, X_k]$ let (A) denote the ideal in $F[X_0, X_1, \ldots, X_k]$ generated by A. Define inductively,

$$M_0 = \{f(X_1), \ldots, f(X_k), s(f)\}, \quad \text{and}$$

$$M_{n+1} = \begin{cases} M_n \cup \{g_n\} & \text{if } 1 \notin (M_n) + (g_n), \\ M_n & \text{if } 1 \in (M_n) + (g_n). \end{cases}$$

Let $M_\infty = \bigcup_{n<\omega} M_n$ and let M be the ideal generated by M_∞. The set M_∞ is α_*-semidecidable by Hermann's Theorem 2.3.8 and hence M is an α_*-semidecidable ideal. After having shown that M is maximal we conclude that M is α_*-decidable since

$$g \notin M \Leftrightarrow 1 \in M + (g).$$

To prove that M is maximal, the nontrivial part is showing that the ideal generated by M_0 is proper. Suppose this were not the case. Then there are polynomials

$$r_0, r_1, \ldots, r_k \in F[X_0, X_1, \ldots, X_k]$$

so that

$$1 = r_0 s(f) + r_1 f(X_1) + \cdots + r_k f(X_k).$$

Let L be the splitting field of f over F and let $a_1, \ldots, a_k$ be the distinct roots of f in L. If we evaluate the right hand side of the above equation using the vector

$$\left(-\frac{1}{\prod_{1 \leqslant i < j \leqslant k}(a_i - a_j)}, a_1, \ldots, a_k \right) \in L^{k+1}$$

we get 0. Thus we have $1 = 0$ in L, which is false. This completes the construction of the splitting field K_f of f as a computable extension field of (F, α). For later reference we record our construction.

3.1.13. PROPOSITION. *Let (F, α) be a computable field. Then for each polynomial $f \in F[X]$, the splitting field K_f of f over F is a computable extension field of (F, α), with a numbering obtained uniformly from f and α.*

Rabin's Theorem (Corollary 3.1.11) can now easily be proved by constructing a computable directed system of splitting fields K_f for $f \in F[X]$ and then taking the direct limit, which will be computable by the extension of Lemma 2.4.5 to computable directed index sets. In fact, this works for any semidecidable set $C \subseteq F[X]$

of polynomials, i.e. the splitting field of any semidecidable set C of polynomials is a computable field extension with a numbering obtained uniformly from C.

3.2. *Splitting algorithms*

A computable ring (R,α) is said to have a *splitting algorithm* if the set

$$Irr(R[X]) = \{f \in R[X]\colon\ f \text{ is irreducible}\}$$

of irreducible polynomials over R is an α_*-decidable subset of $R[X]$. Of course, (R,α) has a splitting algorithm if, and only if, we can compute the irreducible factors of each polynomial in $R[X]$. The study of splitting algorithms is interesting in its own right and important for the theory of computable field extensions. As we shall see, splitting algorithms play an essential part in determining the computability of Steinitz's construction of algebraic closures of fields.

It is an easy matter to construct a computable field with no splitting algorithm. Let A be an r.e. set which is not recursive and let $p_0, p_1, \ldots$ be the increasing enumeration of the prime numbers. Let $K = \mathbb{Q}(\sqrt{p_n}\colon\ n \in A)$. Clearly K is a computable field, in fact it is a canonical extension of $\mathbb{Q}$ by Proposition 3.1.3. Furthermore

$$n \in A \Leftrightarrow X^2 - p_n \notin Irr(K[X])$$

and hence K has no splitting algorithm.

Clearly the set $Irr(F[X])$ is cosemidecidable when F is computable and hence its Turing degree is r.e. The splitting algorithm can take any r.e. Turing degree in the following two senses.

3.2.1. THEOREM.

(i) *For each r.e. Turing degree* **a** *there is a computable field F such that for each computable numbering α of F, the set $Irr(F[X])$ under the standard numbering α_* is in* **a**.

(ii) *There is a computable field F such that for each r.e. Turing degree* **a** *there is a computable numbering α of F such that the set $Irr(F[X])$ under the standard numbering α_* is in* **a**.

There are several notions equivalent to the existence of a splitting algorithm. We say that a computable field (F,α) has a *root algorithm* if the set

$$\{f \in F[X]\colon\ (\exists a \in F)(f(a) = 0)\}$$

is α_*-decidable.

3.2.2. PROPOSITION. *Let (F,α) be a computable field. Then the following are equivalent.*

(i) *(F, α) has a splitting algorithm.*
(ii) *(F, α) has a root algorithm.*
(iii) *The function $\phi : F[X] \to \omega$ defined by*

$$\phi(f) = \left|\{a \in F:\ f(a) = 0\}\right| \text{ (the number of roots of } f\text{)}$$

is α_*-computable.

The proof of the non-trivial implication (iii) $\Rightarrow$ (i) makes use of the theory of symmetric polynomials. Proposition 3.2.2 is uniform in that there are effective procedures which given any algorithm for any one of (i), (ii) or (iii) provide an algorithm for any one of the others.

We now consider the existence of splitting algorithms. First of all it is trivial that the ring of integers $\mathbb{Z}$ has a splitting algorithm. It follows that the field of rationals $\mathbb{Q}$ has a splitting algorithm. In fact, if R is a computable principal ideal domain such that the irreducible elements of R is a decidable set then R has a splitting algorithm if, and only if, the computable quotient field K of R has a splitting algorithm. This fact is also helpful when extending the splitting algorithm to simple transcendental extensions. For supposing the field (F, α) has a splitting algorithm and t is transcendental over F, we need only prove that the computable *ring* $F[t]$ has a splitting algorithm. The latter is rather straight-forward and follows the argument in Kronecker [1882]. It easily extends to countably infinite purely transcendental extensions since a factorisation occurs in a finite part of the extension. Here it is important to consider a standard numbering of the extension.

3.2.3. THEOREM.
(i) *Each prime field has a splitting algorithm.*
(ii) *Let (F, α) be a computable field and let (K, α_*) be a computable purely transcendental extension of (F, α) with a standard numbering α_*. If (F, α) has a splitting algorithm then so does (K, α_*).*

In fact, the α_* and the splitting algorithm for (K, α_*) is obtained uniformly from α, the splitting algorithm for (F, α) and the transcendence degree of K over F.

We now turn our attention to algebraic extensions. First we consider when a splitting algorithm is preserved by a simple algebraic extension.

3.2.4. THEOREM. *Let (F, α) be a computable field with a splitting algorithm. Let K be an extension field of F and suppose $a \in K$ is separable over F. Then the computable extension field $F(a)$ of (F, α) has a splitting algorithm.*

PROOF. By Proposition 3.2.2 it suffices to show that $F(a)$ has a root algorithm. Suppose we are given a non-constant polynomial $f \in F(a)[X]$. Construct the computable splitting field K_f of f over $F(a)$ using Proposition 3.1.13 and its uniformity, and then compute all the distinct roots of f in K_f, say $b_1, \ldots, b_n$. By the theorem on the primitive element, for each i compute $c_i \in K$ using effective search such that $F(a, b_i) = F(c_i)$. Note that the search terminates since a is assumed to be separable over F. Using a splitting algorithm for F compute the irreducible polynomial of each c_i over F, or, equivalently, compute $[F(c_i) : F]$. But

$$\left[F(c_i):F\right]=\left[F(a,b_i):F\right]=\left[F(a,b_i):F(a)\right]\cdot\left[F(a):F\right]$$

so we can compute $[F(a, b_i) : F(a)]$ for each i. Thus f has a root in $F(a)$ if, and only if, for some i, $[F(a, b_i) : F(a)] = 1$. □

It follows from the proof that a splitting algorithm for $F(a)$ is obtained uniformly from a splitting algorithm for F and the irreducible polynomial of a over F. If (K, β) is a computable extension of (F, α) and $a \in K$ is separable over F then a splitting algorithm for $F(a)$ is obtained uniformly from a splitting algorithm for F and a β-index for a.

The separability condition in Theorem 3.2.4 cannot be dropped. We have the following characterisation of when a splitting algorithm extends to finite inseparable extension fields.

A computable field (K, β) of characteristic $p > 0$ is said to have a *pth root algorithm* if the set $\{a \in K\colon (\exists b \in K)(b^p = a)\}$ is β-decidable.

3.2.5. PROPOSITION (Fröhlich and Shepherdson [1956]). *Let (F, α) be a computable field of characteristic $p > 0$ with a splitting algorithm. Let (K, β) be a finite computable algebraic extension field of (F, α). Then (K, β) has a splitting algorithm if, and only if, (K, β) has a pth root algorithm.*

A splitting algorithm always extends finitely for *perfect fields*:

3.2.6. THEOREM. *Let (F, α) be a perfect computable field with a splitting algorithm and let (K, β) be a computable extension field of (F, α) with an algebraic dependence algorithm over (F, α). Then for each $a_1, \ldots, a_n \in K$, the computable subfield $F(a_1, \ldots, a_n)$ of (K, β) has a splitting algorithm obtained uniformly from (β-indices of) $a_1, \ldots, a_n$.*

PROOF. Given $a_1, \ldots, a_n \in K$, compute the transcendence degree r of $F(a_1, \ldots, a_n)$ over F using the algebraic dependence algorithm. Then effectively re-index the elements $a_1, \ldots, a_n$ such that $a_1, \ldots, a_r$ are algebraically independent, and a_i is separable over $F(a_1, \ldots, a_{i-1})$ for $i = r + 1, \ldots, n$. Using Theorems 3.2.3 and 3.2.4 we obtain a splitting algorithm for $F(a_1, \ldots, a_n)$ uniformly in $a_1, \ldots, a_n$. □

Each prime field is perfect and hence we have:

3.2.7. COROLLARY (Krull [1953a, 1953b]). *Every field which is a finitely generated extension of its prime subfield has a splitting algorithm.*

We now consider invariance of computable field extensions under the assumption that the ground field possesses a splitting algorithm. Our first observation is that if the field (F,α) has a splitting algorithm and $C \subseteq F[X]$ is a semidecidable set of separable polynomials then the splitting field K of C over F is a canonical computable extension of (F,α). Let $f_1, f_2, \ldots$ be an effective enumeration of C and let $K_0 = F$ with numbering α. Having defined K_n with numbering α_n we let K_{n+1} be the splitting field of f_{n+1} over K_n and α_{n+1} the numbering obtained using Proposition 3.1.13. From the roots of f_{n+1} in K_{n+1} we compute a_{n+1} such that $K_{n+1} = K_n(a_{n+1})$ using separability. Finally we let $K = \bigcup_n K_n = F(a_1, a_2, \ldots)$. Each K_n has a splitting algorithm, uniformly in n. Using this it is not hard to compute the function ϕ of Proposition 3.1.3 to conclude that K is a canonical computable extension of (F,α).

3.2.8. THEOREM. *Let (F,α) be a computable field with a splitting algorithm. Suppose $C \subseteq F[X]$ is a semidecidable set of separable polynomials and let K be the splitting field of C over F. Then K is a canonical computable extension of (F,α).*

3.2.9. COROLLARY. *Let (F,α) be a perfect computable field with a splitting algorithm.*

(i) *If K is a normal computable extension field of (F,α) then K is computably autostable over (F,α).*

(ii) *The algebraic closure $\overline{F}$ of F is a canonical computable extension of, and is computably autostable over, (F,α).*

PROOF. Statement (i) follows by considering the set $C = \{f \in F[X]\colon f$ irreducible & $(\exists a \in K)(f(a) = 0)\}$; and (ii) follows by letting $C = F[X]$ and using Theorem 3.1.9. □

To strengthen part (i) of the corollary to all computable fields we work in two steps. Assume that (F,α) is a computable field with a splitting algorithm and let (K,β) be a normal computable extension of (F,α). Let K_s be the separable closure of F in K. Then K_s is a β-decidable subfield of K since

$$a \in K_s \Leftrightarrow (\exists f \in F[X])(f(a) = 0 \;\&\; f \text{ irreducible and separable})$$

and the latter is decidable by hypothesis and Lemma 3.1.12. Since K is normal over F it is also the case that K_s is normal over F. It follows that K_s is autostable over F. The field K is a purely inseparable extension of K_s. Assuming $K \neq K_s$, we know that for each $a \in K - K_s$ there is a least integer $e \geqslant 1$ such that $a^{p^e} \in K_s$ and a is the unique p^eth root of a^{p^e}. But then an automorphism θ_1 on K_s extends uniquely to an automorphism θ on K by

$$\theta(a) = \text{unique } p^e\text{th root of } \theta_1\big(a^{p^e}\big).$$

The computability of θ follows from the decidability of K_s by computing the least e such that $a^{p^e} \in K_s$.

3.2.10. THEOREM (Fröhlich and Shepherdson [1956]). *Let (F, α) be a computable field with a splitting algorithm. If K is a normal computable extension field of (F, α) then K is computably autostable over (F, α).*

Neither of the hypotheses of normality and existence of a splitting algorithm can be dropped in the statement of the theorem.

3.3. *Algebraically closed fields*

With the tools now at our disposal we can give a complete analysis of the computability of the Steinitz construction of the algebraic closure $\overline{F}$ of a field F, under the assumption that F is a perfect field. For this purpose recall our previous discussion, where an effective version of Steinitz's construction of an extension field K of F was identified with K being a canonical computable extension of F. The first theorem shows that the Steinitz construction is computable if, and only if, the ground field has a splitting algorithm.

3.3.1. THEOREM (Fröhlich and Shepherdson [1956]). *Let (F, α) be a perfect computable field. Then the algebraic closure $\overline{F}$ of F is a canonical computable extension field of (F, α) if, and only if, (F, α) has a splitting algorithm.*

PROOF. The "if direction" is Corollary 3.2.9. For the "only if direction" let $(\overline{F}, \beta)$ be a canonical computable extension of (F, α). Then by the Characterisation Theorem 3.1.6, F is a β-decidable subfield of $\overline{F}$. Given a polynomial $f \in F[X]$, compute the number of roots of f in $\overline{F}$ using Lemma 3.1.12. Then, by searching through $\overline{F}$ β-effectively, compute all the roots of f, say $a_1, \ldots, a_n$. By the β-decidability of F we can determine whether or not $a_i \in F$ for some i. It follows that F has a root algorithm and hence a splitting algorithm. □

Recall Corollary 3.2.9(ii) which states that for perfect computable fields (F, α), if (F, α) has a splitting algorithm then $\overline{F}$ is computably autostable over (F, α).

Steinitz proved, using his construction, that the algebraic closure $\overline{F}$ of F is unique up to an F-isomorphism. Thus to complete the link between the computable Steinitz construction and the existence of a splitting algorithm we need to establish the computable uniqueness of the algebraic closure precisely when the ground field has a splitting algorithm, i.e. we need a converse of Corollary 3.2.9(ii). This is obtained in the following beautiful theorem.

3.3.2. THEOREM (Metakides and Nerode [1979]). *Let (F, α) be a computable perfect field. Then the algebraic closure $\overline{F}$ of F is computably autostable over (F, α) if, and only if, (F, α) has a splitting algorithm.*

The proof of the remaining direction builds an algebraic closure using Artin's construction, i.e. first constructing the polynomial ring $F[X]$ where $X = X_0, X_1, \ldots$ are infinitely many indeterminates and then constructing a maximal ideal M of $F[X]$ in stages M_s for $s \in \omega$ such that $\overline{F} = F[X]/M$. The construction is done in such a way that if $\overline{F}$ is computably autostable over (F, α) then there exists a computable map $\psi : F_s[t] \to \omega$, where $F_s[t]$ is the set of separable polynomials in the indeterminate t, which given $f \in F_s[t]$ calculates a stage $\psi(f)$ in the construction of M such that the irreducibility of f can be decided from $M_{\psi(f)}$. Let $(\overline{F}, \beta)$ be a computable extension of (F, α). Let α_* be a standard numbering of $F[X]$ obtained from α. Then each partial recursive function with index e induces a partial (α_*, β)-computable function $\phi_e : F[X] \to \overline{F}$. At the completion of the construction of M it will be the case that each computable isomorphism $\Phi : F[X]/M \to \overline{F}$ is induced by some ϕ_e. The basic strategy is to "destroy" as many potential computable isomorphisms as possible by preventing ϕ_e from being lifted to an isomorphism Φ_e. Our failure to destroy all will provide us with the desired splitting algorithm. For details of the proof we refer to the original Metakides and Nerode [1979] or Smith [1981a].

We summarise our results as a theorem giving an account of the computable content of the Steinitz construction.

3.3.3. THEOREM. *Let (F, α) be a perfect computable field and let $\overline{F}$ be the algebraic closure of F. Then the following are equivalent.*

(i) *(F, α) has a splitting algorithm.*
(ii) *$\overline{F}$ is a canonical computable extension of (F, α).*
(iii) *$\overline{F}$ is computably autostable over (F, α).*

It remains to characterise absolute autostability for computable algebraically closed fields. Each countable algebraically closed field can be constructed as the algebraic closure of a countable (finite or infinite) purely transcendental extension of its prime subfield. It follows that *each countable algebraically closed field is computable.*

3.3.4. THEOREM (Ershov [1977a]). *A countable algebraically closed field K is computably autostable if, and only if, K is of finite transcendence degree over its prime subfield.*

PROOF. To prove sufficiency let K be an algebraically closed field of finite transcendence degree over its prime subfield P. Let $t_1, \ldots, t_n \in K$ be algebraically independent over P so that $K = \overline{P(t_1, \ldots, t_n)}$. Then K is computable and the subfield $P(t_1, \ldots, t_n)$ is a computable (though not necessarily decidable) subfield of K with a splitting algorithm (Theorem 3.2.3). The field K is a normal extension of $P(t_1, \ldots, t_n)$ and hence autostable over $P(t_1, \ldots, t_n)$. But $P(t_1, \ldots, t_n)$ is computably stable over P so, by transitivity, K is autostable.

Now suppose K is of infinite transcendence degree over its prime field P. Then K is isomorphic to $\overline{P(t_1, t_2, \ldots)}$, the algebraic closure of the field of rational functions

$P(t_1, t_2, \ldots)$ in the indeterminates $t_1, t_2, \ldots$. Thus there is a numbering β making (K, β) into a canonical computable extension of P, i.e. (K, β) has an algebraic dependence algorithm over P. On the other hand, there is a computable extension field (L, γ) of P with no algebraic dependence algorithm. By necessity L must have infinite transcendence degree over P. Let $(\overline{L}, \delta)$ be the computable algebraic closure of (L, γ). If $(\overline{L}, \delta)$ had an algebraic dependence algorithm over P then this would provide one for (L, γ) by courtesy of the computable embedding. But $K \cong \overline{L}$ and hence K is not autostable. □

3.4. *Some undecidable problems*

There is a large number of undecidability results in the literature to do with fields and other related structures. We have already met two such results, viz. the existence of computable fields with no splitting algorithm and with no algebraic dependence algorithm. Here we will briefly mention three more examples of some algebraic interest and illustrate the proof methods.

The technique for proving undecidability results here follows a general pattern. First some local decidability results are needed. Then these are used in order to carry out a global construction effectively, where there is "enough room" to implant undecidability either via a direct embedding of an r.e. non-recursive set or via a priority construction.

We start with an example using a direct embedding of an r.e. non-recursive set.

3.4.1. THEOREM (Stoltenberg-Hansen and Tucker [1980]). *For any recursively enumerable set A, there exists a computable field (F, α), a subfield of the complex numbers, whose set of roots of unity $U(F) = \{x \in F\colon (\exists n \geqslant 1)(x^n = 1)\}$ is of the Turing degree of A, i.e. $A \equiv_T \alpha^{-1}(U(F))$.*

Here is an outline of the proof. Let (R, α) be a computable ring and a subring of the complex number field $\mathbb{C}$. We say that (R, α) has a *transcendence algorithm* if the set $T(R) = \{x \in R\colon x$ is transcendental (over $\mathbb{Q}$)$\}$ is α-decidable.

3.4.2. LEMMA. *If (R, α) has a transcendence algorithm then its roots of unity problem is decidable.*

PROOF. Given $x \in R$. If $x \in T(R)$ then $x \notin U(R)$. So suppose x is algebraic. Computably search $\mathbb{Z}[X]$ for a polynomial p having x as a zero. Now if x were a root of unity of order n then since the nth cyclotomic polynomial Φ_n is irreducible we would have Φ_n dividing p and $\deg(\Phi_n) \leqslant \deg(p)$. It is not hard to see that there is a recursive function f such that if $\deg(\Phi_n) \leqslant m$ then $n \leqslant f(m)$. This yields a recursive bound as $x \in U(R)$ if, and only if, at least one of the list $x, x^2, \ldots, x^k$ is 1 where $k = f(\deg(p))$. □

The following local decidability result now follows. Let $p_1, \ldots, p_k$ be distinct primes and let $n \geqslant k$. Then $U(R)$ is decidable for $R = \mathbb{Q}[X_1, \ldots, X_n]/(\Phi_1, \ldots, \Phi_k)$, where $\Phi_i(X_i)$ is the p_ith cyclotomic polynomial, uniformly in n and $p_1, \ldots, p_k$. The same result holds for the quotient field of R.

Theorem 3.4.1 is proved by letting $R = \mathbb{Q}[X_1, X_2, \ldots]/I$ for some suitable prime ideal I and then taking the quotient field. There are two kinds of requirements to be satisfied given an r.e. set A. The positive requirements are to code A into $U(R)$; this is done by adding appropriate cyclotomic polynomials as generators for I. On the other hand, there is need for some restraint in order to obtain R computable or, equivalently, to make I into a decidable ideal. The latter is achieved by choosing the generators for I of sufficiently high degree. I is constructed in stages. At each stage the pth cyclotomic polynomial Φ_p is added as a generator for I, where the variable of Φ_p does not appear in the previously added generators and p is a prime larger than the degrees of those generators and sufficiently large to satisfy the negative condition. The local decidability result is used to prove that $U(R)$ has complexity at most that of A.

The second undecidability result concerns the computability of Galois theory. Our previous results show that the Galois theory of finite extensions of computable fields is computable. By this we mean that every finite Galois extension K of a computable field (F, α) is a computable extension and every F-automorphism of K is computable. However, the Galois theory of infinite extensions may be totally non-effective.

Suppose K is a countably infinite Galois extension of the field F, that is K is a normal separable extension of F. It is a theorem of Krull [1928] that the cardinality of the Galois group

$$Gal(K/F) = \{\phi \in Aut(K) \colon (\forall x \in F)(\phi(x) = x)\}$$

is that of the continuum. In contrast we have

3.4.3. THEOREM (Metakides and Nerode [1979]). *There are computable fields (F, α) and (K, β) such that (K, β) is a computable infinite Galois extension of (F, α) and the only β-computable element of $Gal(K/F)$ is the identity.*

Here is an outline of the proof. Let $K = \mathbb{Q}(\sqrt{p_0}, \sqrt{p_1}, \ldots)$ where $p_0, p_1, \ldots$ is the recursive enumeration of the prime numbers in increasing order and let β be a computable numbering of K. K is a normal extension of $\mathbb{Q}$ since K is obtained as a union of a sequence of quadratic extensions. Thus (K, β) is a canonical computable extension of $\mathbb{Q}$ and hence every finite extension of $\mathbb{Q}$ in K is β-decidable uniformly in its generators. The theorem is proved by constructing a computable subfield (F, α) of (K, β) such that $[K : F] = \infty$ and such that the only β-computable F-automorphism is the identity. Observe that if (K, β) were a non-trivial canonical computable extension of (F, α) then the latter could not possibly hold. In particular, (F, α) must not have a splitting algorithm.

First we observe an algebraic fact. Assume that F is a subfield of $K = \mathbb{Q}(\sqrt{p_0}, \sqrt{p_1}, \ldots)$ and let $a_0, a_1, \ldots$ be a subsequence of $\sqrt{p_0}, \sqrt{p_1}, \ldots$. Assume further that the following two conditions hold:

(a) $K = F(a_0, a_1, \ldots)$, and

(b) for each e, $a_e \notin F(a_0, \ldots, a_{e-1})$.

Let ϕ be an F-automorphism different from the identity. Then for each e there is $x \in K$ such that

(1) $\phi(x) \neq x$,

(2) $[F(x) : F] = 2$, and

(3) $x \notin F(a_0, \ldots, a_{e-1})$.

The proof is simple. Since ϕ is not the identity there is i such that $\phi(a_i) \neq a_i$ and hence $\phi(a_i) = -a_i$. If there is such $i \geqslant e$ let $x = a_i$. Otherwise $e > i$ and then let $x = a_i a_e$.

When constructing F there are two requirements which need be satisfied. The first is a negative requirement, namely that $[K : F] = \infty$. It is negative in the sense that certain elements need to be kept out of F so as to obtain an infinite extension. The second is a positive requirement. It says that for each $e \in \omega$, if $\Phi_e \in Gal(K/F)$ then $\Phi_e = \mathrm{id}_K$, where Φ_e is the partial function on K induced by β from the eth partial recursive function.

The requirements are divided into instances for each $e \in \omega$ as follows:

N_e: $[K : F] \geqslant e$.

P_e: If Φ_e is an automorphism on K and Φ_e is not the identity then $\Phi_e(x) \neq x$ for some $x \in F$.

The field F is obtained by constructing a β-semidecidable set $A \subseteq K$ and setting $F = \mathbb{Q}(A)$. The set A is constructed in stages. At the beginning of stage s a finite part of A, denoted by A^s, has been constructed. Furthermore we have a sequence $a_0^s, a_1^s, \ldots$, a β-computable subsequence of $\sqrt{p_0}, \sqrt{p_1}, \ldots$ such that the following two conditions hold:

(a) $K = F(A^s)(a_0^s, a_1^s, \ldots)$, and

(b) for each i, $a_i^s \notin F(A^s)(a_0^s, \ldots, a_{i-1}^s)$.

Suppose this is the case for each s. Then to satisfy N_e it suffices that

$$\lim_{s \to \infty} a_i^s = a_i \quad \text{exists for } i = 0, \ldots, e-1,$$

since then $[K : F] \geqslant 2^e$ by (b). To satisfy P_e assume that $\Phi_e \neq \mathrm{id}_K$ and that Φ_e looks like a possible F-automorphism at stage s. Then, by the algebraic fact above and for sufficiently large s, we add an element x to A^s for which $\Phi_e(x) \neq x$ and which does not destroy any negative requirements N_i for $i < e$, thus ensuring that Φ_e will not be an F-automorphism. Obviously the addition of x to A^s may destroy condition (b) above, but only for finitely many i. Then we redefine the sequence $a_0^{s+1}, a_1^{s+1}, \ldots$ by simply removing, in turn, those finitely many a_i^s for which (b) fails when x is added to A^s.

As a final example we consider a strong undecidability result about algebraic independence. Recall that the set $AD(K/F)$, defined in connection with Definition 3.1.4, can take on arbitrary r.e. Turing degree. It follows from Theorem 3.1.5 that there are computable extension fields K of F with no semidecidable transcendence basis. Here is a strengthening of this result.

3.4.4. THEOREM (Metakides and Nerode [1979]). *Let (F, α) be a computable field. Then there is a computable extension field (K, β) of (F, α) of infinite transcendence degree such that every β-semidecidable subset of K that is algebraically independent over F, is finite.*

The argument is again of the priority type. However, as with previous priority arguments considered in this section, it is of a rather simple kind. It seems that with the added structure in the algebraic setting, at least in connection with rings and fields, it is easier to satisfy requirements once and for all than is the case in similar arguments for r.e. sets.

4. Computable Noetherian rings

In a Noetherian ring and module every ideal and submodule is finitely generated. The Noetherian condition is a finiteness condition that applies to fields and their polynomial rings. It should be expected that such a condition will have a substantial effect on the theory of computability. In this section we investigate and survey some of these effects in the theory of computable Noetherian rings and modules.

If (R, α) is a semicomputable Noetherian ring then its set $I(R)$ of ideals has a canonical numbering obtained from α, simply by considering finite sets of generators. Furthermore $I(R)$ with the operations of addition and multiplication of ideals is a semicomputable algebra.

We consider two fundamental concepts in the theory of computable Noetherian rings and modules. The first is that of *computable coherence*. Every ideal in a Noetherian ring is finitely related (see Section 4.1). A computable Noetherian ring R is said to be computably coherent if there is an algorithm which from a finite set of generators of an ideal I in R computes a finite set of relations for I. Equivalently, there are algorithms which compute intersections and quotients of ideals in R, that is, $I(R)$ expanded with the operations of intersection and quotient is a semicomputable algebra.

The second concept concerns the ideal membership problem (2.3.7). It is the ability to determine computably whether an element of the computable ring R is a member of an ideal *uniformly* in a given finite set of generators for the ideal. When it holds we say that R has an *ideal membership algorithm*. The equality relation between ideals is then decidable, that is, $I(R)$ is a computable algebra.

We show that important classes of computable Noetherian rings are computably coherent and have ideal membership algorithms. As corollaries we obtain old results for polynomial rings over computable fields, due essentially to Hermann [1926], necessary for the theory of computable fields discussed in Section 3.

4.1. *Noetherian rings and modules*

Recall that by a ring we will always mean a *commutative ring with* 1. We are also concerned with *unitary* modules over such rings. A ring R is *Noetherian* if every ideal is finitely generated or, equivalently, if every ascending chain of ideals is stationary. R is *Artinian* if every descending chain of ideals is stationary. Every Artinian ring is Noetherian. An R-module is *Noetherian* if every submodule is finitely generated. The basic algebraic theory of Noetherian rings and modules is covered in many standard texts, e.g., Atiyah and MacDonald [1969] and Cohn [1977].

4.1.1. DEFINITION. Let (R, α) be a semicomputable ring. An R-module M is a *computable* or *semicomputable* (R, α)*-module* if there is a numbering β of M such that

(i) (M, β) is a computable or semicomputable abelian group, and

(ii) the ring action $R \times M \to M$ is $(\alpha \times \beta, \beta)$-computable, that is, there exists a recursive function f such that for each $m \in \Omega_\alpha$ and $n \in \Omega_\beta$, $\beta f(m, n) = \alpha(m)\beta(n)$.

Below we have as a *standing assumption* that (R, α) is a semicomputable Noetherian ring.

When considering an R-module M we use the notation $\langle x_1, \ldots, x_k \rangle_R$ for the submodule of M generated by $x_1, \ldots, x_k \in M$. The ideal in R generated by $a_1, \ldots, a_n \in R$ is denoted by $(a_1, \ldots, a_n)$. Note that an ideal in R is a semicomputable (R, α)-module.

Observe that every finitely generated R-module M is a semicomputable Noetherian (R, α)-module. For suppose $M = \langle m_1, \ldots, m_k \rangle_R$. Let $h : R^k \to M$ be the epimorphism defined by

$$(a_1, \ldots, a_k) \mapsto \sum_{i=1}^{k} a_i m_i.$$

Then, by the Homomorphism Theorem, $M \cong R^k / ker(h)$, where the submodule $ker(h)$ is finitely generated, say by $x_1, \ldots, x_n$, since R and hence R^k are Noetherian R-modules. But then, in the standard numbering α^k of R^k (see Section 2.4), $ker(h)$ is semidecidable because

$$y \in ker(h) \Leftrightarrow (\exists r_1, \ldots, r_n \in R)(y = r_1 x_1 + \cdots + r_n x_n).$$

It follows that (M, β) is a semicomputable (R, α)-module, where $\beta = h\alpha^k$ is the quotient numbering obtained from α^k relative to $ker(h)$.

It is a simple fact that the above numbering β obtained for the finitely generated R-module M is a unique semicomputable numbering of M as an (R, α)-module up to recursive equivalence; that is, M is *computably stable* over (R, α). Thus when considering a finitely generated (R, α)-module it is not necessary to make the semicomputable numbering of the module explicit.

In general, suppose that M is an R-module and let $m_1, \ldots, m_k \in M$. Define

$$Z_R(m_1, \ldots, m_k) = \left\{ (a_1, \ldots, a_k) \in R^k \colon \sum_{i=1}^{k} a_i m_i = 0 \right\}.$$

Then $Z_R(m_1, \ldots, m_k)$ is the kernel of the homomorphism given by

$$(a_1, \ldots, a_k) \mapsto \sum_{i=1}^{k} a_i m_i.$$

It follows that k and a finite set of generators for $Z_R(m_1, \ldots, m_k)$ is a finite presentation of $\langle m_1, \ldots, m_k \rangle_R$. The elements in $Z_R(m_1, \ldots, m_k)$ are the *relations* for $m_1, \ldots, m_k$. Thus if R is Noetherian then every finitely generated submodule $\langle m_1, \ldots, m_k \rangle_R$ of M is *finitely related* in the sense that $Z_R(m_1, \ldots, m_k)$ is finitely generated. The ability to compute a finite generating set for $Z_R(m_1, \ldots, m_k)$ is crucial for the effective theory of Noetherian rings and modules.

Let M and N be semicomputable (R, α)-modules and suppose M is finitely generated. Then every R-module homomorphism $h : M \to N$ is computable since h is completely determined by its values on a finite set of generators for M.

It turns out (Theorem 4.3.7) that in case (R, α) is a semicomputable Noetherian ring and M is a finitely generated R-module then every submodule N of M is decidable, though *not* necessarily uniformly in its generators. In particular, every semicomputable Noetherian ring is computable.

4.2. *Computable coherence*

A ring is said to be *coherent* if every finitely generated ideal is finitely related. The notion of coherence is defined for arbitrary rings (see Bourbaki [1961]) and is a generalisation of the Noetherian condition. It is not a completely satisfactory generalisation in the critical sense that the Hilbert Basis Theorem does not lift, as shown by Soublin [1970]. Soublin constructed a coherent ring R for which the polynomial ring $R[X]$ is not coherent. However, in connection with computability, coherence does capture the essence of the Noetherian condition, as will be apparent below.

4.2.1. DEFINITION. Let (R, α) be a semicomputable Noetherian ring. A semicomputable (R, α)-module M is *computably coherent* if there is an algorithm which given $x_1, \ldots, x_k \in M$ computes a finite set of generators for the module

$$Z_R(x_1, \ldots, x_k) = \left\{ (a_1, \ldots, a_k) \in R^k \colon \sum_{i=1}^{k} a_i x_i = 0 \right\}.$$

We call the algorithm a *coherence algorithm*.

The concept of computable coherence has been studied in Baumslag, Cannonito and Miller [1981], Hingston [1981], Lazard [1976], Jacobsson and Stoltenberg-Hansen [1985], Stoltenberg-Hansen and Tucker [1988], and in a constructive setting in Richman [1974] and Seidenberg [1974a, 1974b]. However, already G. Hermann, in her fundamental paper Hermann [1926], essentially proved that the polynomial ring $F[X_1, \ldots, X_n]$ over a computable field F is computably coherent.

Below we show that every semicomputable Artinian ring (and hence field) is computably coherent. It is routine to verify that computable coherence is preserved under the quotient ring construction $R \mapsto R/I$ and the ring of fractions construction $R \mapsto S^{-1}R$. Furthermore it is preserved under the polynomial ring construction (Theorem 4.4.2) giving Hermann's theorem as a corollary.

Computable coherence corresponds precisely with the ability to compute on ideals. Let (R, α) be a semicomputable Noetherian ring. The set $I(R)$ of ideals in R has a canonical numbering $\alpha^* : \omega \to I(R)$ defined by

$$\alpha^*(e) = \text{ideal generated by } \alpha(D_e \cap \Omega_\alpha),$$

where D_e is the finite set of numbers with canonical index e.

4.2.2. PROPOSITION. *Let (R, α) be a semicomputable Noetherian ring. Then (R, α) is computably coherent if, and only if, the operations of intersection and quotient on $I(R)$ are α^*-computable.*

PROOF. Assume (R, α) is computably coherent and let $I = (x_1, \ldots, x_n)$ and $J = (y_1, \ldots, y_m)$ be given ideals of R. Compute a finite set of generators for $Z_R(x_1, \ldots, x_n, -y_1, \ldots, -y_m)$, say $\{(a_{i1}, \ldots, a_{in}, b_{i1}, \ldots, b_{im})\colon i = 1, \ldots, r\}$. Then

$$I \cap J = \left(\sum_{j=1}^{n} a_{1j} x_j, \ldots, \sum_{j=1}^{n} a_{rj} x_j \right).$$

For the quotient we have $(I : J) = (I : (y_1)) \cap \cdots \cap (I : (y_m))$ so we need only consider the case $m = 1$. Compute a finite set of generators for $Z_R(y, -x_1, \ldots, -x_n)$, say $\{(a_{i0}, \ldots, a_{in})\colon i = 1, \ldots, r\}$. Then $(I : (y)) = (a_{10}, \ldots, a_{r0})$.

For the converse, the ability to compute intersections of ideals in R and the annihilator of each element $x \in R$ suffices. The computation of generators for $Z_R(x_1, \ldots, x_n)$ proceeds by induction on n, the case $n = 1$ being just the annihilator. □

Clearly each computable field is computably coherent as well as the ring $\mathbb{Z}$ of integers. We now proceed to show that each semicomputable Artinian ring is computably coherent.

Let (R, α) be a semicomputable Noetherian ring. A short exact sequence of (R, α)-modules

$$0 \to M' \xrightarrow{\iota} M \xrightarrow{\pi} M'' \to 0$$

is *computable* if the modules M, M' and M'' are semicomputable (R, α)-modules and the homomorphisms ι and π are computable.

4.2.3. LEMMA. *Let* $0 \to M' \xrightarrow{\iota} M \xrightarrow{\pi} M'' \to 0$ *be a computable exact sequence of finitely generated* (R, α)*-modules. Then* M *is computably coherent if, and only if,* M' *and* M'' *are both computably coherent.*

PROOF. Suppose M' and M'' are computably coherent, and let $x_1, \ldots, x_k \in M$. In order to compute generators of $Z_R(x_1, \ldots, x_k)$ we first calculate generators $(a_{11}, \ldots, a_{1k}), \ldots, (a_{l1}, \ldots, a_{lk})$ of $Z_R(\pi(x_1), \ldots, \pi(x_k))$. Thus for each $1 \leqslant j \leqslant l$

$$0 = \sum_{i=1}^{k} a_{ji}\pi(x_i) = \pi\left(\sum_{i=1}^{k} a_{ji}x_i\right),$$

so $\sum_{i=1}^{k} a_{ji}x_i \in ker(\pi) = \iota(M')$ by the exactness condition. Now compute generators for

$$Z_R\left(\sum_{i=1}^{k} a_{1i}\iota^{-1}(x_i), \ldots, \sum_{i=1}^{k} a_{li}\iota^{-1}(x_i)\right)$$

say $(b_{11}, \ldots, b_{1l}), \ldots, (b_{n1}, \ldots, b_{nl})$. It is a tedious but simple exercise to verify that

$$\left(\sum_{j=1}^{l} b_{1j}a_{j1}, \ldots, \sum_{j=1}^{l} b_{1j}a_{jk}\right), \ldots, \left(\sum_{j=1}^{l} b_{nj}a_{j1}, \ldots, \sum_{j=1}^{l} b_{nj}a_{jk}\right)$$

generate $Z_R(x_1, \ldots, x_k)$.

The proof of the converse is similar. □

An analysis of the full proof shows (i) that M is computably coherent uniformly in coherence algorithms for M' and M'', and (ii) that M'' is computably coherent uniformly in a coherence algorithm for M and a finite set of generators of M'.

4.2.4. COROLLARY. *Let* (R, α) *be a semicomputable Noetherian ring. If* (R, α) *is computably coherent, then*

(i) R^n *is a computably coherent* (R, α)*-module uniformly in n, and*

(ii) *every finitely generated* (R, α)*-module is computably coherent, uniformly in a finite presentation.*

PROOF. (i) follows by induction on n since $0 \to R \to R^{n+1} \to R^n \to 0$ is a computable exact sequence. Then (ii) follows from (i) and the above remarks on uniformities since a presentation of a finitely generated R-module M provides us with a representation $M \cong R^n/N$ and hence a computable exact sequence $0 \to N \to R^n \to R^n/N \to 0$. □

We now easily obtain the main result of this subsection.

4.2.5. THEOREM. *Let (R, α) be a semicomputable Noetherian ring.*

(i) *If M is an (R, α)-module of finite length then M is computably coherent.*

(ii) *If R is Artinian then (R, α) is computably coherent.*

(iii) *If R is Artinian and M is a finitely generated (R, α)-module then M is computably coherent, uniformly in a finite presentation of M.*

PROOF. Clearly (ii) follows from (i), and (iii) follows from (ii) and Corollary 4.2.4. To prove (i), assume M is an R-module of finite length. Let

$$0 = M_0 \subset M_1 \subset \cdots \subset M_n = M \ (\subset \text{ denotes strict inclusion})$$

be a maximal composition series. Then for each $i = 1, \ldots, n$ there is a maximal ideal $\mathbf{m}_i$ such that $M_{i+1}/M_i \cong R/\mathbf{m}_i$. Thus

$$0 \to M_i \to M_{i+1} \to R/\mathbf{m}_i \to 0$$

is a computable exact sequence. Now $R/\mathbf{m}_i$ is a computable field and hence trivially computably coherent as a ring. It follows by an easy argument that $R/\mathbf{m}_i$ is also computably coherent as an R-module. Using Lemma 4.2.3 we conclude by induction that each M_i is computably coherent. □

Note that the construction of the composition series in the above proof was not uniform in a presentation of M. For uniformity we need some extra condition on (R, α), for example, that the set of maximal ideals is semidecidable. So the construction is uniform for Artinian rings (since they contain only finitely many maximal ideals) or for $\mathbb{Z}$.

4.3. *The ideal membership relation*

We now consider the problem of deciding membership in ideals of semicomputable Noetherian rings or, more generally, membership in submodules of finitely generated modules over semicomputable Noetherian rings.

Below we assume throughout that (R, α) is a semicomputable Noetherian ring. (R, α) is said to have an *ideal membership algorithm* if the relation $a \in (b_1, \ldots, b_k)$ is decidable for $a, b_1, \ldots, b_k \in R$, uniformly in k. Our first result gives a useful characterisation of the existence of an ideal membership algorithm for computably coherent R.

4.3.1. LEMMA. *Suppose (R,α) is computably coherent. Then (R,α) has an ideal membership algorithm if, and only if, the relation $1 \in (b_1,\dots,b_k)$ is α-decidable uniformly in k.*

PROOF. To prove the nontrivial direction, note that

$$\begin{aligned} a \in (b_1,\dots,b_k) &\Leftrightarrow (\exists r_1,\dots,r_k \in R)(a = r_1b_1 + \cdots + r_kb_k) \\ &\Leftrightarrow (\exists r_1,\dots,r_k \in R)\big((1,r_1,\dots,r_k) \in Z_R(a,b_1,\dots,b_k)\big). \end{aligned}$$

Using the coherence algorithm for (R,α), compute a set of generators of $Z_R(a,b_1,\dots,b_k)$, say $(d_j,e_{j1},\dots,e_{jk})$, $j=1,\dots,n$. Thus

$$\begin{aligned} a \in (b_1,\dots,b_k) &\Leftrightarrow (\exists r_1,\dots,r_k \in R)\big((1,r_1,\dots,r_k) \in Z_R(a,b_1,\dots,b_k)\big) \\ &\Leftrightarrow (\exists c_1,\dots,c_n,r_1,\dots,r_k \in R)\left(\left(1 = \sum_{j=1}^{n} c_jd_j\right) \& \right. \\ &\qquad \left. \left(r_1 = \sum_{j=1}^{n} c_je_{j1}\right) \& \cdots \& \left(r_k = \sum_{j=1}^{n} c_je_{jk}\right)\right) \\ &\Leftrightarrow (\exists c_1,\dots,c_n \in R)(1 = c_1d_1 + \cdots + c_nd_n) \\ &\Leftrightarrow 1 \in (d_1,\dots,d_n). \end{aligned}$$

□

Kudinov has given an example of a computable Noetherian integral domain (R,α) with no ideal membership algorithm but with the relation $1 \in (b_1,\dots,b_k)$ decidable uniformly in k, thus providing an example of a computable Noetherian ring which is not computably coherent.

It follows from the lemma that if (R,α) is computably coherent and there is an α^*-semidecidable family of proper ideals which includes all maximal ideals then (R,α) has an ideal membership algorithm. In particular we have

4.3.2. COROLLARY (Baur [1974]). *If (R,α) is a semicomputable Artinian ring then (R,α) has an ideal membership algorithm.*

PROOF. (R,α) is computably coherent by Theorem 4.2.5. Furthermore, R has only finitely many maximal ideals. Thus the condition of the theorem holds since

$$1 \notin (b_1,\dots,b_k) \Leftrightarrow (\exists \text{ maximal ideal } \mathbf{m})(b_j \in \mathbf{m} \text{ for } j = 1,\dots,k).$$

□

We extend the corollary to semicomputable local rings. Recall that a local ring is a ring with a unique maximal ideal. Of course, if the local ring is computably coherent then it trivially has an ideal membership algorithm.

4.3.3. LEMMA. *Let* (R,α) *be a semicomputable Noetherian local ring with maximal ideal* $\mathbf{m}$. *Then* $R/\mathbf{m}^n$ *is computably coherent, uniformly in* n.

PROOF. $R/\mathbf{m}^n$ is Artinian and hence computably coherent. We prove the claimed uniformity. Note that

$$0 \to \mathbf{m}^n/\mathbf{m}^{n+1} \to R/\mathbf{m}^{n+1} \to R/\mathbf{m}^n \to 0$$

is a computable exact sequence of semicomputable (R,α)-modules, uniformly in n. Assuming inductively that $R/\mathbf{m}^n$ is computably coherent (the induction base is trivial), it suffices to show by Lemma 4.2.3 and the remarks there on uniformity, that $\mathbf{m}^n/\mathbf{m}^{n+1}$ is computably coherent, uniformly in n. Let

$$f(n) = \text{the length of } \mathbf{m}^n/\mathbf{m}^{n+1}.$$

By the Hilbert–Serre theory for Poincaré series (see Chapter 11 in Atiyah and MacDonald [1969]) f is a polynomial except for finitely many n. In particular, f is a recursive function. Furthermore, the annihilator $\mathrm{Ann}(\mathbf{m}^n/\mathbf{m}^{n+1}) \supseteq \mathbf{m}$, so $\mathbf{m}^n/\mathbf{m}^{n+1}$ may be considered as a semicomputable vector space over the residue field $k = R/\mathbf{m}$, and $f(n) = \dim_k(\mathbf{m}^n/\mathbf{m}^{n+1})$. But then, since we can compute generators of $\mathbf{m}^n/\mathbf{m}^{n+1}$, we can effectively find $x_1,\ldots,x_{f(n)} \in \mathbf{m}^n$ generating $\mathbf{m}^n/\mathbf{m}^{n+1}$. Letting $M_i = R\bar{x}_1 + \cdots + R\bar{x}_i$, where as usual $\bar{x}_j = x_j + \mathbf{m}^{n+1}$, we obtain a composition series

$$0 = \mathbf{m}^{n+1}/\mathbf{m}^{n+1} = M_0 \subset M_1 \subset \cdots \subset M_{f(n)} = \mathbf{m}^n/\mathbf{m}^{n+1}.$$

Thus $M_{i+1}/M_i \cong R/\mathbf{m}$, and we can inductively construct coherence algorithms for each M_i; in particular for $M_{f(n)} = \mathbf{m}^n/\mathbf{m}^{n+1}$. □

Using Lemma 4.3.1 we obtain

4.3.4. COROLLARY. *Let* (R,α) *be a semicomputable Noetherian local ring with maximal ideal* $\mathbf{m}$. *Then* $R/\mathbf{m}^n$ *has an ideal membership algorithm, uniformly in* n.

Lemma 4.3.3 can also be used to show that every computable regular local ring of dimension 1 is computably coherent, thus somewhat extending Theorem 4.2.5.

4.3.5. THEOREM (Stoltenberg-Hansen and Tucker [1988]). *Every semicomputable Noetherian local ring* (R,α) *has an ideal membership algorithm.*

PROOF. It suffices to show that the relation $a \notin I$ is semidecidable uniformly in generators of the ideal I. Let $\mathbf{m}$ be the maximal ideal in R. Since R is local, by Krull's theorem,

$$I = \bigcap_{n=1}^{\infty} (I + \mathbf{m}^n)$$

for each ideal I. Thus we have

$$a \notin I \Leftrightarrow \exists n\big(a \notin I + \mathbf{m}^n\big) \Leftrightarrow \exists n\big(a + \mathbf{m}^n \notin I + \mathbf{m}^n \text{ in } R/\mathbf{m}^n\big).$$

And the latter is semidecidable uniformly in generators of I, by Corollary 4.3.4. □

It is trivially true that every semicomputable field is computable. We now show that this extends to all Noetherian rings. The following critical lemma is a trivial consequence of Theorem 4.3.5.

4.3.6. LEMMA. *Let (R, α) be a semicomputable Noetherian ring. Then each prime ideal in R is decidable.*

PROOF. Let $\mathfrak{p}$ be a prime ideal in R and let $\mathbf{m} \supseteq \mathfrak{p}$ be a maximal ideal. Let $R_{\mathbf{m}}$ be the localisation of R at $\mathbf{m}$. Then the local ring $R_{\mathbf{m}}$ is semicomputable, since $\mathbf{m}$ is maximal, with a numbering obtained in a canonical fashion from α so that the natural homomorphism $\lambda : R \to R_{\mathbf{m}}$ given by $x \mapsto x/1$ is computable. It follows from Theorem 4.3.5 that each ideal in $R_{\mathbf{m}}$ is decidable. There is an exact correspondence between prime ideals contained in $\mathbf{m}$ and prime ideals in $R_{\mathbf{m}}$ given by $\mathfrak{p} \leftrightarrow S^{-1}\mathfrak{p}$, where $S = R - \mathbf{m}$. In fact we have

$$a \in \mathfrak{p} \Leftrightarrow \frac{a}{1} \in S^{-1}\mathfrak{p},$$

showing the decidability of $\mathfrak{p}$. □

4.3.7. THEOREM (Baur [1974]). *Let (R, α) be a semicomputable Noetherian ring and let M be a finitely generated R-module. Then M is a computable (R, α)-module. In particular, (R, α) is computable.*

PROOF. Let M be a finitely generated R-module. Then there is a finite chain of submodules $0 = M_0 \subseteq M_1 \subseteq \cdots \subseteq M_n = M$ such that for each i, $M_{i+1}/M_i \cong R/\mathfrak{p}_i$, where $\mathfrak{p}_i$ is a prime ideal. Each M_i is a semicomputable (R, α)-module since it is finitely generated. By induction on i we prove that each M_i is in fact computable. Consider the computable short exact sequence

$$0 \to M_i \xrightarrow{\iota} M_{i+1} \xrightarrow{\pi} R/\mathfrak{p}_i \to 0.$$

For $x \in M_{i+1}$ we have

$$x \neq 0 \Leftrightarrow \pi(x) \neq 0 \text{ or } (\exists y \in M_i)(y \neq 0 \;\&\; \iota(y) = x).$$

It follows from Lemma 4.3.6 that each M_i is computable. In particular, M is a computable (R, α)-module. Letting $M = R$ we have that (R, α) is a computable ring. □

4.3.8. COROLLARY. *Let (R, α) be a computable Noetherian ring and let M be a finitely generated (R, α)-module. Then every submodule of M is decidable. In particular, every ideal of R is decidable.*

PROOF. Let $N \subseteq M$ be a submodule. Then M/N is a computable (R, α)-module by the theorem, that is, N is decidable. □

The proof of the corollary does not provide a uniform method for deciding ideals in computable Noetherian rings. This is in general impossible.

4.3.9. EXAMPLE (Baur [1974]). *A computable Noetherian ring with no ideal membership algorithm.*

Let P be an r.e. set of prime numbers and let S be the multiplicatively closed set generated by P as a subset of the ring of integers $\mathbb{Z}$. Then the ring $R = S^{-1}\mathbb{Z}$ is a computable Noetherian ring. It may be thought of as a subring of the rationals $\mathbb{Q}$. For a given ideal I in R the contracted ideal $I^c = \{x \in \mathbb{Z}\colon x/1 \in I\}$ is a principal ideal in $\mathbb{Z}$, say $I^c = (a)$. Then clearly $I = (a/1)$. Note that for primes p,

$$1 \in (p/1) \Leftrightarrow p \in P.$$

For in case $1 \in (p/1)$ then p divides s for some $s \in S$. It follows that if R has an ideal membership algorithm then the set $\{p \in \mathbb{Z}\colon p \text{ prime} \;\&\; 1 \in (p/1)\}$ is decidable, that is, P is recursive.

4.4. *Polynomial rings*

If (R, α) is a computable Noetherian ring then the polynomial ring $R[X]$ is computable with a standard numbering α_* obtained from α, and it is Noetherian by the Hilbert Basis Theorem. In this subsection we first show that computable coherence is preserved under the polynomial ring construction. Secondly, we prove that the existence of an ideal membership algorithm is preserved under the polynomial ring construction, *provided* that (R, α) is computably coherent.

The proof of the preservation of computable coherence is based on Seidenberg's proof in a constructive setting (Seidenberg [1974b]): Given $r_1, \ldots, r_k \in R[X]$, we effectively construct $s_1, \ldots, s_l \in R[X]$ generating the same ideal but having sufficient closure properties to allow the computation of a finite generating set for $Z_{R[X]}(s_1, \ldots, s_l)$. Then a finite generating set for $Z_{R[X]}(r_1, \ldots, r_k)$ is computed using Lemma 4.4.1, which is based on a result in Richman [1974] in a constructive setting.

4.4.1. LEMMA. *Let (R, α) be a computable Noetherian ring and let M be a finitely generated (R, α)-module. Then there is an algorithm which given a finite presentation of M and an epimorphism $g : R^m \to M$ computes a finite generating set for $ker(g)$.*

PROOF. Let $M \cong R^n/N$ be the given finite presentation of M, that is, we are given n and a finite generating set for N. Suppose $g : R^m \to M$ is the given epimorphism. Then $L = ker(g)$ is a semidecidable submodule of R^m. We are to compute a finite generating set for L.

From the given data we obtain the following diagram of computable exact sequences.

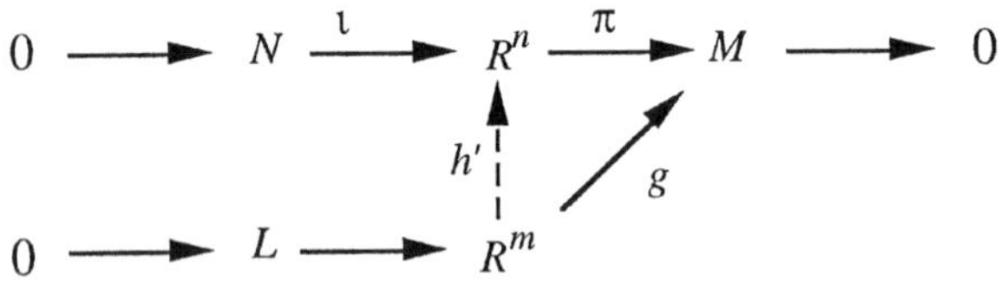

Effectively, using search, we find a homomorphism $h' : R^m \to R^n$ such that the above diagram commutes. Then let $h : R^m \oplus N \to R^n$ be defined by $h(a \oplus s) = h'(a) + s$. It is easily verified that h is an epimorphism and that we obtain the following computable commutative diagram

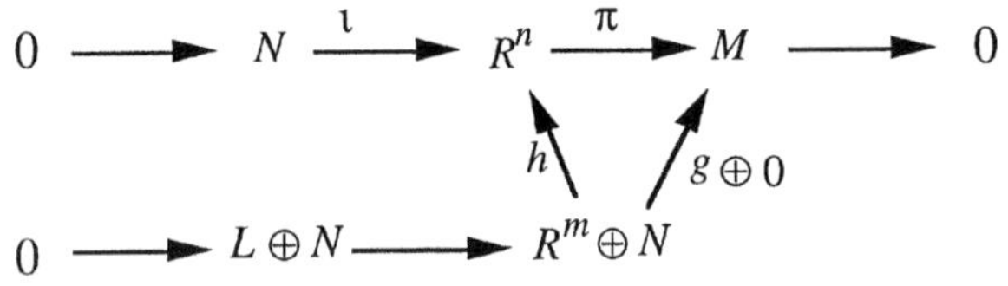

and that the sequence

$$0 \longrightarrow L \oplus N \longrightarrow R^m \oplus N \xrightarrow{g \oplus 0} M \longrightarrow 0$$

is exact.

To compute a finite generating set for L it suffices to compute a finite generating set for $L \oplus N$. Using the standard basis of R^n we effectively find a finite generating set for the submodule $F \subseteq R^m \oplus N$ for which $F \cong R^n$ via h. Then

$$R^m \oplus N = F + ker(h)$$

and the sum is direct. Thus we are able to compute a finite generating set for $ker(h)$ from the computed generating sets for $R^m \oplus N$ and F.

Let $H = \{x \in F\colon \pi h(x) = 0\}$. Using the generating set for N, we compute a finite generating set for H via the isomorphism $F \cong R^n$. Now it is easily verified that

$$L \oplus N = ker(h) + H$$

where the sum is direct. But then we easily obtain a finite generating set for $L \oplus N$ which, as already remarked, enables us to compute a finite generating set for L. □

We now prove our main theorems, proved in Seidenberg [1974b] in a constructive setting.

4.4.2. THEOREM. *Let (R, α) be a computable Noetherian ring. If (R, α) is computably coherent then so is $R[X]$ with the standard numbering α_* obtained from α.*

PROOF. Let $r_1, \dots, r_k \in R[X]$ and consider the ideal $I = (r_1, \dots, r_k)$ in $R[X]$. Let $n = \max\{\deg(r_i)\colon i = 1, \dots, k\}$. We are to compute a finite generating set for $Z_{R[X]}(r_1, \dots, r_k)$.

First (considering $R + RX + \cdots + RX^n$ as an R-module) we compute a finite generating set for the R-module $M = I \cap (R + RX + \cdots + RX^n)$. To compute M let $M_1 = \langle r_1, \dots, r_k \rangle_R$. Having computed a finite generating set for the R-module M_j, let

$$M_{j+1} = (XM_j + M_j) \cap \bigl(R + RX + \cdots + RX^n\bigr).$$

By Proposition 4.2.2, trivially extended to modules, and Corollary 4.2.4 compute a finite generating set for M_{j+1}. We obtain an ascending chain of R-modules

$$M_1 \subseteq M_2 \subseteq \cdots$$

which eventually stabilises since $R + RX + \cdots + RX^n$ is a Noetherian R-module. From the computed generating sets we effectively find j such that $M_{j+1} = M_j$. Then, as desired,

$$M = M_j = I \cap \bigl(R + RX + \cdots + RX^n\bigr).$$

Let $M' = M \cap (R + RX + \cdots + RX^{n-1})$. Now compute a finite set of polynomials S consisting of the computed generators for M and M' and closed as follows: if $f \in S$ and $\deg(f) < n$ then $Xf \in S$. Let S contain l elements, say $S = \{s_1, \dots, s_l\}$. Thus if $\deg(s_i) < n$ then there is a unique index $j(i)$ such that $Xs_i = s_{j(i)}$.

By Lemma 4.4.1 it suffices to compute a finite generating set for $Z_{R[X]}(s_1, \dots, s_l)$. For each i such that $\deg(s_i) < n$ let $e_i \in R[X]^l$ be a vector inducing the relation $Xs_i = s_{j(i)}$. Let F be the computed set of generators for $Z_R(s_1, \dots, s_l)$, and let

$$G = F \cup \bigl\{e_i\colon \deg(s_i) < n\bigr\}.$$

We claim that G generates $Z_{R[X]}(s_1,\ldots,s_l)$. Clearly $G \subseteq Z_{R[X]}(s_1,\ldots,s_l)$.

Suppose $(f_1,\ldots,f_l) \in Z_{R[X]}(s_1,\ldots,s_l)$. By induction on $m = \max\{\deg(f_i)\colon i = 1,\ldots,l\}$ we prove that $(f_1,\ldots,f_l) \in \langle G\rangle_{R[X]}$. It is clearly true for $m = 0$. So suppose $m > 0$ and write $f_i = f_i' + a_i X^m$, where $a_i \in R$ and $\deg(f_i') < m$. Thus,

$$0 = \sum_{i=1}^{l} f_i' s_i + X^m \sum_{i=1}^{l} a_i s_i$$

and we conclude that $\deg(\sum a_i s_i) < n$, that is, $\sum a_i s_i \in M'$. Using generators in S for M' there are $b_1,\ldots,b_l \in R$ such that

$$\sum_{i=1}^{l} a_i s_i = \sum_{i=1}^{l} b_i s_i,$$

where $b_i = 0$ if $\deg(s_i) = n$. Using generators in G for $Z_R(s_1,\ldots,s_l)$ we conclude that it is sufficient to show that $(f_1' + b_1 X^m,\ldots,f_l' + b_l X^m) \in \langle G\rangle_{R[X]}$.

Let $c_j = b_i$ if $s_j = X s_i$ for some i and let $c_j = 0$ otherwise. Then

$$\left(f_1' + c_1 X^{m-1},\ldots,f_l' + c_l X^{m-1}\right) \in Z_{R[X]}(s_1,\ldots,s_l)$$

and hence, by the induction hypothesis, $(f_1' + c_1 X^{m-1},\ldots,f_l' + c_l X^{m-1}) \in \langle G\rangle_{R[X]}$. But

$$\begin{aligned}\left(f_1' + b_1 X^m,\ldots,f_l' + b_l X^m\right) &= \left(f_1' + c_1 X^{m-1},\ldots,f_l' + c_l X^{m-1}\right)\\ &\quad + X^{m-1}(b_1 X - c_1,\ldots,b_l X - c_l),\end{aligned}$$

and the latter term also belongs to $\langle G\rangle_{R[X]}$ by the choice of the generators e_j. □

Now we consider the existence of an ideal membership algorithm and show that this is preserved in the polynomial ring construction provided the original ring is computably coherent.

4.4.3. THEOREM. *Let (R,α) be a computable Noetherian ring and assume that (R,α) is computably coherent. If (R,α) has an ideal membership algorithm then so does the polynomial ring $R[X]$ under the standard numbering α_* obtained from α.*

PROOF. Assume that (R,α) has an ideal membership algorithm. By an argument similar to that of Proposition 4.3.1, the R-module $R + RX + \cdots + RX^n$ has a *submodule membership algorithm*, uniformly in n.

Let $I = (r_1,\ldots,r_k)$ be an ideal in $R[X]$ and let $f \in R[X]$. Compute n, the maximum of the degrees of f and r_i, $i = 1,\ldots,k$. As in the first part of the proof of Theorem 4.4.2, compute a finite generating set for the R-module $M =$

$I \cap (R + RX + \cdots + RX^n)$. Then $f \in I \Leftrightarrow f \in M$, which is decidable uniformly in the computed generators of M. □

G. Hermann's fundamental theorems on polynomial rings, put in a recursion theoretic context, are special cases of the above.

4.4.4. THEOREM (Hermann [1926]). *Let (F, α) be a computable field. Then the polynomial rings $F[X_1, \ldots, X_n]$ are computably coherent and have ideal membership algorithms in their standard numberings obtained from α, uniformly in n.*

PROOF. (F, α) is obviously computably coherent and has an ideal membership algorithm since fields have a trivial ideal structure. Thus, inductively, $F[X_1, \ldots, X_n]$ is computably coherent by Theorem 4.4.2 and has an ideal membership algorithm by Theorem 4.4.3. □

5. Further reading

In this final section we look at a little of the literature on Effective Algebra, and on a few related topics. Our aim is to sketch a map of the subject that will be useful to a range of readers.

5.1. *Computable groups*

Algorithmic decision problems in group theory have been studied in considerable detail in combinatorial group theory. Combinatorial group theory is primarily about finite and recursively enumerable presentations of groups. The discovery of group presentations, in von Dyck [1882], and of the word, conjugacy and related decision problems for group presentations, starting in Dehn [1910, 1911], were made in connection with problems in geometry and the emerging subject of algebraic topology (see also the selected works Dehn [1987], and the historical studies Chandler and Magnus [1982] and Wüssing [1984]). The algorithmic unsolvability of the word problem for finitely presented groups was established independently and published in Novikov [1955] and in Boone [1959].

There is an excellent literature on decision problems in combinatorial group theory, containing many fine undecidability results and characterisation theorems. Convenient short surveys are Stillwell [1982] and Miller [1992]; comprehensive accounts are the monographs Magnus, Karass and Solitar [1976] and Lyndon and Schupp [1977]. An introduction to the word problem is contained in the text-book Rotman [1995]. The proceedings Boone et al. [1973], Adian et al. [1980], and Baumslag and Miller [1992] are convenient and rich sources of further information.

Now it is easy to define the notions of *computable* and *semicomputable groups* (adapting the definition in Section 1.1) and to start a theory of computable groups. What is its connection with decision problems in combinatorial group theory?

THEOREM. *A group is semicomputable if, and only if, it has a recursively enumerable presentation. A group is computable if, and only if, it has a recursively enumerable presentation with decidable word problem. In particular, a finitely presented group is computable if, and only if, it has a decidable word problem.*

This relationship was observed in Rabin [1960a, 1960b] and, in its general form for universal algebras, in Mal'cev [1961]. The result is based on a correspondence between computable numberings and recursively enumerable presentations. The definition of a recursively enumerable presentation requires standard numberings of the free group; and the invariance of the decidability of the word problem with respect to presentations is related to computable stability (recall 2.2.5 (iv); for details see Mal'cev [1961] and Stoltenberg-Hansen and Tucker [1995]).

There are unsolvability results for finitely presented groups involving the commonly used degrees. The fact that any r.e. Turing degree can be realised as the word problem of a finitely presented group was established independently in Fridman [1962], Clapham [1964] and Boone [1966, 1971]. This is sharpened to unbounded truth table degrees in Collins [1971]. In Ziegler [1976] it is shown that not every bounded truth table degree, and hence not every many one degree, is realised by a finitely presented group.

Let us give some examples of important characterisation theorems expressed in the terminology of computable groups:

THEOREM (Higman [1961]). *A finitely generated group is semicomputable if, and only if, it is embeddable in a finitely presented group.*

A theorem in Kuznetsov [1958] on decidability of the term problem in finitely presented simple universal algebras suggested the following:

THEOREM (Boone and Higman [1974]). *A finitely generated group is computable if, and only if, it is embeddable in a simple subgroup of a finitely presented group.*

THEOREM (MacIntyre [1972] and Neumann [1973]). *A finitely generated group is computable if, and only if, it is embeddable in every countable algebraically closed group.*

Details of these results can be found in Lyndon and Schupp [1977].

Although the formulations using presentations and numberings are equivalent, and the combinatorial theory is well established, there are some advantages to using the theory of computable groups. For example, it is suited to investigating infinitely generated groups where the computability properties of infinite presentations are complex. Commonly, the computable numberings of an infinitely generated group are not stable and, hence, the spectrum of numberings of such a group is a complex object of interest.

The group G is computably stable if $|C(G)/{\sim}| = 1$, that is if its computable dimension is 1. If G is a computable group with uncountably many automorphisms then the computable dimension is uncountable. The following remarkable theorem is from Goncharov [1981]:

THEOREM. *For each* $n \geqslant 1$, *there exists a computable metabelian group* G *such that* $|C(G)/{\sim}| = n$. *In detail: there exist* n *computable numberings* $\alpha_1, \ldots, \alpha_n$ *of* G *such that*

(i) *for each* $1 \leqslant i, j \leqslant n$, $\alpha_i \sim \alpha_j \Leftrightarrow i = j$; *and*
(ii) *for any computable numbering* α *of* G,

$$\alpha \sim \alpha_1 \text{ or } \ldots \text{ or } \alpha \sim \alpha_n.$$

The spectra of numberings of groups under recursive autoequivalence $\approx$ have also been studied. The group G is computably autostable if $|C(G)/{\approx}| = 1$. Since the group provided by Goncharov is rigid (i.e. $Aut(G) = 1$) the result about finitely many numberings up to recursive equivalence implies the corresponding result about finitely many numberings up to autoequivalence.

Secondly, the theory of numberings is suited to studying the computability of groups arising in applications, where combinatorial presentations may not be readily available. A simple illustration showing the advantages of the latter point is this result, announced in Rabin [1960a, 1960b]:

THEOREM. *Any finitely generated group of matrices over any field is computable.*

Note that the field is *not* assumed to be computable and that the result applies to the real numbers $\mathbb{R}$ and complex numbers $\mathbb{C}$.

PROOF. Let G be a finitely generated subgroup of the general linear group $GL(n, F)$ over the field F. Take the m matrix generators $(a^1_{ij}), \ldots, (a^m_{ij})$ and extend the prime subfield P of F by the entries in the matrices to form a field $K = P(a^k_{ij} \mid 1 \leqslant k \leqslant m, 1 \leqslant i, j \leqslant n)$. By Theorem 3.1.1, K is a computable field. It is easy to check that $GL(n, K)$ is computable group and that hence G is computable (since every finitely generated subgroup of a computable group is computable). □

To illustrate further the freedom to investigate effectivity provided by the methods:

COROLLARY. *Every finitely generated subgroup of a compact Lie group is computable.*

This is because compact Lie groups have faithful linear representations (see Chevalley [1946]).

The theorem can be generalised to various theorems based on different forms of universal algebra built from the affine spaces F^n over a field F (and, to a lesser extent, over arbitrary commutative rings) which have further applications (see Tucker [1976, 1980a, 1980b]).

Finitely generated abelian groups are finitely presented and computable; this follows from their classification theorem (see Rotman [1995]). Infinitely generated abelian groups have been studied in Mal'cev [1971b], Lin [1981a, 1981b] and Smith [1981b].

The classification of interesting subsets of a computable algebra require notions other than decidable and semidecidable subsets. A subset may be higher up in the arithmetic hierarchy, or not be arithmetically definable at all. We can illustrate the use of higher-order notions from analytic definability theory using Abelian groups.

An Abelian group A is *divisible* if for each $a \in A$ and $n \in \mathbb{N}$ there is a $b \in A$ such that $nb = a$. The additive group $\mathbb{Q}$ of rationals and, for p prime, the groups $\mathbb{Z}(p^\infty)$ are divisible. Any divisible Abelian group is a direct sum of such groups. For any Abelian group A the union of all its divisible subgroups forms a divisible subgroup $Div(A)$. Furthermore, $A \cong Div(A) \oplus N$ where N has no non-trivial divisible subgroups.

THEOREM (Feferman [1975]). *There is a computable abelian 2-group such that $Div(A)$ is not hyperarithmetically definable.*

Decision problems are also important in combinatorial semigroup theory. One can trace this subject to the work of A. Thue on the word problem for Thue systems which are combinatorial systems similar to semigroup presentations (see Thue [1914] and the selected works Thue [1977]). Thue systems and their applications are explained in Davis [1977]. The undecidability of the word problem for finitely presented semigroups was established by Markov and Post in 1947. Turing established the undecidability of the word problem for cancellation semigroups in 1950. A useful reference is Lallement [1979].

5.2. *Linear algebra*

A computable vector space V over a computable field F is a computable F-module (Section 4.1). Finite and countably infinite dimensional vector spaces over F are computable. Their theory is attractive and useful; it concerns linear dependence, decomposition of spaces, linear transformations and automorphisms, matrix theory, and so on. Let V_∞ be a vector space over F of countably infinite dimension; it is computable but not computably stable. All vector spaces V_n over F of finite dimension n embed in V_∞ and so it serves as a universal space for the study of countable dimensional vector spaces up to isomorphism. The set of semidecidable vector subspaces of V_∞ over F forms the lattice $L(V_\infty)$. The study of this lattice $L(V_\infty)$ began in Metakides and Nerode [1977] and has been the subject of an extensive and successful programme of investigation. It is not possible to do justice to the subject here, only to comment briefly on its achievements and working methodology; the reader is referred to the survey Nerode and Remmel [1985] for a comprehensive account.

Consider the subspace membership problem. Let V be a vector space over F and $v_0, v_1, \ldots, v_k \in V$ then $\langle v_1, \ldots, v_k\rangle_F$ is the vector subspace generated by the elements $v_1, \ldots, v_k$; and $v_0 \in \langle v_1, \ldots, v_k\rangle_F$ if, and only if, the set $\{v_0, v_1, \ldots, v_k\}$ is linearly dependent. Define

$$D(V) = \{(v_0, v_1, \ldots, v_k) \in V^* \mid k \geqslant 1 \text{ and } v_0, v_1, \ldots, v_k \text{ are linearly dependent}\}$$

and the subsets $D_k(V) = D(V) \cap V^{k+1}$ for $k = 1, 2, \ldots$. If V is semicomputable then $D(V)$ and $D_k(V)$ are semicomputable. Here is a remarkable undecidability result for the subspace membership problem based on Turing degrees:

THEOREM (Shore [1978]). *Let $A_0, A_1, \ldots$ be any recursive sequence of r.e. sets such that*

(i) $A_i \leqslant_T A_{i+1}$ *uniformly in* $i = 1, 2, \ldots$; *and*
(ii) $A_i \leqslant_T A_0$ *uniformly in* $i = 1, 2, \ldots$.

Then there exists a computable vector space V over an infinite computable field F such that

$$D(V) \equiv_T A_0 \quad \textit{and} \quad D_k(V) \equiv_T A_k \quad \textit{for} \quad k = 1, 2, \ldots.$$

The systematic study of the lattice of subspaces and their associated decision problems was motivated and shaped technically by the theory of the lattice of r.e. sets. Lattice-theoretic and, especially, recursion-theoretic notions about sets and their various degree classifications are used to describe the structure of $L(V_\infty)$. Ideas about r.e. sets suggest concepts about semidecidable subspaces of V_∞; techniques of the priority method are applied and adapted to the construction of semidecidable subspaces with special properties. There are results about the existence of creative subspaces (e.g., Metakides and Nerode [1977]), super-maximal subspaces (e.g., Kalantari and Retzlaff [1977] for which, incidentally, a version of Shore's Theorem is proved in Nerode and Remmel [1983]); there are analogues and extensions of splitting theorems based on the direct sums of spaces (e.g., Retzlaff [1979], Shore [1978], and Ash and Downey [1984]); results about lattice automorphisms of $L(V_\infty)$ (e.g., Guichard [1983]) and the undecidability of the theory of the lattice $L(V_\infty)$ (Nerode and Smith [1982]). The theory of $L(V_\infty)$ demonstrates that the diverse phenomena that has been discovered in the theory of the r.e. sets is relevant and necessary to the complete understanding of the semidecidable subspaces of $L(V_\infty)$.

A great deal of information on the lattice is now available; and there are a number of results of general interest.

The lattice $L(V_\infty)$ of r.e. subspaces has much in common with the lattice $L(F_\infty)$ of subfields of a so called computable countable universal field F_∞ (i.e. an algebraically closed field of countably infinite transcendence degree over its prime subfield, into which all countable fields of a given characteristic may be embedded). The properties of algebraic independence are similar to those of linear independence. For example, in Kudinov [1993] the analogue of Shore's theorem for $L(F_\infty)$ is proved. A general framework for dealing with both types of dependency has been developed and computability properties explored: see Nerode and Remmel [1985].

5.3. *Other computable structures*

Much is known about the computability of many other types of algebraic structure. The literature is scattered and we will not attempt a survey. We will comment on models of Peano arithmetic and Boolean algebras.

The study of models of Peano arithmetic (PA) and its fragments has been influenced by the following fact:

THEOREM (Tennenbaum [1959]). *The only model of PA that is computable is the standard model* $(\mathbb{N}, 0, n+1, n+m, n\times m, n<m)$.

For which subtheories of PA can computable non-standard models exist? Fragments of PA that have computable non-standard models were first identified in Shepherdson [1964, 1965]. The fact that the subtheory $I\Delta_0$, in which the induction scheme involves Δ_0 formulae only, has no computable non-standard models was established in McAloon [1982]. A good account of this topic and its applications is in Kaye [1991].

A good deal is known about the computability of Boolean algebras and related topological structures. Some basic material is in Alton and Madison [1973], and the lattice of substructures has been well studied after the fashion of linear algebra and field theory; see Nerode and Remmel [1985]. A survey of the subject is Remmel [1989].

5.4. *Computable universal algebras*

The theories of different kinds of algebraic structures have a great deal in common. Basic concepts and results about subalgebras, homomorphisms, factor algebras, products, limits, free algebras, equations are unified in the theory of universal algebras (see Grätzer [1979], Cohn [1981], Burris and Sankappanavar [1981] and McKenzie et al. [1987]). In particular, algorithmic properties are unified in the basic paper Mal'cev [1961] which started the theory of computable universal algebras.

A.I. Mal'cev studied computable universal algebras of the form Definition 1.1.1. Most of the notions we use are adaptations of those of Mal'cev (see his selected works Mal'cev [1971a]). Thorough mathematical accounts of computable algebras and their numberings are to be found in the paper Ershov [1977a] and book Ershov [1977b]. An introduction to computable universal algebras is contained in Stoltenberg-Hansen and Tucker [1995].

5.5. *Algebraic theory of data in computer science*

In computer science, many sorted universal algebra is used to provide a general theory of data. Data is represented in software by constructs called *data types*. Algebras are used to define and analyse many new forms of data types, classify data representations, and perform the modularisation of computing systems. The forms of modularisation developed for data types have been generalised through software constructs called *objects*, and algebras can be used in giving a semantics to objects.

The algebraic theory of data has the following basic features. An algebra is a representation or implementation of a data type. An abstract data type is a class of algebras closed under isomorphism. Abstract data types are specified axiomatically using equations and conditional equations. Standard models of data types are constructed as initial algebras and final algebras of equational theories. The initial algebras of equational theories are supported directly by a rich and practical theory of term rewriting. To define a data type, modelled by an algebra up to isomorphism, it may be necessary to choose and add hidden operations, i.e. operations used in the axiomatisation but not present in the algebra. There is a considerable literature available on the algebraic theory of abstract data types and its applications in software and hardware engineering; it can be accessed through survey works such as Wirsing [1990], Wechler [1992], and Meinke and Tucker [1992].

Computable many sorted universal algebra has an obvious role in this theory of data types: it characterises those data types implementable on a computer. Motivated by the need to establish the scope and limits of equational specification methods for abstract data types, a systematic theory of computable many sorted universal algebras has been developed in, for example, Bergstra and Tucker [1980a, 1980b, 1983, 1987, 1995a, 1995b]. Several characterisations of computable, semicomputable and cosemicomputable algebras have been produced motivated by the needs of a theory of data.

There is an algebraic characterisation of the computable data types in terms of equational hidden function specifications that are *complete term rewriting systems*. A complete term rewriting system is one whose reductions or rewrites satisfy the *Church–Rosser* or *confluence property* and are *strongly terminating* or *Noetherian*.

THEOREM (First Completeness Theorem) (Bergstra and Tucker [1980a, 1995b]). *Let A be a finitely generated minimal Σ algebra. Then the following are equivalent*:

(1) *A is computable.*

(2) *There is a finite equational specification* (Σ_0, E_0) *such that*

  (i) $Sort(\Sigma) = Sort(\Sigma_0)$ *and* $\Sigma \subseteq \Sigma_0$;

  (ii) (Σ_0, E_0) *is a complete term rewriting system*;

  (iii) $T(\Sigma_0, E_0)|_\Sigma \cong A$.

Furthermore, the (Σ_0, E_0) *can be taken to be an orthogonal term rewriting system, whose size is independent of the algebra A, depending only on the signature Σ. More specifically, there is a fixed bound for the number of hidden functions needed in* Σ_0, *but there is no fixed bound for the number of equations needed in* E_0.

The fact that (2) implies (1) is straightforward and is a principal reason for the usefulness of the complete term rewriting system. The connections between many sorted equational term rewriting and computable and semicomputable algebras has been further developed in Bergstra and Tucker [1995b].

Another important method of specifying a data type is to use final algebra semantics. Here, an equational specification (Σ_0, E_0) can sometimes be given that has a non-trivial final object $F(\Sigma_0, E_0)$ in $Alg^*(\Sigma_0, E_0)$, which is $Alg(\Sigma_0, E_0)$ with the unit algebras removed.

THEOREM (Second Completeness Theorem) (Bergstra and Tucker [1983]). *Let A be a finitely generated minimal Σ algebra. Then the following are equivalent:*

(1) *A is computable.*

(2) *There is a finite equational specification (Σ_0, E_0) such that*
  - (i) *$Sort(\Sigma) = Sort(\Sigma_0)$, $|Sort(\Sigma)| = n$, and $\Sigma \subseteq \Sigma_0$;*
  - (ii) *Σ_0 has $3(n+1)$ hidden functions, i.e. functions that are not in Σ;*
  - (iii) *E_0 has $2(n+1)$ equations; and*
  - (iv) *both the reduct $I(\Sigma_0, E_0)|_\Sigma$ of the initial algebra $I(\Sigma_0, E_0)$, and the reduct $F(\Sigma_0, E_0)|_\Sigma$ of the final algebra $F(\Sigma_0, E_0)$ are isomorphic with A.*

In particular, the size of (Σ_0, E_0) is independent of A and, indeed, the number of hidden functions and equations is independent of the number of constants and operations in the signature Σ of A.

Characterisations of the semicomputable algebras (Bergstra and Tucker [1987, 1995b]) and cosemicomputable algebras (Bergstra and Tucker [1982] and Moss, Meseguer and Goguen [1992]) have been discovered.

Convenient surveys are contained in Meseguer and Goguen [1985], Wirsing [1990], and Stoltenberg-Hansen and Tucker [1995]. Some simple material is in the text-book Sperschneider and Antonio [1991].

5.6. *Theory of numberings*

The Gödel numberings of formulae and of the partial recursive functions provide motivations for a study of effective numberings of sets. Such a study will also clarify properties of numberings of algebras. A rich theory of effective numberings for sets has been created by Yu. Ershov and his colleagues, starting from some observations of A.I. Mal'cev. Details of this theory have been presented in Ershov [1973, 1975, 1977a], the book Ershov [1977b] and Ershov's chapter in this *Handbook*.

5.7. *Computable numbers*

The study of Analysis on the computable real and complex numbers is particularly important to the development of applications of computability theory in mathematics, and in science and engineering generally. Pour-El and Richards [1989] is an important comprehensive text on the subject and has that purpose – for a survey, see also Pour-El's chapter in this Handbook.

Of interest are the sets of computable real and complex numbers and their algebraic relationship with the fields of real and complex numbers. The computable reals form a real closed subfield of $\mathbb{R}$ and the computable complex numbers form an algebraically closed subfield of $\mathbb{C}$. The numbers we encounter through physical formulae and equation solving, such as the algebraic numbers and the transcendentals π and e, are computable. The fields of computable numbers are not computable algebras, however. Introductions to the study of the computable reals and other topological algebras are contained in Aberth [1980], Weihrauch [1987], Pour-El and Richards [1989] and Stoltenberg-Hansen and Tucker [1995].

The theory of ordered fields involves the real and complex numbers in many ways and, hence, issues concerning computable numbers. For example, there is the following theorem in Madison [1970]:

THEOREM (Madison [1970]). *Let F be an Archimedian ordered field. If F is a computable ordered field then F is isomorphic to a subfield of the recursive real numbers.*

See also Madison [1971] and Lachlan and Madison [1970].

5.8. *General algebraic framework for computations on algebras*

Let us look more closely at the structure of the concept of a computable algebra in order to see how to restrict and to generalise it. The strategy we used is as follows.

(i) Choose an algebra B as a basis for computation.

(ii) Choose a model M of computation that defines sets and functions on B.

(iii) Build algebras R from B that are computable with respect to model of computation M over B.

(iv) To measure computability in A, throw the computability of the algebra R onto A by means of a surjective homomorphism $\alpha : R \to A$; in general, we call α a *representation*.

In these circumstances, we have $A \cong R/\equiv_\alpha$ by the Homomorphism Theorem and computable aspects of the algebra A can be established with respect to the model of computation M over the algebra B.

In our treatment of computable algebra we choose B to be the algebra $(\mathbb{N}; 0, n+1)$ of natural numbers and the model M to be the μ-recursion schemes or register machines. It is possible to give equivalent definitions of a computable algebra, using representations based on other fundamental data sets that are comparable with the natural numbers, for example the set $\{0, 1\}^*$ of strings over 0, 1 and different types of Turing machines. Furthermore, it is also possible reduce or enlarge the class of algebras discussed

(i) by taking a basis and model of computation that is *weaker* than the recursive functions, e.g., primitive recursion on $\mathbb{N}$, or the polynomial time functions on $\{0, 1\}^*$; and

(ii) by taking a basis and model that is *stronger* than the recursive functions, e.g., type 1 objects over $\mathbb{N}$ and recursive type 2 functionals.

A reason for enlarging the class of algebras is to include examples of uncountable algebras; these algebras are not computable but have many algorithmic properties in need of precise analysis. For example, how do we study the effectiveness of the field $\mathbb{R}$ of real numbers? Often, we can find several ways of studying a specific structure like that of the real numbers, but what we would like to have available are systematic methods, as general as those for computable algebras. To investigate algorithmic properties of more general classes of algebras we must consider either

(a) more general models of computation for computing *exactly* in an appropriate algebra R that represents the algebra A of interest; or

(b) general methods for computing *approximately* in such R with available models of computation that results in a method of studying approximate computation in A.

We will begin by commenting on restricting the class of algebras studied by means of weaker models of computation.

5.9. *Primitive recursive algebra*

It is not difficult to define refinements of the notion of a computable algebra to investigate the complexity of computations. Consider first the idea of a *primitive recursive algebra*. We may replace the recursive sets and functions in the Definition 1.1.1 by primitive recursive sets and functions. The notion of the *primitive recursive equivalence* of two numberings, and other adaptations are needed, of course, but the form of the basic theory is similar. In fact Mal'cev [1961] defined the idea of a primitive recursive algebra and presented some other basic concepts and results, simultaneously with those of computable algebras.

The reduction to primitive recursion has the following technical features for computing in an algebra A:

(i) global search of an algebra A is no longer possible, only a limited form of bounded search is possible;

(ii) the theory is independent of representations of the set of natural numbers;

(iii) the special algebraic and logical properties of primitive recursion make the model simpler and it is easy to use the theory of primitive recursive algebras in the specification and verification of data types.

To give an example of the effect of (i) note that if a computable ring is a field then it is a computable field. However, if a primitive recursive ring is a field then it is not necessarily a primitive recursive field.

In the study of Noetherian rings and modules, primitive recursive field extensions, primitive recursive local rings and primitive recursive coherence have been used to prove the following:

THEOREM (Jacobsson and Stoltenberg-Hansen [1985]). *Let R be a local commutative Noetherian ring with maximal ideal N and let $k = R/N$ be the residue field. Let M be a finitely generated R-module. Then the Poincaré–Betti series*

$$P_R^M(z) = \sum_{n \in \mathbb{N}} \dim_k \left(Tor_i^R(M,k)\right) z^i,$$

is primitive recursive.

In their turn the primitive recursive functions can be refined by various hierarchies and corresponding hierarchies of primitive recursive algebras defined.

An early study of primitive recursion and the Grzegorczyk hierarchy in groups is Cannonito [1966, 1973]; this is developed in various papers such as Cannonito and Gatterdam [1973] and Gatterdam [1973].

5.10. *Polynomial-time algebras*

The further refinement of the class of primitive recursive algebras to classes of algebras computable in polynomial-time and other resource bounds is a subtle matter. Some basic concepts can be developed along the lines of computable algebra, but close attention must be paid to the nature of the representations of the algebras. Indeed results seem to depend heavily on representations and there is a need for a systematic abstract theory to help interpret the interesting concepts and results that are currently known.

A general study of polynomial time and space implementations of universal algebras with the theory of data types in mind is Asveld and Tucker [1982]. Universal algebras are represented by terms and (thence) strings, and algebras with polynomial growth are defined and studied. The $P = NP$ problem is generalised and Savitch's theorem proved.

A generalisation of the notion of computable algebra to that of a polynomial time algebra was made in Nerode and Remmel [1987], along with notions based on nondeterministic polynomial time, exponential and nondeterministic exponential time, with and without oracles. Roughly speaking, in Definition 1.1.1, the natural numbers are replaced by subsets of $\{0, 1\}^*$ and the functions are polynomial time operations on strings; they also use another string representation, the tally representation based on $\{0\}^*$, which is not polynomial time equivalent. Intriguing results on vector spaces over finite fields are presented in Nerode and Remmel [1987], and on finite and infinite fields in Nerode and Remmel [1990]. Results on Boolean algebras are presented in Nerode and Remmel [1987]. Further developments are Cenzer and Remmel [1991, 1992].

Just as decision problems in combinatorial group theory were (and continue to be) studied independently of a general approach to computability of algebraic structures (recall Section 5.1) so several results on the complexity of group-theoretic decision problems have been found without reference to any systematic approach to algebra.

In the early paper Lipton and Zalcstein [1977] Rabin's result on the computability of linear groups was analysed (recall the theorem in Section 5.1). There is an outline of a proof that the word problem of a finitely generated group over a field of characteristic 0 is solvable in *log* space. A consequence is that the word problem of a free group is solvable in *log* space; at present, this seems to be the only way of establishing this fact. The case of linear groups over fields of non-zero characteristic is considered in Simon [1979]; and parallel complexity models are applied to linear groups in Waack [1991].

There are a number of results showing that the word problems of certain groups can be efficiently solved (see, e.g., Domanski and Anshel [1985]). Any preassigned space complexity can be realised as the word problem of a finitely presented group: see Avenhaus and Madlener [1977], Waack [1981] and Tretkoff [1988]. However, at present, there is no natural class of groups for which the word problem has been proven to be NP-complete.

Higman's embedding theorem (recall Section 5.1) has been analysed from the point of view of complexity in Valiev [1975], Avenhaus and Madlener [1977], and Waack [1983].

Other algorithmic problems for groups have been examined in Hentzel and Jacobs [1990], Avenhaus and Madlener [1984a, 1984b], and Stewart [1991, 1992].

5.11. *Generalised computability theories and exact computation in uncountable algebras*

The uncountable algebraic systems that dominate mathematics and its applications can be investigated by stronger abstract generalised computability theories. With reference to the framework mentioned in Section 5.8, let us consider three ways in which we could compute "exactly" in an uncountable algebra A, via some representation algebra R and map $\alpha : R \to A$, by means of some generalised recursion theories on R (such as those described in Barwise [1975] and Fenstad [1980]).

(a) *Type 2 Recursion Theory.* Here we can replace the set $\mathbb{N}$ and the total recursive functions on $\mathbb{N}$ by the Baire space $[\mathbb{N} \to \mathbb{N}]$ and the total recursive functionals on $[\mathbb{N} \to \mathbb{N}]$. This approach to effective algebra has been pursued in unpublished work of J.P. Cleave (in 1972) and, independently, by K. Weihrauch who has developed a good generalisation that has interesting applications (e.g., to the real numbers). See Weihrauch [1985, 1987, 1993, 1996]; see also Kreitz and Weihrauch [1984].

(b) *Admissible Recursion Theory.* Here we can replace the set $\mathbb{N}$ and the total recursive functions on $\mathbb{N}$ by an admissible ordinal α and the α-recursive functions (see Sacks [1990]), or more generally by an admissible set. Results known for computable rings and fields (in the sense above) can be lifted to this very general setting. For example, Theorems 3.1.11 and 3.3.2 become the following:

THEOREM. *Let F be a perfect field with cardinality κ that is κ-computable. Then the algebraic closure A of F is κ-computable and, furthermore, A is κ-computably unique if, and only if, F has a κ-computable splitting algorithm.*

(c) *Recursion Theory over Algebras.* Here we can replace the set $\mathbb{N}$ and the total recursive functions on $\mathbb{N}$ by *any* universal algebra B and the functions on B that are computable by means of finite deterministic models of computation, such as **while**-*programs with arrays*, or one of its many equivalents. This generalisation of finite computability is closely connected with programming language theory and the classical theory of the recursive functions, and has a large literature with early contributions, starting in the 1950s, by Y. Ianov, L. Kaloujnine, H. Thiele, E. Engeler, H. Friedman and Y. Moschovakis. Most recently, the computability theory for reals in Blum, Shub and Smale [1989] and Blum, Cucker, Shub and Smale [1998] is a special case of this theory.

The abstract theory of finite computations on universal algebras can be surveyed, for example, by consulting Shepherdson [1985], Tucker [1980b, 1991], Tucker and Zucker [1988, 1994, 1999].

Although much is known about these three recursion theories, there is a great deal to do in developing the generalisations for analysing computation in algebras.

5.12. *Computable approximations of topological algebras*

The methods for the computable approximation of the real and complex numbers seem to be specific to the special nature of the numbers. Computable approximations are needed for other topological algebras. For example, some inverse limits of algebraic systems give rise to topological algebras that are ultrametric algebras. Examples include power series rings, completions of local rings, and profinite groups, which play a key role in the Galois theory of infinite extension fields.

The effective properties of these topological algebras have been studied independently, for example: in Suter [1973] and Stoltenberg-Hansen and Tucker [1988] on local rings and by Lin [1981b] on profinite groups. These and several other studies have many similarities in results; in particular, substructures consisting of computable elements are studied. Can they be unified? Some obvious general questions are:

QUESTION. What is involved in approximating the elements of an arbitrary topological algebra by means of computable elements? What are the algebraic and effective properties of the set of computable elements?

In Stoltenberg-Hansen and Tucker [1995], a method for the systematic study of effective approximations of uncountable topological algebras is presented. It is based on representing topological algebras using algebras built from *domains* and applying the *theory of effective domains*. This method of applying domain theory to approximation problems in mathematics was first developed for topological algebras and used on completions of local rings in Stoltenberg-Hansen and Tucker [1985, 1988] and further developed for universal algebras in Stoltenberg-Hansen and Tucker [1991, 1993, 1995]; see also Stoltenberg-Hansen et al. [1994], Chapter 8.

Suppose that A is a topological algebra, i.e. an algebra whose carrier set is a topological space and whose operations are continuous. The idea is to build an algebra R that represents A by means of the continuous representation map $\upsilon : R \to A$. How do we computably approximate R?

We imagine building R from a set P of approximating data that is a computable structure. Each datum in R is approximated by some sequence $(a_i)_{i \in I}$ of data from P. More specifically, R is a topological space obtained from P by some form of completion process in which the set P is dense in R. The key feature of this approach is that, since P is computable, some of the approximating sequences are computable. Let R_k be the set of elements in R that are computably approximable. This set is the basis of the computable approximation of R and hence of A. We usually use a special type of approximating structure P called a *conditional upper semilattice* (*cusl*) and a completion process called *ideal completion*. This process yields an *algebraic domain*. The method effectively approximates a large class of examples: ultrametric algebras, locally compact Hausdorff algebras (Stoltenberg-Hansen and Tucker [1995]), and complete metric algebras (Blanck [1996]).

The method is related to Weihrauch's generalised computability theory; see also Weihrauch and Schreiber [1981]. Similar ideas have been used in Edalat [1995a, 1995b], applying continuous domains to analytical questions, such as integration and measure; see also the survey Edalat [1997].

These general approaches, involving algebraic domains, continuous domains and type two recursion on Baire space, are complemented by more specific theories such as effective metric spaces (Moschovakis [1964]) and Banach spaces (Pour-El and Richards [1989]). In Stoltenberg-Hansen and Tucker [1999] these approaches are proved to be equivalent in commonly occurring circumstances.

5.13. *Constructive algebra*

The development of logical theories of constructive existence and of intuitionistic algebra is also necessary to examine in order to complete the picture: see Troelstra [1988, 1991], Troelstra and van Dalen [1988a, 1988b], and Beeson [1985]. Work on constructive aspects of algebra since the time of Kronecker is somewhat scattered and includes Molk [1885] and Vandiver [1936]. The subject is partly surveyed in Bridges and Richman [1987] and Mines, Richman and Ruitenburg [1988].

5.14. *Other matters*

There are a number of other subjects that might form part of this review but we do not consider (for example: the application of the theory of isols to algebra, and differential algebra). In addition, the field of computer algebra and symbolic computation is well developed and can be accessed through Becker and Weispfenning [1993] and Adams and Loustaunau [1994].

6. Concluding remarks

The theory of computable rings, fields and modules, and some of the other theories reviewed in the Further Reading, are fundamental subjects. There are many open problems in these areas and, indeed, some of the areas are barely developed. There is evidence that these subjects can be connected but that we are some way from having smooth and convenient links between the worlds of computable discrete algebra and effective continuous algebra. There are also bridges to be made between effective continuous algebra and the world of applications in science and engineering.

Effective Algebra offers its students theoretical depth and scope, many areas of application, and scientific longevity. We hope this chapter has provided an introduction that is satisfying, stimulating and pleasurable.

References

N. H. ABEL

[1824] *Mémoire sur les Équationes Résolubles Algébriques, ou l'on Démontre l'Impossibilité de la Résolution de l'Équation Générale du Cinquième Degré*, Christiania.

[1881] Mémoire sur les équationes résolubles algébriques, ou l'on démontre l'impossibilité de la résolution de l'équation générale du cinquième degré, in: *Ouevres Complètes*, L. Sylow and S. Lie, eds., Cristiania, pp. 28–33.

O. ABERTH

[1980] *Computable Analysis*, McGraw-Hill, New York.

W. W. ADAMS AND P. LOUSTAUNAU

[1994] *An Introduction to Gröbner Bases*, Graduate Studies in Mathematics, Vol. 3, Amer. Math. Soc., Providence, RI.

S. I. ADIAN, W. W. BOONE AND G. HIGMAN (EDS.)

[1980] *Word Problems II*, North-Holland, Amsterdam.

D. ALTON AND E. W. MADISON

[1973] Computability of Boolean algebras and their extensions, *Ann. Math. Logic*, 6, pp. 95–128.

E. ARTIN AND O. SCHREIER

[1927] Algebraischen Konstruktion reeler Körper, *Abh. Math. Sem. Univ. Hamburg*, 5, pp. 85–99.

C. J. ASH AND R. G. DOWNEY

[1984] Decidable subspaces and recursively enumerable subspaces, *J. Symbolic Logic*, 49, pp. 1137–1145.

P. R. J. ASVELD AND J. V. TUCKER

[1982] Complexity theory and the operational structure of algebraic programming systems, *Acta Inform.*, 17, pp. 451–476.

M. ATIYAH AND I. MACDONALD

[1969] *Introduction to Commutative Algebra*, Addison-Wesley, New York.

J. AVENHAUS AND K. MADLENER

[1977] Subrekursive Komplexität bei Gruppen, I. Gruppen mit vorgeschriebener Komplexität, *Acta Inform.*, 9, pp. 87–104.

[1978] Subrekursive Komplexität bei Gruppen, II. Der Einbettungssatz von Higman für entscheidbare Gruppen, *Acta Inform.*, 9, pp. 183–193.

[1984a] The Nielsen reduction and P-complete problems in free groups, *Theoret. Comput. Sci.*, 32, pp. 61–76.

[1984b] On the complexity of intersection and conjugacy problems in free groups, *Theoret. Comput. Sci.*, 32, pp. 279–295.

C. W. AYOUB

[1983] On constructing bases for ideals in polynomial rings over the integers, *J. Number Theory*, 17, pp. 204–225.

C. BABBAGE

[1989] *Collected Works. Volume III: Analytical Engine*, W. Pickering, London.

K. J. BARWISE

[1975] *Admissible Sets and Structures*, Perspectives in Mathematical Logic, Springer, Berlin.

[1977] (ed.) *Handbook of Mathematical Logic*, Studies in Logic, Vol. 90, North-Holland, Amsterdam.

G. BAUMSLAG, F. B. CANNONITO AND C. F. MILLER III

[1981] Computable algebra and group embeddings, *J. Algebra*, 69, pp. 186–212.

G. BAUMSLAG AND C. F. MILLER III (EDS.)

[1992] *Algorithms and Classification in Combinatorial Group Theory*, Springer, Berlin.

W. BAUR

[1974] Rekursive Algebren mit Kettenbedingungen, *Zeits. Math. Logik Grundl. Math.*, 20, pp. 37–46.

T. BECKER AND V. WEISPFENNING

[1993] *Gröbner Bases*, Springer, Berlin.

M. J. BEESON

[1985] *Foundations of Constructive Mathematics*, Springer, Berlin.

J. A. BERGSTRA AND J. V. TUCKER

[1980a] A characterisation of computable data types by means of a finite equational specification method, in: *Automata, Languages and Programming (ICALP), 7th Colloquium, Noordwijkerhout*, J. W. de Bakker and J. van Leeuwen, eds., Lecture Notes in Comput. Sci., Vol. 81, Springer, Berlin, pp. 76–90.

[1980b] A natural data type with a finite equational final semantics specification, but no effective equational initial specification, *Bulletin of the EATCS*, 11, pp. 23–33.

[1982] The completeness of the algebraic specification methods for data types, *Inform. and Control*, 54, pp. 186–200.

[1983] Initial and final algebra semantics for data type specifications: two characterisation theorems, *SIAM J. Comput.*, 12, pp. 366–387.

[1987] Algebraic specifications of computable and semicomputable data types, *Theoret. Comput. Sci.*, 50, pp. 137–181.

[1993] Equational specifications for computable data types: 6 hidden functions suffice and other sufficiency bounds, in: *Many Sorted Logic and its Applications*, K. Meinke and J. V. Tucker, eds., Wiley, pp. 89–102.

[1995a] The data type variety of stack algebras, *Ann. Pure Appl. Logic*, 73, pp. 11–36.
[1995b] Equational specifications, complete term rewriting systems, and computable and semicomputable algebras, *J. ACM*, 42, pp. 1194–1230.

J. BLANCK
[1996] Domain representability of metric spaces, *Ann. Pure Appl. Logic*, 83, pp. 225–247.

L. BLUM, F. CUCKER, M. SHUB AND S. SMALE
[1998] *Complexity and Real Computation*, Springer, Berlin.

L. BLUM, M. SHUB AND S. SMALE
[1989] On a theory of computation and complexity over the real numbers: NP-completeness, recursive functions, and universal machines, *Bull. Amer. Math. Soc.*, 21, pp. 1–46.

W. W. BOONE
[1959] The word problem, *Ann. of Math.*, 70, pp. 207–265.
[1966] Word problems and recursively enumerable degrees of unsolvability, A sequel on finitely presented groups, *Ann. of Math.*, 84, pp. 49–84.
[1971] Word problems and recursively enumerable degrees of unsolvability. An emendation, *Ann. of Math.*, 94, pp. 389–391.

W. W. BOONE, F. B. CANNONITO AND R. C. LYNDON (EDS.)
[1973] *Word Problems*, North-Holland, Amsterdam.

W. W. BOONE AND G. HIGMAN
[1974] Algebraic characterization of the solvability of the word problem, *J. Austral. Math. Soc.*, 18, pp. 41–53.

N. BOURBAKI
[1961] *Algèbre Commutative: I. Modules Plats*, Hermann, Paris.

D. BRIDGES AND F. RICHMAN
[1987] *Varieties of Constructive Mathematics*, London Mathematical Society Lecture Notes Series, Vol. 97, Cambridge Univ. Press, Cambridge.

S. BURRIS AND H. P. SANKAPPANAVAR
[1981] *A Course in Universal Algebra*, Springer, Berlin.

F. B. CANNONITO
[1966] Hierarchies of computable groups and the word problem, *J. Symbolic Logic*, 31, pp. 376–392.
[1973] The algebraic invariance of the word problem in groups, in: *Word Problems*, W. W. Boone, F. B. Cannonito and R. C. Lyndon, eds., Studies in Logic, Vol. 71, North-Holland, Amsterdam, pp. 349–364.

F. B. CANNONITO AND R. W. GATTERDAM
[1973] The computability of group constructions, part I, in: *Word Problems*, W. W. Boone, F. B. Cannonito and R. C. Lyndon, eds., Studies in Logic, Vol. 71, North-Holland, Amsterdam, pp. 365–400.

D. CENZER AND J. B. REMMEL
[1991] Polynomial time versus recursive models, *Ann. Pure Appl. Logic*, 54, pp. 17–58.
[1992] Polynomial time abelian groups, *Ann. Pure Appl. Logic*, 56, pp. 313–363.

B. CHANDLER AND W. MAGNUS
[1982] *The History of Combinatorial Group Theory: A Case Study in the History of Ideas*, Springer, New York.

C. CHEVALLEY
[1946] *Theory of Lie Groups*, Princeton Univ. Press, Princeton, N.J.

C. R. J. Clapham

[1964] Finitely presented groups with word problems of arbitrary degrees of insolubility, *Proc. London Math. Soc. (3)*, 14, pp. 633–676.

P. M. Cohn

[1977] *Algebra*, Vol. 2, Wiley, Chichester.

[1981] *Universal Algebra*, D. Reidel, Dordrecht.

D. J. Collins

[1971] Truth table degrees and the Boone groups, *Ann. of Math.*, 94, pp. 392–396.

N. J. Cutland

[1980] *Computability: An Introduction to Recursive Function Theory*, Cambridge Univ. Press, Cambridge.

M. Davis

[1977] Unsolvable problems, in: *Handbook of Mathematical Logic*, K. J. Barwise, ed., Studies in Logic, Vol. 90, North-Holland, Amsterdam, pp. 567–594.

R. Dedekind

[1888] *Was sind und was sollen die Zahlen?*, Vieweg, Braunschweig, 1888. Reprinted in: *Essays on the Theory of Numbers*, Dover, New York, 1963.

M. Dehn

[1910] Über die Topologie des dreidimensionalen Raumes, *Math. Ann.*, 69, pp. 137–168.

[1911] Über unendliche diskontinuerliche Gruppen, *Math. Ann.*, 71, pp. 116–144.

[1987] *Papers on Group Theory and Topology*, Springer, New York.

V. P. Dobrica

[1981] Constructivizable abelian groups, *Sibirsk. Mat. Zh.*, 22, pp. 208–213.

V. P. Dobrica, A. T. Nurtazin and N. G. Khisamiev

[1978] Constructive periodic abelian groups, *Sibirsk. Mat. Zh.*, 19, pp. 1260–1265.

B. Domanski and M. Anshel

[1985] The complexity of Dehn's algorithm for word problems in groups, *J. Algorithms*, 6, pp. 543–549.

W. Dyck

[1882] Gruppentheoretische Studien, *Math. Ann.*, 20, pp. 1–44.

A. Edalat

[1995a] Domain theory and integration, *Theoret. Comput. Sci.*, 151, pp. 163–193.

[1995b] Dynamical systems, measures and fractals via domain theory, *Inform. and Comput.*, 120, pp. 32–48.

[1997] Domains for computation in mathematics, physics and exact real arithmetic, *Bull. Symbolic Logic*, 3, pp. 401–452.

H. M. Edwards

[1984] *Galois Theory*, Graduate Texts in Mathematics, Vol. 101, Springer, Berlin.

H. B. Enderton

[1977] Elements of recursion theory, in: *Handbook of Mathematical Logic*, K. J. Barwise, ed., Studies in Logic, Vol. 90, North-Holland, Amsterdam, pp. 527–566.

Yu. L. Ershov

[1973] Theorie der Numerierungen I, *Zeits. Math. Logik Grundl. Math.*, 19, pp. 289–388.

[1975] Theorie der Numerierungen II, *Zeits. Math. Logik Grundl. Math.*, 21, pp. 473–584.

[1977a] Theorie der Numerierungen III, *Zeits. Math. Logik Grundl. Math.*, 23, pp. 289–371.

[1977b] *Theory of Numberings* (in Russian), Nauka, Moscow.

S. T. FEDORYAEV

[1993] Countability of widths of algebraic reducibility structures for models in some classes, *Siberian Adv. Math.*, 3, pp. 81–102.

S. FEFERMAN

[1975] Impredicativity of the existence of the largest divisible sub-group of an abelian p-group, in: *Model Theory and Algebra*, D. H. Saracino and V. B. Weispfenning, eds., Lecture Notes in Mathematics, Vol. 498, Springer, Berlin, pp. 117–130.

J. E. FENSTAD

[1980] *General Recursion Theory: An Axiomatic Approach*, Springer, Berlin.

A. A. FRIDMAN

[1962] Degrees of unsolvability of identity in finitely presented groups, *Soviet Math.*, 3, pp. 1733–1737.

A. FRÖHLICH AND J. C. SHEPHERDSON

[1955] On the factorization of polynomials in a finite number of steps, *Math. Z.*, 62, pp. 331–334.

[1956] Effective procedures in field theory, *Philos. Trans. Royal Soc. London Ser. A*, 248, pp. 407–432.

E. GALOIS

[1846] Mémoire sur les conditions de résolubilité des équations par radicaux (dated 6 January 1831, edited by J. Liouville), *Journal de Mathématiques*, 11, pp. 381–444.

R. GANDY

[1988] The confluence of ideas in 1936, in: *The Universal Turing Machine*, R. Herkin, ed., Oxford Univ. Press, Oxford, pp. 55–111.

R. W. GATTERDAM

[1973] The Higman theorem for primitive recursive groups – a preliminary report, in: *Word Problems*, W. W. Boone, F. B. Cannonito and R. C. Lyndon, eds., Studies in Logic, Vol. 71, North-Holland, Amsterdam, pp. 421–425.

S. S. GONCHAROV

[1980a] Computable univalent numerations, *Algebra i Logika*, 19, pp. 507–551.

[1980b] The problem of the number of nonautoequivalent constructivisations, *Algebra i Logika*, 19, pp. 621–639.

[1981] Groups with a finite number of constructivisations, *Soviet Math. Doklady*, 25, pp. 58–61.

R. L. GOODSTEIN

[1961] *Recursive Analysis*, North-Holland, Amsterdam.

G. GRÄTZER

[1979] *Universal Algebra*, Springer, Berlin.

D. GUICHARD

[1983] Automorphisms of substructure lattices in recursive algebra, *Ann. Pure Appl. Logic*, 25, pp. 47–58.

L. A. HARRINGTON, M. D. MORLEY, A. SCEDROV AND S. G. SIMPSON (EDS.)

[1985] *Harvey Friedman's Research on the Foundations of Mathematics*, North-Holland, Amsterdam.

K. HENSEL

[1908] *Theorie der Algebraischen Zahlen*, Leipzig.

I. R. HENTZEL AND D. P. JACOBS

[1990] Complexity and unsolvability properties of nilpotency, *SIAM J. Comput.*, 19, pp. 32–43.

K. HENZELT

[1922] Zur Theorie den Polynomideale und Resultanten (Bearbeitet von E. Noether), *Math. Ann.*, 88, pp. 53–79.

G. HERMANN

[1926] Die Frage der endlichen vielen Schritte in der Theorie der Polynomideale, *Math. Ann.*, 95, pp. 736–788.

G. HIGMAN

[1961] Subgroups of finitely presented groups, *Proc. Royal Soc. London Ser. A*, 262, pp. 455–474.

P. HINGSTON

[1981] Effective decomposition in Noetherian rings, in: *Aspects of Effective Algebra* (*Proceedings of a Conference at Monash University, Australia, 1–4 August, 1979*), J. N. Crossley, ed., Upside Down A Book Company, Steel's Creek, Australia, pp. 122–127.

C. JACOBSSON AND V. STOLTENBERG-HANSEN

[1985] Poincaré–Betti series are primitive recursive, *J. London Math. Soc.*, 31, pp. 1–9.

I. KALANTARI AND A. RETZLAFF

[1977] Maximal vector spaces under automorphisms of the lattice of recursively enumerable vector spaces, *J. Symbolic Logic*, 42, pp. 481–491.

R. KAYE

[1991] *Models of Peano Arithmetic*, Oxford Logic Guide, Vol. 15, Oxford Univ. Press, Oxford.

N. G. KHISAMIEV

[1987] Nonconstructivizibility of some ordered fields of real numbers, *Sibirsk. Mat. Zh.*, 28, pp. 193–195.

[1992] Constructive abelian p-groups, *Siberian Adv. Math.*, 2, pp. 68–113.

S. C. KLEENE

[1981] Origins of recursive function theory, *Ann. Hist. Comput.,* 3, pp. 52–67.

C. KREITZ AND K. WEIHRAUCH

[1984] A unified approach to constructive and recursive analysis, in: *Computation and Proof Theory*, E. Börger, W. Oberschelp, M. M. Richter, B. Schinzel and W. Thomas, eds., Lecture Notes in Mathematics, Vol. 1104, Springer, Berlin, pp. 259–278.

L. KRONECKER

[1882] Grundzüge einer arithmetischen Theorie der algebraischen Grössen, *J. Reine Angew. Math.*, 92, pp. 1–122.

W. KRULL

[1928] Galoische Theorie der unendlichen algebraischen Erweiterungen, *Math. Ann.*, 100, pp. 687–698.

[1953a] Über Polynomzerlegung mit endlich vielen Schritten I, *Math. Z.*, 59, pp. 57–60.

[1953b] Über Polynomzerlegung mit endlich vielen Schritten II, *Math. Z.*, 59, pp. 296–298.

[1954] Über Polynomzerlegung mit endlich vielen Schritten III, *Math. Z.*, 60, pp. 109–111.

O. V. KUDINOV

[1993] Algebraic dependences and reducibilities of constructivizations in universal domains, *Siberian Adv. Math.*, 3, pp. 121–128.

[1996] *An integral domain with finite algorithmic dimension*, Manuscript.

A. V. KUZNETSOV

[1958] Algorithms as operations in algebraic systems, *Uspekhi Mat. Nauk*, 13, pp. 240–241.

G. KÖNIG

[1903] *Einleitung in die Allgemeine Theorie der Algebraischen Größen*, Leipzig.

A. H. LACHLAN AND E. W. MADISON
[1970] Computable fields and arithmetically definable ordered fields, *Proc. Amer. Math. Soc.*, 24, pp. 803–807.

J. L. LAGRANGE
[1772] Réflexions sur la résolution algébrique des équations, *Nouv. Mém. Acad. Berlin, pour les annés 1770/71*, Berlin, pp. 203–421.

G. LALLEMENT
[1979] *Semigroups and Combinatorial Applications*, Wiley, Chichester.

S. LANG
[1965] *Algebra*, Addison-Wesley, Reading, MA.

D. LAZARD
[1976] Algorithmes fondamentaux en algèbre commutative, *Asterisque*, 38–39, pp. 131–138.

C. LIN
[1981a] The effective content of Ulm's theorem, in: *Aspects of Effective Algebra* (*Proceedings of a Conference at Monash University, Australia, 1–4 August, 1979*), J. N. Crossley, ed., Upside Down A Book Company, Steel's Creek, Australia, pp. 147–160.
[1981b] Recursively presented Abelian groups: Effective p-group theory I, *J. Symbolic Logic*, 46, pp. 617–624.

R. J. LIPTON AND Y. ZALCSTEIN
[1977] Word problems solvable in logspace, *J. Assoc. Comput. Mach.*, 24, pp. 522–526.

R. C. LYNDON AND P. E. SCHUPP
[1977] *Combinatorial Group Theory*, Springer, Berlin.

M. MACHTEY AND P. R. YOUNG
[1978] *An Introduction to the General Theory of Algorithms*, Elsevier, Amsterdam.

A. MACINTYRE
[1972] On algebraically closed groups, *Ann. of Math.*, 96, pp. 53–97.

E. W. MADISON
[1970] A note on computable real fields, *J. Symbolic Logic*, 35, pp. 239–241.
[1971] Some remarks on computable (non-archimedian) ordered fields, *J. London Math. Soc.*, 4, pp. 304–308.

W. MAGNUS, A. KARASS AND D. SOLITAR
[1976] *Combinatorial Group Theory*, Dover.

A. I. MAL'CEV
[1961] Constructive algebra I, *Russian Math. Surv.*, 16, pp. 77–129.
[1970] *Algorithms and Recursive Functions*, Wolters-Noordhoff, Groningen.
[1971a] *The Metamathematics of Algebraic Systems. Collected Papers: 1936–1967*, North-Holland, Amsterdam.
[1971b] Constructive algebras I, in: *The Metamathematics of Algebraic Systems. Collected Papers: 1936–1967*, B. F. Wells III, ed., North-Holland, Amsterdam, pp. 148–212.
[1971c] Recursive abelian groups, in: *The Metamathematics of Algebraic Systems. Collected Papers: 1936–1967*, B. F. Wells III, ed., North-Holland, Amsterdam, pp. 282–286.

A. A. MARKOV
[1947] On the impossibility of certain algorithms in the theory of associative systems, *Dokl. Akad. Sci. USSR*, 55, pp. 583–586.

K. MCALOON
[1982] On the complexity of models of arithmetic, *J. Symbolic Logic*, 47, pp. 403–415.

R. N. McKenzie, G. F. McNulty and W. F. Taylor
[1987] *Algebras, Lattices, Varieties*, Vol. 1, Wadsworth and Brooke/Cole, Monterey.

K. Meinke and J. V. Tucker
[1992] Universal algebra, in: *Handbook of Logic in Computer Science, Vol. 1: Mathematical Structures*, S. Abramsky, D. Gabbay and T. S. E. Maibaum, eds., Oxford Univ. Press, Oxford, pp. 189–411.

J. Meseguer and J. A. Goguen
[1985] Initiality, induction and computability, in: *Algebraic Methods in Semantics*, M. Nivat and J. Reynolds, eds., Cambridge Univ. Press, Cambridge, pp. 459–541.

G. Metakides and A. Nerode
[1977] Recursively enumerable vector spaces, *Ann. Math. Logic*, 11, pp. 147–171.
[1979] Effective content of field theory, *Ann. Math. Logic*, 17, pp. 289–320.
[1980] Recursion theory on fields and abstract dependence, *J. Algebra*, 65, pp. 36–59.
[1982] The introduction of non-recursive methods into mathematics, in: *The L. E. J. Brouwer Centenary Symposium,* A. S. Troelstra and D. van Dalen, eds., North-Holland, Amsterdam, pp. 319–335.

C. F. Miller III
[1971] *On Group-Theoretic Decision Problems and their Classification*, Princeton Univ. Press.
[1992] Decision problems for groups. Surveys and reflections, in: *Algorithms and Classification in Combinatorial Group Theory*, G. Baumslag and C. F. Miller III, eds., MSRI Publications, Vol. 23, Springer, Berlin.

R. Mines, F. Richman and W. Ruitenburg
[1988] *A Course in Constructive Algebra*, Springer, New York.

J. Moldestad, V. Stoltenberg-Hansen and J. V. Tucker
[1980a] Finite algorithmic procedures and inductive definability, *Math. Scand.*, 46, pp. 62–76.
[1980b] Finte algorithmic procedures and computation theories, *Math. Scand.*, 46, pp. 77–94.

J. Molk
[1885] Sur une notion qui comprend celle de la divisibilité et sur la théorie de l'elimination, *Acta Math.*, 6, pp. 1–166.

Y. N. Moschovakis
[1964] Recursive metric spaces, *Fund. Math.*, 55, pp. 215–238.

L. Moss, J. Meseguer and J. A. Goguen
[1992] Final algebras, cosemicomputable algebras, and degrees of unsolvability, *Theoret. Comput. Sci.*, 100, pp. 267–302.

A. Nerode and J. B. Remmel
[1983] Recursion theory on matroids II, in: *Proceedings of the Singapore Logic Symposium. Southeast Asian Conference on Logic*, C. T. Chong and M. S. Wicks, eds., North-Holland, Amsterdam, pp. 133–184.
[1985] A survey of r.e. substructures, in: *Recursion Theory*, A. Nerode and R. A. Shore, eds., Proceedings of Symposia in Pure Mathematics 42, Amer. Math. Soc., Providence, RI, pp. 323–375.
[1987] Complexity theoretic algebra I: Vector spaces over finite fields, in: *Proceedings of the 2nd Annual Conference on Structures in Complexity Theory*, IEEE, Silver Spring, pp. 218–239.
[1989] Complexity theoretic algebra II: Boolean algebras, *Ann. Pure Appl. Logic*, 44, pp. 71–99.
[1990] Complexity-theoretic algebra: Vector space bases, in: *Feasible Mathematics*, S. Buss and P. Scott, eds., Springer, Berlin, pp. 293–319.

A. NERODE AND R. SMITH

[1982] The undecidability of the lattice of r.e. subspaces, in: *Proceedings of the Third Brazilian Conference on Mathematical Logic*, A. I. Arruda, N. C. A. DiCosta and A. M. Sette, eds., pp. 245–252.

B. H. NEUMANN

[1973] The isomorphism problem for algebraically closed groups, in: *Word Problems*, W. W. Boone, F. B. Cannonito and R. C. Lyndon, eds., North-Holland, Amsterdam, pp. 553–562.

P. S. NOVIKOV

[1955] On the algorithmic unsolvability of the word problem for group theory, in: *Trudy Mat. Inst. Steklov*, Vol. 44, p. 143 (in Russian).

L. NOVY

[1973] *Origins of Modern Algebra*, Noordhoff, Leiden.

P. ODIFREDDI

[1989] *Classical Recursion Theory*, Studies in Logic, Vol. 125, North-Holland, Amsterdam.

G. PEACOCK

[1830] *A Treatise on Algebra*, J. & J. J. Deighton, Cambridge.

I. C. C. PHILLIPS

[1992] Recursion theory, in: *Handbook of Logic in Computer Science, Vol. 1: Mathematical Structures*, S. Abramsky, D. Gabbay and T. S. E. Maibaum, eds., Oxford Univ. Press, Oxford, pp. 79–187.

E. L. POST

[1947] Recursive unsolvability of a problem of Thue, *J. Symbolic Logic*, 12, pp. 1–11.

M. B. POUR-EL AND J. I. RICHARDS

[1989] *Computability in Analysis and Physics*, Perspectives in Mathematical Logic, Springer, Berlin.

M. O. RABIN

[1960a] Computable algebra, general theory and theory of computable fields, *Trans. Amer. Math. Soc.*, 95, pp. 341–360.

[1960b] Computable algebraic systems, in: *Summer Institute for Symbolic Logic, Cornell University, 1957*, 2nd ed., Institute for Defence Analyses, pp. 134–138.

C. REID

[1970] *Hilbert*, Springer, Berlin.

J. B. REMMEL

[1989] Recursive Boolean algebras, in: *Handbook of Boolean Algebra, Vol. III*, J. D. Monk, ed., North-Holland, Amsterdam, pp. 1099–1165.

A. RETZLAFF

[1979] Direct summands of r.e. vector spaces, *Zeits. Math. Logik Grundl. Math.*, 25, pp. 363–372.

F. RICHMAN

[1974] Constructive aspects of Noetherian rings, *Proc. Amer. Math. Soc.*, 44, pp. 436–441.

H. ROGERS

[1967] *Theory of Recursive Functions and Effective Computability*, McGraw-Hill, New York.

J. ROTMAN

[1995] *An Introduction to the Theory of Groups*, Allyn and Bacon, Boston.

P. RUFFINI

[1779] *Teoria generale delle equazioni, in cui si dimostra impossibile la soluzione algebraica delle equazioni generali di grado superiore alo quarto*, 1779. Reprinted in: *Opere matematiche*, E. Bortolotti, ed., Tomo primo, Palermo, pp. 1–324.

G. E. SACKS

[1990] *Higher Recursion Theory*, Perspectives in Mathematical Logic, Springer, Berlin.

A. SEIDENBERG

[1974a] Constructions in algebra, *Trans. Amer. Math. Soc.*, 197, pp. 273–313.

[1974b] What is Noetherian?, *Rendiconti del Seminario Matematico e Fisico di Milano*, 44, pp. 55–61.

[1985] Survey of constructions in Noetherian rings, in: *Recursion Theory*, A. Nerode and R. A. Shore, eds., Proceedings of Symposia in Pure Mathematics, Vol. 42, Amer. Math. Soc., Providence, RI, pp. 377–386.

J. C. SHEPHERDSON

[1964] A non-standard model for a free variable fragment of number theory, *Bull. Acad. Polon. Sci., Sér. Sci., Math., Astron. Phys.*, 12, pp. 79–86.

[1965] Non-standard models for fragments of number theory, in: *Theory of Models*, J. W. Addison, ed., North-Holland, Amsterdam, pp. 342–358.

[1985] Algorithmic procedures, generalised Turing algorithms, and elementary recursion theory, in: *Harvey Friedman's Research on the Foundations of Mathematics*, L. A. Harrington, M. D. Morley, A. Scedrov and S. G. Simpson, eds., North-Holland, Amsterdam, pp. 285–308.

R. SHORE

[1978] Controlling the dependence degree of a recursively enumerable vector space, *J. Symbolic Logic*, 43, pp. 13–22.

H. U. SIMON

[1979] Word problems for groups and contextfree recognition, in: *Proceedings of FCT'79*, Akademie-Verlag, Berlin, pp. 417–422.

R. L. SMITH

[1981a] Effective valuation theory, in: *Aspects of Effective Algebra* (*Proceedings of a Conference at Monash University, Australia, 1–4 August*, 1979), J. N. Crossley, ed., Upside Down A Book Company, Steel's Creek, Australia, pp. 232–245.

[1981b] Two theorems on autostability in p-groups, in: *Logic Year 1979–80*, M. Lerman, J. H. Schmerl and R. I. Soare, eds., Lecture Notes in Mathematics, Vol. 859, Springer, Berlin, pp. 302–311.

R. SOARE

[1987] *Recursively Enumerable Sets and Degrees*, Perspectives in Mathematical Logic, Springer, Berlin.

V. SPERSCHNEIDER AND G. ANTONIO

[1991] *Logic. A Foundation for Computer Science*, Addison-Wesley.

E. STEINITZ

[1910] Algebraische Theorie der Körper, *Journal für Mathematik*, 137, pp. 167–309.

I. STEWART

[1973] *Galois Theory*, Chapman and Hall.

I. A. STEWART

[1991] Complete problems for symmetric logspace involving free groups, *Inform. Process. Lett.*, 40, pp. 263–267.

[1992] Refining known results on the generalized word problem for free groups, *Internat. J. Algebra Comput.*, 2, pp. 221–236.

J. STILLWELL

[1982] The word problem and the isomorphism problem for groups, *Bull. Amer. Math. Soc.*, 6, pp. 33–56.

V. STOLTENBERG-HANSEN, I. LINDSTRÖM AND E.R. GRIFFOR

[1994] *Mathematical Theory of Domains*, Cambridge Tracts in Theoretical Computer Science, Vol. 22, Cambridge Univ. Press.

V. STOLTENBERG-HANSEN AND J. V. TUCKER

[1980] Computing roots of unity in fields, *Bull. London Math. Soc.*, 12, pp. 463–471.

[1985] *Complete Local Rings as Domains*, Centre for Theoretical Computer Science Reports, University of Leeds, Report 1.85, Leeds.

[1988] Complete local rings as domains, *J. Symbolic Logic*, 53, pp. 603–624.

[1991] Algebraic equations and fixed-point equations in inverse limits, *Theoret. Comput. Sci.*, 87, pp. 1–24.

[1993] Infinite systems of equations over inverse limits and infinite synchronous concurrent algorithms, in: *Semantics – Foundations and Applications*, J. W. de Bakker, G. Rozenberg, and W. P. de Roever, eds., Lecture Notes in Comput. Sci., Vol. 666, Springer, pp. 531–562.

[1995] Effective algebras, in: *Handbook of Logic in Computer Science, Vol. IV: Semantic Modelling*, S. Abramsky, D. M. Gabbay and T. S. E. Maibaum, eds., Oxford Univ. Press, Oxford, pp. 357–526.

[1999] Concrete models of computation for topological algebras, *Theoret. Comp. Sci.*, to appear.

C.-F. STURM

[1835] Mémoire sur la résolution des équations numériques, *Annales de Mathématiques Pures et Appliquées*, 6, pp. 271–318.

J.-P. SOUBLIN

[1970] Anneaux et modules cohérents, *J. Algebra*, 15, pp. 455–472.

G. H. SUTER

[1973] Recursive elements and constructive extensions of computable local integral domains, *J. Symbolic Logic*, 38, pp. 272–290.

S. TENNENBAUM

[1959] Non-archimedian models for arithmetic, *Notices Amer. Math. Soc.*, 6, p. 270.

A. THUE

[1914] Probleme über Veränderungen von Zeichenreihen nach gegebenen Regeln, *Skr. Vid. Kristiania, I Mat. Naturv. Klasse*, 10, p. 34.

[1977] *Axel Thue's Selected Works*, University of Oslo Press, Oslo.

C. TRETKOFF

[1988] Complexity, combinatorial group theory and the language of palutators, *Theoret. Comput. Sci.*, 56, pp. 253–275.

A. S. TROELSTRA

[1988] *On the early history of intuitionistic logic*, University of Amsterdam, ITLI preprint ML-88–04.

[1991] *History of constructivism in the twentieth century*, University of Amsterdam, ITLI preprint ML-91–05.

A. S. TROELSTRA AND D. VAN DALEN

[1988a] *Constructivism in Mathematics. An Introduction, Vol. I*, Studies in Logic, Vol. 121, North-Holland, Amsterdam.

[1988b] *Constructivism in Mathematics. An Introduction, Vol. II*, Studies in Logic, Vol. 123, North-Holland, Amsterdam.

J. V. TUCKER

[1976] *Computability as an Algebraic Property*, Ph.D. Thesis, School of Mathematics, University of Bristol.

[1980a] Computability and the algebra of fields: Some affine constructions, *J. Symbolic Logic*, 45, pp. 103–120.

[1980b] Computing in algebraic systems, in: *Recursion Theory, its Generalisations and Applications*, F. R. Drake and S. S. Wainer, eds., London Mathematical Society Lecture Note Series, Vol. 45, Cambridge Univ. Press, Cambridge, pp. 215–235.

[1991] Theory of computation and specification over abstract data types and its applications, in: *Logic, Algebra and Computation*, F. L. Bauer, ed., Springer, Berlin, pp. 215–235.

J. V. TUCKER AND J. I. ZUCKER

[1988] *Program Correctness over Abstract Data Types with Error-State Semantics*, North-Holland, Amsterdam.

[1991] Examples of semicomputable sets of real and complex numbers, in: *Constructivity in Computer Science*, J. P. Myers Jr. and M. J. O'Donnell, eds., Lecture Notes in Comp. Sci., Vol. 613, Springer, Berlin, pp. 179–198.

[1993] Provable computable selection functions on abstract structures, in: *Proof Theory*, P. Aczel, H. Simmons and S. S. Wainer, eds., Cambridge Univ. Press, Cambridge, pp. 277–306.

[1994] Computable functions on stream algebras, in: *Proof and Computation*, H. Schwichtenberg, ed., Springer, Berlin, pp. 341–383.

[1999] Computable functions and semicomputable sets on many sorted algebras, in: *Handbook of Logic in Computer Science, Vol. V*, S. Abramsky, D. Gabbay and T. S. E. Maibaum, eds., Oxford Univ. Press, in preparation.

A. M. TURING

[1950] The word problem in semigroups with cancellation, *Ann. of Math.*, 52, pp. 491–505.

M. K. VALIEV

[1969] On a theorem of G. Higman, *Algebra and Logic*, 8, pp. 93–128.

[1975] On polynomial reducibility of the word problem under embedding of recursively presented groups in finitely generated groups, in: *Mathematical Foundations of Computer Science 1975*, J. Becvar, ed., Lecture Notes in Comp. Sci., Vol. 32, Springer, Berlin, pp. 432–438.

B. L. VAN DER WAERDEN

[1930a] *Moderne Algebra*, 1st ed., Julius Springer, Berlin.

[1930b] Eine Bemerkung über die Unzerlegbarkeit von Polynomen, *Math. Ann.*, 102, pp. 738–739.

H. S. VANDIVER

[1936] On the ordering of real algebraic numbers by constructive methods, *Ann. of Math.*, 37, pp. 7–16.

YU. G. VENTSOV

[1994] Computable classes of constructivizations of models of infinite algorithmic dimension, *Algebra and Logic*, 33, pp. 22–45.

S. WAACK

[1981] Tape complexity of word problems, in: *Fundamentals of Computation Theory, Proceedings FCT'81*, F. Gécseg, ed., Lecture Notes in Comp. Sci., Vol. 117, Springer, Berlin, pp. 467–471.

[1983] *Raumkomplexität von Wortproblemen endlicher Gruppenpräsentationen*, Dissertation A, Berlin.

[1991] On the parallel complexity of linear groups, *RARIO Inform. Théor. Appl.*, 25, pp. 323–354.

H. WEBER

[1893] Untersuchungen über die allgemeinen Grundlagen der Galois'schen Gleichungstheorie, *Math. Ann.*, 43, pp. 521–549.

W. WECHLER

[1992] *Universal Algebra for Computer Scientists*, EATCS Monographs on Theoretical Computer Science, Vol. 25, Springer, Berlin.

K. WEIHRAUCH

[1985] Type 2 recursion theory, *Theoret. Comput. Sci.*, 38, pp. 17–33.

[1987] *Computability*, EATCS Monographs on Theoretical Computer Science, Vol. 9, Springer, Berlin.

[1993] Computability on computable metric spaces, *Theoret. Comput. Sci.*, 113, pp. 191–210.

[1996] *A foundation of computable analysis*, Fern Universität Hagen, manuscript.

K. WEIHRAUCH AND U. SCHREIBER

[1981] Embedding metric spaces into cpo's, *Theoret. Comput. Sci.*, 16, pp. 5–24.

M. WIRSING

[1990] Algebraic specifications, in: *Handbook of Theoretical Computer Science. Vol. B: Formal Models and Semantics*, J. van Leeuwen, ed., North-Holland, pp. 675–788.

H. WÜSSING

[1984] *The Genesis of the Abstract Group Concept*, MIT Press, Cambridge.

M. ZIEGLER

[1976] Ein rekursiv aufzahlbarer btt-Grad, der nicht zum Wort problem einer Gruppe gehört, *Zeits. Math. Logik Grundl. Math.*, 22, pp. 165–168.

CHAPTER 13

The Structure of Computability in Analysis and Physical Theory: An Extension of Church's Thesis

Marian Boykan Pour-El

School of Mathematics, University of Minnesota, Minneapolis, MN 55455, USA

Contents

1. Introduction . . . 450
2. Part I: a primer for computable analysis . . . 452
   2.1. Introduction . . . 452
   2.2. Computable reals and sequences of reals . . . 453
   2.3. Computability for continuous functions . . . 455
   2.4. Computability and physical theory: two examples, wave propagation and heat dissipation 459
   2.5. L^p-computability . . . 460
3. Part II: computability on a Banach space . . . 462
   3.1. Introduction . . . 462
   3.2. The computability structure . . . 462
   3.3. The First Main Theorem, statement and applications . . . 466
   3.4. The Second Main Theorem, the Eigenvector Theorem and related results . . . 468
4. Addendum: open problems . . . 470
References . . . 470

HANDBOOK OF COMPUTABILITY THEORY
Edited by E.R. Griffor

1. Introduction

This paper is devoted to a survey of computability in analysis and physical theory. Since computers are playing an ever increasing role in solving problems in these and related fields, it is useful to know – at least theoretically – which processes are computable and which are not.

A major consequence of the research described here is the extension of the notion of computability to computability on a Banach space. Recall that Banach space theory is a fundamental tool in analysis, physics and engineering – from the solutions of differential equations in the classical theory to the study of quantum theory. As examples of Banach spaces which are particularly useful we have Hilbert space (more generally L^p-spaces), Sobolev spaces, and others. The Banach space $C[a, b]$, of continuous functions on $[a, b]$, is of particular interest to logicians. Computable continuous functions of a real variable were defined and studied in the 1950's by Grzegorczyk [1957] and Lacombe [1955a, 1955b, 1957]. Thus, the work presented here may be regarded as a generalization of the work begun in the 1950's. It is worth remarking that Grzegorczyk and Lacombe gave several definitions for the computability of continuous functions of a real variable. All of these were proved to be equivalent. The proofs were complicated. We will see that the equivalence of these definitions is an immediate consequence of the notion of a "computability structure" on a Banach space. No additional proof is necessary.

The notion of a *computability structure* on a Banach space is of central importance in this paper. It accords well with the intuitive notion of computability. The definition is given axiomatically. The axioms provide for the interaction of elementary recursion theory with the basic tenets of Banach space theory. The concept which is axiomatized is "computable sequence of elements" of the Banach space. A point x in the Banach space is computable if the constant sequence $x, x, x, \ldots$ is. *It is both necessary and natural to work with sequences of elements rather than individual elements. There are two reasons for this – topological and recursion – theoretic.* Since a Banach space is a metric space, the topology can be given by sequences. In recursion theory one of basic notions is the notion of a recursively enumerable set – a set whose elements can be arranged in a computable sequence.

The computability structure is characterized by three axioms. This is not surprising, since the Banach space, itself, is defined by three properties. It is a *linear space, with a norm, which is complete in the norm.* It is then shown that, under general conditions which in practice are always satisfied, the computability structure on a Banach space is unique. Thus, we are able to achieve the intrinsic quality we associate with the notion of computability. The situation is reminiscent of the one in ordinary recursion theory, when the various definitions, proposed by Turing, Herbrand/Gödel, Church, Post and others, all intuitively convincing, were proved to be equivalent. The notion of a computability structure acts as a unifying concept, since seemingly different definitions of computability, are, in fact, equivalent because of this unicity. Thus, we have a "Church's Thesis" for the given Banach space.

It should be remarked that the notion of computability which appears in the axioms for a computability structure depends ultimately on elementary recursive function theory – i.e. the notion of a recursive function mapping N into N and of a recursively enumerable set. These are combined with the basic facts of Banach space theory to produce the axioms referred to above. No prior knowledge of Banach space theory is presupposed in this paper. Any facts which are required – and they are all quite elementary – will be stated precisely.

It may be useful to comment further on the computability structure. As remarked above, the structure is characterized by three axioms. There is one axiom for each of the basic concepts of Banach space theory – linearity, norm and limit. The axiom provides for the interaction between the associated concept and recursive function theory. When viewed in this light the computability structure is minimal – just sufficient for the fundamental notions of recursive function theory to interact with the basic concepts of Banach space theory. As stated earlier, it will be shown that under very natural conditions the computability structure is not merely minimal, but also maximal – in fact, unique.

We will return to a more detailed discussion of the computability structure and its applications later. We now give a summary of the contents of this paper.

The paper is divided into two parts. Part I is, in essence, a primer on computable analysis. Part II contains the main results. It is suggested that the reader glance briefly at Part I, and go directly to Part II.

Part I begins with the definition of a computable real. We note that the computable reals form a field which contains all algebraic numbers and also the well-known transcendentals – e.g., π and e. This is followed by an account of computability for $C[a,b]$, where a and b are computable reals. (Recall that $C[a,b]$ is the set of continuous functions on $[a,b]$. It is a Banach space with $\|f\| = \sup_{a \leqslant x \leqslant b} |f(x)|$.) The discussion includes integration, differentiation, the max.–min. theorem and the intermediate value theorem, all from the viewpoint of computability. Of particular interest is the relation of computability to physical theory. For example, the propagation of waves need not be computable, even if the initial conditions which determine the wave propagation uniquely are computable. However, heat dissipation is always computable if the initial conditions are.

Part I concludes with a section on computability for L^p-spaces.

We now turn to Part II and the principal results of this work. They are contained in three general theorems: The First Main Theorem, The Second Main Theorem, and the Eigenvector Theorem. The notion of a computability structure plays a fundamental role in the formulation of these theorems, as we now explain.

Why is it the case that wave propagation can be noncomputable even if the initial conditions which uniquely determine the propagation are computable? However, the dissipation of heat is always computable whenever the initial conditions are. The answer is given by the First Main Theorem. Roughly the First Main Theorem states: under general conditions which in practice are always satisfied

Bounded linear operators from one Banach space with a computability structure to another preserve computability, unbounded linear operators do not.

Wave propagation is associated with an unbounded linear operator, whereas the dissipation of heat is associated with a bounded operator. Thus waves can propagate noncomputably even though the initial conditions are computable. However, heat will always dissipate computably.

The First Main Theorem can be applied to a host of examples: integration, differentiation, Fourier series, Fourier transform and many others. We will discuss this briefly in Part II. The fact that the First Main Theorem is so widely applicable is a consequence of two other facts. First, the Theorem is easy to apply. Second, essentially all of the well-known operators in analysis are linear. Indeed, mathematicians are just beginning to study nonlinearity.

We now turn to the Second Main Theorem. It is concerned with the computability/noncomputability of eigenvalues. Since eigenvalues are the quantities which are measured in experiments, it is of some interest to determine whether or not they are computable.

The setting for the Second Main Theorem is self-adjoint linear operators on Hilbert space. Thus the eigenvalues are real numbers. The operators may be bounded or unbounded. We are led to the following questions: are the eigenvalues – i.e. the quantities which are measured – computable real numbers? Can the eigenvalues be arranged in a computable sequence? The Second Main Theorem answers these questions and the answers are quite easy to state. Under mild side conditions which in practice are always satisfied, each eigenvalue is a computable real number. However, in general, the eigenvalues cannot be arranged in a computable sequence.

Incidentally, linear operators which satisfy these mild side conditions are referred to as "effectively determined". All of the standard operators of analysis and physics are effectively determined.

The Second Main Theorem has many corollaries. Some of these will be discussed in Part II. The Theorem can be extended to bounded normal operators.

The Eigenvector Theorem, our third major result, is concerned with the computability/noncomputability of eigenvectors. Recall that eigenvectors have some physical significance: they are associated with the "state of the system". Hence it is of interest to determine whether or not they are computable. The Eigenvector Theorem asserts that there exists an effectively determined, compact, self-adjoint operator such that 0 is an eigenvalue. However, none of the eigenvectors corresponding to zero is computable.

We now turn to Part I.

2. Part I: a primer for computable analysis

2.1. *Introduction*

Part I is a systematic account of elementary computable analysis. It begins in Section 2.2 with the definition of a computable real and of a computable sequence of reals. This is followed by Section 2.3, an account of computability for continuous functions on $[a, b]$, where a and b are computable reals. We give one definition and then indicate why it is that all reasonable definitions are equivalent. Also included is a discussion of the basic theorems of $C[a, b]$, and their relation to computability – integration, differentiation, the max.–min. theorem, the intermediate value theorem.

Section 2.4 gives a brief discussion of the relation of computability to physical theory. In particular we note that wave propagation can be noncomputable even though the initial conditions which determine the propagation uniquely are computable. This cannot happen for the heat equation.

Part I concludes with Section 2.5, a brief discussion of computability on L^p-space, where p is recursive and $1 \leqslant p < \infty$. We note that an element of $L^p[a, b]$ need not be continuous on $[a, b]$ in order to be computable. For example, any step function with rational jump points and rational values is L^p-computable.

2.2. *Computable reals and sequences of reals*

We begin with the definition of a *computable real.* Recall that classically a real is the limit of a Cauchy sequence of rationals. Hence in order for a real number x to be computable, two conditions must be satisfied: (1) there must exist a sequence of rationals which is computable and (2) this sequence must converge effectively to x. Thus we are led to the following definitions.

DEFINITION 1. The sequence $\{r_n\}$ of rationals is *computable* if there exist three recursive functions b, c, s such that $c(n) \neq 0$ for all n and

$$r_n = (-1)^{s(n)} \frac{b(n)}{c(n)}.$$

DEFINITION 2. The sequence $\{r_n\}$ of rationals *converges effectively* to a limit x if there is a recursive function e such that

$$m \geqslant e(n) \quad \text{implies} \quad |r_m - x| < \frac{1}{2^n}.$$

DEFINITION 3. A real number x is computable if there is a computable sequence of rationals which converges effectively to x.

We now give some elementary properties of computable reals. It is not hard to show that if x and y are computable reals then $x + y$ and $x \cdot y$ are also computable. In fact, the computable reals form a field. Furthermore this field is real closed. It contains all algebraic numbers and also the well-known transcendental numbers – e.g., π and e.

Clearly the cardinality of the set of computable reals is $\aleph_0$. Hence most reals are not computable. We now give a specific example of a noncomputable real number.

EXAMPLE (*Noncomputable real*). Let $a : N \to N$ be a $1-1$ recursive function enumerating a recursively enumerable nonrecursive set A. Define x by

$$x = \sum_{n=0}^{\infty} 2^{-a(n)}.$$

Then x is a noncomputable real.

This example dates back to the work of Rice [1954].

Although the proof that x is not computable is quite simple, we will not give it in detail. Instead we make some general remarks.

The proof does require some care. Note that the sequence of partial sums $\{s_k\}$, where

$$s_k = \sum_{n=0}^{k} 2^{-a(n)}$$

forms a computable sequence of rationals. It can be shown that this sequence does not converge effectively. For if it did, the set A would be recursive. However, we cannot conclude immediately that the limit x is noncomputable. There might exist another computable sequence $\{t_k\}$ of rationals which converges effectively to x. That this cannot happen follows from the properties of computable sequences of reals. (See Pour-El and Richards [1989, pp. 17–18, 20].) As remarked earlier, the notion of a computable *sequence* is of great importance in this field.

Alternative definitions of computable real. All of the well-known definitions of a real number effectivize. They include:

(1) Cauchy sequence (the definition we have just discussed).
(2) Nested Interval.
(3) Dedekind Cut.
(4) Expansion to base b, where b is an integer > 1.

All of these effectivizations lead to the same class of reals. This result was proved by R. M. Robinson [1951].

Extension to complex numbers. This is quite straightforward. A complex number is computable if both its real and imaginary parts are computable.

We now turn to a discussion of *computable sequences of reals*. As remarked earlier, the notion of a computable sequence is of fundamental importance.

A sequence of real numbers need not be computable even if each of its elements is a computable real. The sequence must satisfy a stronger requirement. In order for it to be computable, there must exist a master program M, which on input n, effectively approximates the nth element, x_n, of the sequence. The definition is an obvious modification of the notion of a computable real. We require the existence of a computable double sequence of rationals $\{r_{nk}\}$ which converges to $\{x_n\}$ as $k \to \infty$, effectively in n and k. The details may be found in Pour-El and Richards [1989], p. 18.

The extension of the notion of computability to sequences of complex numbers is straightforward. Thus the sequence $\{z_n\}$, where $z_n = x_n + \mathrm{i}y_n$, is computable if the sequences $\{x_n\}$ and $\{y_n\}$ are both computable sequences of reals.

2.3. *Computability for continuous functions*

Let a and b be computable reals. Recall that $C[a, b]$ is the set of continuous functions on $[a, b]$.

We first define computability for elements and sequences of elements of $C[a, b]$. Next we note that all reasonable definitions are equivalent. As we will see later, this is because $C[a, b]$ is a Banach space with a computability structure satisfying some simple conditions. (A more complete description of the computability structure is postponed until Part II.) Finally we give the theorems on integration, differentiation, the max.–min. theorem, the intermediate value theorem and their relation to computability.

2.3.1. *The definition of computability for* $C[a, b]$. Let $P(x)$ be any polynomial with rational coefficients. For example, let

$$P(x) = \frac{1}{3}x^4 + 3x^3 - \frac{7}{5}x + 25.$$

Any reasonable definition ought to require that $P(x)$ be computable on $[a, b]$.

Now let $\varphi(x)$ be an arbitrary element of $C[a, b]$. To define computability for φ we proceed as follows. First we give a seemingly special definition based on an effectivization of the Weierstrass Approximation Theorem (Definition 5, below). Then we indicate why this definition is not specialized at all!

We begin with the statement of the well-known (classical) Weierstrass Approximation Theorem for continuous functions on $[a, b]$.

THEOREM (Weierstrass Approximation Theorem). *Suppose* $\varphi \in C[a, b]$. *Then for every* n *there exists a polynomial* $P_n(x)$ *such that*

$$\left|\varphi(x) - P_n(x)\right| < \frac{1}{2^n}.$$

What does the polynomial $P_n(x)$ look like? One can assume without loss of generality, that it has rational coefficients. Hence

$$P_n(x) = \sum_{j=0}^{d(n)} (-1)^{s(n,j)} \frac{b(n, j)}{c(n, j)} x^j, \tag{1}$$

where $b(n, j)$, $s(n, j)$ and $d(n)$ are nonnegative integers, and $c(n, j)$ is a positive integer. Thus b, c, s are functions from $N \times N \to N$ and d is a function from N to N.

We now turn to the effective version of the Weierstrass Approximation Theorem. The first step is to define "computable sequence of polynomials". Let $P_n(x)$ be as in Eq. (1) above.

DEFINITION 4. The sequence $\{P_n(x)\}$ is a computable sequence of polynomials if s, b, c, d are recursive functions.

It is now very easy to define computability for elements of $C[a, b]$.

DEFINITION 5. Let $\varphi \in C[a, b]$ where a and b are computable reals. Then φ is computable if there is a computable sequence of polynomials $\{P_n(x)\}$ and a recursive function e so that

$$\left|\varphi(x) - P_n(x)\right| < \frac{1}{2^k} \quad \text{whenever} \quad n \geqslant e(k).$$

To summarize: the above definition states that

> Suppose that $\varphi(x)$ is continuous on $[a, b]$, where a and b are computable reals. *Then $\varphi(x)$ is computable on $[a, b]$ if and only if $\varphi(x)$ is in the "effective closure" of the polynomials with rational coefficients.*
> *Note: one does not have to know what value the function assigns to a specific argument. The x appearing in the definition is a "dummy variable".*

This definition can be extended easily to cover computable sequences. Since sequences play a major role we record the definition.

DEFINITION 6. Let $\{\varphi_m\}$ be a sequence of functions in $C[a, b]$, where a and b are computable reals. Then $\{\varphi_m\}$ is a computable sequence if there exists a computable double sequence of polynomials $\{P_{m,n}(x)\}$[1] with rational coefficients and a recursive function e such that

$$\left|\varphi_m(x) - P_{mn}(x)\right| < \frac{1}{2^k} \quad \text{if } n \geqslant e(m, k).$$

EXAMPLES. The following sequences are computable:

(a) $1, x, x^2, x^3, \ldots;$

(b) $\sin x, \cos x, \sin 2x, \cos 2x, \ldots, \sin mx, \cos mx, \ldots.$

---

[1] I.e. $P_{m,n}(x) = \sum_{j=0}^{d(m,n)} (-1)^{s(m,n,j)} b(m, n, j) x^j / c(m, n, j)$, where s, b, c, d are recursive functions, $c(m, n, j) > 0$.

2.3.2. *Other definitions. All reasonable definitions are equivalent.* The definition of "computable continuous function of a real variable" can be formulated in many different ways. It turns out that all of these definitions are equivalent.

One approach is via computable functionals Φ of various sorts. We say that the function $f : N \to N$ represents the real number c if

$$\left| c - \frac{f(n)}{n+1} \right| < \frac{1}{n+1}.$$

Now suppose $\varphi(x)$ is a continuous function defined on $[a, b]$, where a and b are computable reals. Then $\varphi(x)$ is computable if there is a computable functional Φ such that

if f represents the real number c, then $\Phi\langle f\rangle : N \to N$ represents the real $\varphi(c)$.

Many kinds of computable functionals have been used. Grzegorczyk and Lacombe took this route in the 1950's (Grzegorczyk [1957], Lacombe [1955a, 1955b, 1957]; Grzegorczyk's definition, which was so influential in the 1950's, makes use of Kleene's recursive functionals).

Another approach is to recall that a continuous function on $[a, b]$ is determined by two properties:

- the value it gives to the rationals;
- the fact that the function is continuous and hence uniformly continuous on $[a, b]$.

One can obtain a definition of computable continuous function by effectivizing both of these properties.

It turns out that all of these definitions – and any other reasonable definition one may propose – will be equivalent. In Part II we will see that there is a simple explanation for this. On each of these definitions the sequence of functions

$$1, x, x^2, x^3, x^4, \ldots$$

is a computable sequence. This fact completely determines the class of computable continuous functions on $[a, b]$. Actually, the results are more general. As remarked earlier, the proper way to view the problem is via Banach space theory. (Recall that $C[a, b]$ is a Banach space with $\|f\| = \sup_{a \leqslant x \leqslant b} |f(x)|$.) In Part II, when we define a "computability structure" on a Banach space, we will indicate that, under general conditions which in practice are always satisfied, the computability structure is unique. For $C[a, b]$ the condition that $1, x, x^2, \ldots, x^n, \ldots$ is a computable sequence suffices. It is the *unicity of the computability structure for $C[a, b]$ which is responsible for the equivalence of the definitions referred to above.*

One important consequence of the equivalence of the definitions is the fact that we can dispense with the use of computable functionals. Functionals were used in the early results in recursive analysis. (Recall that the domain and range of these functionals are included in the set of functions f mapping N into N.) The fact that we do not need functionals is a considerable advantage. They are rather messy to work with, presumably because the natural topologies of both the reals and of $C[a,b]$ are suppressed. For example, the fact that e^x, $\sin x$, $\cos x$, etc. are computable follows immediately from the "effective Weierstrass" definition – i.e. Definition 5 above. (Use the sequence of partial sums of the power series expansion of each of these functions to obtain the computable sequence of polynomials which converges effectively.) Now try to prove that each of these functions is computable using a suitable computable functional!

The history of computable analysis also indicates the difficulty in obtaining results directly from a definition based on functionals. We cite one example due to Grzegorczyk. There are others. Recall that Grzegorczyk's definition was based on the recursive functionals of S.C. Kleene. Grzegorczyk left open the following question.

QUESTION. Suppose $\varphi \in C[a,b]$ is computable. Suppose further that φ' is continuous. Is φ' computable?

The answer to this question is *no*. A counterexample was given by Myhill fourteen years later – in 1971 (Myhill [1971]). Actually, as we will see later, this result is very easy to prove using a general theorem concerning an unbounded linear operator on a Banach space with a computability structure. This theorem, the First Main Theorem, is discussed in Part II.

2.3.3. *Basic theorems on computability for* $C[a,b]$. We state four theorems: a theorem on integration, a theorem on differentiation, the max.–min. theorem, and the intermediate value theorem. Further details, as well as many additional theorems may be found in Pour-El and Richards [1989], Chapters 0 and 1.

Recall that, in the definition of computability for elements of $C[a,b]$, a and b are assumed to be computable reals.

THEOREM 1. *If* φ *is computable, then*

$$\int_a^x \varphi(x)\,\mathrm{d}x$$

is computable.

THEOREM 2 (Myhill [1971]). *There exists a* C^1 *function* φ, *defined on* $[0,1]$ *such that* φ *is computable but* φ' *is not computable.*

(A slight strengthening of this theorem appears in Pour-El and Richards [1989].)

THEOREM 3 (Max.–Min. Theorem). *Let* $\{\varphi_n\}$ *be a computable sequence of elements of* $C[a,b]$. *Then the maximum values*

$$s_n = \max\{\varphi_n(x)\colon\ x \in [a,b]\}$$

form a computable sequence of reals.

THEOREM 4 (Intermediate Value Theorem). *Let φ be a computable element of $C[a,b]$. Suppose that $\varphi(a) < 0$ and $\varphi(b) > 0$. Then there is a computable element c in $[a,b]$ such that $\varphi(c) = 0$.*

REMARK 1. It can be shown that Theorem 4 cannot be extended to computable sequences (Pour-El and Richards [1989, p. 42]). More precisely, one can define a computable sequence of functions of a real variable $\{\varphi_n(x)\}$ on [0, 1] such that

$$\varphi_n(0) = -1,$$
$$\varphi_n(1) = 1.$$

However, there is no computable sequence $\{c_n\}$ such that $\varphi_n(c_n) = 0$.

This counterexample has an interesting consequence. There is no effective proof of Theorem 4!

REMARK 2. Although Theorems 1 and 2 can be proved directly, it is worth noting that they are both special cases of a very general theorem, the First Main Theorem (See Part II, Section 3.3). Roughly this theorem says that on a Banach space, bounded linear operators preserve computability, unbounded linear operators do not. Integration is a bounded operator, differentiation is unbounded.

Before concluding Section 2.3, we remark that the definition of computability for continuous functions may be extended still further. All of these extensions are easy to obtain.

The first extension concerns computability for $C(\mathbb{R})$, the set of continuous functions on $\mathbb{R}$. The second deals with functions of more than one variable. For functions of n-variables there are two cases to consider. The first case deals with computability on I^n, an n-dimensional rectangle, with computable coordinates for the "corners". The second case is concerned with computability for elements and sequences of elements of $C(\mathbb{R}^n)$. The details, which are straightforward, may be found in Pour-El and Richards [1989].

2.4. *Computability and physical theory: two examples, wave propagation and heat dissipation*

Do waves propagate computably? Does heat dissipate computably? Let us consider each question separately.

Consider the propagation of a wave $u(x, y, z, t)$ governed by the wave equation, a well known partial differential equation, and satisfying the initial conditions

$$u(x, y, z, 0) = f(x, y, z),$$
$$\frac{\partial u}{\partial t}(x, y, z, 0) = 0.$$

There exists a computable – and hence continuous – function $f(x, y, z)$ such that the solution $u(x, y, z, t)$ is continuous but not computable. In fact $u(0, 0, 0, 1)$ is a noncomputable real number. So waves do *not* necessarily propagate computably.

Similarly heat propagation is governed by the heat equation, a partial differential equation satisfying the initial condition

$$u(x, y, z, 0) = f(x, y, z).$$

Here again the question is: if $f(x, y, z)$ is computable, is $u(x, y, z, t)$ computable? However, in this case the answer is: yes.

There is a good reason for the fact that the two answers are different. The explanation will be given in detail in Part II, where we consider computability on a Banach space. Roughly the wave equation is associated with an unbounded linear operator, whereas the operator associated with the heat equation is bounded. For an unbounded operator, one can *always* obtain computable data such that the solution is not computable. By contrast, bounded operators always take computable data into a computable solution. (Further details may be found in Part II, Section 3.3. See also Pour-El and Richards [1989, pp. 115–119].)

2.5. L^p-*computability*

Must a function of a real variable be continuous in order for it to be computable? For intuitionists and other constructivists this appears to be the case. However there are discontinuous phenomena (e.g., shock waves), and mathematicians do perform computations on these phenomena. There is a sense in which computable functions can be discontinuous.

Modeling of physical phenomena is often carried out using the theory of L^2-functions (or, more generally, L^p-functions, $1 \leqslant p < \infty$). Recall that $L^p[a, b]$ is the space of all functions φ such that

$$\int_a^b |\varphi(x)|^p \, \mathrm{d}x < \infty.$$

It is a Banach space, where the norm of φ, $\|\varphi\|_p$ is defined by

$$\|\varphi\|_p = \left(\int_a^b |\varphi(x)|^p \, \mathrm{d}x \right)^{1/p}.$$

The definition of computability for $L^p[a, b]$ is analogous to that for $C[a, b]$. (See Section 2.3, above.) We require that p, a, and b be computable reals and that $1 \leqslant p < \infty$. There are two parts to the definition

(1) A polynomial with rational coefficients is computable.

(2) A function $\varphi(x)$ is L^p-computable if there exists a computable sequence of polynomials $\{P_n(x)\}$ and recursive function $e(n)$ such that

$$\left\| \varphi(x) - P_n(x) \right\|_p < \frac{1}{2^k} \quad \text{if } n \geqslant e(k).$$

(The definition of a computable sequence of polynomials was spelled out earlier-Definition 4, above.)

In summary: the above definition states that

Suppose $\varphi \in L^p[a,b]$. Then φ is L^p-computable if and only if φ is in the "effective closure" in L^p-norm of the polynomials with rational coefficients.

Here again all reasonable definitions for a function φ to be L^p-computable are equivalent. As was the case for computability on $C[a,b]$, there is a simple explanation. For each of these definitions, the sequence of functions

$$1, x, x^2, x^3, \ldots$$

is L^p-computable. The class of L^p-computable functions is determined by this. As remarked earlier, it is fruitful to view the problem via Banach space theory.

The definition of L^p-computability for functions of a real variable can be extended to cover elements of $L^p(\mathbb{R})$. It can also be extended to cover functions of n-variables. For functions of n-variables we again have two cases: computability for $L^p(B)$, where B is an n-dimensional cube with computable coordinates for the corners, and computability for $\mathbb{R}^n$. The extensions are straightforward (see Pour-El and Richards [1989] for details).

We remark in passing that there are *$L^p[a,b]$-computable functions that are not continuous*. In fact any step function with computable "jump-points" and computable values is $L^p[a,b]$-computable. It can be proved that these are the only step-functions which are L^p-computable (Pour-El and Richards [1989]).

The definition of L^p-computability can be extended to sequences in exactly the same way as was done for sequences of elements of $C[a,b]$. (See Definition 6.) Thus:

Suppose $\{\varphi_m\}$ is a sequence of functions in $L^p[a,b]$. Then $\{\varphi_m\}$ is an $L^p[a,b]$-computable sequence if there is a computable double sequence of polynomials $\{P_{m,n}(x)\}$ with rational coefficients and a recursive function e such that

$$\left\| \varphi_m(x) - P_{mn}(x) \right\|_p < \frac{1}{2^k} \quad \text{if } n \geqslant e(m,k).$$

This concludes Part I. We now turn to Part II.

3. Part II: computability on a Banach space

3.1. *Introduction*

We turn now to a discussion of the principal definitions and results. As remarked in the introduction to the paper, this work is motivated by a desire to solve problems in analysis, physics and engineering. Since these problems are often formulated using the theory of Banach spaces, and solved with the help of a computer, it is natural to investigate the relation between computability and Banach space theory.

We first extend the notion of computability to computability on a Banach space. This is done in Section 3.2, below. Then in Sections 3.3 and 3.4 we give the major results: the First Main Theorem, the Second Main Theorem and the Eigenvector Theorem.

The First Main Theorem is concerned with the question: which processes in analysis and physics take computable data into a computable solution and which do not? The answer turns out to be exceedingly simple as we will see in Section 3.3. In particular it explains why wave propagation can be noncomputable even if the initial conditions which determine the wave propagation are computable. The theorem has numerous applications – to Fourier series, Fourier transform, Laplace transform, and a host of others.

The Second Main Theorem and the Eigenvector Theorem are discussed in Section 3.4. The Second Main Theorem is concerned with the computability/noncomputability of eigenvalues. (Recall that eigenvalues are quantities which are measured in experiments.) This theorem has many corollaries which are of independent interest. Some of these are considered in Section 3.4. The Eigenvector Theorem deals with the computability/noncomputability of eigenvectors.

We turn now to Section 3.2, computability on a Banach space.

3.2. *The computability structure*

In Part I we gave some examples of computability on a Banach space. They included computability for $C[a,b]$, the Banach space of continuous functions on $[a,b]$, and computability for $L^p[a,b]$, where a, b, and p are computable reals, $1 \leqslant p < \infty$. We turn now to the definition of computability on an arbitrary Banach space. To avoid complete triviality we assume that the Banach space contains at least one nonzero element.

One note of caution before proceeding further. We do *not* define a computable Banach space. *Throughout this paper the definition of a Banach space is the usual classical one.* (*It is this classical definition which is useful in theory and applications*!) *We merely add an extra structure to a pre-existing Banach space. This structure is called a computability structure.*

As is well-known, there are two approaches to any definition of a mathematical concept, no matter what the field of specialization may be. The first is “intrinsic”.

This is exemplified by the work of the founding fathers of recursion theory – Turing, Post, Herbrand–Gödel, etc. Each gave his definition of computability by answering the question: "What is a computation?" The fact that all the definitions proved to be equivalent gave credence to the belief that computability for functions from N to N was well understood.

The second approach is axiomatic. This is useful in classifying a variety of models under a single definition. The axioms for group theory provide an example of this approach.

The definition of computability on a Banach space will be given axiomatically. We then indicate how the axiomatic approach leads to an *intrinsic characterization*.

As remarked in the introduction to the paper, the concept to be axiomatized is "computable sequence of elements of the Banach space". There are two reasons why this is both natural and fruitful. The first is topological. Since a Banach space is a metric space, the topology can be given by sequences of points. The second is recursion-theoretic. A basic notion of ordinary recursion theory is the notion of a recursively enumerable set, a set whose elements can be arranged in a computable sequence.

We proceed as follows. Given a Banach space X, we define a collection $\mathcal{S}$ of sequences, the "computable sequences". $\mathcal{S}$ is characterized axiomatically. There are three axioms. This is because a Banach space is defined by three properties:

- It is a linear space over the reals or the complex numbers.
- It has a norm.
- It is complete in the norm.

The three axioms provide for the interaction of the basic concepts of Banach space theory with the fundamental notions of computability.

REMARK (*Some obvious notions*). The following obvious notions will be useful in stating the axioms. Suppose that $\mathcal{S}$ is a collection of sequences.

(i) A double sequence $\{x_{nk}\} \in \mathcal{S}$ if it is mapped onto a sequence $\{y_n\} \in \mathcal{S}$ by one of the usual recursive pairing functions from $N \times N$ onto N

(ii) An element $x \in \mathcal{S}$ if the constant sequence $x, x, x, \ldots \in \mathcal{S}$

(iii) The double sequence $\{x_{nk}\}$ of elements of X converges to the sequence $\{x_n\}$ as $k \to \infty$ effectively in k and n, if there exists a recursive function e such that

$$\|x_{nk} - x_n\| < \frac{1}{2^N}$$

for all $k \geqslant e(n, N)$.

We now give the definition of a computability structure $\mathcal{S}$ on a Banach space X. To avoid triviality we assume that $\mathcal{S}$ is a collection of sequences of elements of X which contains at least one member distinct from the sequence $0, 0, 0, 0, \ldots$.

DEFINITION 7. Let X be a Banach space. Let $\mathcal{S}$ be a collection of sequences of elements of X. Then $\mathcal{S}$ is a computability structure for X if $\mathcal{S}$ satisfies the following axioms.

AXIOM 1 (*Linearity*). Let $\{x_n\}$ and $\{y_n\}$ be members of $\mathcal{S}$. Let $\{\alpha_{nk}\}$ and $\{\beta_{nk}\}$ be computable double sequences of real (or complex) numbers. Let $d: N \to N$ be a recursive function. Then the sequence $\{s_n\}$, where

$$s_n = \sum_{k=0}^{d(n)} (\alpha_{nk} x_k + \beta_{nk} y_k)$$

is in $\mathcal{S}$.

AXIOM 2 (*Norm*). Let $\{x_n\} \in \mathcal{S}$. Then the sequence of norms $\{\|x_n\|\}$ form a computable sequence of real numbers.

AXIOM 3 (*Limit*). Let $\{x_{nk}\}$ be an element of $\mathcal{S}$ such that $\{x_{nk}\}$ converges to $\{x_n\}$ as $k \to \infty$ effectively in k and n. Then $\{x_n\} \in \mathcal{S}$.

Banach spaces with computability structures are very common. For example, we have:

$C[a,b]$ with the collection $\mathcal{S}_C$ of computable sequences defined in Part I, Section 2.3.

$L^p[a,b]$ with the collection $\mathcal{S}_p$ of computable sequences defined in Part I, Section 2.5.

3.2.1. *Unicity of the computability structure: the Effective Density Lemma and the Stability Lemma.* In Part I, Section 2.3, we discussed several alternative definitions for computability for a function $\varphi \in C[a,b]$. They included:

(1) A definition based on an effective version of the Weierstrass Approximation Theorem.

(2) A definition using "effective functionals" – e.g., Kleene's general recursive functionals.

(3) A definition based on two properties:

(a) effective uniform continuity;

(b) the mapping of the computable sequences of rationals into a computable sequence of reals.

We now show that all of these definitions – and, indeed, all other reasonable definitions of computability on $C[a,b]$ – are equivalent. This is a consequence of two lemmas, the Effective Density Lemma and the Stability Lemma. Both of these lemmas make use of the notion of an "effective generating set", which we now define. In a nutshell, a computability structure with a given effective generating set is uniquely determined.

DEFINITION 8. Let $(X, \mathcal{S})$ be a Banach space with a computability structure $\mathcal{S}$. Let $\{e_n\} \in \mathcal{S}$. Then $\{e_n\}$ is an effective generating set for $(X, \mathcal{S})$ if the set of finite linear combinations of the e_i's with rational coefficients is dense in the space X.

REMARK. This definition includes the case in which the field of scalars of the Banach space X is the set of complex numbers. In that case, the term "rational" refers to a number of the form $a + b\,\mathrm{i}$ where a and b are rational numbers in the usual sense.

Examples of effective generating sets for a Banach space with a computability structure abound. Consider $C[a, b]$ where a and b are computable reals, with the usual computability structure, $\mathcal{S}_c$, given in Part I. Then $1, x, x^2, \ldots$ is an effective generating set. Another effective generating set for $C[a, b]$ is the collection of all continuous piecewise linear functions of $C[a, b]$ with rational coordinates for the "corners". Both of these sequences are also effective generating sets for $L^p[a, b]$ with its computability structure $\mathcal{S}_p$.

We now state the Effective Density Lemma and the Stability Lemma. The Stability Lemma is an immediate corollary of the Effective Density Lemma. It gives the unicity of the computability structure.

We turn to the Effective Density Lemma. Roughly it says that if $\{e_n\}$ is an effective generating set for $(X, \mathcal{S})$, then $\mathcal{S}$ is the effective closure of the set of rational linear combinations of the e_i. More precisely

THEOREM 5 (Effective Density Lemma). *Suppose that $(X, \mathcal{S})$ is a Banach space with a computability structure. Suppose also, that $\{e_n\} \in \mathcal{S}$ is an effective generating set for $(X, \mathcal{S})$. Then the sequence $\{x_n\}$, $x_n \in X$, is a member of $\mathcal{S}$ if and only if there is a computable double sequence $\{p_{nk}\}$ so that*

$$p_{nk} = \sum_{j=0}^{d(n,k)} \alpha_{nkj} e_j,$$

where $d(n, k)$ is a recursive function and $\{\alpha_{nkj}\}$ is a computable triple sequence of rationals/complex rationals – such that

$$p_{nk} \to x_n \quad \text{as } k \to \infty$$

effectively in k and n.

The Stability Lemma makes the situation clearer. It allows us to conclude that, if two computability structures $\mathcal{S}_1$ and $\mathcal{S}_2$ on a given Banach space X share an effective generating set, then $\mathcal{S}_1 = \mathcal{S}_2$.

THEOREM 6 (Stability Lemma). *Let $\mathcal{S}_1$ and $\mathcal{S}_2$ be computability structures for the Banach space X. Let $\{e_n\} \in \mathcal{S}_1$ and $\{e_n\} \in \mathcal{S}_2$ be an effective generating set for both of these computability structures. Then $\mathcal{S}_1 = \mathcal{S}_2$.*

3.2.2. *Application of the computability structure to prove equivalence of definitions.* It now becomes clear that the three types of definitions for computability on $C[a, b]$, the Weierstrass Approximation definition, the definitions based on computable functionals, and the definition involving effective uniform continuity (see Section 2.3 of Part I) are all equivalent. The sequence $1, x, x^2, \ldots$ is a computable sequence on *all* of these definitions. (This is not hard to see.) Hence the polynomials with rational coefficients can be arranged in a computable sequence. (See Axiom 1 above.) Since these polynomials are dense in $C[a,b]$, they form an effective generating set. We now apply the Stability Lemma to obtain the desired result.

Analogous results hold for $L^p[a,b]$.

3.2.3. *Further applications of the computability structure.* In addition to providing a quick and easy proof that all reasonable definitions of computability on a Banach space are equivalent, the computability structure can be applied in another way. It allows us to obtain effective versions of classical approximation theorems merely from the statement of the classical theorem. It is not necessary to effectivize a proof of the classical result! We mention two. There are many others.

(1) Effective Weierstrass Approximation Theorem, dimension $n \geqslant 1$.

(2) Effective Wiener Tauberian Theorem.

Each of these theorems is an immediate consequence of the statement of the classical theorem together with the Effective Density Lemma. For further details, see Pour-El and Richards [1989, pp. 86–87].

3.3. *The First Main Theorem, statement and applications*

Why is it that wave propagation can take computable initial data into a noncomputable solution, whereas heat dissipation is always computable if the initial conditions are? On a more elementary level, suppose that a and b are computable reals. Why is it that we can find a $\varphi \in C[a,b]$ with continuous derivative φ' such that φ is computable, but φ' is not? By contrast the integral $\int_a^x \varphi(t)\,dt$ of a computable function $\varphi(x)$ is always computable. The answers to these questions, and to many others, are given by the First Main Theorem.

The First Main Theorem is motivated by the question: which processes in analysis and physics take computable data into a computable solution? We first make this question precise, and then we answer it.

It is easy to make the question precise. We need only to define "computability" and "process".

The notion of computability is "computability on a Banach space", which has already been defined.

What shall we mean by the term "process"? Among the processes we wish to include are differentiation, integration, wave propagation, heat dissipation, Fourier series, Fourier transform, and a host of others. Note that each of these processes is associated with a linear operator mapping one Banach space into another. In fact, essentially all of the processes studied in analysis and physics are linear. Mathematicians are just beginning to study nonlinear ones. Thus by a "process" we will mean a "linear operator mapping one Banach space into another".

REMARK. Recall that there are two types of linear operators – bounded and unbounded. A linear operator T – with domain a dense subset of the Banach space X and range included in the Banach space Y – is *bounded* if there exists a positive integer M such that

$$\|Tu - Tv\|_Y \leqslant M\|u - v\|_X$$

for all u, v in the domain of T. A linear operator is *unbounded* if it is not bounded.

Recall also, that in practice the linear operators T of analysis and physics are all *closed* – i.e. the graph of T is a closed subset of $X \times Y$.

The terms have now been defined. The question becomes: suppose T is a linear operator mapping one Banach space with a computability structure into another. Under what conditions does T map computable elements (computable sequences) into computable elements (computable sequences)? The answer is given by the First Main Theorem. A rough statement is:

Bounded operators preserve computability.
Unbounded operators do not.

The precise statement is:

FIRST MAIN THEOREM. *Let X and Y be Banach spaces with computability structures. Let $\{e_n\}$ be a computable sequence in X whose linear span is dense in X (i.e. an effective generating set for X). Let $T : X \to Y$ be a closed linear operator whose domain $D(T)$ contains $\{e_n\}$ and such that the sequence $\{Te_n\}$ is computable in Y. Then T maps every computable element of its domain into a computable element of Y if and only if T is bounded.*

COMPLEMENT. *If T is bounded then more can be asserted. T maps every computable sequence in X into a computable sequence.*

The First Main Theorem is very easy to apply. This is because, in practice, all the hypotheses are automatically satisfied: the linear operator is closed and the domain space has an effective generating set $\{e_n\}$ such that $\{Te_n\}$ is computable in Y. *Thus to determine if T preserves computability we need only determine if T is bounded. But whether or not a linear operator is bounded is a classical fact which is well-known. We are not using any effective version of this fact.*

The theorem gives immediate answers to the questions raised in the beginning of this section. Wave propagation and differentiation are each associated with an unbounded operator, whereas the operators associated with heat dissipation and integration are bounded.

The First Main Theorem has also been applied to Fourier series, Fourier transforms, and a host of other constructs in analysis. One interesting sidelight is that we obtain an Effective Plancherel Theorem and an Effective Riemann–Lebesgue Lemma as immediate corollaries. Additional applications include operators associated with

Laplace's equation and sundry other phenomena (Pour-El and Richards [1989, pp. 104–111, 115–120]).

3.4. *The Second Main Theorem, the Eigenvector Theorem and related results*

We now turn our attention to the computability/noncomputability of eigenvalues and eigenvectors. Recall that in applications, the eigenvalues are the quantities which are measured. Hence it is of interest to determine whether they are computable. The eigenvectors represent the "state of the system".

An appropriate framework for the work in this section is Hilbert space. Recall that a Hilbert space is a Banach space in which the norm is given by an inner product – viz. $\|x\| = (x, x)^{1/2}$. For the most part we assume that our linear operators on the Hilbert space are self-adjoint. We further suppose that they are "effectively determined". (The precise definition of this concept is given below.) All of the well-known operators in analysis and physics are effectively determined. We emphasize the fact that the term "effectively determined", applies to the operator, itself, and not to the concepts derived from it, such as eigenvalue or spectrum.

Before proceeding further, it might be useful to review some well-known concepts. Let H be a Hilbert space. Let $T : H \to H$ be a linear operator. We assume that T is a *closed operator* – i.e. the domain of T is dense in H, and the graph of T is a closed subset of $H \times H$. The scalar λ is an *eigenvalue* of T if there is a nonzero vector $\boldsymbol{x}$ such that $T\boldsymbol{x} = \lambda\boldsymbol{x}$. The vector $\boldsymbol{x}$ is an *eigenvector* of T. The scalar λ is an element of the *spectrum* of T if $T - \lambda I$ has no bounded inverse. Note that an eigenvalue is always an element of the spectrum. However the converse does not hold. In general there are elements of the spectrum which are not eigenvalues.

We now turn our attention to the basic definition of this section, the definition of an effectively determined operator.

DEFINITION 9. A closed operator $T : H \to H$ is *effectively determined* if there is a computable sequence $\{e_n\}$ in H such that the sequence of pairs $\{\langle e_n, Te_n\rangle\}$ is an effective generating set for the graph of T.

We now turn to the Second Main Theorem. This theorem deals with the computability of eigenvalues. We begin with a statement of the theorem. Then we give some corollaries.

Roughly, the Second Main Theorem says that each eigenvalue of an effectively determined self-adjoint operator, whether bounded or unbounded, is computable. However the sequence of eigenvalues need not be computable. Thus, in general, there is no way of arranging the eigenvalues in a sequence $\{\lambda_n\}$ so that for some master program M, $M(n)$ effectively approximates the nth element of the sequence. Actually the Second Main Theorem says quite a bit more. The precise statement is:

SECOND MAIN THEOREM. *Let $T : H \to H$ be an effectively determined (bounded or unbounded) self-adjoint operator. Then there exists a computable sequence of real numbers $\{\lambda_n\}$ and a recursively enumerable set A of natural numbers such that:*

(i) *Each $\lambda_n \in$ spectrum (T), and the spectrum of T is the closure of $\{\lambda_n\}$.*

(ii) *The set of eigenvalues of T is the set $\{\lambda_n\colon\ n \in \mathbb{N} - A\}$. In particular, each eigenvalue of T is computable.*

(iii) *Conversely every set which is the closure of $\{\lambda_n\}$ as in* (i) *above is the spectrum of an effectively determined self-adjoint operator.*

(iv) *Likewise, every set $\{\lambda_n\colon\ n \in \mathbb{N} - A\}$ as in* (ii) *above is the set of eigenvalues of some effectively determined self-adjoint operator T. If the set $\{\lambda_n\}$ is bounded, then T can be chosen to be bounded.*

We now take up some theorems which are immediate consequences of the Second Main Theorem. Although the proofs are not difficult we will not give them. Theorem 8 below, the theorem on operator norm, which states that the norm of an effectively determined bounded self-adjoint operator can be noncomputable, may seem surprising.

THEOREM 7 (Sequence of eigenvalues). *There exists an effectively determined bounded self-adjoint operator $T : H \to H$ such that the sequence of eigenvalues is not computable.*

This theorem follows easily from (iv) of the Second Main Theorem.

We next consider the theorem on operator norm mentioned above.

THEOREM 8 (Norm). *There exists an effectively determined, bounded, self-adjoint operator $T : H \to H$ whose norm is not a computable real.*

THEOREM 9 (Compact operators). *Let $T : H \to H$ be an effectively determined compact self-adjoint operator. Then the set of eigenvalues can be arranged in a computable sequence.*

The Second Main Theorem and its corollaries depend heavily on the fact that the operator is self-adjoint. Analogous results hold for bounded normal operators. The following theorem shows that the Second Main Theorem fails for operators which are neither self-adjoint nor normal.

THEOREM 10 (Noncomputable eigenvalue). *There exists a bounded operator T which is effectively determined (but not self-adjoint or normal) which has a noncomputable real as an eigenvalue.*

This concludes our discussion of the Second Main Theorem. We now take up our third result, the Eigenvector Theorem. This theorem is concerned with the computability/noncomputability of eigenvectors associated with a linear operator. Recall that eigenvectors represent the "state of the system".

EIGENVECTOR THEOREM. *Let $H = L^2[0, 1]$ with the usual computability structure (See Part* I, *Section* 2.5). *There exists an effectively determined, compact, self-adjoint operator $T : H \to H$ such that*

(1) $\lambda = 0$ *is an eigenvalue of* T.

(2) *None of the eigenvectors associated with* $\lambda = 0$ *is computable.*

Thus the "state of the system" need not be computable.

4. Addendum: open problems

There are many open problems. Some were discussed in Pour-El and Richards [1989, pp. 192–194]. In recent years progress has been made in two of the seven problem areas listed in Pour-El and Richards [1989].

The first is concerned with Problem 5, the recursive topology for $\mathbb{R}^n$. The notion of an r.e. open set is well-known,. Using the definitions of an r.e. closed set and of a recursive closed set formulated by Pour-El and Richards in the 1980's, Qing Zhou investigated the recursive topology of $\mathbb{R}^n$ in his Ph.D. thesis (Zhou Qing [1992]). One result of the thesis is the Effective Riemann Mapping Theorem.

The second is associated with Problem 2, the fine structure for noncomputable processes via the theory of the degrees of unsolvability. Some results have been obtained by Anthony Dunlop and the author in two directions: degree theory for real numbers and degree theory for elements of $C[a, b]$, where a and b are recursive reals. These results will appear in Dunlop's Ph.D. thesis. Much more remains to be done in this area. Dunlop and Pour-El continue to work on this problem.

A third problem, not mentioned in Pour-El and Richards [1989], is the problem of finding analogs for the notion of a "computability structure" for spaces which are not Banach spaces. For example, can one characterize the notion of computability on an arbitrary metric space in a meaningful way? The work of Washihara and Yasugi on computability for a Fréchet space may be useful in this regard (Washihara [1995]; Washihara and Yasugi [ta]). The work of Weihrauch [1987, 1993] on computability and computable metric spaces should also be useful.

Recently Pour-El has written a paper (Pour-El [ta]) in which five additional open problems are formulated. Two problems are concerned with the computability structure on a Banach space. The remaining three can be viewed more generally.

The problems discussed above together with the other problems discussed in Pour-El and Richards [1989] represent a sampling of some of the open problems in recursive analysis. Much remains to be done in this field.

References

A. GRZEGORCZYK

[1957] On the definitions of computable real continuous functions, *Fund. Math.*, 44, pp. 61–71.

D. LACOMBE

[1955a] Extension de la notion de fonction récursive aux fonctions d'une ou plusieurs variables réelles I, *C. R. Acad. Sci. Paris*, 240, pp. 2478–2480.

[1955b] Extension de la notion de fonction récursive aux fonctions d'une ou plusieurs variables réelles II, III, *C. R. Acad. Sci. Paris*, 241, pp. 13–14, 151–153.

[1957] Quelques propriétés d'analyse récursive, *C. R. Acad. Sci. Paris*, 244, pp. 838–840, 996–997.

G. METAKIDES, A. NERODE AND R. A. SHORE

[1983] Recursive limits on the Hahn–Banach theorem, in: *E. Bishop – Reflection on Him and Research, 1983, San Diego*, M. Rosenblatt, ed., Contemp. Math., Vol. 39, Amer. Math. Soc., Providence, RI, pp. 85–91.

J. MYHILL

[1971] A recursive function, defined on a compact interval and having a continuous derivative that is not recursive, *Michigan Math. J.*, 18, pp. 97–98.

M. B. POUR-EL

[ta] From axiomatics to intrinsic characterization: Some open problems in computable analysis, *Theoret. Comput. Sci. A,* to appear.

M. B. POUR-EL AND I. RICHARDS

[1981] The wave equation with computable initial data such that its unique solution is not computable, *Adv. Math.*, 39, pp. 215–239.

[1989] *Computability in Analysis and Physics*, Springer, Berlin.

H. G. RICE

[1954] Recursive real numbers, *Proc. Amer. Math. Soc.*, 5, pp. 784–791.

R. M. ROBINSON

[1951] Review of Péter, R.: "Rekursive Funktionen" Akad. Kiado, Budapest, 1951, *J. Symbolic Logic*, 16, pp. 280–282.

R. SOARE

[1987] *Recursively Enumerable Sets and Degrees*, Springer, Berlin.

M. WASHIHARA

[1995] Computability and Fréchet spaces, *Math. Japonica*, 42, 1, pp. 1–13.

M. WASHIHARA AND M. YASUGI

[ta] Computability and metrics in a Fréchet space, to appear.

K. WEIHRAUCH

[1987] *Computability*, Springer, Berlin.

[1993] Computability on computable metric spaces, *Theoret. Comput. Sci.*, 113, pp. 191–210.

QING ZHOU

[1992] *Computability on Open and Closed Subsets of Euclidean Space*, Ph.D. Thesis, University of Minnesota.

CHAPTER 14

Theory of Numberings*

Yuri L. Ershov

Novosibirsk State University, Mechanics and Mathematics Department, Novosibirsk, 630090 Russia

Contents

1. Basic notions 477
2. Computable numberings 486
3. Selected topics 495
   3.1. Index sets 495
   3.2. Creativity and m-universality 499
   3.3. Numbered sets with approximation and the problem P 500
References 502

*Supported by Russian Foundation for Fundamental Research, project code 93-11-16014.

HANDBOOK OF COMPUTABILITY THEORY
Edited by E.R. Griffor

The present survey is essentially an introduction to the problems of the theory of numberings, a developing branch of the algorithm theory. To the author's knowledge, for the first time the idea of a systematical study of numbered sets was proposed by A.N. Kolmogorov in mid-fifties. At that time V.A. Uspenskii [1955a, 1955b] began the work along these lines on computable numberings. Simultaneously, a number of other mathematicians (Rice, Myhill, Friedberg, Lachlan, Lacombe, Pour-El, and others) were involved in developments related to computable numberings.

Independently, an attempt was made (Fröhlich and Shepherdson, Rabin) to study constructive (numbered) algebras which revealed interesting numbering-related peculiarities.

A survey by A.I. Mal'tsev [1961], was the first attempt to synthesize both directions of development and to formulate basic notions of the theory of numberings.

By now elements of the theory of numberings are represented to varying extent in a series of monographs (Mal'tsev [1965], Odifreddi [1989], Weihrauch [1987], etc.). The monograph by Ershov [1977] is devoted entirely to the theory of numberings and contains almost exhaustive information about the research on the theory of numberings up to the year 1977.

In the present survey we formulate basic notions and results contained already in the book and also make an attempt to describe modern trends and achievements.

As an introduction we supply a translation of the first section of introductory chapter from Ershov [1977] which preserves its significance to date.

Introduction

The most "invariant" part of all existing nowadays in mathematics refinements of the notion of algorithm is the class of partial recursive arithmetic functions (a partial arithmetic function is a partial mapping of a finite Cartesian power of the set of all natural numbers N into N), that is those partial arithmetic functions which are computable under any refinement of algorithmic computability that was suggested. Essentially, this invariance enable us to consider all the refinements equivalent and strengthen our belief that the class of all partial recursive functions coincides with the class of all partial arithmetic functions admitting effective (in an intuitive sense) computation.

Thus it is desirable that all investigations in the algorithm theory and its applications were carried out on the "common ground" of the class of partial recursive functions. One of the methods of reduction to natural numbers and arithmetic functions which has already been successfully used time and again in the algorithm theory and mathematical logic is the use of an appropriate numbering, i.e. a mapping of some subset of N onto the class of constructive objects (formulas, words, matrices, etc.) in question. The most brilliant example of the use of a numbering is Gödel's proof of his famous incompleteness theorems.

The theory of numberings is a branch of the algorithm theory dedicated to a solution of problems arising in the above mentioned reduction to the "common ground" on a basis of the notion of a numbered set.

In the theory of numberings the necessary system of notions is being developed; natural questions are being arisen and corresponding answers are being found. Those are questions of dependency of various properties of a set on the choice of a numbering, an existence (uniqueness) of a numbering of a set possessing certain properties, etc. And as usual, in development of the theory the inherent problems arise leading to unexpected beautiful and difficult theorems.

Results obtained in the theory of numberings proved to be important for an understanding of some of the difficulties encountered in modern computer science. For instance, an important problem of contemporary programming is the problem of effective construction of a program to compute a function on a particular computer, given a program which computes the same function on the other computer. Practical realization of these translations for a pair of universal computers proves to be very difficult and often is not realized yet. To clarify that these difficulties could be of the fundamental nature we consider the following model for the situation in question.

By "universal computer" we will mean a computer that calculates a binary function Ψ which is universal for the class of all unary partial recursive functions, and by "a program to compute an unary partial recursive function φ" we will mean its number (one of its numbers), i.e. a number $n \in N$ such that $\varphi = \lambda x \Psi(n, x)$.

Thus, given two universal computers (i.e. two universal functions Ψ_0, Ψ_1), the "translation problem" may be formulated as the problem of existence of an unary recursive function f such that

$$\lambda n \lambda x \Psi_0(n, x) = \lambda n \lambda x \Psi_1\big(f(n), x\big).$$

However, as an investigation of computable numberings revealed, there exist universal functions Ψ_0, Ψ_1 such that there is no desirable function f. Moreover, there does exist Ψ_0, Ψ_1 such that translation from Ψ_0 to Ψ_1 is impossible as well as backward translation.

However, in the class of thus defined universal computers there exists "the most universal" one (which apparently is the only one deserving the name "universal") in a sense that a translation to the language of this computer is possible from any other universal computer.

If we restrict ourselves to the computers that computes only (total) recursive functions from a sufficiently rich class, the situation becomes even worse. For any such a computer there exists another one which computes the same class of functions while both translation problems are unsolvable.

This approach to explanation of difficulties in translation can be used to define a notion of complexity of a class of functions (by means of the semilattice of computable numberings of this class). Apparently this notion of complexity of a class of functions could be more useful in many respects than various notions of complexity of individual functions which are studied now.

A semilattice of (classes of equivalent) computable numberings proved to be a very interesting algebraic object related to numberings, and can serve to distinguish various internal structural properties of classes of recursively enumerable sets and partial recursive functions.

It is precisely the use of an appropriate (principal) computable numbering that allowed S. Kleene to find the most general existence theorems in the theory of recursive functions (Recursion Theorem) which are comparable in significance and the place occupied with main existence theorems in the theory of differential equations.

A notion of a numbering adapted in the book is somewhat restricted; namely, a numbering is a mapping of the whole set N, and not of a subset, onto the corresponding set. This restriction is imposed by the fact that, on one hand, even the restricted notion is sufficient for the investigation of computable numberings, finitely generated algebras, and numerous other interesting and important objects. On the other hand, for the most extended notion of a numbering there is no unambiguously defined notion of reducibility.

Thus, due to this restriction, some important branches of the theory fall out of consideration. Those are systems of notation for ordinals in the theory of recursive functions, some aspects of constructive mathematics, and some other.

A broader viewpoint enables us to consider the theory of numberings as a kind of a modern form of Pythagorism; more precisely, as one of possible answers to the question: what can be constructed in mathematics, given only a notion of a natural number? The most rich construction over the set of natural numbers N is that of a factor set, since finite Cartesian powers of N, for instance, can be naturally identified with N itself. Mappings of factor sets of N into each other which can be realized in arithmetics are exactly those mappings which were called morphisms in the theory of numberings. Indeed, the mappings should be realized by some mappings on indices (natural numbers) by means of mappings (functions) "definable" in arithmetic. But functions which are definable in elementary arithmetic under natural sharpening of this "definability" turn out to be exactly partial recursive functions. Any constructive restrictions on equivalence relations on N are not justified since the study of a class of objects of any kind by means of their names (numbers) together with an equivalence relation (which could be not known in its entirety and be arbitrarily complex) defined by the predicate "is a name of the same object" is a frequently encountered methodological procedure in cognition.

Usually we follow traditional mathematical notation but made occasional use of the next designations:

N – the set of all natural numbers;
$\mathfrak{W}$ – the set of all recursively enumerable subsets of N;
$\mathfrak{W}^n$ – the set of all n-ary recursively enumerable predicates on N;
$\mathfrak{PR}$ – the set of all unary partial recursive functions on N;
$\mathfrak{R}$ – the set of all unary recursive functions on N.

For a (partial) function g we denote by δg and ρg the domain and range of g respectively.

1. Basic notions

Any mapping ν of the set of all natural numbers N onto a set S ($\nu : N \overset{\text{onto}}{\longrightarrow} S$) is called a *numbering of the set* S. A pair $\mathfrak{S} = (S, \nu)$, where ν is a numbering of a set S, is called a *numbered set*.

EXAMPLES.

(1) The identity mapping $\mathrm{id}_N : N \to N$ is a numbering; denote by $\boldsymbol{N}$ the corresponding numbered set (N, id_N).

(2) If $P_\omega(N)$ is the set of all finite subsets of N, then a *canonical numbering* of $P_\omega(N)$ is a numbering $\gamma : N \to P_\omega(N)$ defined by its inverse mapping $\gamma^{-1} : P_\omega(N) \to N$ in the following way:

$$\gamma^{-1}(\emptyset) \leftrightharpoons 0; \qquad \gamma^{-1}(\{a_0 < \cdots < a_k\}) \leftrightharpoons \sum_{i \leqslant k} 2^{a_k}.$$

Denote by $\boldsymbol{\Gamma}$ the numbered set $(P_\omega(N), \gamma)$.

(3) The mapping $c : N^2 \to N$ defined by

$$c(x, y) \leftrightharpoons \frac{(x + y)^2 + 3x + y}{2}$$

is one-to-one and surjective. The inverse mapping $c^{-1} : N \to N^2$ is a numbering of the set N^2. Denote by r and l unary (recursive) functions such that $c^{-1}(x) = \langle l(x), r(x) \rangle$ for $x \in N$.

Let $S_0 \subseteq S$, ν be a numbering of S, and ν_0 be a numbering of S_0. The numbering ν_0 *is reducible to* ν ($\nu_0 \leqslant \nu$), if there exists an unary recursive function f such that $\nu_0 = \nu f$. If ν and ν_0 are numberings of S, $\nu_0 \leqslant \nu$, and $\nu \leqslant \nu_0$, then ν_0 and ν are called *equivalent* ($\nu_0 \equiv \nu$). If $S_0 \subseteq S$, $\nu_0 : N \to S_0$, $\nu : N \to S$ are numberings and $\nu_0 \leqslant \nu$, then we will use the notation $\mathfrak{S}_0 \leqslant \mathfrak{S}$ (for $\mathfrak{S}_0 = (S_0, \nu_0)$, $\mathfrak{S} = (S, \nu)$).

Denote by $Nu(S)$ the set of all numberings of a set S. The relation $\equiv$ on elements of $Nu(S)$ is an equivalence. Denote by $L(S)$ the factor set $Nu(S)/_\equiv$. The reducibility relation $\leqslant$ induces an order on $L(S)$ which will be denoted also by $\leqslant$.

The ordered set $\langle L(S), \leqslant \rangle$ *is an upper semilattice* as the following considerations imply:

Let $\nu_0, \nu_1 \in Nu(S)$. Define a numbering $\nu_0 \oplus \nu_1$ of the set S by

$$(\nu_0 \oplus \nu_1)(2x) \leftrightharpoons \nu_0(x); \quad (\nu_0 \oplus \nu_1)(2x + 1) \leftrightharpoons \nu_1(x), \qquad x \in N.$$

It is easy to verify that:

(1) $\nu_0 \leqslant \nu_0 \oplus \nu_1$, $\nu_1 \leqslant \nu_0 \oplus \nu_1$;

(2) if $\nu \in Nu(S)$, $\nu_0 \leqslant \nu$, $\nu_1 \leqslant \nu$, then $\nu_0 \oplus \nu_1 \leqslant \nu$.

These properties imply that if we denote by $[\nu]$ the class of all numberings equivalent to ν, then $[\nu_0 \oplus \nu_1]$ is the least upper bound of $[\nu_0]$ and $[\nu_1]$ in $\langle L(S), \leqslant\rangle$.

A numbering $\nu : N \to S$ induces a *numbering equivalence* η_ν on N:

$$\eta_\nu \leftrightharpoons \{\langle x, y\rangle \mid x, y \in N,\ \nu(x) = \nu(y)\}.$$

A numbering ν is called *decidable* (*positive, negative*), if η_ν is a recursive (recursively enumerable, complement to recursively enumerable) subset of N^2. A numbering ν is called *single-valued*, if $\eta_\nu = \{\langle x, x\rangle \mid x \in N\}$.

If ν is a positive numbering of S, then $[\nu]$ is a minimal element of $\langle L(S), \leqslant\rangle$.
An upper semilattice $\langle L(S), \leqslant\rangle$ has the least element iff S is finite.

If S is infinite there are continuum many minimal elements in $\langle L(S), \leqslant\rangle$, namely those which contain single-valued numberings.

If S is finite and contains more than one element, or S is countably infinite, then there is no biggest element in $\langle L(S), \leqslant\rangle$.

It is obvious if S_0 and S_1 are countably infinite, then upper semilattices $\langle L(S_0), \leqslant\rangle$ and $\langle L(S_1), \leqslant\rangle$ are isomorphic. (Every one-to-one mapping φ of S_0 onto S_1 induces bijection $\nu \mapsto \varphi\nu$, $\nu \in Nu(S_0)$, of the set $Nu(S_0)$ onto $Nu(S_1)$, which in turn induces an isomorphism of $\langle L(S_0), \leqslant\rangle$ and $\langle L(S_1), \leqslant\rangle$.)

Far less obvious is the following:

If S_0 and S_1 are finite and contain more than one element, then $\langle L(S_0), \leqslant\rangle$ and $\langle L(S_1), \leqslant\rangle$ are isomorphic.

This follows from the explicit description of the upper semilattice $\langle L(S), \leqslant\rangle$ for a finite S ($|S| \geqslant 2$) established in Ershov [1975] (see also Ershov [1977]). Note also that if $|S| = 2$, then $\langle L(S), \leqslant\rangle$ is naturally isomorphic to the upper semilattice $\langle L_m, \leqslant\rangle$ of m-degrees of proper subsets of N.

Sometimes it is convenient to assume that the empty set $\emptyset$ has some (unique) numbering 0 which is reducible to any numbering of any set. Denote by $Nu^*(S)$ the set of all numberings of all subsets (including $\emptyset$) of S. The reducibility relation $\leqslant$ is a preorder on $Nu^*(S)$. Let $L^*(S)$ be the factor set $Nu^*(S)/_\equiv$ and $\leqslant$ be the order on $L^*(S)$ induced by the reducibility relation.

For $\nu_0, \nu_1 \in Nu^*(S)$ define

$$\nu_0 \oplus \nu_1 = \nu_1 \oplus \nu_0 = \nu_0 \quad \text{if } \nu_1 = 0,$$
$$(\nu_0 \oplus \nu_1)(2x) \leftrightharpoons \nu_0(x); \quad (\nu_0 \oplus \nu_1)(2x+1) \leftrightharpoons \nu_1(x), \quad x \in N,$$

if $\nu_0 \neq 0$, $\nu_1 \neq 0$. Note that if ν_0 is a numbering of $S_0 \subseteq S$ and ν_1 is a numbering of $S_1 \subseteq S$, then $\nu_0 \oplus \nu_1$ is a numbering of $S_0 \cup S_1 \subseteq S$.

The element $[\nu_0 \oplus \nu_1]$ of the ordered set $\langle L^*(S), \leqslant\rangle$ is the least upper bound of the elements $[\nu_0]$ and $[\nu_1]$.

The upper semilattice $\langle L^*(S), \leqslant\rangle$ satisfies the following distributivity property:

If $[\nu] \leqslant [\nu_0 \oplus \nu_1]$, *then there exist* $\nu_0', \nu_1' \in Nu^*(S)$ *such that* $\nu_0' \leqslant \nu_0$, $\nu_1' \leqslant \nu_1$, *and* $\nu \equiv \nu_0' \oplus \nu_1'$.

For finite S the upper semilattice $\langle L(S), \leqslant\rangle$ is also distributive.

One of the main methodological problems of the theory of numberings is to select the "proper" numbering for the sets in observation.

One of the general approaches is to select the principal numbering.

We demonstrate it initially in the following environment: let $\mathfrak{S} = (S, \nu)$ be a numbered set and $S_0 \subseteq S$. A numbering $\nu_0 : N \to S_0$ is called *ν-computable* or *computable relative to* $\mathfrak{S}$, if $\nu_0 \leqslant \nu$, i.e. ν_0 is reducible to ν. A subset $S_0 \subseteq S$ is called *ν-computable subset* of $\mathfrak{S}$, if there exists a ν-computable numbering of S_0. A ν-computable numbering $\nu_0 : N \to S_0$ is called *principal computable numbering relative to* $\mathfrak{S}$, if for every ν-computable numbering $\nu_0' : N \to S_0$ we have $\nu_0' \leqslant \nu_0$. If ν_0 is a principal numbering of S_0, then we call S_0 a *principal subset of* $\mathfrak{S}$, and $\mathfrak{S}_0 = (S_0, \nu_0)$ the *principal numbered subset of* $\mathfrak{S}$.

REMARK. If there exists a principal numbering ν_0 for the subset $S_0 \subseteq S$, one may assume that ν_0 is the natural numbering of S_0 induced by the numbering $\nu : N \to S$ of the set S.

If $S_0 \subseteq S$ and $\nu_0 : N \to S_0$ is a (principal) numbering computable relative to $\mathfrak{S} = (S, \nu)$, then we will write $\mathfrak{S}_0 \leqslant \mathfrak{S}$ ($\mathfrak{S}_0 \leqslant_p \mathfrak{S}$).

If $\mathfrak{S}_0 \leqslant_p \mathfrak{S}$, $S_1 \subseteq S_0$, *and* $\nu_1 : N \to S_1$ *is a computable relative to* $\mathfrak{S}$ *numbering, then* $\nu_1 \leqslant \nu_0$.

Indeed, $\nu_0 \oplus \nu_1$ is a ν-computable numbering of S_0; hence, $\nu_0 \oplus \nu_1 \leqslant \nu_0$ and $\nu_1 \leqslant \nu_0$.

There exist interesting sufficient conditions for a subset $S_0 \subseteq S$ to be principal in $\mathfrak{S} = (S, \nu)$.

(1) S_0 is called a *wn-subset* of $\mathfrak{S}$, if there exists a partial recursive function g such that $\delta g \supseteq \nu^{-1}(S_0)$, $\nu g(n) \in S_0$ for all $n \in \delta g$, and if $n \in \nu^{-1}(S_0)$, then $\nu(n) = \nu g(n)$.

(2) S_0 is called a *n-subset* of $\mathfrak{S}$, if there exists total recursive function g satisfying the properties stated in 1.

(3) S_0 is called a *e-subset* of $\mathfrak{S}$, if $\nu^{-1}(S_0)$ is recursively enumerable.

It follows easily from the definitions above that

If S_0 *is a n- or e-subset of* $\mathfrak{S}$, *then* S_0 *is also wn-subset of* $\mathfrak{S}$.
If S_0 *is an wn-subset of* $\mathfrak{S}$, *then* S_0 *is a principal subset of* $\mathfrak{S}$.

We use the notation $\mathfrak{S}_0 \leqslant_{wn} \mathfrak{S}$ ($\mathfrak{S}_0 \leqslant_n \mathfrak{S}$, $\mathfrak{S}_0 \leqslant_e \mathfrak{S}$) if S_0 is a *wn*-subset (*n*-subset, *e*-subset) of $\mathfrak{S}$ and ν_0 is a principal numbering of S_0 ($\mathfrak{S}_0 = (S_0, \nu_0) \leqslant_p \mathfrak{S}$).

Let $\mathfrak{S}_0 = (S_0, \nu_0)$ and $\mathfrak{S}_1 = (S_1, \nu_1)$ be numbered sets. A *morphism* from $\mathfrak{S}_0$ to $\mathfrak{S}_1$ is any mapping $\mu : S_0 \to S_1$ such that there exists a recursive function f such that $\mu\nu_0 = \nu_1 f$; the notation $\mu : \mathfrak{S}_0 \to \mathfrak{S}_1$ assumes that μ is a morphism from $\mathfrak{S}_0$ to $\mathfrak{S}_1$.

The following simple observation can serve as a ground of the definition of some natural topology on an arbitrary numbered set:

If $\mu : \mathfrak{S}_0 \to \mathfrak{S}_1$ is a morphism and $S \subseteq S_1$ is a e-subset of $\mathfrak{S}_1$, then $\mu^{-1}(S) \subseteq S_0$ is a e-subset of $\mathfrak{S}_0$.

Since intersection of two e-subsets of $\mathfrak{S}$ is again an e-subset of $\mathfrak{S}$, the family $e(\mathfrak{S})$ of all e-subsets of $\mathfrak{S}$ can be taken as a basis of some topology (e-topology) on S. Now the proposition implies

Every morphism $\mu : \mathfrak{S}_0 \to \mathfrak{S}_1$ is a continuous mapping in e-topologies on S_0 and S_1.

One can also define a preorder $\leqslant_\nu$ on S by

$$s \leqslant_\nu s' \leftrightharpoons \text{for every } e\text{-subset } S_0 \subseteq S, \text{ if } s \in S_0, \text{ then } s' \in S_0.$$

Now we have:

If $\mu : \mathfrak{S}_0 \to \mathfrak{S}_1$ is a morphism and $s \leqslant_{\nu_0} s'$, then $\mu(s) \leqslant_{\nu_1} \mu(s')$.

A numbered set $\mathfrak{S}$ is called *separated*, if the preorder $\leqslant_\nu$ is a partial order.

A morphism $\mu : \mathfrak{S}_0 \to \mathfrak{S}_1$ is called a *monomorphism* (*epimorphism*), if μ is an injective (surjective) mapping.

A morphism $\mu : \mathfrak{S}_0 \to \mathfrak{S}_1$ is called an *isomorphism* if there exists a morphism $\lambda : \mathfrak{S}_1 \to \mathfrak{S}_0$ such that $\lambda\mu = \mathrm{id}_{S_0}$ and $\mu\lambda = \mathrm{id}_{S_1}$.

An epimorphism $\mu : \mathfrak{S}_0 \to \mathfrak{S}_1$ is called a *factorization* if for any epimorphism $\mu' : \mathfrak{S}_0 \to \mathfrak{S}'$ and any monomorphism $\mu_1 : \mathfrak{S}' \to \mathfrak{S}_1$ such that $\mu_1\mu' = \mu$, μ_1 is an isomorphism.

Let $\mathfrak{S} = (S, \nu)$ be a numbered set, η be an equivalence relation on S. Define $\mathfrak{S}/_\eta \leftrightharpoons (S/_\eta, \nu_\eta)$, where

$$\begin{aligned} S/_\eta &\leftrightharpoons \{[s]_\eta \mid s \in S\}; \\ [s]_\eta &\leftrightharpoons \{s' \mid s' \in S, \langle s, s'\rangle \in \eta\}, \quad s \in S; \\ \nu_\eta(n) &\leftrightharpoons [\nu(n)]_\eta, \quad n \in N. \end{aligned}$$

The mapping $\varphi_\eta : s \mapsto [s]_\eta$, $s \in S$, is a factorization $\varphi_\eta : \mathfrak{S} \to \mathfrak{S}/_\eta$.

If $\mathfrak{S} = (S, \nu) \neq 0$, then the numbering ν is a factorization $\nu : \boldsymbol{N} \to \mathfrak{S}$ and $\boldsymbol{N}/_{\eta_\nu}$ is isomorphic to $\mathfrak{S}$.

Any morphism $\mu : \mathfrak{S}_0 \to \mathfrak{S}_1$ can be presented as a composition $\mu = \overline{\mu}\varphi_\mu$ of the factorization $\varphi_\mu \leftrightharpoons \varphi_{\eta_\mu} : \mathfrak{S}_0 \to \mathfrak{S}_0/_{\eta_\mu}$, where

$$\eta_\mu \leftrightharpoons \{\langle s, s'\rangle \mid s, s' \in S_0,\ \mu(s) = \mu(s')\},$$

and the monomorphism $\overline{\mu} : \mathfrak{S}_0/_{\eta_\mu} \to \mathfrak{S}_1$, which is defined by

$$\overline{\mu}([s]_{\eta_\mu}) \leftrightharpoons \mu(s), \quad s \in S_0.$$

This representation is unique in the following sense: if $\varphi' : \mathfrak{S}_0 \to \mathfrak{S}'$ is a factorization and $\mu' : \mathfrak{S}' \to \mathfrak{S}_1$ is a monomorphism such that $\mu'\varphi' = \mu$, then there exists an isomorphism $\lambda : \mathfrak{S}_0/_{\eta_\mu} \to \mathfrak{S}'$ such that $\lambda\varphi_\mu = \varphi$ and $\mu'\lambda = \overline{\mu}$.

The class of all numbered sets and the morphisms that we introduced constitute the category $\mathfrak{N}$ of numbered sets. Many notions of the theory of numberings that we introduce later are quite natural from the point of view of the category theory.

If $\mathfrak{S}_0 = (S_0, \nu_0)$ and $\mathfrak{S}_1 = (S_1, \nu_1)$ are numbered sets different from 0, then their *direct product* is the numbered set $\mathfrak{S}_0 \times \mathfrak{S}_1 \leftrightharpoons (S_0 \times S_1, \nu_0 \times \nu_1)$, where the numbering $\nu_0 \times \nu_1 : N \to S_0 \times S_1$ is defined by

$$(\nu_0 \times \nu_1)(x) \leftrightharpoons \langle \nu_0 l(x), \nu_1 r(x) \rangle, \quad x \in N.$$

If $\mathfrak{S}_0 = 0$ or $\mathfrak{S}_1 = 0$, then

$$\mathfrak{S}_0 \times \mathfrak{S}_1 \leftrightharpoons 0.$$

If $\mathfrak{S}_0 = (S_0, \nu_0)$ and $\mathfrak{S}_1 = (S_1, \nu_1)$ are numbered sets different from 0, then their *direct sum* is the numbered set

$$\mathfrak{S}_0 + \mathfrak{S}_1 \leftrightharpoons (S'_0 \cup S'_1, \nu'_0 \oplus \nu'_1),$$

where

$$\begin{aligned} S'_0 &\leftrightharpoons S_0 \times \{0\}, \qquad S'_1 \leftrightharpoons S_1 \times \{1\}; \\ \nu'_0(x) &\leftrightharpoons \langle \nu_0(x), 0 \rangle, \quad x \in N; \\ \nu'_1(x) &\leftrightharpoons \langle \nu_1(x), 1 \rangle, \quad x \in N. \end{aligned}$$

(If $S_0 \cap S_1 = \emptyset$, there is a simpler definition of direct sum: $(S_0 \cup S_1, \nu_0 \oplus \nu_1)$.)

If $\mathfrak{S}_0 = 0$, then $\mathfrak{S}_0 + \mathfrak{S}_1 \leftrightharpoons \mathfrak{S}_1$; if $\mathfrak{S}_1 = 0$, then $\mathfrak{S}_0 + \mathfrak{S}_1 \leftrightharpoons \mathfrak{S}_0$.

There exist naturally defined monomorphisms

$$\mu_i : \mathfrak{S}_i \to \mathfrak{S}_0 + \mathfrak{S}_1, \quad i = 0, 1.$$

Now we introduce two most interesting classes of numbered sets (numberings) which are defined by properties similar to injectivity in the category $\mathfrak{N}$. It is shown in [Ershov, 1977] that other possible injectivity properties in the category $\mathfrak{N}$ lead to trivial objects.

A numbered set $\mathfrak{S}$ is called *completely numbered* (*complete*), if for any $\mathfrak{S}_0 \leqslant_e \mathfrak{S}_1$ and a morphism $\mu_0 : \mathfrak{S}_0 \to \mathfrak{S}$ there exists a morphism $\mu_1 : \mathfrak{S}_1 \to \mathfrak{S}$ such that $\mu_1 \restriction S_0 = \mu_0$.

A numbered set $\mathfrak{S}$ is called *precompletely numbered* (*precomplete*), if for any $\mathfrak{S}_0 \leqslant \boldsymbol{N}$ and a morphism $\mu_0 : \mathfrak{S}_0 \to \mathfrak{S}$ there exists a morphism $\mu_1 : \boldsymbol{N} \to \mathfrak{S}$ such that $\mu_1 \restriction S_0 = \mu_0$.

Since $\mathfrak{S}_0 \leqslant N$ implies $\mathfrak{S}_0 \leqslant_e N$ (verify!), every completely numbered set is precomplete.

If $\mathfrak{S} = (S, \nu)$ is a completely (precompletely) numbered set, then the numbering ν is called *complete (precomplete)*.

Now we indicate the most essential properties (see Ershov [1977]) of complete and precomplete numberings (numbered sets).

(1) *A numbered set $\mathfrak{S} = (S, \nu)$ is precomplete iff for any unary partial recursive function g there exists an unary total recursive function h such that $\nu g(k) = \nu h(k)$ for $k \in \delta g$.*

(2) *If $\mathfrak{S}_0 \leqslant_{wn} \mathfrak{S}$ and $\mathfrak{S}_0$ is precompletely numbered, then $\mathfrak{S}_0 \leqslant_n \mathfrak{S}$.*

(3) (The Rice Theorem) *If $\mathfrak{S} = (S, \nu)$ is precomplete, $S_0 \subseteq S$, and $\nu^{-1}(S_0)$ is recursive, then $S_0 = \emptyset$ or $S_0 = S$.*

(4) *If $\mathfrak{S}_0 \leqslant_n \mathfrak{S}$ and $\mathfrak{S}$ is precomplete, then $\mathfrak{S}_0$ is also precomplete.*

(5) *If $\mu : \mathfrak{S} \to \mathfrak{S}_0$ is a factorization and $\mathfrak{S}$ is (pre)complete, then $\mathfrak{S}_0$ is also (pre)complete.*

(6) *Direct product of (pre)completely numbered sets is also (pre)completely numbered.*

(7) (Recursion Theorem) *If $\mathfrak{S} = (S, \nu)$ is precomplete, then for any $n \in N$ and any $(n+1)$-ary partial recursive function G there exists n-ary recursive function Q such that for any $x_1, \ldots, x_n \in N$ if $\langle Q(x_1, \ldots, x_n), x_1, \ldots, x_n\rangle \in \delta G$, then $\nu Q(x_1, \ldots, x_n) = \nu G(Q(x_1, \ldots, x_n), x_1, \ldots, x_n)$.*

(8) (Fixed-Point Theorem) *If $\mathfrak{S}$ is a precompletely numbered set and $\mu : \mathfrak{S} \to \mathfrak{S}$ is a morphism, then there exists $s \in S$ such that $\mu(s) = s$.*

(9) *If $\nu : N \to S$ is a precomplete numbering, $\nu_0 : N \to S_0 \subset S$ and $\nu_1 : N \to S_1 \subset S$ are numberings of subsets of S and $\nu \equiv \nu_0 \oplus \nu_1$, then $\nu \equiv \nu_0$ or $\nu \equiv \nu_1$.*

(10) *If $\mathfrak{S} = (S, \nu)$ is precompletely numbered and $\mu : \mathfrak{S} \to \mathfrak{S}_0 = (S_0, \nu_0)$ is a morphism such that $|\mu(S)| \geqslant 2$, then there exists an injective recursive function f such that $\mu\nu = \nu_0 f$.*

(11) *If $\nu : N \to S$ is a precomplete numbering, $\nu_0 \in Nu(S)$, and $\nu \equiv \nu_0$, then there exists a recursive permutation f such that $\nu_0 = \nu f$.*

A characteristic property of complete numberings is the presence of a special element.

Let $\nu : N \to S$ be a numbering. An element $s_0 \in S$ is called *special* (for the numbering ν), if for every unary partial recursive function g there exists unary recursive function h such that for any $k \in N$

$$\text{if } k \in \delta g, \text{ then } \nu g(k) = \nu h(k),$$
$$\text{if } k \notin \delta g, \text{ then } \nu h(k) = s_0.$$

(12) *A numbering $\nu : N \to S$ is complete iff it has a special element.*

Note that if $s_0 \in S$ is a special element of a complete numbered set $\mathfrak{S}$, $S_0 \subseteq S$ is a e-subset, and $s_0 \in S_0$, then $S_0 = S$. It follows that if $\mathfrak{S}$ is separated, then a special element is the least element (and hence, is unique) with respect to the order $\leqslant_\nu$.

REMARK. A complete numbered set may contain more than one special element (Denisov and Lavrov [1970]).

Now we show the existence of sufficiently many complete numbered sets. We will use unary partial recursive function u with the following property: for any unary partial recursive function g there exists a recursive function h such that $g = uh$. Functions with this property do exist and are called *universal* (see next section for further information).

Let u be a universal function, $\mathfrak{S} = (S, \nu)$ be an arbitrary numbered set different from 0, and assume $* \notin S$. Put $S^* \leftrightharpoons S \cup \{*\}$ and define $\nu^* : N \to S^*$ by

$$\text{if } k \in \delta u, \text{ then } \nu^*(k) \leftrightharpoons \nu u(k),$$
$$\text{if } k \notin \delta u, \text{ then } \nu^*(k) \leftrightharpoons *.$$

Then

$\mathfrak{S}^* \leftrightharpoons (S^*, \nu^*)$ *is a complete numbered set with unique special element* $*$, *and* $\mathfrak{S} \leqslant_e \mathfrak{S}^*$.

We can also construct an interesting example of a precomplete numbered set with the help of a universal function u.

Let η_* be the least equivalence relation on N which contains the graph $\boldsymbol{\Gamma}_u$ of a universal function u. Then η_* is recursively enumerable and hence the numbered set $\mathfrak{S}_* \leftrightharpoons \boldsymbol{N}/_{\eta_*}$ is positive.

The following assertions holds for the numbered set $\mathfrak{S}_*$:

(a) *there exists* $\mathfrak{S}_0 \leqslant_n \boldsymbol{\Pi}$ *isomorphic to* $\mathfrak{S}_*$ (see the next section for the definition of $\boldsymbol{\Pi}$);

(b) $\mathfrak{S}_*$ *is a precomplete numbered set*;

(c) *for every precomplete numbered set* $\mathfrak{S}_1$ *there exists a factorization* $\mu : \mathfrak{S}_* \to \mathfrak{S}_1$;

(d) *for every positive numbered set* $\mathfrak{S}_1$ *there exists* $\mathfrak{S}_0 \leqslant \mathfrak{S}_*$ *isomorphic to* $\mathfrak{S}_1$ (*note that positiveness of* $\mathfrak{S}_*$ *and* $\mathfrak{S}_0 \leqslant \mathfrak{S}_*$ *imply* $\mathfrak{S}_0 \leqslant_e \mathfrak{S}_*$).

Note that $\mathfrak{S}_*$ is not complete since $\mathfrak{S}_*$ is separated and the order $\leqslant_{\nu^*}$ coincides with identity relation.

A. Lachlan [1987] proved the following uniqueness theorem:

THEOREM. *If* $\mathfrak{S}$ *is a positive precomplete numbered set containing at least two elements, then* $\mathfrak{S}$ *and* $\mathfrak{S}_*$ *are isomorphic.*

Now we proceed to the search of natural numberings of morphisms. Let $\mathfrak{S}_0$, $\mathfrak{S}_1$ be numbered sets and $M(\mathfrak{S}_0, \mathfrak{S}_1)$ be the set of all morphisms from $\mathfrak{S}_0$ to $\mathfrak{S}_1$. Let $S \subseteq M(\mathfrak{S}_0, \mathfrak{S}_1)$ and $\nu : N \to S$ be a numbering. We associate with ν a natural mapping $\mu_\nu : N \times S_0 \to S_1$ defined by:

$$\mu_\nu(n, s_0) \leftrightharpoons \big[\nu(n)\big](s_0), \quad n \in N,\ s_0 \in S_0.$$

We call a numbering $\nu : N \to S \subseteq M(\mathfrak{S}_0, \mathfrak{S}_1)$ of a family S of morphisms from $\mathfrak{S}_0$ to $\mathfrak{S}_1$ *computable*, if the mapping $\mu_\nu : N \times S_0 \to S_1$ is a morphism from $\boldsymbol{N} \times \mathfrak{S}_0$ to $\mathfrak{S}_1$.

A numbering $\nu : N \to S \subseteq M(\mathfrak{S}_0, \mathfrak{S}_1)$ is computable iff there exists a binary recursive function g such that $[\nu(n)](\nu_0(m)) = \nu_1 g(n, m)$ for all $n, m \in N$.

Now naturally arises the problem of existence (for a pair of numbered sets $\mathfrak{S}_0, \mathfrak{S}_1$) of a principal computable numbering of the family $M(\mathfrak{S}_0, \mathfrak{S}_1)$, or *the problem P for a pair* $\mathfrak{S}_0, \mathfrak{S}_1$ for short.

If the problem P for a pair $\mathfrak{S}_0, \mathfrak{S}_1$ is solvable and $\nu : N \to M(\mathfrak{S}_0, \mathfrak{S}_1)$ is a principal computable numbering, then the numbered set $(M(\mathfrak{S}_0, \mathfrak{S}_1), \nu)$ will be denoted by $\boldsymbol{M}(\mathfrak{S}_0, \mathfrak{S}_1)$. Note that a principal computable numbering ν (if exists) is unique up to equivalence.

Now we cite a theorem (Ershov [1971, 1977]) that provides category-theoretical justification for this definition.

THEOREM. *The following conditions are equivalent for any pair* $\mathfrak{S}_0, \mathfrak{S}_1$ *of numbered sets*:

(a) *the problem P for the pair* $\mathfrak{S}_0, \mathfrak{S}_1$ *is solvable*;

(b) *the functor* $\lambda\mathfrak{S} M(\mathfrak{S} \times \mathfrak{S}_0, \mathfrak{S}_1) : \mathfrak{N} \to$ *Set is representable.*

If the above conditions are satisfied, then the numbered set $\boldsymbol{M}(\mathfrak{S}_0, \mathfrak{S}_1)$ is a solution to the problem of representation of this functor, i.e. the functors $\lambda\mathfrak{S} M(\mathfrak{S} \times \mathfrak{S}_0, \mathfrak{S}_1)$ and $\lambda\mathfrak{S} M(\mathfrak{S}, \boldsymbol{M}(\mathfrak{S}_0, \mathfrak{S}_1))$ are naturally equivalent.

The book Ershov [1977] contains many results concerning the solvability of the problem P. In Section 3 we formulate a sufficient condition for the solvability of the problem P which is used to define the class of partial computable functionals of finite types. Here we consider only a natural question of solvability of the problem P for the pair $\boldsymbol{N}, \mathfrak{S}$, where $\mathfrak{S}$ is an arbitrary numbered set.

It is easy to verify that

> *If* $\mathfrak{S}$ *is a precomplete numbered set, then the problem* P *is solvable for the pair* $\boldsymbol{N}, \mathfrak{S}$ *(and* $\boldsymbol{M}(\boldsymbol{N}, \mathfrak{S})$ *is precomplete).*

The converse is nontrivial and was proved by S. Dvornikov [1979].

> *If the problem* P *is solvable for a pair* $\boldsymbol{N}, \mathfrak{S}$*, then* $\mathfrak{S}$ *is a precomplete numbered set.*

The definition of reducibility of numbering reveals close connection to the notion of m-reducibility. This connection becomes even more apparent in the development related to the notion of multiple m-reducibility.

Let Λ be an arbitrary nonempty set. A *Λ-sequence* is an arbitrary mapping $\boldsymbol{A} : \Lambda \to P(N)$ of Λ into the set $P(N)$ of all subsets of N. We use the following notation for Λ-sequences:

$$\boldsymbol{A} \leftrightharpoons \{A_\lambda \mid \lambda \in \Lambda\}, \quad A_\lambda \leftrightharpoons \boldsymbol{A}(\lambda) \subseteq N, \quad \lambda \in \Lambda.$$

A Λ-sequence $\boldsymbol{A}$ is *m-reducible to* a Λ-sequence $\boldsymbol{B}$ ($\boldsymbol{A} \leqslant_m \boldsymbol{B}$), if there exists a recursive function f such that for any $\lambda \in \Lambda$, $n \in N$

$$n \in A_\lambda \Leftrightarrow f(n) \in B_\lambda .$$

We associate with a Λ-sequence $\boldsymbol{A}$ a family $S_{\boldsymbol{A}} \subseteq P(\Lambda)$ of subsets of Λ and its numbering as follows:

$$\begin{aligned} S_A &\leftrightharpoons \big\{\Lambda_0 \mid \Lambda_0 \subseteq \Lambda, \text{ there exists } n \in N \text{ such that} \\ &\qquad \Lambda_0 = \{\lambda \mid n \in A_\lambda,\ \lambda \in \Lambda\}\big\}; \\ \nu_{\boldsymbol{A}} &: n \mapsto \{\lambda \mid n \in A_\lambda,\ \lambda \in \Lambda\}, \quad n \in N. \end{aligned}$$

We call the numbered set $(S_{\boldsymbol{A}}, \nu_{\boldsymbol{A}})$ *dual* to $\boldsymbol{A}$ and denote by $\widehat{\boldsymbol{A}}$.

The following proposition establishes the connection between m-reducibility of Λ-sequences and reducibility of numberings.

Let $\boldsymbol{A}$ *and* $\boldsymbol{B}$ *be* Λ-*sequences. Then* $\boldsymbol{A} \leqslant_m \boldsymbol{B}$ *iff* ($S_{\boldsymbol{A}} \subseteq S_{\boldsymbol{B}}$ *and*) $\nu_{\boldsymbol{A}} \leqslant \nu_{\boldsymbol{B}}$.

If $S \subseteq P(\Lambda)$ is a nonempty denumerable family of subsets of Λ and $\nu : N \to S$ is a numbering, we define a Λ-sequence $\boldsymbol{A}(\boldsymbol{A}_{\mathfrak{S}})$ *dual to the numbered set* $\mathfrak{S} = (S, \nu)$ as follows:

$$A_\lambda \leftrightharpoons \big\{n \mid \lambda \in \nu(n),\ n \in N\big\}, \quad \lambda \in \Lambda.$$

We denote it also by $\widehat{\mathfrak{S}}$.

The following propositions are obvious.

(1) *For any* Λ-*sequence* $\boldsymbol{A}$ *we have* $\widehat{\widehat{\boldsymbol{A}}} = \boldsymbol{A}$.

(2) *For any numbered set* $\mathfrak{S} = (S, \nu)$, $S \subseteq P(\Lambda)$, *we have* $\widehat{\widehat{\mathfrak{S}}} = \mathfrak{S}$.

Now we establish an interesting link between an m-universality of a Λ-sequence $\boldsymbol{A}$ in an appropriate class Q of Λ-sequences and a completeness of the numbered set $\widehat{\boldsymbol{A}}$.

A class Q of Λ-sequences is called *closed*, if for any $\boldsymbol{A} \in Q$ and any partial recursive function g the Λ-sequence $g^{-1}(\boldsymbol{A})$ defined by

$$g^{-1}(\boldsymbol{A})(\lambda) \leftrightharpoons g^{-1}(A_\lambda), \quad \lambda \in \Lambda,$$

is contained in Q.

A Λ-sequence $\boldsymbol{A}$ is called *m-universal for* (*in*) a class Q of Λ-sequences, if for any $\boldsymbol{B} \in Q$ we have $\boldsymbol{B} \leqslant_m \boldsymbol{A}$ (and $\boldsymbol{A} \in Q$).

The following propositions hold:

If a Λ-*sequence* $\boldsymbol{A}$ *is* m-*universal in a closed class* Q, *then* $\emptyset \in S_{\boldsymbol{A}}$ *and* $\widehat{\boldsymbol{A}}$ *is a complete numbered set with a special element* $\emptyset$.

If a Λ-*sequence* $\boldsymbol{A}$ *is such that* $\emptyset \in S_{\boldsymbol{A}}$ *and* $\widehat{\boldsymbol{A}}$ *is a complete numbered set with special element* $\emptyset$, *then* $\mathbf{A}$ *is* m-*universal in a closed class* $\{g^{-1}(\mathbf{A}) \mid g$ *is a partial recursive function*$\}$.

Now we formulate a sufficient condition for $\emptyset \in S_{\mathbf{A}}$ and $\widehat{\mathbf{A}}$ to be complete with special element $\emptyset$. These notions are interesting from a general recursion-theoretic point of view.

Let $\boldsymbol{A}$ and $\boldsymbol{B}$ be Λ-sequences. We say that $\boldsymbol{A}$ is pm-reducible to $\boldsymbol{B}$ ($\boldsymbol{A} \leqslant_{pm} \boldsymbol{B}$) if there exists a partial recursive function g such that $\bigcup_{\lambda \in \Lambda} A_\lambda \subseteq \delta g$ and for any $\lambda \in \Lambda$, $n \in \delta g$

$$n \in A_\lambda \Leftrightarrow g(n) \in B_\lambda.$$

This is equivalent to an assertion $\boldsymbol{A} = g^{-1}(\boldsymbol{B})$.

If u is a universal unary partial recursive function, then for any Λ-sequence $\boldsymbol{A}$ we define a Λ-sequence $\boldsymbol{A}^{pm}$ by $\boldsymbol{A}^{pm} \leftrightharpoons u^{-1}(\boldsymbol{A})$ and call it a *p-cylindrification* of $\boldsymbol{A}$. The following properties hold:

(1) $\boldsymbol{A} \leqslant_m \boldsymbol{A}^{pm}$;
(2) $\boldsymbol{A}^{pm} \leqslant_{pm} \mathbf{A}$;
(3) $\boldsymbol{B} \leqslant_{pm} \mathbf{A} \Leftrightarrow \boldsymbol{B} \leqslant_m \boldsymbol{A}^{pm}$;
(4) *For any* Λ-*sequence* $\mathbf{A}$ *the following conditions are equivalent*:
   (a) $\emptyset \in S_{\mathbf{A}}$ *and* $\widehat{\mathbf{A}}$ *is complete with a special element* $\emptyset$;
   (b) $\boldsymbol{A}^{pm} \leqslant_m \mathbf{A}$.

2. Computable numberings

If S is not an arbitrary set but is a family of "constructive" objects, it is natural to study not arbitrary numbering of S but "computable" numberings.

If $S \subseteq \mathfrak{W}$ is a family of recursively enumerable sets, then a numbering $\nu : N \to S$ is called *computable* if a set of pairs

$$G_\nu \leftrightharpoons \{\langle x, y\rangle \mid x, y \in N,\ y \in \nu(x)\}$$

is recursively enumerable. Let $Nu^0(S)$ be the family of all computable numberings of S. Then $Nu^0(S) \subseteq Nu(S)$ and if $\nu_0 \in Nu^0(S)$, $\nu \in Nu(S)$, $\nu \leqslant \nu_0$, then $\nu \in Nu^0(S)$. Denote the image of $Nu^0(S)$ in $L(S)$ by $L^0(S)$.

$\langle L^0(S), \leqslant\rangle$ *is a subsemilattice and an ideal in* $\langle L(S), \leqslant\rangle$.

A family $S \subseteq \mathfrak{W}$ is called *computable* if there exists a computable numbering of S (i.e. $Nu^0(S) \neq \emptyset$).

If $S \subseteq \mathfrak{W}^n$ is a family of recursively enumerable n-ary predicates ($n \geqslant 1$), then a numbering $\nu : N \to S$ is *computable* if

$$G_\nu \leftrightharpoons \{\langle x_0, x_1, \ldots, x_n\rangle \mid \langle x_1, \ldots, x_n\rangle \in \nu(x_0)\}$$

is a (($n+1$)-ary) recursively enumerable predicate.

Identifying functions with their graphs we may assume that $\mathfrak{R} \subseteq \mathfrak{PR} \subseteq \mathfrak{W}^2$; then for $S \subseteq \mathfrak{PR}$ a numbering $\nu : N \to S$ is computable iff a binary function $g_\nu \leftrightharpoons \lambda x \lambda y[\nu(x)](y)$ is partial recursive.

A computable numbering ν of a family $S \subseteq \mathfrak{W}(\mathfrak{PR})$ is called *principal* if $[\nu]$ is a greatest element in $\langle L^0(S), \leqslant\rangle$, or, equivalently, ν is principal if any computable numbering $\nu_0 : N \to S$ is reducible to ν ($\nu_0 \leqslant \nu$).

Note that if ν is a principal computable numbering of a family S and ν_0 is a computable numbering of any subfamily $S_0 \subseteq S$, then $\nu_0 \leqslant \nu$.

The families $\mathfrak{W}$ and $\mathfrak{PR}$ have principal computable numberings.

We start with a class $\mathfrak{PR}$ of all unary partial recursive functions. It is well-known that there exists a binary partial recursive function T universal for the class $\mathfrak{PR}$, i.e. T is such that for any $f \in \mathfrak{PR}$ there exists $k \in N$ such that $f = \lambda x T(k, x)$.

Any binary partial recursive function T which is universal for $\mathfrak{PR}$ defines a computable numbering ν_T of this class as follows:

$$\nu_T(k) \leftrightharpoons \lambda x T(k, x), \quad k \in N.$$

Conversely, if $\nu : N \to \mathfrak{PR}$ is a computable numbering, then

$$T_\nu \leftrightharpoons \lambda x \lambda y\big[\nu(x)\big](y)$$

is a binary partial recursive function which is universal for $\mathfrak{PR}$.

The proof given below of the fact that the existence of any computable numbering of $\mathfrak{PR}$ implies rather formally the existence of a principal computable numbering was originally suggested by A.I. Mal'tsev (see Mal'tsev [1961, 1965]).

Let T be an arbitrary binary partial recursive function universal for the class $\mathfrak{PR}$. Put

$$K \leftrightharpoons \lambda x \lambda y T\big(l(x), c\big(r(x), y\big)\big),$$

where c, l, r are the functions defined in the previous section which define the numbering $c^{-1} : N \to N^2$, $c^{-1}(x) = \langle l(x), r(x)\rangle$, $x \in N$.

The function K is a binary partial recursive function. Show that K defines a principal computable numbering of $\mathfrak{PR}$. Let U be an arbitrary binary partial recursive function (which defines a computable numbering of some subclass of $\mathfrak{PR}$). Then $g \leftrightharpoons \lambda x U(l(x), r(x))$ is an unary partial recursive function and hence there exists $k \in N$ such that $g = \lambda x T(k, x)$. Let $h \leftrightharpoons \lambda x c(k, x)$. Verify that $U = \lambda x \lambda y k(h(x), y)$. Indeed,

$$\begin{aligned} K\big(h(x), y\big) &= T\big(l\big(h(x)\big), c\big(r\big(h(x)\big), y\big)\big) = T\big(k, c(x, y)\big) = g\big(c(x, y)\big) \\ &= U\big(l\big(c(x, y)\big), r\big(c(x, y)\big)\big) = U(x, y). \end{aligned}$$

The principal numbering of $\mathfrak{PR}$ defined by the function K is denoted by $\varkappa$ and is called *Kleene numbering*. The numbering $\varkappa$ is used to define the numbering $\pi : N \to \mathfrak{W}$ which is a composition of $\varkappa : N \to \mathfrak{PR}$ and $\delta : \mathfrak{PR} \to \mathfrak{W}$, where δf is the domain of f, $f \in \mathfrak{PR}$. It is easy to verify that π, which is called *Post numbering*, is a principal computable numbering of $\mathfrak{W}$. Define

$$\boldsymbol{K} \leftrightharpoons (\mathfrak{PR}, \varkappa), \quad \boldsymbol{\Pi} \leftrightharpoons (\mathfrak{W}, \pi).$$

The numberings $\varkappa$ and π are complete.

This is a consequence of the following two facts.

For any unary partial recursive function g there exists a recursive function h such that $\varkappa(h(n))$ is a nowhere defined function if $n \notin \delta g$ and $\varkappa(h(n)) = \varkappa(g(n))$ if $n \in \delta g$.

Define a binary partial recursive function G by

$$G(n, x) \leftrightharpoons \begin{cases} \varkappa(g(n))(x) & \text{if } n \in \delta g,\ x \in \delta\varkappa(g(n)), \\ \text{undefined} & \text{otherwise.} \end{cases}$$

Since G defines a computable numbering $\nu_G : n \mapsto \lambda x G(n, x)$ of the family $\{\lambda x G(n, x) \mid n \in N\}$ of unary partial recursive functions, ν_G is reducible to $\varkappa$, i.e. there exists a recursive function h such that $\nu_G(n) = \varkappa(h(n))$ for all $n \in N$. The function h is a function in question.

The mapping $\delta : \mathfrak{PR} \to \mathfrak{W}$ is a factorization $\delta : \boldsymbol{K} \to \boldsymbol{\Pi}$.

Since π is principal we have

If $S \subseteq \mathfrak{W}$, then a numbering $\nu : N \to S$ is computable iff ν is π-computable (or computable relative to $\boldsymbol{\Pi}$).

Hence,

A family $S \subseteq \mathfrak{W}$ has a principal computable numbering iff S is a principal subset of $\boldsymbol{\Pi}$.

It is useful to note that

A finite family $S \subseteq \mathfrak{W}$ is a wn-subset of $\boldsymbol{\Pi}$ and hence a principal subset of $\boldsymbol{\Pi}$.

A finite family $S \subseteq \mathfrak{W}$ is a n-subset of $\boldsymbol{\Pi}$ iff S has the least element with respect to inclusion.

Another useful fact is

If $S \subseteq \mathfrak{W}$ and $\nu : N \to S$ is a principal computable numbering, then the preorder $\leqslant_\nu$ on S is a partial order and coincides with inclusion relation $\subseteq$ on elements of S.

This is a consequence of the following observations:

If $S \subseteq \mathfrak{W}$, $\nu : N \to S$ is a computable numbering, and $F \subseteq N$ is a finite subset of N, then $S_F \leftrightharpoons \{R \mid R \in S,\ F \subseteq R\}$ is a e-subset of $\mathfrak{S} = (S, \nu)$.

Let $R_0 \subset R_1 \in \mathfrak{W}$ and R be a recursively enumerable nonrecursive set. Then a numbering $\nu_R : N \to \{R_0, R_1\}$ defined by $\nu_R(n) \leftrightharpoons R_1$ if $n \in R$ and $\nu_R(n) \leftrightharpoons R_0$ if $n \notin R$ is computable and $R_0 <_{\nu_R} R_1$.

It implies

If $\nu : N \to S \subseteq \mathfrak{W}$ is a principal computable numbering of S, S_0 is an e-subset of $\mathfrak{S} = (S, \nu)$, $R_0 \in S_0$, $R_0 \subseteq R_1 \in S$, then $R_1 \in S_0$.

We describe, following Rice, all e-subsets of $\boldsymbol{\Pi}$. To this end we use a numbered set $\boldsymbol{\Gamma}$, $\boldsymbol{\Gamma} \leqslant \boldsymbol{\Pi}$. If $S \subseteq \mathfrak{W}$ is a e-subset of $\boldsymbol{\Pi}$, then $S_0 \leftrightharpoons S \cap P_\omega(N)$ is a e-subset of $\boldsymbol{\Gamma}$. The observation above implies

$$S_1 \leftrightharpoons \{R \mid R \in \mathfrak{W},\ \text{there exists } F \in S_0 \text{ such that } F \subseteq R\} \subseteq S.$$

Now S_1 coincides with S as follows from

Let $S \subseteq \mathfrak{W}$ be a e-subset of $\boldsymbol{\Pi}$, $\nu : N \to S' \subseteq \mathfrak{W}$ be a computable numbering such that $\nu(0) \subseteq \nu(1) \subseteq \cdots \subseteq \nu(n) \subseteq \cdots$ and

$$\nu(n) \neq R \leftrightharpoons \bigcup_{k \in N} \nu(k) \quad \textit{for all } n \in N.$$

If $R \in S$, then $\nu(n) \in S$ for an appropriate $n \in N$.

Thus we obtain the following criterion.

A family $S \subseteq \mathfrak{W}$ is an e-subset of $\boldsymbol{\Pi}$ iff the family $S_0 \leftrightharpoons S \cap P_\omega(N)$ is $\boldsymbol{\Gamma}$-computable and

$$S = \{R \mid R \in \mathfrak{W},\ \textit{there exists } F \in S_0 \textit{ such that } F \subseteq R\}.$$

Note that there exists an analogous description of e-subsets of the numbered set $\boldsymbol{K}$. Now we give a nontrivial necessary condition for the set $S \subseteq \mathfrak{W}$ to be principal.

If $S \subseteq \mathfrak{W}$ is a principal subset of $\boldsymbol{\Pi}$, $\nu : N \to S' \subseteq S$ is a computable numbering and the family S' is directed with respect to inclusion, then

$$R \leftrightharpoons \bigcup_{n \in N} \nu(n) \in S.$$

Let $\boldsymbol{R} = R_0, R_1, \ldots, R_n, \ldots,\ n \in N$, be a sequence of recursively enumerable sets. It is called *computable* if the set $s(\boldsymbol{R}) \leftrightharpoons \{c(n, m) \mid m \in R_n\}$ is recursively enumerable. Note that $\boldsymbol{R}$ is computable iff the numbering $\nu : n \mapsto R_n$, $n \in N$, of the set $\{\boldsymbol{R}\} = \{R_0, R_1, \ldots, R_n, \ldots\}$ is computable.

For $S \subseteq \mathfrak{W}$ denote by ∂S the family of all computable sequences $\boldsymbol{R} = R_0, R_1, \ldots, R_n, \ldots$ such that $R_n \in S$, $n \in N$. A numbering $\nu : N \to S \subseteq \partial\mathfrak{W}$ is called *computable* if the set $\{c(m, c(n, k)) \mid k \in R_{\nu(m),n}\}$ is recursively enumerable. Here

$$\nu(m) = \boldsymbol{R}_{\nu(m)} = R_{\nu(m),0}, R_{\nu(m),1}, \ldots, R_{\nu(m),n}, \ldots, \qquad n \in N.$$

The mapping $s : \boldsymbol{R} \mapsto s(\boldsymbol{R})$, $\boldsymbol{R} \in \partial\mathfrak{W}$, is a one-to-one mapping of $\partial\mathfrak{W}$ onto $\mathfrak{W}$. Indeed, there exists an inverse mapping $\Delta : \mathfrak{W} \to \partial\mathfrak{W}$ defined by

$$\Delta(R)_n \leftrightharpoons \{m \mid c(n, m) \in R\}, \quad n \in N;$$
$$\Delta(R) \leftrightharpoons \Delta(R)_0, \Delta(R)_1, \ldots, \Delta(R)_n, \ldots, \quad n \in N,\ R \in \mathfrak{W}.$$

The mapping s possesses the property: for any numbering $\nu : N \to S \subseteq \partial\mathfrak{W}$ the numbering $s\nu : N \to s(S) \subseteq \mathfrak{W}$ is computable iff ν is computable. It follows that the numbering $\Delta\pi : N \to \partial\mathfrak{W}$ is a principal computable numbering of $\partial\mathfrak{W}$ and $\partial\boldsymbol{\Pi} \leftrightharpoons (\partial\mathfrak{W}, \Delta\pi)$ is naturally isomorphic to $\boldsymbol{\Pi}$.

With the help of the notions just introduced we can give the following characterization of principal subsets of $\boldsymbol{\Pi}$.

$S \subseteq \mathfrak{W}$ is a principal subset of $\boldsymbol{\Pi}$ iff the family of sequences $\partial(S \cup \{\emptyset\})$ is computable.

We accompany this statement by the following

If $S \subseteq \mathfrak{W}$ is a principal (wn-) subset of $\boldsymbol{\Pi}$, then $S \cup \{\emptyset\}$ is a principal (n-) subset of $\boldsymbol{\Pi}$ and $S\backslash\{\emptyset\}$ is a principal (wn-) subset of $\boldsymbol{\Pi}$.

If $\boldsymbol{R} = R_0, R_1, \ldots$ is a computable sequence ($\boldsymbol{R} \in \partial\mathfrak{W}$), then the corresponding (computable) numbering $\nu : n \mapsto R_n$, $n \in N$, defines a morphism $\mu_R : \boldsymbol{N} \to \boldsymbol{\Pi}$ and, conversely, any morphism $\mu : \boldsymbol{N} \to \boldsymbol{\Pi}$ defines a computable sequence $\boldsymbol{R}_\mu \leftrightharpoons \mu(0), \mu(1), \ldots$. Thus there exists a bijection between $\partial\mathfrak{W}$ and $M(\boldsymbol{N}, \boldsymbol{\Pi})$ (the set of all morphisms from $\boldsymbol{N}$ to $\boldsymbol{\Pi}$) which defines an isomorphism of numbered sets $\partial\boldsymbol{\Pi}$ and $\boldsymbol{M}(\boldsymbol{N}, \boldsymbol{\Pi})$.

Define an involution $* : \partial\mathfrak{W} \to \partial\mathfrak{W}$ on the set $\partial\mathfrak{W}$ by: for $\boldsymbol{R} \in \partial\mathfrak{W}$ the sequence $\boldsymbol{R}^* = R_0^*, R_1^*, \ldots$ is defined by

$$R_n^* \leftrightharpoons \{k \mid n \in R_k\}, \quad n \in N.$$

It is easy to verify that $\boldsymbol{R}^{**} = \boldsymbol{R}$ and $*$ is a morphism from $\partial\boldsymbol{\Pi}$ to $\partial\boldsymbol{\Pi}$.

Denote by $P_m(N)$, $m \in N$, the family of all subsets of N containing not more than m elements. It is easily seen that

$P_m(N)$ is a n-subset of $\boldsymbol{\Pi}$.

Let $\pi_m : N \to P_m(N)$ be a principal computable numbering of $P_m(N)$ and

$$\boldsymbol{\Pi}_m \leftrightharpoons (P_m(N), \pi_m) \leqslant_n \boldsymbol{\Pi}.$$

Now we pay more attention to the case $m = 1$. If $\nu : N \to S \subseteq P_1(N) \subseteq \mathfrak{W}$ is a computable numbering, then $P_\nu \leftrightharpoons \{\langle k, l\rangle \mid l \in \nu(k)\}$ is (by definition) a recursively enumerable predicate. Since $|\nu(k)| \leqslant 1$ for all $k \in N$, the set P_ν is a graph of some partial recursive function which we denote by f_ν. Thus we obtain a mapping $\nu \mapsto f_\nu$ from $Nu^0(P_1(N))$ to $\mathfrak{PR}$ which is easily seen to be a bijection. If $f \in \mathfrak{PR}$, we denote by ν_f the numbering in $Nu^0(P_1(N))$ such that $\nu_f(k) \leftrightharpoons \emptyset$ for $k \notin \delta f$ and $\nu_f(k) \leftrightharpoons \{f(k)\}$ for $k \in \delta f$. The mapping $f \mapsto \nu_f$, $f \in \mathfrak{PR}$, is the inverse mapping to $\nu \mapsto f_\nu$, $\nu \in Nu^0(P_1(N))$.

A partial recursive function u is called *universal* if the numbering ν_u is a principal computable numbering of the family $P_1(N)$.

We list the properties of universal functions:

(1) *The following properties are equivalent for any partial recursive function u:*

(a) *u is universal;*

(b) *for any partial recursive function g there exists a recursive function h such that $g = uh$;*

(c) *there exists a binary recursive function g such that the binary partial recursive function ug is universal for the class of unary partial recursive functions;*

(d) *there exists a binary recursive function g such that $ug = K^2$, the Kleene binary universal function.*

(2) *If u is universal and g is a recursive permutation, then gu and ug are universal functions.*

(3) (H. Rodgers [1958, 1967]) *If u_0 and u_1 are universal functions, then there exists a recursive permutation g such that $u_0 = g^{-1}u_1 g$.*

Kleene numbering $\varkappa$ can be used to obtain the following characterizations of the notions of precomplete numbering and wn-subset.

The first statement shows that precomplete numbered sets are characterized by the effective fixed-point property.

> *A numbered set $\mathfrak{S} = (S, \nu)$ is precomplete iff there exists a recursive function g such that $\nu g(n) = \nu\varkappa_n(g(n))$ for all n such that $\varkappa_n$ is recursive.*

The second statement shows that wn-subsets are characterized by the property of being effectively principal.

> *Let $\mathfrak{S}_0 \leqslant_p \mathfrak{S}$. The set S_0 is a wn-subset of $\mathfrak{S}$ iff there exists an unary partial recursive function g such that for any $n \in N$ if $\varkappa_n$ is recursive and $\nu\varkappa_n$ is a numbering of S_0 then $n \in \delta g$, $\varkappa_{g(n)}$ is recursive, and $\nu\varkappa_n = \nu_0\varkappa_{g(n)}$.*

There exists a huge amount of publications studying the properties of the upper semilattice $L^0(S)$ for a computable family $S \subseteq \mathfrak{W}$. It is not possible to survey all these investigations here. A kind of survey of related publications up to the year 1977 is contained in Ershov [1977]. Here we present only sample results which either have a general qualitative nature or reveal a kind of brilliance (in author's opinion).

Let $S \subseteq \mathfrak{W}$ be an arbitrary computable family. The following qualitative properties of the upper semilattice $\langle L^0(S), \leqslant\rangle$ hold:

(1) (A.B. Khutoretski [1971]) *$L^0(S)$ is either one-element or infinite.*

(2) *If $L^0(S)$ is infinite, then there exist sequences $a_0, a_1, \ldots; b_0, b_1, \ldots$ of elements in $L^0(S)$ such that $a_0 < a_1 < \cdots$ and b_i, b_j are incomparable for $i < j \in N$.*

Call a sequence $c_0, c_1, \ldots$ of elements in $L^0(S)$ *computable* if there exists a sequence $\mu_0, \mu_1, \ldots$ of morphisms from $\boldsymbol{N}$ to $\boldsymbol{\Pi}$ such that $\mu_n(N) = S$ and $c_n = [\mu_n]$ for all $n \in N$, and the numbering $\nu : n \mapsto \mu_n$, $n \in N$, of the family $\{\mu_0, \mu_1, \ldots, \mu_n, \ldots\} \subseteq M(\boldsymbol{N}, \boldsymbol{\Pi})$ is computable.

Thus, one can state the existence of computable sequences $a_0, a_1, \ldots$ and $b_0, b_1, \ldots$ in the previous property.

(3) *If $a_0 < a_1 < \cdots < a_n < \cdots$ is a computable sequence of elements in $L^0(S)$. Then the set $\{a_0, a_1, \ldots, a_n, \ldots\}$ does not have the least upper bound in $\langle L^0(S), \leqslant\rangle$.*

(4) *If $b_0, b_1, \ldots, b_n, \ldots$ is a computable sequence of incomparable elements in $L^0(S)$, then there exists $b \in L^0(S)$ which is incomparable with all b_n, $n \in N$.*

(5) (V.L. Selivanov [1976]) *If $L^0(S)$ is infinite, then $\langle L^0(S), \leqslant\rangle$ is not a lattice, i.e. there exist $a, b \in L^0(S)$ having no greatest lower bound in $\langle L^0(S), \leqslant\rangle$.*

(6) (A.B. Khutoretski [1971]) *If $L^0(S)$ is infinite and has the greatest element a, then for any $b < a$ there exists c such that $b < c < a$.*

Consider some examples.

(1) Let $S \leftrightharpoons \{\emptyset, \{0\}\}$. Then $\langle L^0(S_0), \leqslant\rangle$ is naturally isomorphic to the upper semilattice L^0_m of m-degrees of proper recursively enumerable subsets of N.

Indeed, if $\nu \in Nu^0(S_0)$, then

$$R_\nu \leftrightharpoons \nu^{-1}(\{0\}) = \{x \mid 0 \in \gamma(x)\}$$

is recursively enumerable (and different from $\emptyset$, N). If $\emptyset \neq R \subset N$ is recursively enumerable, then

$$\nu_R(n) \leftrightharpoons \begin{cases} \{0\}, & n \in R, \\ \emptyset, & n \notin R \end{cases}$$

is a computable numbering of S_0. Moreover, if $\nu_0, \nu_1 \in Nu^0(S_0)$, then

$$\nu_0 \leqslant \nu_1 \Leftrightarrow R_{\nu_0} \leqslant_m R_{\nu_1}.$$

(2) Let $n > 0$, $S_n \leftrightharpoons \{\emptyset, \{0\}, \{1\}, \ldots, \{n\}\}$. A profound result is given by the following theorem:

THEOREM (S.D. Denisov [1978]). *For any $n > 1$ the upper semilattice $\langle L^0(S_n), \leqslant\rangle$ is isomorphic to the upper semilattice L^0_m.*

(3) Let $\mathcal{P}$ be the family of all unary primitive recursive functions.

For any computable numbering ν of the family $\mathcal{P}$ there exists a single-valued computable numbering ν_0 of $\mathcal{P}$ which is not reducible to ν.

This implies that $\langle L^0(\mathcal{P}), \leqslant\rangle$ has neither least nor greatest element.

Many publications deal with a problem of finding conditions sufficient for the existence of a single-valued computable numbering of a family $S \subseteq \mathfrak{W}$. A profound and difficult result having important applications for the constructive (recursive) model theory was obtained by S.S. Goncharov:

THEOREM (Goncharov [1980]). *For any $n > 1$ there exists a family $G_n \subseteq \mathfrak{W}$ having exactly n mutually nonequivalent single-valued computable numberings.*

The notions of multiple reducibility, m-universality, etc., treated in broad generality in the previous section gained considerable attention of investigators in the case of computable (or finite) sequences of recursively enumerable sets also.

Thus, for instance, m-universal (= effectively inseparable) pairs of disjoint recursively enumerable sets found interesting applications in the statement of main theorems of mathematical logic (Smullyan [1962]). Sequences of mutually disjoint (embedded) sets were also considered: $R_i \cap R_j = \emptyset,\ i < j \in N$ ($R_i \supseteq R_j, i, j \in N$).

If $\Lambda, \Lambda' \subseteq N$, let

$$S(\Lambda, \Lambda') \leftrightharpoons \left\{ \boldsymbol{R} \mid \boldsymbol{R} \in \partial\mathfrak{W}, \bigcap_{i\in\Lambda} R_i \subseteq \bigcup_{j\in\Lambda'} R_j \right\}.$$

A family $S \subseteq \partial\mathfrak{W}$ is *structurally definable* if there exists a family of pairs (Λ_i, Λ_i'), $i \in I$, such that

$$S = \bigcap_{i\in I} S(\Lambda_i, \Lambda_i').$$

For instance, the condition $R_n \cap R_m = \emptyset$ is equivalent to

$$\bigcap_{i\in\{n,m\}} R_i \subseteq \bigcup_{j\in\emptyset} R_j,$$

and the condition $R_0 \cap R_1 = R_2 \cup R_3$ is equivalent to a conjunction of

$$\begin{gathered} R_0 \cap R_1 \subseteq R_2, \quad R_0 \cap R_1 \subseteq R_3, \\ R_2 \subseteq R_0, \quad R_2 \subseteq R_1, \quad R_3 \subseteq R_0, \quad R_3 \subseteq R_1. \end{gathered}$$

Note also that if a family $S \subseteq \partial\mathfrak{W}$ is structurally defined then there exists a family $\Lambda_i,\ i \in I$, of subsets of N such that

$$S = \bigcap_{i\in I} S(\Lambda_i, N\backslash\Lambda_i).$$

If a family $S \subseteq \mathfrak{W}$ is structurally defined, then:

(1) *S is closed, i.e. if $\boldsymbol{R} \in S$ and $g \in \mathfrak{PR}$, then $g^{-1}(\mathfrak{R}) \in S$;*

(2) *If* $\boldsymbol{R}, \boldsymbol{R}' \in S$, *then* $\boldsymbol{R} \oplus \boldsymbol{R}' \in S$, *where*

$$\boldsymbol{R} \oplus \boldsymbol{R}' \leftrightharpoons R_0 \oplus R_0', \quad R_1 \oplus R_1', \ldots, R_n \oplus R_n', \ldots .$$

Now we give a necessary and sufficient condition for the existence of a m-universal sequence in a structurally defined subset of $\partial\mathfrak{W}$.

Let $S \subseteq \partial\mathfrak{W}$ be a structurally defined family. Then S contains a m-universal (for the sequences in S) sequence iff the family

$$U(S^*) \leftrightharpoons \{R \mid \textit{there exists } R^* \in S^* \textit{ and } n \in N \textit{ such that } R = R_n^*\}$$

is a principal subset of $\mathfrak{W}$.

We give a sketch of the proof.

(1) It is easily verifiable that for any structurally defined family $S \subseteq \partial\mathfrak{W}$ the following assertion holds:

$$\boldsymbol{R} \in S \Leftrightarrow \text{for any } n \in N,\ R_n^* \in U(S^*).$$

(2) If $\boldsymbol{R}, \boldsymbol{R}' \in S$ and $\boldsymbol{R} \leqslant_m \boldsymbol{R}'$, then $\nu_{\boldsymbol{R}} \leqslant \nu_{\boldsymbol{R}'}$, where the numbering $\nu_{\boldsymbol{R}}$ ($\nu_{\boldsymbol{R}'}$) of some subfamily of $U(S^*)$ is defined by $\nu_{\boldsymbol{R}} : n \mapsto R_n^*$ ($\nu_{\boldsymbol{R}'} : n \mapsto R_n'^*$), $n \in N$. On the other hand, $\nu_{\boldsymbol{R}} \leqslant \nu_{\boldsymbol{R}'}$ implies $\boldsymbol{R} \leqslant_m \boldsymbol{R}'$. Moreover, $\boldsymbol{R}$ is m-reducible to $\boldsymbol{R}'$ via a recursive function f iff $\nu_{\boldsymbol{R}} = \nu_{\boldsymbol{R}'} f$.

(3) If $\nu : N \to U(S^*)$ is a principal computable numbering, then the sequence $R^\nu = R_0^\nu, R_1^\nu, \ldots$, where

$$R_n^\nu \leftrightharpoons \{k \mid k \in N,\ n \in \nu(k)\}, \quad n \in N,$$

is contained in S and is m-universal for S. (Note also that $\nu_{\boldsymbol{R}^\nu} = \nu$.)

The theorem follows from (1)–(3).

Now we give an easy application of this theorem.

If for a structurally defined family $S \subseteq \partial\mathfrak{W}$ there exists $n \in N$ such that for any $\boldsymbol{R} \in S$, $R_m = \emptyset$ for $m \geqslant n$, then S contains an m-universal sequence.

Indeed, in this case the elements of $U(S^*)$ are only subsets of $\{0, 1, \ldots, n-1\}$, i.e. $U(S^*)$ is finite and hence is a principal subset of $\mathfrak{W}$.

There is no m-universal sequence in a structurally defined family

$$S \leftrightharpoons \left\{\boldsymbol{R} \,\middle|\, \bigcap_{i \in N} R_i = \emptyset\right\}.$$

It is easy to verify that $U(S^*) = \{R \mid R \in \mathfrak{W}, R \neq N\}$. This family is not principal since it contains, for instance, a computable sequence of sets

$$\{0\} \subseteq \{0, 1\} \subseteq \cdots \subseteq \{0, 1, \ldots, n\} \subseteq \cdots$$

having the union ($= N$) which is not contained in $U(S^*)$. This contradicts a theorem on page 17.

A useful supplement is the following:

Let $S \subseteq \mathfrak{W}$ be closed and such that for any $\boldsymbol{R}, \boldsymbol{R}' \in S$ there exists $\boldsymbol{R}'' \in S$ such that $\boldsymbol{R} \leqslant_m \boldsymbol{R}''$, $\boldsymbol{R}' \leqslant_m \boldsymbol{R}''$. If $U(S^*)$ is a principal subset of $\mathfrak{W}$, then S is structurally defined (and hence contains a m-universal sequence).

REMARK. In the next section we establish a connection between m-universal sequences and a notion of co-productiveness (creativity).

3. Selected topics

The present section is devoted to three separate themes related to the theory of numberings. They are interesting in themselves; moreover, a series of useful and promising notions arise in a discussion.

3.1. *Index sets*

Since the numbered sets $\boldsymbol{\Pi}$ and $\boldsymbol{K}$ occupy a particularly important place, there arises a natural problem of characterization of index sets in $\boldsymbol{\Pi}(\boldsymbol{K})$ for various families of recursively enumerable sets (partial recursive functions). The exact definition follows:

Let $\mathfrak{S} = (S, \nu)$ be a numbered set and $S_0 \subseteq S$. The set $\nu^{-1}(S_0)$ ($\subseteq N$) is called an *index set of S_0 in $\mathfrak{S}$*.

The Rice theorem (Rice [1953]) may be considered as the first result in this direction. It asserts:

There is no nontrivial family $S \subseteq \mathfrak{W}$ with a recursive index set in $\boldsymbol{\Pi}$ (nontrivial means that $S \neq \emptyset, \mathfrak{W}$).

Numerous publications are devoted to determining of an exact place in Kleene–Mostowski hierarchy of particular families of recursively enumerable sets (partial recursive functions). (See, for instance, Rodgers [1967].)

L. Hay [1966] proved that there exist exactly three recursively nonisomorphic index sets of one-element subsets of $\mathfrak{W}$ ($\mathfrak{PR}$).

Yu. Ershov [1968a] and L. Hay [1969] gave independently an exhaustive investigation of index sets of finite subsets of $\mathfrak{W}$.

For the purpose of this investigation in Ershov [1968a] (see also Ershov [1968b, 1970a, 1973]) a new hierarchy was introduced. Namely, the hierarchy Σ_n^m, $n \in N$, of classes of subsets of N contained in the class Δ_2^0 of the Kleene–Mostowski hierarchy (in Ershov [1968b, 1970a] this hierarchy was extended along a system of notation for ordinals so that the union of all classes $\Sigma_{(\alpha)}^m$, where (α) is a notation for an ordinal, is Δ_2^0). Here we present a description of the hierarchy Σ_n^m which uses a notion of pm-reducibility introduced in Section 1 and a notion of m-jump which has also an independent significance.

We say that B ($\subseteq N$) is *m-enumerable in A* ($\subseteq N$) if $B \leqslant_{pm} A \oplus (N \backslash A)$; for a set $A \subseteq N$ the *m-jump of A* is the set

$$mj(A) \leftrightharpoons (A \oplus (N \backslash A))^{pm}.$$

The following properties hold:

(1) $A \leqslant_1 mjA$, $N \backslash A \leqslant_1 mjA$;

(2) $mjA \approx mj(N \backslash A)$;

(3) B is m-enumerable in $A \Leftrightarrow B \leqslant_1 mjA$;

(4) $mjA \nleqslant_m A$.

Define a sequence of sets $L_0, L_1, \ldots, L_n, \ldots,\ n \in N$, by:

$$L_0 \leftrightharpoons N \oplus \emptyset (= \{2k \mid k \in N\});$$
$$L_{n+1} \leftrightharpoons mjL_n, \quad n \in N,$$

and let

$$\Sigma_n^m \leftrightharpoons \{A \mid A \subseteq N,\ A \leqslant_m L_n\},$$
$$\Pi_n^m \leftrightharpoons \{A \mid N \backslash A \in \Sigma_n^m\}, \quad n \in N.$$

Now we have

$$\Sigma_0^m = \Pi_0^m \textit{ is a family of all recursive subsets of } N.$$

HIERARCHY THEOREM.

$$\Sigma_n^m \cup \Pi_n^m \subseteq \Sigma_{n+1}^m \cap \Pi_{n+1}^m, \quad n \in N;$$
$$\Sigma_n^m \subseteq \Delta_2^0, \quad n \in N.$$

The application of this hierarchy to index sets may be formulated as follows:

If $\emptyset \neq S \subseteq \mathfrak{W}$ is finite, then $\pi^{-1}(S)$ is m-universal in one of the classes

$$\Pi_2^0,\ \Pi_{2n+1}^m,\ \Sigma_{2n+2}^m, \quad n \in N.$$

A more precise information on the class in which the set $\pi^{-1}(S)$ is contained may be found in Ershov [1968a], Hay [1969].

The following qualitative results on partially ordered sets of m-degrees of all index sets in $\boldsymbol{\Pi}(\boldsymbol{K})$ are rather curious.

Let $\langle S, \leqslant \rangle$ be a partially ordered set. If S_0 is a subset of S, then a finite subset T of S is called a *generalized upper bound of* S_0 if the following conditions hold:

$$\forall s_0 \in S_0\ \forall t \in T\ (s_0 \leqslant t);$$
$$\forall s \in S\ \big(\forall s_0 \in S_0 (s_0 \leqslant s) \to \exists t \in T (t \leqslant s)\big).$$

A generalized lower bound is defined analogously.

A partially ordered set $\langle S, \leqslant \rangle$ is called a *generalized discrete upper semilattice of rank* k (> 0) if for any $\emptyset \neq S_0 \in P_\omega(S)$ there exists $T_0 \in P_\omega(S)$ such that $0 < |T_0| \leqslant k$, T_0 is a generalized upper bound of S_0, S_0 is a generalized lower bound of T_0, and k is least possible.

Let

$$J_m \leftrightharpoons \big\{d_m\big(\pi^{-1}(S)\big) \mid S \subseteq \mathfrak{W}\big\}$$

be a family of m-degrees of all index sets in $\boldsymbol{\Pi}$ $(J_m^* \leftrightharpoons \{d_m(\varkappa^{-1}(S)) \mid S \subseteq \mathfrak{P}\mathfrak{R}\})$ and $\leqslant$ be a partial order on J_m (J_m^*) induced by the order on m-degrees. Then the following two propositions hold:

The partially ordered set $\langle J_m, \leqslant \rangle$ *is a generalized discrete upper semilattice of rank* 4 (Selivanov [1979]).
The partially ordered set $\langle J_m^*, \leqslant \rangle$ *is a generalized discrete upper semilattice of rank* 2 (Kuz'mina [1981]).

An interesting problem is the one of studying the multiple reducibility and multiple isomorphism of index sets of finite families in $\mathfrak{W}$.

The most simple answer was obtained for the case of infinite sets.

Let $\langle R_0, \ldots, R_n \rangle$, $\langle R'_0, \ldots, R'_n \rangle$ be two sequences of mutually different infinite recursively enumerable sets. Then

$$\big\langle \pi^{-1}(R_0), \ldots, \pi^{-1}(R_n)\big\rangle \leqslant_m \big\langle \pi^{-1}\big(R'_0\big), \ldots, \pi^{-1}\big(R'_n\big)\big\rangle,$$
$$\big(\big\langle \pi^{-1}(R_0), \ldots, \pi^{-1}(R_n)\big\rangle \approx \big\langle \pi^{-1}\big(R'_0\big), \ldots, \pi^{-1}\big(R'_n\big)\big\rangle\big)$$

iff

$$R_i \subseteq R_j \Rightarrow R'_i \subseteq R'_j, \quad i, j \leqslant n,$$
$$\big(R_i \subseteq R_j \Leftrightarrow R'_i \subseteq R'_j, \quad i, j \leqslant n\big).$$

For the case of finite sets the answer is more complicated.

Let $\langle R_0, \ldots, R_n \rangle$, $\langle R'_0, \ldots, R'_n \rangle$ *be two families of mutually different finite sets. Then*

$$\big\langle \pi^{-1}(R_0), \ldots, \pi^{-1}(R_n)\big\rangle \leqslant_m \big\langle \pi^{-1}\big(R'_0\big), \ldots, \pi^{-1}\big(R'_n\big)\big\rangle$$

iff the following conditions hold:
(1) $R_i \neq \emptyset \Rightarrow R'_i \neq \emptyset,\ i \leqslant n$;
(2) $R_i \subseteq R_j \Rightarrow R'_i \subseteq R'_j,\ i, j \leqslant n$;
(3) *for any* $i \leqslant n$ *and* $I \subseteq \{0, \ldots, n\}$ *if there exists* $R \notin \{R_0, \ldots, R_n\}$ *such that*

$$R_i \subseteq R \subseteq \bigcap_{j \in I} R_j,$$

then there exists $R' \notin \{R'_0, \ldots, R'_n\}$ *such that*

$$R'_i \subseteq R' \subseteq \bigcap_{j \in I} R'_j.$$

Under the assumptions of this proposition, the sequences are multiply recursive isomorphic if the converse of conditions (1)–(3) also hold.

As a consequence we get the following qualitative result:

For any n there exists only finitely many multiple recursive isomorphism types of $\langle \pi^{-1}(R_0), \ldots, \pi^{-1}(R_n)\rangle$ *for sequences which either contain only finite or contain only infinite sets* $R_0, \ldots, R_n$.

The combined case turned out to be much more difficult and is not studied yet for $n > 2$. But in the case $n = 2$ there already appear infinitely many multiple isomorphism types.

Consider the case of $R_0 = R'_0 = \emptyset$, R_1, R_2, R'_1, R'_2 are infinite recursively enumerable sets such that $R_1 \cap R_2 = R'_1 \cap R'_2 = \emptyset$. To find a necessary and sufficient condition for $\langle \pi^{-1}(R_0), \pi^{-1}(R_1), \pi^{-1}(R_2)\rangle$ to be m-reducible to $\langle \pi^{-1}(R'_0), \pi^{-1}(R'_1), \pi^{-1}(R'_2)\rangle$ we introduce a relation $\leqslant_{sm}$ on pairs of disjoint subsets of N.

Let $\langle A_0, B_0\rangle$ and $\langle A_1, B_1\rangle$ be two pairs of disjoint subsets of N. We say that $\langle A_0, B_0\rangle$ is *m-reducible by separation* (*sm-reducible*) to $\langle A_1, B_1\rangle$ ($\langle A_0, B_0\rangle \leqslant_{sm} \langle A_1, B_1\rangle$) if there exist $A_2, B_2 \subseteq N$ such that $A_0 \subseteq A_2$, $B_0 \subseteq B_2$, $\langle A_2, B_2\rangle \leqslant_m \langle A_1, B_1\rangle$.

If $\langle A_0, B_0\rangle \leqslant_{sm} \langle A_1, B_1\rangle$ and $\langle A_1, B_1\rangle \leqslant_{sm} \langle A_0, B_0\rangle$ then the pairs $\langle A_0, B_0\rangle$ and $\langle A_1, B_1\rangle$ are called *sm-equivalent* ($\langle A_0, B_0\rangle \equiv_{sm} \langle A_1, B_1\rangle$).

Under the assumptions about the triples $\langle R_0, R_1, R_2\rangle$ and $\langle R'_0, R'_1, R'_2\rangle$ stated above the following two propositions are equivalent:
(a) $\langle \pi^{-1}(R_0), \pi^{-1}(R_1), \pi^{-1}(R_2)\rangle \leqslant_m \langle \pi^{-1}(R'_0), \pi^{-1}(R'_1), \pi^{-1}(R'_2)\rangle$;
(b) $\langle R_1, R_2\rangle \leqslant_{sm} \langle R'_1, R'_2\rangle$.

REMARK. The factor set L_{sm} of the family of all disjoint pairs of subsets of N with respect to sm-equivalence coupled with the induced order is a distributive lattice (see Dvornikov and Rybina [1990]; in Ershov [1970b, 1977] it was erroneously stated that L_{sm} is not a lattice).

The sublattice L^0_{sm} formed by equivalence classes which contain a pair of recursively enumerable sets is infinite (Ershov [1970b, 1977]).

3.2. *Creativity and m-universality*

A problem of characterization of m-universal Λ-sequences in various classes has a certain interest. The first and unexpected characterization of m-universal recursively enumerable sets as creative sets was obtained in a paper by Myhill [1955]. It was followed by a number of publications attempted to extend the notion of creativity (productiveness) to (disjoint) pairs, n-tuples, etc.

We present this theory of m-universality/creativity in a somewhat final form. This form has a strong flavour of the theory of numberings.

Let Q be a closed class of Λ-sequences containing a m-universal one. Such a class is called a *C-class*. Given any m-universal Λ-sequence $\mathbf{A}$ in Q, we define a numbering $\nu : N \to Q$ by $\nu(n) \leftrightharpoons \varkappa_n^{-1}(\mathbf{A})$, $n \in N$. This numbering will be called *canonical*; it does not depend (up to the equivalence) on the choice of a m-universal sequence $\mathbf{A}$ in a C-class Q.

For $A, B \subseteq N$ denote by $A \bigtriangledown B$ the set $(A \cap B) \cup ((N\backslash A) \cap (N\backslash B))$ (which is a complement of the symmetric difference $A \bigtriangleup B$).

A Λ-sequence $\mathbf{A}$ is called *co-productive for* a C-class Q if there exists a recursive function h such that for any $n \in N$

$$h(n) \in \bigcap_{\lambda \in \Lambda} \left(A_\lambda \bigtriangledown \big(\nu(n)\big)_\lambda\right), \qquad (*)$$

where $\nu : N \to Q$ is the canonical numbering. The function h is called *co-productive for* $\mathbf{A}$.

A co-productive for Q Λ-sequence $\mathbf{A}$ is called *creative* if additionally $\mathbf{A} \in Q$.

The following theorem holds:

Let Q be a C-class of Λ-sequences and $\nu : N \to Q$ a canonical numbering. For any Λ-sequence $\mathbf{A}$ the following conditions are equivalent:

(1) *$\mathbf{A}$ is m-universal for Q;*

(2) *$\mathbf{A}$ is co-productive for Q;*

(3) *there exists a recursive function g such that*

$$\nu_{\mathbf{A}}\big(h(n)\big) = \nu_{\nu(n)}\big(h(n)\big)$$

for any $n \in N$.

COROLLARY. *A Λ-sequence $\mathbf{A}$ is m-universal in Q iff it is creative.*

REMARK. Actually the proof of the theorem provides the following additional information on co-productive functions.

(1) If $\boldsymbol{A}$ is m-universal for a C-class Q, then there exists an injective recursive co-productive function h for $\boldsymbol{A}$.

(2) In the definition of a co-productive sequence it is sufficient to require the existence of a partial recursive function h which satisfies either of the conditions:

If $n \in N$ is such that for any $\lambda \in \Lambda$ either $\nu(n)_\lambda = N$ or $\nu(n)_\lambda = \emptyset$, then $n \in \delta h$ and $(*)$ holds;
If $n \in N$ is such that there exists $m \in N$ such that for any $\lambda \in \Lambda$ either $\nu(n)_\lambda = \{m\}$ or $\nu(n)_\lambda = \emptyset$ (which is equivalent to $|\bigcup_{\lambda\in\Lambda} \nu(n)_\lambda| \leqslant 1$), then $n \in \delta h$ and $(*)$ holds.

In the second section we considered a problem of existence of m-universal sequences in structurally defined classes Q of (N-) sequences for $Q \subseteq \partial\mathfrak{W}$.

Note that if a structurally defined (hence, closed) class $Q \subseteq \partial\mathfrak{W}$ contains a m-universal sequence (i.e. is a C-class), then

A canonical numbering of Q is a principal computable numbering of Q.

When a C-class Q is contained in $\partial\mathfrak{W}$ one can use the numbering $\Delta\pi$ instead of a canonical numbering of Q to formulate the notion of co-productiveness. A sequence $\boldsymbol{A} = \{A_n \mid n \in N\}$ is called *co-productive for Q relative to $\Delta\pi$* if there exists a partial recursive function g (*co-productive for* $\boldsymbol{A}$) such that for any $n \in N$ such that $\Delta\pi(n) \in Q$ we have $n \in \delta g$ and $g(n) \in \bigcap_{k\in N}(A_k \bigtriangledown (\Delta\pi(n))_k)$.

REMARK. Similarly to the definition of a co-productive function it is sufficient for the function g to be defined and satisfy (*) for such n that ($\Delta\pi(n) \in Q$ and) for any $k \in N$, $\Delta\pi(n)_k = \emptyset$ or $\Delta\pi(n)_k = N$; or for such n that ($\Delta\pi(n) \in Q$ and) $|\bigcup_{k\in N} \Delta\pi(n)_k| \leqslant 1$.

If, moreover, $\boldsymbol{A} \in Q$, then $\boldsymbol{A}$ is called *creative*.

Any co-productive for Q relative to $\Delta\pi$ sequence is m-universal for Q.

When, finally, a structurally defined C-class Q ($\subseteq \partial\mathfrak{W}$) contains a creative (relative to $\Delta\pi$) sequence? An answer is given by the following theorem.

A structurally defined C-class $Q \subseteq \partial\mathfrak{W}$ contains a creative relative to $\Delta\pi$ sequence iff $U(Q^*)$ is a n-subset of $\boldsymbol{\Pi}$.

REMARK. There exists a principal subset $S \subseteq \mathfrak{W}$ such that $\emptyset \in S$ but S is not a n-subset of $\boldsymbol{\Pi}$. Hence $\partial S \subseteq \partial\mathfrak{W}$ is an example of a C-class of sequences containing a m-universal but no creative relative to $\Delta\pi$ sequence.

3.3. *Numbered sets with approximation and the problem P*

We introduce a number of notions which primarily serve to formulate extensive sufficient conditions for the problem P to be solvable but have also an independent significance.

Let $\mathfrak{S}$ be a separated numbered set; $\mathfrak{S}_0 \leqslant \mathfrak{S}$ is called an *approximation for* $\mathfrak{S}$ if the following two conditions are satisfied:

(1) The predicate $\{\langle x, y\rangle \mid \nu_0 x \leqslant_\nu \nu y\}$ is recursively enumerable;

(2) If S', S'' are e-subsets of $\mathfrak{S}$ and $S' \nsubseteq S''$, then there exists $s_0 \in S_0$ such that $s_0 \in S' \backslash S''$.

The following properties hold:

If $\mathfrak{S}_0 \leqslant \mathfrak{S}$ is an approximation for $\mathfrak{S}$, then

(1) *$\mathfrak{S}_0$ is a positively numbered set and the predicate $\{\langle x, y\rangle \mid \nu_0 x \leqslant_\nu \nu_0 y\}$ is recursively enumerable*;

(2) *If S' is an e-subset of $\mathfrak{S}$, then $S'_0 \leftrightharpoons S' \cap S_0$ is an e-subset of $\mathfrak{S}_0$ such that if $s_0 \in S'_0$ and $s_0 \leqslant_\nu s_1$ $(s_1 \in S_0)$, then $s_1 \in S'_0$, and*

$$S' = \{s \mid \textit{there exists } s_0 \in S'_0 \textit{ such that } s_0 \leqslant_\nu s\};$$

(3) *If, moreover, $\mathfrak{S}_1 \leqslant \mathfrak{S}$ is an approximation for $\mathfrak{S}$, then $S_1 = S_0$ and $\nu_1 \equiv \nu_0$.*

Let $\mathfrak{S}_0 \leqslant \mathfrak{S}$ be an approximation for $\mathfrak{S}$. We call $\mathfrak{S}$ *complete over approximation* if for all $\mathfrak{S}'$ such that $\mathfrak{S}_0 \leqslant \mathfrak{S}'$ and $\mathfrak{S}_0$ is an approximation for $\mathfrak{S}'$ such that $\leqslant_\nu \restriction S_0^2 = \leqslant_{\nu'} \restriction S_0^2$ there exists (unique mono-) morphism $\mu : \mathfrak{S}' \to \mathfrak{S}$ such that $\mu \restriction S_0 = \mathrm{id}_{S_0}$.

If $\mathfrak{S}_0$ is a positively numbered set and $\leqslant$ is a partial order on S_0 such that the set $\{\langle x, y\rangle \mid \nu_0 x \leqslant \nu_0 y\}$ is recursively enumerable, then there exists a numbered set $\mathfrak{S}$ such that $\mathfrak{S}_0 \leqslant \mathfrak{S}$ is an approximation for $\mathfrak{S}$, $\leqslant_\nu \restriction S_0^2 = \leqslant$, and $\mathfrak{S}$ is complete over approximation.

A pair $\langle \mathfrak{S}_0, \leqslant\rangle$ is called a *constructive partial upper semilattice*, or a *constructive parus* (from PARtial Upper Semilattice), if $\mathfrak{S}_0$ is positive, $\langle S_0, \leqslant\rangle$ is a partial upper semilattice, the predicate

$$C_\leqslant \leftrightharpoons \{\langle x, y\rangle \mid \nu_0 x \text{ and } x_0 y \text{ are compatible} \\ \text{(that is there exists } s_0 \in S_0 \text{ such that } \nu_0 x \leqslant s_0,\ \nu_0 y \leqslant s_0)\}$$

is recursive, and there exists a binary partial recursive function $\mathfrak{S}$ such that $\delta g = C_\leqslant$ and for $x, y \in N$ if $\langle x, y\rangle \in C_\leqslant = \delta g$, then $\nu_0 g(x, y)$ is the least upper bound of $\nu_0 x$ and $\nu_0 y$ in $\langle S_0, \leqslant\rangle$.

Define the classes $\boldsymbol{C}_2$ and $\boldsymbol{C}^*_{20}$ of numbered sets as follows:

$\mathfrak{S} \in \boldsymbol{C}_2$ if $\mathfrak{S}$ has an approximation $\mathfrak{S}_0 \leqslant \mathfrak{S}$ such that $\langle \mathfrak{S}_0, \leqslant_\nu \restriction S_0^2\rangle$ is a constructive parus;

$\mathfrak{S} \in \boldsymbol{C}^*_{20}$ if $\mathfrak{S}$ has an approximation $\mathfrak{S}_0 \leqslant \mathfrak{S}$ such that $\langle \mathfrak{S}_0, \leqslant_\nu \restriction S_0^2\rangle$ is a constructive parus with a least element and g is complete over approximation.

EXAMPLES 1. (1) $\boldsymbol{N} \in \boldsymbol{C}_2 \backslash \boldsymbol{C}^*_{20}$;

(2) $\boldsymbol{\Pi}_1 \in \boldsymbol{C}^*_{20}$.

If $\mathfrak{S}_0, \mathfrak{S}_1 \in \boldsymbol{C}_2$ ($\boldsymbol{C}^*_{20}$), then $\mathfrak{S}_0 \times \mathfrak{S}_1 \in \boldsymbol{C}_2$ ($\boldsymbol{C}^*_{20}$).

The main theorem on solvability of the problem P for pairs of numbered sets which may be used to define a class of partial computable functionals of finite types is the following:

If $\mathfrak{S}_0 \in \boldsymbol{C}_2$ *and* $\mathfrak{S}_1 \in \boldsymbol{C}_{20}^*$, *then the problem* P *is solvable for the pair* $\mathfrak{S}_0, \mathfrak{S}_1$ *and* $\boldsymbol{M}(\mathfrak{S}_0, \mathfrak{S}_1) \in \boldsymbol{C}_{20}^*$.

The following theorem can be found in Ershov [1972]. It may be considered as establishing the existence of a big family of constructive models for the λ-calculus.

For any $\mathfrak{S} \in \boldsymbol{C}_{20}^*$ *there exist* $\mathfrak{S}_0 \in \boldsymbol{C}_{20}^*$, $\mathfrak{S} \leqslant \mathfrak{S}_0$, *and a morphism* $\mu : \mathfrak{S}_0 \to \mathfrak{S}$ *such that* $\mu(s) = s$ *for* $s \in S$, $\mu(s_0) \leqslant_{\nu_0} s_0$ *for* $s_0 \in S_0$, *and* $\mathfrak{S}_0$ *is isomorphic to* $\boldsymbol{M}(\mathfrak{S}_0, \mathfrak{S}_0)$.

In Ershov [1985, 1986] an attempt is made to use similar ideas in construction of classes of partial Σ-predicates and Σ-functionals of finite types in any admissible set.

References

S. D. Denisov

[1978] Structure of the upper semilattice of recursively enumerable m-degrees, *Algebra i Logika*, 17, 6, pp. 643–683.

S. D. Denisov and I. A. Lavrov

[1970] Complete numerations with infinitely many singular elements, *Algebra i Logika*, 9, 5, pp. 503–509.

S. G. Dvornikov

[1979] Precompletely enumerated sets, *Sibirsk. Matem. Zh.*, 20, 6, pp. 1303–1306.

S. G. Dvornikov and T. V. Rybina

[1990] Denseness properties of a lattice of separation-degrees, *Sibirsk. Matem. Zh.*, 31, 1, pp. 64–69.

Yu. L. Ershov

[1968a] On a hierarchy of sets. I, *Algebra i Logika*, 7, 1, pp. 47–74.

[1968b] On a hierarchy of sets. II, *Algebra i Logika*, 7, 4, pp. 15–47.

[1970a] On a hierarchy of sets. III, *Algebra i Logika*, 9, 1, pp. 34–51.

[1970b] On index sets, *Sibirsk. Matem. Zh.*, 11, 2, pp. 326–342.

[1971] Computable numerations of morphisms, *Algebra i Logika*, 10, 3, pp. 247–308.

[1972] Continuous lattices and A-space, *Dokl. Akad. Nauk SSSR*, 207, 3, pp. 523–526.

[1973] Hierarchies of sets of class Δ_2^0, in: *Proceedings IV ICLMPS* (*Bucharest, 1971*), North-Holland, Amsterdam, pp. 69–76.

[1975] The upper semilattice of numerations of a finite set, *Algebra i Logika*, 14, 3, pp. 258–283. (E. A. Palyutin, Supplement to Yu. L. Ershov's article "The upper semilattice of numerations of a finite set", *ibid.*, pp. 284–287.)

[1977] *The Theory of Numberings*, Nauka, Moscow (in Russian).

[1985] Σ-predicates of finite types over an admissible set, *Algebra i Logika*, 24, 5, pp. 499–536.

[1986] f_A-spaces, *Algebra i Logika*, 25, 5, pp. 533–543.

S. S. Goncharov

[1980] Computable single-valued numerations, *Algebra i Logika*, 19, 5, pp. 507–551.

L. HAY

[1966] Isomorphism types of index sets partial recursive functions, *Proc. Amer. Math. Soc.*, 17, pp. 106–110.

[1969] Index sets of finite classes of recursively enumerable sets, *J. Symbolic Logic*, 34, pp. 39–44.

A. B. KHUTORETSKII

[1971] On the cardinality of the upper semilattice of computable enumerations, *Algebra i Logika*, 10, 5, pp. 561–569.

T. M. KUZ'MINA

[1981] Structure of the m-degrees of the index sets of families of partial recursive functions, *Algebra i Logika*, 20, 1, pp. 55–68.

A. H. LACHLAN

[1987] A note on positive equivalence relations, *Zeits. Math. Logik Grundl. Math.*, 33, 1, pp. 43–46.

D. LACOMBE

[1966] *Propriétés Récursives des Structures Enumeérées*, Publ. l'Inst. B. Pascal.

A. I. MAL'TSEV

[1961] Constructive algebras. I, *Uspekhi Mat. Nauk*, 16, 3, pp. 3–60.

[1963] Completely enumerable sets, *Algebra i Logika*, 2, 2, pp. 4–29.

[1965] *Algorithms and Recursive Functions*, Nauka, Moscow (in Russian); English transl.: Wolters-Noordhoff Publishing, Groningen.

J. MYHILL

[1955] Creative sets, *Zeits. Math. Logik Grundl. Math.*, 1, 2, pp. 97–108.

P. ODIFREDDI

[1989] *Classical Recursion Theory*, North-Holland, Amsterdam.

H. G. RICE

[1953] Classes of recursively enumerable sets and their decision problems, *Trans. Amer. Math. Soc.*, 74, 2, pp. 358–366.

H. RODGERS, JR.

[1958] Gödel numberings of partial recursive functions, *J. Symbolic Logic*, 23, 3, pp. 331–341.

[1967] *Theory of Recursive Functions and Effective Computability*, McGraw-Hill, New York.

V. L. SELIVANOV

[1976] Two theorems on computable numberings, *Algebra i Logika*, 15, 4, pp. 470–484.

[1979] Structures of the degrees of unsolvability of index sets, *Algebra i Logika*, 18, 4, pp. 463–480.

R. M. SMULLYAN

[1962] *Theory of Formal Systems*, Princeton Univ. Press.

V. A. USPENSKII

[1955a] On computable operations, *Dokl. Akad. Nauk SSSR*, 103, pp. 773–776.

[1955b] Systems of enumerable sets and their enumerations, *Dokl. Akad. Nauk SSSR*, 105, pp. 1155–1158.

[1960] *Lectures on Computable Functions*, Fizmatgiz, Moscow.

K. WEIHRAUCH

[1987] *Computability*, Springer, Berlin.

Part 5
Logic and Computability Theory

CHAPTER 15

Pure Recursive Model Theory

Terrence S. Millar

Department of Mathematics, University of Wisconsin–Madison, Van Vleck Hall, 480 Lincoln Drive, WI 53706-1388, USA

Contents

1. Introduction 508
2. A computational approach to classical model theoretic results 509
   2.1. Strong similarities 509
   2.2. Weak similarities 512
   2.3. Differences 514
3. Computational hierarchies of model theoretic domains 517
   3.1. Elementary embeddings 518
   3.2. Homogeneous models revisited 518
   3.3. Ehrenfeucht theories 520
4. New perspectives from recursive model theory 521
   4.1. Almost homogeneity 521
   4.2. Ehrenfeucht theories and stability 522
   4.3. Omitting types revisited 524
   4.4. Number of countable models 525
   4.5. An Ash legacy 529
References 530

HANDBOOK OF COMPUTABILITY THEORY
Edited by E.R. Griffor

1. Introduction

Recursive model theorists study the computable properties of model theoretic objects and constructions. Their work is flanked by classical model theory – the study of the properties and relations between syntax and semantics in mathematics, and classical recursion theory – the study of the properties of computable functions and related sets. A fundamental result in classical model theory was Gödel's Completeness theorem (Gödel [1930]): a first order theory is consistent if and only if it has a model. The foundations of recursion theory were partially shaped by another result of Gödel, the Incompleteness theorem (Gödel [1931]). Implicit in the Incompleteness theorem is an early result in recursive model theory – a fragment of arithmetic provides an example of a theory that, in the terminology that will be introduced below, is axiomatizable and has a computably presented model, but has no decidable model.

Recursive model theory is one of several areas of interaction between recursion theory and model theory. Researchers in both of these classical disciplines benefit from the tools and perspectives of the other. For instance, the cardinality and automorphisms of a structure, and the categoricity and existence of prime and saturated models of a theory are common model theoretic concerns. One of the first questions in recursion theory concerned the cardinality of the upper semi-lattice of recursively enumerable degrees – "Does the semi-lattice have more than two elements?" Post's [1944] examples of more complex model-theoretic questions are whether the theory of that structure is $\aleph_0$-categorical (Lerman et al. [1984]), whether or not the structure has any non-trivial automorphisms (Simpson [1977]), and whether or not the theory has a prime model. Conversely, model theorists, for example, find recursively saturated models a very useful tool (Schlipf [1978]).

During the last twenty-five years, recursive model theorists broadened the connection between these two central disciplines. Sometimes they restrict the domain of model theoretic inquiry to computable objects and effective arguments, in order to investigate which corresponding classical results hold. Alternatively, they start with model theoretic objects and relations, and then determine the hierarchy of computational complexity among them. And finally, they investigate new areas and results that are suggested by a recursive model theoretic perspective. I will discuss and give some examples of these three approaches. The chapter is not intended to be comprehensive, and therefore many important contributions are not cited. A more comprehensive review of results in related areas can be found in Harizanov [ta]. Some of the important definitions are indented.

This chapter assumes some familiarity with model theory (Chang and Keisler [1990]) and recursion theory (Soare [1987]), although most of the discussion should be accessible without extensive knowledge in these two disciplines. Historically the underlying domain of recursion theory was the set ω of natural numbers. However, this is easily generalized to other effective object types. For the most part, this chapter is written in an "object type independent" fashion, since the particular formalizations one adopts are not important to most of the arguments. Throughout this chapter, all languages and models are assumed countable. Historically the phrases "computable functions" and "recursive functions" have been used synonymously and refer to the class of functions that are, intuitively speaking, effectively computable. There are many equivalent formal descriptions of this class of functions (Kleene [1982]).

- A set is said to be *computable* (*recursive*) if its characteristic function is a computable function.
- A set is said to be *computably enumerable* (*recursively enumerable*) if it is the range of a computable function.
- A function f is said to be *computable in* (*recursive in*) a function g (or a set A) just if f is computable using an oracle for g (or A).

In this paper we will use the terms "computable" and "computably enumerable", since many of the results can be adapted to other notions of "computable" other than "recursive". Similarly, we will talk about relative computability and, strictly speaking, we will mean with respect to Turing reducibility (Kleene [1982]). Again, other reducibilities also could be investigated in a "computable" model theoretic spirit.

2. A computational approach to classical model theoretic results

I start with examples of the first approach – cataloguing the effectiveness of results in classical model theory from a computational perspective.

2.1. *Strong similarities*

A number of results have translations that are also theorems. There will be two groups of examples. This first group represents results that have more or less direct translations. The second group represents results where the translation is more problematic.

2.1.1. *The Completeness theorem.* The Completeness theorem proves that a theory in first order predicate logic is syntacticly consistent if and only if it has a model. What happens if the first order theory is computable? First we must decide what "computable" means for first order theories and models. Certainly the syntax should be reasonable.

Thus the language must have a countable, effectively presented set of logical, variable, constant, relation, and function symbols. Traditionally a first order language $\mathcal{L}$ is thought of as a tuple $\langle \mathcal{R}, \mathcal{F}, \mathcal{C}, \mathcal{V}, \mathcal{I} \rangle$, where the domains of $\mathcal{R}$, $\mathcal{F}$, and the three sets $\mathcal{C}, \mathcal{V}, \mathcal{I}$ are pairwise disjoint. $\mathcal{R}$ is a function whose domain is a set of symbols referred to as the relation symbols, $\mathcal{F}$ is a function whose domain is a set of symbols referred to as the function symbols, $\mathcal{C}$ is a set of constant symbols, $\mathcal{V}$ is a countable set of variable symbols, and $\mathcal{I}$ is a set of logical symbols. If $R \in \mathcal{R}$ ($F \in \mathcal{F}$), then $\mathcal{R}(R)$ ($\mathcal{F}(F)$) is a positive natural number referred to as the *arity* of R (F). An effectively presented language is one where the domains of $\mathcal{R}$, $\mathcal{F}$, and the three sets $\mathcal{C}, \mathcal{V}, \mathcal{I}$ also are computable sets, and the two functions $\mathcal{R}$ and $\mathcal{F}$ are computable. For the remainder of the paper, all languages are assumed effectively presented.

While there are several possibilities for the definition of a computable first order theory, a natural candidate is that the theory be

- *decidable* – that is, a theory whose set of consequences is a computable set of sentences.

There are also several possibilities for the definition of a computable model. One option is that the model is

- *computably presented*, that is, the universe of the model and the interpretations of constant, function, and relation symbols are uniformly computable objects (again, "recursively presented" is commonly used for this notion).

An equivalent formulation is that the atomic diagram of the model is a computable set of sentences.

- The *atomic diagram* of a model is the set of atomic and negated atomic sentences true in an extension of the model to a language in which every element of the model is named, i.e. interprets a constant symbol, in a canonical fashion.

For example, the standard model of arithmetic is computably presented. However, by the Incompleteness theorem, the set of first order sentences true in this model is not computable. Thus, a natural and stronger condition for the definition of a computable model is that the elementary diagram of the model be a computable set of sentences.

- The *elementary diagram* of a model is the set of all first order sentences true in the model, in the expanded language mentioned above.

If a model satisfies this stronger version, then it is *decided*. Obviously a decided model is computably presented. The standard model of arithmetic is computably presented but not decided. Since isomorphic models usually are identified in mathematics, there are two related formulations –

- a model is *computably presentable* if it is isomorphic to a computably presented model, and similarly a *decidable* model is a structure that is isomorphic to a decided model.

A fundamental result in recursive model theory now follows from two observations. The first observation is that the proof of the Completeness theorem is effective. The second is that the proof actually uniformly effectively produces, given a consistent theory, not only the atomic diagram of a finite or countable model for the theory, but also the elementary diagram of that model. So the corresponding theorem in recursive model theory is that a consistent and decidable theory has a decidable model. The converse of this result fails because of intrinsic features of recursion theory. For example, consider the theory of one unary predicate symbol that is true of infinitely many pairwise distinct constant symbols. This theory is consistent and decidable, but it has $2^{\aleph_0}$ pairwise non-equivalent subtheories, and therefore they can not all be decidable, even though a decidable model of the original theory also is a model for each subtheory.

Although it is true that a consistent theory has a decidable model if and only if it has a consistent decidable extension, a sufficient intrinsic condition for a theory to have a computable model has not been forthcoming (Lerman and Schmerl [1979]). For instance, it is not sufficient that the theory in question be axiomatizable.

– A theory is *axiomatizable* if there is a computably enumerable set of sentences that is semantically equivalent to the theory.

(Two sets of sentences are semantically equivalent just if they have the same models.)

A simple example of this pathology is the theory T whose axioms are

$$\{U(c_i) \mid i \in A\} \cup \{\neg U(c_i) \mid i \in B\},$$

where A and B are computably enumerable, computably inseparable sets.

– Two sets are *computably inseparable* if they are disjoint and there is no computable set containing all of one set and none of the other.

Since A and B are computably enumerable, the theory T is axiomatizable. For any model $\mathcal{D}$ of this theory, the set

$$\{i \mid \mathcal{D} \models U(c_i)\}$$

is computable from the atomic diagram of $\mathcal{D}$, and separates the sets A and B – therefore $\mathcal{D}$ cannot be computably presented. There is a price that must be paid to exhibit this pathology – an axiomatizable consistent theory that does not have a decidable model must be incomplete, and in fact must have $2^{\aleph_0}$ complete extensions.

2.1.2. *The omitting types theorem.* Another example of a classical argument that has a "smooth" translation is the omitting types theorem (Chang and Keisler [1990]) – every non-principal type of a theory is omitted in some countable model of the theory.

– A *type* of a theory is a set of formulas in finitely many free variables consistent with the theory.
– A *principal type* is a type such that for some generating formula $\theta(\overline{x})$ consistent with the theory, and for every formula $\gamma(\overline{x})$ of the type, the theory proves $\forall\overline{x}(\theta(\overline{x}) \rightarrow \gamma(\overline{x}))$. A *non-principal type* is a type that is not principal. A type is *realized* in a model if some tuple of elements from the model satisfies all the formulas in the type; otherwise the type is *omitted* from the model.
– A *complete* type is a type that is maximal.
– A *complete formula* is a generating formula for a complete principal type.
– $\mathbf{S(T)}$ denotes the set of complete types of T; $\mathbf{S^r(T)}$ denotes the set of computable complete types; $\mathbf{S_p(T)}$ denotes the set of principal complete types; $\mathbf{Typ}(\mathcal{A}, \bar{\mathbf{a}})$ denotes the complete type realized by $\bar{a}$ in $\mathcal{A}$; and $\mathbf{TySp}(\mathcal{A}) \subseteq S(T)$ is the set of complete types realized in the model $\mathcal{A}$.

Again the classical omitting types argument is effective and not only produces a model that omits the type, but also produces the elementary diagram of that model. The corresponding theorem in recursive model theory is that every non-principal computable type of a decidable theory is omitted in some decidable model of that theory.

2.1.3. *Craig's Interpolation theorem.* Some model theoretic results immediately lift to corresponding recursive model theoretic results because of their form, and the relation of that form to prior fundamental results. One of those fundamental results is that although the set of valid sentences of first order predicate logic is not usually computable (if, for example, the language has at least one binary predicate symbol), the set is computably enumerable (Kleene [1982]). Now consider the Craig Interpolation theorem (Chang and Keisler [1990]): if the language L_0 (with equality, for simplicity) has two otherwise disjoint extensions L_1 and L_2, θ_1, θ_2 are sentences of L_1, L_2, respectively, and $(\theta_1 \rightarrow \theta_2)$ is valid, then there is a sentence θ_0 of L_0 such that $(\theta_1 \rightarrow \theta_0)$ and $(\theta_0 \rightarrow \theta_2)$ are valid also. The translated version in recursive model theory says that there is a uniformly computable method that, given θ_1 and θ_2 as above, computes a corresponding θ_0. This translated version is true. A procedure for finding θ_0 is as follows: first note, by the classical result, that given θ_1 and θ_2, where $(\theta_1 \rightarrow \theta_2)$ is valid, such a θ_0 does exist. Therefore, start computably enumerating all valid sentences in L_1 and L_2, respectively. Search the expanding enumerations for a pair of sentences γ_1 in L_1 and γ_2 in L_2 that have the forms, for some θ_0 in L_0, of $(\theta_1 \rightarrow \theta_0)$ and $(\theta_0 \rightarrow \theta_2)$, respectively. Since such a θ_0 and the corresponding γ_1 and γ_2 exist, and since they must eventually be enumerated, this is an effective method for computing such a desired θ_0.

2.2. *Weak similarities*

Many results have less obvious translations because of the nature of the objects addressed in the classical model theoretic setting. Classical model theorists move comfortably through a diverse collection of set theoretic objects. Translations to recursive model theory then depend on how the notions of effectiveness are interpreted. For example, fix a complete theory T and consider $S(T)$. Since the language of T is effective, the set of formulas $Frm_{\mathcal{L}(T)}$ of the language is a computable domain. Thus the computable complexity of a set of formulas, say for instance a complete type of T, can be discussed with the object type being formula. However, for infinitary objects, such as the set $S(T)$ of complete types of T, the picture is more difficult. The problem is that $S(T)$ can be very complex from a recursion theoretic point of view (for example, it may not be countable). Thus there is not automatically a natural domain in which to discuss the computable complexity of subsets of $S(T)$. But for subsets of the set $S^r(T)$ of computable complete types of T, there is another sensible route. Each element $\Gamma \in S^r(T)$ can be identified with its computable characteristic function $\chi_\Gamma : Frm_{\mathcal{L}(T)} \rightarrow \{0, 1\}$. A computable function is a manageable object, since it can be identified with one of its finite programs. This provides a natural way to discuss the complexity of a set of computable functions – in terms of the complexity of some corresponding set of finite programs. In turn say that $\mathcal{G} \subset S^r(T)$ is computable (in A) just if a corresponding set of finite programs for $\mathcal{G}$ is computable (in A). Of course such approaches lead to some ambiguity, since for instance there are different sets of finite programs that represent the same characteristic functions, such that the different sets have very different computational properties.

2.2.1. *Elementary chains.* An example of a similar kind of complication involves the notion of uniformity and can be seen in the elementary chain theorem (Chang and Keisler [1990]). The classical result is that if each model of an elementary chain is a model of a fixed theory, then so is the union of the elementary chain. One possible translated version is that if there is an elementary chain of decidable models of a theory T, then the union of the elementary chain is also a decidable model of the theory T. This version is false in general. It is one thing for each model to be computable, and quite another for the set of models to be uniformly computable. In other words, it is important whether or not there is a single computable process that simultaneously can represent the decidability of all the models in the chain. If that additional assumption is made as part of the translation of the classical result, then the lifted version follows easily.

Here is an example that shows that some additional assumption is necessary. Consider the theory T in the language with infinitely many unary predicate symbols $U_i, i \in \omega$, and infinitely many constant symbols c^i_j, $i, j \in \omega$, whose axioms are:

$$\begin{aligned} &c^i_j \neq c^i_k, \quad j < k;\ i, j, k \in \omega, \\ &U_i\big(c^i_j\big), \quad i, j \in \omega, \\ &\forall x\big[U_i(x) \rightarrow \neg U_j(x)\big], \quad i < j;\ i, j \in \omega. \end{aligned}$$

This theory is complete, decidable, and consistent. For any $X \subseteq \omega$ there is a countable model $\mathcal{A}_X$ of T such that for those $n \in \omega - X$, the only elements of $(U_n)_{\mathcal{A}_X}$ are the $(c^n_j)_{\mathcal{A}_X}$'s, and such that for those $n \in X$ there is exactly one additional element a^n_X, other than the $(c^n_j)_{\mathcal{A}_X}$'s, in $(U_n)_{\mathcal{A}_X}$. If $X \neq Y$, then the models $\mathcal{A}_X, \mathcal{A}_Y$ are not isomorphic. For each such X and for every $k \in \omega$, let $\mathcal{A}^k_X$ be the elementary submodel of $\mathcal{A}_X$ whose universe is $|\mathcal{A}_X| - \{a^n_X \mid k < n, n \in \omega\}$. Then it follows that for $m < n$, $\mathcal{A}^m_X$ is an elementary submodel of $\mathcal{A}^n_X$. Also the union of the elementary chain $\{\mathcal{A}^n_X \mid n \in \omega\}$ is just $\mathcal{A}_X$. It is easy to check that for every such X and n, $\mathcal{A}^n_X$ is decidable. Since there are $2^{\aleph_0}$ subsets X of ω, not all of the $\mathcal{A}_X$ can be decidable, even though each one is the union of an elementary chain of decidable models.

2.2.2. *Prime and saturated models.* As another such example, consider the result from model theory that every consistent complete theory T with only countably many complete types has both a prime model and a countable saturated model (Chang and Keisler [1990]).

- A *prime model* of a complete theory T is a countable model that is elementarily embeddable in every model of T. Or equivalently a prime model $\mathcal{A}$ is a countable model that realizes only principal types, i.e. $TySp(\mathcal{A}) = S_p(T)$.
- A countable *saturated* model $\mathcal{B}$ is a countable model of T that realizes every type, i.e. $TySp(\mathcal{B}) = S(T)$, and is "homogeneous".

– A countable model is *homogeneous* if every two pairs of finite tuples of elements realizing the same complete type are in the same orbit (two tuples of elements are in the same orbit of a model if there is an automorphism of the model taking one tuple to the other).

If the condition of this result is translated only as “a decidable theory T such that $S(T) = S^r(T)$”, then both results fail. The problem is similar to the one in the last example, but in this case the question of uniformity applies to the set $S^r(T)$. If the translation is strengthened to include the conditions that $S(T) = S^r(T)$ and $S(T)$ is computable, then the results lift – in other words such a theory has a decidable prime model and a decidable saturated model. These two examples will be considered in more detail below.

The translation issue can become even more complex. There are two equivalent classical conditions that are necessary and sufficient for the existence of the prime model for a complete consistent theory:

(i) for every formula $\phi(\overline{x})$ consistent with the theory, there is a complete formula $\theta(\overline{x})$ such that $\forall\overline{x}[\theta(\overline{x}) \to \phi(\overline{x})]$ is a consequence of the theory;

(ii) for every formula $\phi(\overline{x})$ consistent with the theory, there is a principal type $\Gamma(\overline{x})$ of the theory such that $\phi(\overline{x}) \in \Gamma(\overline{x})$.

Reasonable translations of these conditions would be:

(i) there is a computable process that, given a formula $\phi(\overline{x})$ consistent with the theory, produces a complete formula $\theta(\overline{x})$ such that $\forall\overline{x}[\theta(\overline{x}) \to \phi(\overline{x})]$ is a consequence of the theory;

(ii) there is a computable process that, given a formula $\phi(\overline{x})$ consistent with the theory, produces a principal type $\Gamma(\overline{x})$ such that $\phi(\overline{x}) \in \Gamma(\overline{x})$. (It is not difficult to see that $S_p(T) \subset S^r(T)$ for a decidable theory T. Therefore a principal type Γ can be “produced”, as discussed above, by providing a finite program for the characteristic function of Γ.)

Both conditions are sufficient for the existence of a decidable prime model of a decidable complete theory. However, these conditions are not equivalent for all complete decidable theories. A moment’s reflection reveals that for such theories condition (i) implies condition (ii); but only condition (ii) is equivalent to the existence of a decidable prime model for a complete decidable theory (Goncharov and Nurtazin [1973], Harrington [1974]).

2.3. *Differences*

Some model theoretic results entirely fail to lift to corresponding results in recursive model theory because they rely on properties whose possible lifted versions fail in the new setting. I will give three examples of this behavior.

2.3.1. *Vaught’s theorem.* Vaught [1961] proved that no complete theory has exactly two countable models up to isomorphism. This result depends on several regularities in the classical domain:

(i) a complete theory that has more than one countable model up to isomorphism must have a complete non-principal type;

(ii) any complete type can be realized in a countable model of its theory;

(iii) a non-principal type of a complete theory is omitted by some countable model of the theory;

(iv) if a complete theory has more than one countable model up to isomorphism, then no extension of the theory (in a larger language) has exactly one countable model up to isomorphism.

The first three results from above do lift directly:

(i) a complete decidable theory that has more than one decidable model must have a computable non-principal complete type;

(ii) any computable complete type can be realized in a decidable model of its decidable theory;

(iii) a computable non-principal type can be omitted by some decidable model of its decidable theory.

But property (iv) fails – there is a complete decidable theory that has more than one decidable model such that each relevant (in the sense of the classical proof) complete extension of the theory has only one decidable model, up to isomorphism. This is the key to producing a decidable complete theory that has exactly two decidable models up to isomorphism (Kudaibergenov [1979], Millar [1979]). This failure can in fact be realized without losing control of the number of countable models of the theory – there is a complete theory with only countably many countable models up to isomorphism, such that exactly two of them are decidable (Millar [1981b]).

Before continuing with the next example, a remark is in order concerning the method used to construct many of the examples cited in this chapter. If a theory is to be constructed such that some object or property associated with the theory is not computable, then the coding of the required recursion theoretic complexity must be reflected in the axioms of the theory. Fortunately, this coding and the associated model theoretic constraints often can be partially separated from some of the other model theoretic problems of constructing a complete theory. Specifically, the coding usually can be done with a set of universal axioms in such a way that the desired theory is the model completion of a set of universal axioms that include those axioms that do the coding.

– T is the *model completion* of T' just if:

(i) Every model of T is a model of T';

(ii) Every model of T' is a submodel of a model of T;

(iii) For every model $\mathcal{A}$ of T', the union of T with the atomic diagram of $\mathcal{A}$ is a complete theory.

A nice feature of model completions of universal theories, when they exist, is that they allow elimination of quantifiers (Sacks [1972]). Therefore it is useful to have a general criteria for when a universal, consistent, decidable theory has a decidable, complete, model completion. In fact there is a straightforward set of syntactic conditions for a universal theory that are equivalent to the existence of such a model completion (Millar [1981a]). An application of the technique is contained in the next example.

2.3.2. *Robinson's Consistency theorem.* By the Robinson Consistency theorem (Chang and Keisler [1990]), if L_1 and L_2 are countable languages whose intersection is the language L_0, T_1, T_2 are L_1, L_2 theories, respectively, that are individually consistent, and the restrictions of T_1 and T_2 to L_0 are identical and complete, then $T_1 \cup T_2$ has a model. However, even if T_1 and T_2 are in addition individually complete and decidable, the resulting $T_1 \cup T_2$ does not necessarily have a decidable model. This is another consequence of the model theory being sufficiently flexible to encode a recursion theoretic pathology.

To construct an example, fix computably enumerable, computably inseparable sets A and B as before. Fix computable f, g whose ranges are A, B, respectively. Start the construction of such a counterexample with the universal theory T_0' in the language L_0 with infinitely many unary predicate symbols U_i, V_i, $i \in \omega$, and infinitely many binary predicates W_i, $i \in \omega$, with the universal axioms, for all $s \in \omega$:

$$\forall x \forall y \left[\bigwedge_{i<s} [U_i(x) \wedge V_i(y)] \rightarrow W_{f(n)}(x, y) \right], \quad n < s;$$

$$\forall x \forall y \left[\bigwedge_{i<s} [U_i(x) \wedge V_i(y)] \rightarrow \neg W_{g(n)}(x, y) \right].$$

By the methods in Millar [1981a] it is easy to verify that T_0' has a complete decidable model completion T_0. Now let the languages L_1, L_2 be L_0 augmented by one addition constant symbol d_1, d_2, respectively. Let the universal axioms for T_1' be those of T_0' together with, for $i \in \omega$

$$U_i(d_1) \wedge \neg V_i(d_1) \wedge W_i(d_1, d_1),$$

and similarly let the universal axioms for T_2' be those of T_0' together with, for $i \in \omega$

$$\neg U_i(d_2) \wedge V_i(d_2) \wedge W_i(d_2, d_2).$$

Then, again using the methods of Millar [1981a], it is routine to check that T_1', T_2' have complete decidable model completions T_1, T_2, respectively. However, there is no computably presentable model $\mathcal{D}$ of $T_1 \cup T_2$, since otherwise, as before, the computable set

$$\{n \mid \mathcal{D} \models W_n(d_1, d_2)\}$$

would then separate A and B. Again, one can remove the pathology by restricting to those T_1, T_2 such that in addition T_1, T_2 are axiomatizable and $T_1 \cup T_2$ has at most countably many completions. In that case, $T_1 \cup T_2$ must then have a decidable model.

2.3.3. *Homogeneous models.* The third example is the classical theorem that every consistent countable theory has a countable homogeneous model (Chang and Keisler [1990]). It is false that every consistent decidable theory has a decidable homogeneous model, or even a computably presentable homogeneous model (Goncharov [1980]). One version of a proof of the classical result is to form an ω-ordered elementary chain, where each model is "homogenized" over elements and types of the model at the previous step. The desired model is then the union of the elementary chain. Since the required elementary chain theorem does lift, where does the proof in the effective domain break down? The answer has to be in the "homogenizing" step – managing infinitary information in a computable construction has its limits. The specific problem occurs when attempting to consistently amalgamate infinitary extensions (from $S(T)$) that are needed to homogenize the desired model. Although such amalgamations must exist by the classical argument, the amalgamations may not be computable, or even if computable, they may not be uniformly computable.

3. Computational hierarchies of model theoretic domains

I turn now to the second approach mentioned in the introduction – starting with model theoretic objects, and then classifying the relative computable complexity of those and related objects and constructions. As mentioned earlier, for the purposes of this exposition relative computability will be with respect to Turing reduction. The computable sets and functions occupy the bottom of this complexity hierarchy. Another important level in the hierarchy is characterized by the

- *Halting Problem* – the set of $\langle \mathcal{P}, n \rangle$, where $\mathcal{P}$ is a finite program in some fixed programming language, n is a natural number, and the program $\mathcal{P}$ with input n halts in finitely many steps.

It is a classical result that this set is computably enumerable (Kleene [1982]). The degree of this set (let us take the *degree* of a set to be the set of subsets of the natural numbers that are mutually computable with the given set) is often denoted $\mathbf{O}'$. If the previous definition is "relativized" to the set of $\langle \mathcal{Q}, n \rangle$, where $\mathcal{Q}$ is a program with oracle $\mathbf{O}'$, then the resulting degree is referred to as $\mathbf{O}''$ or $\mathbf{O}^{(2)}$. By taking computable unions at computable limit ordinals, this process can be iterated effectively through all computable ordinals – the hierarchy of those sets computable from those sets at these levels is referred to as the "hyperarithmetic hierarchy" (Kleene [1982]). For instance, the resulting $\mathbf{O}^{(\omega)}$ includes a set that codes the set of true statements for the standard model for arithmetic.

- A set is *hyperarithmetic* if it computable in any set of degree $\mathbf{O}^{(\alpha)}$ for some computable ordinal α.
- A set is *arithmetic* if it is computable in any set of degree $\mathbf{O}^{(n)}$ for some natural number n.

3.1. *Elementary embeddings*

The presentation of the second perspective starts with a simple example that involves elementary embeddings, homogeneous models, and prime models. A classical result is that for countable models $\mathcal{A}$ and $\mathcal{B}$, if $TySp(\mathcal{A}) \subseteq TySp(\mathcal{B})$, and $\mathcal{B}$ is homogeneous, then $\mathcal{A}$ can be elementarily embedded into $\mathcal{B}$ (Chang and Keisler [1990]). The strongest form of the corresponding result in recursive model theory is false – there are decidable models $\mathcal{A}$ and $\mathcal{B}$ as above such that $\mathcal{A}$ is not elementarily embeddable into $\mathcal{B}$ by any computable function. But here the nature of the failure suggests itself – any elementary embedding of $\mathcal{A}$ into $\mathcal{B}$ must take elements in $\mathcal{A}$ to elements in $\mathcal{B}$ that realize the same complete type. This information is, in general, infinitary. On the other hand, there always is such an embedding that is relatively tame from a computational point of view – that is, there is always such an embedding which is computable relative to $\mathbf{O}'$. This result follows from the constructive nature of a possible proof of the classical result, as long as there is an oracle to indicate when two finite tuples of elements realize the same complete type.

The general embedding problem is computationally complex. Coding computational complexity into the embedding problem is made possible by the model theoretic complexity that can arise in structures through their inhomogeneity, as reflected in their Scott sentences. For countable structures with simple Scott sentences, the embedding problem is correspondingly simpler. For example, as we saw above, if $\mathcal{B}$ is homogeneous (and therefore has a low Scott rank) and decidable, then the embedding problem is only as complicated as the Halting Problem, independent of the model theoretic complexity of $\mathcal{A}$ (as long as $\mathcal{A}$ is decidable).

Alternatively, if $\mathcal{A}$ is prime and decidable, then again the embedding problem into decidable models can only be as complex as the Halting Problem, although for a different reason. The embedding of the prime model is actually finitary in nature, depending only on which complete formulas finite tuples satisfy. For a complete decidable theory the complete formulas are always computable relative to the Halting Problem. This can be used to show that the prime model always embeds elementarily in an $\mathbf{O}'$ fashion into any other decidable model of the theory.

3.2. *Homogeneous models revisited*

The second example of the second approach is the characterization of the possible computational complexity of homogeneous countable models under various restrictions. It is known that homogeneous structures are model theoretically simple objects; for instance, each is uniquely determined by its cardinality and type spectrum. Is this model theoretic regularity reflected in their possible computational complexity? A classical theorem is that every consistent countable theory has a countable homogeneous model. Since a model can be no simpler computationally than its complete theory, and since there are $2^{\aleph_0}$ complete theories in any non-trivial language, clearly there are arbitrarily computationally complex countable homogeneous models. Therefore restrict the inquiry to decidable complete theories.

As noted above, a countable homogeneous model is uniquely determined by its type spectrum. Clearly the computational complexity of a model bounds the computational complexity of the set of complete types realized in the model. Since there are decidable theories with $2^{\aleph_0}$ complete types, and since every type is realized in a countable homogeneous model, there is thus no bound to the possible complexity of homogeneous models, even for decidable theories. Therefore consider only decidable theories T such that $S(T)$ is countable. It is then a result from recursion theory and the definitional form of a complete type for a decidable theory that the computational complexity of the types is hyperarithmetic (Millar [1978]). But this still does not give a bound to the possible complexity of countable homogeneous models of such a theory, since there are such theories with $2^{\aleph_0}$ pairwise non-isomorphic countable homogeneous models (they each have a unique type spectrum).

Therefore assume a further restriction to complete decidable theories that have only countably many countable homogeneous models (this is the only alternative, since if a theory has uncountably many countable homogeneous models, then it has $2^{\aleph_0}$ such models). Then of course there can only be countably many distinct type spectra of homogeneous models. It again follows by recursion model theory and the form of the definition of a type spectrum that all countable homogeneous models of such a theory must be hyperarithmetic (Millar [1982a]).

Is there a better bound then this? No – there is a class of examples of complete decidable theories T such that $S(T) = S^r(T)$, each T has only countably many countable models, and yet the computable complexity of the homogeneous models of these theories is unbounded in the hyperarithmetic hierarchy (Millar [1989]). Such theories must have very complex relations between their non-principal types. On the one hand, there must be enough flexibility that there are homogeneous models $\mathcal{A}$ whose $TySp(\mathcal{A})$ is complex from a recursion theoretic point of view. On the other hand, there must be sufficient cohesiveness that the number of possible type spectra for each T is countable. If the number of countable models of the theory is assumed to be finite, then it does follow that all homogeneous countable models of such theories ($S(T) = S^r(T)$) are decidable (Millar [1982b]). I will elaborate on this further below.

Alternatively, if one looks at very special homogeneous models then better bounds also are possible. This is true for the prime and saturated models of a decidable theory, if such models exist. If a decidable theory has a prime model, then the prime model is always computable relative to the Halting Problem. However, there are decidable theories that have prime models that are not decidable. In fact there are finitely axiomatizable, decidable theories with prime models that are not decidable (Peretyat'kin [1982]). If a decidable theory has a countable saturated model, then such a model is always hyperarithmetic (Millar [1982a]). This is the best possible general bound. If $S(T) = S^r(T)$ then, like the prime model, the saturated model of T is always computable relative to the Halting Problem (Millar [1978]). And again, there are examples of decidable theories T with $S(T) = S^r(T)$ such that the countable saturated model is not decidable. There are in fact such examples where the theory in question only has countably many countable models – in other words even this added regularity does not guarantee that the saturated model is decidable (Millar [1984]). The corresponding question for the prime model is open – it is not known if there is a complete decidable theory with only countably many countable models such that the prime model of the theory is not decidable. I will return to this point.

3.3. *Ehrenfeucht theories*

The third example of the second approach that I shall elaborate concerns

– *Ehrenfeucht theories* – complete theories that have more than one but only finitely many countable models up to isomorphism.

One of the first examples of such a theory was a dense linear order without end-points with an ω-ordered set of constant symbols (Vaught [1961]). This theory is the model completion of the following axioms:

$$\forall x[x \not< x],$$
$$\forall x \forall y[x < y \vee y < x \vee x = y],$$
$$\forall x \forall y \forall z[x < y \wedge y < z \rightarrow x < z],$$
$$c_i < c_{i+1}, \quad i \in \omega.$$

Up to isomorphism, the three countable models of T are:

(i) The prime model $\mathcal{A} = \langle \mathcal{Q}, <_{\mathcal{A}}, (c_i)_{\mathcal{A}} \rangle_{i \in \omega}$, where $\mathcal{Q}$ is the set of rationals, $<_{\mathcal{A}}$ is the usual order relation on the rationals, and $(c_i)_{\mathcal{A}} = i$ (the c_i interpretations are cofinal);

(ii) The weakly saturated model $\mathcal{B} = \langle \mathcal{Q}, <_{\mathcal{B}}, (c_i)_{\mathcal{B}} \rangle_{i \in \omega}$, where $<_{\mathcal{B}} = <_{\mathcal{A}}$, and $(c_i)_{\mathcal{A}} = 1 - 2^{-i}$ (the c_i interpretations have a least upper bound of 1);

(iii) The saturated model $\mathcal{C} = \langle \mathcal{Q}, <_{\mathcal{C}}, (c_i)_{\mathcal{C}} \rangle_{i \in \omega}$, where $<_{\mathcal{C}} = <_{\mathcal{A}}$, and $(c_i)_{\mathcal{A}} = \sum_{j<i+1} 1/j!$ (the c_i interpretations are bounded above, but there is no least upper bound).

This example has the property that the theory and all of its countable models are decidable. This raises an obvious question: are all countable models of decidable Ehrenfeucht theories decidable? In other words, is the relationship between an Ehrenfeucht theory and its models sufficiently tame that if the theory is decidable, then all of its models must be decidable? If one considers an $\aleph_0$ – categorical theory, then since every consistent decidable theory has a decidable model, this produces a class of decidable theories all of whose countable models are decidable. If one restricts down to decidable theories with fewer than five models, then although the theory may have a countable model that is not decidable, all of its models must be decidable relative to a rather tame oracle – $\mathbf{O}''$ [Millar, 1982b] (in fact there is a decidable complete theory with exactly three countable models, where two of the models are not decidable [Peretyat'kin, 1973]). The corresponding result for arbitrary Ehrenfeucht theories with more than four countable models is false. It is not too difficult to show that all countable models for decidable Ehrenfeucht theories must be hyperarithmetic (Sacks [1981]). But there is no bound in the hyperarithmetic hierarchy for the computable complexity of countable models of decidable Ehrenfeucht theories. In fact, there is a unbounded class of such theories each of which has only five countable models (Reed [1991]).

Another question that arises in this context is whether the computable complexity that is possible in countable models of decidable Ehrenfeucht theories is related to other objects associated with the theories and their models. Until a recent announcement by Goncharov, for the examples known the answer was yes – every example of a decidable Ehrenfeucht theory that failed to have all of its models decidable also had computationally complex complete types realized in the complex models. Goncharov has now shown that this is not necessary. In other words, Goncharov has an example of a decidable Ehrenfeucht theory with $S(T) = S^r(T)$ such that not all of the countable models are decidable. A partial result in the other direction is that if an arithmetic Ehrenfeucht theory (1) has only arithmetic complete types; and (2) has no complete extension in finitely many additional constant symbols which fails to be Ehrenfeucht, then it has only arithmetic countable models (Ash and Millar [1983]).

4. New perspectives from recursive model theory

4.1. *Almost homogeneity*

This topic provides a natural transition into the third approach mentioned in the introduction. For example, homogeneity is important to the question of whether or not every countable model of a decidable Ehrenfeucht theory realizing only computable types must be decidable. There are complete theories all of whose models are homogeneous – a non-$\aleph_0$-categorical example would be the theory of infinitely many distinct constant symbols. It is a classical result that every Ehrenfeucht theory must have at least one model that is not homogeneous (Benda [1974]) – for the original example mentioned above, it is the second model where the interpretations of the ω-chain of constant symbols have a least upper bound of 1. In that case, 1 has the same 1-type as larger elements, but it is the only element in its orbit.

However, all known examples of countable models of Ehrenfeucht theories are almost homogeneous. In other words, there are finitely many points in the model that, once named by constant symbols in an expanded language, produce an expanded model that is homogeneous. In the case of the model mentioned above, naming 1 results in a homogeneous model (which is, in fact, prime). Results about the decidability of homogeneous models make this interesting from a recursive model theoretic perspective. Although not all countable homogeneous models $\mathcal{A}$ whose *TySp*($\mathcal{A}$) is computable are decidable, it is true that if, for a complete theory T, $S^r(T)$ is computable, then a countable homogeneous model $\mathcal{A}$ of the theory is decidable if and only if *TySp*($\mathcal{A}$) is computable (Goncharov [1978], Millar [1980], Peretyat'kin [1978]). Since every computable type is realized in some decidable model, it is not hard to see that for a decidable Ehrenfeucht theory T, $S^r(T)$ must be computable. The same kind of observation shows that $TySp(\mathcal{A})$ must be computable for any $\mathcal{A}$ such that $TySp(\mathcal{A}) \subset S^r(T)$ (with T Ehrenfeucht). Thus every countable homogeneous model of a decidable Ehrenfeucht theory T with $S(T) = S^r(T)$ must be decidable.

This raises a model theoretic question – is every model of an Ehrenfeucht theory almost homogeneous? If the answer were yes, then the computable complexity of models of decidable Ehrenfeucht theories, all of whose complete extensions by finitely many constant symbols were also Ehrenfeucht and decidable, would be known – all countable models of such theories would be decidable (even in this context, however, the question is open whether or not the additional assumption that all complete extensions in finitely many constant symbols are Ehrenfeucht is necessary).

4.2. *Ehrenfeucht theories and stability*

The behavior of expansions of models of Ehrenfeucht theories when realizations of non-principal types are named is thus related to the decidability of the models of those theories. Another model theoretic concept about Ehrenfeucht theories that arose from recursive model theoretic investigations is the notion of being "projectively finitary". This concept has a bearing on the well-known open problem in model theory of whether there is an Ehrenfeucht theory that is stable.

- A theory is *stable* just in case for some infinite cardinal κ and all models $\mathcal{A}$ of the theory of size κ, the number of complete types in the theory of $\langle \mathcal{A}, a \rangle_{a \in |\mathcal{A}|}$ is κ (Chang and Keisler [1990]).

Let $B_n(T)$ be the set of equivalence classes of formulas of the language of T in the free variables $\{x_1, \ldots, x_n\}$ (where, as usual, two formulas are equivalent if they are provably equivalent in T). Recall that the Ryll-Nardzewski theorem shows that a complete consistent theory T is $\aleph_0$-categorical just if for every $n \in \omega$, $B_n(T)$ is finite (Chang and Keisler [1990]).

Fix a complete type $\Gamma(x_1, \ldots, x_n)$ of a theory T. With this n fixed, we will write $\bar{u}$ to denote the tuple $\langle u_1, \ldots, u_n \rangle$. Now for any $m \in \omega$, and formulas $\theta_1(\bar{u}_1, \ldots, \bar{u}_m), \theta_2(\bar{u}_1, \ldots, \bar{u}_m)$ we will write

$$\theta_1 \equiv_\Gamma \theta_2$$

to mean that

$$\bigcup_{1 \leqslant i \leqslant m} \Gamma(\bar{u}_i) \vdash \big[\theta_1(\bar{u}_1, \ldots, \bar{u}_m) \leftrightarrow \theta_2(\bar{u}_1, \ldots, \bar{u}_m)\big].$$

It is easy to check that this is an equivalence relation. We will write B_m^Γ to denote the set of equivalence classes of such formulas for Γ and m.

DEFINITION 1. A type Γ is *finitary* just in case B_m^Γ is finite for every $m \in \omega$.

For instance, every type that has only finitely many realizations in any model is finitary. In the Ehrenfeucht theory above, every type is finitary.

DEFINITION 2. A complete theory is *projectively finitary* just if some non-principal type of the theory is finitary.

So again, the Ehrenfeucht theory above is projectively finitary: for the non-principal 1-type Γ and a fixed $m > 0$, each formula represented in B_m^Γ is equivalent to a Boolean combination of formulas from the set

$$\{u_i < u_j,\ u_i = u_j \mid 1 \leqslant i, j \leqslant m\}.$$

In fact, all known Ehrenfeucht theories are projectively finitary, though there are Ehrenfeucht theories that have non-principal types that are not finitary.

Fix a structure $\mathcal{A}$ and a complete type $\Sigma(x_1, \ldots, x_n)$ of the theory of $\mathcal{A}$. Let $\mathcal{A}_\Sigma$ be the set of tuples $\bar{a}$ such that $Typ(\mathcal{A}, \bar{a}) = \Sigma$. If $\mathcal{A}$ and $\mathcal{B}$ are two elementarily equivalent structures and Σ is a type of their theory, then we write $\mathcal{A} \cong_\Sigma \mathcal{B}$ just if there is a 1–1 onto map

$$f : \mathcal{A}_\Sigma \Rightarrow \mathcal{B}_\Sigma$$

such that for all $m < \omega$, all $\bar{a}_0, \ldots, \bar{a}_{m-1} \in \mathcal{A}_\Sigma$, and all formulas $\theta(\bar{x}_0, \ldots, \bar{x}_{m-1})$,

$$\begin{aligned} &\langle \mathcal{A}, \bar{a}_i \rangle_{i<m} \models \theta(\bar{a}_0, \ldots, \bar{a}_{m-1}) \quad \text{iff} \\ &\langle \mathcal{B}, f(\bar{a}_i) \rangle_{i<m} \models \theta(f(\bar{a}_0), \ldots, f(\bar{a}_{m-1})). \end{aligned}$$

For a theory T and type Σ, $\cong_\Sigma$ is an equivalence relation on the class of models of T. Say that a complete type Σ of the theory T is $\aleph_0$-*categorical* just if there are only finitely many equivalence classes with respect to $\cong_\Sigma$ represented in $\{\mathcal{A} \mid \mathcal{A}$ is a countable model of $T\}$. For example, it is not difficult to see that every complete type of an Ehrenfeucht theory is $\aleph_0$-categorical (the concepts of finitary and $\aleph_0$-categorical are independent). It then turns out that every theory that has a complete, $\aleph_0$-categorical, finitary, non-principal type is unstable. Since every complete type of an Ehrenfeucht theory is $\aleph_0$-categorical, this shows that every projectively finitary Ehrenfeucht theory is unstable. This result for types can be worded informally as follows: if a non-principal type (for which there is always a model that has infinitely many realizations) has only finitely many finite or countable isomorphism types (is $\aleph_0$-categorical) and satisfies a condition similar to the syntactic condition in the Ryll-Nardzewski theorem (B_n^Γ finite for all n), then the theory of the type must have a model with a definable, infinite, linear order. The proof of this result is similar to proofs of related results in the literature (Millar [1981c], Pillay [1980]). However, as mentioned, this result is applicable to all known examples of Ehrenfeucht theories, whereas former results are not similarly general. As a corollary of this result, any Ehrenfeucht theory that allows elimination of quantifiers in a language whose only non-logical symbols are constant symbols, unary predicate symbols, and finitely many binary predicate symbols must be unstable. The reason is that in this case every non-principal type is finitary, since an n-type is completely determined by its 1-type projections and finitely many atomic and negated atomic formulas involving the binary predicate symbols.

4.3. *Omitting types revisited*

Another group of examples of the third approach concerns the omitting types theorem. As mentioned above, every computable non-principal type of a decidable theory is omitted from some decidable model of the theory. The standard proof of the classical result can be modified easily to omit countably many non-principal types. That modification is effective, and so we have the result that every computable sequence of non-principal types of a decidable theory is omitted from some decidable model of the theory. On the other hand, what happens if there is a computable list of types of a complete theory, some of which are principal and some of which are non-principal? Well of course one can not hope to omit the principal types in any model of a complete theory. However, can one omit the non-principal types from the sequence in some decidable model of the theory?

This is an example where one returns to the proof of the classical theorem and tries for a more subtle mastery of the construction. In the Henkin style proof of the omitting types theorem for say a single 1-type, the idea is that the elementary diagram of the desired model is constructed by increasing finite approximations. Parsed into the normal steps of a Henkin construction are infinitely many actions designed to omit the 1-type from the model. Since the 1-type will be omitted from the model if and only if no element of the model satisfies the type, it is enough to make sure that for each element of the model there is a stage of the construction where one guarantees that that element satisfies a formula whose negation is in the designated 1-type. Of course this guarantee is made by putting such a sentence into the next approximation of the elementary diagram of the desired model. Since only finitely much about that element will have been "said" in the approximation to the elementary diagram to that point, our ability to find such a formula that is consistent with that approximation and the theory is a consequence of the definition of a non-principal type.

In the effective version of this theorem, each step of the construction is computable in a uniform way. Consistency with the theory can be checked effectively by using the finite approximations to the elementary diagram, since the theory is decidable. The required formulas can be found effectively as above, because the type is computable. But what happens in a sequence of computable types, some of which may be principal? Then it would not be wise to start a search for a formula to make sure some constant does not realize the type, since the type may be principal and the construction already may have committed that element to the type by putting the complete formula about that constant into the elementary diagram.

A solution to this problem is to abandon the aggressive "omit as you go" strategy, and replace it by an "omit as you can" strategy. Imagine that making sure a particular constant does not realize a type in the sequence, as long as the type turns out to be non-principal, has highest priority among the goals past some stage of the construction. If the elementary diagram of the model at that stage already commits that element to realize some principal type, then the possibility that that constant could realize a non-principal type is eliminated. If the element is not committed yet to realizing a principal type, then at some point later in the construction, a sentence will be considered such that neither what is said about the constant by the construction so far together with the information in the new sentence, nor the negation of that augmented information, is implied by what will have been said so far about the constant. This then would provide the opportunity to make sure that that element does not realize the type in question. So this is the start for obtaining a sharper effective theorem. In fact by using standard techniques in recursion theory applied to the model theoretic setting, one can show that for a decidable theory and any sequence of computable complete types that is computably enumerable in the Halting Problem, there is a decidable model of the theory that omits every non-principal type from the sequence (Millar [1983]). This result plays a role in our next example.

4.4. *Number of countable models*

Sacks showed that Vaught's conjecture holds for a theory if the Scott rank of each countable model $\mathcal{A}$ of the theory is less than its maximum possible value $(\omega_{CK}^{\mathcal{A}}+1)$ (Sacks [1981]). That is, any such theory that has more than countably many countable models has $2^{\aleph_0}$ countable models. Similarly, recursive model theoretic aspects of a theory are related to the possible number of countable models of the theory. A very simple example of this phenomenon is that if a decidable theory has a countable model that is not decidable, then the theory has more than one countable model. I will pursue this perspective.

It is useful here to introduce a common refinement of the hierarchy of computable sets. Suppose that an effective object type is specified, such as natural number, partial computable function, or formula of an effective language. A subset of such a domain is said to be Σ_2 just in case it is computably enumerable in $\mathbf{O}'$. A set that is the complement of such a set is said to be Π_2. Similarly, a subset computably enumerable in $\mathbf{O}^{(n)}$ is said to be Σ_{n+1} and its complement Π_{n+1}. For instance the subset of the partial computable functions that are total is a Π_2 set, whereas the subset of the partial computable functions whose domains are co-finite is Σ_3. Similarly, $S^r(T)$ is itself no worse than Π_2, since the set of total computable functions $f: Frm_{\mathcal{L}(T)} \to \{0,1\}$ is Π_2, and such an f is a characteristic function for an element of $S^r(T)$ just in case, for some $n \in \omega$:

(i) for all finite subsets $\Theta \subset Frm_{\mathcal{L}(T)}$, $T \cup \{\theta \in \Theta \mid f(\theta)=1\}$ is consistent;
(ii) if $f(\theta)=1$ then the free variables of θ are included in $\{x_1, x_2, \ldots, x_n\}$;

(iii) for each θ whose free variables are included in $\{x_1, x_2, \ldots, x_n\}$, either $f(\theta) = 1$ or $f(\neg\theta) = 1$.

As another example, the set $S_p(T)$ of principal complete types of such a T is no worse than Σ_2, essentially because the subset of complete formulas is computable in $\mathbf{O}'$.

Now consider a language $\mathcal{L}$ with unary predicate symbols P_α for each $\alpha \in 2^{<\omega}$. Let T be the $\mathcal{L}$-theory whose axioms are:

$$\forall x[P_\alpha(x) \to P_\gamma(x)], \quad \gamma \subseteq \alpha \in 2^{<\omega};$$
$$\forall x[P_\alpha(x) \to \neg P_\gamma(x)], \quad \alpha \nsubseteq \gamma, \gamma \nsubseteq \alpha, \quad \alpha, \gamma \in 2^{<\omega};$$
$$\exists x\, P_\alpha(x), \quad \alpha \in 2^{<\omega};$$
$$\forall x \bigvee_{\alpha \in 2^n} P_\alpha(x), \quad n \in \omega.$$

This theory is complete, consistent, decidable, and has $2^{\aleph_0}$ 1-types. In fact, for each $f : \omega \to 2$, the set $\{P_\alpha(x) \mid \alpha \subset f; \alpha \in 2^{<\omega}\}$ is contained in a unique complete 1-type of T. All 1-types of T are non-principal. In particular T does not have a prime model. By the remarks above, the set of complete computable 1-types of T is Π_2. But what happens if $S^r(T')$ of a complete decidable theory T' is less complex than Π_2? It is a theorem that then T' must have a prime model, and in fact the prime model is decidable [Millar, 1982a]! For example, if T has a decidable model whose type spectrum is $S^r(T)$, then $S^r(T)$ is Σ_1 and thus T has a decidable prime model.

Recall from above that the set of principal types of a complete decidable theory is always Σ_2. Goncharov and Nurtazin, and Harrington proved that if such a T has a prime model, then the prime model is decidable if and only if $S_p(T)$ is Σ_1 (Goncharov and Nurtazin [1973], Harrington [1974]). So if a decidable complete theory T has a prime model that is not decidable, then T has more than one countable model, the set $S_p(T)$ is not Σ_1, and the set $S^r(T)$ is not Σ_2. Using the omitting types theorem described earlier, one can prove that if there is no $\Gamma(\overline{x}) \in S^r(T)$ such that $\Gamma(\overline{c})$ has a decidable prime model, then T has $2^{\aleph_0}$ countable models (Millar [1983]).

As another example of the technique, I will use the omitting types theorem to prove that if a decidable theory T has a prime model that is not decidable, and the theory has, for instance, only countably many countable models, then it has other computationally complex structural features. Notice by the result mentioned at the end of the previous paragraph, if the subset $\mathcal{DP}(T) \subset S(T)$ defined by ($\Gamma(\overline{x}) \in \mathcal{DP}(T)$ iff $\Gamma(\overline{c})$ has a decidable prime model) is empty, then T has $2^{\aleph_0}$ countable models. Therefore for such T, $\mathcal{DP}(T)$ can not be empty. In fact we will see that $\mathcal{DP}(T)$ can not be Σ_2.

For a complete theory T and a model $\mathcal{A}$ of T, $TySp(\mathcal{A})$ is sometimes uniquely determined by its containment of a particular type and its omission of finitely many others types.

– $TySp(\mathcal{A})$ has the *finite basis property* just if there is a type $\Gamma \in TySp(\mathcal{A})$ and a finite $\mathcal{O} \subset S(T)$, such that for all models $\mathcal{B}$ of T, $TySp(\mathcal{A}) = TySp(\mathcal{B})$ iff

(i) $\Gamma \in TySp(\mathcal{B})$; and
(ii) $TySp(\mathcal{B}) \cap \mathcal{O} = \emptyset$.

– The pair $\langle \Gamma, \mathcal{O} \rangle$ is a *basis* – two bases are *equivalent* if they characterize the same type spectrum.

(Notice that any finite set of types that are realized in $\mathcal{A}$ can be represented by one type that is realized in $\mathcal{A}$.)

Let $\Gamma \in S(T)$ be the unique 1-type that describes an element that is larger than all of the interpretations of the constant symbols in the Ehrenfeucht theory T described above. Then $\langle T, \{\Gamma\} \rangle$ is a basis for the prime model, and $\langle \Gamma, \{\ \} \rangle$ is a basis for the other two models. No model of the theory mentioned above in the signature P_α, $\alpha \in 2^{<\omega}$, has a basis. If a decidable theory has a prime model that is not decidable, then since the omitting types theorem is effective, the prime model fails to have a basis. Let $\mathcal{B}(T) \subseteq S^r(T)$ be the set of computable types Γ such that for some finite $\mathcal{O} \subset (S(T) - S_p(T))$, either $\langle \Gamma, \mathcal{O} \rangle$ is a basis, or, for all decidable models $\mathcal{A}, \mathcal{B}$ that realize Γ and omit $\mathcal{O}$, $TySp(\mathcal{A}) = TySp(\mathcal{B})$. Thus for a decidable theory that has a prime model that is not decidable, $S_p(T) \cap \mathcal{B}(T) = \emptyset$.

THEOREM. *Suppose T is a decidable, consistent, complete theory. Then sufficient conditions for T to have a decidable prime model include*:
(i) *$S^r(T)$ is Σ_2*;
(ii) *$S_p(T)$ is Σ_1 and T has a prime model*;
(iii) *$\mathcal{DP}(T)$ is Σ_2 and $\{TySp(\mathcal{A}) \mid \mathcal{A}$ models $T\}$ is countable*;
(iv) *$\mathcal{B}(T)$ is Σ_2, and $\{TySp(\mathcal{A}) \mid \mathcal{A}$ models $T\}$ is countable.*

PROOF. (i) follows immediately from the more general omitting types theorem mentioned above – there is a decidable model $\mathcal{A}$ that omits all non-principal types in $S^r(T)$. Since $\mathcal{A}$ is decidable, $TySp(\mathcal{A}) \subset S^r(T)$. But that means that $\mathcal{A}$ can realize only principal types and is thus prime. (ii) is Goncharov, Nurtazin, and Harrington's theorem (Goncharov and Nurtazin [1973], Harrington [1974]). I will describe a construction that proves both (iii) and (iv). The proof is by contradiction. Since both conditions assume that $\{TySp(\mathcal{A}) \mid \mathcal{A}$ models $T\}$ is countable, $S(T)$ is therefore countable and so it follows that for every $\Gamma(\overline{x}) \in S(T)$, $\Gamma(\overline{c})$ has a prime model. We will define $\Gamma_\alpha \in S^r(T)$ for $\alpha \in 2^{<\omega}$ such that:

(i) $\Gamma_\alpha(\overline{x}) \subseteq \Gamma_\gamma(\overline{x}, \overline{y})$ for $\alpha \subseteq \gamma$;

(ii) The prime model of $\Gamma_\alpha(\overline{c})$ omits Γ_β for every $\beta | \alpha$ such that for the least i such that $\beta(i) \neq \alpha(i)$, $\beta(i) < \alpha(i)$;

(iii) $\Gamma_\alpha \notin \mathcal{DP}(T)$ in case (iii), and $\Gamma_\alpha \notin \mathcal{B}(T)$ in case (iv).

Assume for the moment that Γ_α, $\alpha \in 2^{<\omega}$, is defined. Let $\langle \mathcal{A}_\alpha, \overline{c}_\alpha \rangle$ be the prime model of $\Gamma_\alpha(\overline{d})$, where $\mathcal{A}_\alpha$ is a model of T. For each $f : \omega \to 2$, define

$$\mathcal{A}_f = \bigcup_{\alpha \subset f} \mathcal{A}_\alpha .$$

By (i) $\mathcal{A}_f$ is the union of an elementary chain of models of T and is thus a model of T. By (ii), for $f, g : \omega \to 2$, $f \neq g$, $TySp(\mathcal{A}_f) \neq TySp(\mathcal{A}_g)$. But that means that $\mathcal{A}_f$ and $\mathcal{A}_g$ are countable models that do not have the same type spectrum. Thus T does not have countably many type spectra, which is the desired contradiction.

Write $\beta \prec \alpha$ just in case for the least i (and it exists) such that $\beta(i) \neq \alpha(i)$, $\beta(i) < \alpha(i)$. Let $\Gamma_{\langle\rangle}$ be any principal 1-type of T. Assume inductively that Γ_β has been defined for all $\beta \prec \alpha$ such that length(β) $\leqslant$ length(α), and for all $\beta \subset \alpha$. Let $\alpha' \subset \alpha$ have length one less than α. In addition, assume that for all $\beta \prec \alpha$, $\Gamma_\beta(x)$ is a non-principal type of the theory $\Gamma_{\alpha'}(c)$ and similarly that all types in $\mathcal{DP}(T)$ ($\mathcal{B}(T)$) are non-principal in the theory $\Gamma_{\alpha'}(c)$. For case (iii) of the theorem, the omitting types theorem of Millar [1983] must be stated with more precision:

THEOREM (Effective Omitting Types Theorem). *Let T be a complete decidable consistent theory, let $\Phi \subset S^r(T)$ be a Σ_2 set of types of T, and let Ψ be a Σ_2 set of non-principal types of T, perhaps incomplete. Then T has a decidable model that omits all the types in Ψ and all non-principal types in Φ.*

The theorem will be applied to the theory $\Gamma_{\alpha'}(\bar{c})$. The difficult case is when $\alpha = \alpha'^\frown\langle 1\rangle$. So assume we are in that case and let $\alpha_0 = \alpha^\frown\langle 0\rangle$ and let $\mathcal{A}$ be any decidable model of T that realizes Γ_{α_0}. Then it is not difficult to see that the set E of types $\Delta \in TySp(\mathcal{A})$ such that

$$\Gamma_{\alpha'}(\bar{x}) \subset \Delta(\bar{x}, \bar{y})$$

is a Σ_2 set. The Φ of the omitting types theorem will be

$$\Phi = \{\Delta(\bar{c}, \bar{y}) \mid \Delta \in E\}.$$

Let Ψ of the omitting types theorem be

$$\Psi = \{\Gamma_\beta \mid \beta \prec \alpha\} \cup \mathcal{DP}(T).$$

Now apply the theorem to $\Gamma_{\alpha'}(\bar{c})$ to obtain a decidable model $\langle \mathcal{B}, \bar{d}\rangle$ that omits all the types in Ψ and all the non-principal types in Φ. Then since $\Gamma_{\alpha'} \notin \mathcal{DP}(T)$, $\langle \mathcal{B}, \bar{d}\rangle$ must realize a computable non-principal type $\Sigma(\bar{c}, \bar{y})$ of $\Gamma_{\alpha'}(\bar{c})$ (where $\Sigma(\bar{x}, \bar{y}) \in S(T)$). Thus $\Sigma(\bar{c}, \bar{y}) \notin \Phi$, i.e. $\Sigma(\bar{x}, \bar{y}) \notin TySp(\mathcal{A})$. Define $\Gamma_\alpha = \Sigma(\bar{x}, \bar{y})$. Note that the induction hypothesis is maintained by the omitting types argument.

Assume next that (iv) of the theorem holds. In that case, since $\Gamma_{\alpha'} \notin \mathcal{B}(T)$, it follows that

$$\langle \Gamma_{\alpha'}, \{\Gamma_\beta \mid length(\beta) \leqslant length(\alpha); \beta \prec \alpha\}\rangle$$

is not a basis. In fact we can fix two decidable models $\mathcal{A}, \mathcal{B}$ realizing $\Gamma_{\alpha'}$ and omitting $\{\Gamma_\beta \mid length(\beta) \leqslant length(\alpha); \beta \prec \alpha\}$ such that

$$TySp(\mathcal{A}) \neq TySp(\mathcal{B}).$$

Let $\Sigma'(\overline{y})$ be any type in the symmetric difference of those two sets, and let $\Gamma_\alpha(\overline{x}, \overline{y})$ be a type in one of the two sets such that

$$\Gamma_{\alpha'}(\overline{x}), \Sigma'(\overline{y}) \subset \Gamma_\alpha(\overline{x}, \overline{y}).$$

□

4.5. *An Ash legacy*

I will finish the third approach and the chapter by touching on a set of related topics that has drawn considerable attention during the last ten years. Out of this work has come a new and powerful priority argument construction introduced by Ash. Ash's method of labelling systems is introduced in Ash [1986a, 1986b]. The technique is used to solve problems of recursive model theory, e.g., Ash [1986a, 1986b], Ash and Knight [1990], Barker [1988]. The technique is then simplified and (slightly) generalized in Ash [1990], and further developed in Ash and Knight [1994]. Although developed specifically to solve a series of problems in recursive model theory, the method also will have applications in classical recursion theory. A common feature of these topics is they concern isomorphic computable structures.

- A relation $X \subset |\mathcal{A}|^n$ on a computably presented model $\mathcal{A}$ is *intrinsically computably enumerable* just if for every computably presented model $\mathcal{B}$ and relation Y such that $\langle \mathcal{A}, X\rangle$ is isomorphic to $\langle \mathcal{B}, Y\rangle$, Y is computably enumerable. (So note that if $X = R_{\mathcal{A}}$ for some relation symbol in the language of the model, then automatically X is intrinsically computably enumerable.)
- A computably presentable model is *computably stable* just if every isomorphism between computably presented models in the isomorphism class is computable.
- A computably presentable model is *computably categorical* just if for every pair of computably presented models in the isomorphism class, there is a computable isomorphism between them.

For example, if the underlying model is the natural numbers with the successor function, then the set of even numbers is intrinsically computably enumerable, and the model is computably stable. If instead the underlying model is the set of natural numbers with the "less than" relation, then the set of even numbers is not intrinsically computably enumerable, and the model is not computably stable. Although not computably stable, the rationals with the "less than" relation is computably categorical.

In Ash and Nerode [1981] the condition of intrinsic computable enumerability is proven equivalent, with some additional assumptions on the model, to a natural syntactic condition that can be expressed in $\mathcal{L}_{\omega_1,\omega}$. Generalizations and variations of this are pursued through the hyperarithmetic hierarchy in many different papers, for example, Ash [1986a, 1986b, 1990], Barker [1988], Harizanov [1991a, 1991b], Chisolm [1990a, 1990b]. Similar strategies and techniques arise in the pursuit of syntactic characterizations of computable stability and categoricity. Again, many papers (Ash [1986a, 1986b], Ash and Knight [1990], Ash and Goncharov [1985]) have been written on these and related subjects.

References

C. J. ASH
[1986a] Stability of recursive structures in arithmetical degrees, *Ann. Pure Appl. Logic*, 32, pp. 113–135.
[1986b] Recursive labelling systems and stability of recursive structures in hyperarithmetic degrees, *Trans. Amer. Math. Soc.*, 298, pp. 497–514.
[1990] Labelling systems and r.e. structures, *Ann. Pure Appl. Logic*, 47, pp. 99–119.

C. J. ASH AND S. S. GONCHAROV
[1985] Strong Δ_2^0-categoricity, *Algebra and Logic*, 24, pp. 471–476.

C. J. ASH AND T. S. MILLAR
[1983] Persistently finite, persistently arithmetic theories, *Proc. Amer. Math. Soc.*, 89, pp. 487–492.

C. J. ASH AND A. NERODE
[1981] Intrinsically recursive relations, in: *Aspects of Effective Algebra*, Crossley, ed., Steel's Creek, Australia, pp. 26–41.

C. J. ASH AND J. F. KNIGHT
[1990] Pairs of recursive structures, *Ann. Pure Appl. Logic*, 46, pp. 211–234.
[1994] Mixed systems, *J. Symbolic Logic*, 59, pp. 1383–1399.

E. BARKER
[1988] Intrinsically Σ_α^0 relations, *Ann. Pure Appl. Logic*, 39, pp. 105–130.

M. BENDA
[1974] Remarks on countable models, *Fund. Math.*, 81, pp. 107–119.

C. C. CHANG AND H. J. KEISLER
[1990] *Model Theory*, 3rd ed., North-Holland, Amsterdam.

J. CHISOLM
[1990a] Effective model theory vs. recursive model theory, *J. Symbolic Logic*, 55, pp. 1168–1191.
[1990b] The complexity of intrinsically r.e. subsets of existentially decidable models, *J. Symbolic Logic*, 55, pp. 1213–1232.

K. GÖDEL
[1930] Die Vollständigkeit der Axiome des logischen Funktionenkalküls, *Monatsh. Math. Phys.*, 37, pp. 349–360.
[1931] Über formal unentscheidbare Sätze der Principia Mathematica und verwandter Systeme, I, *Monatsh. Math. Phys.*, 38, pp. 173–198.

S. S. GONCHAROV
[1978] Strongly constructivizability of homogeneous models, *Algebra and Logic*, 17, pp. 247–263.
[1980] Totally transcendental decidable theory without constructivizable homogeneous models, *Algebra and Logic*, 19, pp. 85–93.

S. S. GONCHAROV AND A. T. NURTAZIN
[1973] Constructive models of complete solvable theories, *Algebra and Logic*, 12, pp. 67–77.

L. HARRINGTON
[1974] Recursively presentable prime models, *J. Symbolic Logic*, 39, pp. 305–309.

V. S. HARIZANOV
[1991a] Uncountable degree spectra, *Ann. Pure Appl. Logic*, 54, pp. 255–263.
[1991b] Some effects of Ash–Nerode and other decidability conditions on degree spectra, *Ann. Pure Appl. Logic*, 55, pp. 51–65.
[ta] Pure recursive model theory, in: *Handbook of Recursive Algebra*, A. Nerode, J. B. Remmel, Y. Ershov and S. Goncharov, eds.

S. C. KLEENE

[1982] *Introduction to Metamathematics*, 8th reprint, Wolters-Noordhoff Publishing Co., Groningen and North-Holland, Amsterdam.

K. ZH. KUDAIBERGENOV

[1979] A theory with two strongly constructivizable models, *Algebra and Logic*, 18, pp. 111–117.

M. LERMAN AND J. H. SCHMERL

[1979] Theories with recursive models, *J. Symbolic Logic*, 44, pp. 59–76.

M. LERMAN, R. A. SHORE AND R. I. SOARE

[1984] The elementary theory of the recursively enumerable degrees is not $\aleph_0$-categorical, *Adv. in Math.*, 53, pp. 301–320.

T. S. MILLAR

[1978] Foundations of recursive model theory, *Ann. Math. Logic*, 13, pp. 305–320.

[1979] A complete, decidable theory with two decidable models, *J. Symbolic Logic*, 44, pp. 307–312.

[1980] Homogeneous models and decidability, *Pacific J. Math.*, 91, pp. 407–418.

[1981a] Counterexamples via model completions, in: *Lecture Notes in Mathematics*, M. Lerman, J. H. Schmerl, R. I. Soare, eds., Springer, Berlin, pp. 215–229.

[1981b] Vaught's theorem recursively revisited, *J. Symbolic Logic*, 46, pp. 397–411.

[1981c] Stability, complete extensions, and the number of countable models, in: *Aspects of Effective Algebra*, J. N. Crossley, ed., Upside Down A Book Company, Yarra Glen, pp. 196–205.

[1982a] Type structure complexity and decidability, *Trans. Amer. Math. Soc.*, 271, pp. 73–81.

[1982b] Decidable Ehrenfeucht theories, in: *Recursion Theory*, Nerode and Shore, ed., Proc. Symp. Pure Math., Vol. 42, pp. 311–321.

[1983] Omitting types, type spectrums, and decidability, *J. Symbolic Logic*, 48, pp. 171–181.

[1984] Decidability and the number of countable models, *Ann. Pure Appl. Logic*, 27, pp. 137–153.

[1989] Tame theories with hyperarithmetic homogeneous models, *Proc. Amer. Math. Soc.*, 105, pp. 712–726.

M. G. PERETYAT'KIN

[1973] On complete theories with a finite number of denumerable models, *Algebra and Logic*, 12, pp. 310–326.

[1978] Criterion for strong constructivizability of a homogeneous model, *Algebra and Logic*, 17, pp. 290–301.

[1982] Turing machine computations in finitely axiomatizable theories, *Algebra and Logic*, 21, pp. 272–295.

A. PILLAY

[1980] Instability and theories with a few models, *Proc. Amer. Math. Soc.*, 80, pp. 461–468.

E. L. POST

[1944] Recursively enumerable sets of positive integers and their decision problems, *Bull. Amer. Math. Soc.*, 50, pp. 284–316.

C. R. REED

[1991] A decidable Ehrenfeucht theory with exactly two hyperarithmetic models, *Ann. Pure Appl. Logic*, 53, pp. 135–168.

G. E. SACKS

[1972] *Saturated Model Theory*, W. A. Benjamin, Reading, MA.

[1981] On the number of countable models, in: *Southeast Asian Conference on Logic*, Barwise, Kaplan, Keisler, Suppes, Troelstra, eds., Studies in Logic and the Foundation of Mathematics, Vol. 111, pp. 185–195.

J. SCHLIPF

[1978] Toward model theory through recursive saturation, *J. Symbolic Logic*, 43, pp. 183–206

S. G. SIMPSON

[1977] Degrees of unsolvability: a survey of results, in: *Handbook of Mathematical Logic*, J. Barwise, ed., Studies in Logic and the Foundations of Mathematics, Vol. 90, pp. 631–652.

R. I. SOARE

[1987] *Recursively Enumerable Sets and Degrees*, Springer, Berlin.

R. L. VAUGHT

[1961] Denumerable models of complete theories, in: *Logic, Methodology, and Philosophy of Science*, Y. Bar-Hillel, ed., North-Holland, Amsterdam, pp. 390–401.

CHAPTER 16

Classifying Recursive Functions

Helmut Schwichtenberg

University of Munich, Institute of Mathematics, Theresienstrasse 39, D-80333 Munich, Germany

Contents

1. Introduction 534
2. Collapsing results 536
3. The extended Grzegorczyk hierarchy 540
4. Partial continuous functionals 542
5. Computability in higher types 558
6. Bounded fixed point operators 569
7. Elimination of detours through higher types by transfinite recursion 576
References 582

HANDBOOK OF COMPUTABILITY THEORY
Edited by E.R. Griffor

1. Introduction

Ever since the recursive functions have been identified there was a challenge to measure their inherent computational complexity, or in Kleene's [1958] words to "classify the recursive functions into a hierarchy, according to some general principle". The constructive ordinals of Church and Kleene lend themselves as an obvious scale for such classification attempts.

The most natural way to measure the "complexity" of a recursive function by constructive ordinals is to consider recursion along recursive well-orderings of larger and larger order types. A positive result in this direction is due to Rósza Péter, who has shown that the class of functions definable with (substitution and) n-fold nested recursion grows with n increasing. One can view n-fold nested recursion as recursion on a special, "natural" well-ordering of order type ω^n. In Section 2 we will prove a theorem due to Myhill [1953] and Routledge [1953], which says that any recursive function can be defined using (Csillag–Kalmár) elementary operations and just one recursion on an elementary well-ordering of type ω. Myhill refers to this fact as "a stumblingblock in constructive mathematics"; in particular it implies that Péter's result essentially depends on the special form of the well-orderings used. With a similar method we prove that such a "collapse" will also occur if we allow transfinite recursion for the definition of 0-1-valued functions only. Finally we prove a rather general result, which says that it is impossible to index the recursive functions in any "reasonable" way by means of a Π^1_1-path through Kleene's system $\mathcal{O}$ of notations for constructive ordinals.

Therefore it seems necessary to look out for a notion of a "standard" well-ordering of the natural numbers. But this turns out to be a very difficult task. Although an interesting attempt has been made by Zemke [1977] and later been extended by Buchholz, Cichon and Weiermann [1994], it seems fair to say that up to now no satisfactory such notion has been found. Therefore in this paper we restrict ourselves to "canonical" well-orderings of concrete order types like ε_0. In this case the Cantor normal form can be used to define a canonical coding; moreover, one also has a canonical choice of fundamental sequences approximating the limit ordinals. Using these, a quite natural sequence F_α, $\alpha < \varepsilon_0$, of fast growing functions can be defined. Then one defines $\mathcal{E}_\alpha$ as the elementary closure of F_α; the $\mathcal{E}_\alpha$ form the *extended Grzegorczyk hierarchy*.

For the classes $\mathcal{E}_\alpha$ quite a number of interesting characterizations are known. First of all, $\mathcal{E}_\alpha$ is the class of functions computable by a register machine with time (i.e. number of computation steps) bounded by a finite iteration of the function F_α. Moreover, the functions in $\mathcal{E}_\alpha$ can be characterized by counting the number of recursions used and their order types, or else by transfinitely iterating the process of extending an effectively generated class of functions by enumeration (cf. Schwichtenberg [1971]). So we have a quite satisfactory theory here.

However, by definition our smallest class $\mathcal{E}_0$ consists of the class of Csillag–Kalmár elementary functions and hence already contains exponentially growing functions. If one is interested in analyzing more "feasible" notions of computation, then it is necessary to take into account the fact that numbers are represented as strings of numerals. This leads to a notion of "bounded recursion on notation" introduced by Cobham [1965]. We do not attempt to deal with this theory here, but refer the reader to the excellent survey of Clote [1999] in the present Handbook.

So as a measure of complexity for recursive functions we use ordinals. We consider a recursive function to be given by a computation method or algorithm; hence closely connected to such a definition is a termination proof for the algorithm. Therefore it is to be expected that methods from proof theory are of central importance for the subject. More precisely, there is not much difference between classifying proofs of $\forall\exists$-theorems and definitions of recursive functions. For instance, by a classic result of Ackermann [1940] and Kreisel [1952] the ε_0-recursive functions are just those functions which are provably recursive in Peano arithmetic (cf., e.g., Fairtlough and Wainer [1998] or Schwichtenberg [1977]), and the primitive recursive functions are just those definable by the subsystem of Peano arithmetic where induction is restricted to Σ_1-formulas; this result is due to Kreisel, Parsons, Mints [1973] and Takeuti [1987]. Since there is a recent survey on new developments along these lines in the paper by Fairtlough and Wainer [1998], we do not go into this subject here.

One very interesting phenomenon needs to be mentioned: the role of ε_0 in classifying the provably recursive functions of Peano arithmetic is not as mandatory as one might think. It has been shown by Girard that the so-called slow growing hierarchy G_α, $\alpha <$ the Bachmann–Howard ordinal, yields the same growth as the fast growing hierarchy up to ε_0. This is interesting because a natural description of the Bachmann–Howard ordinal seems to require the least uncountable ordinal Ω. In Schwichtenberg and Wainer [1995] an attempt has been made to explain why uncountable ordinals play a role here.

The main subject of this article is the study of the interplay between recursion in higher types and transfinite recursion. To make the paper sufficiently self-contained we include a short introduction to the theory of partial continuous functionals, based on Scott's notion of an information system. The partial continuous functionals are central for any analysis of higher type computability which is based on the rather natural assumption that any computation ought to be finite. The reason is simply that they form the mathematically appropriate domain of a computable functional. The total continuous functionals of Kleene–Kreisel (Kleene [1959a], Kreisel [1959]) can then be singled out from the partial continuous ones (cf. Ershov [1974], Berger [1993] and also Dag Normann's paper [Normann, 1999] in the present Handbook), and it seems best to define them that way.

Since any partial continuous functional Φ is the limit of finite approximations to it, it is straightforward to define computability: Φ is computable if and only if it is the limit of a computably enumerable set of finite approximations. This is an externally defined notion of computability, and the question arises whether there is an internal characterization. This is indeed the case: we will prove a result due to Plotkin [1977] that a partial continuous functional is computable if and only if it can be defined explicitly from fixed point operators introduced into this framework by

Platek [1966], the parallel conditional pcond and a continuous approximation $\exists$ to the existential quantifier. We call such functionals recursive in pcond and $\exists$.

The term language with a fixed point operator leads us back to our original question of classifying recursive functions. To demonstrate the relevance of higher types for our subject, we first show that higher type iteration operators suffice to define all F_α, $\alpha < \varepsilon_0$, by application alone. We then show that it is possible to combine the power and elegance of the general fixed point operator with the desire to have better control over computational complexity by introduction of a bounded version of the fixed point operator. Also for $\exists$ one can define a bounded version; together with (partial) primitive recursion this leads to subrecursive hierarchies over the partial continuous functionals, developed by Niggl [1993, 1995]. Niggl [1994] also studies another restricted notion of computability, generalizing Cook and Kapron's (Cook [1992]) typed while programs to partial continuous functionals.

We finally establish some kind of converse of the definability of the F_α, $\alpha < \varepsilon_0$, by means of higher type iteration operators. We show that finite types are not only sufficient to obtain the ε_0-recursive functions, but that their use can also be eliminated at the expense of higher ordinals. So we have a certain trade-off between the two concepts. The result itself is not new; however, here we sketch a different proof, by adapting a method of Buchholz.

The paper is organized as follows. Section 2 contains the collapsing results, and Section 3 discusses the extended Grzegorczyk hierarchy. Section 4 gives an introduction to partial continuous functionals, and Section 5 then contains some general material concerning computability in higher types, including definability of the F_α, $\alpha < \varepsilon_0$, by means of higher type iteration operators, and Plotkin's definability theorem. The discussion of bounded fixed point operators takes place in Section 6, and the final Section 7 concerns elimination of detours through higher types by transfinite recursion.

2. Collapsing results

To set the stage, we prove the failure of some natural attempts to classify the recursive functions. Here we assume knowledge of certain introductory material from recursion theory, e.g., the Csillag–Kalmár definition of the elementary functions, Minsky's register machines and also Kleene's system $\mathcal{O}$ of ordinal notations, Π^1_1-sets and paths through $\mathcal{O}$.

2.1. *Register machines.* Minsky [1961] has introduced (see also Shepherdson and Sturgis [1963]) a type of idealized computing machines now called register machines. These machines allow a rather direct and perspicuous proof that all recursive functions are computable. In particular one can prove in the well-known way the following theorem.

THEOREM (Normal form theorem). *Let f be a unary recursive function, p be the Gödel number of a register machine computing f and $s_f(x)$ the number of steps performed by this machine when computing $f(x)$. Then f can be written in the form*

$$f(x) = D\big(C(p, x, s_f(x))\big)$$

with elementary functions D ("decoding function") and C ("configuration function").

2.2. Now we can prove our first collapse result, due to Myhill [1953] and Routledge [1953].

THEOREM. *For any recursive function f we can find an elementary well-ordering $\prec$ of the natural numbers with order type ω and a recursive function h such that f is elementary in h, and h can be defined in the form*

$$\begin{aligned} h(0) &= 0, \\ h(u) &= 1 + h(g(u)) \quad \textit{for } u \neq 0, \end{aligned}$$

with g an elementary function such that $g(u) \prec u$ for $u \neq 0$.

PROOF. It clearly suffices to prove the theorem for unary functions f. Let p be the Gödel number of a register machine computing f and let $s_f(x)$ be the number of steps performed by this machine when computing $f(x)$; we may assume $s_f(x) \geqslant 1$ for all x. Note first that the relation $t < s_f(x)$ is elementary, since

$$t < s_f(x) \leftrightarrow C(p, x, t) \neq C(p, x, t + 1)$$

with C the configuration function from 2.1. The pairs (t, x) with $t < s_f(x)$ can be well-ordered by

$$(t_1, x_1) \ll (t_2, x_2) \leftrightarrow x_1 < x_2 \vee (x_1 = x_2 \wedge t_1 > t_2).$$

Let B be the image of all these pairs under the mapping $\lambda t \lambda x(\pi(t, x) + 1)$, i.e.

$$B(u) :\leftrightarrow \exists t \leqslant u \exists x \leqslant u\big(t < s_f(x) \wedge \pi(t, x) + 1 = u\big).$$

Here π is an elementary pairing function with elementary inverses π_1, π_2. Clearly B is elementary. The "induced" well-ordering on B can be extended easily to an elementary well-ordering $\prec$ on all natural numbers with least element 0 (note that $\neg B(0)$, $B(1)$). Let

$$\begin{aligned} b(u) &:= \mu v \leqslant u\big[B(v) \wedge \forall w \leqslant u\big(B(w) \to w \leqslant v\big)\big], \\ t(u) &:= \pi_1\big(b(u) \dot{-} 1\big), \\ x(u) &:= \pi_2\big(b(u) \dot{-} 1\big). \end{aligned}$$

Then we define $\prec$ by

$$\begin{aligned} u \prec v \leftrightarrow\ & [u = 0 \wedge v \neq 0] \vee \\ & \big[u \neq 0 \wedge \big(x(u) < x(v) \vee \\ & \qquad \big(x(u) = x(v) \wedge t(u) > t(v)\big) \vee \\ & \qquad \big(x(u) = x(v) \wedge t(u) = t(v) \wedge u < v\big)\big)\big]. \end{aligned}$$

Now consider the function

$$g(u) := \begin{cases} \pi(t(u)+1, x(u))+1 & \text{if } B(u) \text{ and } t(u)+1 < s_f(x(u)), \\ 0 & \text{otherwise.} \end{cases}$$

Then clearly we have $g(u) \prec u$ for all $u \neq 0$, and from $\pi(0,x)+1$ we come to 0 by exactly $s_f(x)$ applications of g. Moreover, g is elementary. So if we define h by

$$\begin{aligned} h(0) &= 0, \\ h(u) &= 1 + h\big(g(u)\big) \quad \text{for } u \neq 0, \end{aligned}$$

then we have

$$s_f(x) = h\big(\pi(0,x)+1\big).$$

Now the claim follows, since by the normal form Theorem 2.1 we have

$$f(x) = D\big(C\big(p, x, s_f(x)\big)\big).$$

□

2.3. It is tempting to try to avoid this "collapse" by allowing bounded recursion only. This may seem particularly promising, since Rósza Péter has proved that bounded multiple recursion does not lead out of the primitive recursive functions (Peter [1957, p. 94]). However, the collapse cannot be avoided in this simple way.

THEOREM. *For any recursive* 0-1-*valued function* f *we can find an elementary well-ordering* $\prec$ *of the natural numbers with order type* ω *and a recursive* 0-1-*function h such that* f *is elementary in h, and h can be defined in the form*

$$\begin{aligned} h(0) &= 0, \\ h(u) &= g_0\big(u, h\big(g(u)\big)\big) \quad \textit{for } u \neq 0, \\ h(u) &\leqslant 1, \end{aligned}$$

with g, g_0 *elementary functions such that* $g(u) \prec u$ *for* $u \neq 0$.

PROOF. We proceed as in the previous proof, up to and including the definition of g. Then let

$$h(0) := 0,$$
$$h(u) := \begin{cases} h(g(u)) & \text{if } g(u) \neq 0, \\ \mathsf{sg}(D(C(p, x(u), t(u)+1))) & \text{otherwise.} \end{cases}$$

Then $h(u) \leqslant 1$ and we have $f(x) = h(\pi(0, x) + 1)$ and hence the claim. □

2.4. *A general collapse result.* We now prove a rather general result, which says that it is impossible to index the recursive functions in any "reasonable" way by means of a Π^1_1-path through Kleene's system $\mathcal{O}$ of notations for constructive ordinals. I have learned this result from Stan Wainer, who in turn attributed it to some unpublished work of Yiannis Moschovakis.

We will need in this section some knowledge of Kleene's system $\mathcal{O}$, Π^1_1-sets and the like (cf. Shoenfield [1967], Rogers [1967], Sacks [1990], Hinman [1978] or other papers in the present Handbook). For definiteness, here is a list of what we need. A subset P of $\mathcal{O}$ is called a *path* through $\mathcal{O}$ if

(P1) P is linearly ordered by $<_{\mathcal{O}}$,
(P2) $b \in P$ and $a <_{\mathcal{O}} b$ implies $a \in P$, and
(P3) any constructive ordinal is denoted by a $b \in P$.

We will make use of the following facts.

(F1) There is a recursively enumerable relation $<'_{\mathcal{O}}$ extending $<_{\mathcal{O}}$, such that for any $b \in \mathcal{O}$ we have

$$a <_{\mathcal{O}} b \iff a <'_{\mathcal{O}} b.$$

(F2) (Feferman, Spector). There is a Π^1_1-path through $\mathcal{O}$.
(F3) There is no Σ^1_1-path through $\mathcal{O}$.

Let us now formulate what we mean by a "reasonable" hierarchy. First, it should be indexed by a path P through $\mathcal{O}$, i.e. be of the form $(f_a)_{a \in P}$. Moreover, the property $R(a, x, y)$ that the a-th function f_a at argument x has value y should not be too complex. By this we mean that at least it should be inductively definable from arithmetical (or even Π^1_1-) relations, hence R itself should be a Π^1_1-relation. So let us assume that we have a Π^1_1-relation R satisfying

$$\forall a \in P \forall x \exists! y\, R(a, x, y),$$

and define $f_a(x)$ to be the unique y such that $R(a, x, y)$. We then require the following properties of our hierarchy $(f_a)_{a \in P}$.

(H1) $\forall a, b \in P(a \neq b \to f_a \neq f_b)$.
(H2) $\forall a \in P(f_a \in \mathrm{REC})$.
(H3) $\forall f \in \mathrm{REC} \exists a \in P(f = f_a)$.

Here REC denotes the set of all unary total recursive functions. So (H1)–(H3) expresses that our assumed hierarchy $(f_a)_{a\in P}$ provides a unique indexing of all $f \in$ REC by means of the path P through $\mathcal{O}$.

We now show that from all these assumptions we can derive a Σ^1_1-definition of P, in the form

$$a \in P \leftrightarrow \exists\{e\}{\in}\mathrm{REC}\forall b{\in}P\big[\forall x R\big(b, x, \{e\}(x)\big) \to a <'_{\mathcal{O}} b\big]. \qquad (*)$$

Clearly the right hand side has Σ^1_1-form, since $\{e\} \in$ REC is arithmetical, the Π^1_1-relations P and R appear as premises and $<'_{\mathcal{O}}$ in the conclusion is recursively enumerable by (F1). Since by (F3) there is no Σ^1_1-definition of P, we have the desired contradiction.

It remains to prove (∗). (⇒). Let $a \in P$. Then by (P3) there is a $b \in P$ such that $a <_{\mathcal{O}} b$, hence $f_b \in$ REC by (H2). Pick e such that $f_b = \{e\}$. Then by (H1) b is uniquely determined by $f_b = \{e\}$, i.e. by the Π^1_1-relation $\forall x R(b, x, \{e\}(x))$. Since $b \in P \subseteq \mathcal{O}$, we also have $a <'_{\mathcal{O}} b$ by (F1).

(⇐). Let $\{e\} \in$ REC such that $\forall b{\in}P[f_b = \{e\} \to a <'_{\mathcal{O}} b]$. By (H3) there is a $b \in P$ such that $f_b = \{e\}$, and again by (H1) b is uniquely determined. Since $b \in P \subseteq \mathcal{O}$, from $a <'_{\mathcal{O}} b$ we can conclude $a <_{\mathcal{O}} b$ by (F1), and hence $a \in P$ by (P2).

3. The extended Grzegorczyk hierarchy

From now on we restrict attention to "canonical" well-orderings of concrete order types like ε_0, Γ_0 or the Bachmann–Howard ordinal. We assume some knowledge of such ordinals, in particular of their Cantor normal form.

3.1. The *fast growing* functions F_α, $\alpha < \varepsilon_0$, are defined as follows.

$$\begin{aligned} F_0(x) &:= 2^x, \\ F_{\alpha+1}(x) &:= F_\alpha^{(x)}(x) \quad (F_\alpha^{(x)}\ x\text{-th iterate of } F_\alpha), \\ F_\lambda(x) &:= F_{\lambda[x]}(x). \end{aligned}$$

Here the fundamental sequence $\lambda[x]$ for limit numbers $\lambda < \varepsilon_0$ and $x \in \mathbb{N}$ is defined in a quite natural way. First note that any such limit number can be written uniquely in the form $\lambda = \omega^{\alpha_n} + \cdots + \omega^{\alpha_0}$ with $\lambda > \alpha_n \geqslant \cdots \geqslant \alpha_0 > 0$. Then we let

$$\lambda[x] := \begin{cases} \omega^{\alpha_n} + \cdots + \omega^{\alpha_1} + \omega^{\alpha_0 - 1} \cdot x & \text{if } \alpha_0 \text{ is a successor,} \\ \omega^{\alpha_n} + \cdots + \omega^{\alpha_1} + \omega^{\alpha_0[x]} & \text{if } \alpha_0 \text{ is a limit.} \end{cases}$$

There are many variants of the fast growing functions, which have essentially the same rate of growth. We only mention the *Hardy*-functions [Hardy, 1904], which are defined by

$$\begin{aligned} H_0(x) &:= x, \\ H_{\alpha+1}(x) &:= H_\alpha(x+1), \\ H_\lambda(x) &:= H_{\lambda[x]}(x). \end{aligned}$$

A detailed comparison of closely related functions can be found in Fairtlough and Wainer [1998].

3.2. We now define the *extended Grzegorczyk hierarchy*. Let $\mathcal{E}_\alpha$ be the elementary closure of F_α, i.e. the least class of functions containing the F_α and some initial functions ($U^i_n = \lambda x_1, \ldots, x_n.x_i$ (for $1 \leqslant i \leqslant n$), $C^i_n = \lambda x_1, \ldots, x_n.i$ (for $n \geqslant 0$, $i \geqslant 0$), $\lambda x, y.x + y$ and $\lambda x, y.x \dot{-} y$), that is closed against (simultaneous) substitution and bounded sums and products. The ε_0-*recursive functions* are defined to be the functions in

$$\bigcup_{\alpha<\varepsilon_0} \mathcal{E}_\alpha.$$

One of the classic results of the subject is the identification due to Ackermann [1940] and Kreisel [1952] of the ε_0-recursive functions as the provably recursive functions of arithmetic. In fact, there is quite an intimate relationship between proof theory and subrecursive hierarchies. However, since there is a recent survey on new developments along these lines in the paper by Fairtlough and Wainer [1998], we do not go into this subject here. Another quite interesting approach can be found in Friedman and Sheard [1995].

Also for the particular classes $\mathcal{E}_\alpha$ quite a number of interesting characterizations are known. First of all, $\mathcal{E}_\alpha$ is the class of functions computable by a register machine with time (i.e. number of computation steps) bounded by a finite iteration of the function F_α. Moreover, the functions in $\mathcal{E}_\alpha$ can be characterized by counting the number of recursions used and their order types, or else by transfinitely iterating the process of extending an effectively generated class of functions by enumeration (cf. Schwichtenberg [1971]). Also proof theoretic characterizations of these classes are known (e.g., Schwichtenberg [1972] and Takeuti [1994]). So we have a quite satisfactory theory here.

3.3. One might ask to what extent ε_0 is typical for the set of recursive functions considered, i.e. the ε_0-recursive functions and hence the functions provably recursive in arithmetic. It seems that the results mentioned provide strong evidence that this is the case. However, this impression is not correct: Girard [1981] has shown that one might as well associate a far bigger ordinal with the functions provably recursive in arithmetic, the Bachmann–Howard ordinal. To this end, Girard has introduced the so-called slow growing hierarchy G_α, $\alpha <$ the Bachmann–Howard ordinal:

$$G_0(x) := x,$$

$$G_{\alpha+1}(x) := G_\alpha(x) + 1,$$
$$G_\lambda(x) := G_{\lambda[x]}(x).$$

This hierarchy catches up with the F_α only at the Bachmann–Howard ordinal. However, the natural description of this ordinal seems to need the least uncountable ordinal Ω, and one might wonder why Ω can play a role in a characterization of the provably recursive functions of arithmetic. In Schwichtenberg and Wainer [1995] an attempt has been made to explain why this is so.

3.4. The subrecursive classifications in terms of the fast growing hierarchy are due to Grzegorczyk [1953] for $\alpha < \omega$, to Robbin [1965] and the author (Schwichtenberg [1969]) for $\alpha < \omega^\omega$ and generally to Löb and Wainer [1970a, 1970b], Wainer [1970] and the author (Schwichtenberg [1971]). From the many other more recent contributions to the subject we only mention Rose's book [1984] and the papers by Buchholz and Wainer [1987], Fairtlough and Wainer [1998], Friedman and Sheard [1995] and Weiermann [1995].

4. Partial continuous functionals

In the rest of this paper we will consider what happens when higher types are taken into account. This means that as arguments not only numbers are allowed, but also functions and even functionals of any finite type, where for now we take types to be built from the type ι of natural numbers by means of an operation $\rho \to \sigma$ to form function types. So for any type ρ a set D_ρ of objects of type ρ must be given. The precise definition of D_ρ depends on the particular approach to higher type computability, but in most cases we have $D_0 = \mathbb{N}$ and $D_{\rho\to\sigma}$ is a set of (possibly partial) functions from D_ρ into D_σ. In the present section we give a short exposition of the theory of partial continuous functionals, in order to make the paper sufficiently self-contained.

NOTATION. We will normally use $f, g, F, G, \dots$ for total and $\varphi, \psi, \Phi, \Psi, \dots$ for partial functions and functionals.

4.1. *Historical comments.* Let us first review Kleene's notion of partial recursive functionals, introduced in Kleene [1959b]. In his approach, D_ρ is the set HT_ρ of all "hereditarily total" functionals of type ρ:

$$\mathsf{HT}_0 := \mathbb{N},$$
$$\mathsf{HT}_{\rho\to\sigma} := \{f\colon \mathsf{HT}_\rho \to \mathsf{HT}_\sigma \mid f \text{ total}\}.$$

Let $\mathsf{HT} := \bigcup_\rho \mathsf{HT}_\rho$. Kleene only considered the special cases $\sigma_0 := \iota$ and $\sigma_{n+1} := \sigma_n \to \iota$, but this is not an essential restriction; Gandy and Hyland [1977] have developed Kleene's approach for the more general types considered here.

Kleene gave an inductive definition of the *partial recursive functionals* on these domains, by means of certain schemata (S1), ..., (S9) (cf. Normann's paper [1999] in the present Handbook). He defined a *recursive functional* as a partial recursive functional which happens to be total. The class of partial recursive functionals has been studied in detail by Kleene and others. For type level 1 one obtains the ordinary partial recursive functions, and for type level 2 the functionals computable by Turing machines with input functions available as oracles.

If one uses the schemata (S1), ..., (S9) to compute $\Phi(x_1, \ldots, x_n)$, then one thinks of the higher type arguments as being given by oracles. This means that one is questioning the x_i's with certain computed functionals as arguments. Now the only scheme which properly uses an argument of a level $\geqslant 2$ is (S8). It says that for any recursive functional Ψ of an appropriate type also the following functional Φ is partial recursive.

$$\Phi(x_1, \ldots, x_n) :\simeq x_1\bigl(\lambda y\, \Psi(y, x_1, \ldots, x_n), x_2, \ldots, x_n\bigr). \tag{S8}$$

Here we require that Φ is defined only for arguments $x_1, \ldots, x_n$ such that the functional $\lambda y\, \Psi(y, x_1, \ldots, x_n)$ is total; this must be assumed since x_1 is only defined for hereditarily total arguments. But this requirement has the undesired consequence that the partial recursive functionals are not closed under substitution. More precisely we have:

LEMMA. *There are partial recursive functionals Ψ, F such that F is total and*

$$\varphi(x_1, \ldots, x_n) := \Psi\bigl(\lambda y\, F(y, x_1, \ldots, x_n), x_1, \ldots, x_n\bigr)$$

is not partial recursive.

PROOF. Let

$$\begin{aligned}
\chi(n, p) &:= \begin{cases} 0 & \text{if } \neg T(n, n, p), \\ \text{undefined} & \text{otherwise.} \end{cases} \\
\Psi(I, n) &:= I(\lambda p\, \chi(n, p)) \quad \text{by (S8)}. \\
\varphi(n) &:= \Psi\bigl(0^2, n\bigr) \quad \text{with } 0^2(\alpha) := 0.
\end{aligned}$$

Then

$$\begin{aligned}
\varphi(n)\!\downarrow &\leftrightarrow \Psi\bigl(0^2, n\bigr)\!\downarrow \\
&\leftrightarrow 0^2\bigl(\lambda p\, \chi(n, p)\bigr)\!\downarrow \\
&\leftrightarrow \forall p\, \chi(n, p)\!\downarrow \\
&\leftrightarrow \forall p\, \neg T(n, n, p).
\end{aligned}$$

Hence φ is not partial recursive. □

Restriction to recursive functionals leads to a stable class of functionals, which in particular is closed under substitution. For the recursive functionals many characterizations are known, including some using abstract machines (see, e.g., Kleene [1962]). However, one serious problem remains: Due to the scheme (S8) computation trees are in general infinite. The resulting theory is therefore more a definability theory than a theory of computation, since in the latter the requirement that a computation is finite seems to be essential.

The pathology apparent in the lemma suggests that one should extend the domains HT_ρ in Kleene's theory, i.e. should also allow partial arguments. Platek [1966] (see also Moldestad [1977]) has developed such a theory. However, the problem mentioned above about the infinite nature of the resulting computation trees remains.

In fact, this problem was identified quite early in the development of higher type recursion theory. Kreisel [1959, p. 104] mentions it explicitly. Consequently, he introduced the *continuous functionals*, and in the same volume Kleene [1959a] proposed a notion of *countable functionals*, using so-called *associates*. Both notions turned out to be essentially equivalent. However, it took considerably more time and effort to find the "right" mathematical context for dealing with continuous functionals: this seems to be what is known today as Scott–Ershov domain theory (Scott [1970, 1982], Ershov [1972, 1977]). In this setting the appropriate *domains* of computable functionals have been identified as the partial continuous functionals. Moreover, the countable functionals of Kleene–Kreisel have been singled out as *total* objects by Ershov [1974]. Note that they are defined now on *all* partial continuous functionals. An abstract, domain theoretic characterization of totality has been given by Berger [1990, 1993]; this turned out to be quite useful for further generalizations, e.g., of the density theorem and the Kreisel–Lacombe–Shoenfield theorem. We refer to Stoltenberg-Hansen, Griffor and Lindström [1994] as an excellent textbook on domain theory; in Chapter 8.3 it contains an exposition of Berger's approach.

4.2. To set the stage for our introduction to the theory of partial continuous functionals, we begin with a discussion of certain general principles which follow from an analysis of higher type computation.

We take it as a basic assumption that any computation has to be finite. Hence, if a partial function φ is an input for a computable functional Φ, then the computation can only make use of a finite subfunction φ_0 of φ. Similarly, if a functional Φ is used as an argument of a computable type three functional $\mathcal{H}$, then Φ can only be called upon finitely many times, and each time Φ's argument must be presented to Φ in an explicit form, e.g., as a finite set of argument-value pairs (n, m). Hence $\mathcal{H}$ can only make use of a "finite approximation" of its argument Φ. Such a notion of a finite approximation can be defined in a similar fashion as we move up through the types, and using it we can formulate our first general principle as follows.

PRINCIPLE OF FINITE SUPPORT. *If $\mathcal{H}(\Phi)$ is defined with value n, then there is a finite approximation Φ_0 of Φ such that $\mathcal{H}(\Phi_0)$ is defined with value n.*

If $\Phi(\varphi_0)$ is defined with value n and φ_0' extends the finite function φ_0, then clearly also $\Phi(\varphi_0')$ should be defined with the same value n. This notion of an extension can be carried up through the types: e.g., a finite approximation Φ_0' extends Φ_0 if for any pair (φ_0, n) in the latter there exists a pair (φ_0', n) in the former such that φ_0 extends φ_0' (note the reversed order). The notion of an extension also makes sense for arbitrary functionals, so we require.

MONOTONICITY PRINCIPLE. *If $\mathcal{H}(\Phi)$ is defined with value n and Φ' extends Φ, then also $\mathcal{H}(\Phi')$ is defined with value n.*

It is a consequence of these two principles that for a computable functional a finite approximation to its arguments suffices to find the value, and moreover, that the functional itself is completely determined by its finite approximations. Hence it seems natural to require that

(i) the domain of a computable functional consists of all functionals that can be represented as limits of finite approximations (these are called partial continuous functionals), and

(ii) a functional is computable if and only if it is the limit of a computably enumerable set of finite approximations.

We postpone a discussion of the second requirement to the next section; in the present section we give an abstract, axiomatic formulation of the above principles, in terms of the so-called information systems of Scott [1970, 1982]. From these we will define the notion of a continuous functional of arbitrary finite type over $\mathbb{N}$.

4.3. *Information systems.* The basic idea of information systems is to provide an axiomatic setting to describe approximations of abstract objects (like functions or functionals) by concrete, finite ones. We do not attempt to analyze the notion of "concreteness" or finiteness here, but rather take an arbitrary countable set A of "bits of data" or "tokens" as a basic notion to be explained axiomatically. In order to use such data to build approximations of abstract objects, we need a notion of "consistency", which determines when the elements of a finite set of tokens are consistent with each other. We also need an "entailment relation" between consistent sets X of data and single tokens a, which intuitively expresses the fact that the information contained in X is sufficient to compute the bit of information a. The axioms below are a minor modification of Scott's, due to Larsen and Winskel [1991].

DEFINITION. An *information system* is a structure $(A, \text{CON}, \vdash)$ where A is a countable set (the *tokens*), CON is a nonempty set of finite subsets of A (the *consistent* sets), and $\vdash$ is a subset of $\text{CON} \times A$ (the *entailment relation*) which satisfy:

(1) If $X \subseteq Y \in \text{CON}$ then $X \in \text{CON}$;

(2) If $a \in A$ then $\{a\} \in \text{CON}$;

(3) If $X \vdash a$ then $X \cup \{a\} \in \text{CON}$;

(4) If $X \in \text{CON}$ and $a \in X$ then $X \vdash a$;

(5) If $X, Y \in \text{CON}$ and $\forall b \in Y (X \vdash b)$ and $Y \vdash c$, then $X \vdash c$.

We shall write $X \vdash Y$ to mean $\forall b \in Y (X \vdash b)$. Lower case x, y, z will denote subsets of A, and upper case X, Y, Z will normally be used for finite subsets of A.

DEFINITION. The *objects* or *ideals* of an information system $\mathbf{A} = (A, \mathrm{CON}, \vdash)$ are defined to be those subsets z of A which satisfy:

(1) $X \subseteq^{\mathrm{fin}} z$ implies $X \in \mathrm{CON}$ (z is "consistent");

(2) $X \subseteq^{\mathrm{fin}} z$ and $X \vdash a$ implies $a \in z$ (z is "deductively closed").

The set of all objects of $\mathbf{A}$ is written $|\mathbf{A}|$.

EXAMPLES. Any countable set A can be turned into a *flat* information system $\mathbf{A}$ by letting the set of tokens be A, and

$$\mathrm{CON} := \{\emptyset\} \cup \{\, \{a\} \mid a \in A \,\} \quad \text{and} \quad X \vdash a :\equiv a \in X.$$

For $A = \mathbb{N}$ we have the following picture of the CON-sets.

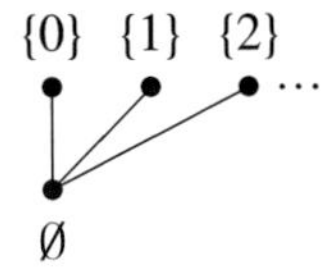

In this case the objects are just the elements of CON. Note that if instead we took CON to be the set of all finite subsets of A, then the objects would be all subsets of A.

Another rather important example is the following, which concerns approximations of functions from a countable set A into a countable set B. The tokens are the pairs (a, b) with $a \in A$ and $b \in B$, and

$$\mathrm{CON} := \big\{ \{(a_0, b_0), \dots, (a_{k-1}, b_{k-1})\} \mid \forall i, j < k\, (a_i = a_j \to b_i = b_j) \big\},$$
$$X \vdash (a, b) :\equiv (a, b) \in X.$$

It is not difficult to verify that this defines an information system whose objects are (the graphs of) all partial functions from A to B.

A final example is provided by any fixed partial functional Φ. A token should now be a pair (φ_0, n) where φ_0 is a finite function and $\Phi(\varphi_0)$ is defined with value n. Thus if we take CON to be the set of all finite sets of tokens and for $X := \{\, (\varphi_i, n_i) \mid i = 1, \dots, k \,\}$ define $X \vdash (\varphi_0, n)$ iff φ_0 extends some φ_i, then this structure becomes an information system. The objects in this case are all sets x of tokens with the property that, whenever (φ_0, n) belongs to x, then also all (φ_0', n) with φ_0' extending φ_0 belong to x.

REMARK. Suppose $\mathbf{A} = (A, \mathrm{CON}, \vdash)$ is an information system. Then it is easy to see that:

(1) $\emptyset \in \mathrm{CON}$.

(2) If $X \in \text{Con}$, $Y \subseteq X$ and $Y \vdash a$, then $X \vdash a$.

(3) If $X \in \text{Con}$ and $Y \subseteq^{\text{fin}} A$ and $X \vdash a$ for every $a \in Y$, then $X \cup Y \in \text{Con}$ by axiom 3; hence $Y \in \text{Con}$ by axiom 1.

(4) The intersection of any number of ideals is an ideal again.

4.4. *Function spaces.* We now define the "function space" $\mathbf{A} \to \mathbf{B}$ between two information systems $\mathbf{A}$ and $\mathbf{B}$.

DEFINITION. Let $\mathbf{A} = (A, \text{Con}_A, \vdash_A)$ and $\mathbf{B} = (B, \text{Con}_B, \vdash_B)$ be information systems. Then the structure $\mathbf{A} \to \mathbf{B} = (C, \text{Con}, \vdash)$ is defined as follows:

(1) The set C of tokens is $C = \text{Con}_A \times B$.

(2) The set Con consists of of all finite sets $\{(X_1, b_1), \ldots, (X_n, b_n)\} \subseteq C$ such that for every index set $I \subseteq \{1, \ldots, n\}$,

$$\bigcup_{i \in I} X_i \in \text{Con}_A \implies \{b_i \mid i \in I\} \in \text{Con}_B.$$

(3) For the definition of the entailment relation $\vdash$ it is helpful first to define the notion of an *application* of $W := \{(X_1, b_1), \ldots, (X_n, b_n)\} \in \text{Con}$ to $X \in \text{Con}_A$:

$$WX = \{(X_1, b_1), \ldots, (X_n, b_n)\}X := \{b_i \mid X \vdash_A X_i\}.$$

From the definition of Con we know that this set is in Con_B. Clearly application is *monotone in the second argument*, in the sense that $X \vdash_A X'$ implies $WX' \subseteq WX$, hence $WX \vdash_B WX'$. Now define $W \vdash (X, b)$ by

$$W \vdash (X, b) :\equiv WX \vdash_B b.$$

In fact, application is also *monotone in the first argument*, i.e., $W \vdash W'$ implies $WX \vdash_B W'X$. To see this let $W' = \{(X_1, b_1), \ldots, (X_n, b_n)\}$ and observe that

$$WX \vdash_B \bigcup\{WX_i \mid X \vdash_A X_i\} \vdash_B \{b_i \mid X \vdash_A X_i\} = W'X.$$

LEMMA. *If* $\mathbf{A}$ *and* $\mathbf{B}$ *are information systems, then so is* $\mathbf{A} \to \mathbf{B}$ *defined as above.*

PROOF. Let $\mathbf{A} = (A, \text{Con}_A, \vdash_A)$ and $\mathbf{B} = (B, \text{Con}_B, \vdash_B)$. The axioms 1, 2 and 4 are clearly satisfied. For axiom 3, suppose

$$\{(X_1, b_1), \ldots, (X_n, b_n)\} \vdash (X, b), \quad \text{i.e.} \quad \{b_j \mid X \vdash_A X_j\} \vdash_B b.$$

We have to show that $\{(X_1, b_1), \ldots, (X_n, b_n), (X, b)\} \in \text{Con}$. So let $I \subseteq \{1, \ldots, n\}$ and suppose

$$X \cup \bigcup_{i \in I} X_i \in \text{Con}_A.$$

We must show that $b \cup \{ b_i \mid i \in I \} \in \mathrm{CON}_B$. Let $J \subseteq \{1, \ldots, n\}$ consists of those j with $X \vdash_A X_j$. Then also

$$X \cup \bigcup_{i \in I} X_i \cup \bigcup_{j \in J} X_j \in \mathrm{CON}_A.$$

Since

$$\bigcup_{i \in I} X_i \cup \bigcup_{j \in J} X_j \in \mathrm{CON}_A,$$

from the consistency of $\{(X_1, b_1), \ldots, (X_n, b_n)\}$ we can conclude that

$$\{b_i \mid i \in I\} \cup \{b_j \mid j \in J\} \in \mathrm{CON}_B.$$

But $\{ b_j \mid j \in J \} \vdash_B b$ by assumption. Hence

$$\{b_i \mid i \in I\} \cup \{b_j \mid j \in J\} \cup \{b\} \in \mathrm{CON}_B.$$

For axiom 5, suppose

$$W \vdash \big\{(X_1, b_1), \ldots, (X_n, b_n)\big\} \quad \text{and} \quad \big\{(X_1, b_1), \ldots, (X_n, b_n)\big\} \vdash (X, b).$$

We have to show that $WX \vdash_B b$. Note that from $X \vdash_A X_i$ we can conclude $WX \vdash_B WX_i$ by the monotonicity of application in the second argument. Hence

$$WX \vdash_B \bigcup \{WX_i \mid X \vdash_A X_i\} \vdash_B \{b_i \mid X \vdash_A X_i\} \vdash_B b.$$

□

We shall now give two alternative characterizations of the function space: firstly as "approximable maps", and secondly as continuous maps with respect to the so-called Scott topology.

4.5. *Approximable maps.* We want to study "information respecting" maps from **A** into **B**. Such a map is given by a relation r between CON_A and B, where $r(X, b)$ intuitively means that whenever we are given the information $X \in \mathrm{CON}_A$, then we know that at least the token b appears in the value.

DEFINITION. Let $\mathbf{A} = (A, \mathrm{CON}_A, \vdash_A)$ and $\mathbf{B} = (B, \mathrm{CON}_B, \vdash_B)$ be information systems. A relation $r \subseteq \mathrm{CON}_A \times B$ is an *approximable map* if it satisfies:

(1) If $r(X, b_1), \ldots, r(X, b_n)$, then $\{b_1, \ldots, b_n\} \in \mathrm{CON}_B$;
(2) If $r(X, b_1), \ldots, r(X, b_n)$ and $\{b_1, \ldots, b_n\} \vdash_B b$, then $r(X, b)$;
(3) If $r(X', b)$ and $X \vdash_A X'$, then $r(X, b)$.

We write $r : \mathbf{A} \to \mathbf{B}$ to mean r is an approximable map from **A** to **B**.

THEOREM. *Let* **A** *and* **B** *be information systems. Then the objects of* $\mathbf{A} \to \mathbf{B}$ *are exactly the approximable maps from* **A** *to* **B**.

PROOF. Let $\mathbf{A} = (A, \mathrm{CON}_A, \vdash_A)$ and $\mathbf{B} = (B, \mathrm{CON}_B, \vdash_B)$. If $r \in |\mathbf{A} \to \mathbf{B}|$ then $r \subseteq \mathrm{CON}_A \times B$ is consistent and deductively closed. We have to show that r satisfies the axioms for approximable maps.

(1) Let $r(X, b_1), \dots, r(X, b_n)$. We must show that $\{b_1, \dots, b_n\} \in \mathrm{CON}_B$. But this clearly follows from the consistency of r.

(2) Let $r(X, b_1), \dots, r(X, b_n)$ and $\{b_1, \dots, b_n\} \vdash_B b$. We must show that $r(X, b)$. But

$$\{(X, b_1), \dots, (X, b_n)\} \vdash (X, b)$$

by the definition of the entailment relation $\vdash$ in $\mathbf{A} \to \mathbf{B}$, hence $r(X, b)$ by the deductive closure of r.

(3) Let $X \vdash_A X'$ and $r(X', b)$. We must show that $r(X, b)$. But

$$\{(X', b)\} \vdash (X, b)$$

since $\{(X', b)\}X = \{b\}$ (which follows from $X \vdash_A X'$), hence again $r(X, b)$ by the deductive closure of r.

For the other direction suppose that $r : \mathbf{A} \to \mathbf{B}$ is an approximable map. We must show that $r \in |\mathbf{A} \to \mathbf{B}|$.

Consistency of r: suppose $r(X_1, b_1), \dots, r(X_n, b_n)$ and

$$X = \bigcup \{X_i \mid i \in I\} \in \mathrm{CON}_A \text{ for some } I \subseteq \{1, \dots, n\}.$$

We must show that $\{b_i \mid i \in I\} \in \mathrm{CON}_B$. Now from $r(X_i, b_i)$ and $X \vdash_A X_i$ we obtain $r(X, b_i)$ by axiom 3 for any $i \in I$, and hence $\{b_i \mid i \in I\} \in \mathrm{CON}_B$ by axiom 1.

Deductive closure of r: suppose $r(X_1, b_1), \dots, r(X_n, b_n)$ and

$$W := \{(X_1, b_1), \dots, (X_n, b_n)\} \vdash (X, b).$$

We must show $r(X, b)$. By definition of $\vdash$ for $\mathbf{A} \to \mathbf{B}$ we have $WX \vdash_B b$, which is $\{b_i \mid X \vdash_A X_i\} \vdash_B b$. Further by our assumption $r(X_i, b_i)$ we know $r(X, b_i)$ by axiom 3 for all i with $X \vdash_A X_i$. Hence $r(X, b)$ by axiom 2. □

4.6. *Continuous maps.* We now introduce the Scott topology.

DEFINITION. Suppose $\mathbf{A} = (A, \mathrm{CON}, \vdash)$ is an information system and $X \in \mathrm{CON}$. Define $\overline{X}$, the *deductive closure* of X, by

$$\overline{X} := \{a \in A \mid X \vdash a\}.$$

Define also $U_X \subseteq |\mathbf{A}|$ by

$$U_X := \{x \in |\mathbf{A}| \mid X \subseteq^{\mathsf{fin}} x\}.$$

Note that, since the objects $x \in |\mathbf{A}|$ are deductively closed, if $x \in U_X$ then $\overline{X} \subseteq x$.

LEMMA. *The system of all U_X with $X \in$ CON forms the basis of a topology on $|\mathbf{A}|$, called the Scott topology.*

PROOF. Suppose $X, Y \in$ CON and $x \in U_X \cap U_Y$. We have to find $Z \in$ CON such that $x \in U_Z \subseteq U_X \cap U_Y$. Choose $Z = X \cup Y$. □

LEMMA. *Let $\mathbf{A}$ be an information system and $U \subseteq |\mathbf{A}|$. Then the following are equivalent.*

(1) *U is open in the Scott topology.*
(2) *U satisfies:*
- *If $x \in U$ and $x \subseteq y$, then $y \in U$ (Alexandrov condition).*
- *If $x \in U$ then $\overline{X} \in U$ for some $X \subseteq^{\mathsf{fin}} x$ (Scott condition).*

(3) $U = \bigcup_{\overline{X} \in U} U_X$.

So open sets U may be seen as those determined by a (possibly infinite) system of finitely observable properties, namely all X such that $\overline{X} \in U$.

PROOF. (1) ⇒ (2) If U is open, then U is the union of some U_X's, $X \in$ CON. Since each U_X is upward closed, also U is; this proves the Alexandrov condition. For the Scott condition assume $x \in U$. Then $x \in U_X \subseteq U$ for some $X \in$ CON. Hence $X \subseteq^{\mathsf{fin}} x$, and $\overline{X} \in U$ since $\overline{X} \in U_X$.

(2) ⇒ (3) Assume that $U \subseteq |\mathbf{A}|$ satisfies the Alexandrov and Scott conditions. Let $x \in U$. By the Scott condition, $\overline{X} \in U$ for some $X \subseteq^{\mathsf{fin}} x$, so $x \in U_X$ for this X. Conversely, let $x \in U_X$ for some $\overline{X} \in U$. Then $\overline{X} \subseteq x$. Now $x \in U$ follows from $\overline{X} \in U$ by the Alexandrov condition.

(3) ⇒ (1) The U_X's are the basic open sets of the Scott topology. □

We now give some simple characterizations of the continuous functions $f : |\mathbf{A}| \to |\mathbf{B}|$. Call f *monotone* if $x \subseteq y$ implies $f(x) \subseteq f(y)$. Call $D \subseteq |\mathbf{A}|$ *directed* if for any $x, y \in D$ there is a $z \in D$ such that $x \subseteq z$ and $y \subseteq z$; note that then $\bigcup D \in |\mathbf{A}|$. An important example of a directed set is the set $\{\overline{X} \mid X \subseteq^{\mathsf{fin}} x\}$ of *finitely generated* or *compact* approximations (f.g. approximations for short) of a given $x \in |\mathbf{A}|$. Any $x \in |\mathbf{A}|$ can be written as the union of its finitely generated approximations:

$$x = \bigcup \{\overline{X} \mid X \subseteq^{\mathsf{fin}} x\}.$$

PROPOSITION. *Let $\mathbf{A}$ and $\mathbf{B}$ be information systems and $f : |\mathbf{A}| \to |\mathbf{B}|$. Then the following are equivalent.*

(1) *f is continuous with respect to the Scott topology;*

(2) *f is monotone and satisfies the "Principle of finite support" PFS: if $b \in f(x)$, then $b \in f(\overline{X})$ for some $X \subseteq^{\mathsf{fin}} x$;*

(3) *f is monotone and commutes with directed unions: for every directed $D \subseteq |\mathbf{A}|$*

$$f\Big(\bigcup_{x\in D} x\Big) = \bigcup_{x\in D} f(x).$$

Note that in (3) the set $\{f(x) \mid x \in D\}$ is directed by monotonicity of f, hence its union is indeed an object in $|\mathbf{B}|$. Note also that from PFS it follows immediately that if $Y \subseteq^{\mathsf{fin}} f(x)$, then $Y \subseteq^{\mathsf{fin}} f(\overline{X})$ for some $X \subseteq^{\mathsf{fin}} x$.

So continuous maps $f : |\mathbf{A}| \to |\mathbf{B}|$ are those that can be completely described from the point of view of finite approximations of the abstract objects $x \in |\mathbf{A}|$ and $f(x) \in |\mathbf{B}|$: Whenever we are given a finite approximation Y to the value $f(x)$, then there is a finite approximation X to the argument x such that already $f(\overline{X})$ contains the information in Y; note that by monotonicity $f(\overline{X}) \subseteq f(x)$.

PROOF. (1) $\Rightarrow$ (2) Let f be continuous. Then for any basic open set $U_Y \in |\mathbf{B}|$ (so $Y \in \mathrm{CON}_B$) the set $f^{-1}[U_Y] = \{x \mid Y \subseteq^{\mathsf{fin}} f(x)\}$ is open in $|\mathbf{A}|$. To prove monotonicity assume $x \subseteq y$; we must show $f(x) \subseteq f(y)$. So let $b \in f(x)$, i.e. $\{b\} \subseteq^{\mathsf{fin}} f(x)$. The open set $\{z \mid \{b\} \subseteq^{\mathsf{fin}} f(z)\}$ satisfies the Alexandrov condition, so from $x \subseteq y$ we can infer $\{b\} \subseteq^{\mathsf{fin}} f(y)$, i.e. $b \in f(y)$. To prove PFS assume $b \in f(x)$. The open set $\{z \mid \{b\} \subseteq^{\mathsf{fin}} f(z)\}$ satisfies the Scott condition, so for some $X \subseteq^{\mathsf{fin}} x$ we have $\{b\} \subseteq^{\mathsf{fin}} f(\overline{X})$.

(2) $\Rightarrow$ (1) Assume that f satisfies monotonicity and PFS. We must show that f is continuous, i.e. that for any fixed $Y \in \mathrm{CON}_B$ the set $f^{-1}[U_Y] = \{x \mid Y \subseteq^{\mathsf{fin}} f(x)\}$ is open. We prove

$$\big\{x \mid Y \subseteq^{\mathsf{fin}} f(x)\big\} = \bigcup \big\{U_X \mid X \in \mathrm{CON}_A \text{ and } Y \subseteq^{\mathsf{fin}} f\big(\overline{X}\big)\big\}.$$

Let $Y \subseteq^{\mathsf{fin}} f(x)$. Then by PFS $Y \subseteq^{\mathsf{fin}} f(\overline{X})$ for some $X \in \mathrm{CON}_A$ such that $X \subseteq^{\mathsf{fin}} x$, and $X \subseteq^{\mathsf{fin}} x$ implies $x \in U_X$. Conversely, let $x \in U_X$ for some $X \in \mathrm{CON}_A$ such that $Y \subseteq^{\mathsf{fin}} f(\overline{X})$. Then $\overline{X} \subseteq x$, hence $Y \subseteq^{\mathsf{fin}} f(x)$ by monotonicity.

For (2) $\Leftrightarrow$ (3) assume that f is monotone. Let f satisfy PFS, and $D \subseteq |\mathbf{A}|$ be directed.

$$f\Big(\bigcup_{x\in D} x\Big) \supseteq \bigcup_{x\in D} f(x)$$

follows from monotonicity. For the reverse inclusion let $b \in f(\bigcup_{x\in D} x)$. Then by PFS $b \in f(\overline{X})$ for some $X \subseteq^{\mathsf{fin}} \bigcup_{x\in D} x$. From the directedness and the fact that X is finite we obtain $X \subseteq^{\mathsf{fin}} z$ for some $z \in D$. From $b \in f(\overline{X})$ and monotonicity we infer

$b \in f(z)$. Conversely, let f commute with directed unions, and assume $b \in f(x)$. Then

$$b \in f(x) = f\Big(\bigcup_{X \subseteq^{\text{fin}} x} \overline{X}\Big) = \bigcup_{X \subseteq^{\text{fin}} x} f(\overline{X}),$$

so $b \in f(\overline{X})$ for some $X \subseteq^{\text{fin}} x$. □

Clearly the identity and constant functions are continuous, and also the *composition* $g \circ f$ of continuous functions $f : |\mathbf{A}| \to |\mathbf{B}|$ and $g : |\mathbf{B}| \to |\mathbf{C}|$.

THEOREM. *Let* $\mathbf{A}$ *and* $\mathbf{B} = (B, \mathrm{CON}_B, \vdash_B)$ *be information systems. Then the objects of* $\mathbf{A} \to \mathbf{B}$ *are in a natural bijective correspondence with the continuous functions from* $|\mathbf{A}|$ *to* $|\mathbf{B}|$, *as follows:*

- *With any approximable map* $r : \mathbf{A} \to \mathbf{B}$ *we can associate a continuous function* $|r| : |\mathbf{A}| \to |\mathbf{B}|$ *by*

$$|r|(z) := \{b \in B \mid r(X, b) \textit{ for some } X \subseteq^{\text{fin}} z\}.$$

- *Conversely, with any continuous function* $f : |\mathbf{A}| \to |\mathbf{B}|$ *we can associate an approximable map* $\hat{f} : \mathbf{A} \to \mathbf{B}$ *by*

$$\hat{f}(X, b) :\equiv b \in f(\overline{X}).$$

These assignments are inverse to each other, i.e. $f = |\hat{f}|$ *and* $r = \widehat{|r|}$.

PROOF. Let r be an object of $\mathbf{A} \to \mathbf{B}$; then by the theorem just proved r is an approximable map. We first show that $|r|$ is well-defined. So let $z \in |\mathbf{A}|$.

$|r|(z)$ is consistent: let $b_1, \dots, b_n \in |r|(z)$. Then there are $X_1, \dots, X_n \subseteq^{\text{fin}} z$ such that $r(X_i, b_i)$. Hence $X := X_1 \cup \dots \cup X_n \subseteq^{\text{fin}} z$ and $r(X, b_i)$ by axiom 3 of approximable maps. Now from axiom 1 we can conclude that $\{b_1, \dots, b_n\} \in \mathrm{CON}_B$.

$|r|(z)$ is deductively closed: let $b_1, \dots, b_n \in |r|(z)$ and $\{b_1, \dots, b_n\} \vdash_B b$. We must show $b \in |r|(z)$. As before we find $X \subseteq^{\text{fin}} z$ such that $r(X, b_i)$. Now from axiom 2 we can conclude $r(X, b)$ and hence $b \in |r|(z)$.

To prove continuity of $|r|$ let $Y \in \mathrm{CON}_B$; we must show that $|r|^{-1}[U_Y]$ is open. Now for every $z \in |\mathbf{A}|$

$$\begin{aligned} z \in |r|^{-1}[U_Y] &\equiv |r|(z) = \{b \in B \mid r(X, b) \text{ for some } X \subseteq^{\text{fin}} z\} \in U_Y \\ &\equiv Y \subseteq^{\text{fin}} \{b \in B \mid r(X, b) \text{ for some } X \subseteq^{\text{fin}} z\} \\ &\equiv \forall b \in Y \exists X [X \subseteq^{\text{fin}} z \land r(X, b)] \\ &\equiv \forall b \in Y \exists X [z \in U_X \land r(X, b)] \\ &\equiv z \in \bigcap_{b \in Y} \bigcup \{U_X \mid r(X, b)\}. \end{aligned}$$

Since Y is finite, this implies that $|r|^{-1}[U_Y]$ is open.

Now let $f : |\mathbf{A}| \to |\mathbf{B}|$ be continuous. It is easy to verify that $\hat{f}$ is indeed an approximable map. Furthermore

$$\begin{aligned} b \in |\hat{f}|(z) &\equiv \hat{f}(X, b) \text{ for some } X \subseteq^{\mathsf{fin}} z \\ &\equiv b \in f(\overline{X}) \text{ for some } X \subseteq^{\mathsf{fin}} z \\ &\equiv b \in f(z) \text{ by monotonicity and PFS.} \end{aligned}$$

Finally, for any approximable map $r : \mathbf{A} \to \mathbf{B}$ we have

$$\begin{aligned} r(X, b) &\equiv \exists Y \subseteq^{\mathsf{fin}} \overline{X}\, r(Y, b) \text{ by axiom 3 for approximable maps} \\ &\equiv b \in |r|(\overline{X}) \\ &\equiv \widehat{|r|}(X, b). \end{aligned}$$

This completes the proof. □

REMARK. From now on we will usually write $r(z)$ for $|r|(z)$, and similarly $f(X, b)$ for $\hat{f}(X, b)$. It should always be clear from the context where the mods and hats should be inserted.

4.7. *Cartesian products.* We define an information system $\mathbf{A} \times \mathbf{B} = (C, \mathrm{CON}, \vdash)$ from $\mathbf{A}$ and $\mathbf{B}$ in such a way that the inclusion ordering $\subseteq$ on the objects of $|\mathbf{A} \times \mathbf{B}|$ is isomorphic to the Cartesian product of the inclusion orderings on $|\mathbf{A}|$ and $|\mathbf{B}|$. Without loss of generality we may assume that A and B are disjoint. But then any pair (x, y) of elements $x \in |\mathbf{A}|$ and $y \in |\mathbf{B}|$ can be approximated in each component separately. Hence we choose as tokens in $\mathbf{A} \times \mathbf{B}$ simply the union $C := A \cup B$. Consistency and entailment is inherited in the expected way from $\mathbf{A}$ and $\mathbf{B}$:

$$\begin{aligned} X \in \mathrm{CON} &:\equiv X \cap A \in \mathrm{CON}_A \text{ and } X \cap B \in \mathrm{CON}_B, \\ X \vdash c &:\equiv (c \in A \implies X \cap A \vdash_A c) \text{ and } (c \in B \implies X \cap B \vdash_B c). \end{aligned}$$

It is then obvious that we have

LEMMA. *If* $\mathbf{A}$ *and* $\mathbf{B}$ *are information systems with* $A \cap B = \emptyset$, *then so is* $\mathbf{A} \times \mathbf{B}$ *defined as above. The objects of* $\mathbf{A} \times \mathbf{B}$ *are exactly the unions* $x \cup y$ *of the objects of* $x \in |\mathbf{A}|$ *and of* $y \in |\mathbf{B}|$.

When using both $\to$ and $\times$ to build information systems, we assume that $\times$ has a higher precedence that $\to$, hence, e.g., $\mathbf{A} \times \mathbf{B} \to \mathbf{C}$ means $(\mathbf{A} \times \mathbf{B}) \to \mathbf{C}$.

REMARK. Clearly the pairs $(x, y) \in |\mathbf{A}| \times |\mathbf{B}|$ are in a natural bijective correspondence with the objects $x \cup y$ of $|\mathbf{A} \times \mathbf{B}|$. Therefore from now on we will usually write x, y for $x \cup y$, to increase readability. It should be clear from the context where a comma needs to be replaced by a union.

Clearly the *projections* $\pi_i : \mathbf{A}_1 \times \mathbf{A}_2 \to \mathbf{A}_i$ defined by $\pi_i(x_1, x_2) := x_i$ for $i = 1, 2$ are continuous.

LEMMA (Universal property of Cartesian products). *Let* **A**, **B** *and* **C** *be information systems such that* $A \cap B = \emptyset$. *Then for any pair* $f : \mathbf{C} \to \mathbf{A}$ *and* $g : \mathbf{C} \to \mathbf{B}$ *of continuous functions there is a unique continuous function* $h : \mathbf{C} \to \mathbf{A} \times \mathbf{B}$ *such that* $f = \pi_1 \circ h$ *and* $g = \pi_1 \circ h$.

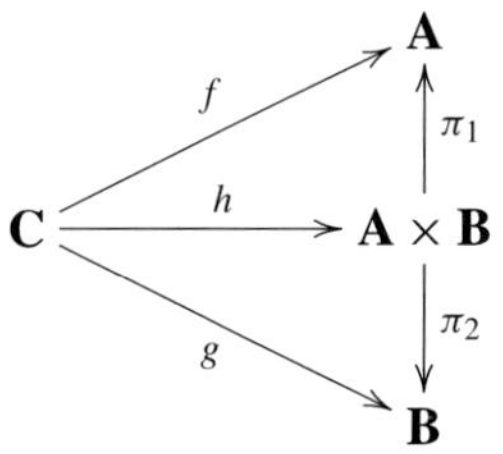

PROOF. Uniqueness follows from the fact that any such h must satisfy $h(x) = (f(x), g(x))$ for $x \in |\mathbf{C}|$. To prove existence, define $h(x) := (f(x), g(x))$ for $x \in |\mathbf{C}|$. We must show that this h is continuous. Monotonicity is obvious; for PFS assume $b \in (f(x), g(x))$. Since $A \cap B = \emptyset$, the token b must be in exactly one component, say B. So $b \in g(x)$, hence $b \in g(\overline{X})$ for some $X \subseteq^{\text{fin}} x$ and therefore $b \in h(\overline{X}) = (f(\overline{X}), g(\overline{X}))$. □

As an application let us construct the *product* $f \times g$ of two continuous functions $f : \mathbf{A} \to \mathbf{C}$ and $g : \mathbf{B} \to \mathbf{D}$ (with $A \cap B = \emptyset$).

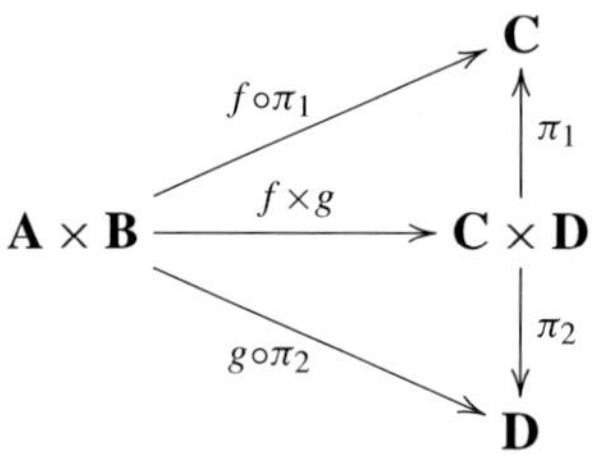

$f \times g$ satisfies $(f \times g)(x, y) = (f(x), g(y))$ for $x \in |\mathbf{A}|$ and $y \in |\mathbf{B}|$.

LEMMA. *Let* **A**, **B** *and* **C** *be information systems with* $A \cap B = \emptyset$. *A function* $f : \mathbf{A} \times \mathbf{B} \to \mathbf{C}$ *is continuous iff it is continuous in each argument separately, i.e. all its sections* $f_1^y : \mathbf{A} \to \mathbf{C}$, $f_1^y(x) := f(x, y)$ *for fixed* $y \in |\mathbf{B}|$ *and similarly* $f_2^x : \mathbf{B} \to \mathbf{C}$, $f_2^x(y) := f(x, y)$ *for fixed* $x \in |\mathbf{A}|$ *are continuous.*

PROOF. The function sending x to (x, y) for fixed y clearly is continuous, hence by composition so is f_1^y; for f_2^x the argument is similar. Conversely, assume that all f_1^y and f_2^x are continuous. To prove monotonicity of f, assume $x \subseteq x'$ and $y \subseteq y'$ with

$x, x' \in |\mathbf{A}|$ and $y, y' \in |\mathbf{B}|$. Then $f(x, y) \subseteq f(x', y) \subseteq f(x', y')$ by monotonicity of the sections. To prove PFS for f, assume $c \in f(x, y)$. Then $c \in f_1^x(y)$, hence

$$c \in f_1^x(\overline{Y}) = f(x, \overline{Y}) = f_2^{\overline{Y}}(x)$$

for some $Y \subseteq^{\mathsf{fin}} y$ by PFS for f_1^x, and also

$$c \in f_2^{\overline{Y}}(\overline{X}) = f(\overline{X}, \overline{Y}) = f(\overline{X \cup Y})$$

for some $X \subseteq^{\mathsf{fin}} x$ by PFS for $f_2^{\overline{Y}}$. □

4.8. *Evaluation and currying.* We now show that the information systems together with the approximable mappings form a "Cartesian closed category".

LEMMA. *Let* **A** *and* **B** *be information systems. Then the function*

$$\mathsf{eval}\colon (\mathbf{A} \to \mathbf{B}) \times \mathbf{A} \to \mathbf{B},$$
$$\mathsf{eval}(f, x) := f(x) \quad \textit{for } f \in |\mathbf{A} \to \mathbf{B}|,\ x \in |\mathbf{A}|$$

is continuous.

PROOF. To avoid confusion we shall use the more appropriate (but less suggestive) letter r instead of f, and also use the correct notation $|r|(x)$ instead of $r(x)$.

By the previous lemma it suffices to prove continuity in each component separately. For the second component this is obvious. For the first component monotonicity follows from the definition of $|r|$. To prove PFS assume $b \in \mathsf{eval}(r, x) = |r|(x)$. Then $r(X, b)$ for some $X \subseteq^{\mathsf{fin}} x$, hence for $W := \{(X, b)\}$ we have $W \subseteq^{\mathsf{fin}} r$ and

$$\mathsf{eval}\big(\overline{W}, x\big) = \big|\overline{W}\big|(x) = \big\{c \in B \mid \overline{W}(Y, c) \text{ for some } Y \subseteq^{\mathsf{fin}} x\big\} \ni b.$$

□

LEMMA. *Let* **A**, **B** *and* **C** *be information systems with* $A \cap B = \emptyset$. *Then the function*

$$\mathsf{curry}\colon (\mathbf{A} \times \mathbf{B} \to \mathbf{C}) \to (\mathbf{A} \to (\mathbf{B} \to \mathbf{C})),$$
$$\mathsf{curry}(f)(x)(y) := f(x, y) \quad \textit{for } f \in |\mathbf{A} \times \mathbf{B}|,\ x \in |\mathbf{A}| \textit{ and } y \in |\mathbf{B}|$$

is well-defined and continuous.

PROOF. We again use the letter r instead of f, and also use the correct notation $|r|(x)$ instead of $r(x)$.

Fix $r \in |\mathbf{A} \times \mathbf{B} \to \mathbf{C}|$. Then for fixed $x \in |\mathbf{A}|$ the function $k := |r|_2^x$ sending $y \in |\mathbf{B}|$ to $|r|(x, y) \in |\mathbf{C}|$ is continuous, since it is a section of $|r|$. We now show that the function h sending $x \in |\mathbf{A}|$ into $\hat{k} \in |\mathbf{B} \to \mathbf{C}|$ is continuous. For monotonicity assume

$x \subseteq x'$. Let $\hat{k}' = h(x')$. We must show $\hat{k} \subseteq \hat{k}'$. It suffices to show that $\hat{k}(Y, c)$ implies $\hat{k}'(Y, c)$, i.e. $c \in k(\overline{Y})$ implies $c \in k'(\overline{Y})$. But $k(\overline{Y}) \subseteq k'(\overline{Y})$ by monotonicity of $|r|$. For PFS assume $(Y, c) \in h(x) = \hat{k}$. We want $X \subseteq^{\mathsf{fin}} x$ such that $(Y, c) \in h(\overline{X})$. Now $(Y, c) \in \hat{k}$, so $c \in k(\overline{Y}) = |r|(x, \overline{Y})$. PFS for the section $|r|_1^{\overline{Y}}$ yields an $X \subseteq^{\mathsf{fin}} x$ such that $c \in |r|(\overline{X}, \overline{Y}) = |h(\overline{X})|(\overline{Y})$, i.e. $(Y, c) \in h(\overline{X})$.

It remains to show that the function g sending r into $\hat{h}$ is continuous. Monotonicity follows from the definition of $|r|$. For PFS assume $(X, (Y, c)) \in g(r)$, for some $X \in \mathrm{C{\small ON}}_A$, $Y \in \mathrm{C{\small ON}}_B$ and $c \in C$. Then it is easy to verify that $(X, (Y, c)) \in g(\overline{W})$ for $W := \{(X \cup Y, c)\} \subseteq^{\mathsf{fin}} r$. □

Another way to say this is that for any continuous $f : \mathbf{A} \times \mathbf{B} \to \mathbf{C}$ there is a continuous $c_f : \mathbf{A} \to (\mathbf{B} \to \mathbf{C})$ such that $\mathsf{eval} \circ (c_f \times \mathsf{id}) = f$:

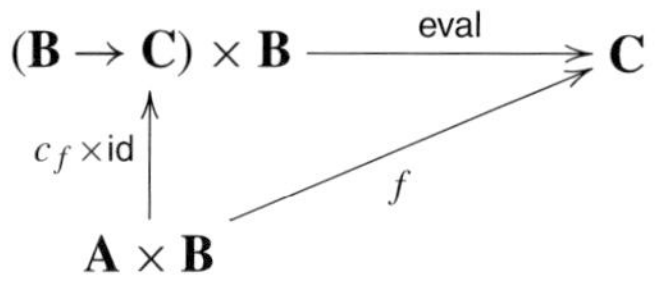

Moreover, such a c_f clearly is uniquely determined. This together with the fact that we can form Cartesian products $\mathbf{A} \times \mathbf{B}$ (satisfying the universal property above) and the existence of a "terminal" information system $(\emptyset, \{\emptyset\}, \emptyset)$ (such that for every $\mathbf{A}$ there is a unique $f : \mathbf{A} \to (\emptyset, \{\emptyset\}, \emptyset)$), establishes that the information systems with continuous maps between them form a *Cartesian closed category*.

4.9. *Typed lambda terms.* An important consequence of working in a Cartesian closed category is that every "explicit definition" involving higher types actually defines an object. We first explain precisely what makes up such an explicit definition.

(*Finite* or *simple*) *types* $\rho, \sigma, \tau, \ldots$ are built from a symbol ι for the type of natural numbers by *function type* formation $\rho \to \sigma$ and *product type* formation $\rho \times \sigma$. The *level* $\mathsf{lev}(\rho)$ of a type ρ is defined inductively by

$$\begin{aligned} \mathsf{lev}(\iota) &:= 0, \\ \mathsf{lev}(\rho \to \sigma) &:= \max\bigl(\mathsf{lev}(\rho) + 1, \mathsf{lev}(\sigma)\bigr), \\ \mathsf{lev}(\rho \times \sigma) &:= \max\bigl(\mathsf{lev}(\rho), \mathsf{lev}(\sigma)\bigr). \end{aligned}$$

The type constructor $\to$ is understood to be associative to the right, so $\rho_1 \to \cdots \to \rho_{n-1} \to \rho_n$ means $(\rho_1 \to \cdots \to (\rho_{n-1} \to \rho_n) \ldots)$. Similarly $\times$ is understood to be associative to the left. We also let $\times$ have a higher precedence than $\to$, so $\rho \times \sigma \to \tau$ means $(\rho \times \sigma) \to \tau$. Note that any type without $\times$ can be written uniquely in the form $\rho_1 \to \cdots \to \rho_n \to \iota$ and its level is $\max(\mathsf{lev}(\rho_1), \ldots, \mathsf{lev}(\rho_n)) + 1$ if $n > 0$, and 0 otherwise. For simplicity, we assume that there is only one ground type ι, the type of natural numbers.

Typed λ-terms are built up from typed variables $x^\rho, y^\rho, z^\rho, \dots$ (countably many of each type) and constants denoted by c^ρ by means of the operations of *application* $(M^{\rho\to\sigma} N^\rho)^\sigma$ and *λ-abstraction* $(\lambda x^\rho M^\sigma)^{\rho\to\sigma}$. The type superscripts will be omitted whenever the typing is clear from the context or else immaterial. The variable x in $\lambda x M$ is considered to be bound; with this in mind the set $\mathsf{FV}(M)$ of variables free in M is defined in the expected way.

DEFINITION. We define the information system $\mathbf{C}_\rho = (C_\rho, \mathrm{CON}_\rho, \vdash_\rho)$ by induction on the type ρ:

$$\begin{aligned}
&\mathbf{C}_\iota := \mathbb{N} \quad \text{(viewed as a flat information system)},\\
&\mathbf{C}_{\rho\to\sigma} := \mathbf{C}_\rho \to \mathbf{C}_\sigma,\\
&\mathbf{C}_{\rho\times\sigma} := \mathbf{C}_\rho \times \mathbf{C}_\sigma.
\end{aligned}$$

The *partial continuous functionals* (over $\mathbb{N}$) of type $\rho \to \sigma$ are the continuous functions from $|\mathbf{C}_\rho|$ to $|\mathbf{C}_\sigma|$.

LEMMA. *For every λ-term M^ρ and every list $x_1^{\rho_1}, \dots, x_n^{\rho_n}$ of variables containing all variables free in M we have $[\![M]\!]\colon \mathbf{C}_{\rho_1} \times \cdots \times \mathbf{C}_{\rho_n} \to \mathbf{C}_\rho$ such that for all $u_1 \in |\mathbf{C}_{\rho_1}|, \dots, u_n \in |\mathbf{C}_{\rho_n}|$*

$$\begin{aligned}
[\![x_i]\!](u_1,\dots,u_n) &= u_i,\\
[\![MN]\!](u_1,\dots,u_n) &= [\![M]\!](u_1,\dots,u_n)\big([\![N]\!](u_1,\dots,u_n)\big),\\
[\![\lambda x M]\!](u_1,\dots,u_n)(v) &= [\![M]\!](u_1,\dots,u_n,v).
\end{aligned}$$

PROOF. By induction on the term M. In case M is a variable x_i, let $[\![x_i]\!] := \pi_i^n$, where π_i^n is the projection of an n-tuple to its i-th component. Formally we define π_i^n from the projections π_1, π_2 (introduced above) as follows (recall that $A \times B \times C := (A \times B) \times C$): $\pi_1^1 = \mathsf{id}$, $\pi_{n+1}^{n+1} = \pi_2$, $\pi_i^{n+1} = \pi_i^n \circ \pi_1$ for $i = 1, \dots, n$.

For an application term, define $[\![MN]\!] := \mathsf{eval} \circ ([\![M]\!] \times [\![N]\!])$. This is a continuous function by what was proved above, and

$$\begin{aligned}
[\![MN]\!](u_1,\dots,u_n) &= \mathsf{eval}\big([\![M]\!] \times [\![N]\!]\big)(u_1,\dots,u_n)\\
&= \mathsf{eval}\big([\![M]\!](u_1,\dots,u_n), [\![N]\!](u_1,\dots,u_n)\big)\\
&= [\![M]\!](u_1,\dots,u_n)\big([\![N]\!](u_1,\dots,u_n)\big).
\end{aligned}$$

Finally for a λ-abstraction, let $[\![\lambda x M]\!] := \mathsf{curry}([\![M]\!])$. Then by the properties of curry

$$\begin{aligned}
[\![\lambda x M]\!](u_1,\dots,u_n)(v) &= \mathsf{curry}\big([\![M]\!]\big)(u_1,\dots,u_n)(v)\\
&= [\![M]\!](u_1,\dots,u_n,v).
\end{aligned}$$

□

4.10. *Coherence.* The information systems $\mathbf{C}_\rho$ enjoy the rather pleasant property of *coherence*, which amounts to the possibility to locate inconsistencies in two-element sets of data objects. Generally, an information system $\mathbf{A} = (A, \text{CON}, \vdash)$ is called *coherent* if for any set $X \subseteq^{\text{fin}} A$ we have

$$X \in \text{CON} \equiv \forall a, b \in X(\{a, b\} \in \text{CON}).$$

Clearly any flat information system is coherent, and moreover we have

LEMMA. *Let* $\mathbf{A}$ *and* $\mathbf{B}$ *be information systems.*

(1) *If* $\mathbf{B}$ *is coherent, then so is* $\mathbf{A} \to \mathbf{B}$.

(2) *If* $\mathbf{A}$ *and* $\mathbf{B}$ *are coherent, then so is* $\mathbf{A} \times \mathbf{B}$.

PROOF. Let $\mathbf{A} = (A, \text{CON}_A, \vdash_A)$ and $\mathbf{B} = (B, \text{CON}_B, \vdash_B)$.

(1) Let $\mathbf{A} \to \mathbf{B} = (\text{CON}_A \times B, \text{CON}, \vdash)$ and $\{(X_1, b_1), \dots, (X_n, b_n)\} \subseteq \text{CON}_A \times B$. Assume

$$\forall i, j(1 \leqslant i < j \leqslant n \to \{(X_i, b_i), (X_j, b_j)\} \in \text{CON}). \tag{$*$}$$

We have to show that $\{(X_1, b_1), \dots, (X_n, b_n)\} \in \text{CON}$. So assume $I \subseteq \{1, \dots, n\}$ and $\bigcup_{i \in I} X_i \in \text{CON}_A$. We must show that $\{b_i \mid i \in I\} \in \text{CON}_B$. Now since $\mathbf{B}$ is coherent by assumption, it suffices to show that $\{b_i, b_j\} \in \text{CON}_B$ for all $i, j \in I$. So let $i, j \in I$. By assumption we have $X_i \cup X_j \in \text{CON}_A$ and hence by $(*)$ and the definition of CON also $\{b_i, b_j\} \in \text{CON}_B$.

(2) Let $\mathbf{A} \times \mathbf{B} = (A \cup B, \text{CON}, \vdash)$ (with $A \cap B = \emptyset$). Let $a_1, \dots, a_n \in A$ and $b_1, \dots, b_m \in B$ such that any two-element subset of $\{a_1, \dots, a_n, b_1, \dots, b_m\}$ is consistent. Then $\{a_1, \dots, a_n\} \in \text{CON}_A$ and $\{b_1, \dots, b_m\} \in \text{CON}_B$ since $\mathbf{A}$ and $\mathbf{B}$ are coherent. Therefore $\{a_1, \dots, a_n, b_1, \dots, b_m\} \in \text{CON}$ by definition of CON. □

COROLLARY. *The information systems* $\mathbf{C}_\rho$ *are all coherent.*

5. Computability in higher types

We now use the theory presented in the previous section to set up a framework for computability in our iterated function spaces $\mathbf{C}_\rho$. But first we want want to make it clear that higher types are relevant for our subject of classifying recursive functions: as observed in Schwichtenberg [1973] (cf. also Fortune, Leivant and O'Donnell [1983]), the functions F_α, $\alpha < \varepsilon_0$, of the fast growing hierarchy can be defined quite easily from higher type iteration functionals.

As already mentioned in 4.1, a functional is taken to be computable if it is the limit of a computably enumerable set of finite approximations. This is an "external" notion of computability, and the question arises whether it can be characterized internally, as a class of functionals generated from initial functionals by certain schemata. Such

a characterization has been given by Plotkin [1977]: a partial continuous functional is computable if and only if it can be defined explicitly from
- fixed point operators introduced into this framework by Platek in his thesis [1966],
- the parallel conditional pcond of type $\iota \to \iota \to \iota \to \iota$, and
- a continuous approximation $\exists$ of type $(\iota \to \iota) \to \iota$ to the existential quantifier.

We shall give a detailed proof of this result below.

5.1. *Iteration functionals.* We first extend the definition of the functions F_α into higher types. It is convenient here to introduce *integer types* ρ_n:

$$\begin{aligned} \rho_0 &:= \iota, \\ \rho_{n+1} &:= \rho_n \to \rho_n. \end{aligned}$$

So if $x_0, \dots, x_{n+1}$ are of integer types $\rho_0, \dots, \rho_{n+1}$, then we can form $x_{n+1}(x_n)$ (which is of type ρ_n) and hence finally $x_{n+1}(x_n) \cdots (x_0)$; the latter will be abbreviated by $x_{n+1}(x_n, \dots, x_0)$. Note that $\mathsf{lev}(\rho_n) = n$. Now we define F_α^{n+1} of type ρ_{n+1} for $\alpha < \varepsilon_0$, by the following induction on α.

$$\begin{aligned} F_0^{n+1}(x_n, \dots, x_0) &:= \begin{cases} 2^{x_0} & \text{if } n = 0, \\ x_n^{(x_0)}(x_{n-1}, \dots, x_0) & \text{otherwise,} \end{cases} \\ F_{\alpha+1}^{n+1}(x_n, \dots, x_0) &:= \big(F_\alpha^{n+1}\big)^{(x_0)}(x_n, \dots, x_0), \\ F_\lambda^{n+1}(x_n, \dots, x_0) &:= F_{\lambda[x_0]}^{n+1}(x_n, \dots, x_0). \end{aligned}$$

Here $x_n^{(y)}(x_{n-1}, \dots, x_0)$ denotes $I(y, x_n, \dots, x_0)$ with an iteration functional I of the type $\iota, \rho_n, \rho_{n-1}, \dots, \rho_0 \to \rho_0$ defined by

$$\begin{aligned} I(0, y, z) &:= z, \\ I(x+1, y, z) &:= y\big(I(x, y, z)\big). \end{aligned}$$

In the following lemma, the condition $\beta + \omega^\alpha = \beta \mathbin{\#} \omega^\alpha$ expresses that in the Cantor normal form of β the last summand ω^{β_0} (if it exists) has an exponent $\beta_0 \geqslant \alpha$.

LEMMA. $F_\alpha^{n+1}(F_\beta^n) = F_{\beta+\omega^\alpha}^n$ *if* $\beta + \omega^\alpha = \beta \mathbin{\#} \omega^\alpha$.

PROOF. We use induction on α. $\alpha = 0$:

$$\begin{aligned} F_0^{n+1}\big(F_\beta^n, x_{n-1}, \dots, x_0\big) &= \big(F_\beta^n\big)^{(x_0)}(x_{n-1}, \dots, x_0) \\ &= F_{\beta+1}^n(x_{n-1}, \dots, x_0); \end{aligned}$$

α successor:

$$F_\alpha^{n+1}\big(F_\beta^n, x_{n-1}, \dots, x_0\big) = \big(F_{\alpha-1}^{n+1}\big)^{(x_0)}\big(F_\beta^n, x_{n-1}, \dots, x_0\big)$$

$$
\begin{aligned}
&= F^n_{\beta+\omega^{\alpha-1}\cdot x_0}(x_{n-1},\dots,x_0) \\
&\quad \text{by induction hypothesis} \\
&= F^n_{(\beta+\omega^\alpha)[x_0]}(x_{n-1},\dots,x_0) \\
&= F^n_{\beta+\omega^\alpha}(x_{n-1},\dots,x_0);
\end{aligned}
$$

α limit:

$$
\begin{aligned}
F^{n+1}_\alpha\big(F^n_\beta, x_{n-1},\dots,x_0\big) &= F^{n+1}_{\alpha[x_0]}\big(F^n_\beta, x_{n-1},\dots,x_0\big) \\
&= F^n_{\beta+\omega^{\alpha[x_0]}}(x_{n-1},\dots,x_0) \\
&\quad \text{by induction hypothesis} \\
&= F^n_{(\beta+\omega^\alpha)[x_0]}(x_{n-1},\dots,x_0) \\
&= F^n_{\beta+\omega^\alpha}(x_{n-1},\dots,x_0).
\end{aligned}
$$

□

Hence each F^{n+1}_α, $\alpha < \varepsilon_0$, and therefore in particular the functions $F_\alpha = F^1_\alpha$ for $\alpha < \varepsilon_0$ can be built from the simple functionals F^{n+1}_0, which are essentially iteration functionals, by application alone. Note that in the resulting representation of the functions F_α, $\alpha < \varepsilon_0$, we do not need the fundamental sequences $\lambda[x]$ any more.

5.2. *Abstract computability.* With the preparations done it is now rather straightforward to define computability in our iterated function spaces $\mathbf{C}_\rho$ (based on $\mathbb{N}$ viewed as a flat information system). The tokens and finite sets of tokens are encodable by integers using sequence-coding. It is easy to see that the notions $X \in \mathrm{CON}_\rho$ of consistency and $X \vdash_\rho a$ of entailment correspond to recursive (in fact elementary) relations.

DEFINITION. A partial continuous functional φ of type ρ is said to be *computable* if – when viewed as a set of (codes of) tokens – it is Σ^0_1-definable.

LEMMA. *For all types ρ, σ, τ the functionals*

$$
\begin{aligned}
&\mathsf{eval}_{\rho,\sigma}\colon (\mathbf{C}_\rho \to \mathbf{C}_\sigma) \times \mathbf{C}_\rho \to \mathbf{C}_\sigma, \\
&\mathsf{curry}_{\rho,\sigma,\tau}\colon (\mathbf{C}_\rho \times \mathbf{C}_\sigma \to \mathbf{C}_\tau) \to (\mathbf{C}_\rho \to (\mathbf{C}_\sigma \to \mathbf{C}_\tau))
\end{aligned}
$$

are computable.

PROOF. The tokens of $\mathsf{eval}_{\rho,\sigma}$ are of the form (W, X, a) with $W \in \mathrm{CON}_{\rho\to\sigma}$, $X \in \mathrm{CON}_\rho$ and $a \in C_\sigma$, and we have

$$(W, X, a) \in \mathsf{eval} \equiv a \in \overline{W}\big(\overline{X}\big) \equiv a \in \overline{WX} \equiv WX \vdash a.$$

Here WX is the application of finite approximations introduced in 4.4, and we have made use of $\overline{WX} = \overline{W}(\overline{X})$, which can be proved easily. Thus we have a Σ_1^0-definition of $\mathsf{eval}_{\rho,\sigma}$. For curry we similarly have

$$\begin{aligned}(W, X, Y, a) \in \mathsf{curry} &\equiv a \in \overline{W}\big(\overline{X \cup Y}\big) \equiv a \in \overline{W(X \cup Y)} \\ &\equiv W(X \cup Y) \vdash a,\end{aligned}$$

which again is a Σ_1^0-definition. This completes the proof. □

LEMMA.
(1) *If* $\varphi : \mathbf{C}_\rho \to \mathbf{C}_\sigma$ *and* $\psi : \mathbf{C}_\sigma \to \mathbf{C}_\tau$ *are computable, then so is* $\psi \circ \varphi$.
(2) *If* $\Phi : \mathbf{C}_\rho \to \mathbf{C}_\sigma$ *and* $\varphi \in |\mathbf{C}_\rho|$ *are computable, then so is* $\Phi(\varphi)$.
(3) *If* $\varphi : \mathbf{C}_\rho \to \mathbf{C}_\sigma$ *and* $\psi : \mathbf{C}_\rho \to \mathbf{C}_\tau$ *are computable, then so is* $\varphi \times \psi$.
(4) *The projections* $\pi_1 : \mathbf{C}_\rho \times \mathbf{C}_\sigma \to \mathbf{C}_\rho$ *and* $\pi_2 : \mathbf{C}_\rho \times \mathbf{C}_\sigma \to \mathbf{C}_\sigma$ *are computable.*

PROOF. (1) The tokens of $\psi \circ \varphi$ are of the form (X, b) with $X \in \mathrm{CON}_\rho$ and $b \in C_\tau$, and we have

$$(X, b) \in \psi \circ \varphi \equiv \exists Y \subseteq^{\mathsf{fin}} \varphi\big[(Y, b) \in \psi \wedge \forall a \in Y\, (X, a) \in \varphi\big].$$

This is a Σ_1^0-definition, since φ and ψ are computable.
(2) Since eval is continuous, we have from PFS

$$\begin{aligned}a \in \Phi(\varphi) &\equiv \exists W \subseteq^{\mathsf{fin}} \Phi \exists X \subseteq^{\mathsf{fin}} \varphi\big(a \in \overline{W}\big(\overline{X}\big)\big) \\ &\equiv \exists W \subseteq^{\mathsf{fin}} \Phi \exists X \subseteq^{\mathsf{fin}} \varphi(WX \vdash a).\end{aligned}$$

This again is a Σ_1^0-definition.
(3) and (4) are proved similarly. □

LEMMA. *For every* λ*-term* M^ρ *and every list* $x_1^{\rho_1}, \ldots, x_n^{\rho_n}$ *of variables containing all variables free in* M *the functional* $[\![M]\!] : \mathbf{C}_{\rho_1} \times \cdots \times \mathbf{C}_{\rho_n} \to \mathbf{C}_\rho$ *defined above is computable.*

PROOF. The construction of $[\![M]\!]$ (by induction on M) in 4.9 is such that the previous lemma implies immediately its computability. □

5.3. *Least fixed points.* Since the least fixed point operator makes sense for arbitrary information systems, we for a moment re-introduce that level of generality. So let $\mathbf{A}$ be an information system and $f : \mathbf{A} \to \mathbf{A}$. An object $x \in |\mathbf{A}|$ is a *fixed point* of f if $f(x) = x$. It is called the least fixed point of f if for any other $y \in |\mathbf{A}|$ with $f(y) = y$ we have $x \subseteq y$.

THEOREM (Fixed Point Theorem). *Let* $\mathbf{A}$ *be an information system and* $f\colon \mathbf{A} \to \mathbf{A}$. *Then* f *has a least fixed point* $Y(f)$ *given by*

$$Y(f) := \bigcup_{n\in\mathbb{N}} f^{(n)}(\overline{\emptyset}).$$

Moreover, Y is continuous.

PROOF. From $\overline{\emptyset} \subseteq f(\overline{\emptyset})$ we get $f^{(n)}(\overline{\emptyset}) \subseteq f^{(n+1)}(\overline{\emptyset})$ by induction, since f is monotone. Hence $x := \bigcup_n f^{(n)}(\overline{\emptyset}) \in |\mathbf{A}|$. Moreover, by the characterization of continuity in part 3 of the proposition in 4.6

$$f(x) = f\Big(\bigcup_n f^{(n)}(\overline{\emptyset})\Big) = \bigcup_n f^{(n+1)}(\overline{\emptyset}) = x.$$

If y is another fixed point, from $\overline{\emptyset} \subseteq y$ we get $f^{(n)}(\overline{\emptyset}) \subseteq y$ by induction and hence

$$x = \bigcup_n f^{(n)}(\overline{\emptyset}) \subseteq y.$$

Finally we show that Y is continuous. Monotonicity follows immediately from its definition. For PFS assume $b \in Y(f)$, i.e. $b \in f^{(m)}(\overline{\emptyset})$ for some m. Now since evaluation is continuous, for fixed m the function $f \mapsto f^{(m)}(\overline{\emptyset})$ is continuous as well. Hence by PFS, for our $b \in f^{(m)}(\overline{\emptyset})$ we find a $W \subseteq^{\text{fin}} f$ such that

$$b \in \overline{W}^{(m)}(\overline{\emptyset}) \subseteq Y(\overline{W}).$$

□

Let Y_ρ be the least fixed point operator associated with the information system $\mathbf{C}_\rho$, so $Y_\rho : (\mathbf{C}_\rho \to \mathbf{C}_\rho) \to \mathbf{C}_\rho$. We now show that Y_ρ is computable.

LEMMA. Y_ρ *is computable.*

PROOF. We need to show that Y_ρ is Σ^0_1-definable when viewed as a set of tokens. From the definitions of approximable maps, their correspondence with continuous functions, the definition of Y_ρ above, and the easily verified fact that $\overline{X}(\overline{Z}) = \overline{XZ}$, we obtain

$$\begin{aligned}
(X, a) \in Y_\rho &\equiv (X', a) \in Y_\rho \quad \text{for some } X' \subseteq^{\text{fin}} \overline{X} \\
&\equiv a \in Y_\rho(\overline{X}) \\
&\equiv \exists n \big(a \in \overline{X}^{(n)}(\overline{\emptyset})\big) \\
&\equiv \exists n \big(a \in \overline{X^{(n)}\emptyset}\big) \\
&\equiv \exists n \big(X^{(n)}\emptyset \vdash a\big).
\end{aligned}$$

Thus Y_ρ is Σ^0_1-definable if the relation $X^{(n)}\emptyset \vdash a$ is. But note that

$$X^{(n)}\emptyset \vdash a \equiv \exists \vec{Z}\big[\vec{Z} \text{ has length } n{+}1 \wedge Z_0 = \emptyset \wedge \\ \forall i{<}n(Z_{i+1} = XZ_i) \wedge Z_n \vdash a\big].$$

Now the Z's are finite and encodable as integers using sequence-coding. Hence finite sequences of Z's are also encodable as integers. Furthermore, the relation $Z' = XZ$ as defined in 4.4 is Σ^0_1-definable from a positive occurrence of the entailment relation $Z \vdash b$ at a type level lower than that of X. Therefore $X^{(n)}\emptyset \vdash a$ is Σ^0_1-definable from the entailment relation at this lower level, which in turn is then Σ^0_1-definable from an even lower level of entailment relation, etc. Therefore the relation $(X, a) \in \widehat{Y_\rho}$ can eventually be unravelled into an outright Σ^0_1-definition. □

5.4. *Functionals recursive in* pcond *and* $\exists$. We first define some particular computable functionals. Recall that $\mathbb{N}$ is the flat information system whose tokens are the natural numbers, whose CON-sets are just the singletons $\{n\}$ and $\emptyset$, and whose entailment relation is simply $\ni$ (so $\{n\} \vdash m$ iff $n = m$, and $\emptyset \nvdash m$). We shall often write $\bot$ for the empty set, and simply n for the singleton $\{n\}$ where no confusion should occur.

The *parallel conditional* pcond of type $\iota \to \iota \to \iota \to \iota$ is given by

$$\mathsf{pcond}(p, x, y) := \begin{cases} x & \text{if } p > 0, \\ y & \text{if } p = 0, \\ x & \text{if } x = y, \\ \bot & \text{otherwise.} \end{cases}$$

From it we can define the *sequential conditional* cond of the same type by

$$\mathsf{cond}(p, x, y) := \mathsf{pcond}\big(p, \mathsf{pcond}(p, x, \bot), \mathsf{pcond}(p, \bot, y)\big).$$

Then

$$\mathsf{cond}(p, x, y) = \begin{cases} x & \text{if } p > 0, \\ y & \text{if } p = 0, \\ \bot & \text{if } p = \bot. \end{cases}$$

We can also define from pcond the *parallel or* of type $\iota \to \iota \to \iota$ given by

$$\vee(p, q) := \mathsf{pcond}(p, 1, q).$$

Then

$$\vee(p, q) = \begin{cases} 1 & \text{if } p > 0, \\ 1 & \text{if } q > 0, \\ 0 & \text{if } p = 0 \text{ and } q = 0, \\ \bot & \text{otherwise.} \end{cases}$$

Because these are at the lowest type level, it is necessary only for them to be monotone in order to be continuous, since PFS is trivial over the flat system $\mathbb{N}$. But their monotonicity is clear from their definitions. It is equally clear that each of these examples is computable. We only need to show that as relations (i.e. sets of tokens of the appropriate type) they are Σ^0_1-definable. But this is immediate: in the case of pcond the tokens are all 4-tuples (p, x, y, z) of numbers such that either $p > 0$ and $z = x$, or $p = 0$ and $z = y$, or $x = y = z$.

The *continuous existential quantifier* $\exists$ of type $(\iota \to \iota) \to \iota$ is defined as follows:

$$\exists(\varphi) := \begin{cases} 0 & \text{if } \varphi(\bot) = 0, \\ 1 & \text{if } \varphi(n) > 0 \text{ for some } n, \\ \bot & \text{otherwise.} \end{cases}$$

Clearly $\exists$ is monotone. Furthermore, if $\exists(\varphi)$ has a defined value by either of the first two clauses, then it obviously has that same value on a finite subfunction of φ (only one bit of data is needed). Therefore $\exists$ satisfies PFS and hence is continuous.

Observe on the other hand, that the full existential quantifier, where the first line is replaced by "0 if $\varphi(n) = 0$ for every n", is *not* continuous. For if it were, then by PFS its value (namely 0) on the total constant-0 function would also be attained on some finite subfunction; this is impossible.

The continuous $\exists$-functional is computable since its tokens are all pairs (X, n) such that either $(\emptyset, 0) \in X$ and $n = 0$ or else there exists an m and a $k > 0$ such that $(\{m\}, k) \in X$ and $n = 1$. Clearly this is another Σ^0_1 condition.

DEFINITION. A partial continuous functional Φ of type $\rho_1 \to \cdots \to \rho_p \to \iota$ is said to be *recursive in* pcond *and* $\exists$ if it can be defined explicitly, for all arguments $\varphi_1, \ldots, \varphi_p$, by an equation

$$\Phi(\varphi_1, \ldots, \varphi_p) = M(\varphi_1, \ldots, \varphi_p),$$

where M is a λ-term built up from the variables $\varphi_1, \ldots, \varphi_p$ together with the constants 0, successor, predecessor, the fixed point operators Y_ρ, and pcond and $\exists$.

THEOREM (Plotkin). *A partial continuous functional is computable if and only if it is recursive in* pcond *and* $\exists$.

PROOF. The examples above together with the facts proved in 5.2 that

- (pure) λ-terms are computable, and
- computable functionals are closed under application

show that every functional recursive in pcond and $\exists$ is computable. For the converse, we restrict attention to the case of a composed type $\rho_1 \to \cdots \to \rho_p \to \iota$; the details concerning product types are left as an exercise. Let Φ be computable of this type. Then Φ is a primitive recursively enumerable set of tokens

$$\Phi = \{(X^1_{f_1(n)}, \ldots, X^p_{f_p(n)}, g(n)) \mid n \in \mathbb{N}\},$$

where for each type ρ_j, $(X_i^j)_{i\in\mathbb{N}}$ is an enumeration of CON_{ρ_j}, and $f_1, \dots, f_p$ and g are fixed primitive recursive functions. Henceforth, we will drop the superscripts from the X's.

Let $\vec{\varphi} = \varphi_1, \dots, \varphi_p$ be arbitrary continuous functionals of types $\rho_1, \dots, \rho_p$, respectively. We show that Φ is definable by the equation

$$\Phi(\vec{\varphi}) = Y(w)(0),$$

with w of type $(\iota \to \iota) \to (\iota \to \iota)$ given by

$$\begin{aligned} w(\psi)(n) := \mathsf{pcond}\big(&\mathsf{incons}(\varphi_1, f_1(n)) \vee \dots \\ &\vee \mathsf{incons}(\varphi_p, f_p(n)), \psi(n+1), g(n)\big), \end{aligned}$$

where the incons_{ρ_i}'s of type $\rho_i \to \iota \to \iota$ are the continuous functionals given by

$$\mathsf{incons}(\varphi, n) := \begin{cases} 1 & \text{if } \varphi \cup X_n \text{ is inconsistent,} \\ 0 & \text{if } \varphi \supseteq X_n, \\ \bot & \text{otherwise.} \end{cases}$$

We will prove in the lemma below that they are recursive in pcond and $\exists$; their definition will involve the functional $\exists$.

First suppose $\Phi(\vec{\varphi})$ is defined. By the definition of Φ this means that there is an n such that $\varphi_i \supseteq X_{f_i(n)}$ for every i. Let n_0 be the least such n. Then again from the definition of Φ, $\Phi(\vec{\varphi}) = g(n_0)$. We claim that

$$w^{k+1}(\lambda x \bot)(n_0 - k) = g(n_0) \quad \text{for every } k \leqslant n_0.$$

This is proved by induction on k. The base case $k = 0$ is immediate since $\varphi_i \supseteq X_{f_i(n_0)}$ hence $\mathsf{incons}(\varphi_i, f_i(n_0)) = 0$ for every i, so by definition of pcond, $w(\lambda x \bot)(n_0) = g(n_0)$. For the step case $k \to k+1$

$$\begin{aligned} w^{k+2}(\lambda x \bot)(n_0-k-1) &= w\big(w^{k+1}(\lambda x \bot)\big)(n_0-k-1) \\ &= \mathsf{pcond}\big(\dots, w^{k+1}(\lambda x \bot)(n_0-k), g(n_0-k-1)\big) \\ &= \mathsf{pcond}\big(\dots, g(n_0), g(n_0-k-1)\big) \quad \text{by ind. hyp.} \\ &= g(n_0), \end{aligned}$$

with $\dots = \mathsf{incons}(\varphi_1, f_1(n_0-k-1)) \vee \dots \vee \mathsf{incons}(\varphi_p, f_p(n_0-k-1))$. The last step in this chain needs some explanation. Note that the first argument in pcond cannot have value 0, because this contradicts the minimality of n_0. There are two remaining cases: either the first argument of pcond is 1, in which case the value is the second argument, namely $g(n_0)$; or else the first argument of pcond is $\bot$, in which case $\varphi_i \cup X_{f_i(n_0-k-1)}$ is consistent for every i, and therefore

$$g(n_0) = \Phi\big(\overline{\varphi_i \cup X_{f_i(n_0-k-1)}}\big) = g(n_0-k-1).$$

Hence the value of the pcond-term is again $g(n_0)$, and this completes the proof of the claim.

From the claim for $k = n_0$ we obtain $w^{n_0+1}(\lambda x\bot)(0) = g(n_0)$, and since $Y(w)$ is the union of all the iterates $w^k(\lambda x\bot)$, it follows that $Y(w)(0) = g(n_0)$. We have therefore proved that if $\Phi(\vec{\varphi})$ is defined, then its value is $Y(w)(0)$. It remains to show the converse.

So suppose $Y(w)(0)$ is defined. Then there must be an n such that $\varphi_i \supseteq X_{f_i(n)}$ for every i. Otherwise we would have $w(\lambda x\bot) = \lambda x\bot$ by definition of w, since the first argument of pcond is never 0, so the value must be $\bot$; therefore $Y(w) = \lambda x\bot$, a contradiction. Let n_0 be the least such n. Then $\Phi(\vec{\varphi}) = g(n_0)$ by definition of Φ, and the above claim again applies to give $Y(w)(0) = g(n_0)$. Thus if $Y(w)(0)$ is defined then its value is $\Phi(\vec{\varphi})$. □

Note that from all fixed point operators Y_ρ we only made use of $Y_{\iota\to\iota}$, to construct the least fixed point of $w\colon (\iota\to\iota)\to(\iota\to\iota)$. The proof of the lemma below (which is all we need to finish the proof of Plotkin's theorem) will not make any essential use of Y (it only uses $Y_{\iota\to\iota}$ again to take care of primitive recursion).

LEMMA. *The following functionals are recursive in* pcond *and* $\exists$:

(1) en_ρ *of type* $\iota\to\iota\to\rho$ *such that* $\mathsf{en}(m)$ *enumerates all finitely generated extensions of* $\overline{X_m}$, *thus*

$$\mathsf{en}(m,\bot) = \overline{X_m},$$
$$\mathsf{en}(m,n) = \overline{X_n} \quad \text{if } \overline{X_n} \supseteq \overline{X_m}.$$

(2) incons_ρ *of type* $\rho\to\iota\to\iota$, *defined by*

$$\mathsf{incons}(\varphi,n) := \begin{cases} 1 & \text{if } \varphi\cup X_n \text{ is inconsistent,}\\ 0 & \text{if } \varphi\supseteq X_n,\\ \bot & \text{otherwise.}\end{cases}$$

PROOF. We show (1) and (2) by simultaneous induction on ρ. Again we restrict attention to the case of a composed type $\rho_1\to\cdots\to\rho_p\to\iota$; the details concerning product types are left as an exercise.

(1) We first prove that en_ρ is recursive in pcond and $\exists$. For its definition we need to look in more detail into the definition of the sets X_m of type ρ.

For any type ρ, fix an enumeration $(X_n^\rho)_{n\in\mathbb{N}}$ of CON_ρ such that $X_0 = \emptyset$ and the following relations are primitive recursive:

$$X_n \subseteq X_m,$$
$$X_n \cup X_m \in \mathrm{CON}_\rho,$$
$$X_n^{\rho\to\sigma} X_m^\rho = X_k^\sigma,$$
$$X_n \cup X_m = X_k \quad (\text{with } k = 0 \text{ if } X_n\cup X_m \notin \mathrm{CON}_\rho).$$

We also assume an enumeration $(b_i^\rho)_{i\in\mathbb{N}}$ of the set of tokens of type ρ.

Note that any primitive recursive function f can be regarded as a continuous functional of type $\iota \to \cdots \to \iota \to \iota$ if we identify it with its strict extension. It is easy to see that any primitive recursive function can be represented in this way by a term of the simply typed λ-calculus involving 0, successor, predecessor, the least fixed point operators $Y_{\iota\to\iota}$ and the sequential conditional cond_ι. For instance, addition can be written as

$$m + n = Y_{\iota\to\iota}(\lambda\varphi\lambda x[\textbf{if } x = 0 \textbf{ then } m \textbf{ else } \varphi(x-1)+1 \textbf{ fi}])n.$$

Let $\rho = \rho_1 \to \cdots \to \rho_p \to \iota$ and j, k and h be primitive recursive functions such that

$$X_m = \{(X_{j(m,1,l)}, \ldots, X_{j(m,p,l)}, k(m,l)) \mid l < h(m)\}.$$

en_ρ will be defined from an auxiliary functional Ψ of type $\rho_1 \to \cdots \to \rho_p \to \iota \to \iota \to \iota$ given by

$$\begin{aligned} \Psi(\vec{\varphi}, m, d, 0) &:= d, \\ \Psi(\vec{\varphi}, m, d, l+1) &:= \mathsf{pcond}(p_l, \Psi(\vec{\varphi}, m, d, l), k(m,l)), \end{aligned}$$

where p_l denotes $\mathsf{incons}_{\rho_1}(\varphi_1, j(m,1,l)) \vee \cdots \vee \mathsf{incons}_{\rho_p}(\varphi_p, j(m,p,l))$. Hence

$$p_l = \begin{cases} 1 & \text{if } \varphi_i \cup X_{j(m,i,l)} \text{ is inconsistent for some } i = 1, \ldots, p, \\ 0 & \text{if } \varphi_i \supseteq X_{j(m,i,l)} \text{ for all } i = 1, \ldots, p, \\ \bot & \text{otherwise.} \end{cases}$$

Let

$$X_m^0 := \emptyset, \quad X_m^{l+1} := X_m^l \cup \{(X_{j(m,1,l)}, \ldots, X_{j(m,p,l)}, k(m,l))\},$$

hence $X_m = X_m^{h(m)}$. We first show that

$$\Psi(\vec{\varphi}, m, \bot, l) = \overline{X_m^l}(\vec{\varphi}). \tag{$*$}$$

This is proved by induction on l. For $l = 0$ both sides $= \bot$. In the step $l \to l+1$ we distinguish three cases according to the possible values 1, 0 and $\bot$ of p_l. In case $p_l = 1$ by the definition of Ψ, the induction hypothesis and the fact that $p_l = 1$ implies that $\varphi_i \cup X_{j(m,i,l)}$ is inconsistent for some $i = 1, \ldots, p$ we obtain

$$\Psi(\vec{\varphi}, m, \bot, l+1) = \Psi(\vec{\varphi}, m, \bot, l) = \overline{X_m^l}(\vec{\varphi}) = \overline{X_m^{l+1}}(\vec{\varphi}).$$

In case $p_l = 0$ by the definition of Ψ and the fact that $p_l = 0$ implies that $\varphi_i \supseteq X_{j(m,i,l)}$ for all $i = 1, \ldots, p$ we obtain

$$\Psi\big(\vec{\varphi}, m, \bot, l+1\big) = k(m,l) = \overline{X_m^{l+1}}\big(\vec{\varphi}\big).$$

In case $p_l = \bot$ first assume that $\Psi(\vec{\varphi}, m, \bot, l) = \bot$. Then by induction hypothesis $\overline{X_m^l}(\vec{\varphi}) = \bot$, hence also $\overline{X_m^{l+1}}(\vec{\varphi}) = \bot$ (since we do not have $\varphi_i \supseteq X_{j(m,i,l)}$ for all $i = 1, \ldots, p$), and $\Psi(\vec{\varphi}, m, \bot, l+1) = \bot$ by definition of pcond. If on the other hand $\Psi(\vec{\varphi}, m, \bot, l)$ is defined, then by induction hypothesis $\overline{X_m^l}(\vec{\varphi})$ is defined as well. Hence it must be $= k(m,l)$ (since $\varphi_i \cup X_{j(m,i,l)}$ is consistent for all $i = 1, \ldots, p$, and X_m^{l+1} is consistent), hence by definition of pcond

$$\Psi\big(\vec{\varphi}, m, \bot, l+1\big) = k(m,l) = \overline{X_m^{l+1}}\big(\vec{\varphi}\big).$$

This completes the proof of $(*)$.

We now show

$$\Psi\big(\vec{\varphi}, m, d, l\big) = d, \qquad (**)$$

provided d is the common value of all $k(m,l')$ such that $p_{l'} = 0$ or $= \bot$ (i.e., $\varphi_i \cup X_{j(m,i,l')}$ is consistent for all $i = 1, \ldots, p$). Again, the proof is by induction on l. For $l = 0$ we have $\Psi(\vec{\varphi}, m, d, 0) = d$ by definition. In the step $l \to l+1$ we again distinguish cases according to the possible values of p_l. In case $p_l = 1$ we know that $\varphi_i \cup X_{j(m,i,l)}$ is inconsistent for some $i = 1, \ldots, p$, hence we have

$$\Psi(\vec{\varphi}, m, d, l+1) = \Psi(\vec{\varphi}, m, d, l) = d$$

by induction hypothesis. In case $p_l = 0$ or $= \bot$ we know that that $\varphi_i \cup X_{j(m,i,l)}$ is consistent for all $i = 1, \ldots, p$. Then $k(m,l) = d$ by the choice of d, hence $\Psi(\vec{\varphi}, m, d, l+1) = d$ by induction hypothesis and the definition of pcond. This completes the proof of $(**)$.

We can now proceed with the proof of (1). Let

$$\begin{aligned} \Phi\big(\vec{\varphi}, m, d\big) &:= \Psi\big(\vec{\varphi}, m, d, h(m)\big), \\ \mathsf{en}\big(m, n, \vec{\varphi}\big) &:= \Phi\big(\vec{\varphi}, m, \Phi\big(\vec{\varphi}, n, \bot\big)\big) \end{aligned}$$

with the understanding that n is also allowed to take the value $\bot$. Note that $\Phi(\vec{\varphi}, n, \bot) = \overline{X_n}(\vec{\varphi})$ by $(*)$. The first property of en is now obvious, since

$$\mathsf{en}\big(m, \bot, \vec{\varphi}\big) = \Phi\big(\vec{\varphi}, m, \Phi\big(\vec{\varphi}, \bot, \bot\big)\big) = \Phi\big(\vec{\varphi}, m, \bot\big) = \overline{X_m}\big(\vec{\varphi}\big).$$

For the second property let $\overline{X_n} \supseteq \overline{X_m}$ and $\vec{\varphi}$ be given, and $d := \overline{X_n}(\vec{\varphi})$. Then by definition $\mathsf{en}(m, n, \vec{\varphi}) = \Phi(\vec{\varphi}, m, d)$, and $\Phi(\vec{\varphi}, m, d) = d$ follows from $(**)$.

(2) Let $\rho = \sigma \to \tau$ and fix primitive recursive functions f and g such that the i-th token at type ρ is $b_i^\rho = (X^\sigma_{f(i)}, b^\tau_{g(i)})$. We will define the functionals incons_ρ from similar functionals ic_ρ of type $\rho \to \iota \to \iota$ given by

$$\mathsf{ic}(\varphi, i) := \begin{cases} 1 & \text{if } \varphi \cup \{b_i\} \text{ is inconsistent,} \\ 0 & \text{if } \varphi \supseteq \{b_i\}, \\ \bot & \text{otherwise.} \end{cases}$$

In order to show that the ic's are recursive in pcond and $\exists$ we carry out the following computations:

$$\begin{aligned} \mathsf{ic}_\rho(\varphi, i) = 1 &\equiv \varphi \cup \{b_i\} \text{ is inconsistent} \\ &\equiv \varphi \cup \{(X_{f(i)}, b_{g(i)})\} \text{ is inconsistent} \\ &\equiv \exists n \big(\overline{X_n} \supseteq X_{f(i)} \text{ and } \varphi\big(\overline{X_n}\big) \cup \{b_{g(i)}\} \text{ is inconsistent}\big) \\ &\equiv \exists n \big(\varphi\big(\mathsf{en}_\sigma(f(i), n)\big) \cup \{b_{g(i)}\} \text{ is inconsistent}\big) \\ &\equiv \exists n \big(\mathsf{ic}_\tau\big(\varphi\big(\mathsf{en}_\sigma(f(i), n)\big), g(i)\big) = 1\big) \end{aligned}$$

and

$$\begin{aligned} \mathsf{ic}_\rho(\varphi, i) = 0 &\equiv b_i \in \varphi \\ &\equiv (X_{f(i)}, b_{g(i)}) \in \varphi \\ &\equiv b_{g(i)} \in \varphi\big(\overline{X_{f(i)}}\big) \\ &\equiv \mathsf{ic}_\tau\big(\varphi\big(\mathsf{en}_\sigma(f(i), \bot)\big), g(i)\big) = 0. \end{aligned}$$

Hence we can define

$$\mathsf{ic}_\rho(\varphi, i) := \exists\big(\lambda n.\mathsf{ic}_\tau\big[\varphi\big(\mathsf{en}_\sigma(f(i), n)\big), g(i)\big]\big).$$

We still have to define incons_ρ from ic_ρ. Let

$$\begin{aligned} \mathsf{ic}^*(\varphi, n, 0) &:= 0, \\ \mathsf{ic}^*(\varphi, n, l+1) &:= \mathsf{ic}^*(\varphi, n, l) \vee \mathsf{ic}\big(\varphi, j(n, l)\big), \end{aligned}$$

where $j(n, l)$ is defined by $X_n^\rho = \{b_{j(n,l)} \mid l < h(n)\}$. It is now easy to see that

$$\mathsf{incons}_\rho(\varphi, n) = \mathsf{ic}^*_\rho\big(\varphi, n, h(n)\big).$$

Note that one needs the coherence of $\mathrm{C{\small ON}}_\rho$ here (cf. Corollary 4.10). Note also that we do need the parallel or in the definition of ic^*. □

6. Bounded fixed point operators

Having seen that there is a quite nice internal definition of the computable functionals over the partial continuous functionals, we now come back to our subject of classifying recursive functions. The idea is simply to combine the power and elegance of treating recursion by means of the fixed point operator with the advantages of having a higher order term language with more "controlled" forms of higher type recursion, like Gödel's [1958] system T for the primitive recursive functionals. Moreover, we want to have conversion rules that actually carry out the computations, and are terminating and confluent. To achieve this goal, we introduce bounded versions of the fixed point operators, depending on a given well-ordering. We essentially follow Schwichtenberg and Wainer [1995] here, but – in order to simplify our arguments in Section 7 – we do not allow recursion with respect to a measure function.

6.1. Let a well-ordering $\prec$ of the natural numbers be given. We want the bounded fixed point operator $[\![\mathsf{Y}_{\rho,\prec}]\!]$ to satisfy, for $x \neq \emptyset$,

$$[\![\mathsf{Y}_{\rho,\prec}]\!](G, x) := G\left(\lambda y, \vec{y}.\begin{cases} [\![\mathsf{Y}_{\rho,\prec}]\!](G, y, \vec{y}) & \text{if } y \prec x \\ \emptyset & \text{otherwise} \end{cases}, x\right).$$

This equality, when viewed as a conversion rule from left to right, clearly gives rise to an infinite reduction sequence: just expand the subterm corresponding to $[\![\mathsf{Y}_{\rho,\prec}]\!](G, y)$ by the same rule again, and carry on doing that. Hence in order to achieve termination we have to make use of the fact that $[\![\mathsf{Y}_{\rho,\prec}]\!](G, y)$ occurs in a "guarded" context only. Therefore we restrict our conversion rules and allow conversion of a term $\mathsf{cond}MNL$ only if M is a constant (i.e. a numeral $\mathbf{n}$ or $\perp$). Then clearly the infinite reduction sequence above can not be formed any more, but still we can expect that any closed term of ground type reduces to a numeral.

It is shown below that every closed term of ground type involving only bounded fixed point operators is strongly normalizable. Hence the object "undefined" in the model can be viewed as a finite error. This is in contrast to the term language PCF of Plotkin [1977], where the value undefined is to be interpreted as a nonterminating computation. The addition of constants for finite errors to a term language has also been considered by Dosch [1992]; however, Dosch only deals with strict functionals.

6.2. Let us now carry out this program. We fix a well-ordering $\prec$ of the natural numbers. First we have to extend the term language by

- the sequential conditional cond of type $\iota \to \iota \to \iota \to \iota$,
- $c_\prec$ and $c_=$ of type $\iota \to \iota \to \iota$,
- for each type ρ, the constant $\mathsf{Y}_{\rho,\prec}$ of type $[(\iota \to \rho) \to \iota \to \rho] \to \iota \to \rho$.

Let us call the resulting term language $\text{PCF}_\prec$.

Recall that the sequential conditional cond was definable in our original language. However, we must introduce cond as a separate constant now, since it will be given a special status in our operational semantics, which makes it possible to use it as a "guard" blocking infinite reduction sequences.

NOTATION. In this section we have been (and will continue to be) slightly sloppy in not distinguishing notationally a constant of the term language $\text{PCF}_\prec$ (e.g., cond) from the partial continuous functional it denotes (e.g., $[\![\mathsf{cond}]\!]$). However, it should always be clear from the context what is meant. Recall also that we normally write $\perp$ for the empty set, and simply n for the singleton $\{n\}$ where no confusion should occur.

For any of the new constants $c_{\prec}, c_{=}, \mathsf{Y}_{\rho,\prec}$, we define its value as follows:

$$c_{\prec}(x,y) := \begin{cases} 1 & \text{if } x \prec y \text{ (and } x, y \text{ are both defined)}, \\ 0 & \text{if } x \not\prec y \text{ (and } x, y \text{ are both defined)}, \\ 0 & \text{if either } x \text{ of } y \text{ is undefined (i.e.} = \perp), \end{cases}$$

and similarly for $c_{=}$.

$\mathsf{Y}_{\rho,\prec}$ is to be an object of type $[(\iota \to \rho) \to \iota \to \rho] \to \iota \to \rho$. So let $\rho = \rho_1 \to \cdots \to \rho_n \to \iota$ and objects G of type $(\iota \to \rho) \to \iota \to \rho$, x of type ι and $\vec{x}$ of type $\vec{\rho}$ be given. We define $\mathsf{Y}_{\rho,\prec}(G, x, \vec{x})$ by (transfinite) $\prec$-recursion on x, as follows. If x is undefined, then $\mathsf{Y}_{\rho,\prec}(G, x, \vec{x})$ is undefined. Otherwise, assume that for all y with $y \prec x$ and all $\vec{y}$ the value $\mathsf{Y}_{\rho,\prec}(G, y, \vec{y})$ is already known. Then let

$$\mathsf{Y}_{\rho,\prec}\big(G, x, \vec{x}\big) := G\big([\mathsf{Y}_{\rho,\prec}(G)]_{\prec x}, x, \vec{x}\big),$$

where for any F of type $\iota \to \rho$ the restriction $[F]_{\prec x}$ of F below x is defined by

$$[F]_{\prec x}\big(y, \vec{y}\big) := \begin{cases} F(y, \vec{y}) & \text{if } y \prec x, \\ \emptyset & \text{otherwise.} \end{cases}$$

6.3. *Operational semantics.* The operational semantics is given by the following conversion rules. Recall that $\mathbf{m}$ is the m-th numeral, so $[\![\mathbf{m}]\!] = m$.

(i) $(\lambda x^{\rho} M)N \to M[N/x]$.

(ii) $\dfrac{M \to M'}{MN \to M'N}$.

(iii) $\mathsf{Y}_{\rho,\prec} N K \vec{N} \to \mathsf{cond}(c_{=} K K)[N(\lambda y, \vec{y}.\mathsf{cond}(c_{\prec} y K)(\mathsf{Y}_{\rho,\prec} N y \vec{y})\perp) K \vec{N}]\perp$.

(iv) $\mathsf{cond}\mathbf{k} M N \to M$ if $k > 0$; $\quad \mathsf{cond}0 M N \to N$; $\quad \mathsf{cond}\perp M N \to \perp$.

(v) $\dfrac{M \to M'}{\mathsf{cond}M \to \mathsf{cond}M'}$.

(vi) $\mathsf{pcond}\mathbf{k} M N \to M$ if $k > 0$; $\quad \mathsf{pcond}0 M N \to N$; $\quad \mathsf{pcond} M N N \to N$; $\mathsf{pcond}\perp M \perp \to \perp$; $\quad \mathsf{pcond}\perp\perp N \to \perp$; $\quad \mathsf{pcond}\perp c d \to \perp$ if c and d are different constants or numerals.

(vii) $\dfrac{M \to M'}{\mathsf{pcond}M \to \mathsf{pcond}M'}$; $\quad \dfrac{N \to N'}{\mathsf{pcond}MN \to \mathsf{pcond}MN'}$;
$\dfrac{L \to L'}{\mathsf{pcond}MNL \to \mathsf{pcond}MNL'}$.

(viii) $c_{\prec}\mathbf{m}\mathbf{n} \to \begin{cases} 1 & \text{if } m \prec n \\ 0 & \text{otherwise} \end{cases}$; $\quad c_{\prec}\perp N \to \perp$; $\quad c_{\prec} M \perp \to \perp$; $\quad$ similarly for $c_{=}$.
$\mathsf{S}\perp \to \perp$.

(ix) $\dfrac{M \to M'}{c_{\prec} M \to c_{\prec} M'}$; $\quad \dfrac{N \to N'}{c_{\prec} M N \to c_{\prec} M N'}$; $\quad$ similarly for $c_{=}$ and S.

It might seem that the simpler rule
(iii)′ $\mathsf{Y}_{\rho,\prec}NK\vec{N} \to N(\lambda y, \vec{y}.\mathsf{cond}(c_{\prec}yK)(\mathsf{Y}_{\rho,\prec}Ny\vec{y})\bot)K\vec{N}$
for the $\mathsf{Y}_{\rho,\prec}$-operator would suffice. However, it does not in case K evaluates to $\bot$. For then the outcome of the simplified rule (iii′) depends on the form of N and might be defined, which would be inconsistent with our denotational semantics.

Note that we do *not* have

- $\dfrac{N \to N'}{MN \to MN'}$,
- $\dfrac{M \to M'}{\lambda x M \to \lambda x M'}$,

for then following an application of rule (iii) for the bounded fixed point operator we could easily construct an infinite reduction sequence.

6.4. *Normalization via computability predicates.* We now prove that every closed term M of ground type can be computed by the rules of 6.3, i.e. that any reduction sequence starting with M terminates. It clearly follows from the form of our conversion rules that the final term in any reduction sequence must be a constant or a numeral. Note that it is only in the range of the constants other than the sequential conditional cond that we have a choice of where to reduce.

For the proof we use W.W. Tait's method of computability predicates. So for every type ρ we define what it means for a term of type ρ to be computable. The definition is by induction on types.

(1) A closed term M of ground type is computable if any reduction sequence starting with M terminates.
(2) A closed term M of type $\rho \to \sigma$ is computable if for every closed term N of type ρ, if N is computable, then so is MN.
(3) A term M containing at least one free variable is computable if every type respecting substitution of closed computable terms for the free variables in M yields a computable term.

THEOREM. *Every term M is computable.*

PROOF. By induction on M.

Case x. Trivial.

Case MN. Let $\vec{L}$ be a sequence of closed computable terms. We have to show that $M[\vec{L}]N[\vec{L}]$ is computable. But this holds, since by induction hypothesis we know that $M[\vec{L}]$ as well as $N[\vec{L}]$ are computable.

Case $\lambda x M$. Let $\vec{L}$ be a sequence of closed computable terms. We have to show that $\lambda x M[\vec{L}]$ is computable. So let $N, \vec{M}$ be closed computable terms. We must show that $(\lambda x M[\vec{L}])N\vec{M}$ is computable, i.e. that every reduction sequence starting with that term terminates. But from the rules of our operational semantics it follows that the second member of any such sequence must be $M[N, \vec{L}]\vec{M}$, which is computable by induction hypothesis.

Case $\mathsf{Y}_{\rho,\prec}$. Let $N, K, \vec{N}$ be closed computable terms. We have to show that $\mathsf{Y}_{\rho,\prec}NK\vec{N}$ is computable, i.e. that every reduction sequence starting with that term terminates. Since K is computable and of type ι, we know that K reduces to a constant or a numeral c; we use an auxiliary (transfinite) $\prec$-induction on (the value of) c. If $c = \bot$, then every reduction sequence starting with $\mathsf{Y}_{\rho,\prec}NK\vec{N}$ must, after its first two steps, work on reducing $c_{=}KK$ until this term has been reduced to a constant or a numeral; this follows from the form of the conversion rules for the sequential conditional. Now since by assumption K reduces to $\bot$ we must get $\bot$, and hence $\mathsf{Y}_{\rho,\prec}NK\vec{N}$ reduces to $\bot$. So assume $c = \mathbf{n}$. Then as before we can see that every reduction sequence starting with $\mathsf{Y}_{\rho,\prec}NK\vec{N}$ must come across

$$N\big(\lambda y, \vec{y}.\mathsf{cond}(c_{\prec}yK)\big(\mathsf{Y}_{\rho,\prec}Ny\vec{y}\big)\bot\big)K\vec{N}.$$

Since N is computable by assumption, it suffices to show that the closed term

$$\lambda y, \vec{y}.\mathsf{cond}(c_{\prec}yK)\big(\mathsf{Y}_{\rho,\prec}Ny\vec{y}\big)\bot$$

is computable. So let $L, \vec{L}$ be closed computable terms. We must show that

$$\mathsf{cond}(c_{\prec}LK)\big(\mathsf{Y}_{\rho,\prec}NL\vec{L}\big)\bot$$

is computable. Since L is computable by assumption, we know that L reduces to a constant or a numeral d. If $d = \bot$, then $c_{\prec}LK$ must reduce to $\bot$ and hence the whole term must reduce to $\bot$. So assume $d = \mathbf{m}$. Then the whole term must reduce to

$$\mathsf{cond}(c_{\prec}\mathbf{mn})\big(\mathsf{Y}_{\rho,\prec}NL\vec{L}\big)\bot.$$

If $m \prec n$ holds, then $c_{\prec}\mathbf{mn} \to \mathbf{k}$ for some $k > 0$ and hence the whole term must reduce to $\mathsf{Y}_{\rho,\prec}NL\vec{L}$, which by hypothesis of our auxiliary $\prec$-induction is computable. If $m \prec n$ does not hold, then $c_{\prec}\mathbf{mn} \to 0$ and hence the whole term reduces to $\bot$.

(*Case* cond, pcond) Let M, N, L be closed computable terms. We have to show that $\mathsf{cond}MNL$ and $\mathsf{pcond}MNL$ are computable, i.e. that every reduction sequence starting with either of them terminates. But this clearly holds, since otherwise one of the terms M, N, L would have an infinite reduction sequence.

For the other constants of $\text{PCF}_{\prec}$ the argument is similar. □

Note that this proof involves computability predicates of arbitrary types and hence is not formalizable in Peano arithmetic in all finite types. It is well-known that this is necessarily so for every proof of a result implying the existence of a normal form for every term M.

6.5. *Reduction of* $\mathsf{Y}_{\rho,<}$ *to higher order primitive recursion.* We first consider the case where the well-ordering $\prec$ is the standard ordering $<$ of the natural numbers. Then the bounded fixed point operator is definable by means of primitive recursion, as follows.

Define a functional H_ρ of type $[(\iota \to \rho) \to \iota \to \rho] \to \iota \to \iota \to \rho$ by primitive recursion, as in Gödel's [1958] system T:

$$\begin{aligned} H_\rho(G, \bot) &:= \lambda y, \vec{y}\,\bot, \\ H_\rho(G, 0) &:= \lambda y, \vec{y}\,\bot, \\ H_\rho(G, n+1) &:= K_\rho\big(G, n, H_\rho(G, n)\big), \end{aligned}$$

where K_ρ of type $[(\iota \to \rho) \to \iota \to \rho] \to \iota \to (\iota \to \rho) \to \iota \to \rho$ is defined by

$$\begin{aligned} &K_\rho\big(G, x, z, y, \vec{y}\big) \\ &\quad := \mathsf{cond}\big(c_=(y, x), G\big(z, x, \vec{y}\big), \mathsf{cond}\big(c_<(y, x), z\big(y, \vec{y}\big), \bot\big)\big). \end{aligned}$$

Then we have, by an easy induction on n.

LEMMA.

$$H_\rho(G, n, m) = \begin{cases} \mathsf{Y}_{\rho,<}(G, m) & \text{if } m < n, \\ \lambda \vec{y}\,\bot & \text{otherwise.} \end{cases}$$

PROOF. It clearly suffices to prove this for $n+1$. In case $m = n$ we have

$$\begin{aligned} H\big(G, n+1, n, \vec{y}\big) &= K\big(G, n, H(G, n), n, \vec{y}\big) \\ &= G\big(H(G, n), n, \vec{y}\big) \\ &= G\big([\mathsf{Y}_{\rho,<}(G)]_{<n}, n, \vec{y}\big) \quad \text{by ind. hyp. for } n \\ &= \mathsf{Y}_{\rho,<}\big(G, n, \vec{y}\big), \end{aligned}$$

and in case $m < n$

$$\begin{aligned} H\big(G, n+1, m, \vec{y}\big) &= K\big(G, n, H(G, n), m, \vec{y}\big) \\ &= H\big(G, n, m, \vec{y}\big) \\ &= \mathsf{Y}_{\rho,<}\big(G, m, \vec{y}\big) \quad \text{by ind. hyp. for } n. \end{aligned}$$

In case $n < m$ both sides are undefined. □

Hence $\mathsf{Y}_{\rho,<}$ is explicitly definable from H_ρ, by

$$\mathsf{Y}_{\rho,<}(G, x) = H_\rho\big(G, \mathsf{S}(x), x\big).$$

6.6. *Reduction of higher order primitive recursion to* $\mathsf{Y}_{\rho,\prec}$. Let us now consider a general well-ordering $\prec$. We show that, under some mild assumptions on $\prec$, higher order primitive recursion can be reduced to the bounded fixed point operator $\mathsf{Y}_{\rho,\prec}$. The assumptions on $\prec$ say that $<$ can be embedded by means of a primitive recursive embedding-projection pair into $\prec$, i.e. that there are primitive recursive functions h, h' such that

$$x < y \to h(x) \prec h(y), \tag{$\prec$1}$$

$$h'(h(x)) = x. \tag{$\prec$2}$$

From now on let us assume that all well-orderings $\prec$ considered satisfy ($\prec$1) and ($\prec$2).

LEMMA (Reduction of higher order primitive recursion to $\mathsf{Y}_{\rho,\prec}$). *Let G of type ρ and H of type $\rho \to \iota \to \rho$ be given. Define*

$$\begin{aligned} G_1(u, y) &:= \begin{cases} G & \text{if } y = h(0), \\ H(u(h(x)), x) \text{ with } x = h'(y) - 1 & \text{otherwise,} \end{cases} \\ F_1 &:= \mathsf{Y}_{\rho,\prec}(G_1), \\ F(x) &:= F_1\big(h(x)\big). \end{aligned}$$

Then F satisfies the primitive recursion equations with respect to G and H, i.e.

$$\begin{aligned} F(\bot) &= \lambda\vec{y}\,\bot, \\ F(0) &= G, \\ F(n+1) &= H(F(n), n). \end{aligned}$$

PROOF. $F(0) = F_1(h(0)) = G_1([F_1]_{\prec h(0)}, h(0)) = G$, and

$$\begin{aligned} F(n+1) &= F_1\big(h(n+1)\big) \\ &= G_1\big([F_1]_{\prec h(n+1)}, h(n+1)\big) \\ &= H\big([F_1]_{\prec h(n+1)}\big(h(n)\big), n\big) \\ &= H\big(F_1\big(h(n)\big), n\big) \\ &= H\big(F(n), n\big). \end{aligned}$$

□

6.7. *Constants introduced by bounded recursion.* In Section 7 it will be more convenient to work with an equivalent version of the bounded fixed point operators, namely with functionals introduced by bounded recursion. By this we mean a definition of a functional F from a given functional G, of the form

$$F(x) := \begin{cases} \lambda\vec{y}\,\bot & \text{if } x = \bot, \\ G([F]_{\prec x}, x) & \text{otherwise,} \end{cases}$$

with $[F]_{\prec x}$ as in 6.2. Then clearly $F = \mathsf{Y}_{\rho,\prec}(G)$; hence the fixed point operators are strong enough to allow this form of $\prec$-recursive definition. In fact, a fragment of the term language $\mathrm{PCF}_{\prec}$ suffices, where $\mathsf{Y}_{\rho,\prec}$ is only allowed to occur in a context $\mathsf{Y}_{\rho,\prec}N$ with N a *closed* term.

We now prove that this restriction of the use of fixed point operators is not essential.

LEMMA. *$\mathsf{Y}_{\rho,\prec}$ is definable by explicit definitions and one $\prec$-recursion yielding a functional of the same type level as $\mathsf{Y}_{\rho,\prec}$.*

PROOF. Let

$$\begin{aligned} G_0(u,x,G) &:= G(\lambda y.u(y,G),x), \\ F(x) &:= \begin{cases} \lambda\vec{y}\,\bot & \text{if } x=\bot, \\ G_0([F]_{\prec x},x) & \text{otherwise,} \end{cases} \\ H(G,x) &:= F(x,G). \end{aligned}$$

Then $H(G,\bot)=F(\bot,G)=\lambda\vec{y}\,\bot$ and for $x\neq\bot$

$$\begin{aligned} H(G,x) &= G_0\big([F]_{\prec x},x,G\big) \\ &= G\big(\lambda y.[F]_{\prec x}(y,G),x\big) \\ &= G\left(\lambda y\begin{cases} F(y,G) & \text{if } y\prec x \\ \lambda\vec{y}\,\bot & \text{otherwise} \end{cases},\ x\right) \\ &= G\left(\lambda y\begin{cases} H(G,y) & \text{if } y\prec x \\ \lambda\vec{y}\,\bot & \text{otherwise} \end{cases},\ x\right) \\ &= G\big([H(G)]_{\prec x},x\big). \end{aligned}$$

Hence $H=\mathsf{Y}_{\rho,\prec}$. □

Note that a $\prec$-recursion of the form $F(x)=G([F]_{\prec x},x)$ might be a "nested" (see Tait [1961]) recursion; e.g., $G(f,x)=f(f(x))$ is allowed.

7. Elimination of detours through higher types by transfinite recursion

We have seen in 5.1 that a detour through higher types (specifically the integer types ρ_n) makes it possible to give a rather perspicuous definition of the generating functions F_α, $\alpha<\varepsilon_0$, of the extended Grzegorczyk hierarchy: every F_α can be built from iteration functionals by application alone, in a way determined by the Cantor normal form of α to base ω. Now our goal is to establish some kind of converse to this result, namely that detours through higher types can be eliminated at the expense of raising the ordinals of the transfinite recursions involved. The first result of this kind is due to Kreisel [1952]; here we discuss a generalization that first appeared in Schwichtenberg [1973, 1975]. We will also sketch a new proof of this result, by adapting a method of Buchholz [1991].

7.1. In order to formulate the result we first need some preparations. In what follows it will be cumbersome to construct explicitly the well-orderings to which our norms need to refer. But since we have to work with "canonical" well-orderings of concrete order types like ε_0, Γ_0 or the Bachmann–Howard ordinal anyway (cf. the introduction to Section 3), and since such ordinals can be coded by natural numbers in a straightforward way, we may think as well of norms as mappings directly into the ordinals, e.g., those $<\varepsilon_0$. This will be done in the rest of the paper.

Similarly, instead of $\mathsf{Y}_{\rho,\prec}$, $\prec$-recursion and $\mathrm{PCF}_\prec$ we will speak of $\mathsf{Y}_{\rho,\alpha}$, α-recursion and PCF_α, where α (e.g., $<\varepsilon_0$) is thought of as representing a standard well-ordering of this order type.

7.2. THEOREM (Trade-off theorem). *Let M be a closed term of* PCF_α *of type level $n+1$ such that every $\mathsf{Y}_{\rho,\alpha}$ appearing in M has $\mathsf{lev}(\rho) \leqslant n+m+1$ with $m \geqslant 1$. Then one can find $\beta < 2_{m+1}(\alpha+\omega)$ and a closed term M' of* PCF_β *involving just one $\mathsf{Y}_{\sigma,\beta}$ such that $[\![M]\!] = [\![M']\!]$ and $\mathsf{lev}(\sigma) \leqslant n+1$.*

Here $2_m(\gamma)$ is the m-fold iteration of the ordinal function $2^{\cdot}$ to γ, i.e. $2_0(\gamma) := \gamma$, $2_{m+1}(\gamma) := 2^{2_m(\gamma)}$.

7.3. We now sketch the proof given in Schwichtenberg [1973, 1975]. First, Tait [1965] has observed in that every α-recursive functional or – as we should say here – every term M in PCF_α can be represented by an infinite term t_M, built up from typed variables and constants by application, λ-abstraction and the formation of sequences $\langle t_i \rangle_{i\in\mathbb{N}}$ of type $\iota \to \iota$ with terms t_i of type ι.

For instance, if F is defined by α-recursion at type ρ, i.e. $F = Y_{\rho,\alpha} N$ with N a closed term of type $(\iota \to \rho) \to \iota \to \rho$, then one can represent $F\zeta$ by

$$t_\zeta := N\langle t_{\zeta\eta}\rangle_\eta \quad \text{with } t_{\zeta\eta} := \begin{cases} t_\eta & \text{if } \eta < \zeta, \\ \lambda\vec{x}\,\bot & \text{otherwise.} \end{cases}$$

So F can be represented by $t_F := \langle t_\zeta \rangle_\zeta$; here a sequence $\langle s_i \rangle_{i\in\mathbb{N}}$ of terms s_i of a non ground type is an abbreviation for $\lambda x\vec{x}\,\langle s_i\vec{x}\rangle_{i\in\mathbb{N}}x$.

Now let M be a closed term of PCF_α of type level $n+1$ such that every $\mathsf{Y}_{\rho,\alpha}$ appearing in M has level $\leqslant n+m+1$ with $m \geqslant 1$. We may assume that M is β-normal, and moreover (by 6.7) that every $Y_{\rho,\alpha}$ in M appears in a context $Y_{\rho,\alpha}N$ with N closed. Now expand every subterm $Y_{\rho,\alpha}N$ as explained above. One can see easily that the resulting infinite term t_M has depth $|t_M| < \alpha+\omega$ and rank $\mathsf{Rk}(t_M) \leqslant n+m+2$; here the rank $\mathsf{Rk}(t)$ of an infinite term t is defined to be the supremum of the type levels of all subterms λxs in a context $(\lambda xs)r$. One can define a reduction relation (essentially as in Tait [1965], but using β-conversions only) such that every infinite term t of $\mathsf{Rk}(t) \leqslant k+1$ can be reduced to a t' with $\mathsf{Rk}(t') \leqslant k$ and $|t'| \leqslant 2^{|t|}$. Hence, the above M of type level $n+1$ can be represented by an infinite term t^*_M with $\mathsf{Rk}(t^*_M) \leqslant n+1$ and depth $|t^*_M| < 2_{m+1}(\alpha+\omega)$.

We consider now finite numbers or codes for infinite terms, and define valuation terms $\mathrm{VAL}^{\alpha,S}_\rho$ with the following property. Let $\ulcorner t \urcorner$ be a code of a closed infinite term

t of type ρ and depth $|t| < \alpha$, all of whose subterms have types from a finite set S. Then $\mathrm{VAL}_\rho^{\alpha,S}\ulcorner t \urcorner$ represents the same functional as t. The definition of the $\mathrm{VAL}_\rho^{\alpha,S}$, $\rho \in S$, is by simultaneous α-recursion.

Now let M be as before. We first obtain a number $\ulcorner t_M \urcorner$ of an infinite term representing M, such that $\mathrm{VAL}_\rho^{\alpha+k,S}\ulcorner t_M \urcorner$ for suitable k, S represents the same functional as M. Corresponding to the reduction of the term t_M of $\mathsf{Rk}(t_M) \leqslant n+m+2$ and with depth $|t_M| < \alpha + \omega$ to a term t_M^* with $\mathsf{Rk}(t_M^*) \leqslant n+1$ and with depth $|t_M^*| < 2_{m+1}(\alpha+\omega)$ we construct a function RED^* such that

$$\mathrm{VAL}_\rho^{\alpha+k,S}\ulcorner t_M \urcorner = \mathrm{VAL}_\rho^{\beta,S_{n+1}}\left(\mathrm{RED}^*\ulcorner t_M \urcorner\right)$$

with $\beta < 2_{m+1}(\alpha+\omega)$ and a finite set S_{n+1} of types with levels $\leqslant n+1$. The function RED^* turns out to be primitive recursive; a similar situation occurs in the theory of Kleene's $\mathcal{O}$, where $+_{\mathcal{O}}$ can be chosen primitive recursive; cf. Kleene [1958]. Since $\mathrm{VAL}_\rho^{\beta,S_{n+1}}$ is obtained using β-recursion but without auxiliary functionals of type level $> n+1$ we have the desired result.

7.4. In the rest of the paper we sketch a different proof of the trade-off theorem, by adapting a method of Buchholz [1991] originally developed for well-founded sequent style derivations in ω-arithmetic. In that paper, the basic idea is to introduce a primitive recursive notation system for well-founded ω-derivations in the same way as one usually introduces an ordinal notation system as a system of terms generated from constants for particular ordinals by function symbols corresponding to certain ordinal functions. A less-than relation between ordinal notations is then derived from the properties of the denoted ordinals and ordinal functions. Instead of ordinals Buchholz considers well-founded derivations in the standard system $\mathbf{Z}^\infty$ of ω-arithmetic, with an unrestricted ω-rule. Each derivation in Peano-arithmetic $\mathbf{Z}$ can be viewed as a notation for a particular $\mathbf{Z}^\infty$-derivation; so $\mathbf{Z}$-derivations can play the role of constants here. The role of ordinal functions is taken over by the standard operators for cut elimination in infinite derivations, as introduced in Schütte [1951], Tait [1968] and Mints [1978, 1992]. Buchholz then introduces a system $\mathbf{Z}^*$ of *finite* terms built from derivations in $\mathbf{Z}$ by these function symbols. Each term (or notation) $h \in \mathbf{Z}^*$ then clearly denotes an infinite derivation $[\![h]\!] \in \mathbf{Z}^\infty$.

The main technical achievement of Buchholz' paper is that without explicit use of indices for (primitive) recursive functions in a coding of infinite derivations he obtains rather elegantly defined *primitive* recursive functions $\cdot[\cdot]$ and $o(\cdot)$, such that $h[n]$ denotes the nth premise of the infinite derivation $[\![h]\!]$ denoted by h, and $o(h)$ gives an ordinal bound for $[\![h]\!]$. It is quite clear that one can give *recursive* definitions of these functions, since the corresponding operators on infinite derivations in $\mathbf{Z}^\infty$ are defined by transfinite recursion. The trick to obtain primitive recursive such functions is the usual one, namely to introduce some delay operators, in the present case the repetition rule introduced by Mints [1978].

Here we transfer Buchholz' approach to the somewhat more flexible setting of infinite (typed) terms. We then have to deal with β-contraction and hence with substitutions, and this will cause some additional complication. However, one benefit is that this approach allows applications in recursion theory as well as in proof theory, since by the well-known Curry–Howard correspondence every natural (infinite) deduction can be viewed as an (infinite) term with formulas as types. In the present paper we restrict ourselves to just one application of the method in recursion theory, namely the trade-off theorem mentioned above. More details including another application can be found in Schwichtenberg [1998].

7.5. *Infinite terms with the repetition rule.* Infinite terms are defined inductively, as in Tait [1965]. They are built from

- typed variables x^ρ by
- constant application $c\vec{t}$,
- λ-abstraction $\lambda x^\rho t$,
- application ts,
- sequence formation $\langle t_i^\iota \rangle_{i \in \mathbb{N}}$, and
- repetition $\text{REP}(t)$.

$\mathsf{Tp}(t)$ denotes the type of t. We assume that finitely many constants c denoting primitive recursive functions are given, where each such function expects $n \geqslant 0$ arguments of ground types $\iota_1, \dots, \iota_n$ and yields a value of ground type ι; in addition, for every $k \in \mathbb{N}^\perp$ we have a constant $\mathbf{k}$. The *level* of a term is the level of its type.

The *rank* of an infinite term is defined to be the supremum of the levels of all subterms of the form λxt or $\text{REP}(t)$ in a context $(\lambda xt)s$ or $(\text{REP}(t))s$. Substitution, β-conversion and a function $\text{RED}(t)$ reducing the rank of t by 1 are defined as usual. However, at certain points of the inductive definition one has to insert applications of the repetition rule; this is needed to allow later a primitive recursive recognition which rule had been applied at a given node. It turns out that the depth $|\text{RED}(t)|$ can only be estimated by $3^{|t|}$ (not $2^{|t|}$); this leads to slightly weaker bounds in the trade-off theorem.

7.6. *Notation systems for infinite terms.* We now introduce notation systems for infinite terms. From each notation a of a term t we want to be able to read off the type of t (as $\mathsf{tp}(a)$), the rule applied last to form t (as $\mathsf{rl}(a)$) and for each $n < \mathsf{arity}(\mathsf{rl}(a))$ the n-th predecessor of t (as $a[n]$). We will also introduce a bound function fv, which for every notation a of a term t gives an upper bound $\mathsf{fv}(a)$ on the variables free in t. This information will be needed when we define substitution, for then a certain renaming of bound variables will be necessary.

A *notation system* $\mathcal{T} = (T, \mathsf{tp}, \mathsf{fv}, \mathsf{rl}, \cdot[\cdot])$ is given by a primitive recursive nonempty set $T \subseteq \mathbb{N}$ together with four functions

$$\mathsf{tp}: T \to \text{(codes of) types},$$

$$\mathsf{fv}: T \to \text{(codes of) finite sets of variables},$$

$$\mathsf{rl}: T \to \text{(codes of) rules},$$

$$\cdot[\cdot]: T \times \mathbb{N} \to T \cup \{0\} \quad \text{(we assume } 0 \notin T\text{)}$$

such that

$$\forall a \in T \forall n \in \mathbb{N}\big(a[n] \in T \leftrightarrow n < \mathsf{arity}(\mathsf{rl}(a))\big).$$

A notation system $\mathcal{T} = (T, \mathsf{tp}, \mathsf{fv}, \mathsf{rl}, \cdot[\cdot])$ is *correct* if for every notation a and its predecessors the assigned types and sets of free variables satisfy some obvious conditions. An α*-norm* for a notation system $\mathcal{T}$ is a function o from T into the ordinals $<\alpha$ such that for all $a \in T$ and all $n < \mathsf{arity}(\mathsf{rl}(a))$ we have $o(a[n]) < o(a)$. A *rank* function for a notation system $\mathcal{T}$ is a function $\mathsf{rk}\colon T \to \mathbb{N}$ from T into the natural numbers such that (i) for all $a \in T$ and all $n < \mathsf{arity}(\mathsf{rl}(a))$ we have $\mathsf{rk}(a[n]) \leqslant \mathsf{rk}(a)$, and also (ii) $\mathsf{lev}(\mathsf{tp}(a[0])) \leqslant \mathsf{rk}(a)$ if $\mathsf{rl}(a)$ is a "critical" application, i.e., not an application starting with a variable or a sequence.

7.7. *The extension $\mathcal{T}^*$ of a notation system $\mathcal{T}$.* Let $\mathcal{T} = (T, \mathsf{tp}, \mathsf{fv}, \mathsf{rl}, \cdot[\cdot])$ be a notation system. We extend $\mathcal{T}$ by adding for every variable y a constant a_y, for every $n \geqslant 1$ an $(n+2)$-ary function symbol sub for substitution (written $\mathsf{sub}_{\vec{x}}(a, \vec{b})$, where $\vec{x}$ is a (code for a) list of n distinct variables $x_1, \ldots, x_n$), and also a binary function symbol β for beta-conversion and a unary function symbol red for reduction. Let T^* be the closure of T generated by these function symbols. Using some obvious coding machinery we may clearly assume that T^* is a primitive recursive subset of $\mathbb{N}$. The main point is that we can now extend the functions tp, fv, rl and $\cdot[\cdot]$ from T to T^*, in a primitive recursive way. It is here where the insertion of repetition rules in 7.5 is used. One can show

- If $\mathcal{T}$ is a correct notation system, then so is $\mathcal{T}^*$.
- Every norm on $\mathcal{T}$ can be extended to a norm on $\mathcal{T}^*$ by defining

$$\begin{aligned} o(a_y) &:= 0, \\ o\big(\mathsf{sub}_{\vec{x}}(a, \vec{b})\big) &:= \big(\max o(\vec{b})\big) + o(a), \\ o\big(\beta(a, b)\big) &:= o(b) + o(a) + 1, \\ o\big(\mathsf{red}(a)\big) &:= 4^{o(a)}. \end{aligned}$$

- Every rank function on $\mathcal{T}$ can be extended to a rank function on $\mathcal{T}^*$ by defining

$$\begin{aligned} \mathsf{rk}(a_y) &:= 0, \\ \mathsf{rk}\big(\mathsf{sub}_{\vec{x}}(a, \vec{b})\big) &:= \max\big(\mathsf{lev}(\vec{\rho}^{\circ}), \mathsf{rk}(\vec{b}), \mathsf{rk}(a)\big), \\ &\quad\ \text{where } \vec{\rho}^{\circ} \text{ consists of those} \\ &\quad\ \rho_i \text{ from } \vec{\rho} := \mathsf{Tp}(\vec{x}) \text{ with } \mathsf{rl}(b_i) \\ &\quad\ \text{an abstraction or R}\text{EP}, \\ \mathsf{rk}\big(\beta(a, b)\big) &:= \max\big(\mathsf{lev}(\mathsf{tp}(b)), \mathsf{rk}(a), \mathsf{rk}(b)\big), \\ \mathsf{rk}\big(\mathsf{red}(a)\big) &:= \mathsf{rk}(a) \dot{-} 1. \end{aligned}$$

7.8. *Embedding of* PCF_α. The next step is to translate the PCF_α-terms into infinite terms; this is done essentially by Tait's method, as in 7.3. We obtain a correct notation system $\mathcal{P}$ for PCF_α-terms with primitive recursive rank function and $\alpha \cdot \omega$-norm.

As described in 7.7, we can now extend $\mathcal{P}$ to a correct notation system $\mathcal{P}^* = (P^*, \dots)$ with primitive recursive rank function $\mathsf{rk}\colon P^* \to \mathbb{N}$ and norm $o\colon P^* \to \varepsilon(\alpha)$, where $\varepsilon(\alpha)$ is the least ε-number bigger than α.

7.9. *Valuation functionals.* Consider a notation system $\mathcal{P}$ with α-norm o. Let $a \in P$ be a notation of a closed term t with type ρ. Then we want to define the value $\mathrm{VAL}_\rho(a)$ of t. However, for the definition of the VAL_ρ it is necessary to allow free variables in t. Therefore, we introduce an additional argument w which serves to code an assignment $\vec{x} \mapsto \vec{u}$ for these variables.

So before we can give a definition of the VAL_ρ, we have to introduce such codes for assignments. Since we are interested ultimately in closed terms and have to consider free variables only for their inductive construction, it is sufficient to restrict ourselves to terms whose free variables have types from an appropriate finite set $S = \{\tau_1, \dots, \tau_m\}$. So let an assignment of functionals $u_1, \dots, u_m$ to variables $x_1, \dots, x_m$ be given, both of types $\tau_1, \dots, \tau_m$. Let $a_1, \dots, a_m \in P$ be notations for $x_1, \dots, x_m$ (i.e. $\mathsf{rl}(a_i) = \mathrm{VAR}_{x_i}$). First, transform all the u_i to a common type $\tau = \tau_S$ by means of transformation functionals $\mathsf{Tr}^{\tau}_{\tau_i}$ with inverses $\mathsf{Tr}^{\tau_i}_{\tau}$ (i.e. $\mathsf{Tr}^{\tau_i}_{\tau}(\mathsf{Tr}^{\tau}_{\tau_i}(u)) = u$). With these $\mathsf{Tr}^{\tau}_{\tau_i}(u_i)$ build a function w of type $\iota \to \tau$ such that $w(a_i) = \mathsf{Tr}^{\tau}_{\tau_i}(u_i)$. From w one can then read off the functionals u_i assigned to x_i by $u_i = \mathsf{Tr}^{\tau_i}_{\tau}(w(a_i))$. With these tools one can define VAL_ρ ($\rho \in S$) with respect to a fixed finite set S of types and our given notation system $\mathcal{P}$ with α-norm o. Note that for $a \in P$ with norm $o(a) < \beta \leqslant \alpha$ we have a definition by simultaneous β-recursion of the finitely many functionals VAL_ρ for $\rho \in S$, where all auxiliary functionals (in particular $\cdot[\cdot]$) are primitive recursive.

7.10. *Second proof of the trade-off theorem.* We now can assemble the tools discussed and give a second proof of the trade-off Theorem 7.2. However, due to the slightly weaker estimates in 7.5 we obtain $4_m(\alpha \cdot \omega)$ as a bound for β (not $2_{m+1}(\alpha + \omega)$ as above).

The proof is as follows. By 6.7 we may assume that every $\mathsf{Y}_{\sigma,\alpha}$ in M appears in a context $\mathsf{Y}_{\sigma,\alpha} N$ with closed N, and that M is β-normal. Construct a correct notation system $\mathcal{P}$ for PCF_α as in 7.8, and also a primitive recursive rank function and an $\alpha \cdot \omega$-norm for $\mathcal{P}$. Now extend $\mathcal{P}$ to a correct notation system $\mathcal{P}^*$, as described above, again with a primitive recursive rank function and an $\varepsilon(\alpha)$-norm. In $\mathcal{P}^*$ we can form $\mathsf{red}^m(M)$. Since we have $\mathsf{rk}(M) = \mathsf{lev}(\rho) \leqslant n + m + 1$ in $\mathcal{P}$, we know that $\mathsf{rk}(\mathsf{red}^m(M)) \leqslant n + 1$ in $\mathcal{P}^*$, hence all bound variables in the term denoted by $\mathsf{red}^m(M)$ have levels $\leqslant n$. Concerning the norm in $\mathcal{P}$ we have $o(M) < \alpha \cdot k < \alpha \cdot \omega$ for some $k < \omega$, hence in $\mathcal{P}^*$

$$o\big(\mathsf{red}^m(M)\big) < 4_m(\alpha \cdot k) =: \beta < 4_m(\alpha \cdot \omega).$$

Let S be the finite set of all types used in the term M, and $S_k := \{\rho \in S \mid \mathsf{lev}(\rho) \leqslant k\}$. As described in 7.9 we can now define $\mathrm{VAL}_\rho(a, w)$ for $\rho \in S_{n+1}$, with $a \in P^*$ of norm $o(a) < \beta$ and w of type $\tau := \tau_{S_n}$, by simultaneous β-recursion. But

$$M' := \mathrm{VAL}_{\mathsf{Tp}(M)}\big(\mathsf{red}^m(M), w_0\big),$$

– where w_0 codes the empty assignment – denotes the same functional as M. Hence we have found the closed PCF_β-term of we have looked for.

Acknowledgements

I would like to thank Peter Clote and Stan Wainer for their useful comments on various drafts of this paper.

References

W. Ackermann

[1940] Zur Widerspruchsfreiheit der Zahlentheorie, *Math. Ann.*, 117, pp. 162–194.

U. Berger

[1990] *Totale Objekte und Mengen in der Bereichstheorie*, Ph.D. Thesis, Mathematisches Institut der Universität München.

[1993] Total sets and objects in domain theory, *Ann. Pure Appl. Logic*, 60, pp. 91–117.

W. Buchholz

[1991] Notation systems for infinitary derivations, *Arch. Math. Logic*, 30, pp. 277–296.

W. Buchholz, A. Cichon and A. Weiermann

[1994] A uniform approach to fundamental sequences and hierarchies, *Math. Logic Quart.*, 40, pp. 273–286.

W. Buchholz and S. Wainer

[1987] Provably computable functions and the fast growing hierarchy, in: *Logic and Combinatorics*, Contemporary Mathematics, Vol. 65, Amer. Math. Soc., Providence, RI, pp. 179–198.

P. Clote

[1999] Computation models and function algebras, in: *Handbook of Computability Theory*, E. R. Griffor, ed., Elsevier, Amsterdam, pp. XXX–XXX.

A. Cobham

[1965] The intrinsic computational difficulty of functions, in: *Logic, Methodology and Philosophy of Science II*, Y. Bar-Hillel, ed., North-Holland, Amsterdam, pp. 24–30.

S. A. Cook

[1992] Computability and complexity of higher type functions, in: *Logic from Computer Science, Proceedings of a Workshop, held November 13–17, 1989*, Y. N. Moschovakis, ed., MSRI Publications, Vol. 21, Springer, Berlin, pp. 51–72.

W. Dosch

[1992] Reduction relations in strict applicative languages, in: *ISTCS '92*, Lecture Notes in Comput. Sci., Vol. 601, Springer, Berlin, pp. 55–66.

YU. L. ERSHOV
[1972] Computable functionals of finite types, *Algebra i Logika*, 11 (4), pp. 367–437.
[1974] Maximal and everywhere defined functionals, *Algebra i Logika*, 13 (4), pp. 374–397.
[1977] Model *C* of partial continuous functionals, in: *Logic Colloquium 1976*, R. Gandy and M. Hyland, eds., North-Holland, Amsterdam, pp. 455–467.

M. V. H. FAIRTLOUGH AND S. S. WAINER
[1998] Hierarchies of provably recursive functions, in: *Handbook of Proof Theory*, S. Buss, ed., North-Holland, Amsterdam, pp. 149–207.

S. FORTUNE, D. LEIVANT AND M. O'DONNELL
[1983] The expressiveness of simple and second-order type structures, *J. Assoc. Comput. Machin.*, 30 (1), pp. 151–185.

H. FRIEDMAN AND M. SHEARD
[1995] Elementary descent recursion and proof theory, *Ann. Pure Appl. Logic*, 71, pp. 1–45.

R. O. GANDY AND J. M. E. HYLAND
[1977] Computable and recursively countable functions of higher type, in: *Logic Colloquium 76*, North-Holland, Amsterdam, pp. 407–438.

J.-Y. GIRARD
[1981] Π^1_2-logic. Part I: Dilators, *Ann. Math. Logic*, 21, pp. 75–219.

K. GÖDEL
[1958] Über eine bisher noch nicht benützte Erweiterung des finiten Standpunkts, *Dialectica*, 12, pp. 280–287.

A. GRZEGORCZYK
[1953] *Some Classes of Recursive Functions*, Rozprawy Matematyczne, Vol. IV, Warszawa.

G. H. HARDY
[1904] A theorem concerning the infinite cardinal numbers, *Quart. J. Math.*, 35, pp. 87–94.

P. G. HINMAN
[1978] *Recursion-Theoretic Hierarchies*, Springer, Berlin.

S. C. KLEENE
[1958] Extension of an effectively generated class of functions by enumeration, *Colloq. Math.*, 7, pp. 67–78.
[1959a] Countable functionals, in: *Constructivity in Mathematics*, A. Heyting, ed., North-Holland, Amsterdam, pp. 81–100.
[1959b] Recursive functionals and quantifiers of finite types I, *Trans. Amer. Math. Soc*, 91, pp. 1–52.
[1962] Turing machine computable functionals of finite types I, in: *Proc. of the Congress for Logic, Methodology and the Philosophy of Science*, E. Nagel et al., eds., Stanford, pp. 38–45.

G. KREISEL
[1952] On the interpretation of non-finitist proofs II, *J. Symbolic Logic*, 17, pp. 43–58.
[1959] Interpretation of analysis by means of constructive functionals of finite types, in: *Constructivity in Mathematics*, A. Heyting, ed., North-Holland, Amsterdam, pp. 101–128.

K. G. LARSEN AND G. WINSKEL
[1991] Using information systems to solve recursive domain equations, *Information and Computation*, 91, pp. 232–258.

M. H. LÖB AND S. S. WAINER
[1970a] Hierarchies of number-theoretic functions I, *Arch. Math. Logik Grundlagenforschung*, 13, pp. 39–51.
[1970b] Hierarchies of number-theoretic functions II, *Arch. Math. Logik Grundlagenforschung*, 13, pp. 97–113.

M. L. MINSKY

[1961] Recursive unsolvability of Post's problem of "Tag" and other topics in the theory of Turing machines, *Ann. Math.*, 74, pp. 437–455.

G. E. MINTS

[1973] Quantifier-free and one-quantifier systems, *J. Soviet Math.*, 1, pp. 71–84.

[1978] Finite investigations of transfinite derivations, *J. Soviet Math.*, 10, pp. 548–596. Translated from: Zap. Nauchn. Semin. LOMI 49 (1975).

[1992] *Selected Papers in Proof Theory*, Studies in Proof Theory, Bibliopolis Napoli and North-Holland, Amsterdam.

J. MOLDESTAD

[1977] *Computations in Higher Types*, Lecture Notes in Mathematics, Vol. 811, Springer, Berlin.

J. MYHILL

[1953] A stumblingblock in constructive mathematics (abstract), *J. Symbolic Logic*, 18, p. 190.

K.-H. NIGGL

[1993] Subrecursive hierarchies on Scott domains, *Arch. Math. Logic*, 32, pp. 239–257.

[1994] *Subrecursive Hierarchies on the Partial Continuous Functionals*, Ph.D. Thesis, Universität München, Fakultät für Mathematik.

[1995] Towards the computational complexity of PR^{ω}-terms, *Ann. Pure Appl. Logic*, 75, pp. 153–178.

D. NORMANN

[1999] The continuous functionals, in: *Handbook of Computability Theory*, E. R. Griffor, ed., Elsevier, Amsterdam, pp. 251–275.

R. PÉTER

[1957] *Rekursive Funktionen*, Akademie-Verlag, Berlin.

R. A. PLATEK

[1966] *Foundations of Recursion Theory*, Ph.D. Thesis, Department of Mathematics, Stanford University.

G. D. PLOTKIN

[1977] LCF considered as a programming language, *Theoret. Comput. Sci.*, 5, pp. 223–255.

J. W. ROBBIN

[1965] *Subrecursive Hierarchies*, Ph.D. Thesis, Princeton University.

H. ROGERS, JR.

[1967] *Theory of Recursive Functions and Effective Computability*, McGraw Hill.

H. E. ROSE

[1984] *Subrecursion: Functions and Hierarchies*, Oxford Logic Guides, Vol. 9, Clarendon Press, Oxford.

N. A. ROUTLEDGE

[1953] Ordinal recursion, *Proc. Cambridge Phil. Soc.*, 49, pp. 175–182.

G. E. SACKS

[1990] *Higher Recursion Theory*, Perspectives in Mathematical Logic, Springer, Berlin.

K. SCHÜTTE

[1951] Beweistheoretische Erfassung der unendlichen Induktion in der Zahlentheorie, *Math. Ann.*, 122, pp. 369–389.

H. SCHWICHTENBERG

[1969] Rekursionszahlen und die Grzegorczyk-Hierarchie, *Arch. Math. Logik Grundlagenforschung*, 12, pp. 85–97.

[1971] Eine Klassifikation der ε_0-rekursiven Funktionen, *Zeits. Math. Logik Grundl. Math.*, 17, pp. 61–74.

[1972] Beweistheoretische Charakterisierung einer Erweiterung der Grzegorczyk-Hierarchie, *Arch. Math. Logik Grundlagenforschung*, 15, pp. 129–145.

[1973] *Einige Anwendungen von unendlichen Termen und Wertfunktionalen*, Habilitationsschrift, Mathematisches Institut der Universität Münster.

[1975] Elimination of higher type levels in definitions of primitive recursive functionals by means of transfinite recursion, *Logic Colloquium '73*, in: H. E. Rose and J. C. Shepherdson, eds., North-Holland, Amsterdam, pp. 279–303.

[1977] Proof theory: Some applications of cut-elimination. in: *Handbook of Mathematical Logic*, J. Barwise, ed., Studies in Logic and the Foundations of Mathematics, Vol. 90, North-Holland, Amsterdam, pp. 867–895.

[1998] Finite notations for infinite terms, *Ann. Pure Appl. Logic*, 94, pp. 201–222.

H. SCHWICHTENBERG AND S. S. WAINER

[1995] Ordinal bounds for programs, in: *Feasible Mathematics II*, P. Clote and J. Remmel, eds., Birkhäuser, Boston, pp. 387–406.

D. SCOTT

[1970] *Outline of a Mathematical Theory of Computation*, Technical Monograph PRG-2, Oxford University Computing Laboratory.

[1982] Domains for denotational semantics, in: *Automata, Languages and Programming*, E. Nielsen and E. M. Schmidt, eds., Lecture Notes in Comput. Sci., Vol. 140, Springer, Berlin, pp. 577–613. A corrected and expanded version of a paper prepared for ICALP'82, Aarhus, Denmark.

J. C. SHEPHERDSON AND H. E. STURGIS

[1963] Computability of recursive functions, *J. Assoc. Comput. Machin.*, 10, pp. 217–255.

J. R. SHOENFIELD

[1967] *Mathematical Logic*, Addison-Wesley, Reading, MA.

V. STOLTENBERG-HANSEN, E. GRIFFOR AND I. LINDSTRÖM

[1994] *Mathematical Theory of Domains*, Cambridge Tracts in Theoretical Computer Science, Cambridge Univ. Press.

W. W. TAIT

[1961] Nested recursion, *Math. Ann.*, 143, pp. 236–250.

[1965] Infinitely long terms of transfinite type, in: *Formal Systems and Recursive Functions*, J. Crossley and M. Dummett, eds., North-Holland, Amsterdam, pp. 176–185.

[1968] Normal derivability in classical logic, in: *The Syntax and Semantics of Infinitary Languages*, J. Barwise, ed., Lecture Notes in Mathematics, Vol. 72, Springer, Berlin, pp. 204–236.

G. TAKEUTI

[1987] *Proof Theory*, 2nd edn., North-Holland, Amsterdam.

[1994] Grzegorcyk's hierarchy and $iep\sigma_1$, *J. Symbolic Logic*, 59 (4), pp. 1274–1284.

S. S. WAINER

[1970] A classification of the ordinal recursive functions, *Arch. Math. Logik Grundlagenforschung*, 13, pp. 136–153.

A. WEIERMANN
[1995] Investigations on slow versus fast growing: how to majorize slow growing functions nontrivially by fast growing ones, *Arch. Math. Logic*, 34, pp. 313–330.

F. ZEMKE
[1977] P.R.-regulated systems of notation and the subrecursive hierarchy equivalence property, *Trans. Amer. Math. Soc.*, 234, pp. 89–118.

Part 6
Computer Science and Computability Theory

CHAPTER 17

Computation Models and Function Algebras

P. Clote*

Institut für Informatik, Ludwig-Maximilians-Universität München, Oettingenstrasse 67, D-80538 München, Germany

E-mail: clote@informatik.uni-muenchen.de

Contents

1. Introduction . . . 590
2. Machine models . . . 592
   2.1. Turing machines . . . 592
   2.2. Parallel machine model . . . 602
   2.3. Circuit families . . . 605
3. Some recursion schemes . . . 608
   3.1. An algebra for the logtime hierarchy LH . . . 609
   3.2. Bounded recursion on notation . . . 622
   3.3. Bounded recursion . . . 625
   3.4. Bounded minimization . . . 632
   3.5. Divide and conquer, course-of-values and miscellaneous . . . 638
   3.6. Safe recursion . . . 646
4. Type 2 functionals . . . 656

References . . . 670

List of symbols . . . 678

*A preliminary, abbreviated version of this paper appears in the proceedings of *Logic and Computational Complexity*, edited by D. Leivant, *Lecture Notes in Computer Science*, Vol. 960 (1995), pp. 98–130. This research partially supported by NSF CCR-9408090, by US–Czechoslovak Science and Technology Program Grant 93–025 and the Volkswagen Foundation.

HANDBOOK OF COMPUTABILITY THEORY
Edited by E.R. Griffor

1. Introduction

The modern digital computer, a force which has shaped the latter part of the 20-th century, can trace its origins back to work in mathematical logic concerning the formalization of concepts such as *proof* and *computable function*. Numerous examples support this assertion. For instance, in his development of the universal Turing machine, A.M. Turing seems to have been the first, along with J. von Neumann, to have understood the potential of memory-stored programs executed by a universal computational device. Moreover, certain function classes and proof systems can be viewed as prototypes of programming languages: LISP was developed from the Church–Kleene λ-calculus; PROLOG was developed from resolution (Gentzen sequent calculus); polymorphic programming languages such as ML were inspired by J.-Y. Girard's system **F**; imperative programming languages such as PASCAL and C can be viewed as an implementation of S.C. Kleene's μ-recursive functions.

One recurring theme in recursion theory is that of a *function algebra* – i.e. a smallest class of functions containing certain initial functions and closed under certain operations (especially substitution and primitive recursion).[1] In 1904, G.H. Hardy [1904] used related concepts to define sets of real numbers of cardinality $\aleph_1$. In 1923, Th. Skolem [1923] introduced the primitive recursive functions, and in 1925, as a technical tool in his claimed sketch proof of the continuum hypothesis, D. Hilbert [1925] defined classes of higher type functionals by recursion. In 1928, W. Ackermann [1928] furnished a proof that the diagonal function $\varphi_a(a, a)$ of Hilbert [1925], a variant of the Ackermann function, is not primitive recursive. In 1931, K. Gödel [1931] defined the primitive recursive functions, there calling them "rekursive Funktionen", and used them to arithmetize logical syntax via Gödel numbers for his incompleteness theorem. Generalizing Ackermann's work, in 1936 R. Péter [1936] defined and studied the k-fold recursive functions. The same year saw the introduction of the fundamental concepts of Turing machine (A.M. Turing [1936–37]), λ-calculus (A. Church [1936]) and μ-recursive functions (S.C. Kleene [1936a]). By restricting the scheme of primitive recursion to allow only limited summations and limited products, the *elementary functions* were introduced in 1943 by L. Kalmár [1943]. In 1953, A. Grzegorczyk [1953] studied the classes $\mathcal{E}^k$ obtained by closing certain fast growing "diagonal" functions under composition and *bounded primitive recursion* or *bounded minimization.*

H. Scholz's [1952] question concerning the characterization of *spectra* $\{n \in \mathbb{N}: (\exists \text{ model } M \text{ of } n \text{ elements}) (M \models \phi)\}$ of first order sentences ϕ, which was shown in 1974 by N. Jones and A. Selman [1974] to equal NTIME$(2^{O(n)})$, was the starting point for J.H. Bennett's work [1962]. Among other results, Bennett introduced the key notions of *positive extended rudimentary* and *extended rudimentary* (equivalent to the notions of nondeterministic polynomial time NP and the polynomial time hierarchy PH), characterized the spectra of sentences of higher type logic as exactly

[1] Hilbert [1925] stated that "*substitution* (i.e. replacement of an argument by a new variable or function) and *recursion* (the scheme of deriving the function value for $n+1$ from that of n)" are "the elementary operations for the construction of functions".

the Kalmár elementary sets, and proved that *rudimentary* coincides with Smullyan's notion of *constructive arithmetic* (those sets definable in the language $\{0, 1, +, \cdot, \leqslant\}$ of arithmetic by first order bounded quantifier formulas). Only much later in 1976 did C. Wrathall [1976] connect these concepts to computer science by proving that the linear time hierarchy LTH coincides with rudimentary, hence constructive arithmetic, sets. In 1963 R.W. Ritchie [1963] proved that Grzegorczyk's class $\mathcal{E}^2$ is the collection of functions computable in linear space on a Turing machine. In 1965, A. Cobham [1965] characterized the polynomial time computable functions as the smallest function algebra closed under Bennett's scheme of *bounded recursion on notation*.[2] These arithmetization techniques led to a host of characterizations of computational complexity classes by machine-independent function algebras in the work of D.B. Thompson [1972] on polynomial space, of K. Wagner [1979] on general time complexity classes. Function algebra characterizations of *parallel* complexity classes were given more recently by P. Clote [1990] and B. Allen [1991], while certain small *boolean circuit* complexity classes were treated by P. Clote and G. Takeuti [1995]. Higher type analogues of certain characterizations were given in 1976 by K. Mehlhorn [1976], in 1991 by S. Cook and B. Kapron [1996], Kapron and Cook [1990] for sequential computation, and in 1993 by Clote, A. Ignjatovic, B. Kapron [1993] for parallel computation. In 1995 H. Vollmer and K. Wagner [1996] Valiant's class $\#P$. Though distinct, the arithmetization techniques of function algebras are related to those used in proving numerous results like (i) NP equals generalized first order spectra (R. Fagin [1974]), (ii) the characterization of complexity classes via finite models (the program of *descriptive complexity theory* investigated by R. Fagin [1990], N. Immerman [1987, 1989], Y. Gurevich and S. Shelah [1986], and others).

From this short historical overview, it clearly emerges that *function algebras* and *computation models* are intimately related as the software (class of programs) and hardware (machine model) counterparts of each other. Historically, these notions are among the central concepts of recursion theory, proof theory and theoretical computer science. Perhaps this is the reason that K. Gödel [1975] claimed in 1975 that the most important open problem in recursion theory is the classification of all total recursive functions, presumably in a hierarchy of function algebras determined by admitting more and more complex operations. While much work characterizing ever larger subrecursive hierarchies has been done by W. Buchholz, J.-Y. Girard, G.E. Sacks, K. Schütte, H. Schwichtenberg, G. Takeuti, S.S. Wainer and others, in this paper we concentrate principally on subclasses of the primitive recursive functions and their relations to computational complexity. For primitive recursive functions, ε_0-functions, etc. and strong higher type functionals, see the articles of H. Schwichtenberg and D. Normann in this volume.

Apart from its interest as part of recursion theory, there are applications of function algebras to proof theory, especially in the study of theories T of first and second order arithmetic, whose *provably total* functions (having suitably definable graphs)

[2] According to Mehlhorn [1976], K. Weihrauch independently proved a similar characterization in 1972.

coincide with those of a particular function algebra. Using such techniques, for instance, G. Takeuti [1995] provided a simpler proof of the existence of an alternating logtime algorithm for the boolean formula evaluation problem, a result first proved by S. Buss [1987, 1993] (see Theorem 2.11). For a further discussion of such applications, see the recent monograph by J. Krajíček [1995].

Historically, Cobham's machine independent characterization of the polynomial time computable functions was the start of modern complexity theory, indicating a robust and mathematically interesting field. As outlined in Section 4, current work on type 2 and higher type function algebras suggests directions for the extension of complexity theory to higher type computation. The development of function algebras is potentially important in computer science for programming language design. New kinds of operations used in defining function algebras could possibly be incorporated in *small*, non-universal programming languages for dedicated purposes. All the function algebras defined in this paper could be used to define free variable equational calculi. For instance, S. Cook's [1971] system PV comes from Theorem 3.19, P. Clote's [1992a, 1993] systems AV, ALV, ALV' come from Theorems 3.26 and 3.27, J. Johannsen's [1996] systems TV, $A2V$ come from Theorem 3.16, while M. O'Donnell [1985] has proposed equational calculus as a programming language.

In this paper, we will survey a selection of results which illustrate the arithmetization techniques used in characterizing certain computation models by function algebras.

2. Machine models

Despite the immense diversity of abstract machine models and complexity classes (see for instance P. van Emde Boas [1990] or K. Wagner and G. Wechsung [1986]), only the most natural and robust models and classes will be treated in this paper. Many of the following machine models are familiar. Nevertheless, definitions are given in sufficient detail to provide an idea of the required initial functions and closure operations which permit function algebra characterizations of complexity classes.

2.1. *Turing machines*

In proving the recursive unsolvability of Hilbert's *Entscheidungsproblem* (independently established as well by A. Church [1936] using the λ-calculus), A.M. Turing [1936–37] introduced the Turing machine, largely motivated by his attempt to make precise the notion of computable (real) number, i.e. "those whose decimals which are calculable by finite means". Considering the "computer" as an idealized human clerk, Turing argued that the "behavior of the computer at any moment is determined by the symbols which he is observing, and his 'state of mind' at that moment", and specified that the number of "states of mind" should be finite, since "human memory is necessarily limited". Formally, we have the following.

DEFINITION 2.1. A multitape Turing machine (TM) M is specified by $(Q, \Sigma, \Gamma, \delta, q_0, k)$ where $k \in \mathbb{N}$,

- Q is a finite set of *states* containing the accept and reject states q_A, q_R, as well as the start state q_0,
- Σ [resp. Γ] is a finite read-only input [resp. read-write work] tape alphabet not containing the blank symbol B,
- δ is the transition function and maps

  $$\left(Q - \{q_A, q_R\}\right) \times \left(\Sigma \cup \{B\}\right) \times \left(\Gamma \cup \{B\}\right)^k$$

  into

  $$Q \times \left(\Gamma \cup \{B\}\right)^k \times \{-1, 0, 1\}^{k+1}.$$

A Turing machine is assumed to have a one-way infinite input tape and k one-way infinite work tapes. The work tapes are initially blank, while on input $w = w_1 \cdots w_n$ with $w_i \in \Sigma$, the initial input tape is of the form below.

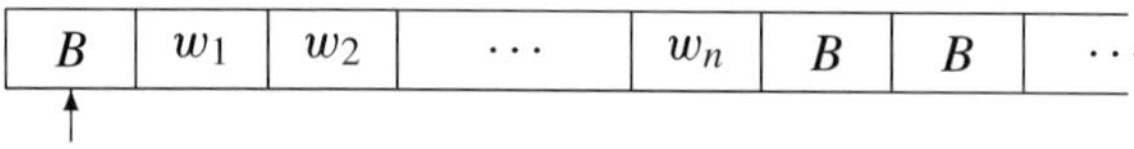

Each work tape has a tape head (above indicated by an arrow) capable of reading the symbol in the currently scanned square, writing a symbol in that square and remaining stationary or moving one square left or right. The leftmost cell is the 0-th cell. Since the input tape is read-only, the input tape head can scan a tape cell and remain stationary or move one square left or right. A *configuration* is a member of $Q \times (\Sigma \cup \{B\})^* \times (\Gamma \cup \{B\})^{*k} \times \mathbb{N}^{k+1}$, and indicates the current state, tape contents, and head positions. Alternately, a configuration can be abbreviated by underscoring the symbols currently scanned by a tape head, in order to indicate the current tape head position. For instance, $(q, Bab\underline{a}B, Bb\underline{b}B)$ abbreviates the configuration of a TM in state q, with an input tape, whose head currently scans an a, and one work tape, whose head currently scans a b. A halted configuration is one whose state is q_A or q_R.

Let

$$\begin{aligned}\alpha &= (q, BxB, \alpha_1, \ldots, \alpha_k, n_0, n_1, \ldots, n_k),\\ \beta &= (r, BxB, \beta_1, \ldots, \beta_k, m_0, m_1, \ldots, m_k)\end{aligned}$$

be configurations for M on input x. Then β is the *next configuration* after α in M's computation on x, denoted $\alpha \vdash_M \beta$, if the following conditions are satisfied:

(1) the n_0-th cell of the input tape BxB contains symbol a,

(2) for $1 \leqslant i \leqslant k$ the following hold:

  (a) $\sigma_i, \tau_i \in \Gamma \cup \{B\}$ and $u_i, v_i, w_i \in (\Gamma \cup \{B\})^*$,

(b) $\alpha_i = u_i \sigma_i v_i$ and $\beta_i = u_i \tau_i w_i$,

(c) $|u_i| = n_i$ (Recall that the leftmost cell is the 0-th cell, so the n-th cell has n cells to its left. This implies that σ_i [resp. τ_i] is the contents of the n_i-th cell of the i-th tape in configuration α [resp. β].)

(3) $\delta(q, a, \sigma_1, \ldots, \sigma_k) = (r, \tau_1, \ldots, \tau_k, m_0 - n_0, m_1 - n_1, \ldots, m_k - n_k)$, where for $1 \leqslant i \leqslant k$:

(a) $m_i < |\beta_i|$,

(b) either $v_i = w_i$ or $v_i = \lambda$ (the empty word), $w_i = B$, and $m_i = n_i + 1$.

The reflexive, transitive closure of $\vdash_M$ is denoted by $\vdash_M^*$, and a configuration C is said to *yield* a configuration D *in n-steps*, denoted $C \vdash_M^n D$, if there are $C_1, \ldots, C_n$ such that $C = C_1 \vdash_M C_2 \vdash_M \cdots \vdash_M C_n = D$, while C *yields* D if $C \vdash_M^* D$. A Turing machine M *accepts* a language $L \subseteq \Sigma^*$, denoted by $L = L(M)$, if L is the collection of words w such that the *initial configuration* $(q_0, \underline{B}wB, \underline{B}, \ldots, \underline{B})$ yields $(q_A, \underline{B}wB, \underline{B}, \ldots, \underline{B})$; a word w is *accepted* in n steps if $(q_0, \underline{B}wB, \underline{B}, \ldots, \underline{B}) \vdash_M^n (q_A, \underline{B}wB, \underline{B}, \ldots, \underline{B})$. The machine M accepts $L \subseteq \Sigma^*$ in *time* $T(n)$ (resp. *space* $S(n)$) if $L = L(M)$ and for each word $w \in L(M)$ of length n, w is accepted in at most $T(n)$ steps (resp. the maximum number of cells visited on each of M's work tapes is $S(n)$). A language $L \subseteq \Sigma^*$ is *decided* by M in *time* $T(n)$ (resp. *space* $S(n)$) if L [resp. $\Sigma^* - L$] is the collection of words for which M halts in state q_A [resp. q_R], and for each word $w \in \Sigma^*$ of length n, M halts in at most $T(n)$ steps (resp. the maximum number of cells visited on each of M's work tapes is $S(n)$). This article concerns complexity classes, so for the most part we identify the notions of acceptance and decision (for most of the complexity classes here considered, machines of a certain complexity class can be clocked so as to reject a word if they don't accept it).

Recall that

$$\begin{aligned} \mathrm{O}(f) &= \big\{g\colon (\exists c > 0)(\exists n_0)(\forall n \geqslant n_0)\big[g(n) \leqslant c \cdot f(n)\big]\big\}, \\ \Omega(f) &= \big\{g\colon (\exists c > 0)(\exists n_0)(\forall n \geqslant n_0)\big[f(n) \leqslant c \cdot g(n)\big]\big\}, \\ \Theta(f) &= \mathrm{O}(f) \cap \Omega(f) \end{aligned}$$

so that $n^{\mathrm{O}(1)}$ denotes the set of all polynomially bounded functions. If T, S are one-place functions, then

$$\begin{aligned} \mathrm{DTIME}(T(n)) &= \big\{L \subseteq \Sigma^*\colon L \text{ accepted by a TM in time } \mathrm{O}\big(T(n)\big)\big\}, \\ \mathrm{DSPACE}(S(n)) &= \big\{L \subseteq \Sigma^*\colon L \text{ accepted by a TM in space } \mathrm{O}\big(S(n)\big)\big\}, \\ \mathrm{PTIME} &= \mathrm{P} = \mathrm{DTIME}\big(n^{\mathrm{O}(1)}\big), \\ \mathrm{PSPACE} &= \mathrm{DSPACE}\big(n^{\mathrm{O}(1)}\big), \\ \mathrm{ETIME} &= \bigcup_{c \geqslant 1} \mathrm{DTIME}\big(2^{c \cdot n}\big) = \mathrm{DTIME}\big(2^{\mathrm{O}(n)}\big), \\ \mathrm{EXPTIME} &= \bigcup_{c \geqslant 1} \mathrm{DTIME}\big(2^{n^c}\big) = \mathrm{DTIME}\big(2^{n^{\mathrm{O}(1)}}\big). \end{aligned}$$

Finally, DTIMESPACE($T(n)$, $S(n)$) is defined as

$$\{L \subseteq \Sigma^*\colon L \text{ accepted by a TM in time } \mathrm{O}(T(n)) \text{ and space } \mathrm{O}(S(n))\}.$$

DEFINITION 2.2. A nondeterministic multitape Turing machine (NTM) M is specified by $(Q, \Sigma, \Gamma, \Delta, q_0, k)$ where $Q, \Sigma, \Gamma, q_0, k$ are as in Definition 2.1 and the *transition relation* Δ is contained in

$$\big((Q - \{q_A, q_R\}) \times (\Sigma \cup \{B\}) \times (\Gamma \cup \{B\})^k\big) \times \big(Q \times (\Gamma \cup \{B\})^k \\ \times \{-1, 0, 1\}^{k+1}\big).$$

If α, β are configurations in the computation of the nondeterministic Turing machine (NTM) M on input x, then write $\alpha \vdash_M \beta$ if

$$(q, a, \sigma_1, \ldots, \sigma_k, r, \tau_1, \ldots, \tau_k, m_0 - n_0, m_1 - n_1, \ldots, m_k - n_k) \in \Delta,$$

where $\sigma_i, \tau_i, a, n_i, m_i$ are as in the deterministic case.

With this change, the notions of configuration, yield and acceptance are analogous to the previously defined notions. A nondeterministic computation corresponds to a *computational tree* whose root is the initial configuration, whose leaves are halted computations, and whose internal nodes α have as children those configurations β obtained in one step from α, $\alpha \vdash_M \beta$. A word $w \in \Sigma^*$ is *accepted* if there is an accepting path in the computation tree, though many non-accepting paths may exist. A NTM M accepts a word of length n in *time* $T(n)$ [resp. *space* $S(n)$] if the depth of the associated computation tree is at most $T(n)$ [resp. for each configuration α in the computation tree the number of cells used on each work tape is at most $S(n)$]. NTIME($T(n)$) [resp. NSPACE($S(n)$)] is the collection of languages $L \subseteq \Sigma^*$ accepted by a NTM in time $\mathrm{O}(T(n))$ [resp. space $\mathrm{O}(S(n))$]; NP = NTIME($n^{\mathrm{O}(1)}$).

Similarly, NTIMESPACE($T(n)$, $S(n)$) is the set of languages $L \subseteq \Sigma^*$ accepted by a NTM in time $\mathrm{O}(T(n))$ and space $\mathrm{O}(S(n))$.

With the previous definitions, any computation depending on all bits of the input requires at least linear time, the minimum amount of time taken to scan the input. However, by allowing a TM to access its input bitwise via pointers or random access, sublinear runtimes can be achieved, as shown by Chandra et al. [1981].

DEFINITION 2.3. A Turing machine M with *random access* (RATM) is given by a finite set Q of states, an input tape having no tape head, k work tapes, an *index query* tape and an *index answer* tape. To permit random access, the alphabet Γ is always assumed to contain the symbols 0, 1. Except for the input tape, all other tapes have a tape head. M contains a distinguished input query state q_I, in which state M writes into the leftmost cell of the index answer tape that symbol which appears in the k-th input tape cell, where $k = \sum_{i<m} k_i \cdot 2^i$ is the integer whose binary representation is given by the contents

B	k_{m-1}	k_{m-2}	$\cdots$	k_0	B	$\cdots$

of the query index tape. Unlike the oracle Turing machine in Definition 2.5, the query index tape is not automatically erased after making an input bit query. A logtime RATM runs in time $O(\log n)$, where n is the length of the input.[3]

Logtime on a RATM is not so weak, and can compute certain simple functions, as shown by the next result. In the following, the function value $f(u) = v$ is computed by a logtime Turing machine in the sense that on input (k, u), the machine outputs the k-th bit of v in time logarithmic in the length of the input.

FACT 2.4 (Barrington, Immerman and Straubing [1990]). *Given an input of length n, a deterministic logtime* RATM *can*
(i) *compute the length of its input,*
(ii) *add and subtract integers of* $O(\log n)$ *bits,*
(iii) *decode a simple pairing function on strings of length* $O(n)$.

PROOF. Since a RATM has no output tape, we adopt the convention that M computes the function $f : \Sigma^* \to \Sigma^*$ if $|f(x)|$ is bounded by a polynomial in $|x|$, and for all bits, the i-th bit of $f(x)$ is a iff M accepts (x, i, a). The proof of (i) uses binary search, and according to Buss [1987], appears to have been first noticed by M. Dowd. The proof of (ii) is clear, since addition and subtraction take time linear in the input length. In (iii), for $u, v \in \Sigma^*$ the pair (u, v) can be encoded by $\tau(|u|)11\tau(|v|)11uv$, where τ replaces each 0 [resp. 1] by 00 [resp. 01]. Decoding can then be done by using addition and random access. □

A.M. Turing [1936–37] introduced the notion of *relative computation* using an *oracle Turing machine.*

DEFINITION 2.5. Let $B \subseteq \Gamma^*$. An oracle Turing machine (OTM) with oracle B is a Turing machine M which in addition to a read-only input tape, a distinguished output tape and finitely many work tapes, has a one-way infinite *oracle query tape.* The machine M has oracle answer states q_{yes}, q_{no} as well as a special oracle query state $q_?$ in which it queries whether the current contents of the oracle query tape belongs to oracle B. The transition function δ of M is a mapping from

$$(Q - \{q_A, q_R, q_?\}) \times (\Sigma \cup \{B\}) \times (\Gamma \cup \{B\})^{k+1}$$

[3] Logarithms are with respect to base 2.

into

$$Q \times \left(\Gamma \cup \{B\}\right)^{k+1} \times \{-1,0,1\}^{k+2}.$$

A computation is defined as previously, except that if M is in state $q_?$ then the machine queries whether the word given by the current contents of the oracle query tape belongs to B. Dependent on the outcome of the oracle query, M goes into state q_{yes} or q_{no}, and simultaneously erases the query tape and places the oracle tape head at the leftmost square. This entire sequence of events takes place in one step. Finally, nondeterministic oracle Turing machines are analogously defined by adding the oracle apparatus to the NTM model.

For $A \subseteq \Sigma^*$ and $B \subseteq \Gamma^*$, write $A \leqslant_T B$ if A can be decided by an oracle Turing machine with oracle B. Similarly write $A \leqslant_T^P B$ [resp. $A \leqslant_T^{NP} B$] if A can be computed by a deterministic [resp. nondeterministic] oracle Turing machine with oracle B in polynomial time. Let $\Sigma_0^P = P$ and Σ_{n+1}^P be

$$\left\{A \colon \left(\exists B \in \Sigma_n^P\right)\left(A \leqslant_T^{NP} B\right)\right\}.$$

A. Chandra, D. Kozen and L. Stockmeyer [1981] introduced the *alternating Turing machine* (ATM), a model suitable for formalizing divide and conquer algorithms.[4] When used with random access, this model allows sublinear runtimes and can be viewed as a kind of parallel computational device; in particular, uniform boolean circuit families, another parallel computation model, can be related to ATM's.

DEFINITION 2.6. An alternating multitape Turing machine (ATM) M is specified by $(Q, \Sigma, \Gamma, \Delta, q_0, k, \ell)$ where $\ell : (Q - \{q_A, q_R\}) \to \{\wedge, \vee\}$ and $Q, \Sigma, \Gamma, \Delta, q_0, k$ are as in Definition 2.2 of a nondeterministic machine.

The function ℓ labels non-halting states as *universal* ($\wedge$) and *existential* ($\vee$). An *accepting computation tree* T is a subtree of the computation tree of M on x such that for any configuration $\alpha \in T$,

- the root of T is the initial configuration of M on x,
- if α is a leaf of T, then α is an *accepting* configuration,
- if α is universal, then for all β, $\alpha \vdash_M \beta \Rightarrow \beta \in T$, and
- if α is existential, then there exists $\beta \in T$ for which $\alpha \vdash_M \beta$.

The ATM M *accepts* input x if there is a non-empty accepting computation tree of M on x; otherwise x is rejected. $L(M)$ denotes the set of $x \in \Sigma^*$ accepted by M. The language $L(M)$ is accepted by M in *time* $T(n)$ [resp. *space* $S(n)$] if for each $w \in L(M)$ of length n, there is an accepting computation tree T of depth at most $T(n)$ [resp. in which at most $S(n)$ cells are used for each of the work tapes and index

[4] Divide and conquer algorithms are generally space efficient. The *parallel computation thesis* states that sequential space equals parallel time (see [Borodin, 1973]). In this sense, ATM's provide a parallel computation model.

tapes at any node in the tree T]. The number of *alternations* M makes in an accepting computation tree T is defined to be the maximum number of alternations between existential and universal nodes in a path from the root to a leaf.

CONVENTION 2.7. From now on, unless otherwise indicated, for any sublinear runtime $T(n) = o(n)$, the intended Turing machine model is RATM, while for runtimes $T(n) = \Omega(n)$, the intended Turing machine model is the conventional TM. This convention applies to deterministic, nondeterministic, and alternating Turing machines. While it is a simple exercise to show that PTIME is the same class, regardless of model, it appears to be an open problem to determine the relationship between DTIME($T(n)$) on TM and RATM, for $T(n) = \Omega(n)$.

DEFINITION 2.8.

$$\begin{aligned}
\mathrm{ATIME}(T(n)) &= \{L \subseteq \Sigma^*\colon L \text{ accepted by an ATM in time } \mathrm{O}(T(n))\},\\
\mathrm{ASPACE}(S(n)) &= \{L \subseteq \Sigma^*\colon L \text{ accepted by an ATM in space } \mathrm{O}(S(n))\},\\
\mathrm{ALOGTIME} &= \mathrm{ATIME}(\mathrm{O}(\log n)),\\
\mathrm{APOLYLOGTIME} &= \bigcup_{k \geqslant 1} \mathrm{ATIME}(\mathrm{O}(\log^k n)),\\
\mathrm{ALINTIME} &= \mathrm{ATIME}(\mathrm{O}(n)).
\end{aligned}$$

The *logtime hierarchy* LH [resp. the *linear time hierarchy* LTH, resp. the *polynomial time hierarchy* PH] is the collection of languages $L \subseteq \Sigma^*$, for which L is accepted by an ATM in time $\mathrm{O}(\log n)$ [resp. $\mathrm{O}(n)$, resp. $n^{\mathrm{O}(1)}$] with at most $\mathrm{O}(1)$ alternations.[5] Σ_k-TIME($T(n)$) is the collection of languages accepted by an ATM in time $\mathrm{O}(T(n))$ with at most k alternations, beginning with an existential state.

The class ALOGTIME is surprisingly powerful. It is not difficult to see that it contains all of the regular languages. To see this, recall that a finite state automaton M is a 5-tuple $(Q, q_0, \Sigma, \delta, F)$, where Q is a finite set of states, q_0 is the initial state, Σ a finite alphabet, $\delta : Q \times \Sigma \to Q$ and $F \subseteq Q$. For the empty word λ, let $\delta^*(\lambda) = q_0$, and for $w_1 \cdots w_n \in \Sigma^*$, let $\delta^*(w_1 \cdots w_n) = \delta(\delta^*(w_1 \cdots w_{n-1}), w_n)$. A word $w \in \Sigma^*$ is *accepted* by the finite state automaton M if $\delta^*(w) \in F$. The language L is *accepted* by M, denoted by $L = L(M)$, if it consists of the words accepted by M. Finally, a language L is *regular* if it is accepted by a finite state automaton.

FACT 2.9. *If $L \subseteq \Sigma^*$ is regular, then $L \in$* ALOGTIME.

PROOF. Suppose that L is recognized by a finite state automaton M given by $(Q, q_0, \Sigma, \delta, F)$. For each word w of Σ^*, associate the mapping $f_w : Q \to Q$ ob-

[5] It follows from [Furst, Saxe and Sipser, 1984, 2] that LH is a hierarchy, where the collection of languages accepted by k-alternations is properly contained in the collection of languages accepted by $k+1$-alternations. The question of whether LTH or PH is a proper hierarchy is still open.

tained by repeatedly applying the transition function δ on the letters from w. Formally, if $w = w_1 \cdots w_n$ then

$$f_w(q) = \delta\big(\cdots\delta\big(\delta(q, w_1), w_2\big), \ldots, w_n\big)\cdots\big).$$

When M is the minimal finite state automaton recognizing the regular language L, then the (finite) collection $\{f_w \colon w \in \Sigma^*\}$, constructed as above from M, is called the *syntactic monoid* of L.

Now, given the word $w = w_1 \cdots w_n \in \Sigma^*$, associate $f_{w_n}, \ldots, f_{w_1}$ with the leaves of a binary tree T, and at each internal node of T, compute the composition of two children nodes (the tree's root is at the top). The root of T then contains $f_w = f_{w_n} \circ \cdots \circ f_{w_1}$. It follows that $w \in L$ if and only if $f_w(q_0) \in F$. This construction can be formalized to yield an ALOGTIME algorithm. □

Much more striking are the results of D. Barrington[6] and S. Buss. First, define a language L to be *complete* for ALOGTIME under DLOGTIME reductions, if $L \in$ ALOGTIME and for any $L' \in$ ALOGTIME, there is a logtime many-one function f with the property that $|f(u)|$ is polynomial in $|u|$, and $u \in L'$ iff $f(u) \in L$. Here, the function value $f(u) = v$ is computed by a logtime Turing machine in the sense that on input (k, u), the machine outputs the k-th bit of v in time logarithmic in the length of the input.

THEOREM 2.10 (D. Barrington [1989]). *Let G be any finite non-solvable permutation group (for example S_5). Then the word problem*

$$\big\{(\sigma_1, \ldots, \sigma_n) : \sigma_i \in G,\ \sigma_1 \circ \sigma_2 \circ \cdots \circ \sigma_n = id\big\}$$

for G is complete for ALOGTIME *under* DLOGTIME *reductions.*

SKETCH OF PROOF. If G is a group, then the *commutator* of elements $a, b \in G$ is the element $aba^{-1}b^{-1}$. The *commutator subgroup* of G, denoted by $[G, G]$, is the subgroup of G generated by all the commutators of G. For any group G, define $G^{(0)} = G$, and $G^{(n+1)} = [G^n, G^n]$. By definition, a group G is solvable if there is a finite series $G = G^{(0)} \geqslant G^{(1)} \geqslant \cdots \geqslant G^{(n)} = \{e\}$. If G is finite, then G is non-solvable if and only if $G = G^{(0)} \geqslant G^{(1)} \geqslant \cdots \geqslant G^{(n)} = G^{(n+1)} \neq \{e\}$, i.e. $G^{(n)}$ is non-trivial and equal to its commutator subgroup. For example, the groups A_k, S_k for $k \geqslant 5$ are non-solvable.

Assume now that G is a non-solvable group with series $G = G^{(0)} \geqslant \cdots \geqslant G^{(n)} = H$, and that H is non-trivial and equal to its commutator subgroup $[H, H]$; i.e. there exists m such that every element of H can be expressed as a product $\prod_{i=1}^{m} a_i b_i a_i^{-1} b_i^{-1}$ of m commutators of H. Using this observation, Barrington showed how to represent conjunctions and disjunctions as a word problem over H.

[6] D. Barrington changed his name to D. Mix Barrington, so that some articles appear under the former name and some under the latter name.

Namely, given a non-identity element $g \in H$ and an alternating AND/OR computation tree $T(x)$ for the computation of M on x, describe a word $w_{T(x)}$ in the elements of H such that

$$w_{T(x)} = \begin{cases} e & \text{if } M \text{ accepts } x, \\ g & \text{else.} \end{cases}$$

This is done by induction on depth of node A in $T(x)$. Recall that $H = [H, H]$ and every element of H can be written as the product of m commutators of H. Then Barrington observed that if $A = (B \vee C)$ then

$$w_A(g) = w_{B \vee C}(g) = \prod_{i=1}^{m} w_B(b_i) w_C(c_i) w_B\big(b_i^{-1}\big) w_C\big(c_i^{-1}\big).$$

Similarly, $B \wedge C$ and $\neg B$ can be expressed. Inductively one forms the word $w_{T(x)}$ whose product equals e exactly when M accepts x.

From the above discussion, with a close look at uniformity issues, it follows that the word problem for a finite non-solvable permutation group is hard for ALOGTIME. On the other hand, the word problem is clearly in ALOGTIME, since one can compose n permutations by associating them with the leaves of a binary tree, whose internal nodes compute the composition of their children. □

THEOREM 2.11 (S. Buss [1987, 1993]). *The boolean formula valuation problem*

$$\{\Theta\colon\ \Theta \text{ is a true variable-free propositional logic formula}\}$$

is complete for ALOGTIME *under* DLOGTIME *reductions.*

The proof of Theorem 2.11 is long and difficult, so cannot be sketched here. The results of Barrington and Buss are complementary in the sense that the word problem for S_5 is clearly in ALOGTIME, but not obviously complete, while the boolean formula evaluation problem is clearly complete but not obviously in ALOGTIME.

W. Savitch [1970] proved that $\text{NSPACE}(S(n)) \subseteq \text{DSPACE}(S^2(n))$, for any space constructible $S(n) \geqslant \log n$. The following theorem, due to Chandra, Kozen and Stockmeyer [1981], is in part a generalization of Savitch's result that PSPACE = NPSPACE, and relates alternating time and space to deterministic time and space.

THEOREM 2.12 (Chandra, Kozen, Stockmeyer [1981]). *For $f(n) \geqslant n$,*

$$\begin{aligned} \text{ATIME}\big(f(n)\big) &\subseteq \text{DSPACE}\big(f(n)\big) \subseteq \text{NSPACE}\big(f(n)\big) \\ &\subseteq \bigcup_{c>0} \text{ATIME}\big(c \cdot f(n)^2\big). \end{aligned}$$

For $f(n) \geqslant \log n$,

$$\text{ASPACE}\big(f(n)\big) \subseteq \bigcup_{c>0} \text{DTIME}\big(c^{f(n)}\big).$$

From definitions, it is clear that

$$\text{LH} \subseteq \text{ALOGTIME} \subseteq \text{LOGSPACE} \subseteq \text{PTIME} \subseteq \text{PH} \subseteq \text{PSPACE}$$

and

$$\text{LH} \subseteq \text{LTH} \subseteq \text{ALINTIME} \subseteq \text{DLINSPACE} \subseteq \text{PSPACE}.$$

By Furst, Saxe, Sipser [1984] and Ajtai [1983], integer multiplication does not belong to LH (since multiplication $\times$ is a function, what is meant is that $\times \notin$ FLH, where the latter is the class of functions of polynomial growth rate, whose bitgraph belongs to LH; this is defined later). Note that Buss [1992] has even shown that the graph of multiplication does not belong to LH. Since the graph of integer multiplication belongs to ALOGTIME, the first containment above is proper. With this exception, nothing else is known about whether the other containments are proper.

All the previous machine models concern language recognition problems. Predicates $R \subseteq (\Sigma^*)^k$ can be recognized by allowing input of the form

$$Bx_1Bx_2B \cdots Bx_nB$$

consisting of n inputs $x_i \in \Sigma^*$, each separated by the blank $B \notin \Sigma$. By adding a write-only output tape with a tape head capable only of writing and moving to the right, and by allowing input of the form $Bx_1Bx_2B \cdots Bx_nB$, a TM or RATM can compute an n-place function. In the literature, function classes such as the polynomial time computable functions were so introduced. To provide uniform notation for such function classes, along with newer classes of sublinear time computable functions, we proceed differently.

DEFINITION 2.13. A function $f(x_1, \ldots, x_n)$ has *polynomial growth* resp. *linear growth* resp. *logarithmic growth* if

$$\big|f(x_1, \ldots, x_n)\big| = O\Big(\max_{1 \leqslant i \leqslant n} |x_i|^k\Big), \quad \text{for some } k$$

resp.

$$\big|f(x_1, \ldots, x_n)\big| = O\Big(\max_{1 \leqslant i \leqslant n} |x_i|\Big)$$

resp.

$$\left|f(x_1,\ldots,x_n)\right| = O\Big(\log\Big(\max_{1\leqslant i\leqslant n} |x_i|\Big)\Big).$$

The *graph* G_f satisfies $G_f(\vec{x}, y)$ iff $f(\vec{x}) = y$. The *bitgraph* B_f satisfies $B_f(\vec{x}, i)$ iff the i-th bit of $f(\vec{x})$ is 1. If $\mathcal{C}$ is a complexity class, then $\mathcal{FC}$ [resp. *Lin*$\mathcal{FC}$ resp. *Log*$\mathcal{FC}$] is the class of functions of polynomial [resp. linear resp. logarithmic] growth whose bitgraph belongs to $\mathcal{C}$. In this paper, $\mathcal{GC}$ will abbreviate *Lin*$\mathcal{FC}$. The iteration $f^{(n)}(x)$ is defined by induction on n: $f^{(0)}(x) = x$, $f^{(n+1)}(x) = f(f^{(n)}(x))$. With this notation, the iteration $\log^{(n)} x$ should not be confused with the power $\log^n x = (\log x)^n$.

There are other extensions of the Turing machine model not covered in this survey, such as the *probabilistic* Turing machine (yielding classes such as R and BPP, see [Van Emde Boas, 1990]), the *genetic* Turing machine (defined by P. Pudlák [1994], who showed that polynomial time bounded genetic TM's compute exactly PSPACE), and the *quantum* Turing machine (first introduced by D. Deutsch [1985], and for which P. Shor [1994] proved that integer factorization is computable in bounded error probabilistic quantum polynomial time BQP).

2.2. *Parallel machine model*

> "Having one processor per data element changes the way one thinks."
> W.D. Hillis and G.L. Steele, Jr. [1986]

Emerging around 1976–77 from the work of Goldschlager [1977, 1982], Fortune and Wyllie [1978], and Shiloach and Vishkin [1982], the *parallel random access machine* (PRAM) provides an abstract model of parallel computation for algorithm development. While existent "massively parallel" computers generally require a specific communication network (e.g., hypercube, mesh, etc.) for message passing between processors (and such details are of immense practical importance), the PRAM abstracts out all such processor communication details and postulates a global shared memory. Individual processors of a PRAM additionally have local memory, and while operating synchronously on the same program, are capable of performing arithmetic and logical operations as well as local and global read/write in both direct and indirect addressing mode. Processors may have different data stored in their local memories and have access to their unique processor identity number PID. Thus the effect of an instruction like "add the contents of the PID-th global memory register to local memory register 2 and store in local memory register 7" may be quite different in different processors. Different models of PRAM have been studied, depending on the strength of local arithmetic operations allowed, and whether simultaneous read/write in the same global memory register is allowed by several processors. This yields EREW, CREW, and CRCW models, according to whether exclusive read, exclusive write, concurrent read or concurrent write are allowed. An excellent survey of parallel algorithms and models is R.M. Karp and V. Ramachandran [1990]. The formal development follows.

A *concurrent random access machine* CRAM has a sequence $R_0, R_1, \ldots$ of random access machines which operate in a synchronous fashion in parallel. Each R_i has its own local memory, an infinite collection of registers, each of which can hold an arbitrary non-negative integer. Global memory consists of an infinite collection of registers accessible to all processors, which are used for reading the input, processor message passing, and output. Global registers are designated $M_0^g, M_1^g, M_2^g, \ldots$, and local registers by $M_0, M_1, M_2, \ldots$ – local registers of processor P_i might be denoted $M_{i,0}, M_{i,1}, \ldots$. A global memory register can be read simultaneously by several processors (*concurrent read*, rather than *exclusive read*). In the case where more than one processor may attempt to write to the same global memory register, the lowest numbered processor succeeds (*priority resolution* of write conflict in this *concurrent write* rather than *exclusive write* model). An input x is initially given bitwise in the global registers, the register M_i^g holding the i-th bit of x. All other registers initially contain the blank symbol B (different from 0, 1) which designates that the register is empty. Similarly at termination, the output y is specified in the global memory, the register M_i^g holding the i-th bit of y. At termination of a computation all other global registers contain the blank symbol. [The input/output convention of one integer per global memory register yields an equivalent model for the complexity classes here considered.] Let *res* (result), *op*0, *op*1, *op*2 (operands 0,1,2) be non-negative integers. If any register occurring on the right side of an instruction contains 'B', then the register on the left side of the instruction will be assigned the value 'B' (undefined).

Instructions are as follows.

M_{res} = `constant`

M_{res} = `processor number`

$M_{res} = M_{op1}$

$M_{res} = M_{op1} + M_{op2}$

$M_{res} = M_{op1} \dot{-} M_{op2}$

M_{res} = `MSP`(M_{op1}, M_{op2})

M_{res} = `LSP`(M_{op1}, M_{op2})

$M_{res} = *M_{op1}$

$M_{res} = *M_{op1}^g$

$*M_{res} = M_{op1}$

$*M_{res}^g = M_{op1}$

`GOTO label`

`GOTO label IF` $M_{op1} = M_{op2}$

`GOTO label IF` $M_{op1} \leqslant M_{op2}$

`HALT`

Cutoff subtraction is defined by $x \dotdiv y = x - y$, provided that $x \geq y$, else 0. The shift operators MSP and LSP are defined by
- $\text{MSP}(x, y) = \lfloor x/2^y \rfloor$, provided that $y < |x|$, otherwise 'B',
- $\text{LSP}(x, y) = x - 2^y \cdot (\lfloor x/2^y \rfloor)$, provided that $y \leq |x|$, otherwise 'B'.

The CRAM model is due to N. Immerman [1989], though there slightly different conventions are made.

Instructions with '$*$' concern indirect addressing. The instruction $M_{res} = *M_{op1}$ assigns to local register M_{res} the contents of local register with address given by the value M_{op1}. Similarly, $M_{res} = *M^g_{op1}$ performs an indirect read from global memory into local memory. The instruction $*M_{res} = M_{op1}$ assigns the value of local register M_{op1} to the local register whose address is given by the current contents of the local register M_{res}. Similarly, $*M^g_{res} = M_{op1}$ performs an indirect write into global memory.

In summary, the CRAM has instructions for
(i) local operations – addition, cutoff subtraction, shift,
(ii) global and local indirect reading and writing,
(iii) control instructions – GOTO, conditional GOTO and HALT.

A program is a finite sequence of instructions, where each individual processor of a CRAM has the same program. Each instruction has unit-cost (*uniform time cost*). During the course of a computation, only finitely many *active* processors perform computations. An input x of length n is *accepted* by a CRAM M in time $T(n)$ with $P(n)$ many active processors, if M halts after $T(n)$ time where processors $P_0, \ldots, P_{n-1}$ synchronously execute the program. The class TIMEPROC($T(n), P(n)$) consists of those languages accepted by a CRAM in time $T(n)$ with $P(n)$ many processors.

EXAMPLE 2.14. The following is a CRAM program for computing $|x| = \lceil \log_2(x + 1) \rceil$, where comments begin by '%'.

Let $M_{res} = \text{BIT}(M_{op1}, M_{op2})$ be the instruction which, for $i = M_{op2}$ computes the coefficient of 2^i in the binary representation of the integer stored in M_{op1}, provided that $i < |M_{op1}|$, and otherwise returns the value 'B'.

```
 1  M₁ = processor number
 2  M₂ = *M₁ᵍ  % in Pᵢ, Mᵢ = Mᵢᵍ
 3  if (M₂ = B) then M₀ᵍ = M₁
 4  M₃ = M₀ᵍ  % in Pᵢ, M₃ = least i [Mᵢᵍ = B] = |x|
 5  *M₁ᵍ = B  % erase global memory
 6  M₄ = 1
 7  M₄ = M₁ + 1
 8  M₄ = M₃ ∸ M₄
 9  M₅ = MSP(M₃,M₄)
10  M₆ = MSP(M₅,1)
```

11 $M_6 = M_6 + M_6$

12 $M_4 = M_5 \dot{-} M_6$

13 $*M_1^g = M_4$ `% output placed in global memory`

14 HALT

Processor bound: $P(|x|) = |x|$.

Strictly speaking, line 3 is not syntactically allowed, but can easily be implemented with a few extra lines of code, and will not affect the time or processor bound. Lines 6–12 ensure that $M_4 = \mathrm{BIT}(M_3, M_3 \dot{-} (M_1 + 1))$, so that in processor P_i, $M_4 = \mathrm{BIT}(|x|, |x| \dot{-} (i + 1))$.

To further illustrate the CRAM model, Algorithm 2.15 computes $\max(x_1, \ldots, x_n)$ of n integers in constant time with $O(n^2)$ processors.

ALGORITHM 2.15. Constant time algorithm for maximum.

(1) for all $\binom{n}{2}$ pairs $1 \leqslant i < j \leqslant n$ in parallel do

$$a_{i,j} = \begin{cases} 1 & \text{if } x_i < x_j \\ 0 & \text{else} \end{cases}$$

(2) for $i := 1$ to n in parallel do
  $m_i := 0$

(3) for $1 \leqslant i < j \leqslant n$ in parallel do
  if $a_{i,j} = 1$ then $m_i := 1$

(4) for $i := 1$ to n in parallel do
  if $m_i = 0$ then $m := i$

(5) $\max := x_m$

Time = O(1), Processors = $O(n^2)$

2.3. *Circuit families*

A directed graph G is given by a set $V = \{1, \ldots, m\}$ of vertices (or nodes) and a set $E \subseteq V \times V$ of edges. The *in-degree* or *fan-in* [resp. *out-degree* or *fan-out*] of node x is the size of $\{i \in V\colon (i, x) \in E\}$ [resp. $\{i \in V\colon (x, i) \in E\}$]. A *circuit* C_n is a labeled, directed acyclic graph whose nodes of in-degree 0 are called *input* nodes and are labeled by one of $0, 1, x_1, \ldots, x_n$, and whose nodes v of in-degree $k > 0$ are called *gates* and are labeled by a k-place function from a *basis* set of boolean functions. A circuit has a unique *output* node of out-degree 0.[7] A family $\mathcal{C} = \{C_n\colon n \in \mathbb{N}\}$

[7] The usual convention is that a circuit may have any number of output nodes, and hence compute a function $f: \{0,1\}^n \to \{0,1\}^m$. In this paper, we adopt the convention that a circuit computes a boolean function $f: \{0,1\}^n \to \{0,1\}$. An m-output circuit C computing function $g: \{0,1\}^n \to \{0,1\}^m$ can then be simulated by a circuit computing the boolean function $f: \{0,1\}^{n+m} \to \{0,1\}$ where $f(x_1, \ldots, x_n, 0^{m-i}1^i) = 1$ iff the i-th bit of $g(x_1, \ldots, x_n)$ is 1.

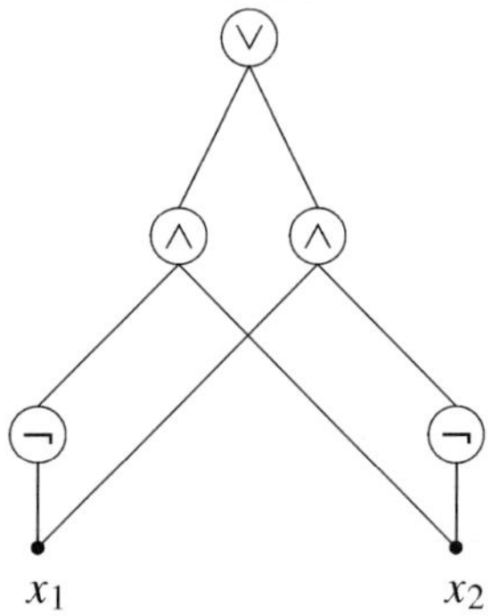

Fig. 1. Exclusive-or.

of circuits has *bounded fan-in* if there exists k, for which all gates of all C_n have in-degree at most k; otherwise C has *unbounded* or *arbitrary* fan-in.

Boolean circuits have basis $\wedge, \vee, \neg$, where $\wedge, \vee$ may have fan-in larger than 2 (as described below, the AC^k [resp. NC^k] model concerns unbounded fan-in [resp. fan-in 2] boolean circuits). A *threshold* gate $\mathrm{TH}_{k,n}$ outputs 1 if at least k of its n inputs is 1. A *modular counting* gate $\mathrm{MOD}_{k,n}$ outputs 1 if the sum of its n inputs is evenly divisible by k. A *parity* gate $\oplus$ outputs 1 if the number of input bits equal to 1 is even, where as for $\wedge, \vee$ the fan-in may be restricted to 2, or arbitrary, depending on context.

An input node v labeled by x_i computes the boolean function $f_v(x_1, \ldots, x_n) = x_i$. A node v having in-edges from $v_1, \ldots, v_m$, and labeled by the m-place function g from the basis set, computes the boolean function $f_v(x_1, \ldots, x_n) = g(f_{v_1}(x_1, \ldots, x_n), \ldots, f_{v_m}(x_1, \ldots, x_n))$. A circuit C_n *accepts* the word $x_1 \cdots x_n \in \{0, 1\}^n$ if $f_v(x_1, \ldots, x_n) = 1$, where f_v is the function computed by the unique output node v of C_n. A family $(C_n\colon n \in \mathbb{N})$ of circuits *accepts* a language $L \subseteq \{0, 1\}^*$ if for each n, $L^n = L \cap \{0, 1\}^n$ consists of the words accepted by C_n.

The *depth* of a circuit is the length of the longest path from an input to an output node, while the *size* is the number of gates. A language $L \subseteq \{0, 1\}^*$ belongs to $\mathrm{SIZEDEPTH}(S(n), D(n))$ over basis B if L consists of those words accepted by a family $(C_n\colon n \in \mathbb{N})$ of circuits over basis B, where $size(C_n) = \mathrm{O}(S(n))$ and $depth(C_n) = \mathrm{O}(D(n))$.

A boolean circuit which computes the function $f(x_1, x_2) = x_1 \oplus x_2$ is as in Figure 1.

EXAMPLE 2.16. The function $\max(a_0, \ldots, a_{n-1})$ of n integers, each of size at most m, can be computed by a boolean circuit as follows. Assume the integers a_i are distinct (a small modification is required for non-distinct integers). Then the k-th bit of $\max(a_0, \ldots, a_{n-1})$ is 1 exactly when

$$(\exists i < n)(\forall j < n)\big(j \neq i \rightarrow a_j \leqslant a_i \wedge \mathrm{BIT}(k, a_i) = 1\big).$$

This bounded quantifier formula is translated into a boolean circuit by

$$\bigvee_{i<n} \bigwedge_{j<n, j\neq i} \bigvee_{\ell<n} \bigwedge_{\ell<p<m} \big(\text{BIT}(p, a_j) = \text{BIT}(p, a_i)\big) \wedge \text{BIT}(\ell, a_j) = 0 \wedge \text{BIT}(\ell, a_i) = 1.$$

Note that by Algorithm 2.15, $\max(a_0, \ldots, a_{n-1})$ is computed by a CRAM in constant time with a polynomial number of processors and by Example 2.16 $\max(a_0, \ldots, a_{n-1})$ is computed by a constant depth polynomial size family of boolean circuits. As this suggests, there is a relation between time/processors for a CRAM and depth/size for a boolean circuit family. The exact relation between the two models is given in Theorem 2.18.

Without a uniformity condition, circuit families of depth 2 and size 1 can accept non-recursive languages (e.g., all inputs are accepted [resp. rejected] if the n-th circuit is of the form $x_1 \vee \neg x_1$ [resp. $x_1 \wedge \neg x_1$]). Various notions of uniformity have been suggested (PTIME-uniformity [Beame, Cook and Hoover, 1986], LOGSPACE-uniformity [Borodin, 1973], U_{E^*}-uniformity [Ruzzo, 1981], etc.), but the most robust (and strictest) appears to be that of LOGTIME-uniformity [Buss, 1986a; Mix Barrington, Immerman and Straubing, 1990], which is adopted in this paper.

DEFINITION 2.17 (*W. Ruzzo* [1981], *also Mix Barrington, Immerman and Straubing* [1990]). The *direct connection language* (DCL) of a circuit family $(C_n\colon n \in \mathbb{N})$ is the set of $(a, b, \ell, 0^n)$, where a is the parent of b in the circuit C_n, and the label of gate a is ℓ. A circuit family is LOGTIME-uniform if its associated DCL belongs to DLOGTIME. For $k \geqslant 0$, AC^k [resp. NC^k] is the class of languages accepted by LOGTIME-uniform $\text{SIZEDEPTH}(n^{O(1)}, O(\log^k n))$ over the boolean basis, where $\wedge$, $\vee$ have arbitrary fan-in [resp. fan-in 2], and $\text{NC} = \bigcup_k \text{AC}^k = \bigcup_k \text{NC}^k$. $\text{ACC}(k)$ is the class of languages accepted by LOGTIME-uniform $\text{SIZEDEPTH}(n^{O(1)}, O(1))$ over the basis $\wedge$, $\vee$, $\neg$, $\text{MOD}_{k,n}$, where $\wedge$, $\vee$ have unbounded fan-in, and $\text{ACC} = \bigcup_{k\geqslant 2} \text{ACC}(k)$. TC^0 is the class of languages in LOGTIME-uniform $\text{SIZEDEPTH}(n^{O(1)}, O(1))$ over the basis $\text{TH}_{k,n}$.[8]

In L. Stockmeyer and U. Vishkin [1984] related PRAM time and processors to boolean circuit depth and size. The LOGTIME-uniform version of that result was proved by N. Immerman [1989] and follows.

THEOREM 2.18. *For $k \geqslant 0$, AC^k equals $\text{TIMEPROC}(O(\log^k n), n^{O(1)})$ on a* CRAM.

The following containments are known:

$$\text{NC}^k \subseteq \text{AC}^k \subseteq \text{NC}^{k+1}$$

[8] In this paper, circuit classes such as AC^k, NC^k, NC and TC^0 sometimes denote both language classes, though more often function classes, where the intended meaning is clear from context. That is, we write NC in place of $\mathcal{F}$NC, etc. *NC* is an acronym for "Nick's Class", as this class was first studied by N. Pippenger. AC^k was studied by W.L. Ruzzo, using the alternating Turing machine model.

and

$$\mathrm{NC}^1 = \mathrm{ALOGTIME} \subseteq \mathrm{LOGSPACE} \subseteq \mathrm{NLOGSPACE} \subseteq \mathrm{AC}^1.$$

None of the inclusions are known to be strict or not. For more information on circuits, see the excellent survey article by R. Boppana and M. Sipser [1990].

From this point on, we will assume that all language and function complexity classes are over the alphabet {0, 1}.

3. Some recursion schemes

Kleene's [1936a] normal form theorem states that for each recursive (partial) function f there is an index e for which $f(\vec{x}) = U(\mu y[T(e, \vec{x}, y) = 0])$, where T, U are primitive recursive. The proof relies on arithmetizing computations via Gödel numbers, a technique introduced by Gödel [1931], and with which Turing computable functions can be shown equivalent to μ-recursive functions. Since then, there have been a number of *arithmetizations* of machine models [Kleene, 1936a, 1936b; Bennett, 1962; Cobham, 1965; Ritchie, 1963; Thompson, 1972; Wagner, 1979; Clote, 1990; Clote and Takeuti, 1995], etc. Key to all of these results is the availability in a function algebra $\mathcal{F}$ of a conditional function, a pairing function, and some string manipulating functions, in order to show that the function $\mathrm{NEXT}_M(x, c) = d$ belongs to $\mathcal{F}$. Here, c, d encode configurations of machine M on input x and d is the configuration obtained in one step from configuration c.

DEFINITION 3.1. An *operator* (here also called *operation*) is a mapping from functions to functions. If $\mathcal{X}$ is a set of functions and OP is a collection of operators, then $[\mathcal{X}; \mathrm{OP}]$ denotes the smallest set of functions containing $\mathcal{X}$ and closed under the operations of OP. The set $[\mathcal{X}; \mathrm{OP}]$ is called a *function algebra*. In a straightforward inductive manner, define *representations* or *names* for functions in $[\mathcal{X}; \mathrm{OP}]$. The *characteristic function* $c_P(\vec{x})$ of a predicate P satisfies

$$c_p(\vec{x}) = \begin{cases} 1 & \text{if } P(\vec{x}), \\ 0 & \text{else,} \end{cases} \tag{1}$$

where P is often written in place of c_P. If $\mathcal{F}$ is a class of functions, then $\mathcal{F}_*$ is the class of predicates whose characteristic function belongs to $\mathcal{F}$.[9]

DEFINITION 3.2. Let $\mathcal{F} = [f_1, f_2, \ldots; O_1, O_2, \ldots]$ be a function algebra. Let O denote operator O_{i_0}, and fix a representation R of $f \in \mathcal{F}$. The rank $rk_{O,R}(f)$ of applications of O in the representation R of $f \in \mathcal{F}$ is defined by induction. If f is an initial function $f_1, f_2, \ldots$ then $rk_{O,R}(f) = 0$. Suppose that f is defined

[9] Grzegorczyk [1953] defined $\mathcal{F}_*$ as the collection of predicates P for which there is a function $f \in \mathcal{F}$ satisfying $P(\vec{x}) \iff f(\vec{x}) = 0$. For the function classes here considered, these are equivalent definitions.

by application of operator O_i to functions $g_1, \dots, g_m$ where $rk_{O,R}(g_j) = r_j$ for $1 \leqslant j \leqslant m$. If $i = i_0$ then $rk_{O,R}(f) = 1 + \max\{r_1, \dots, r_m\}$; otherwise $rk_{O,R}(f) = \max\{r_1, \dots, r_m\}$. The *O-rank* $rk_O(f)$ of a function $f \in \mathcal{F}$ is the minimum of $rk_{O,R}(f)$ over all representations R of f in $\mathcal{F}$.

Operations which have been studied in the literature include composition, primitive recursion, minimization, and their variants including bounded composition, bounded recursion, bounded recursion on notation, bounded minimization, simultaneous recursion, multiple recursion, course-of-values recursion, divide and conquer recursion, safe and tiered recursion, etc.[10] Good surveys of function algebras include the monographs by H. Rose [1984] and K. Wagner and G. Wechsung [1986] (Chapters 2, 10).

Since newer results concerning smaller complexity classes yield older results concerning larger classes as corollaries, we begin with a function algebra introduced by the author for the class $\mathcal{F}$LH of functions in the logtime hierarchy. By [Mix Barrington, Immerman and Straubing, 1990], this class is equal to the class AC^0 of functions computable on a concurrent random access machine in constant parallel time with a polynomial number of processors.

3.1. *An algebra for the logtime hierarchy* LH

DEFINITION 3.3. The *successor* function satisfies $s(x) = x + 1$; the *binary successor* functions s_0, s_1 satisfy $s_0(x) = 2 \cdot x$, $s_1(x) = 2 \cdot x + 1$; the n-place projection functions $I_k^n(x_1, \dots, x_n) = x_k$; I denotes the collection of all projection functions.

DEFINITION 3.4. The function f is defined by *composition* (COMP) from the functions $h, g_1, \dots, g_m$ if

$$f(x_1, \dots, x_n) = h\big(g_1(x_1, \dots, x_n), \dots, g_m(x_1, \dots, x_n)\big).$$

The function f is defined by *primitive recursion* (PR) from functions g, h if

$$\begin{aligned} f(0, \vec{y}) &= g(\vec{y}), \\ f(x+1, \vec{y}) &= h\big(x, \vec{y}, f(x, \vec{y})\big). \end{aligned}$$

The collection $\mathcal{PR}$ of primitive recursive functions is $[0, I, s; \text{COMP}, \text{PR}]$. The function f is defined by *iteration* (ITER) from functions g if

$$\begin{aligned} f(0) &= 0, \\ f(x+1) &= g\big(f(x)\big). \end{aligned}$$

[10] In this paper, for uniformity of notation, a number of operations are introduced as *bounded* instead of *limited* operations. For example, Grzegorczyk's schemes of *limited recursion* and *limited minimization* are here called *bounded recursion* and *bounded minimization*.

THEOREM 3.5 (R.M. Robinson [1947]). *Define the operation* ADD *by*

$$\text{ADD}(f, g)(x) = f(x) + g(x),$$

and let $q(x) = x - \lfloor\sqrt{x}\rfloor^2$. *Let* $\mathcal{PR}_1$ *denote the collection of one-place primitive recursive functions. Then* $\mathcal{PR}_1$ *equals* $[0, s, q;$ COMP, ITER, ADD$]$.

G. Asser [1987] presented a version of the previous theorem for primitive recursive word functions of one variable.

Primitive recursion defines $f(x+1)$ in terms of $f(x)$, so that the computation of $f(x)$ requires approximately $2^{|x|}$ many steps, an exponential number in the length of x. To define smaller complexity classes of functions, Bennett [1962] introduced the scheme of *recursion on notation*, which Cobham [1965] later used to characterize the polynomial time computable functions.

DEFINITION 3.6. Assume that $h_0(x, \vec{y}), h_1(x, \vec{y}) \leqslant 1$. The function f is defined by *concatenation recursion on notation* (CRN) from g, h_0, h_1 if

$$\begin{aligned} &f(0, \vec{y}) = g(\vec{y}),\\ &f\big(s_0(x), \vec{y}\big) = s_{h_0(x,\vec{y})}\big(f(x, \vec{y})\big), \quad \text{if } x \neq 0,\\ &f\big(s_1(x), \vec{y}\big) = s_{h_1(x,\vec{y})}\big(f(x, \vec{y})\big). \end{aligned}$$

This scheme can be written in the abbreviated form

$$\begin{aligned} f(0, \vec{y}) &= g(\vec{y}),\\ f\big(s_i(x), \vec{y}\big) &= s_{h_i(x,\vec{y})}\big(f(x, \vec{y})\big). \end{aligned}$$

The scheme CRN was introduced by Clote [1990], though motivated by a similar scheme due to J. Lind [1974]. If concatenation of the empty string is allowed, or if the condition $h_i(x, \vec{y}) \leqslant 1$ is dropped (i.e. if concatenation of $f(x, \vec{y})$ with an arbitrary string $h_i(x, \vec{y})$ is allowed as in Lind's scheme), then the resulting scheme is provably stronger (i.e. parity $\bigoplus_{i=1}^{n} x_i$ is easily defined using Lind's version, although parity is known not to belong to LH).

DEFINITION 3.7. The length of x in binary satisfies $|x| = \lceil\log(x+1)\rceil$; $||x||$ is defined as $|(|x|)|$, etc.; $\text{MOD2}(x) = x - 2 \cdot \lfloor\frac{x}{2}\rfloor$; the function $\text{BIT}(i, x) = \text{MOD2}(\lfloor\frac{x}{2^i}\rfloor)$ yields the coefficient of 2^i in the binary representation of x; the *smash* function satisfies $x\#y = 2^{|x|\cdot|y|}$. The algebra A_0 is defined to be

$$[0, I, s_0, s_1, \text{BIT}, |x|, \#; \text{COMP}, \text{CRN}].$$

Arbitrary constants belong to A_0. For instance the integer 6 has binary representation 110 and is represented by $s_0(s_1(s_1(0)))$. The auxiliary reverse function $rev0(x, y)$ gives the $|y|$ many least significant bits of x written in reverse. Let

$$\begin{aligned} rev0(x,0) &= 0, \\ rev0(x, s_i(y)) &= s_{\text{BIT}(|y|,x)}(rev0(x,y)). \end{aligned} \tag{2}$$

The reverse of the binary notation for x is given by $rev(x) = rev0(x, x)$. For instance the integer 10 has binary notation 1010 whose reverse is 101, corresponding to the integer 5, so $rev(10) = 5$. The following computation may be helpful, where $\overline{w}$ temporarily denotes the integer having binary representation w.

$$\begin{aligned} rev(10) &= rev(\overline{1010}) \\ &= rev0(\overline{1010}, \overline{1010}) \\ &= s_{\text{BIT}(|\overline{101}|,\overline{1010})}(rev0(\overline{1010}, \overline{101})) \\ &= s_1(s_{\text{BIT}(|\overline{10}|,\overline{1010})}(rev0(\overline{1010}, \overline{10}))) \\ &= s_1 s_0(s_{\text{BIT}(|\overline{1}|,\overline{1010})}(rev0(\overline{1010}, \overline{1}))) \\ &= s_1 s_0 s_1(s_{\text{BIT}(\overline{0},\overline{1010})}(rev0(\overline{1010}, \overline{0}))) \\ &= s_1 s_0 s_1 s_0(0) \\ &= 5. \end{aligned}$$

Let

$$\begin{aligned} ones(0) &= 0, \\ ones(s_i(x)) &= s_1(ones(x)) \end{aligned} \tag{3}$$

so that $ones(x) = 2^{|x|} - 1$ whose binary representation consists of $|x|$ many 1's. Let

$$\begin{aligned} pad(x,0) &= x, \\ pad(x, s_i(y)) &= s_0(pad(x,y)) \end{aligned} \tag{4}$$

so that $pad(x, y) = 2^{|y|} \cdot x$ whose binary representation is that of x with $|y|$ many 0's appended to the right. Kleene's signum functions sg, $\overline{sg}$, which satisfy $sg(x) = \min(x, 1)$ and $\overline{sg}(x) = 1 - sg(x)$, are defined by

$$\begin{aligned} sg(x) &= \text{BIT}(0, ones(x)), \\ \overline{sg}(x) &= \text{BIT}(0, pad(1, x)). \end{aligned} \tag{5}$$

The conditional function, easily defined using stronger recursion schemes,

$$cond(x, y, z) = \begin{cases} y & \text{if } x = 0, \\ z & \text{else} \end{cases} \tag{6}$$

is here defined using the auxiliary functions $cond0$, $cond1$, $cond2$. Define

$$\begin{aligned} &cond0(0, y) = 0, \\ &cond0(s_i(x), y) = s_{\mathrm{BIT}(0,y)}(cond0(x, y)), \end{aligned} \tag{7}$$

$$cond1(x, y) = sg(cond0(x, y)), \tag{8}$$

$$\begin{aligned} &cond2(x, 0) = 0, \\ &cond2(x, s_i(y)) = s_{cond1(x, s_i(y))}(cond2(x, y)) \end{aligned} \tag{9}$$

so that

$$cond0(x, y) = \begin{cases} 0 & \text{if } \mathrm{BIT}(0, y) = 0, \\ 2^{|x|} - 1 & \text{else,} \end{cases} \tag{10}$$

$$cond1(x, y) = \begin{cases} 0 & \text{if } x = 0, \\ \mathrm{BIT}(0, y) & \text{else,} \end{cases} \tag{11}$$

$$cond2(x, y) = \begin{cases} 0 & \text{if } x = 0, \\ y & \text{else.} \end{cases} \tag{12}$$

The concatenation function $x * y = 2^{|y|} \cdot x + y$ is defined by

$$\begin{aligned} &x * 0 = x, \\ &x * s_i(y) = s_i(x * y). \end{aligned} \tag{13}$$

Then the conditional function *cond* is defined by

$$cond(x, y, z) = cond2(\overline{sg}(x), y) * cond2(x, z).$$

With *cond* one can form (characteristic functions of) predicates by applying boolean operations AND, OR, NOT to other predicates. Additionally, using *cond*, one can introduce functions using *definition by cases*

$$f(\vec{x}) = \begin{cases} g_1(\vec{x}) & \text{if } P_1(\vec{x}), \\ g_2(\vec{x}) & \text{if } P_2(\vec{x}), \\ & \vdots \\ g_n(\vec{x}) & \text{if } P_n(\vec{x}) \end{cases} \tag{14}$$

where predicates $P_1, \ldots, P_n$ are disjoint and exhaustive. A *sharply bounded quantifier* is of the form $(\exists x \leqslant |y|)$ or $(\forall x \leqslant |y|)$.

LEMMA 3.8. *$(A_0)_*$ is closed under sharply bounded quantifiers.*

PROOF. Suppose that the predicate $R(x, \vec{z})$ belongs to A_0 and that $P(y, \vec{z})$ is defined by $(\exists x \leqslant |y|)R(x, \vec{z})$. Define

$$f(0, \vec{z}) = 0,$$
$$f(s_i(x), \vec{z}) = s_{R(|x|, \vec{z})}(f(x, \vec{z})).$$

Then $P(y, \vec{z}) = sg(f(s_1(y), \vec{z}))$ belongs to A_0. Bounded universal quantification can be derived from bounded existential quantification using $\overline{sg}$. □

DEFINITION 3.9. The function f is defined by *sharply bounded minimization* (SBMIN) from the function g, denoted by $f(x, \vec{y}) = \mu i \leqslant |x|[g(i, \vec{y}) = 0]$, if

$$f(x, \vec{y}) = \begin{cases} \min\{i \leqslant |x|\colon\ g(i, \vec{y}) = 0\} & \text{if such exists,} \\ 0 & \text{else.} \end{cases}$$

Sharply bounded maximization (SBMAX) is analogously defined.

From Lemma 3.8, it follows that A_0 is closed under SBMIN and SBMAX. Namely, define

$$k(0, \vec{y}) = 0,$$
$$k(s_i(z), \vec{y}) = s_{h(z, \vec{y})}(k(z, \vec{y}))$$

where

$$h(z, \vec{y}) = \begin{cases} 0 & \text{if } (\exists x \leqslant |z|)[g(x, \vec{y}) = 0], \\ 1 & \text{else.} \end{cases}$$

Then $f(x, \vec{y}) = \mu i \leqslant |x|[g(i, \vec{y}) = 0]$ is defined by

$$f(x, \vec{y}) = \begin{cases} 0 & \text{if } g(0, \vec{y}) = 0 \text{ or } \neg(\exists i \leqslant |x|)[g(i, \vec{y}) = 0], \\ |rev(k(s_1(x), \vec{y}))| & \text{else.} \end{cases}$$

The integer x is a beginning of y, denoted xBy, if the binary representation of x is an initial segment (from left to right) of the binary representation of y; formally xBy iff $x = 0$ or $x, y > 0$ and

$$(\forall i \leqslant |x|)[\text{BIT}(i, rev(s_1(x))) = \text{BIT}(i, rev(s_1(y)))].$$

Thus the predicate $B \in A_0$. Similarly, predicates xPy (x is part of y, i.e. a convex subword of y) and xEy (x is an end of y) can be shown to belong to A_0.

To show the closure of A_0 under part-of quantifiers $(\exists x\, By)$, $(\exists x\, Py)$, $(\exists x\, Ey)$, etc. define the *most significant part* function MSP by

$$\begin{aligned} &\mathrm{MSP}(0, y) = 0, \\ &\mathrm{MSP}(s_i(x), y) = s_{\mathrm{BIT}(y, s_i(x))}(\mathrm{MSP}(x, y)) \end{aligned} \tag{15}$$

and the *least significant part* function LSP by

$$\mathrm{LSP}(x, y) \;=\; \mathrm{MSP}(rev(\mathrm{MSP}(rev(s_1(x)), |\mathrm{MSP}(x, y)|)), 1). \tag{16}$$

These functions satisfy $\mathrm{MSP}(x, y) = \lfloor \frac{x}{2^y} \rfloor$ and $\mathrm{LSP}(x, y) = x \bmod 2^y$, where x mod 1 is defined to be 0. For later reference, define the unary analogues *msp*, *lsp* by

$$msp(x, y) = \lfloor x/2^{|y|} \rfloor = \mathrm{MSP}(x, |y|), \tag{17}$$

$$lsp(x, y) = x \bmod 2^{|y|} = \mathrm{LSP}(x, |y|), \tag{18}$$

and note that *lsp* is definable from *msp*, *rev* as follows

$$lsp(x, y) \;=\; msp(rev(msp(rev(s_1(x)), msp(x, y))), 1). \tag{19}$$

Using MSP, LSP together with ideas of the proof of the previous lemma, the following is easily shown.

LEMMA 3.10. *$(A_0)_*$ is closed under part-of quantifiers.*

Using part-of quantification, the inequality predicate $x \leqslant y$ can be defined by

$$\begin{aligned} &|x| < |y| \\ &\quad \text{OR} \\ &|x| = |y| \text{ AND } (\exists u\, Bx)\big[u\, By \wedge \mathrm{BIT}(|x| \dot{-} |u| \dot{-} 1, y) = 1 \wedge \\ &\qquad\qquad\qquad\qquad \mathrm{BIT}(|x| \dot{-} |u| \dot{-} 1, x) = 0\big] \end{aligned}$$

where $|x| < |y|$ has characteristic function $sg(\mathrm{MSP}(y, |x|))$. Note that $|x| \dot{-} |u| \dot{-} 1$ can be expressed by

$$\big|msp(msp(x, u), 1)\big| = \left|\left\lfloor \frac{msp(x, u)}{2} \right\rfloor\right|.$$

Addition $x + y$ can be defined in A_0 by applying CRN to $sum(x, y, z)$, whose value is the $|z|$-th bit of $x + y$. In adding x and y, the $|z|$-th bit of the sum depends whether a *carry* is *generated* or *propagated*. Define the predicates GEN, PROP by having $\mathrm{GEN}(x, y, z)$ hold iff the $|z|$-th bit of both x and y is 1 and $\mathrm{PROP}(x, y, z)$ hold iff the $|z|$-th bit of either x or y is 1. Define $carry(x, y, 0) = 0$ and $carry(x, y, s_i(z))$ to be 1 iff

$$(\exists u Bz)\big[\mathrm{GEN}(x, y, u) \wedge (\forall v Bz)\big[|v| > |u| \rightarrow \mathrm{PROP}(x, y, v)\big]\big].$$

Then $sum(x, y, z) = x \oplus y \oplus carry(x, y, z)$ where the EXCLUSIVE-OR $x \oplus y$ is defined by $cond(x, cond(y, 0, 1), cond(y, 1, 0))$. Using the 2's complement trick, modified subtraction $x \dot{-} y = \max(x - y, 0)$ can be shown to belong to A_0. In order to arithmetize machine computations, pairing and sequence encoding functions are needed. To that end, define the *pairing* function $\tau(x, y)$ by

$$\tau(x, y) = \big(2^{\max(|x|,|y|)} + x\big) * \big(2^{\max(|x|,|y|)} + y\big). \tag{20}$$

Noting that $2^{\max(|x|,|y|)} = cond(msp(x, y), pad(1, y), pad(1, x))$, this function is easily definable from *msp*, *cond*, *pad*, $*$, $+$ hence belongs to A_0. As an example, to compute $\tau(4, 3)$, note that $\max(|4|, |3|) = 3$ and so one concatenates $1\underline{100}$ with $1\underline{011}$, where the underlined portions represent 4 resp. 3 in binary. Define the functions TR [resp. TL] which truncate the rightmost [resp. leftmost] bit: $\mathrm{TR}(x) = \mathrm{MSP}(x, 1) = \lfloor \frac{x}{2} \rfloor$ and $\mathrm{TL}(x) = \mathrm{LSP}(x, |\mathrm{TR}(x)|) = \mathrm{TR}(rev(\mathrm{TR}(rev(s_1(x)))))$, where the latter definition is used later to show that TL belongs to a certain subclass of A_0. The left π_1 and right π_2 projections are defined by

$$\pi_1(z) = \mathrm{TL}\left(\mathrm{MSP}\left(z, \left\lfloor \frac{|z|}{2} \right\rfloor\right)\right), \tag{21}$$

$$\pi_2(z) = \mathrm{TL}\left(\mathrm{LSP}\left(z, \left\lfloor \frac{|z|}{2} \right\rfloor\right)\right) \tag{22}$$

and satisfy $\tau(\pi_1(z), \pi_2(z)) = z$, $\pi_1(\tau(x, y)) = x$ and $\pi_2(\tau(x, y)) = y$. An n-tuple $(x_1, \ldots, x_n)$ can be encoded by $\tau_n(x_1, \ldots, x_n)$, where $\tau_2 = \tau$ and

$$\tau_{k+1}(x_1, \ldots, x_{k+1}) = \tau\big(x_1, \tau_k(x_2, \ldots, x_{k+1})\big).$$

At this point, it should be mentioned that by using the functions so far defined, Turing machine configurations (TM and RATM) are easily expressed in A_0, and even in subalgebras of A_0. A *configuration* of RATM is of the form $(q, u_1, \ldots, u_{k+2}, n_1, \ldots, n_{k+2})$ where $q \in Q$, $u_i \in (\Gamma \cup \{B\})^*$ and $n_i \in \mathbb{N}$. The u_i represent the contents of the k work tapes and of the index query and the index answer tapes, and the n_i represent the head positions on the tapes (the input tape has no head). Since the input is accessed through random access, the input does not form part of the configuration of the RATM. Let ℓ_i [resp. r_i] represent the contents of the left portion [resp. the reverse of the right portion] of the i-th tape (i.e. tape cells of index $\leqslant n_i$ [resp. $> n_i$]). Assuming some simple binary encoding of $\Gamma \cup \{B\}$, a RATM configuration can be represented using the tupling function by

$$\tau_{2k+5}(q, \ell_1, r_1, \ldots, \ell_{k+2}, r_{k+2}).$$

Let $\text{INITIAL}_M(x)$ be the function mapping x to the initial configuration of RATM M on input x. For configurations α, β in the computation of RATM M on x, let predicate $\text{NEXT}_M(x, \alpha, \beta)$ hold iff $(x, \alpha) \vdash_M (x, \beta)$.

If M is a TM with input x, then a configuration can be similarly represented by $\tau_{2k+3}(q, \ell_0, r_0, \ldots, \ell_k, r_k)$ where $initial_M(x)$, $next_M(x, \alpha, \beta)$ are the counterparts for Turing machine computations without random access.

LEMMA 3.11. INITIAL_M, NEXT_M *belong to* $[0, I, s_0, s_1, \text{BIT}, |x|; \text{COMP}, \text{CRN}]$. *Moreover,* $\tau, \pi_1, \pi_2, initial_M, next_M$ *belong to* $[0, I, s_0, s_1, \text{MOD2}, msp; \text{COMP}, \text{CRN}]$.

PROOF. Using s_0, s_1, pad, $*$, $\lfloor x/2 \rfloor$, $cond$, BIT, MSP, LSP, the pairing and tupling functions, etc. it is routine to show that INITIAL_M, NEXT_M are definable in A_0 without use of the smash function. For instance, a move of the first tape head to the right would mean that in the next configuration $\ell_1' = 2 \cdot \ell_1 + \text{MOD2}(r_1)$ and $r_1' = \lfloor r_1/2 \rfloor$.

Temporarily, let $\mathcal{F}$ designate the algebra $[0, I, s_0, s_1, \text{MOD2}, msp; \text{COMP}, \text{CRN}]$. Using MOD2 and msp appropriately, functions from (2) through (14) can be introduced in $\mathcal{F}$. For instance, in (2)

$$rev0(x, s_i(y)) = s_{\text{MOD2}(msp(x,y))}(rev0(x, y)).$$

Part-of quantifiers, the pairing function (20), its left, right projections (21) can be defined in $\mathcal{F}$, by using msp, lsp appropriately in place of MSP, LSP. For instance, to define the projections of the pairing function, define auxiliary functions g, h as follows:

$$\begin{aligned} g(0, x) &= 0, \\ g\big(s_i(z), x\big) &= s_{\text{BIT}(z*z, ones(x))}\big(g(z, x)\big), \\ h(x) &= rev\big(g(x, x)\big). \end{aligned}$$

Then $|h(x)| = \lfloor \frac{|x|}{2} \rfloor$ and for x of even length (i.e. $ones(h(x)) * ones(h(x)) = ones(x)$), the left and right projections of the pairing function are defined by

$$\begin{aligned} \pi_1(x) &= msp\big(x, h(x)\big), \\ \pi_2(x) &= lsp\big(x, h(x)\big). \end{aligned}$$

From this, the function $initial_M$ and predicate $next_M$ are now routine to define. □

We can now describe how short sequences of small numbers are encoded in A_0. To illustrate the idea, what follows is a first approximation to the sequence encoding technique. Generalizing the pairing function, to encode the sequence $(3, 9, 0, 4)$ first compute $\max\{|3|, |9|, |0|, |4|\}$. Temporarily let t denote the integer having binary representation

$$1\underline{0011}1\underline{1001}1\underline{0000}1\underline{0100}$$

where the underlined portions correspond to the binary representations of $3, 9, 0, 4$. Now the length ℓ of sequence $(3, 9, 0, 4)$ is 4, the *block size* BS is 5, and $|t| = \ell \cdot \text{BS}$. Define, as a first approximation, the *sequence number* $\langle 3, 9, 0, 4\rangle$ by $\tau(t, \ell)$.

Given the sequence number $z = \langle 3, 9, 0, 4\rangle$, the Gödel β function decoding the sequence is given by

$$\beta(0, z) = \pi_2(z) = \ell = 4.$$

The blocksize $\text{BS} = \lfloor |\pi_1(z)| / \pi_2(z) \rfloor = \lfloor 20/4 \rfloor = 5$, and for $i = 1, \ldots, 4$

$$\beta(i, z) = \text{LSP}\big(\text{MSP}\big(\pi_1(z), (\ell - i) \cdot \text{BS}\big), \text{BS} - 1\big).$$

Thus $\beta(1, z) = \text{LSP}(\text{MSP}(\pi_1(z), 3 \cdot 5), 4) = 3$, etc. All the above operations belong to A_0, with the exception of multiplication and division (which provably do not belong to A_0). However, multiplication and division by powers of 2 is possible in A_0, so the previously described sequence encoding technique is slightly modified. The sequence $(a_1, \ldots, a_n)$ is encoded by $z = \langle a_1, \ldots, a_n\rangle$ where

$$\begin{aligned} &z = \tau(t, n), \\ &\text{BS} = \max\left\{2^{||a_i||}\colon\ 1 \leqslant i \leqslant n\right\}, \\ &t = h(N) \end{aligned}$$

where

$$\begin{aligned} &|N| = n \cdot \text{BS}, \\ &h(0) = 0, \\ &h\big(s_i(x)\big) = s_{g(x)}\big(h(x)\big) \end{aligned}$$

and

$$g(x) = \begin{cases} 1 & \text{if } |x| \bmod \text{BS} = 0, \\ \text{BIT}((\text{BS} \dot{-} 1) \dot{-} (|x| \bmod \text{BS}), a_{\lfloor |x|/\text{BS} \rfloor + 1}) & \text{else.} \end{cases}$$

Finally define

$$\ell h(z) = \beta(0, z) = \begin{cases} \pi_2(z) & \text{if } z \text{ encodes a pair,} \\ 0 & \text{else} \end{cases} \tag{23}$$

and for $1 \leqslant i \leqslant \beta(0, z)$

$$\beta(i, z) = \text{LSP}\left(\text{MSP}\left(\pi_1(z), (\ell h(z) \dot{-} i) \cdot \left\lfloor \frac{|\pi_1(z)|}{\ell h(z)} \right\rfloor\right), \left\lfloor \frac{|\pi_1(z)|}{\ell h(z)} \right\rfloor \dot{-} 1\right). \tag{24}$$

Suppose that $z = \tau(t, n)$ codes a sequence of length n, where $|t| = \text{BS} \cdot n$ and the block size $\text{BS} = 2^m$ for some m. The exponent m can be computed, since $m = \mu x \leqslant$

$||a||$ [$\text{MSP}(|t|, x) = n$], and A_0 is closed under sharply bounded minimization. Using this observation, it is clear that the β function belongs to A_0. Using the techniques introduced, the following can be proved.

THEOREM 3.12 (Clote [1993]). *If $f \in A_0$ then there exists $g \in A_0$ such that for all x,*

$$g(x, \vec{y}) = \langle f(0, \vec{y}), \ldots, f(|x| - 1, \vec{y}) \rangle.$$

The following two lemmas, together with the sequence encoding machinery of A_0, will allow us soon to establish that $A_0 = \mathcal{F}\text{LH}$.

LEMMA 3.13. *For every $k, m > 1$,*

$$\text{DTIMESPACE}\big(\log^k(n), \log^{1-1/m}(n)\big) \subseteq A_0.$$

PROOF. Let M be a RATM running in time $\log^k(n)$ and space $\log^{1-1/m}(n)$. For each $i \leqslant m \cdot k$, define a predicate $\text{NEXT}_{M,i}$ belonging to A_0 such that

$$\text{NEXT}_{M,i}(x, c, d) \iff d \text{ follows } c \text{ in at most } \log^{i/m}(n) \text{ steps} \tag{25}$$

where c, d are encodings of configurations in the computation of M on input x, and $n = |x|$. By Lemma 3.11, the predicate $\text{NEXT}_{M,0}$ belongs to A_0 and satisfies (25). Assume that $\text{NEXT}_{M,i} \in A_0$ has been defined and satisfies (25). Define the formula $\text{NEXT}_{M,i+1}(x, c, d)$ by

$$\begin{aligned} &\big(\exists s \leqslant |x|^3\big)\big(\forall j < ||x||^{1/m} - 1\big) \\ &\quad \big[s = \langle s_0, \ldots, s_{||x||^{1/m}-1}\rangle \wedge \\ &\quad c = s_0 \wedge d = s_{||x||^{1/m}-1} \wedge \text{NEXT}_{M,i}(s_j, s_{j+1})\big]. \end{aligned} \tag{26}$$

Since for all $j < ||x||^{1/m}$, $|s_j| \leqslant ||x||^{1-1/m}$, $|s| \leqslant (||x||^{1-1/m} + 1) \cdot ||x||^{1/m} \leqslant 2 \cdot ||x||$. This establishes the validity of the bound $s \leqslant |x|^3$ in the definition of $\text{NEXT}_{M,i+1}$. It follows that M accepts input x iff $\text{NEXT}_{M,m\cdot k}(x, c, d)$ holds, where c and d respectively are the initial configuration and the terminal accepting configuration in the computation of M on x. □

The following result for LH is similar.

LEMMA 3.14. *DSPACE$(\log\log(n))$ on a RATM is contained in LH.*

PROOF. Using the logtime computable pairing function from Fact 2.4, one can define a logtime predicate $\text{NEXT}_{M,0}$ which identifies consecutive configurations in the computation of the RATM M running in polylogarithmic time ($\log^{O(1)}(n)$) and simultaneous sublogarithmic space ($\log^{1-\varepsilon}(n)$). As in the preceding lemma, by using

ATM existential and universal branching, the predicate $\mathrm{NEXT}_{M,i}$ can be shown to belong to LH. Thus $\mathrm{DTIMESPACE}(\log^k(n), \log^{1-1/m}(n)) \subseteq \mathrm{LH}$. Since there are only $2^{c \cdot \log\log n} = \log^c(n)$ many possible configurations for some constant c, it follows that DSPACE(O(log log(n))) on RATM is contained in

$$\mathrm{DTIMESPACE}\big(\log^k(n), \log^{1-1/m}(n)\big)$$

on RATM, hence in LH. □

THEOREM 3.15 (P. Clote). $A_0 = \mathcal{F}\mathrm{LH}$.

PROOF. Consider the direction $A_0 \subseteq \mathcal{F}\mathrm{LH}$. It follows from Fact 2.4 that $0, s_0, s_1, |x|$ are computable in logtime. To compute $\mathrm{BIT}(i, x)$, the machine M_1 on input $BiBxB$ writes the bits of i onto a work tape, computes $|i|$ and writes $i + |i| + 1$ onto its input query tape and reads the input answer tape. To compute the i-th bit of $I_k^n(x_1, \ldots, x_n)$, the machine M_2 on input $BiBx_1Bx_2B \cdots Bx_nB$ uses existential and universal states find the locations of the input separators B, computes

$$m = i + k + \sum_{j<i} |x_j|$$

and returns the m-th bit of the input. To compute the the i-th bit of $x\#y = 2^{|x|\cdot|y|}$, the machine M_3 outputs 1 if $i = |x| \cdot |y|$, else 0. Since the product $|x| \cdot |y|$ can be computed in DSPACE(log log n) on a RATM, hence in LH by Lemma 3.14, it follows that $\# \in \mathcal{F}\mathrm{LH}$.

To see that $\mathcal{F}\mathrm{LH}$ is closed under composition, suppose that $f(x)$ equals $g(h_1(x), h_2(x))$, where the bitgraphs of g, h_1, h_2 are computed by the ATM M_g, M_{h_1}, M_{h_2} running in logtime with constantly many alternations. The bitgraph of f is then computed by the ATM M_f obtained from M_g as follows. Recall that M_g expects input of the form By_1By_2B, where $y_1, y_2 \in \{0, 1\}^*$. Whenever M_g requests the i-th bit of its input, M_f computes $|h_1(x)|$, $|h_2(x)|$, and then executes the following code.

```
if i = 0 then
        return B
else if i ≤ |h1(x)| then
        return M_h1(i − 1, x)
else if i = |h1(x)| + 1 then
        return B
else if i ≤ |h1(x)| + 1 + |h2(x)| then
        return M_h2(i − |h1(x)| − 2, x)
else
        return B
```

Inequalities like $i \leqslant |h_1(x)|$ can be decided by checking whether $|i| < ||h_1(x)||$ or $|i| = ||h_1(x)||$ and i precedes $|h_1(x)|$ in lexicographic order (i.e. $i\,B|h_1(x)|$). Values $M_{h_1}(i-1, x)$ can be computed by simulating M_{h_1}, providing bits of input $i-1$ when required, etc. It is similarly easy to see that $\mathcal{F}$LH is closed under CRN. It follows that $A_0 \subseteq \mathcal{F}$LH.

Consider the direction that $\mathcal{F}\text{LH} \subseteq A_0$. A first attempt to arithmetize the computation of the logtime bounded RATM M on input x might be to use Lemma 3.11 together with sequence numbers. However, this encoding of M's computation cannot be done in A_0 because there are $\mathrm{O}(\log n)$ many configurations in M's computation, with each configuration of size $\mathrm{O}(\log n)$, thus requiring sequence numbers of size $\mathrm{O}(\log^2 n)$, and quantification over such values is not sharply bounded. However, the integer s, which encodes the sequence of *instructions* executed (rather than configurations), *is* bounded by a polynomial in n and so can be expressed within the scope of a sharply bounded quantifier. What then remains to be shown is the existence of functions in A_0 which recognize whether a sequence of instructions corresponds to a correct computation.

Suppose that $M = (Q, \Sigma, \Gamma, \Delta, q_0, k+2, \ell)$, is a Σ_m-RATM, running in time $c \cdot |n|$, where $n = |x|$. For notational simplicity, assume $c = 1$ and that $\Sigma = \{0, 1\}$. An instruction of M is of the form

$$(q, a_1, \ldots, a_{k+2}, q', b_1, \ldots, b_{k+2}, d_1, \ldots, d_{k+2})$$

belonging to the transition relation Δ, where q is the current state, $a_1, \ldots, a_{k+2} \in (\Gamma \cup \{B\})$ are the symbols currently read on the k work tapes and input query and answer tapes, q' is the next state, $b_1, \ldots, b_{k+2} \in (\Gamma \cup \{B\})$ are the symbols printed on the work tapes and input query and answer tapes, and $d_i \in \{-1, 0, 1\}$ is the direction of head movement on tape $1 \leqslant i \leqslant k+2$.

Using the earlier sequence encoding, one can code the sequence of $\log n$ instructions by an integer bounded by $|x|^{\mathrm{O}(1)}$. Thus the Σ_m machine M accepts input x iff

$$(\exists y_1 \leqslant |x|^d)(\forall y_2 \leqslant |x|^d)(\exists y_3 \leqslant |x|^d) \cdots (Q y_m \leqslant |x|^d)\Theta(x, y_1, \ldots, y_m),$$

where Q is $\forall$ (resp. $\exists$) if m is even (resp. odd) and $\Theta(x, \vec{y})$ says that

if

(i) for $i = 1, 2, \ldots, m$, each y_i encodes a sequence of instructions from M's program, where the states occurring in y_i are existential (resp. universal) if i is odd (resp. even),

and

(ii) the sequence of instructions coded by $y_1, \ldots, y_m$ determines a correct computation of M,

then

(iii) this computation is accepting.

Using BIT, MSP, LSP, etc. it is not difficult to express (i) and (iii) in A_0. It remains to see how to formulate (ii) in A_0. If $y_1, \ldots, y_m$ encodes a sequence s of $m \cdot \log n$ instructions from M's program, then s corresponds to a correct computation of M, provided that

- The state in the first instruction is q_0, the state in the last instruction is q_A, and for all $0 \leqslant r < m \cdot \log n - 1$, the new state in the r-th instruction is the old state in the $(r+1)$-st instruction.
- For all tape cells, and all $r < m \cdot \log n$, if the r-th instruction is

$$(q, a_1, \ldots, a_{k+2}, q', b_1, \ldots, b_{k+2}, d_1, \ldots, d_{k+2})$$

then for $1 \leqslant j \leqslant k+2$, a_j is the symbol written on the j-th work tape at the last visit of the position p_j, where p_j is the current head position of the j-th work tape, provided that this position has previously been visited, and $a_j = B$ otherwise. Moreover, if q is the input query state q_I, then $b_{k+2} = \text{BIT}(i, x)$, where i is the content of the input query tape.

Note that the function SBBITSUM (sharply bounded bitsum)

$$\text{SBBITSUM}(x, y) = \begin{cases} \sum_{i<|y|} \text{BIT}(i, y) & \text{if } y \leqslant |x|, \\ |x| + 1 & \text{else} \end{cases}$$

is computable in $\log^2 n$ time and log log space, hence by Lemma 3.13, belongs to the algebra A_0. Using SBBITSUM, one can determine whether, given i_0, i_1, j, at instruction i_0 the head of tape j is in the same position as at instruction i_1 in the execution of M on input x. It follows that $\mathcal{F}\text{LH} \subseteq A_0$. □

In Furst et al. [1984], integer multiplication was shown to be AC^0 reducible to MAJ, where $\text{MAJ}(x)$ is 1 if $\sum_{i<|x|} \text{BIT}(i, x) \geqslant \lceil |x|/2 \rceil$, else 0. In Chandra et al. [1984] as refined by Barrington et al. [1990], MAJ was shown to be AC^0 reducible to integer multiplication. The following characterization of polysize, constant depth threshold circuits TC^0 is proved by formalizing these reductions, using the previous techniques.

THEOREM 3.16 (Clote and Takeuti [1995]).

$$\text{TC}^0 = [0, I, s_0, s_1, |x|, \text{BIT}, \times, \#; \text{COMP}, \text{CRN}].$$

REMARK 3.17. Theorem 3.15 was first obtained by combining the author's result [Clote, 1990] that A_0 equals FO definable functions, and the Barrington, Immerman and Straubing [1990] result that FO = LH, an analogue of Bennett's Theorem 3.56.[11] The current proof is direct, influenced by A. Woods' [1986] presentation, and simplifies the argument of [Barrington et al., 1990] by using Lemma 3.13 and Lemma 3.14,

[11] See [Barrington et al., 1990] for definition of FO.

both of which were generalized from Lemma 3.55. Lemma 3.55 was first proved by Nepomnjascii [1970] (a related result proved by Bennett [1962]), though R. Kannan [1981] later rediscovered this result. The idea of encoding a sequence of instructions rather than a sequence of configurations has been repeatedly used by a number of persons.

3.2. *Bounded recursion on notation*

Cobham's original characterization of $\mathcal{F}$PTIME was in terms of functions on words in a finite alphabet, rather than integers. To our knowledge, the first published proof of Cobham's result, additionally formulated for functions on the integers, appeared in [Rose, 1984].

DEFINITION 3.18. The function f is defined by *bounded recursion on notation* (BRN) from g, h_0, h_1, k if

$$\begin{aligned} f(0,\vec{y}) &= g(\vec{y}), \\ f\big(s_0(x),\vec{y}\big) &= h_0\big(x,\vec{y},f(x,\vec{y})\big), \quad \text{if } x \neq 0, \\ f\big(s_1(x),\vec{y}\big) &= h_1\big(x,\vec{y},f(x,\vec{y})\big) \end{aligned}$$

provided that $f(x,\vec{y}) \leqslant k(x,\vec{y})$ for all $x, \vec{y}$.

THEOREM 3.19 (A. Cobham [1965], see H. Rose [1984]).

$$\mathcal{F}\text{PTIME} = [0, I, s_0, s_1, \#; \text{COMP}, \text{BRN}].$$

PROOF. Temporarily denote the algebra $[0, I, s_0, s_1, \#; \text{COMP}, \text{BRN}]$ by $\mathcal{F}$. Consider first the inclusion from left to right. Let M be a TM with input x, running in polynomial time $p(|x|)$. Using BRN the functions MOD2, TR, *msp* can be defined in $\mathcal{F}$ as follows: $\text{MOD2}(0) = 0$, $\text{MOD2}(s_0(x)) = 0$, $\text{MOD2}(s_1(x)) = 1$; $\text{TR}(0) = 0$, $\text{TR}(s_i(x)) = x$; $msp(x,0) = x$, $msp(x, s_i(y)) = \text{TR}(msp(x,y))$, where $\text{MOD2}(x)$, $\text{TR}(x)$, $msp(x,y)$ are bounded by x. It follows that

$$[0, I, s_0, s_1, \text{MOD2}, msp; \text{COMP}, \text{CRN}] \subseteq \mathcal{F}$$

so by Lemma 3.11, $initial_M$, $next_M$ belong to $\mathcal{F}$. By suitably composing $0, s_0, s_1, \#$, there is a function $k \in \mathcal{F}$ satisfying $p(|x|) \leqslant |k(x)|$ for all inputs x. Using BRN, define

$$\begin{aligned} Run_M(x,0) &= initial_M(x), \\ Run_M\big(x, s_i(y)\big) &= next_M\big(x, Run_M(x,y)\big). \end{aligned}$$

Then the value computed by M on input x can be obtained from $Run_M(x, k(x))$ by π_1, π_2.

The inclusion from right to left is proved by an easy induction on term formation in the Cobham algebra. □

Using the same techniques, one can characterize the class $\mathcal{G}\text{TIMESPACE}(n^{O(1)}, O(n))$ of polynomial time linear space computable functions of linear growth as follows. The first assertion is due to D.B. Thompson [1972] (recall that $*$ is concatenation), and the other assertion follows by an alternate function in bounding the recursion on notation.

THEOREM 3.20.

$$\begin{aligned} \mathcal{G}\text{TIMESPACE}\big(n^{O(1)}, O(n)\big) &= [0, I, s_0, s_1, *; \text{COMP}, \text{BRN}] \\ &= [0, I, s_0, s_1, \times; \text{COMP}, \text{BRN}]. \end{aligned}$$

DEFINITION 3.21. The function f is defined from functions g, h_0, h_1, k by *sharply bounded recursion on notation*[12] (SBRN) if

$$\begin{aligned} f(0, \vec{y}) &= g(\vec{y}), \\ f\big(s_0(x), \vec{y}\big) &= h_0\big(x, \vec{y}, f(x, \vec{y})\big), \quad \text{if } x \neq 0, \\ f\big(s_1(x), \vec{y}\big) &= h_1\big(x, \vec{y}, f(x, \vec{y})\big), \end{aligned}$$

provided that $f(x, \vec{y}) \leqslant |k(x, \vec{y})|$ for all $x, \vec{y}$.

J. Lind [1974] characterized $\mathcal{F}\text{LOGSPACE}$ functions on words $w \in \Sigma^*$ as the smallest class of functions containing the initial functions $c_=$ (characteristic function of equality), $*$ (string concatenation) and closed under the operations of explicit transformation, log bounded recursion on notation, and a (provably stronger) version of concatenation on notation. An arithmetic version of Lind's characterization is the following.

THEOREM 3.22.

$$\begin{aligned} \mathcal{F}\text{LOGSPACE} &= [0, I, s_0, s_1, |x|, \text{BIT}, \#; \text{COMP}, \text{CRN}, \text{SBRN}] \\ &= [0, I, s_0, s_1, \text{MOD2}, msp, \#; \text{COMP}, \text{CRN}, \text{SBRN}]. \end{aligned}$$

The first statement appeared in [Clote, 1988; Clote and Takeuti, 1995] and the second can be proved using similar techniques.

Recently, function algebras have been found for small parallel complexity classes. Consider the following variants of recursion on notation.

[12] In [Clote and Takeuti, 1995], this scheme was denoted B_2RN.

DEFINITION 3.23. The function f is defined by *k-bounded recursion on notation* (k-BRN) from g, h_0, h_1 if

$$\begin{aligned} f(0,\vec{y}) &= g(\vec{y}), \\ f\big(s_0(x),\vec{y}\big) &= h_0\big(x,\vec{y},f(x,\vec{y})\big), \quad \text{if } x \neq 0, \\ f\big(s_1(x),\vec{y}\big) &= h_1\big(x,\vec{y},f(x,\vec{y})\big) \end{aligned}$$

provided that $f(x,\vec{y}) \leqslant k$ holds for all $x, \vec{y}$, where k is a constant.

DEFINITION 3.24. The function f is defined by *weak bounded recursion on notation* (WBRN) from g, h_0, h_1, k if $F(x,\vec{y})$ is defined from g, h_0, h_1, k by BRN and $f(x,\vec{y}) = F(|x|,\vec{y})$; i.e.

$$\begin{aligned} F(0,\vec{y}) &= g(\vec{y}), \\ F\big(s_0(x),\vec{y}\big) &= h_0\big(x,\vec{y},F(x,\vec{y})\big) \quad \text{if } x \neq 0, \\ F\big(s_1(x),\vec{y}\big) &= h_1\big(x,\vec{y},F(x,\vec{y})\big), \\ f(x,\vec{y}) &= F\big(|x|,\vec{y}\big) \end{aligned}$$

provided that $F(x,\vec{y}) \leqslant k(x,\vec{y})$ holds for all $x, \vec{y}$.

The characterization of polynomial size, constant depth boolean circuits with parity gates (resp. MOD6 gates) uses sequence encoding techniques of A_0 together with logtime hierarchy analogues of work of Handley, Paris, Wilkie [1984].

THEOREM 3.25 (Clote and Takeuti [1995]).

$$\text{ACC}(2) = [0, I, s_0, s_1, |x|, \text{BIT}, \#; \text{COMP}, \text{CRN}, 1\text{-BRN}], \tag{27}$$

$$\text{ACC}(6) = [0, I, s_0, s_1, |x|, \text{BIT}, \#; \text{COMP}, \text{CRN}, 2\text{-BRN}], \tag{28}$$

$$\text{ACC}(6) = [0, I, s_0, s_1, |x|, \text{BIT}, \#; \text{COMP}, \text{CRN}, 3\text{-BRN}]. \tag{29}$$

The following characterization of $\mathcal{F}$ALOGTIME uses earlier techniques with a formalization of Barrington's [1989] trick in Theorem 2.10 of expressing boolean connectives AND, OR by permutation group word problems.

THEOREM 3.26 (P. Clote [1993]).

$$\mathcal{F}\text{ALOGTIME} = [0, I, s_0, s_1, |x|, \text{BIT}, \#; \text{COMP}, \text{CRN}, 4\text{-BRN}].$$

A natural question arising from work in vectorizing compilers is whether there is a recursive procedure to effectively *parallelize* sequential code (i.e. from sequential code, generate optimal code for a parallel machine). Though I am not aware of details having been worked out, it seems clear that the non-existence of such a procedure must follow from the unsolvability of the halting problem. More importantly, it is not known whether NC is properly contained in PTIME, with modular powering $a^b \bmod m$ being a candidate to separate the classes. Though effective optimal parallelization of sequential code is hopeless, it may seem surprising that certain well-known parallel complexity classes can be characterized in a sequential manner. From the following theorem it follows that NC is characterized by a fragment of the PASCAL language allowing only `for`-loops of the form

```
for i = 1 to |x| if P then y := 2*y
  else y := 2*y+1;
for i = 1 to ||x|| if <statement>;
```

Using repeated squaring (see proof of Theorem 3.53), modular powering is evidently a polynomial time algorithm, yet cannot obviously be written using the above two `for`-loops.

THEOREM 3.27 (P. Clote [1990]).

$$\begin{aligned} \mathrm{NC} &= [0, I, s_0, s_1, |x|, \mathrm{BIT}, \#; \mathrm{COMP}, \mathrm{CRN}, \mathrm{WBRN}], \\ \mathrm{AC}^k &= \{f \in \mathrm{NC} : rk_{\mathrm{WBRN}}(f) \leqslant k\}. \end{aligned}$$

It should be mentioned that independently and at about the same time, B. Allen [1991] characterized NC by a function algebra using a form of *divide and conquer recursion*, and noticed without giving details that over a basis of appropriate initial functions, NC could also be characterized by the scheme of WBRN.[13] A precise statement of Allen's characterization is given later in Theorem 3.77.

Using such techniques, two characterizations of NC^k were given in [Clote, 1990; Clote and Takeuti, 1995]. Levels of a natural time-space hierarchy between $\mathcal{F}$PTIME and $\mathcal{F}$PSPACE were characterized in [Clote, 1992b].

3.3. *Bounded recursion*

A. Grzegorczyk [1953] investigated a hierarchy of subclasses $\mathcal{E}^n$ of primitive recursive functions, defined as the closure of certain initial functions under composition and bounded recursion.

DEFINITION 3.28. The function f is defined by *bounded recursion* (BR) from functions g, h, k if

$$\begin{aligned} f(0, \vec{y}) &= g(\vec{y}), \\ f(x+1, \vec{y}) &= h\big(x, \vec{y}, f(x, \vec{y})\big) \end{aligned}$$

provided that $f(x, \vec{y}) \leqslant k(x, \vec{y})$ holds for all $x, \vec{y}$.

[13] See remark at bottom of p. 13 of [Allen, 1991].

DEFINITION 3.29. Define the following *principal* functions

$$\begin{aligned} f_0(x) &= s(x) = x+1, \\ f_1(x,y) &= x+y, \\ f_2(x,y) &= (x+1)\cdot(y+1), \\ f_3(x) &= 2^x, \\ f_{n+1}(x) &= f_n^{(x)}(1), \quad \text{for } n \geqslant 3. \end{aligned}$$

Let $\mathcal{E}f$ denote $[0, I, s, f; \text{COMP}, \text{BR}]$ and $\mathcal{E}^n$ denote $\mathcal{E}f_n$.

REMARK 3.30. For $n \geqslant 3$, Grzegorczyk's [1953] original functions were defined by $f_{n+1}(0,y) = f_n(y+1, y+1)$, and $f_{n+1}(x+1,y) = f_{n+1}(x, f_{n+1}(x,y))$. The above functions, for $n \geqslant 3$, were taken from [Bel'tyukov, 1982].

A number of characterizations of Grzegorczyk's classes $\mathcal{E}^n$, for $n \geqslant 3$, have been given. A. Meyer and D. Ritchie [1967] characterized $\mathcal{E}^n$ in terms of certain *loop* programming languages, H. Schwichtenberg [1969] investigated the number of nested bounded recursions used in function definitions (the so-called *Heinermann hierarchy*), S.S. Muchnick [1976] investigated *vectorized* Grzegorczyk classes (essentially related to simultaneous bounded recursion schemes), etc. The following theorem is due to H. Schwichtenberg [1969] for $n \geqslant 3$ and to H. Müller [1974] for $n = 2$.[14]

THEOREM 3.31 (Schwichtenberg [1969], Müller [1974]). *Let $\mathcal{H}_n$ be the set*

$$\{f\colon\ f \text{ primitive recursive, } rk_{\text{PR}}(f) \leqslant n\}.$$

Then for $n \geqslant 2$, $\mathcal{H}_n = \mathcal{E}^{n+1}$.

Grzegorczyk [1953] proved that for all $n \geqslant 0$, $\mathcal{E}^n$ is properly contained in $\mathcal{E}^{n+1}$ by demonstrating that $f_{n+1} \notin \mathcal{E}^n$. Concerning the relational classes, he showed that for $n \geqslant 2$, $\mathcal{E}^n_*$ is properly contained in $\mathcal{E}^{n+1}_*$, and asked whether $\mathcal{E}^0_* \subset \mathcal{E}^1_* \subset \mathcal{E}^2_*$. This question remains open. In fact, $\text{LTH} \subseteq \mathcal{E}^0_*$ and $\mathcal{E}^2_* = \text{LINSPACE}$,[15] so Grzegorczyk's question is related to the yet open problem whether the linear time hierarchy is properly contained in linear space. An interesting partial result concerning the containment of the first two relational classes is the following.

THEOREM 3.32 (A. Bel'tyukov [1982]). *For $s \geqslant 1$, let $\beta_s(x) = \max(1, x + \lceil x^{1-1/s} \rceil)$. Then for $s \geqslant 1$, $\mathcal{E}^0_* = (\mathcal{E}\beta_s)_*$. Additionally, $\mathcal{E}^2_* = \mathcal{E}^1_*$ implies $\mathcal{E}^2_* = \mathcal{E}^0_*$.*

[14] It should be mentioned that Schwichtenberg [1969] used slightly different functions f_i; there f_i is the i-th Ackermann branch A_i.

[15] See Corollary 3.37.

To obtain this result, Bel'tyukov introduced the *stack register machine*, a machine model capable of describing $(\mathcal{E} f)_*$. The stack register machine, a variant of the successor register machine, has a finite number of *input registers* and *stack* registers $S_0, \ldots, S_k$ together with a *work* register W. Branching instructions

$$\texttt{if } p(x_1, \ldots, x_m) = q(x_1, \ldots, x_m) \texttt{ then } I_i \texttt{ else } I_j$$

allow to jump to different instructions I_i, I_j depending on the comparison of two polynomials whose variables are current register values. Storage instructions

$$\texttt{W} = S_i$$

allow a value to be saved from a stack register to the work register. Incremental instructions

$$S_i = S_i + 1$$

perform the only computation, and have a side effect of setting to 0 all S_j for $j < i$. A program is a finite list of instructions, where for each i there is at most one incremental instruction for S_i.

Apart from characterizing $\mathcal{E}^2_*$ or LINSPACE, Bel'tyukov characterized the linear time hierarchy LTH. The papers of Paris, Wilkie [1983] and Handley, Paris, Wilkie [1984] study counting classes between LTH and LINSPACE defined by stack register machines. Recent work of Clote [1997a] and of W. Handley [1994b, 1994a] further study the effect of nondeterminism for this model.

LEMMA 3.33 (Grzegorczyk [1953]). *The functions $x \dot{-} y$, $sg(x)$, $\overline{sg}(x)$, $sg(x) \cdot y$, $\overline{sg}(x) \cdot y$ belong to $\mathcal{E}^0$. If $f \in \mathcal{E}^0$ then*

$$\sum_{i \leqslant x} sg(f(i)),\ \sum_{i \leqslant x} \overline{sg}(f(i)),\ \prod_{i \leqslant x} sg(f(i)) \text{ and } \prod_{i \leqslant x} \overline{sg}(f(i))$$

belong to $\mathcal{E}^0$.

DEFINITION 3.34. The function f is defined by *bounded minimization* (BMIN) from the function g, denoted by $f(x, \vec{y}) = \mu i \leqslant x[g(i, \vec{y}) = 0]$, if

$$f(x, \vec{y}) = \begin{cases} \min\{i \leqslant x\colon\ g(i, \vec{y}) = 0\} & \text{if such exists,} \\ 0 & \text{else.} \end{cases}$$

COROLLARY 3.35 (Grzegorczyk [1953]). *For $n \geqslant 0$, $\mathcal{E}^n_*$ is closed under boolean connectives and bounded quantification, and $\mathcal{E}^n$ is closed under bounded minimization.*

PROOF. The predicate $\neg P(\vec{x})$ has characteristic function

$$\overline{sg}\big(c_P(\vec{x})\big),$$

the predicate $P(\vec{x}) \vee Q(\vec{x})$ has characteristic function

$$\overline{sg}\big(\overline{sg}(c_P(\vec{x})) \cdot \overline{sg}(c_Q(\vec{x}))\big),$$

while $(\exists i \leqslant x) R(i, \vec{y})$ has characteristic function

$$sg\left(s(x) \dot{-} \sum_{i \leqslant x} \overline{sg}\big(c_R(i, \vec{y})\big)\right),$$

and $(\forall i \leqslant x) R(i, \vec{y})$ has characteristic function

$$\overline{sg}\left(s(x) \dot{-} \sum_{i \leqslant x} c_R(i, \vec{y})\right).$$

To define $f(x, \vec{y}) = \mu i \leqslant x[g(i, \vec{y}) = 0]$, define the auxiliary function h by

$$\begin{aligned} h(i, \vec{y}) &= \sum_{j \leqslant i} \overline{sg}\big(g(j, \vec{y})\big) \\ &= \text{cardinality of } \big\{j \leqslant i\colon\ g(j, \vec{y}) = 0\big\} \end{aligned}$$

so $\overline{sg}(h(i, \vec{y})) = 1$ iff $(\forall j \leqslant i)(g(j, \vec{y}) \neq 0)$, and

$$\sum_{i \leqslant x} \overline{sg}\big(h(i, \vec{y})\big) = \begin{cases} \mu i \leqslant x[g(i, \vec{y}) = 0] & \text{if } (\exists i \leqslant x)(g(i, \vec{y}) = 0), \\ x + 1 & \text{else.} \end{cases}$$

Then

$$f(x, \vec{y}) = \overline{sg}\left(\left(\sum_{i \leqslant x} \overline{sg}\big(h(i, \vec{y})\big)\right) \dot{-} x\right) \cdot \sum_{i \leqslant x} \overline{sg}\big(h(i, \vec{y})\big).$$

□

The following characterization of LINSPACE in terms of the Grzegorczyk hierarchy was proved by R.W. Ritchie [1963].

THEOREM 3.36. $\mathcal{F}$LINSPACE $= \mathcal{E}^2$.

PROOF. Consider first the direction from right to left. The initial functions of $\mathcal{E}^2$ are computable in LINSPACE, and $\mathcal{F}$LINSPACE is closed under composition and bounded recursion.

Now consider the direction from left to right. We first claim that

$$[0, I, s_0, s_1, \text{MOD2}, msp; \text{COMP}, \text{CRN}] \subseteq \mathcal{E}^2.$$

Clearly $s_0, s_1, cond \in \mathcal{E}^2$ and $\mathcal{E}^2$ is closed under bounded quantification. Now $\overline{sg}(0) = 1$, $\overline{sg}(s(x)) = 0$, $\text{MOD2}(0) = 0$, $\text{MOD2}(s(x)) = \overline{sg}(\text{MOD2}(x))$, so that $\text{MOD2} \in \mathcal{E}^2$. Using BR define the following functions in $\mathcal{E}^2$:

$$\begin{aligned}
\lfloor x/2 \rfloor &= \mu y \leqslant x[y + y = x \vee y + y + 1 = x],\\
\text{MSP}(x, 0) &= x,\\
\text{MSP}(x, i + 1) &= \lfloor \text{MSP}(x, i)/2 \rfloor,\\
\text{BIT}(i, x) &= \text{MOD2}(\text{MSP}(x, i)).
\end{aligned}$$

Temporarily define the auxiliary function h by

$$\begin{aligned}
h(x, 0) &= 0,\\
h(x, i + 1) &= \begin{cases} h(x, i) + 1 & \text{if BIT}(i, x) = 1,\\ h(x, i) & \text{else.}\end{cases}
\end{aligned}$$

Note that $ones(x) = 2^{|x|} - 1$ is defined by

$$\mu y \leqslant s_0(x)\big[(\forall i \leqslant x)\big(\text{BIT}(i, y) = 1 \leftrightarrow (\exists j \leqslant x)(i \leqslant j \wedge \text{BIT}(j, x) = 1)\big)\big].$$

Then $|x| = h(ones(x), x)$ and $msp(x, y) = \text{MSP}(x, |y|)$ belong to $\mathcal{E}^2$.

Suppose that f is defined from g, h_0, h_1 by CRN, where $g, h_0, h_1 \in \mathcal{E}^2$. Then $f(x, \vec{y})$ is $\mu z \leqslant g(\vec{y}) \cdot (2x + 1) + 2x[(30) \vee (31) \vee (32) \vee (33)]$ where

$$|z| = |g(\vec{y})| + |x|, \tag{30}$$

$$\text{MSP}(z, |x|) = g(\vec{y}), \tag{31}$$

$$\begin{aligned}
(\forall i, j < |x|)\big(j = |x| - i - 1 \wedge \text{BIT}(j, x) = 0&\\
\rightarrow&\\
\text{BIT}(j, z) = h_0\big(\text{MSP}(x, j + 1), \vec{y}\big)\big),&
\end{aligned} \tag{32}$$

$$\begin{aligned}
(\forall i, j < |x|)\big(j = |x| - i - 1 \wedge \text{BIT}(j, x) = 1&\\
\rightarrow&\\
\text{BIT}(j, z) = h_1\big(\text{MSP}(x, j + 1), \vec{y}\big)\big).&
\end{aligned} \tag{33}$$

The above bound on f suffices, since $f(x, \vec{y}) \leqslant g(\vec{y}) \cdot 2^{|x|} + 2^{|x|} - 1$, and the latter is at most $g(\vec{y}) \cdot (2x + 1) + 2x$. By Corollary 3.35, $\mathcal{E}^2$ is closed under BMIN, so $f \in \mathcal{E}^2$. It follows that

$$[0, I, s_0, s_1, \text{MOD2}, msp; \text{COMP}, \text{CRN}] \subseteq \mathcal{E}^2.$$

Now let M be a linear space bounded multitape Turing machine computing a function f. By Lemma 3.11, $initial_M$ and $next_M$ belong to $\mathcal{E}^2$. Define the function T by

$$T(x,0) = initial_M(x),$$
$$T(x,y+1) = next_M\big(T(x,y)\big).$$

From the linear space bound, there exists a constant c such that $|T(x,y)| \leqslant c \cdot |x|$, and so $T(x,y) \leqslant (x+1)^c + 1$. Thus T is definable using bounded recursion from functions belonging to $\mathcal{E}^2$. It follows that $\mathcal{F}\text{LINSPACE} \subseteq \mathcal{E}^2$. □

COROLLARY 3.37 (R. Ritchie [1963]). LINSPACE $= \mathcal{E}^2_*$.

The following well-known fact follows from Ritchie's arithmetization techniques together with the observations that a TM with space bound $S(n)$ has time bound $2^{O(S(n))}$ and that $2^x \in \mathcal{E}^k$ for $k \geqslant 3$.

THEOREM 3.38. *For* $k \geqslant 3$,

$$\mathcal{E}^k = \text{DTIME}\big(\mathcal{E}^k\big) = \text{DSPACE}\big(\mathcal{E}^k\big).$$

Similar techniques yield a characterization of PSPACE.

THEOREM 3.39 (D.B. Thompson [1972]).

$$\mathcal{F}\text{PSPACE} = \big[0, I, s, \max, x^{|x|}; \text{COMP}, \text{BR}\big].$$

DEFINITION 3.40. Let k be an integer. The function f is defined by *k-bounded recursion* (k-BR) from functions g, h, k if

$$f(0,\vec{y}) = g(\vec{y}),$$
$$f(x+1,\vec{y}) = h\big(x,\vec{y},f(x,\vec{y})\big)$$

provided that $f(x,\vec{y}) \leqslant k$ holds for all $x, \vec{y}$, where k is a constant.

The following characterization results from the method of the proof of Barrington's Theorem 2.10, arithmetization techniques of this paper, and Theorem 2.12 implying that ATIME$(n^{O(1)})$ = PSPACE. J.-Y. Cai and M. Furst [1988] give a related characterization of PSPACE using *safe-storage* Turing machines, a model related to Bel'tyukov's earlier stack register machines. The next result follows from a characterization of PSPACE by a variant of the stack register machine model.

THEOREM 3.41 (P. Clote [1997a]). *For* $k \geqslant 4$,

$$\text{PSPACE} = [0, I, s_0, s_1, |x|, \text{BIT}, \#; \text{COMP}, \text{CRN}, k\text{-BR}]_*.$$

K. Wagner [1979] extended Ritchie's characterization to more general complexity classes.

THEOREM 3.42 (K. Wagner [1979]). *Let f be an increasing function such that for some $r > 1$ and for all but finitely many x, it is the case that $f(x) \geqslant x^r$. Let $\mathcal{F}$ temporarily denote the algebra $[|f(2^n)|; \text{COMP}]$. Then recalling that $f_2(x, y) = (x+1) \cdot (y+1)$,*

$$\text{DSPACE}(\mathcal{F}) = [0, I, s, \max, f; \text{COMP}, \text{BR}]_* = [0, I, s, f_2, f; \text{COMP}, \text{BR}]_*.$$

Let $|x|_0 = x$, $|x|_{k+1} = ||x|_k|$.

COROLLARY 3.43 (K. Wagner [1979]). *For $k \geqslant 1$,*

$$\text{DSPACE}\big(n \cdot \big(\log^{(k)} n\big)^{\mathrm{O}(1)}\big) = \big[0, I, s, \max, x^{|x|_{k+1}}; \text{COMP}, \text{BR}\big]_*.$$

DEFINITION 3.44 (*K. Wagner* [1979]). The function f is defined by *weak bounded primitive recursion* (WBPR) from g, h, k if $f(x, \vec{y}) = F(|x|, \vec{y})$, where F is defined by bounded primitive recursion, i.e.

$$\begin{aligned} F(0, \vec{y}) &= g(\vec{y}), \\ F(x+1, \vec{y}) &= h\big(x, \vec{y}, F(x, \vec{y})\big), \\ f(x, \vec{y}) &= F\big(|x|, \vec{y}\big) \end{aligned}$$

provided that $F(x, \vec{y}) \leqslant k(x, \vec{y})$ holds for all $x, \vec{y}$.

Provided the proper initial functions are chosen, WBPR is equivalent with BRN. Using this observation, Wagner characterized certain general complexity classes as follows.

THEOREM 3.45 (K. Wagner [1979]). *Let f be an increasing function such that $f(x) \geqslant x^r$ for some $r > 1$, and temporarily let $\mathcal{F}$ denote the algebra $[|f(2^n)|; \text{COMP}]$ and $\mathcal{G}$ denote $\{g^k\colon g \in \mathcal{F}, k \in \mathbb{N}\}$. Then*

$$\begin{aligned} \text{DTIMESPACE}(\mathcal{G}, \mathcal{F}) &= [0, I, s_0, s_1, \max, f; \text{COMP}, \text{BRN}]_* \\ &= [0, I, s, \max, 2 \cdot x, f; \text{COMP}, \text{BRN}]_* \\ &= [0, I, s, +, f; \text{COMP}, \text{BRN}]_* \\ &= [0, I, s, \max, 2 \cdot x, \lfloor x/2 \rfloor, \dot{-}, f; \text{COMP}, \text{WBPR}]_* \\ &= [0, I, s, \lfloor x/2 \rfloor, +, \dot{-}, f; \text{COMP}, \text{WBPR}]_*. \end{aligned}$$

The class DTIMESPACE$(n^{\mathrm{O}(1)}, \mathrm{O}(n))$ of simultaneous polynomial time and linear space can be characterized from the previous theorem by taking $f(x) = x^2$. As referenced in [Wagner and Wechsung, 1986], S.V. Pakhomov [1979] has characterized general complexity classes DTIMESPACE(T, S), DTIME(T), and DSPACE(S)

for suitable classes S, T of unary functions. The class $\mathrm{QL} = \mathrm{DTIME}(n \cdot (\log n)^{O(1)})$ of *quasilinear time* was studied by C.P. Schnorr in [1978]. In analogy, let *quasilinear space* be the class $\mathrm{DSPACE}(n \cdot (\log n)^{O(1)})$. Though Corollary 3.43 characterizes quasilinear space via a function algebra, there appears to be no known function algebra for *quasilinear time*. Y. Gurevich and S. Shelah [1989] studied the class NLT (*nearly linear time*) of functions computable in time $O(n \cdot (\log n)^{O(1)})$ on a random access Turing machine RTM, which is allowed to change its input tape.

DEFINITION 3.46 (*Gurevich and Shelah* [1989]). A RTM is a Turing machine with one-way infinite main tape, address tape and auxiliary tape, such that the head of the main tape is at all times in the cell whose position is given by the contents of the address tape. Instructions of a RTM are of the form

$$(q, a_0, a_1, a_2) \to (q', b_0, b_1, b_2, d_1, d_2)$$

where q, q' are states, $a_i, b_i \in (\Sigma \cup \{B\})$, and $d_i \in \{-1, +1\}$. For such an instruction, if the machine is in state q reading a_0, a_1, a_2 on the main, address and auxiliary tapes, then the machine goes to state q', writes b_0, b_1, b_2 on the respective tapes, and the head of the address [resp. auxiliary] tape goes one square to the right if $d_1 = 1$ [resp. $d_2 = 1$] otherwise one square to the left.

Gurevich and Shelah [1989] show the robustness of NLT by proving the equivalence of this class with respect to different machine models, and give a function algebra for NLT. Their algebra, defined over words from a finite alphabet, is the closure under composition of certain initial functions and weak iterates $f^{(|x|)}(x)$ of certain string manipulating initial functions.

3.4. *Bounded minimization*

Grzegorczyk [1953] considered function classes defined by bounded minimization, defined in Definition 3.34.

DEFINITION 3.47 (*Harrow* [1975, 1979]). For $n = 0, 1, 2$, define $\mathcal{M}^n = [0, I, s, f_n;$ COMP, BMIN]. For $n \geqslant 3$, define $\mathcal{M}^n = [0, I, s, x^y, f_n;$ COMP, BMIN].

Though implicitly asserted in [Grzegorczyk, 1953], the details for the proof of the following result, which follow those in the proof of Theorem 3.65, are given by K. Harrow [1979].[16] The idea of the proof is simply to encode via sequence numbers a definition by bounded primitive recursion and apply the bounded minimization operator.

THEOREM 3.48 (Grzegorczyk [1953], Harrow [1979]). *For* $n \geqslant 3$, $\mathcal{E}^n = \mathcal{M}^n$.

[16] The result was proved in K. Harrow [1979] for $[0, I, s, x^y, f_n;$ COMP, BMIN], but, as previously explained, the exponential is unnecessary.

In the literature, the algebra RF of *rudimentary functions* is sometimes defined by

$$\text{RF} = [0, I, s, +, \times; \text{COMP}, \text{BMIN}].$$

As noticed in [Harrow, 1979], it follows from J. Robinson's [1949] bounded quantifier definition of addition from successor and multiplication that $\mathcal{M}^2 = \text{RF}$. As is well-known, there is a close relationship between (bounded) minimization and (bounded) quantification. This is formalized as follows.

Terms in the first order language of $0, s, +, \cdot, \leqslant$ of arithmetic are defined inductively by: 0 is a term; $x_0, x_1, \ldots$ are terms; if t, t' are terms, then $s(t)$, $t + t'$ and $t \cdot t'$ are terms. Atomic formulas are of the form $t = t'$ and $t \leqslant t'$, where t, t' are terms. The set Δ_0 of bounded quantifier formulas is defined inductively by: if ϕ is an atomic formula, then $\phi \in \Delta_0$; if $\phi, \theta \in \Delta_0$ then $\neg\phi$, $\phi \wedge \theta$, and $\phi \vee \theta$ belong to Δ_0; if $\phi \in \Delta_0$ and t is a term, then $(\exists x \leqslant t)\phi(x, t)$ and $(\forall x \leqslant t)\phi(x, t)$ belong to Δ_0. A k-ary relation $R \subseteq \mathbb{N}^k$ belongs to Δ_0^N if there is a Δ_0 formula ϕ for which $R = \{(a_1, \ldots, a_n)\colon \mathbb{N} \models \phi(a_1, \ldots, a_n)\}$.

DEFINITION 3.49. A predicate $R \subseteq \mathbb{N}^k$ belongs to CA (*constructive arithmetic*), a notion due to R. Smullyan, if there is $\phi(\vec{x}) \in \Delta_0$ such that $R(a_1, \ldots, a_k)$ holds iff $\mathbb{N} \models \phi(a_1, \ldots, a_k)$. Following Definition 2.13, a function $f(\vec{x}) \in \mathcal{G}\text{CA}$ if the bitgraph $B_f \in \text{CA}$ and f is of linear growth.[17]

DEFINITION 3.50. *Presburger arithmetic* (PRES) is the collection of all predicates $R \subseteq \mathbb{N}^k$ for which there exists a first order formula $\phi(\vec{x})$ in the language $0, s, +$ of arithmetic such that $R(a_1, \ldots, a_k)$ holds iff $\mathbb{N} \models \phi(a_1, \ldots, a_k)$.

The following theorem is proved by using quantifier elimination for Presburger arithmetic to show the equivalence between first order formulas and bounded formulas in a richer language allowing congruences, and then exploiting the correspondence between bounded quantification and bounded minimization.

THEOREM 3.51 (Harrow [1975]). *$\mathcal{M}_*^1$ equals the collection of Presburger definable sets .*

From J. Robinson's definition of addition from successor and multiplication, the following easily follows.

PROPOSITION 3.52 (Harrow [1975]). $\mathcal{M}_*^2 = \text{CA}$ *and* $\mathcal{M}^2 = \mathcal{G}\text{CA}$.

While it is obvious that $\text{RF}_* = \text{CA}$, it is non-trivial and surprising that CA equals the linear time hierarchy. J.H. Bennett [1962] showed that the collection of constructive

[17] In the literature, especially in [Paris and Wilkie, 1983], a function f is defined to be Δ_0^N if its graph G_f belongs to Δ_0^N and f is of linear growth. It easily follows from Corollary 3.54 that $f \in \mathcal{G}\text{CA} \iff f \in \Delta_0^N$.

arithmetic sets (Δ_0 definable) is equal to RUD, the class of *rudimentary* sets in the sense of Smullyan [1961]. Later, C. Wrathall [1976] proved that the rudimentary sets are exactly those in the linear time hierarchy LTH.

THEOREM 3.53 (J. Bennett [1962]). *The ternary relation $G(x, y, z)$ for the graph $x^y = z$ of exponentiation is in constructive arithmetic.*

PROOF. Using the technique of *repeated squaring* to compute the exponential $x^y = z$, the idea is to encode the computation in a Δ_0 manner where all quantifiers are bounded by a polynomial in z. Suppose that $x, y > 1$ and that $y = \sum_{i<n} y_i \cdot 2^i$ is the binary representation of y. The following algorithm computes $z = x^y$ by repeatedly applying the fact that $x^{2y} = (x^y)^2$ and $x^{2y+1} = (x^y)^2 \cdot x$. Throughout the rest of the proof, let n denote $|y|$.

```
z = 1
for i = 1 to n
        if y_{n-i} = 0 then z = z^2 else z = z^2· x
```

To encode the computation, for $0 \leqslant i \leqslant n$, let $a_i = \text{MSP}(y, |y| - i)$ and $b_i = x^{a_i}$. The binary representation of a_i consists of the i leftmost bits of y, while b_i equals the value of z at the end of the i-th pass through the above **for**-loop. Except for trivial cases where x, y take values among 0, 1 it follows that $x^y = z$ if and only if there exist sequences $(a_0, \ldots, a_n)$ and $(b_0, \ldots, b_n)$ for which

$$a_0 = 0,\ b_0 = 1,\ a_n = y,\ b_n = z,\ (\forall i < n)(a_{i+1} \in \{2a_i, 2a_i + 1\})$$

and

$$(\forall i < n)\big((a_{i+1} = 0 \to b_{i+1} = b_i^2) \wedge (a_{i+1} = 1 \to b_{i+1} = b_i^2 \cdot x)\big).$$

Thus the graph of exponentiation is Δ_0 provided that sequences $(a_0, \ldots, a_n)$, $(b_0, \ldots, b_n)$ can be encoded in a manner where all quantifiers are bounded by a polynomial in z.

To do this, we will find relatively prime $m_0 < m_1 < \cdots < m_n$ satisfying $a_i, b_i < m_i$ and apply the Chinese remainder theorem to obtain $M = \prod_{i \leqslant n} m_i$ and unique $A, B < M$ for which

$$A \equiv a_i \pmod{m_i},$$
$$B \equiv b_i \pmod{m_i}$$

for $i \leqslant n$. By choosing the m_i to be prime powers of distinct primes, and m_{i+1} to be the smallest prime power divisor of M greater than m_i, one can determine the m_i from M in a Δ_0 manner. Formally, define the predicates PRIME(p), MPP(m, M), I(m, M), N(m, m', M), and F(m, M) as follows.

1. PRIME(p) means that p is prime:

$$2 \leqslant p \wedge (\forall q < p)(q | p \rightarrow q = 1).$$

2. MPP(m, M) means that m is a maximal prime power divisor of M:

$$m|M \wedge (\exists p \leqslant m)\big(\text{PRIME}(p) \wedge p|m \wedge \\ (\forall q < m)\big(q|m \rightarrow q \in \{1, p\}\big) \wedge p \cdot m \nmid M\big).$$

3. I(m, M) means that $m = m_0$ is the *initial* (least) maximal prime power divisor of M:

$$\big(m = 1 \wedge M \in \{0, 1\}\big) \vee \\ \big(\text{MPP}(m, M) \wedge (\forall m' \leqslant M)\big(\text{MPP}(m', M) \rightarrow m \leqslant m'\big)\big).$$

4. N(m, m', M) means that $m' = m_{i+1}$ is the *next* maximal prime power divisor of M after $m = m_i$:

$$\text{MPP}(m, M) \wedge \text{MPP}(m', M) \wedge m < m' \wedge \\ (\forall m'' < m')\big(\text{MPP}(m'', M) \rightarrow m'' \leqslant m\big).$$

5. F(m, M) means that $m = m_n$ is the *final* (greatest) maximal prime power divisor of M:

$$\big(m = 1 \wedge M \in \{0, 1\}\big) \vee \\ \big(\text{MPP}(m, M) \wedge (\forall m' \leqslant M)\big(\text{MPP}(m', M) \rightarrow m' \leqslant m\big)\big).$$

Assume that neither x nor y take values among 0, 1. Then $a_i < 2^i$ and $b_i \leqslant x^{2^i}$ for $i \leqslant n$. Define $m_0, \ldots, m_n$ to be an increasing sequence of prime powers of distinct primes as follows. Let $m_0 = 2$. Given $m_0, \ldots, m_{k-1}$ let $m_k = p^\alpha$, where p is the least prime not dividing $\prod_{i<k} m_i$ and α is the least integer for which $p^\alpha > x^{2^k}$. By the prime number theorem, $p = \mathrm{O}(k \cdot \ln k) < k^2 \leqslant x^{2^k}$, and by choice of α, it is the case that $p^{\alpha-1} < x^{2^k}$, and so

$$p^\alpha = p \cdot p^{\alpha-1} < x^{2^k} \cdot x^{2^k} \leqslant x^{2^{k+1}}.$$

An inductive proof yields that $\prod_{i<k} m_i \leqslant x^{2^{k+1}}$ for all $k > 0$, hence

$$M = \prod_{i \leqslant n} m_i \leqslant x^{2^{n+2}} = \big(x^{2^{n-1}}\big)^8 \leqslant z^8.$$

It now follows that the relation $x^y = z$ is Δ_0 definable. □

The main lines of this proof were influenced by Wilkie's [1983] presentation. See [Hájek and Pudlák, 1992] for other proofs.

COROLLARY 3.54. *The function algebra* $[0, I, s_0, s_1, |x|, \text{BIT}; \text{COMP}, \text{CRN}]$ *is contained in* $\mathcal{M}^2$.

PROOF. Note that $s_0(x) = x + x$, $s_1(x) = x + x + 1$,

$$|x| = \mu y \leqslant x\big[(\exists z \leqslant 2 \cdot x)\big(2^y = z \wedge x < z \wedge \lfloor z/2 \rfloor \leqslant x\big)\big]$$

and that

$$\text{BIT}(i, x) = \mu y \leqslant 1\big[(\exists u, v \leqslant x)\big(|u| = i + 1 \wedge v = 2^{i+1} \wedge v|(x - u)\big)\big]$$

so that $s_0, s_1, |x|$, BIT belong to $\mathcal{M}^2$. Using these functions and bounded minimization, it is easy to show that $\mathcal{M}^2$ is closed under CRN. □

The following is proved in a manner similar to that of Lemma 3.13 and Lemma 3.14.

LEMMA 3.55 (Nepomnjascii [1970]). *For every* $k, m > 1$,

$$\text{NTIMESPACE}(n^k, n^{1-1/m})$$

on a TM *is contained in* CA. *Moreover,* NSPACE(O(log(n))) $\subseteq$ LTH.

THEOREM 3.56. LTH = CA.

PROOF. Consider first the direction from left to right. It follows from Lemmas 3.54 and 3.11 that $initial_M$ and $next_M$ are CA. Now proceed in a similar fashion as in the proof of Theorem 3.15.

The direction from right to left is proved by induction on the length of Δ_0 formulas. Addition, inequality $\leqslant$, and multiplication are computable in LOGSPACE, and $\mathcal{F}$LOGSPACE is closed under composition. By Lemma 3.55 it follows that atomic Δ_0 formulas define relations in LTH. Bounded quantifiers can be handled by existential and universal branching of an alternating Turing machine. □

COROLLARY 3.57. $\mathcal{M}^2_* = \text{LTH}$, *and* $\mathcal{M}^2 = \mathcal{F}\text{LTH}$.

Though the linear time hierarchy equals the bounded arithmetic hierarchy, there is no known exact level-by-level result. The sharpest result we know is due to A. Woods [1986].

If Γ is a class of first order formulas, then Γ^N denotes the collection of predicates definable by a formula in Γ. Let $\Sigma_{0,m}$ denote the collection of bounded quantifier formulas of the form $(\exists \vec{x}_1 \leqslant y)(\forall \vec{x}_2 \leqslant y) \ldots (Q\vec{x}_m \leqslant y)\phi$ where ϕ is a quantifier free formula in the first order language $0, 1, +, \cdot, \leqslant$. Thus $\Sigma_{0,0}$ is the collection of quantifier free formulas. In the following theorem, recall the definition of $\Sigma_n - \text{TIME}(T(n))$ from Definition 2.8.

THEOREM 3.58 (A. Woods [1986]). *For* $m \geqslant 1$, $\Sigma^N_{0,m} \subseteq \Sigma_{m+2} - \text{TIME}(n)$.

SKETCH OF PROOF. The inclusion $\Sigma^N_{0,0} \subseteq \Sigma_2 - \text{TIME}(\text{O}(n))$ is shown as follows. Given an atomic formula $\phi(n)$, suppose that all terms appearing in $\phi(n)$ are bounded by a polynomial in n. By the prime number theorem, there exists a constant c such that the product of the first $c \cdot \ln(n)$ primes is larger than the values of all terms occurring in $\phi(n)$. Using non-determinism guess all terms and subterms appearing in the given quantifier free formula, guess the first $c \cdot \ln(n)$ many prime numbers p and the residues modulo p of the products of subterms occurring in a term, and branching universally, verify that the computations are correct. Now, by the Chinese remainder theorem, the computations are correct iff they are correct modulo the primes.

Since the negation of a quantifier free formula is quantifier free, it follows that

$$\Sigma^N_{0,0} \subseteq \Sigma_2 - \text{TIME}(\text{O}(n)) \cap \Pi_2 - \text{TIME}(\text{O}(n)).$$

Now induct on the number of quantifier blocks. □

By Corollary 3.57 and Theorem 3.36, $\mathcal{M}^2_* = \text{LTH} \subseteq \text{LINSPACE} = \mathcal{E}^2_*$. While LINSPACE is clearly closed under *counting*, this may not be the case for LTH. A typical open question is whether $\pi(x) \in \mathcal{M}^2$, where $\pi(x)$ is the number of primes less than x. J. Paris, A. Wilkie [1983] and later W. Handley, J.B. Paris and A.J. Wilkie [1984] studied the effect of adding k-bounded recursion to LTH. Using the techniques of Barrington, Paris, Wilkie and Handley, together with those of this paper, the following result can be proved.

THEOREM 3.59 (P. Clote [1997a]). *For any* $k \geqslant 4$,

$$\text{ALINTIME} = [0, I, s, +, \dot{-}, \times; \text{COMP}, \text{BMIN}, k\text{-BR}]_*.$$

As in Corollary 3.57, $\mathcal{F}$PH can similarly be characterized.

THEOREM 3.60 (Folklore).

$$\begin{aligned} \mathcal{F}\text{PH} &= [0, I, s, +, \dot{-}, \times, \#; \text{COMP}, \text{BMIN}] \\ &= [0, I, s, +, \dot{-}, \times, \#; \text{COMP}, \text{BRN}, \text{BMIN}] \\ &= [0, I, s_0, s_1, \#; \text{COMP}, \text{BRN}, \text{BMIN}]. \end{aligned}$$

The last assertion of this theorem was sharpened by S. Bellantoni as follows. Following S. Buss [1986a] let $\square^P_i$ denote the class of functions computed in polynomial time on a Turing machine with oracle A, for some set $A \in \Sigma_i^P$. With this notation, $\mathcal{F}\text{PH} = \bigcup_i \square_i^P$.

DEFINITION 3.61 (*S. Bellantoni* [1995]). Let μFP_i denote the algebra

$$\{f \in [0, I, s_0, s_1, \#; \text{COMP}, \text{BRN}, \text{BMIN}] : rk_{\text{BMIN}}(f) \leqslant i\}.$$

THEOREM 3.62 (S. Bellantoni [1995]). *For $i \geqslant 0$, $\square_i^P = \mu FP_i$.*

3.5. *Divide and conquer, course-of-values and miscellaneous*

DEFINITION 3.63 (*Grzegorczyk* [1953], *Constable* [1973]). The function f is defined by *bounded summation* (BSUM) [resp. *bounded product* (BPROD)] from g, k if $f(x, \vec{y})$ equals

$$\sum_{i=0}^{x} g(i, \vec{y}) \qquad \left[\text{resp. } \prod_{i=0}^{x} g(i, \vec{y})\right]$$

provided that $f(x, \vec{y}) \leqslant k(x, \vec{y})$ for all $x, \vec{y}$.

The function f is defined by *sharply bounded summation* (SBSUM) [resp. *sharply bounded product* (SBPROD)] from g, k if $f(x, \vec{y})$ equals

$$\sum_{i=0}^{|x|} g(i, \vec{y}) \qquad \left[\text{resp. } \prod_{i=0}^{|x|} g(i, \vec{y})\right]$$

provided that $f(x, \vec{y}) \leqslant k(x, \vec{y})$ for all $x, \vec{y}$.[18]

The elementary functions were first introduced by Kalmár [1943] and Csillag [1947].

DEFINITION 3.64. The class $\mathcal{E}$ of *elementary* functions is the algebra

$$[0, I, s, +, \dot{-}; \text{COMP}, \text{BSUM}, \text{BPROD}].$$

The elementary functions have many alternate characterizations, among them that $\mathcal{E} = \mathcal{E}^3$.

[18] Sharply bounded summation [resp. product] is called *weak sum* [resp. *product* in [Wagner and Wechsung, 1986], and bounded summation [resp. product] is called *limited sum* [resp. *product* in [Grzegorczyk, 1953].

THEOREM 3.65 (Grzegorczyk [1953]).

$$\begin{aligned}\mathcal{E} &= [0, I, s, \max, f_3; \text{COMP}, \text{BR}] \\ &= [0, I, s, \dot{-}, x^y; \text{COMP}, \text{BMIN}] \\ &= [0, I, s, \dot{-}, \times, x^y; \text{COMP}, \text{BSUM}].\end{aligned}$$

Grzegorczyk asked whether $\mathcal{E}$ had a *finite basis*, i.e. a finite number of functions, whose closure with I under composition equals $\mathcal{E}$. As surveyed in [Wagner and Wechsung, 1986], D. Rödding first gave a positive answer, which was refined by C. Parsons. S.S. Marčenkov [1980] gave a particularly elegant characterization of $\mathcal{E}$ as $[0, I, s, \lfloor x/y \rfloor, x^y, \phi(x, y); \text{COMP}]$, where $\phi(x, y)$ is 0 for $x \leqslant 1$, and otherwise is the least index i for which $a_i = 0$ in the radix x representation of y, i.e. $y = \sum_{i=0}^{\infty} a_i \cdot x^i$, where $0 \leqslant a_i < x$ for all i. In the following theorem, the first statement is due to S.S. Marčenkov [1980], while the second statement to J.P. Jones and Y. Matijasevič [1982].

THEOREM 3.66.

$$\begin{aligned}\mathcal{E}^3_* &= \big([0, I, s, \dot{-}, \lfloor x/y \rfloor, x^y; \text{COMP}]\big)_* \\ &= \big([0, I, +, \dot{-}, \lfloor x/y \rfloor, x!, 2^x; \text{COMP}]\big)_*.\end{aligned}$$

R. Constable [1973] defined the class $\mathcal{K}$ by

$$[0, I, s, +, \dot{-}, \times, \lfloor x/y \rfloor; \text{COMP}, \text{SBSUM}, \text{SBPROD}]$$

a polynomial analogue of the definition of Kalmár elementary functions. The class $\mathcal{K}(f)$ is defined as above, but with f as an additional initial function. Let $FP(f)$ denote the collection of functions polynomial time computable in f ($FP(f)$ can equivalently be defined as the set of type 1 functions in $BFF(f)$; see Definition 4.6).

On p. 118 of [Constable, 1973], the following claim is stated as a theorem without proof.

CLAIM 3.67. *For all non-decreasing f, $K(f) = FP(f)$.*

The statement $\mathcal{F}\text{PTIME} = \mathcal{K}$ was then claimed as a corollary in [Constable, 1973]. This statement was again cited as a theorem (without proof) in [Wagner and Wechsung, 1986].

It now appears that this assertion is doubtful, since $\mathcal{K} \subseteq NC$ and it is currently conjectured that NC is properly contained in $\mathcal{F}\text{PTIME}$. Moreover, using an oracle separation of NC^A from P^A, Clote [1996] provided a counterexample to the previous claim.

THEOREM 3.68 (P. Clote [1996]). *There exists a non-decreasing function f for which $K(f) \neq FP(f)$.*

Nevertheless, Constable's class $\mathcal{K}$ is very natural, suggesting the following question.

QUESTION 3.69. What complexity class corresponds to

$$[0, I, s, +, \dot{-}, \times, \lfloor x/y \rfloor; \text{COMP}, \text{SBSUM}, \text{SBPROD}]?$$

H.-J. Burtschick (personal correspondence) suggested that polynomial size uniform arithmetic circuits could be related to the class $\mathcal{K}$.

Somewhat related is the recent work on counting classes. The class $\#P$, introduced by Valiant [1979], is the set of functions f, for which there exists a nondeterministic polynomial time bounded Turing machine M, such that $f(x)$ is the number of accepting paths in the computation tree of M on input x. Unless $P = NP$, it is unlikely that $\#P$ is closed under composition. Using the arithmetization of boolean formulas from A. Shamir (see [Babai and Fortnow, 1991]), H. Vollmer and K. Wagner gave the following characterization of $\#P$.[19]

THEOREM 3.70 (H. Vollmer, K. Wagner [1996]).

$$\begin{aligned}\#P &= \big[[0, I, S, +, \dot{-}, \times, \lfloor x/y \rfloor, \#; \text{COMP}, \text{SBPROD}]; \text{BSUM}\big]\\ &= \big[[0, I, S, +, \dot{-}, \times, \#; \text{COMP}]; \text{SBPROD}, \text{BSUM}\big].\end{aligned}$$

DEFINITION 3.71. Let $\mathcal{R}_k$ be the smallest class of functions definable from the constant functions $0, \ldots, k$, the projections I, the characteristic functions of the *graphs* of $+, \times, =$, and closed under composition and bounded recursion.

The following result was proved by the Paris–Wilkie modification of Bel'tyukov's stack register machines.

THEOREM 3.72 (Paris and Wilkie [1983]). $(\mathcal{R}_2)_* = (\mathcal{R}_3)_*$.

The next theorem follows from the author's work in [Clote, 1997a] and is based on Barrington's trick.

THEOREM 3.73 (P. Clote [1997a]). *For $n \geqslant 4$,*

$$(\mathcal{R}_n)_* = (\mathcal{R}_{n+1})_* = \text{ALINTIME}.$$

Kutyłowski [1988] considered oracle versions of the Paris–Wilkie work.

[19] In the notation of [Vollmer and Wagner, 1996], the characterization reads $\#P = [[+, \dot{-}, \times, \dot{:}]_{\text{Sub}}, \text{WProd}]_{\text{Sum}}$, and $\#P = [[+, \dot{-}, \times]_{\text{Sub}}]_{\text{WProd,Sum}}$. This formulation is equivalent to that given in Theorem 3.70.

DEFINITION 3.74 (*M. Kutyłowski* [1988]). f is a k-function[20] if for all $x_1, \ldots, x_n$

$$f(x_1, \ldots, x_n) = f\big(\min(x_1, k), \ldots, \min(x_n, k)\big) \leqslant k.$$

For a family $\mathcal{F}$ of functions, $\mathcal{W}_k(\mathcal{F})$ is the smallest class of functions containing I, $\mathcal{F}$, all k-functions and closed under composition and k-bounded recursion. The function f is defined from g, h by *m-counting* if

$$\begin{aligned} f(0, \vec{x}) &= g(\vec{x}), \\ f(n+1, \vec{x}) &= \big(f(n, \vec{x}) + h(n, \vec{x})\big) \pmod{m}. \end{aligned}$$

The class $\mathcal{CW}_k(\mathcal{F})$ is the smallest class of functions containing I, $\mathcal{F}$, all m-functions for $m \in \mathbb{N}$ and closed under composition, k-bounded recursion and arbitrary counting.

THEOREM 3.75 (M. Kutyłowski [1988]). *For every class $\mathcal{F}$ of functions, $\mathcal{W}_2(\mathcal{F})_* = \mathcal{W}_3(\mathcal{F})_*$. For every $k \geqslant 3$, there exists a family $\mathcal{F}$ of functions, such that $\mathcal{W}_k(\mathcal{F})_* \subset \mathcal{W}_{k+1}(\mathcal{F})_*$. For every $k \geqslant 3$, there is a family $\mathcal{F}$ of functions such that $\mathcal{CW}_k(\mathcal{F})_* \subset \mathcal{CW}_{k+1}(\mathcal{F}_*)$.*

Parallel algorithms often employ a divide and conquer strategy. B. Allen [1991] formalized this approach to characterize NC.

DEFINITION 3.76. The *front half* FH(x) is defined by MSP$(x, \lfloor |x|/2 \rfloor)$ and the *back half* BH(x) by LSP$(x, \lfloor |x|/2 \rfloor)$. The function f is defined by *polynomially bounded branching recursion* (PBBR) from functions g, h if there exists a polynomial p such that

$$\begin{aligned} f(0, \vec{y}) &= g_0(\vec{y}), \\ f(1, \vec{y}) &= g_1(\vec{y}), \\ f(x, \vec{y}) &= h\big(x, \vec{y}, f\big(\mathrm{FH}(x), \vec{y}\big), f\big(\mathrm{BH}(x), \vec{y}\big)\big), \quad \text{if } x > 1, \end{aligned}$$

provided that $|f(x, \vec{y})| \leqslant p(\max(|x|, |y_i|))$ for all $x, \vec{y}$. Let $Seq(x) = 0$ if x encodes a sequence[21] else 0. If x encodes a sequence $(x_1, \ldots, x_n)$ and f is a one-place function, then the operation MAP is defined by MAP$(f, x) = \langle f(x_1), \ldots, f(x_n) \rangle$. Define the *bounded shift left function* by SHL$(x, i, y) = x \cdot 2^{\min(i, |y|)}$.

THEOREM 3.77 (B. Allen [1991]). NC *is characterized by the function algebra*

$$[0, I, s, +, \dot{-}, |x|, \mathrm{BIT}, cond, c_{\leqslant}, Seq, \beta, \mathrm{MSP}, \mathrm{SHL}; \mathrm{COMP}, \mathrm{MAP}, \mathrm{PBBR}].$$

[20] What is here called a k-function is called a $k+1$-function in [Kutyłowski, 1988]. As our definition of k-bounded recursion corresponds to Kutyłowski's definition of $k+1$-bounded recursion, the indices of $\mathcal{W}_k(\mathcal{F})$ and $\mathcal{CW}_k(\mathcal{F})$ differ by 1 from [Kutyłowski, 1988].

[21] Here, we use the earlier defined sequence numbers, though Allen [1991] uses a different sequence encoding technique.

Allen explicitly did not attempt to find the smallest set of initial functions, but went on to develop a proof theory for NC functions, similar in spirit to that of S. Buss [1986a]. Independently and at the same time, an equivalent theory of arithmetic for NC functions was given by Clote [1989], later appearing in joint work P. Clote and G. Takeuti [1992].

F. Pitt [1995] considered a variant of B. Allen's polynomial bounded branching recursion, where the function value is bounded by a constant.

DEFINITION 3.78 (*F. Pitt* [1995]). The function f is defined by *k-bounded tree recursion on notation* (k-BTRN) from functions g, h if

$$\begin{aligned} f(0, \vec{y}) &= g_0(\vec{y}), \\ f(1, \vec{y}) &= g_1(\vec{y}), \\ f(x, \vec{y}) &= h\big(x, \vec{y}, f(\mathrm{FH}(x), \vec{y}), f(\mathrm{BH}(x), \vec{y})\big), \quad \text{if } x > 1, \end{aligned}$$

provided that $f(x, \vec{y}) \leqslant k$, for all $x, \vec{y}$. When k is unspecified, the scheme k-BTRN is meant to allow all constants $k \in \mathbb{N}$.

THEOREM 3.79 (F. Pitt [1995]).

$$\mathcal{F}\mathrm{ALOGTIME} = [0, I, s_0, s_1, |x|, \#, \mathrm{MSP}, \mathrm{LSP}; \mathrm{COMP}, \mathrm{CRN}, k\text{-}\mathrm{BTRN}].$$

The theorem is proved by showing that FH, BH can be defined from the initial functions, and then by defining the function TREE in the above function algebra, where TREE is a function evaluating a full binary tree with alternating levels of AND's and OR's, and whose leaves are the bits of a given input x (see Clote [1990, 1992a] for details of definition). Clote [1990] characterized ALOGTIME as $[0, I, s_0, s_1, |x|, \mathrm{BIT}, \#, \mathrm{TREE}; \mathrm{COMP}, \mathrm{CRN}]$, so the proof sketch is complete.

It is often useful to define two or more functions simultaneously. Simultaneous versions of recursion, recursion on notation, k-bounded recursion on notation, etc. are defined in the obvious manner. For example, simultaneous recursion and simultaneous recursion on notation are defined as follows.

DEFINITION 3.80. The functions $f_1, \ldots, f_n$ are defined from functions $g_1, \ldots, g_n$, $h_1, \ldots, h_n$ by *simultaneous recursion* if

$$\begin{aligned} f_i(0, \vec{y}) &= g_i(\vec{y}), \quad \text{for } 1 \leqslant i \leqslant n, \\ f_i(x+1, \vec{y}) &= h_i\big(x, \vec{y}, f_1(x, \vec{y}), \ldots, f_n(x, \vec{y})\big), \quad \text{for } 1 \leqslant i \leqslant n. \end{aligned}$$

If additionally $f_i(x, \vec{y}) \leqslant k_i(x, \vec{y})$ for $1 \leqslant i \leqslant n$, then the f_i are defined by *simultaneous bounded recursion* from $\vec{g}, \vec{h}, \vec{k}$.

The functions $f_1, \ldots, f_n$ are defined from functions $g_1, \ldots, g_n$, $h_1^0, \ldots, h_n^0$ and $h_1^1, \ldots, h_n^1$ by *simultaneous recursion on notation* if for $1 \leqslant i \leqslant n$

$$f_i(0, \vec{y}) = g_i(\vec{y}),$$

$$f_i\big(s_0(x), \vec{y}\big) = h_i^0\big(x, \vec{y}, f_1(x, \vec{y}), \ldots, f_n(x, \vec{y})\big), \quad \text{provided } x \neq 0,$$
$$f_i\big(s_1(x), \vec{y}\big) = h_i^1\big(x, \vec{y}, f_1(x, \vec{y}), \ldots, f_n(x, \vec{y})\big).$$

If additionally $f_i(x, \vec{y}) \leqslant k_i(x, \vec{y})$ for $1 \leqslant i \leqslant n$, then the f_i are defined by *simultaneous bounded recursion on notation* from $\vec{g}$, $\vec{h^0}$, $\vec{h^1}$, $\vec{k}$.

A function algebra $\mathcal{F}$, whose primary closure operation is a certain form of recursion, can often be proved to be closed under the simultaneous version of that form of recursion, by using the pairing function τ and its projections π_1, π_2. For instance, the following is straightforward to establish.

PROPOSITION 3.81 (Kapron and Cook [1996]). *The Cobham algebra* $[0, I, s_0, s_1, \#;$ COMP, BRN] *is closed under simultaneous bounded recursion on notation.*

PROOF. For notational simplicity, suppose that $n = 2$. Define

$$f(0, \vec{y}) = \tau\big(g_1(\vec{y}), g_2(\vec{y})\big),$$
$$f\big(s_i(x), \vec{y}\big) = \tau\Big(h_1^i\big(x, \vec{y}, \pi_1\big(f(x, \vec{y})\big), \pi_2\big(f(x, \vec{y})\big)\big), h_2^i\big(x, \vec{y}, \pi_1\big(f(x, \vec{y})\big), \pi_2\big(f(x, \vec{y})\big)\big)\Big)$$

where $f(x, \vec{y}) \leqslant \tau(k_1(x, \vec{y}), k_2(x, \vec{y}))$. Then $f_1(x, \vec{y})$ is $\pi_1(f(x, \vec{y}))$ and $f_2(x, \vec{y})$ is $\pi_2(f(x, \vec{y}))$. □

The Fibonacci sequence $1, 1, 2, 3, 5, 8, \ldots$ is defined by $Fib(0) = Fib(1) = 1$, and $Fib(n+2) = Fib(n) + Fib(n+1)$. This is a special case of course-of-values recursion.

DEFINITION 3.82. The function f is defined from functions g, h by *course-of-values recursion* (VR) if

$$f(0, \vec{y}) = g(\vec{y}),$$
$$f(x+1, \vec{y}) = h\big(x, \vec{y}, \langle f(0, \vec{y}), \ldots, f(x, \vec{y})\rangle\big).$$

The class $\mathcal{PR}$ of primitive recursive functions is easily seen to be closed under VR.

DEFINITION 3.83. The function f is defined from functions g, h, r, k by *bounded 2-value recursion* (BVR) if

$$f(0, \vec{y}) = g(\vec{y}),$$
$$f(x+1, \vec{y}) = h\big(x, \vec{y}, f(x, \vec{y}), f\big(r(x, \vec{y}), \vec{y}\big)\big)$$

provided that $f(x, \vec{y}) \leqslant k(x, \vec{y})$ and $r(x, \vec{y}) < x$ for all $x, \vec{y}$.

THEOREM 3.84 (Monien [1977]). *Let* $f_2(x,y)=(x+1)\cdot(y+1)$. *Then*

$$\{f\in \text{ETIME}: f \text{ has linear growth rate}\}=[0,I,s,f_2;\text{COMP},\text{BVR}].$$

PROOF. Our exposition follows Wagner and Wechsung [1986]. Temporarily, let $\mathcal{F}$ denote $\{f\in\text{ETIME}: f$ has linear growth rate$\}$ and $\mathcal{G}$ denote $[0,I,s,f_2;\text{COMP},\text{BVR}]$. Consider first the inclusion $\mathcal{F}\subseteq\mathcal{G}$. Suppose that M is a TM which computes the bit-graph B_f of $f:\mathbb{N}^k\to\mathbb{N}$ in time $2^{c\cdot n}$. For notational simplicity, suppose $k=1$ and $|f(x)|\leqslant d\cdot|x|$.

CLAIM. $B_f\in\mathcal{G}$.

Since $\sum_{i<2^{c\cdot n}} i\leqslant 2^{2c\cdot n}$, without loss of generality, M's head movements before halting may be assumed to be of the form

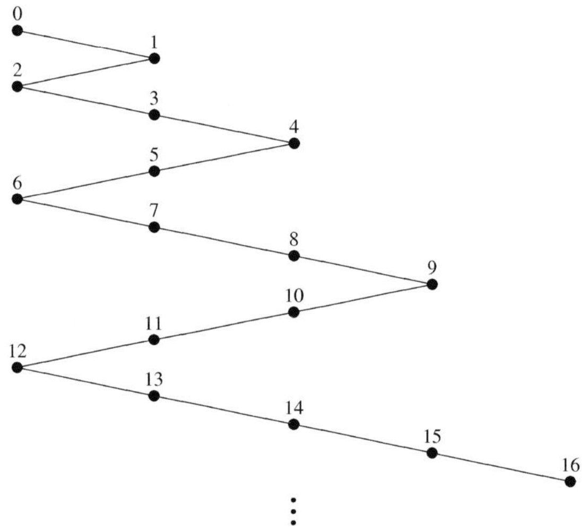

For notational simplicity, assume that M has only one tape, and that the transition function

$$\delta:\big(Q-\{q_A,q_R\}\big)\times\big(\Sigma\cup\{B\}\big)\to Q\times(\Sigma\cup\{B\})\times\{-1,0,1\}$$

satisfies $\delta(q,\sigma)=(state(q,\sigma),symbol(q,\sigma),direction(q,\sigma))$ for suitable functions *state*, *symbol*, *direction*. Let $h(t)$ be M's head position at the *beginning* of step t; let $s(t,x)$ be the state of M at the *completion* of step t on input x; let $a(t,x)$ be the symbol written by M on cell $h(t)$ during step t. Let

$$p_0(t,t')=\begin{cases}\max\{t'': t''\leqslant t'\wedge h(t'')=h(t)\} & \text{if such exists,}\\ t'+1 & \text{else}\end{cases}$$

and $p(t) = p_0(t, t \dot{-} 1)$. Let $sqrt(x) = \lfloor\sqrt{x}\rfloor$. Note that

$$h(t) = \begin{cases} sqrt(x) + sqrt(x)^2 \dot{-} x & \text{if } x \leqslant sqrt(x)^2 + sqrt(x), \\ x \dot{-} sqrt(x) \dot{-} sqrt(x)^2 & \text{else.} \end{cases}$$

Using BMIN, *sqrt* is definable by $sqrt(x) = \mu y \leqslant x[x < (y+1)^2]$ and so $p, h \in \mathcal{M}^2 \subseteq \mathcal{G}$. Define the functions $s(t, x)$ and $a(t, x)$ by

$$\begin{aligned} s(0, x) &= state(q_0, B), \\ s(t+1, x) &= \begin{cases} state(s(t, x), a(p(t+1), x)) \quad \text{if } p(t+1) \leqslant t, \\ state(s(t, x), \text{BIT}(|x| \dot{-} h(t+1), x)) \\ \quad \text{else and } 1 \leqslant h(t+1) \leqslant |x|, \\ state(s(t, x), B) \quad \text{else,} \end{cases} \end{aligned}$$

$$\begin{aligned} a(0, x) &= symbol(q_0, B), \\ a(t+1, x) &= \begin{cases} symbol(s(t, x), a(p(t+1), x)) \\ \quad \text{if } p(t+1) \leqslant t, \\ symbol(s(t, x), \text{BIT}(|x| \dot{-} h(t+1), x)) \\ \quad \text{else and } 1 \leqslant h(t+1) \leqslant |x|, \\ symbol(s(t, x), B) \quad \text{else.} \end{cases} \end{aligned}$$

Instead of defining the functions $s(t, x)$, $a(t, x)$ by simultaneous bounded recursion, define $F(t, x) = \tau(s(t, x), a(t, x))$ by bounded 2-value recursion in the obvious manner. Since $\tau, \pi_1, \pi_2 \in \mathcal{M}^2 \subseteq \mathcal{G}$, it is now routine to complete the proof of the claim that $B_f \in \mathcal{G}$.

Define

$$\begin{aligned} f(x) &= \mu y \\ &\leqslant (x+1)^{d+1} \cdot 2^d \big[\big(\forall i \leqslant d \cdot |x|\big)\big((x, i) \in B_f \leftrightarrow \text{BIT}(i, y) = 1\big)\big]. \end{aligned}$$

Since BR is included in BVR, by the proof of Theorem 3.36, BIT $\in \mathcal{G}$. By Corollary 3.35, $\mathcal{G}$ is closed under bounded quantification and bounded minimization, so it follows that $f \in \mathcal{G}$.

Consider now the inclusion $\mathcal{G} \subseteq \mathcal{F}$. By induction, all functions of $\mathcal{G}$ are of linear growth rate. The functions $0, I_k^n, s, f_2$ are computable in exponential time, and because functions of $\mathcal{F}$ are of linear growth rate, $\mathcal{F}$ is closed under composition. If f is defined by BVR from g, h, r, k, then when computing $f(x+1, \vec{y})$, an exponential time bounded machine M has sufficient space to store the entire sequence $f(0, \vec{y}), \ldots, f(x, \vec{y})$ of previous values on a work tape. It follows that any function of the algebra $\mathcal{G}$ is computable in exponential time. □

A more powerful version of simultaneous recursion was introduced in [Kapron and Cook, 1996].

DEFINITION 3.85. The functions $f_1, \ldots, f_n$ are defined from functions $g_1, \ldots, g_n$, $h_1^0, \ldots, h_n^0, h_1^1, \ldots, h_n^1$ and $k_1, \ldots, k_n$ by *multiple bounded recursion on notation* if the f_i are defined by simultaneous recursion on notation from $\vec{g}_i, \vec{h_i^0}, \vec{h_i^1}$ and moreover

$$f_1(x, \vec{y}) \leqslant k_1(x, \vec{y}),$$
$$f_i(x, \vec{y}) \leqslant k_i\big(x, \vec{y}, f_1(x, \vec{y}), \ldots, f_{i-1}(x, \vec{y})\big), \quad \text{for } 2 \leqslant i \leqslant n.$$

The following non-trivial closure property has an important application in the Kapron and Cook characterization of type 2 polynomial time computations described in the next section.

THEOREM 3.86 (Kapron and Cook [1996]). *The Cobham algebra* $[0, I, s_0, s_1, \#;$ COMP, BRN] *is closed under multiple bounded recursion on notation.*

3.6. *Safe recursion*

All the function algebras from the previous subsection are defined from specific initial functions, using some version of bounded recursion. Without any bound, even schemes such as WBRN can generate all the primitive recursive functions. Recently, certain *unbounded* recursion schemes have been introduced which distinguish between variables as to their position in a function $f(x_1, \ldots, x_n; y_1, \ldots, y_m)$. Variables x_i occurring to the left of the semi-colon are called *normal*, while variables y_j to the right are called *safe*. By allowing only recursions of a certain form, which distinguish between normal and safe variables, particular complexity classes can be characterized. *Normal* values are considered as known in totality, while *safe* values are those obtained by impredicative means (i.e. via recursion). Sometimes, to help distinguish normal from safe positions, the letters $u, v, w, x, y, z, \ldots$ denote normal variables, while $a, b, c, \ldots$ denote safe variables. This terminology, due to Bellantoni and Cook [1992], was chosen to indicate that a *safe* position is one where it is safe to substitute an impredicative value. Related *tiering* notions, though technically different, have occurred in the literature, as in P. Clote and G. Takeuti [1986] (k sorted variables used in defining k-fold multiple exponential time), but most especially in H. Simmons [1988] (*control* variables, i.e. those used for recursion, are distinguished from usual variables; by separating their function, one prevents diagonalization as in the Ackermann function) and in D. Leivant [1989, 1991, 1993, 1994] (stratified polymorphism, second order system $L_2(QF^+)$ corresponding to polynomial time computable functions, stratified functional programs, ramified recurrence over 2 tiered word algebras corresponding to polynomial time). Of these, Simmons [1988] and Leivant [1991] are the most related to the Bellantoni and Cook work described below.

If $\mathcal{F}$ and $\mathcal{O}$ are collections of initial functions and operations which distinguish normal and safe variables, then NORMAL $\cap\, [\mathcal{F}; \mathcal{O}]$ denotes the collection of all functions $f(\vec{x};) \in [\mathcal{F}; \mathcal{O}]$ which have only normal variables. Similarly, $(\text{NORMAL} \cap [\mathcal{F}; \mathcal{O}])_*$ denotes the collection of predicates whose characteristic function $f(\vec{x};)$ has only normal variables and belongs to $[\mathcal{F}; \mathcal{O}]$.

Define the following initial functions by

(0-ary constant)	$0,$
(projections)	$I_j^{n,m}(x_1, \ldots, x_n; a_1, \ldots, a_m) = \begin{cases} x_j & \text{if } 1 \leqslant j \leqslant n, \\ a_{j-n} & \text{if } n < j \leqslant n+m, \end{cases}$
(successors)	$S_0(; a) = 2 \cdot a,\ S_1(; a) = 2 \cdot a + 1,$
(binary predecessor)	$P(; a) = \lfloor a/2 \rfloor,$
(conditional)	$C(; a, b, c) = \begin{cases} b & \text{if } a \bmod 2 = 0, \\ c & \text{else.} \end{cases}$

DEFINITION 3.87 (*Bellantoni and Cook* [1992]). The function f is defined by *safe composition* (SCOMP) from $g, u_1, \ldots, u_n, v_1, \ldots, v_m$ if

$$f(\vec{x}; \vec{a}) = g\big(u_1(\vec{x};), \ldots, u_n(\vec{x};); v_1(\vec{x}; \vec{a}), \ldots, v_m(\vec{x}; \vec{a})\big).$$

If $h(x; y)$ is defined, then SCOMP allows one to define

$$f(x, y;) = h\big(I_1^{2,0}(x, y;); I_2^{2,0}(x, y;)\big) = h(x; y).$$

However, one *cannot* similarly define $g(; x, y) = h(x; y)$.

DEFINITION 3.88. The function f is defined by *safe recursion on notation*[22] (SRN) from the functions g, h_0, h_1 if

$$f(0, \vec{y}; \vec{a}) = g(\vec{y}; \vec{a}),$$
$$f\big(s_0(x), \vec{y}; \vec{a}\big) = h_0\big(x, \vec{y}; \vec{a}, f(x, \vec{y}; \vec{a})\big), \quad \text{provided } x \neq 0,$$
$$f\big(s_1(x), \vec{y}; \vec{a}\big) = h_1\big(x, \vec{y}; \vec{a}, f(x, \vec{y}; \vec{a})\big).$$

The function algebra B is defined by

$$[0, I, S_0, S_1, P, C; \text{SCOMP}, \text{SRN}].$$

THEOREM 3.89 (Bellantoni and Cook [1992]). *The polynomial time computable functions are exactly those functions of B having only normal arguments, i.e.*

$$\mathcal{F}\text{PTIME} = \text{NORMAL} \cap B.$$

The difficult direction of the proof is the inclusion from left to right. By Theorem 3.19 of Cobham, PTIME functions are those in the algebra

$$[0, I, s_0, s_1, \#; \text{COMP}, \text{BRN}].$$

[22] In Bellantoni and Cook [1992] this scheme is called *predicative notational recursion*.

To see the difficulties involved, suppose that f is defined by BRN from g, h_0, h_1 and that $g(\vec{y}) = g'(\vec{y};)$, $h_0(x, \vec{y}, z) = h_0'(x, \vec{y}, z;)$ and $h_1(x, \vec{y}, z) = h_1'(x, \vec{y}, z;)$. In trying to define f' by recursion on notation, one has $f'(0, \vec{y};) = g'(\vec{y};)$ and $f'(s_i(x), \vec{y};) = h_i'(x, \vec{y}, f'(x, \vec{y};);)$. However, this violates the requirement of SRN that the function value $f'(x, \vec{y};)$ be in a safe position in h_i'. For this reason a different approach is necessary.

LEMMA 3.90. *If $f \in \mathcal{F}$PTIME then there exist $f' \in B$ and a monotone increasing polynomial p_f such that $f(\vec{x}) = f'(w; \vec{x})$ for all $|w| \geqslant p_f(|\vec{x}|)$.*

PROOF. Temporarily, let's say that a function f is defined by polynomially bounded recursion on notation (PBRN) from g, h_0, h_1 if f is defined by recursion on notation from these functions, and additionally there exists a polynomial p satisfying $|f(x, \vec{y})| \leqslant p(|x|, |\vec{y}|)$ for all $x, \vec{y}$. Since *pad*, # are easily defined by PBRN [for instance, $0\#y = 1$, $s_i(x)\#y = pad(y, x\#y)$ where $|x\#y| \leqslant |x| \cdot |y| + 1$] it follows from Theorem 3.19 that $\mathcal{F}\text{PTIME} = [0, I, s_0, s_1; \text{COMP}, \text{PBRN}]$. The lemma is now proved by induction on the construction of f in the latter algebra.

If f is $0, I_k^n, s_0, s_1$ then we may take f' to be the corresponding initial function of B and p_f to be 0. Suppose that $f(\vec{x}) = g(h_1(\vec{x}), \ldots, h_n(\vec{x}))$ is defined by composition, where by the induction hypothesis

$$g(y_1, \ldots, y_n) = g'(w; y_1, \ldots, y_n) \quad \text{for } |w| \geqslant p_g(|y_1|, \ldots, |y_n|), \tag{34}$$

$$h_i(\vec{x}) = h_i'(w; \vec{x}) \quad \text{for } |w| \geqslant p_{h_i}(|\vec{x}|). \tag{35}$$

Define

$$f'(w; \vec{x}) = g'\bigl(w; h_1'(w; \vec{x}), \ldots, h_n'(w; \vec{x})\bigr), \tag{36}$$

$$p_f(|\vec{x}|) = p_g\bigl(p_{h_1}(|\vec{x}|), \ldots, p_{h_n}(|\vec{x}|)\bigr) + \sum_{i=1}^{n} p_{h_i}(|\vec{x}|). \tag{37}$$

It follows that $f(\vec{x}) = f'(w; \vec{x})$ for all $|w| \geqslant p_f(|\vec{x}|)$.

Suppose that f is defined from g, h_0, h_1 by PBRN as follows

$$f(0, \vec{y}) = g(\vec{y}),$$

$$f\bigl(s_i(x), \vec{y}\bigr) = h_i\bigl(x, \vec{y}, f(x, \vec{y})\bigr)$$

where $|f(x, \vec{y})| \leqslant q(|x|, |\vec{y}|)$ for some polynomial q. By the induction hypothesis then there exist $g', h_0', h_1', p_g, p_{h_0}, p_h$ satisfying

$$g(\vec{y}) = g'(w; \vec{y}) \qquad \text{for } |w| \geqslant p_g(|\vec{y}|),$$

$$h_i(x, \vec{y}, z) = h_i'(w; x, \vec{y}, z) \quad \text{for } |w| \geqslant p_{h_i}\bigl(|x|, |\vec{y}|, |z|\bigr).$$

Let $E(z, w; x)$ be the initial segment of x obtained by removing from x the $|w| \dot{-} |z|$ lowest order bits. By SRN define F by

$$F(0, w; x, \vec{y}) = 0,$$

$$F\bigl(s_i(z), w; x, \vec{y}\bigr) = \begin{cases} g'(w; \vec{y}) & \text{if Case 1,} \\ h'_0(w; E(z, w; x), \vec{y}, F(z, w; x, \vec{y})) & \text{if Case 2,} \\ h'_1(w; E(z, w; x), \vec{y}, F(z, w; x, \vec{y})) & \text{if Case 3,} \\ F(z, w; x, \vec{y}) & \text{otherwise} \end{cases} \tag{38}$$

where

- Case 1 holds if $|w| - |x| = |s_i(z)| \leqslant |x|$,
- Case 2 holds if $|w| - |x| < |s_i(z)| \leqslant |x|$ and the low order bit of $E(s_i(z), w; x)$ is 0,
- Case 3 holds if $|w| - |x| < |s_i(z)| \leqslant |x|$ and the low order bit of $E(s_i(z), w; x)$ is 1.

To see that $F \in B$, introduce the following functions. The low order bit $M(; a) = a \bmod 2$ is defined by $M(; a) = C(; a, 0, S_1(0))$. The truncation function $T(x; a) = \lfloor a/2^{|x|} \rfloor$ is defined by

$$\begin{aligned} T(0; a) &= a, \\ T\bigl(s_i(x); a\bigr) &= P\bigl(; T(x; a)\bigr). \end{aligned}$$

Let $T'(x, y;) = T(x, y;) = \lfloor y/2^{|x|} \rfloor$, and define the extraction operator

$$E(x, w; a) = T\bigl(T'(x, w;); a\bigr) = \lfloor a/2^{|w| \dot{-} |x|} \rfloor$$

so that $E(x, w; a)$ is the initial segment of a produced by removing from a the $|w| \dot{-} |x|$ lowest order bits. Define the bitwise OR function by

$$\begin{aligned} \vee(0; a) &= M(; a), \\ \vee(s_i(x); a) &= C\bigl(; \vee(x; a), M(; T(s_i(x); a)), 1\bigr). \end{aligned}$$

Note that for $|w| - |x| \leqslant |y| \leqslant |w|$ it follows that $|w| - |x| = |y|$ if and only if $\vee(w; E(y, w; x)) = 0$ and $|w| - |x| < |y|$ iff $\vee(w; E(y, w; x)) = 1$. It is now straightforward, to give a more formal definition of F placing it in B. Now set

$$f'(w; x, \vec{y}) = F(w, w; x, \vec{y}) \tag{39}$$

and

$$\begin{aligned} p_f(|x|, |\vec{y}|) = {} & p_{h_0}\bigl(|x|, |\vec{y}|, q(|x|, |\vec{y}|)\bigr) + p_h\bigl(|x|, |\vec{y}|, q(|x|, |\vec{y}|)\bigr) \\ & + p_g(|\vec{y}|) + |x| + 1. \end{aligned} \tag{40}$$

CLAIM 3.91. *If u satisfies $|w| - |x| \leqslant |u| \leqslant |w|$ and $|w| \geqslant p_f(|x|, |\vec{y}|)$ then $F(u, w; x, \vec{y}) = f(E(u, w; x), \vec{y})$.*

PROOF OF CLAIM. Fix w satisfying $|w| \geqslant p_f(|x|, |\vec{y}|)$. Proceed by induction on $|u|$. First suppose that $|u| = |w| - |x|$. Then $E(u, w; x) = \lfloor x/2^{|w|-|u|} \rfloor = \lfloor x/2^{|x|} \rfloor = 0$. By (40), $|w| \geqslant |x| + 1$, so $|u| \geqslant 1$ and Case 1 applies. It then follows that

$$F(u, w; x, \vec{y}) = g'(w; \vec{y}) = f\big(E(u, w; x), \vec{y}\big).$$

Now suppose that $|w| - |x| < |u| \leqslant |w|$, and that $u = s_0(z)$ or $u = s_1(z)$. In the definition of $F(s_i(z), w; x, \vec{y})$ only case 2 or case 3 can occur.

Case 1. The $(|x| + |u| - |w|)$-th bit of x from the left is 0, or equivalently the $(|w| - |z| - 1)$-st bit of x from the right is 0. Then

$$\begin{aligned}
F\big(s_i(z), w; x, \vec{y}\big) &= h'_0\big(w; E(z, w; x), \vec{y}, F(z, w; x, \vec{y})\big) \\
&= h'_0\big(w; E(z, w; x), \vec{y}, f(E(z, w; x), \vec{y})\big) \\
&\quad \text{induction hypothesis} \\
&= h_0\big(E(z, w; x), \vec{y}, f(E(z, w; x), \vec{y})\big) \\
&\quad \text{by justification below} \\
&= f\big(s_0(E(z, w; x)), \vec{y}\big) \\
&\quad \text{by definition of } f \text{, if } E(z, w; x) \neq 0 \\
&= f\big(E(s_i(z), w; x), \vec{y}\big).
\end{aligned}$$

The last line follows, because in case 1, the low order bit of $E(s_i(z), w; x)$ is 0, so by the definition of E, $E(s_i(z), w; x) = s_0(E(z, w; x))$. The justification for the second line in the above equations is given as follows.

$$\begin{aligned}
|w| &\geqslant p_{h_0}\big(|x|, |\vec{y}|, q(|x|, |\vec{y}|)\big) \\
&\geqslant p_{h_0}\big(|E(z, w; x)|, |\vec{y}|, q(|E(z, w; x)|, |\vec{y}|)\big) \\
&\quad \text{as } |E(z, w; x)| \leqslant |x| \text{ and } q \text{ is monotonic} \\
&\geqslant p_{h_0}\big(|E(z, w; x)|, |\vec{y}|, |f(E(z, w; x), \vec{y})|\big) \\
&\quad \text{as } q \text{ bounds length of } f \\
&\geqslant p_{h_0}\big(|E(z, w; x)|, |\vec{y}|, |F(z, w; x, \vec{y})|\big) \\
&\quad \text{induction hypothesis of claim.}
\end{aligned}$$

Case 2. The $(|w| - |z| - 1)$-st bit of x from the left is 1.
This case in handled similarly to that of case 1, and so the proof of the claim is complete. □

By definition of E, it is clear that for all $|u| = |v|$ we have $E(u, w; x) = E(v, w; x)$. Using this, an easy induction on notation yields that for $|u| \geqslant |x|$ and $|w| \geqslant p_f(|x|, |\vec{y}|)$

$$F(u, w; x, \vec{y}) = F(x, w; x, \vec{y}).$$

For $|w| \geqslant p_f(|x|, |\vec{y}|)$ then

$$\begin{aligned} f'(w; x, \vec{y}) &= F(w, w; x, \vec{y}) && \text{by definition} \\ &= F(x, w; x, \vec{y}) && \text{as } |w| \geqslant |x| + 1 \\ &= f\big(E(x, w; x), \vec{y}\big) && \text{by Claim 3.91} \\ &= f(x, \vec{y}) && \text{by definition of } E. \end{aligned}$$

This completes the proof of the lemma. □

To show all functions of $\mathcal{F}$PTIME are functions of B containing only normal arguments, appropriate bounding functions in B must be defined.

THEOREM 3.92. *If* $f \in \mathcal{F}$PTIME *then* $f(\vec{x};) \in B$.

PROOF. Since f is polynomially bounded, let m, c be such that

$$\big|f(x_1, \ldots, x_n)\big| \leqslant \left(\sum_{i=1}^{n} |x_i|\right)^m + c.$$

Define

$$\begin{aligned} &Pad^2(0; y) = y, \\ &Pad^2\big(s_i(x); y\big) = S_1\big(; Pad^2(x; y)\big), \\ &Pad^{k+1}(x_1, \ldots, x_k; x_{k+1}) = Pad^2\big(x_1; Pad^k(x_2, \ldots, x_k; x_{k+1})\big). \end{aligned}$$

Define

$$\begin{aligned} &Smash(0, x;) = 1, \\ &Smash\big(s_i(y), x;\big) = Pad^2\big(x; Smash(y, x;)\big). \end{aligned}$$

Then $Smash(y, x;) = 2^{|x| \cdot |y|}$. Let $b_0(x;)$ be obtained by composing $Smash(x, x;)$ with itself so as to satisfy $|b_0(x;)| \geqslant |x|^m + c$ and define

$$b(x_1, \ldots, x_n;) = b_0\big(Pad^{n+1}(x_1, \ldots, x_n; 1);\big).$$

Then $|f(x_1, \ldots, x_n)| \leqslant |b(x_1, \ldots, x_n;)|$ and so for f' given by Lemma 3.90, define F by $F(\vec{x};) = f'(b(\vec{x};); \vec{x})$. Then $F \in B$ and $f(\vec{x}) = F(\vec{x};)$. □

For the reverse inclusion, the following bounding lemma is proved by induction on the construction of f in B.

LEMMA 3.93. *Let f belong to B. There is a monotone increasing polynomial q_f such that $|f(\vec{x};\vec{y})| \leqslant q_f(|\vec{x}|) + \max_i |y_i|$ for all $\vec{x}, \vec{y}$.*

THEOREM 3.94 (Bellantoni and Cook [1992]). *If $f(\vec{x};\vec{y}) \in B$ then there is $f'(\vec{x},\vec{y}) \in \mathcal{F}$PTIME such that $f(\vec{x};\vec{y}) = f'(\vec{x},\vec{y})$ for all $\vec{x}, \vec{y}$.*

PROOF. By induction on the construction of f in B. The case for initial functions and composition is straightforward. For monotonic bounding polynomial q_f given by the preceding lemma, there is a function $g \in [0, I, s_0, s_1, \#; \text{COMP}, \text{BRN}]$ satisfying $q_f(|\vec{x}|, |\vec{y}|) \leqslant |g(\vec{x},\vec{y})|$. Thus SRN may be simulated using BRN. □

COROLLARY 3.95.

$$\text{PTIME} = (\text{NORMAL} \cap [0, I, S_0, S_1, P, C; \text{SCOMP}, \text{SRN}])_*.$$

This approach has led to other characterizations of familiar complexity classes using *safe* variants of unbounded recursion schemes.

THEOREM 3.96 (Bellantoni [1992]). *Let $f(\vec{x})$ be a function satisfying $|f(\vec{x})| = O(\log|x|)$. Then $f(\vec{x})$ is computable by a logspace Turing machine iff*

$$f(\vec{x};) \in [0, I, S_1, P, C; \text{SCOMP}, \text{SRN}].$$

Note that the function S_0 does not belong to the above algebra, so the intuition is that for *small* functions (of logarithmic growth rate), LOGSPACE computations are arithmetized by using unary numerals on the work tape along with the same closure operators as for polynomial time. Bellantoni first proves his result for the operation of *simultaneous* safe recursion on notation, and then simulates this simultaneous scheme by SRN, a non-trivial task since the usual pairing function uses S_0. The previous theorem yields a nice characterization of LOGSPACE, to be compared with Corollary 3.95.

COROLLARY 3.97.

$$\text{LOGSPACE} = (\text{NORMAL} \cap [0, I, S_1, P, C; \text{SCOMP}, \text{SRN}])_*.$$

DEFINITION 3.98. The function f is defined by *safe minimization* (SMIN) from the function g, denoted $f(\vec{x};\vec{b}) = s_1(\mu a[g(\vec{x};a,\vec{b}) \bmod 2 = 0)])$, if

$$f(\vec{x};\vec{b}) = \begin{cases} s_1(\min\{a\colon\ g(\vec{x};a,\vec{b}) = 0\}) & \text{if such exists,} \\ 0 & \text{else.} \end{cases}$$

The algebra $\mu B = [0, I, S_0, S_1, P, C; \text{SCOMP}, \text{SRN}, \text{SMIN}]$. Let μB_i denote the set of functions derivable in μB using at most i applications of safe minimization.

THEOREM 3.99 (Bellantoni [1995]).

$$\square_i^P = \{f(\vec{x};): f \in \mu B_i\}.$$

DEFINITION 3.100 (*Bellantoni* [1992]). The function f is defined by *safe recursion*[23] (SR) from the functions g, h if

$$f(0, \vec{y}; \vec{a}) = g(\vec{y}; \vec{a}),$$
$$f(x+1, \vec{y}; \vec{a}) = h(x, \vec{y}; \vec{a}, f(x, \vec{y}; \vec{a})).$$

Define the following initial functions by

$$\text{(successor)} \quad S(; a) = a + 1, \tag{41}$$
$$\text{(predecessor)} \quad \Pr(; a) = a \dot{-} 1, \tag{42}$$
$$\text{(conditional)} \quad K(; a, b, c) = \begin{cases} b & \text{if } a = 0, \\ c & \text{else.} \end{cases} \tag{43}$$

Recall that $\mathcal{E}^2$, the second level of the Grzegorczyk hierarchy, is the collection of linear space computable functions.

THEOREM 3.101 (Bellantoni [1992]).

$$\mathcal{E}^2 = \text{NORMAL} \cap [0, I, S, Pr, K; \text{SCOMP}, \text{SR}].$$

W. Handley (unpublished) and D. Leivant (unpublished) both independently obtained Theorem 3.101. Building on Bellantoni's proof, in her work on linear space reasoning, A.P. Nguyen [1996] gave a slightly different characterization of this class. P. Clote [1997b] gave a *safe* characterization of ETIME functions of linear growth, by adapting the proof of Theorem 3.84.

D. Leivant [1994] gave an alternative formulation of the safe characterizations of polynomial time and of linear space, by introducing a *tiering* notion to arbitrary word algebras. The idea is that one admits various copies or *tiers* $W_0, W_1, \ldots$ of the word algebra W (generated from 0 by s_0, s_1),[24] and defines *ramified recurrence* by

$$f(s_i(x), \vec{y}) = h_i(f(x, \vec{y}), x, \vec{y})$$

where the tier of the first argument $s_i(x)$ is larger than the tier of the value $f(x, \vec{y})$. Comparing with the Bellantoni–Cook notation, tier 0 is safe, whereas tier 1 is normal. Leivant then shows that f is computable by a register machine over algebra A

[23] In [Bellantoni, 1992] this scheme is called *predicative primitive recursion*.

[24] Leivant considers more general algebras defined from finitely many constructors.

in time polynomial in the length of the inputs iff f is definable by explicit definition (corresponding essentially to safe composition) and ramified recurrence over A. This yields that f is polynomial time computable iff f is definable by explicit definition and ramified recurrence over W_0, W_1, whereas f is linear space computable (i.e. in $\mathcal{E}^2$) iff f is definable by explicit definition and ramified recurrence over $\mathbb{N}$, the unary algebra defined from 0, S.

These characterizations of complexity classes in terms of safe operations suggests the following problem.

PROBLEM 3.102. Characterize the classes $\mathcal{M}^n$, for $n = 0, 1, 2$ and for each $n \geqslant 0$, the Grzegorczyk class $\mathcal{E}^n$ via appropriate initial functions, and safe operations. In particular, can one characterize $\mathcal{M}^2$ by $[0, I, S, Pr, K; \text{SCOMP}, \text{SMIN}]$? (Note that the conditional function $cond \in \mathcal{E}^1 - \mathcal{E}^0$.)

Turning to parallel computation, by building on Theorem 3.27, S. Bellantoni [1992] characterizes NC as those functions with normal variables in an algebra built up from 0, I, S_0, S_1, the conditional C, the bit function BIT, the length function $L(; a) = |a|$, a variant $\#'$ of the smash function, and closed under safe composition, concatenation recursion on notation and a *safe* version of weak bounded recursion on notation. Define the *half* function by $H(x) = \lfloor x/(2^{\lceil |x|/2 \rceil}) \rfloor$, and note that the least number of times which H can be iterated on x before reaching 0 is $||x||$. The function f is defined by *safe weak recursion on notation* (SWRN) [25] from the functions g, h if

$$f(0, \vec{y}; \vec{a}) = g(\vec{y}; \vec{a}),$$
$$f(x, \vec{y}; \vec{a}) = h\big(x, \vec{y}; \vec{a}, f(H(x), \vec{y}; \vec{a})\big), \quad \text{provided } x \neq 0.$$

THEOREM 3.103 (S. Bellantoni [1992]).

$$\text{NC} = [0, I, S_0, S_1, C, L, \text{BIT}, \#'; \text{SCOMP}, \text{CRN}, \text{SWRN}].$$

Following Allen [1991], define $\text{BH}(x) = x \bmod 2^{\lceil |x|/2 \rceil}$ and

$$\text{FH}(x) = msp\big(x, \text{BH}(x)\big).$$

The *back half* $\text{BH}(x)$ consists of the $\lceil |x|/2 \rceil$ rightmost bits of x, while the *front half* $\text{FH}(x)$ consists of the $\lfloor |x|/2 \rfloor$ leftmost bits of x. S. Bloch [1994] defines two distinct safe versions of Allen's divide and conquer recursion.

[25] In [Bellantoni, 1992] this scheme is called *log recursion*.

DEFINITION 3.104 (*S. Bloch* [1994]). The function f is defined by *safe divide and conquer recursion* (SDCR) from the functions g, h if

$$f(x,y,\vec{z};\vec{a}) = \begin{cases} g(x,\vec{z};\vec{a}) & \text{if } |x| \leqslant \max(|y|,1), \\ h(x,y,\vec{z};\vec{a}, f(\mathrm{FH}(;x),y,\vec{z};\vec{a}), f(\mathrm{BH}(;x),y,\vec{z};\vec{a})) \\ \quad \text{else.} \end{cases}$$

The function f is defined by *very safe divide and conquer recursion* (VSDCR) from the functions g, h if

$$f(x,y,\vec{z};\vec{a}) = \begin{cases} g(x,\vec{z};\vec{a}) & \text{if } |x| \leqslant \max(|y|,1), \\ h(;x,\vec{z},\vec{a}, f(\mathrm{FH}(;x),y,\vec{z};\vec{a}), f(\mathrm{BH}(;x),y,\vec{z};\vec{a})) \\ \quad \text{else.} \end{cases}$$

Note that in VSDCR the iteration function h has no normal parameters, and hence cannot itself be defined by recursion.

THEOREM 3.105 (S. Bloch [1994]). *There is a collection* BASE *of initial functions, for which*

$$\mathrm{ALOGTIME} = (\mathrm{NORMAL} \cap [\mathrm{BASE};\ \mathrm{SCOMP},\ \mathrm{VSDCR}])_*,$$
$$\mathrm{POLYLOGTIME} = (\mathrm{NORMAL} \cap [\mathrm{BASE};\ \mathrm{SCOMP},\ \mathrm{SDCR}])_*.$$

SKETCH OF PROOF. The collection BASE of initial functions consists of NC^0 computable versions of MSP, LSP, FH, BH, a conditional function, and some string manipulating functions (see S. Bloch [1994] for details).

Only the proof of the first assertion will be sketched. Consider first the inclusion $[\mathrm{BASE};\ \mathrm{SCOMP},\ \mathrm{VSDCR}]_* \subseteq \mathrm{ALOGTIME}$. Show that $\mathrm{BASE} \subseteq \mathrm{NC}^0 \subseteq \mathrm{ALOGTIME}$. Since the iterating function $h(;z,\vec{x},\vec{y},u,v)$ has only safe parameters, h must be obtained from BASE by safe composition, hence belongs to NC^0. Very safe divide and conquer recursion corresponds to the evaluation of a binary tree of logarithmic depth, whose leaves correspond to $g(x,\vec{z};\vec{a})$ for $|x| \leqslant \max(|y|,1)$, and whose internal nodes correspond to the NC^0 function h. Since the resulting circuit is of logarithmic depth, it follows that the function f defined by VSDCR belongs to ALOGTIME.

Now consider the inclusion $\mathrm{ALOGTIME} \subseteq [\mathrm{BASE};\ \mathrm{SCOMP},\ \mathrm{VSDCR}]_*$. Without using VSDCR, define certain string manipulating functions explicitly. Let M be a RATM. The computation tree of M corresponds to a binary branching logdepth tree, all nodes of which are encodings of the current work tape, index tapes, state and tape head positions. Without loss of generality, one may assume that a bit of the input can be queried, using the index tape, only at a leaf configuration. Depending on the current contents (say i) of the index tape, a bit (say the i-th bit) of the input is accessed. Depending on that query, evaluation of the leaves of the computation tree is defined, and evaluation of the internal nodes involves the simple evaluation of an AND-OR tree (minimax strategy). Describe the leaf nodes by a function in

[BASE; SCOMP, VSDCR]. Evaluation of the AND-OR tree is very simply described, using an iterating function having only safe parameters. □

It seems clear that linear time on multitape Turing machines or on random access machines can be characterized using appropriate initial functions, closure under safe composition and some form of simultaneous very safe recursion (simultaneous recursion, since a pairing function apparently cannot be defined from the initial functions using safe composition – such would be necessary for defining NEXT_M). Details have been worked out by S. Bloch [1995] and J. Otto [1994, 1996, 1995], the latter using category theory.

4. Type 2 functionals

Many programming languages allow functions to be passed as parameters to other functions or procedures. For instance, in different limited manners PASCAL and C allow function parameters, while C^{++} supports function templates and ADA, ML admit polymorphism.[26] The oracle Turing machine is a reasonable construct to model function parameter passing, though it has principally been used to study reducibilities $A \leqslant_T B$, $A \leqslant_T^P B$ etc. between *sets*. Nevertheless, higher type *functional* complexity theory is a new area with fundamental open problems. In particular, though various classes have been proposed as candidates for the *feasible* type 2 functionals, there is not yet general agreement about the right notion. For reasons of space, only a few recent directions in higher type functional complexity will be presented. For more information, see the survey by S.A. Cook [1992] and the volume edited by D. Leivant [1995] (higher type complexity theory) and the articles by D. Norman and H. Schwichtenberg in this volume (higher type recursion theory).

DEFINITION 4.1. A *type* 2 functional F of *rank* (k, ℓ) is a total mapping from $(\mathbb{N}^N)^k \times \mathbb{N}^\ell$ into $\mathbb{N}$.

It is worth noting[27] that at type 2, Gödel's question about classification of recursive functions is completely answered. Namely, a rank $(1,1)$ recursive functional F has normal form

$$F(f,x) = U\big(\mu y\big[T_F(x, \overline{f}(y)) = 0\big]\big)$$

where $\overline{f}(y) = \langle f(0), \ldots, f(y-1)\rangle$ and T_F is a well founded tree of height $< \omega_1^{ck}$. Thus type 2 total recursive functionals can be classified with respect to recursive ordinals.

[26] Polymorphism allows function and procedures to abstract over data types – e.g., a generic sorting algorithm for any data type having a comparison function.

[27] This well-known fact, pointed out to the author by S.S. Wainer, is proved in H. Schwichtenberg's article in this volume.

DEFINITION 4.2. A function oracle Turing machine (OTM) is a Turing machine M which in addition to read-only input tape, distinguished output tape and finitely many work tapes, has an *oracle query tape* and *oracle answer tape*, both one-way infinite, for each *function* input. Additionally M has a special oracle query state for each function input.

Note that the previous definition, unlike Definition 2.5, allows function (rather than set) arguments. In order to query a function input f at x, the machine M takes steps to write x in binary on the oracle query tape. When the oracle query tape head is in its leftmost square, M enters a special query state. In the next step, M erases both the oracle query and answer tapes, writes the function value $f(x)$ in binary on the oracle answer tape, and leaves the oracle query and answer tape heads in their leftmost squares. Upon entering the oracle query state, there seem to be two natural measures for the time to complete the function query $f(x)$. The *unit cost*, considered by Mehlhorn [1976], charges unit time, while the *function length cost*, considered by Constable [1973] and later Kapron and Cook [1996], charges $\max\{1, |f(x)|\}$ time. The machine M computes the *rank*(n, m) functional $F(f_1, \ldots, f_n, x_1, \ldots, x_m)$ if M has n oracle query states, query and answer tapes corresponding to $f_1, \ldots, f_n$ and if M outputs the integer $F(f_1, \ldots, f_n, x_1, \ldots, x_m)$ in binary on the output tape, when started in its initial state q_0 with input tape $\underline{B}x_1Bx_2B\cdots Bx_mB$.

DEFINITION 4.3. For any OTM M, for any inputs $f_1, \ldots, f_n, x_1, \ldots, x_m$ and integer t, the query answer set $QA_M(\vec{f}, \vec{x}, t)$ is defined as

$$\{(y, z)\colon\ M \text{ on input } \vec{f}, \vec{x} \text{ queries some } f_i(y) = z \text{ within time } S(t) \text{ steps}\}$$

where $S(t)$ is the least number of steps s for which if M runs s steps then its time complexity is at least t. The *query set* $Q_M(\vec{f}, \vec{x}, t)$ is $\{y\colon (\exists z)[(y, z) \in QA_M(\vec{f}, \vec{x}, t)]\}$ and the *answer set* $A_M(\vec{f}, \vec{x}, t)$ is $\{z\colon (\exists y)[(y, z) \in QA_M(\vec{f}, \vec{x}, t)]\}$.

An OTM M is a *polynomial time oracle Turing machine* (POTM) if M computes a total *rank*(n, m) functional F and there is a polynomial p such that for all input $f_1, \ldots, f_n, x_1, \ldots, x_m$ and times t

$$t \leqslant p\big(\big|\max\big(\{x_1, \ldots, x_m\} \cup A_M(\vec{f}, \vec{x}, t)\big)\big|\big).$$

OPT is the collection of type 2 functionals computable by an oracle polynomial time oracle Turing machine.

EXAMPLE 4.4.

(1) $F(f, x) = \max\{f(y)\colon y \leqslant |x|\}$ belongs to OPT.

(2) $G(f, x) = \max\{f(y)\colon |y| \leqslant |x|\}$ does not belong to OPT.

(3) $H(f, x) = f^{(|x|)}(x)$ belongs to OPT.

K. Mehlhorn [1976] extended Cobham's function algebra to type 2 functionals. A modern presentation of Mehlhorn's definition uses the following schemes.

DEFINITION 4.5 (*Townsend* [1990]). F is defined from $H, G_1, \ldots, G_m$ by *functional composition* if for all $\vec{f}, \vec{x}$,

$$F(\vec{f}, \vec{x}) = H\big(\vec{f}, G_1(\vec{f}, \vec{x}), \ldots, G_m(\vec{f}, \vec{x}), \vec{x}\big).$$

F is defined from G by *expansion* if for all $\vec{f}, \vec{g}, \vec{x}, \vec{y}$,

$$F(\vec{f}, \vec{g}, \vec{x}, \vec{y}) = G(\vec{f}, \vec{x}).$$

F is defined from $G, G_1, \ldots, G_m$ by *functional substitution* if for all $\vec{f}, \vec{x}$,

$$F(\vec{f}, \vec{x}) = H\big(\vec{f}, \lambda y.G_1(\vec{f}, \vec{x}, y), \ldots, \lambda y.G_m(\vec{f}, \vec{x}, y), \vec{x}\big).$$

F is defined from G, H, K by *limited recursion on notation*[28] (LRN) if for all $\vec{f}, \vec{x}, y$,

$$\begin{aligned} F(\vec{f}, \vec{x}, 0) &= G(\vec{f}, \vec{x}), \\ F(\vec{f}, \vec{x}, y) &= H\Big(\vec{f}, \vec{x}, y, F\Big(\vec{f}, \vec{x}, \Big\lfloor \frac{y}{2} \Big\rfloor\Big)\Big), \quad \text{if } y \neq 0 \end{aligned}$$

provided that $F(\vec{f}, \vec{x}, y) < K(\vec{f}, \vec{x}, y)$ holds for all $\vec{f}, \vec{x}, y$.

DEFINITION 4.6 (*Townsend* [1990], *Cook and Kapron* [1990]). Let X be a class of type 2 functionals. The class of *basic feasible functionals* defined from X, denoted BFF(X), is the smallest class of functionals containing $X, 0, s_0, s_1, I, \#$ and the application functional Ap, defined by $Ap(f, x) = f(x)$, and which is closed under functional composition, expansion, and LRN. If $F \in$ BFF(X), then F is basic feasible in X. The class BFF of basic feasible functionals[29] is BFF$(\emptyset)$.

Mehlhorn [1976] introduced the Turing machine model with function oracle, charging unit cost for a function oracle call, independent of the length of the function value returned. Mehlhorn's model has an oracle input tape and an oracle output tape, thus avoiding the situation where m successive iterates of a function $f(f(\ldots f(x)\ldots))$ might take m steps. Using the techniques of low-level arithmetization from the proof of Theorem 3.19, the following result is proved.

THEOREM 4.7 (Mehlhorn [1976]). *For every functional F in* BFF, *there is a unit cost model* OTM *M which computes F, i.e. $M(\vec{f}, \vec{x}) = F(\vec{f}, \vec{x})$, and where the runtime of M on all input $\vec{f}, \vec{x}$ is bounded by $|G(\vec{f}, \vec{x})|$ for some G belonging to* BFF.

[28] This scheme is clearly equivalent to that of bounded recursion on notation BRN for functionals, yet notationally easier to manipulate in the proofs which follow.

[29] Townsend [1990] calls this class **POLY**. Cook and Kapron call this class *basic feasible*, leaving open the possibility that with future research a more natural class of *feasible* functionals may be investigated. The original definition of basic feasible functional required closure under functional substitution, but this can be defined from the remaining schemes, as noted in [Townsend, 1990].

Conversely, if functional F is computed by OTM M*, which on input* $\vec{f}, \vec{x}$ *has runtime at most* $|G(\vec{f}, \vec{x})|$ *for some* G *belonging to* BFF*, then* $F \in$ BFF.

DEFINITION 4.8. A class $\mathcal{F}$ of type 2 functionals has the *Ritchie–Cobham property* if

$$\mathcal{F} = \big\{F\colon \text{there exist } G \in \mathcal{F} \text{ and OTM } M \text{ which on any input } \vec{f}, \vec{x} \text{ computes } F(\vec{f}, \vec{x}) \text{ within time } |G(\vec{f}, \vec{x})|\big\}.$$

With this definition, Theorem 4.7 can be rephrased by the statement that BFF has the Ritchie–Cobham property using unit cost OTM.

It is clear that OPT contains functionals which are not intuitively feasible. In particular, substituting the polytime computable function $\lambda y.y^2$ for f in H, where $H(f, x) = f^{(|x|)}(x)$, above yields $H(\lambda y.y^2, x) = x^{2^{|x|}}$ which is not a polytime computable type 1 function (example due to A. Seth [1992]). The following example, due to S. Cook, provides a functional which belongs to OPT yet not to BFF.

Let $\preceq$ quasi-order $\mathbb{N} \times \mathbb{N}$ by *length first difference*; i.e. $(a, b) \preceq (c, d)$ iff $|a| < |c|$ or $(|a| = |c|$ and $|b| \leqslant |d|)$. Transfer this ordering to $\mathbb{N}$ by a standard polynomial time pairing function. Define the *rank*$(1, 0)$ functional L by $L(f) = \mu i[(\exists j < i)(f(j) \preceq f(i))]$. Note that $\preceq$ defines a well quasi-ordering on $\mathbb{N} \times \mathbb{N}$, so L is well defined.

THEOREM 4.9 (S. Cook [1992]). *The functional* L *belongs to* OPT *yet not to* BFF.

S. Cook [1992] points out that the type-1 section of the closure of OPT with L is just the type-1 section of OPT, i.e. the class of polynomial time computable functions, and so L should be considered a feasible functional. This argument suggests that BFF should not be considered the class of all feasible type-2 functionals. Against this, A. Seth [1992] proves that the type-1 section of the closure of type-2 exponential time with L is not the class of exponential time computable functions, and hence L should not be considered a feasible functional. It is worth noting that Bellantoni [1992, p. 85] showed that if one adds the length function $|x|$ to a modification of class B from [Bellantoni, 1992], and closes under lambda abstraction and application, then the resulting higher type class *does* compute the functional L. S. Bellantoni (private correspondence) has raised the question whether the class obtained by omitting $|x|$ is equivalent to BFF.

Kapron and Cook [1996] lift Cobham's characterization of polynomial time computable functions to functionals of level 2. To state their result, the notion of *length* of a function and that of second order polynomial must be introduced.

DEFINITION 4.10. The length $|f|$ of one-place function f is itself a one-place function defined by

$$|f|(n) = \max_{|x| \leqslant n}\big\{|f(x)|\big\}.$$

Let $f_1, \ldots, f_m$ be variables ranging over $\mathbb{N}^N$ and $x_1, \ldots, x_n$ be variables ranging over $\mathbb{N}$. The collection C of second order polynomials $P(f_1, \ldots, f_m, x_1, \ldots, x_n)$ is defined inductively as follows.

(i) for any integer c, $c \in C$,
(ii) for every $1 \leqslant i \leqslant n$, $x_i \in C$,
(iii) if $P, Q \in C$ then $P + Q \in C$ and $P \cdot Q \in C$,
(iv) if $P \in C$ then $f_i(P) \in C$ for $1 \leqslant i \leqslant m$.

The depth $d(P)$ of a second order polynomial P is defined inductively by $d(c) = 0 = d(x_i)$, $d(P + Q) = d(P \cdot Q) = \max(d(P), d(Q))$, $d(f_i(P)) = 1 + d(P)$. For any f and $Q \subseteq \mathbb{N}$, let f_Q be defined by

$$f_Q(x) = \begin{cases} f(x) & \text{if } x \in Q, \\ 0 & \text{else.} \end{cases}$$

If $Q = Q_M(f, x, t)$ then M on inputs f_Q, x behaves identically to M on inputs f, x.

FACT 4.11. *Suppose that M is an* OTM *and P a second order polynomial such that the runtime of M on inputs f, x is bounded by $P(|f|, |x|)$.*

(i) *Suppose that $Q = Q_M(f, x, t)$ and $t \geqslant P(|f_Q|, |x|)$. Then M halts within t steps.*

(ii) *Suppose that $Q = Q_M(f, x, t)$ and $Q' = Q_M(f, x, P(|f_Q|, |x|))$. Then either M halts within $P(|f_Q|, |x|)$ steps or $Q \subset Q'$.*

The preceding fact is clear, since in (i) M on inputs f, x makes identical moves as M on f_Q, x, and in (ii) if $Q = Q'$, then apply (i) with Q' in place of Q.

THEOREM 4.12 (B. Kapron and S. Cook [1996]). BFF *is the collection of functionals $F(f_1, \ldots, f_n, x_1, \ldots, x_m)$ computable in time $P(|f_1|, \ldots, |f_n|, |x_1|, \ldots, |x_m|)$ for some second order polynomial P on an* OTM *with function length cost.*[30]

PROOF. The inclusion from left to right is straightforward. Consider the inclusion from right to left. Suppose that M computes a functional F of rank $(1, 1)$ and the runtime of M is bounded by the depth d second order polynomial P.

For $1 \leqslant c \leqslant d$, let $P^c_1, \ldots, P^c_{k_c}$ be an enumeration of depth c subpolynomials of P of the form $f(Q)$, where Q is of depth $c - 1$. If $P^c_i = f(Q)$ then denote the associated Q by Q^c_i. Note that $d(P^c_i) = c$ and $d(Q^c_i) = c - 1$. For any Q^c_i there is a first order polynomial q^c_i satisfying

$$\begin{aligned} &q^c_i\big(P^1_1(\vec{f}, \vec{x}), \ldots, P^1_{k_1}(\vec{f}, \vec{x}), \ldots, P^{c-1}_1(\vec{f}, \vec{x}), \ldots, P^{c-1}_{k_{c-1}}(\vec{f}, \vec{x}), \vec{x}\big) \\ &\quad = Q^c_i(\vec{f}, \vec{x}). \end{aligned}$$

30 As noted in [Clote, Kapron and Ignjatovic, 1993], this result holds as well for unit cost.

For any inputs f, x of M and time t, there exist queries $q_1, \ldots, q_d$ in $Q = Q_M(f, x, t)$ such that for $1 \leqslant c \leqslant d$

$$|q_c| \leqslant \max\{Q_1^c(|f_Q|, |x|), \ldots, Q_{k_c}^c(|f_Q|, |x|)\} \tag{44}$$

and

$$|f(q_c)| \geqslant \max\{P_1^c(|f_Q|, |x|), \ldots, P_{k_c}^c(|f_Q|, |x|)\}. \tag{45}$$

For $1 \leqslant c \leqslant d$ there exist first order polynomials q_c satisfying

$$Q_i^c(|f_Q|, |x|) \leqslant q_c(|f(q_1)|, \ldots, |f(q_{c-1})|, |x|).$$

As well there is a first order polynomial p such that

$$P(|f_Q|, |x|) \leqslant p(|f(q_1)|, \ldots, |f(q_d)|, |x|).$$

For every first order polynomial q, there exists a function h_q built up from $0, x_i, s_0, s_1, \#$ using composition, such that

$$q(|x_1|, \ldots, |x_n|) \leqslant |h_q(x_1, \ldots, x_n)|.$$

For example, $|x|^2 \cdot (|y| + 3)$ is bounded by $|(x\#x)\#(s_1(s_1(s_1(y))))|$. It follows that there exist BFF functionals $\overline{Q_c}$, $1 \leqslant c \leqslant d$, and $\overline{P}$ such that

$$Q_i^c(|f_Q|, |x|) \leqslant |\overline{Q_c}(f, q_1, \ldots, q_{c-1}, x)|$$

and

$$P(|f_Q|, |x|) \leqslant |\overline{P}(f, q_1, \ldots, q_d, x)|.$$

For $1 \leqslant c \leqslant d$ define maxquery_M^c of rank (1,2) by

$$\text{maxquery}_M^c(f, x, r) = \mu q_c \in Q(M, f, x, t)[q_c \text{ satisfies (44) and (45)}],$$

where t is the least time satisfying $|Q(M, f, x, t)| \geqslant r$ or M halts in t steps. Define A_{lmax} to be the rank (1,1) functional satisfying

$$|f(A_{lmax}(f, x))| = \max_{y \leqslant |x|}\{|f(y)|\}.$$

Note that A_{lmax} is BFF, since it can be defined by LRN as follows

$$A_{lmax}(f, 0) = 0,$$
$$A_{lmax}(f, x) = \begin{cases} A_{lmax}(f, \lfloor \frac{x}{2} \rfloor) & \text{if } |f(|x|)| \leqslant |f(A_{lmax}(\lfloor \frac{x}{2} \rfloor))| \\ |x| & \text{otherwise} \end{cases}$$

and $A_{lmax}(f,x) \leqslant |x|$ for all f, x.

Assuming that for $1 \leqslant c \leqslant d$, maxquery^c_M is BFF, we can show that F is BFF. Let

$$\begin{aligned} r_1 &= \overline{Q_1}(f,x), \\ T_c &= \overline{P}\big(f, \text{maxquery}^1_{M(f,x,r_c)}, \ldots, \text{maxquery}^d_{M(f,x,r_c)}, x\big) && \text{for } 1 \leqslant c \leqslant d, \\ l_c &= A_{lmax}(f, 2\#T_c) && \text{for } 1 \leqslant c \leqslant d, \\ r_c &= \overline{Q_c}(f, l_1, \ldots, l_{c-1}, x) && \text{for } 2 \leqslant c \leqslant d. \end{aligned}$$

Finally define G by

$$G(f,x) = \max\Big\{\overline{P}(f, l_1, \ldots, l_d, x), \max_{1\leqslant c\leqslant d} T_c\Big\}.$$

CLAIM 4.13. *M halts within $|G(f,x)|$ steps on inputs f, x.*

PROOF OF CLAIM. Suppose that $q_1, \ldots, q_d$ are as in (44) and (45). Then $|q_1| \leqslant |r_1|$ and

$$T_1 = \overline{P}\big(f, \text{maxquery}^1_{M(f,x,r_1)}, \ldots, \text{maxquery}^d_{M(f,x,r_1)}, x\big).$$

If M halts in $|T_1|$ steps, then surely M halts in $|G(f,x)|$ steps. If not, then M must have made more than r_1 queries to the oracle f. Thus

$$l_1 = A_{lmax}(f, 2\#T_1),$$

so

$$\begin{aligned} \big|f(l_1)\big| &= f\big(A_{lmax}(f, 2\#T_1)\big) = \max_{y\leqslant|2\#T_1|} \big|f(y)\big| \\ &= \max_{y\leqslant 2\cdot|T_1|+1} \big|f(y)\big| \geqslant \max_{y\leqslant 2r_1+1} \big|f(y)\big| \\ &\geqslant \max_{|y|\leqslant|r_1|} \big|f(y)\big| = |f|(|r_1|). \end{aligned}$$

The only non-obvious step above relies on the observation that $|T_1| \geqslant r_1$, since when executing M for $|T_1|$ steps, either M halts or M asks more than r_1 oracle queries in that time. As M is assumed not to halt in $|T_1|$ steps, the r_1 oracle queries were posed in $|T_1|$ steps, so $r_1 \leqslant |T_1|$. Since $|q_1| \leqslant |r_1|$, it follows that $|f(q_1)| \leqslant |f|(|r_1|) \leqslant |f(l_1)|$. Now

$$r_2 = \overline{Q_2}(f, l_1, x)$$

and

$$T_2 = \overline{P}\big(f_1, \text{maxquery}^1_M(f,x,r_2), \ldots, \text{maxquery}^d_M(f,x,r_2), x\big).$$

If M halts in $|T_2|$ steps, then M halts in $|G(f,x)|$ steps. Otherwise, $l_2 = A_{lmax}(f, 2\#T_2)$ and a similar argument as before yields

$$\left|f(l_2)\right| \geqslant |f|\left(|r_2|\right).$$

As well,

$$|q_2| \leqslant \max_{1\leqslant i\leqslant k_2} Q_i^2\left(|f|,|x|\right) \leqslant \left|\overline{Q_2}(f,q_1,x)\right| \leqslant |r_2|.$$

Hence $|f(q_2)| \leqslant |f|(|r_2|) \leqslant |f(l_2)|$. Proceeding inductively, if M does not halt in $|T_i|$ steps for some $1 \leqslant i \leqslant d$, then it is the case that

$$|q_i| \leqslant |r_i|, \quad r_i \leqslant |T_i|, \quad \text{and} \quad \left|f(q_i)\right| \leqslant |f|\left(|r_i|\right) \leqslant \left|f(l_i)\right|$$

so that M halts in

$$\begin{aligned} P\left(|f|,|x|\right) &\leqslant p\left(|f(q_1)|,\dots,|f(q_d)|,|x|\right) \\ &\leqslant \left|\overline{P}(f,q_1,\dots,q_d,x)\right| \end{aligned}$$

steps. Thus M halts in $|G(f,x)|$ steps. This establishes the claim. □

Since a BFF functional $Run_M(f,x,T)$ can be defined which arithmetizes the execution of OTM M on inputs f, x for $|T|$ steps, it follows that

$$F(f,x) = Output_M\left(Run_M\left(f,x,G(f,x)\right)\right),$$

where $Output_M$ is an easily defined BFF functional. To establish that F is BFF, it only remains to prove that $\text{maxquery}_M^c(f,x,r)$ is BFF for $1 \leqslant c \leqslant d$. But this follows from the type 2 version of Theorem 3.86. This completes the proof of Theorem 4.12. □

The *oracle concurrent random access machine* (OCRAM), introduced by P. Clote, A. Ignjatovic and B. Kapron [1993] has instructions for (i) local operations – addition, cutoff subtraction, shift, (ii) global and local indirect reading and writing, (iii) control instructions – GOTO, conditional GOTO and HALT, (iv) oracle calls, where in one step, all active processors simultaneously can retrieve

$$f(x_i \cdots x_j) = f\left(\sum_{k=i}^{j} x_k \cdot 2^{j-k}\right)$$

where i, j are current values of local registers, and x_i is the 0,1 value held in the i-th oracle register. If M on argument f, x runs in time $T(|f|,|x|)$ with $P(|f|,|x|)$ processors, then the formal details of the model ensure that $|u| \leqslant T(|f|,|x|) \cdot P(|f|,|x|)$ for every oracle call $f(u)$. If T, P are bounded by second order polynomials, then it follows that there is a second order polynomial Q, such that $|f(u)| \leqslant Q(|f|, |x|)$ for all oracle calls $f(u)$ during the computation of M on input f, x.

The OCRAM is formally defined as follows. For each k-ary function argument f, there are k infinite collections of *oracle registers*, the i-th collection labeled $M_0^{o,i}, M_1^{o,i}, M_2^{o,i}, \ldots$, for $1 \leqslant i \leqslant k$. As with global memory, in the event of a write conflict the lowest numbered processor succeeds in writing to an oracle register. Let *res* (result), $op0$ (operand 0) and $op1$ (operand 1) be non-negative integers, as well as $op2, op3, \ldots, op(2k)$.

In addition to the instructions for the CRAM, the OCRAM has instructions concerning the oracle registers and oracle calls.

$$\begin{aligned}
*M^o_{res} &:= 0,\\
*M^o_{res} &:= 1,\\
M^o_{res} &:= 0,\\
M^o_{res} &:= 1,\\
M_{res} &:= *M^o_{op1},\\
M_{res} &:= f\big([M_{op1} \cdots M_{op2}]_1, [M_{op3} \cdots M_{op4}]_2, \ldots,\\
&\qquad [M_{op(2k-1)} \cdots M_{op(2k)}]_k\big).
\end{aligned}$$

The notation $[M_{op(2i-1)} \cdots M_{op(2i)}]_i$ denotes the integer whose binary notation is given in oracle registers $M^{o,i}_{M_{op(2i-1)}}$ through $M^{o,i}_{M_{op(2i)}}$. In other words,

$$[M_{op(2i-1)} \cdots M_{op(2i)}]_i = \sum_{m=M_{op(2i-1)}}^{M_{op(2i)}} M_m^{o,i} \cdot 2^{op(2i)-m}.$$

The instruction $*M^o_{res} := 0$ sets the contents of the oracle register whose address is given by the current contents of local memory M_{res} to 0. Similarly for the instruction $*M^o_{res} := 1$. The instruction $M_{res} := *M^o_{op1}$ sets the contents of local memory M_{res} to be the current contents of the oracle register whose address is given by the current contents of local memory M_{op1}. With these instructions, it will be the case that oracle registers hold a 0 or 1 but no larger integer. If any register occurring on the right side of an instruction contains 'B' meaning undefined, then the register on the left side of the instruction will be assigned the value 'B' (undefined). For instance, if a unary oracle function f is called in the instruction

$$M_{res} := f([M_{op1} \cdots M_{op2}])$$

and if some register M_i contains 'B', where $op1 \leqslant i \leqslant op2$, then M_{res} is assigned the value 'B'.

In characterizing AC^k in the non-oracle case, Stockmeyer and Vishkin [1984] require a polynomial bound $p(n)$ on the number of active processors on inputs of length n. With the above definition of OCRAM one might hope to characterize the class of type 2 functionals computable in constant parallel time with a second-order polynomial number of processors as exactly the type 2 functionals in the algebra $\mathcal{A}_0$. Using the definitions given so far, this is not true. To rectify this situation, proceed as follows.

DEFINITION 4.14. For every OCRAM M, functions f, g and integers x, t the query set $Q(M, f, x, t, g)$ is defined as

$$\{y\colon \ M \text{ with inputs } f, x \text{ queries } f \text{ at } y \text{ within } t \text{ steps, where for each } i < t \text{ the active processors are those with index } 0, \ldots, g(i) - 1\}.$$

Let M be an OCRAM, P a functional of rank (1,1), f a function and x, t integers. If $Q \subseteq \mathbb{N}$ then define

$$f_Q(u) = \begin{cases} f(u) & \text{if } u \in Q, \\ 0 & \text{else.} \end{cases}$$

Define $\mathcal{M} = \langle M, P\rangle$ to be a *fully specified* OCRAM if for all f, x, t the OCRAM M on input f, x either is halted at step t or executes at step t with active processors $0, \ldots, P(|f_{Q_t}|, |x|) - 1$ where

$$Q_t = Q\big(M, f, x, t, P(|f_{Q_{t-1}}|, |x|)\big)$$

is the collection of queries made by $\mathcal{M}$ before step t.

If $\mathcal{M} = \langle M, P\rangle$ is a fully specified OCRAM with input f, x define

$$Q_{\mathcal{M}}(f, x, t) = \{y\colon \ \mathcal{M} \text{ queries } y \text{ at time } i < t \text{ on input } f, x\}.$$

In place of stating that $\mathcal{M} = \langle M, P\rangle$ is fully specified, usually M is said to run with processor bound P. If $F(\vec{f}, \vec{x})$ abbreviates $F(f_1, \ldots, f_m, x_1, \ldots, x_n)$ and P is a second order polynomial, then $P(|\vec{f}|, |\vec{x}|)$ abbreviates $P(|f_1|, \ldots, |f_m|, |x_1|, \ldots, |x_n|)$.

The type 2 analogue of concatenation recursion on notation is given by the following.

DEFINITION 4.15. F is defined from G, H, K by concatenation recursion on notation (CRN) if for all $\vec{f}, \vec{x}, y$,

$$\begin{aligned} F(\vec{f}, \vec{x}, 0) &= G(\vec{f}, \vec{x}), \\ F\big(\vec{f}, \vec{x}, s_0(y)\big) &= F(\vec{f}, \vec{x}, y) * \mathrm{BIT}\big(H(\vec{f}, \vec{x}, y), 0\big), \quad \text{provided that } x \neq 0, \\ F\big(\vec{f}, \vec{x}, s_1(y)\big) &= F(\vec{f}, \vec{x}, y) * \mathrm{BIT}\big(K(\vec{f}, \vec{x}, y), 0\big) \end{aligned}$$

where $*$ denotes concatenation.

DEFINITION 4.16. The type 2 functional H is defined by weak bounded recursion on notation WBRN from G, H_0, H_1, K if

$$F(\vec{f},\vec{x},0) = G(\vec{f},\vec{x}),$$
$$F(\vec{f},\vec{x},s_0(y)) = H_0(\vec{f},\vec{x},y,F(\vec{f},\vec{x},y)), \quad \text{if } n \neq 0,$$
$$F(\vec{f},\vec{x},s_1(y)) = H_1(\vec{f},\vec{x},y,F(\vec{f},\vec{x},y)),$$
$$H(\vec{f},\vec{x},y) = F(\vec{f},\vec{x},|y|)$$

provided that $F(\vec{f},\vec{x},y) \leqslant K(\vec{f},\vec{x},y)$ holds for all $\vec{f},\vec{x},y$.

DEFINITION 4.17. The algebra $\mathcal{A}_0$ is the smallest class of functionals (of type 1 and 2) containing $0, s_0, s_1, I$, BIT, $|x|, \#, Ap$ and closed under functional composition, expansion, functional substitution and CRN. The algebra $\mathcal{A}$ is the closure of $0, s_0, s_1, I$, BIT, $|x|, \#, Ap$ under functional composition, expansion, functional substitution, CRN and WBRN.

The following theorem is the type 2 analogue of the fact that AC^0 (or equivalently LH) is characterized by the function algebra $\mathcal{A}_0$.

THEOREM 4.18 (Clote, Kapron, Ignjatovic [1993]). *A functional $F(\vec{f},\vec{x})$ belongs to $\mathcal{A}_0$ if and only if it is computable on an* OCRAM *in constant time with at most $P(|\vec{f}|,|\vec{x}|)$ many processors, for some second-order polynomial P.*

To provide some intuition for working with the OCRAM, consider the following program for $Ap(f,x) = f(x)$. Recall that $M_{res} = \text{BIT}(M_{op1}, M_{op2})$ is the easily programmed instruction which, for $i = M_{op2}$, computes the coefficient of 2^i in the binary representation of the integer stored in M_{op1}, *provided* that $i < |M_{op1}|$, and otherwise returns the value 'B'. Define "reverse bit" RBIT, where $\text{RBIT}(x,y) = \text{BIT}(x,|x| \dotdiv (y+1))$ *provided* that $y < |x|$, and 'B' otherwise.

OCRAM program for $Ap(f,x) = f(x)$.

```
1  M₀ = 0
2  M₁ = processor number
3  M₂ = *M₁ᵍ
4  if (M₂ = B) then M₀ᵍ = M₁
5  M₃ = M₀ᵍ ∸ 1 % M₃ = |x| ∸ 1
6  *M₁ᵍ = B % erase global memory
7  *M₁ᵒ = M₂ % in Pᵢ, Mᵢᵒ = Xᵢ
8  M₄ = f([M₀...M₃])
9  M₅ = BIT(M₄, M₁)
10 if (M₅ = B) then M₀ᵍ = M₁
```

M_5 = M_0^g % M_5 = |f(x)|
M_6 = M_5 $\dot{-}$ (M_1 + 1)
M_7 = BIT(M_4, M_6)
*M_1^o = B
*M_1^g = B
*M_1^g = M_7
if ($M_1 \geqslant M_5$) then M_1^g = B % erase trailing 0's
HALT % Now X_i = RBIT(f(x), i)

The type 2 analogue of Theorem 3.27 was established by the author (in preparation), and strengthens the principal result of Clote, Kapron and Ignjatovic [1993].

In his attempted proof of the continuum hypothesis, D. Hilbert [1925] studied classes of higher type functionals defined by the operations of composition and primitive recursion. Hilbert's general scheme [1925, p. 186] was of the form

$$\begin{aligned}&\mathcal{F}(G,H,0)=H,\\&\mathcal{F}(G,H,n+1)=G\big(\mathcal{F}(G,H,n),n\big)\end{aligned}$$

where $\mathcal{F}$, G, H are higher type functionals of appropriate types possibly having other parameters not indicated. Illustrating the power of primitive recursion over higher type objects, Hilbert gave a simple higher type primitive recursive definition of the Ackermann function, known not to be primitive recursive. For example, define

$$\begin{aligned}&F(g,0)=1,\\&F(g,n+1)=g\big(F(g,n)\big).\end{aligned}$$

Then for $n \geqslant 3$, the principal functions f_n from Definition 3.29 satisfy $f_{n+1}(x) = F(f_n, x)$.

DEFINITION 4.19. The set T_ρ of all finite types is defined inductively as follows:
- $0 \in T_\rho$,
- if $\sigma, \tau \in T_\rho$ then $(\sigma \to \tau) \in T_\rho$.

By induction on τ, every type $\rho = (\sigma \to \tau)$ can be written uniquely in the normal form

$$\rho = \rho_1 \to \rho_2 \to \cdots \to \rho_k \to 0$$

when association is to the right and parentheses are dropped. The *level* of a type is defined as follows:
- level $(0) = 0$,
- level $(\rho_1 \to \cdots \to \rho_k \to 0) = 1 + \max_{1 \leqslant i \leqslant k}\{level(\rho_i)\}$.

If F is of type ρ, where $\rho = \rho_1 \to \cdots \to \rho_k \to 0$, then often $F(X_1, \ldots, X_k)$ is written in place of $F(X_1)(X_2)\cdots(X_k)$.

Higher type functional complexity theory is an emerging field. For reasons of space, only references to a few papers will be given. Ker-I Ko [1986] surveyed the theory of sequential complexity theory of real valued functions. H.J. Hoover [1990] investigated parallel computable real valued functions. S. Cook [1992] gave a survey of higher type computational approaches, and proved Theorem 4.9. Cook further proposed that any class C of feasible type 2 functionals must satisfy the following two conditions:

(1) BFF $\subseteq C \subseteq$ OPT,

(2) C is closed under abstraction and application.

A. Seth [1992] defined a class C_2 of type 2 functionals defined by *counter* Turing machines with polynomial bounds, which satisfies the previous conditions, and proved that no recursively presentable class of functionals exists which contains C_2 and satisfies the previous conditions. Seth [1993] further investigated closure conditions for feasible functionals. J. Royer [1994] studied a polynomial time counterpart to the of Kreisel, Lacombe, Shoenfield [1957] theorem.

Complexity theory for functionals of all finite types was initiated by S. Buss [1986b] who introduced a polynomial time analogue of the *hereditarily recursive operations* HRO to define polynomial time functionals of all finite types decorated with runtime bounds. A. Nerode, J. Remmel and A. Scedrov [1989] studied a polynomially graded type system. J.-Y. Girard, A. Scedrov and P. Scott [1990] introduced *bounded linear logic*, and proved a normalization theorem which yielded a characterization of a feasible class of type 2 functionals, whose type 1 section is the class of polytime computable functions. S. Cook and A. Urquhart [1993] introduced an analogue of Gödel's system **T** by admitting a recursor for bounded recursion on notation for type 1 objects. Their system PV^ω provided a natural class of polynomial time higher type functionals (called the *basic feasible functionals of higher type*), whose type-2 section of PV^ω is BFF. V. Harnik [1992] extended Cook–Urquhart's functionals to levels of the polynomial time hierarchy. S. Cook and B. Kapron [1990] characterized the higher type functionals in PV^ω by certain kinds of programming language constructs, *typed while* programs and *bounded loop* programs. This kind of characterization was extended P. Clote, B. Kapron and A. Ignjatovic [1993] to the higher type functionals in NC^ω, relating *bounded loop* programs with higher type parallel complexity classes. A. Seth [1994] extended his definition of *counter* Turing machine to all finite types, thus characterizing PV^ω by a machine model. If one additionally allows dynamic computation of indices of subprograms within this counter Turing machine model, then Seth has conjectured this class to properly contain PV^ω.

In finite model theory, many complexity classes $\mathcal{C}$ have been characterized via word models over a logic as follows: $L \in \mathcal{C}$ iff there is a closed formula ϕ (in a certain logic over a certain signature) for which

$$L = \{w \in \{0,1\}^* \colon w \models \phi\}.$$

As mentioned in the introduction, though techniques are similar in spirit to those surveyed in this paper, for reasons of space we do not present such results here. Another direction of finite model theory is the investigation of function algebras, as interpreted over finite structures, rather than over $\mathbb{N}$. Here Y. Gurevich [1983] showed that LOGSPACE "global" functions can be characterized by primitive recursion over finite structures. A. Goerdt [1992] generalized this to prove that type level $k+1$ recursive definitions over finite structures characterize global functions in the class $\text{DTIME}(\exp_k(n^{O(1)})$ where $\exp_0(n) = n$ and $\exp_{k+1}(n) = 2^{\exp_k(n)}$.

D. Leivant and J.-Y. Marion [1993] gave various characterizations of PTIME by typed λ-calculi with pairing over an algebra $\mathbf{W}$ of words over $\{0, 1\}$. Recently, Leivant and Marion showed how a natural restriction of functional recurrence with substitution generates exactly PSPACE. In a series of papers (see for instance [Leivant, 1994]) D. Leivant investigated various tiering schemes of recursion (extensions of safe recursion) and related complexity classes. Such investigations may have some applicability to programming language design. K.-H. Niggl [1993], building on H. Schwichtenberg [1991], investigated certain subrecursive hierarchies (analogues of primitive recursive) of partial continuous functionals on Scott domains. As evidenced by the articles in the conference proceedings [Leivant, 1995], edited by D. Leivant, higher type functional complexity is an exciting area with many interesting theoretical questions, and the possibility of contributing to new programming language features.

Acknowledgements

Many thanks to T. Altenkirch, S. Bellantoni, M. Hofmann, H. Schwichtenberg, T. Strahm and especially N. Danner and K.-H. Niggl for correcting typos and making various suggestions, though of course, the author is solely responsible for any remaining errors. A special note of thanks is due to Sam Buss, Jan Krajíček, Pavel Pudlák, and Gaisi Takeuti, with whom the author has collaborated either directly or indirectly over a period of years. Thanks to I. Mignani for help in preparing the subject and symbol index. During the preparation of this manuscript, support from the National Science Foundation and Volkswagen Foundation is gratefully acknowledged.

Note added in proof

Professors F. Felix Lara and Alejandro Fernandez Margarit of the University of Sevilla (Spain) have kindly pointed out a missing function from our definition of the Grzegorczyk classes $\mathcal{E}^n$ and $\mathcal{M}^n$, for $n \geqslant 3$. Namely, the function $\max(x, y)$ should be added to the list of initial functions. In fact, F. Felix Lara and A. Fernandez Margarit have shown the following Proposition.

"For all n-ary functions $g \in \mathcal{E}^3$, there exists a unary function $b(x)$ and an index $1 \leqslant i \leqslant n$ such that for all $x_1, \ldots, x_n$ we have $f(x_1, \ldots, x_n) \leqslant b(x_i)$, where b can be obtained by composition of f_3 and successor." A similar omission in the definition of the Grzegorczyk classes appears on p. 33 of Rose [1984], though I have not checked whether this causes a problem in that definition. As referenced in the text, our principal functions in the definition of the Grzegorczyk hierarchy, for $n \geqslant 3$ were taken from Bel'tyukov [1982].

References

W. ACKERMANN
[1928] Zum Hilbertschen Aufbau der reelen Zahlen, *Math. Ann.*, 99, pp. 118–133.

M. AJTAI
[1983] Σ_1^1-formulae on finite structures, *Ann. Pure Appl. Logic*, 24, pp. 1–48.

B. ALLEN
[1991] Arithmetizing uniform NC, *Ann. Pure Appl. Logic*, 53(1), pp. 1–50.

G. ASSER
[1987] Primitive recursive word-functions of one variable, in: *Computation Theory and Logic*, E. Börger, ed., Lecture Notes in Comput. Sci., Vol. 270, Springer, Berlin, pp. 14–19.

L. BABAI AND L. FORTNOW
[1991] Arithmetization: a new method in structural complexity theory, *Computational Complexity*, 1, pp. 41–66.

D. MIX BARRINGTON, N. IMMERMAN AND H. STRAUBING
[1990] On uniformity in NC^1, *J. Computer System Sci.*, 41(3), pp. 274–306.

D. A. BARRINGTON
[1989] Bounded-width polynomial-size branching programs recognize exactly those languages in NC^1, *J. Computer System Sci.*, 38, pp. 150–164.

P. W. BEAME, S. A. COOK AND H. J. HOOVER
[1986] Log depth circuits for division and related problems, *SIAM J. Computing*, 15, pp. 994–1003.

S. BELLANTONI
[1992] *Predicative recursion and computational complexity*, Technical Report 264/92, University of Toronto, Computer Science Department, 164 pages.
[1995] Predicative recursion and the polytime hierarchy, in: *Feasible Mathematics II*, P. Clote and J. Remmel, eds., Birkhäuser, pp. 15–29.

S. BELLANTONI AND S. COOK
[1992] A new recursion-theoretic characterization of the polytime functions, *Computational Complexity*, 2, pp. 97–110.

A. BEL'TYUKOV
[1982] A computer description and a hierarchy of initial Grzegorczyk classes, *J. Soviet Math.*, 20, pp. 2280–2289. Translation from *Zap. Nauchn. Sem. Lening. Otd. Mat. Inst. V. A. Steklova AN SSSR*, Vol. 88, pp. 30–46, 1979.

J. H. BENNETT
[1962] *On Spectra*, PhD thesis, Princeton University, Department of Mathematics.

S. BLOCH
[1994] Function-algebraic characterizations of log and polylog parallel time, *Computational Complexity*, 4(2), pp. 175–205.

S. Bloch, J. Buss and J. Goldsmith

[1995] Sharply bounded alternation within $\mathcal{P}$, to appear in *Proceedings, DMTCS '96*; submitted for journal publication.

R. Boppana and M. Sipser

[1990] The complexity of finite functions, in: *Handbook of Theoretical Computer Science*, Vol. A, J. van Leeuwen, ed., Amsterdam, Elsevier, MIT Press, Cambridge, pp. 759–804.

A. Borodin

[1973] On relating time and space to size and depth, *SIAM J. Computing*, 6, pp. 733–744.

S. Buss

[1986a] *Bounded Arithmetic*, Vol. 3 of *Studies in Proof Theory*, Bibliopolis, 221 pages.

[1986b] The polynomial hierarchy and intuitionistic bounded arithmetic, in: *Structure in Complexity Theory*, A. L. Selman, ed., Lecture Notes in Comput. Sci., Vol. 223, Springer, Berlin, pp. 77–103.

[1987] The boolean formula value problem is in ALOGTIME, in: *Proceedings of the 19th Annual ACM Symposium on Theory of Computing*, pp. 123–131.

S. R. Buss

[1992] The graph of multiplication is equivalent to counting, *Inform. Process. Lett.*, 41, pp. 199–201.

[1993] Algorithms for boolean formula evaluation and for tree contraction, in: *Arithmetic, Proof Theory and Computational Complexity*, P. Clote and J. Krajíček, eds., Oxford Univ. Press, pp. 96–115.

J.-Y. Cai and M. L. Furst

[1988] *PSPACE* survives three-bit bottlenecks, in: *Proceedings of 3th Annual IEEE Conference on Structure in Complexity Theory*, pp. 94–102.

A. Chandra, D. Kozen and L. J. Stockmeyer

[1981] Alternation, *J. Assoc. Comput. Machin.*, 28, pp. 114–133.

A. Chandra, L. J. Stockmeyer and U. Vishkin

[1984] Constant depth reducibility, *SIAM J. Computing*, 13, pp. 423–439.

A. Church

[1936] An unsolvable problem in elementary number theory, *Amer. J. Math.*, 58, pp. 345–363.

P. Clote

[1989] *A first order theory for the parallel complexity class NC*, Technical Report BCCS-89-01, Boston College, Computer Science Department.

[1992a] *ALOGTIME* and a conjecture of S.A. Cook, *Ann. Math. Artif. Intell.*, 6, pp. 57–106.

[1992b] A time-space hierarchy between P and PSPACE, *Math. Systems Theory*, 25, pp. 77–92.

[1993] Polynomial size frege proofs of certain combinatorial principles, in: *Arithmetic, Proof Theory and Computational Complexity*, P. Clote and J. Krajíček, eds., Oxford Univ. Press, pp. 162–184.

[1996] A note on the relation between polynomial time functionals and Constable's class $\mathcal{K}$, in: *Computer Science Logic*, Kleine-Büning, ed., Lecture Notes in Comput. Sci., Springer, Berlin.

[1997a] Nondeterministic stack register machines, *Theoret. Computer Sci. A*, 178, pp. 37–76.

[1997b] A safe recursion scheme for exponential time, in: *Logical Foundations of Computer Science, LFCS'97* (July 6–12, 1997 in Yaroslavl, Russia), L. Beklemishev, ed., Springer.

P. Clote, B. Kapron and A. Ignjatovic

[1993] Parallel computable higher type functionals, in: *Proceedings of IEEE 34th Annual Symposium on Foundations of Computer Science*, Nov. 3–5, 1993. Palo Alto, CA, pp. 72–83.

P. CLOTE AND G. TAKEUTI

[1986] Exponential time and bounded arithmetic, in: *Structure in Complexity Theory*, A. L. Selman, ed., Lecture Notes in Comput. Sci., Vol. 223, Springer, Berlin, pp. 125–143.

[1992] Bounded arithmetics for *NC*, *ALOGTIME*, *L* and *NL*, *Ann. Pure Appl. Logic*, 56, pp. 73–117.

[1995] First order bounded arithmetic and small boolean circuit complexity classes, in: *Feasible Mathematics II*, P. Clote and J. Remmel, eds., Birkhäuser, Boston, pp. 154–218.

P. G. CLOTE

[1988] *A sequential characterization of the parallel complexity class NC*, Technical Report BCCS-88-07, Department of Computer Science, Boston College.

[1990] Sequential, machine-independent characterizations of the parallel complexity classes $ALOGTIME, AC^k, NC^k$ and NC, in: *Feasible Mathematics*, P. J. Scott, S. R. Buss, eds., Birkhäuser, pp. 49–70.

A. COBHAM

[1965] The intrinsic computational difficulty of functions, in: *Logic, Methodology and Philosophy of Science II*, Y. Bar-Hillel, ed., North-Holland, Amsterdam, pp. 24–30.

R. CONSTABLE

[1973] Type 2 computational complexity, in: *5th Annual ACM Symposium on Theory of Computing*, pp. 108–121.

S. COOK

[1992] Computability and complexity of higher type functions, in: *Logic from Computer Science*, Y. N. Moschovakis, ed., Springer, Berlin, pp. 51–72.

S. A. COOK

[1971] The complexity of theorem proving procedures, in: *3rd Annual ACM Symposium on Theory of Computing*, pp. 151–158.

S. A. COOK AND B. M. KAPRON

[1990] Characterizations of the feasible functionals of finite type, in: *Feasible Mathematics*, P. J. Scott and S. R. Buss, eds., Birkhäuser, pp. 71–98.

S. A. COOK AND A. URQUHART

[1993] Functional interpretations of feasibly constructive arithmetic, *Ann. Pure Appl. Logic*, 63(2), pp. 103–200.

P. CSILLAG

[1947] Eine Bemerkung zur Auflösung der eingeschachtelten Rekursion, *Acta Sci. Math. Szeged.*, 11, pp. 169–173.

D. DEUTSCH

[1985] Quantum theory, the Church–Turing principle and the universal quantum computer, *Proc. Royal Soc. London*, A 400, pp. 73–90.

R. FAGIN

[1974] Generalized first–order spectra and polynomial–time recognizable sets, in: *Complexity of Computation*, SIAM–AMS Proceedings, Vol. 7, R. M. Karp, ed., pp. 43–73.

[1990] Finite-model theory – a personal perspective, in: *Proc. 1990 International Conference on Database Theory*, S. Abiteboul and P. Kanellakis, eds., Lecture Notes in Comput. Sci. 470, Springer, Berlin, pp. 3–24. Journal version to appear in *Theoretical Computer Science*.

S. FORTUNE AND J. WYLLIE

[1978] Parallelism in random access machines, in: *10th Annual ACM Symposium on Theory of Computing*, pp. 114–118.

M. FURST, J. B. SAXE AND M. SIPSER

[1984] Parity circuits and the polynomial time hierarchy, *Mathematical Systems Theory*, 17, pp. 13–27. Preliminary version in *Proceedings of the 22nd IEEE Symposium on Foundations of Computer Science*, 1981.

J.-Y. GIRARD, A. SCEDROV AND P. SCOTT

[1990] Bounded linear logic, in: *Feasible Mathematics*, P. J. Scott and S. R. Buss, eds., Birkhäuser, pp. 195–210.

K. GÖDEL

[1931] Über formal unentscheidbare Sätze der Principia Mathematica und verwandter Systeme, *J. Monat. Math. Phys.*, 38, pp. 173 –198.

[1975] *Conversation with G. E. Sacks*, Institute for Advanced Study.

A. GOERDT

[1992] Characterizing complexity classes by general recursive definitions in higher types, *Information and Computation*, 101(2), pp. 202–218.

L. GOLDSCHLAGER

[1977] *Synchronous parallel computation*, Technical Report 114, University of Toronto, 131 pages.

[1982] A unified approach to models of synchronous parallel machines, *J. Assoc. Comput. Machin.*, 29(4), pp. 1073–1086.

A. GRZEGORCZYK

[1953] Some clases of recursive functions, *Rozprawy Matematyczne*, 4.

Y. GUREVICH

[1983] Algebras of feasible functions, in: *Proceedings of 24th IEEE Symposium on Foundations of Computer Science*, pp. 210–214.

Y. GUREVICH AND S. SHELAH

[1986] Fixed-point extensions of first-order logic, *Ann. Pure Appl. Logic*, 32, pp. 265–280.

[1989] Nearly linear time, in: *Symposium on Logical Foundations of Computer Science*, Pereslavl-Zalessky, USSR, Lecture Notes in Comput. Sci., Vol. 363, Springer, Berlin, pp. 108–118.

P. HÁJEK AND P. PUDLÁK

[1992] *Metamathematics of First Order Arithmetic*, Springer, Berlin.

W. HANDLEY, J. B. PARIS AND A. J. WILKIE

[1984] Characterizing some low arithmetic classes, in: *Theory of Algorithms*, Colloquia Societatis Janos Bolyai, Akademie Kyado, Budapest, pp. 353–364.

W. G. HANDLEY

[1994a] LTH plus nondeterministic summation mod M_3 yields ALINTIME, Submitted.

[1994b] Deterministic summation modulo $\mathcal{B}_n$, the semi-group of binary relations on $\{0, 1, \ldots, n-1\}$, Submitted.

G. H. HARDY

[1904] A theorem concerning the infinite cardinal numbers, *Quart. J. Math.*, pp. 87–94.

V. HARNIK

[1992] Provably total functions of intuitionistic bounded arithmetic, *J. Symbolic Logic*, 57(2), pp. 466–477.

K. HARROW

[1975] Small Grzegorczyk classes and limited minimum, *Z. Math. Logik*, 21, pp. 417–426.

[1979] Equivalence of some hierarchies of primitive recursive functions, *Z. Math. Logik*, 25, pp. 411–418.

D. HILBERT
[1925] Über das Unendliche, *Math. Ann.*, 95, pp. 161–190.

W. D. HILLIS AND G. L. STEELE, JR.
[1986] Data parallel algorithms, *Commun. ACM*, 29(12), pp. 1170–1183.

H. J. HOOVER
[1990] Computational models for feasible real analysis, in: *Feasible Mathematics*, S. R. Buss and P. J. Scott, eds., Birkhäuser, pp. 221–238.

N. IMMERMAN
[1987] Languages that capture complexity classes, *SIAM J. Comput.*, 16, pp. 760–778.
[1989] Expressibility and parallel complexity, *SIAM J. Comput.*, 18(3), pp. 625–638.

J. JOHANNSEN
[1996] *Schwache Fragmente der Arithmetik und Schwellwertschaltkreise beschränkter Tiefe*, PhD thesis, Universität Erlangen-Nürnberg.

J. P. JONES AND Y. MATIJASEVIČ
[1982] A new representation for the symmetric binomial coefficient and its applications, *Ann. Sci. Math. Quebec*, 6(1), pp. 81–97.

N. D. JONES AND A. L. SELMAN
[1974] Turing machines and the spectra of first-order formulas, *J. Symbolic Logic*, 39, pp. 139–150.

L. KÁLMAR
[1943] Egyszerü példa eldönthetetlen aritmetikai problémára, *Mate és Fizikai Lapok*, 50, pp. 1–23. (In Hungarian with German abstract.)

R. KANNAN
[1981] Towards separating nondeterministic time from deterministic time, in: *Proceedings of 22nd IEEE Symposium on Foundations of Computer Science*, pp. 235–243.

B. KAPRON AND S. COOK
[1996] A new characterization of type-2 feasibility, *SIAM J. Comput.*, 25(1), pp. 117–132. Preliminary version appeared in *IEEE Symposium on Foundations of Computer Science*, pp. 342–347 (1991).

R. M. KARP AND V. RAMACHANDRAN
[1990] Parallel algorithms for shared-memory machines, in: *Handbook of Theoretical Computer Science*, Vol. A, J. van Leeuwen, ed., Elsevier, Amsterdam, MIT Press, Cambridge, pp. 871–942.

S. C. KLEENE
[1936a] General recursive functions of natural numbers, *Math. Ann.*, 112, pp. 727–742.
[1936b] Lambda-definability and recursiveness, *Duke Math. J.*, 2, pp. 340–353.

KER-I KO
[1986] Applying techniques of discrete complexity theory to numerical computation, in: *Studies in Complexity Theory*, R. V. Book, ed., Wiley, pp. 1–62.

J. KRAJÍČEK
[1995] *Bounded Arithmetic, Propositional Logic, and Complexity Theory*, Cambridge Univ. Press.

G. KREISEL, D. LACOMBE AND J. R. SHOENFIELD
[1957] Partial recursive functionals and effective operations, in: *Constructivity in Mathematics: Proceedings of a colloquium held in Amsterdam*, A. Heyting, ed., North-Holland, Amsterdam, pp. 195–207.

M. KUTYŁOWSKI
[1988] Finite automata, real time processes and counting problems in bounded arithmetics, *J. Symbolic Logic*, 53(1), pp. 243–258.

D. LEIVANT

[1989] Stratified polymorphism, in: *Proceedings of IEEE 4th Annual Symposium on Logic in Computer Science*, pp. 39–47. Journal version: Finitely stratified polymorphism, *Information and Computation*, 93 (1991), pp. 93–113.

[1991] A foundational delineation of computational feasibility, in: *Proceedings of IEEE 6th Annual Symposium on Logic in Computer Science*.

[1993] Stratified functional programs and computational complexity, in: *Conference Record of the Twentieth Annual ACM Symposium on Principles of Programming Languages*.

[1994] Ramified recurrence and computational complexity I: word recurrence and poly-time, in: *Feasible Mathematics II*, P. Clote and J. Remmel, eds., Birkhäuser, pp. 320–343.

[1995] *Logic and Computational Complexity*, Lecture Notes in Comput. Sci., Vol. 960, Springer, Berlin.

D. LEIVANT AND J.-Y. MARION

[1993] Lambda-calculus characterizations of poly-time, *Fundamenta Informaticae*, 19, pp. 167–184.

J. C. LIND

[1974] *Computing in logarithmic space*, Technical Report Project MAC Technical Memorandum 52, Massachusetts Institute of Technology.

S. S. MARČENKOV

[1980] A superposition basis in the class of Kalmar elementary functions, *Mat. Zametki*, 27(3), pp. 321–332. Translation in *Mathematical Notes of the Academy of Sciences of the USSR*, Plenum Publishing Company.

K. MEHLHORN

[1976] Polynomial and abstract subrecursive classes, *J. Computer System Sci.*, 12, pp. 147–178.

A. R. MEYER AND D. RITCHIE

[1967] The complexity of loop programs, in: *Proc. ACM Nat. Conf.*, pp. 465–469.

B. MONIEN

[1977] A recursive and grammatical characterization of exponential time languages, *Theoret. Computer Sci.*, 3, pp. 61–74.

S. S. MUCHNICK

[1976] The vectorized Grzegorczyk hierarchy, *Z. Math. Logik.*, 22, pp. 441–80.

H. MÜLLER

[1974] *Klassifizierungen der primitiv rekursiven Funktionen*, PhD thesis, Universität Münster.

V. A. NEPOMNJASCII

[1970] Rudimentary predicates and turing calculations, *Dokl. Akad. Nauk SSSR*, 195, pp. 29–35. Translated in *Soviet Math. Dokl.*, 11, pp. 1462–1465.

A. NERODE, J. REMMEL AND A. SCEDROV

[1989] Polynomially graded logic I – a graded version of system T, in: *Proceedings of IEEE 4th Annual Symposium on Logic in Computer Science*.

A. P. NGUYEN

[1996] *A formal system for linear space reasoning*, Technical Report 300/96, University of Toronto.

K.-H. NIGGL

[1993] Subrecursive hierarchies on Scott domains, *Arch. Math. Logic*, 32, pp. 239–257.

M. J. O'DONNELL

[1985] *Equational Logic as a Programming Language*, MIT Press.

J. OTTO

[1994] *Tiers, tensors, and Δ_0^0*, Talk at meeting LCC, Indianapolis, organizer D. Leivant, pp. 13–16.

[1995] *Complexity Doctrines*, PhD thesis, Department of Mathematics and Statistics, McGill University.

[1996] *Half tiers and linear space (and time)*, Talk at DIMACS Workshop on Computational Complexity and Programming Languages, organized by B. M. Kapron and J. Royer, July 25–26, 1996.

S. V. PAKHOMOV

[1979] Machine independent description of some machine complexity classes (in Russian), *Issledovanija po Konstrukt. Matemat. i Mat. Logike*, VIII, pp. 176–185, LOMI.

J. B. PARIS AND A. J. WILKIE

[1983] Counting problems in bounded arithmetic, in: *Methods in Mathematical Logic*, C. A. di Prisco, ed., Lecture Notes in Mathematics, Proceedings of Logic Conference held in Caracas, Springer, Berlin, pp. 317–340.

R. PÉTER

[1936] Über die mehrfache Rekursion, *Math. Ann.*, 113, pp. 489–526.

F. PITT

[1995] *$\widehat{N}_0$ and ALOGTIME*, Typeset manuscript.

P. PUDLÁK

[1994] Complexity theory and genetics, in: *Proceedings of 9th Annual IEEE Conference on Structure in Complexity Theory*.

R. W. RITCHIE

[1963] Classes of predictably computable functions, *Trans. Amer. Math. Soc.*, 106, pp. 139–173.

J. ROBINSON

[1947] Primitive recursive functions, *Bull. Amer. Math. Soc.*, 53, pp. 923–943.

[1949] Definability and decision problems in arithmetic, *J. Symbolic Logic*, 14, pp. 98–114.

H. E. ROSE

[1984] *Subrecursion: Function and Hierarchies*, Oxford Logic Guides, Vol. 9, Clarendon Press, Oxford.

J. S. ROYER

[1994] *Semantics vs. Syntax vs. Computation*, Typescript, November 29, 1994.

W. L. RUZZO

[1981] On uniform circuit complexity, *J. Comput. System Sci.*, 22, pp. 365–383.

W. J. SAVITCH

[1970] Relationship between nondeterministic and deterministic tape complexities, *J. Comput. System Sci.*, 4, pp. 177–192.

C. P. SCHNORR

[1978] Satisfiability is quasilinear complete in *NQL*, *J. Assoc. Comput. Machin.*, 25(1), pp. 136–145.

H. SCHOLZ

[1952] Ein ungelöstes Problem in der symbolischen Logik, *J. Symbolic Logic*, 17, p. 160.

H. SCHWICHTENBERG

[1969] Rekursionszahlen und die Grzegorczyk-Hierarchie, *Arch. Math. Logik.*, 12, pp. 85–97.

[1991] Primitive recursion on the partial continuous functionals, in: *Informatik und Mathematik*, M. Broy, ed., Springer, Berlin, pp. 251–259.

A. SETH
[1992] There is no recursive axiomatization for feasible functionals of type 2, in: *Proceedings of IEEE 7th Annual Symposium on Logic in Computer Science*, pp. 286–295.
[1993] Some desirable conditions for feasible functionals of type 2, in: *Proceedings of IEEE 8th Annual Symposium on Logic in Computer Science*.
[1994] Turing machine characterizations of feasible functionals of all finite types, in: *Feasible Mathematics II*, P. Clote and J. Remmel, eds., Birkhäuser, pp. 407–428.

Y. SHILOACH AND U. VISHKIN
[1982] Finding the maximum, merging and sorting in a parallel computation model, *J. Algorithms*, 3, pp. 57–67.

P. SHOR
[1994] Algorithms for quantum computation: discrete log and factoring, in: *Proceedings of IEEE 35th Annual Symposium on Foundations of Computer Science*.

H. SIMMONS
[1988] The realm of primitive recursion, *Arch. Math. Logic*, 27, pp. 177–188.

TH. SKOLEM
[1923] Begründung der elementaren Arithmetik durch die rekurrierende Denkweise ohne Anwendung scheinbarer Veränderlichen mit unendlichem Ausdehnungsbereich, *Skrifter utgit av Videnskapsselskapet, I. Mate. Klasse*, 6, Oslo.

R. SMULLYAN
[1961] *Theory of Formal Systems*, Annals of Mathematical Studies, Vol. 47, Princeton Univ. Press.

L. STOCKMEYER AND U. VISHKIN
[1984] Simulation of parallel random access machines by circuits, *SIAM J. Computing*, 13, pp. 409–422.

G. TAKEUTI
[1995] Frege proof system and TNC^0, in: *Logic and Computational Complexity*, D. Leivant, ed., Lecture Notes in Comput. Sci., Vol. 960, Springer, Berlin, pp. 221–252.

D. B. THOMPSON
[1972] Subrecursiveness: machine independent notions of computability in restricted time and storage, *Math. Systems Theory*, 6, pp. 3–15.

M. TOWNSEND
[1990] Complexity for type-2 relations, *Notre Dame J. Formal Logic*, 31, pp. 241–262.

A. M. TURING
[1936–37] On computable numbers, with an application to the Entscheidungsproblem, *Proc. London Math. Soc., Ser.* 2, 42, pp. 230–265.

L. VALIANT
[1979] The complexity of computing the permanent, *Theoret. Computer Sci.*, 8, pp. 189–201.

P. VAN EMDE BOAS
[1990] Machine models and simulations, in: *Handbook of Theoretical Computer Science*, Vol. A, J. van Leeuwen, ed., Elsevier, Amsterdam, MIT Press, Cambridge, pp. 1–66.

H. VOLLMER AND K. WAGNER
[1996] Recursion theoretic characterizations of complexity classes of counting functions, *Theoret. Computer Sci.*, 163, pp. 245–258.

K. WAGNER
[1979] Bounded recursion and complexity classes, in: *Lecture Notes in Comput. Sci.*, Vol. 74, Springer, Berlin, pp. 492–498.

K. WAGNER AND G. WECHSUNG

[1986] *Computational Complexity*, Reidel Publishing Co.

A. J. WILKIE

[1983] Modèles non standard de l'arithmétique, et complexité algorithmique, in: *Modèles Non Standard en Arithmétique et en Théorie des Ensembles*, A. J. Wilkie and J.-P. Ressayre, eds., Publications Mathématiques de l'Université Paris VII, pp. 5–45.

A. WOODS

[1986] Bounded arithmetic formulas and Turing machines of constant alternation, in: *Logic Coloquium 1984*, J. B. Paris, A. J. Wilkie and G. M. Wilmers, eds., North-Holland, Amsterdam.

C. WRATHALL

[1976] Complete sets and the polynomial time hierarchy, *Theoret. Computer Sci.*, 3, pp. 23–33.

List of symbols

$\Box_i^P$, 637, 638, 653

$*$, 612

$\dot{-}$, 604

$\oplus$, 606

$\leqslant_T$, 656

$\leqslant_T^P$, 656

$\leqslant_T^{NP}$, 597

$\leqslant_T^P$, 597

$\vdash_M$, 595

$\vdash_M^*$, 594

$\vdash_M^n$, 594

$|f|$, 659

$\#P$, 640

$|x|$, 610

$\|x\|$, 610

$[\mathcal{X}; \text{OP}]$, 608

A_0, 610, 619

A_{lmax}, 661

$A_M(\vec{f}, \vec{x}, t)$, 657

$\mathcal{A}$, 666

$\mathcal{A}_0$, 666

AC^k, 607

ACC(2), 624

ACC(6), 624

ACC(k), 607

ADD, 610

ALINTIME, 637, 598, 640

ALOGTIME, 598

Ap, 658

APOLYLOGTIME, 598

ASPACE($S(n)$), 598

ATM, 597

ATIME($n^{O(1)}$), 630

ATIME($T(n)$), 598

$\alpha \vdash_M \beta$, 593

B, 647

B_f, 602

BFF, 658–660

$BFF(f)$, 639

BFF(X), 658

BH, 654

BH(x), 641

BIT, 610

BMIN, 627

BPROD, 638

BR, 625

BRN, 622

BS, 617

BSUM, 638

BVR, 643

β, 617

$\beta(i, z)$, 618

C_2, 668

$c_P(\vec{x})$, 608

C(;A,B,C), 647

CA, 633, 636

COMP, 609

cond, 611, 612

$cond0$, 612
$cond1$, 612
$cond2$, 612
CRAM, 603
CRCW, 602
CREW, 602, 665
CRN, 610
$\mathcal{CW}_k(\mathcal{F})$, 641
$\mathcal{CW}_k(\mathcal{F})_*$, 641

$d(P)$, 660
DCL, 607
DSPACE($n \cdot (\log^{(k)} n)^{O(1)}$), 631
DSPACE($S(n)$), 595
DTIME($T(n)$), 595
DTIMESPACE($n^{O(1)}$, $O(n)$), 631
DTIMESPACE($T(n)$, $S(n)$), 595
DTIMESPACE(T, S), 631
Δ_0, 633
Δ_0 FORMULA, 633

$E(z, w; x)$, 649
EREW, 602
ETIME, 595, 644, 653
EXPTIME, 595
$(\mathcal{E} f)_*$, 627
$\mathcal{E}^2$, 653
$\mathcal{E}^2_*$, 630
$\mathcal{E}^2_* =$ LINSPACE, 626
$\mathcal{E}^3_*$, 639
$\mathcal{E}^k$, 630
$\mathcal{E}^n$, 626, 654
$\mathcal{E}^n_*$, 626
$\mathcal{E} f$, 626

f_0, 626
f_1, 626
f_2, 626
f_3, 626
f_{n+1}, 626
f_Q, 660
$f^{(n)}(x)$, 602
FH, 654
FH(x), 641
$FP(f)$, 639
$\mathcal{F}_*$, 608
$\mathcal{F}$ALOGTIME, 624
$\mathcal{F}$LH, 618, 619
$\mathcal{F}$LINSPACE, 628
$\mathcal{F}$LOGSPACE, 636
$\mathcal{F}$LTH, 636
$\mathcal{F}$PH, 637
$\mathcal{F}$PSPACE, 625, 630, 631
$\mathcal{F}$PTIME, 622, 625, 647, 651
$\mathcal{FC}$, 602

G_f, 602
$\mathcal{G}$CA, 633
GPTIME, 623
$\mathcal{GC}$, 602
Γ^N, 636

$H(x)$, 654
$\mathcal{H}_n$, 626

I, 609
$I^{n,m}_j$, 647
I^n_k, 609
$initial_M$, 616
INITIAL$_M$, 616
ITER, 609

$K(; a, b, c)$, 653
$K(f)$, 639
k-BR, 630
k-BRN, 624
k-BTRN, 642
$\mathcal{K}$, 639
$\mathcal{K}(f)$, 639

$L(M)$, 594
$L_2(QF^+)$, 646
ℓh, 617
LH, 598
$Lin\mathcal{FC}$, 602
LINSPACE, 630
$\log^{(n)} x$, 602
$Log\mathcal{FC}$, 602
LOGSPACE, 636
LRN, 658
lsp, 614
LSP, 614
LSP, 604
LTH, 598, 636
LTH $\subseteq \mathcal{E}^0_*$, 626

m-COUNTING, 641

MAJ, 621
MAP, 641
MAXQUERY^c_M, 661
msp, 614
MSP, 614
MSP, 604
$\text{MOD}_{k,n}$, 606
MOD2, 610
$\mathcal{M}^1_*$, 633
$\mathcal{M}^2$, 633, 636
$\mathcal{M}^2_*$, 633, 636
$\mathcal{M}^n$, 632
μB, 653
μB_i, 653
μFP_i, 638

NC, 607, 641, 654
NC, 639
NC^0, 655
NC^k, 607, 625
NC^ω, 668
$next_M$, 616
NEXT_M, 616
NLT, 632
NORMAL $\cap [\mathcal{F}; \mathcal{O}]$, 646
(NORMAL $\cap [\mathcal{F}; \mathcal{O}])_*$, 647
NP, 595
NSPACE($S(n)$), 595
NSPACE(O(log(n))), 636
NTIME($T(n)$), 595
NTIMESPACE($T(n), S(n)$), 595
NTM, 595

OCRAM, 663, 664, 666
ones, 611
OPT, 657
OTM, 657
$\Omega(f)$, 594

P, 595
P(;A), 647
pad, 611
PBBR, 641
PBRN, 648
PH, 598
PID, 602
POLYLOGTIME, 655
POTM, 657
PR, 609
Pr(; a), 653
PRAM, 602
PRES, 633
PSPACE, 595, 630, 669
PTIME, 595, 652
PV^ω, 668
$\mathcal{PR}$, 609, 643
$\mathcal{PR}_1$, 610
π_1, 615
π_2, 615

$q_?$, 596
q_A, 593
q_R, 593
q_{no}, 596
q_{yes}, 596
$Q(M, f, x, t, g)$, 665
$Q_M(f, x, t)$, 660
$Q_M(\vec{f}, \vec{x}, t)$, 657
$Q_{\mathcal{M}}(f, x, t)$, 665
$QA_M(\vec{f}, \vec{x}, t)$, 657
QL, 632

O(f), 594

RATM, 595
rev, 611
*rev*0, 611
RF, 633
RF_*, 633
rk_O, 609
$rk_{O,R}$, 608
RTM, 632
$(\mathcal{R}_2)_*$, 640
$(\mathcal{R}_3)_*$, 640
$(\mathcal{R}_n)_*$, 640
$\mathcal{R}_k$, 640
$\rho = (\sigma \to \tau)$, 667

$S(; a)$, 653
$S(t)$, 657
$S_0(; a)$, 647
$S_1(; a)$, 647
$s(x)$, 609
$s_0(x)$, 609
$s_1(x)$, 609
SBBITSUM, 621

SBMAX, 613
SBMIN, 613
SBPROD, 638
SBRN, 623
SBSUM, 638
SCOMP, 647
SDCR, 655
sg, $\overline{sg}$, 611
SHL, 641
SIZEDEPTH($S(n)$, $D(n)$), 606
SMIN, 652
SR, 653
SRN, 647
SWRN, 654
Σ_0^P, 597
Σ_{n+1}^P, 597
Σ_n-TIME($T(n)$), 637

T_ρ, 667
TC^0, 607, 621
$\mathrm{TH}_{k,n}$, 606
TIMEPROC($T(n)$, $P(n)$), 604
TL, 615
TM, 593
TR, 615
$\Theta(f)$, 594
$\tau(x, y)$, 615
τ_n, 615

VR, 643

W, 669
WBPR, 631
WBRN, 624, 666
$\mathcal{W}_2(\mathcal{F})_*$, 641
$\mathcal{W}_3(\mathcal{F})_*$, 641
$\mathcal{W}_k(\mathcal{F})$, 641
$\mathcal{W}_k(\mathcal{F})_*$, 641

$x\#y$, 610
xBy, 613
xEy, 613
xPy, 613

CHAPTER 18

Polynomial Time Reducibilities and Degrees

Klaus Ambos-Spies

Mathematisches Institut, Universität Heidelberg, D-69120 Heidelberg, Germany
E-mail: ambos@math.uni-heidelberg.de

Contents

1. Introduction 684
2. Basic definitions and results 685
3. The meet operator and gap languages 688
4. Delayed diagonalization or the looking back technique 690
5. The iterated look-ahead technique 693
6. The theory of the polynomial time degrees 696
7. Other reducibilities and the axiomatic approach 700
References 702

HANDBOOK OF COMPUTABILITY THEORY
Edited by E.R. Griffor

1. Introduction

Just as the recursive reducibilities are fundamental concepts for the classification of the undecidable problems according to their degree of unsolvability, resource-bounded reducibilities can serve as natural tools for the classification of the decidable problems according to their degree of complexity. First examples of subrecursive reducibilities, namely primitive recursive and elementary recursive reducibilities were introduced by Kleene [1958], Axt [1959], Machtey [1974] and others in order to investigate hierarchies of the recursive sets in general and of the primitive recursive sets, respectively. Later, by the influence of theoretical computer science, the interest shifted to reducibilities of lower complexity. In particular, the polynomial time bounded variants of Turing (p-T) and many-one (p-m) reducibilities introduced by Cook [1971], Karp [1972] and Levin [1973] became central concepts in computational complexity theory. There is the widely accepted thesis that feasible computability coincides with computability by a deterministic Turing machine in polynomial time. By this complexity theoretical analogue of Church's thesis, the polynomial time reducibilities formalize feasibly computable reductions, thereby qualifying as the most natural tool for the comparison of the relative complexity of the solvable but intractable problems, i.e. of the problems in $\mathbf{REC} - \mathbf{P}$.

While the main applications of the effective reducibilities are undecidability proofs based on completeness or hardness of a problem for a class containing nonrecursive sets (most frequently for the class of recursively enumerable sets), the polynomial time reducibilities were designed for intractability proofs based on hardness or completeness for hyperpolynomial classes. Here the interest focused on completeness results for the class **NP** of the problems solvable in polynomial time by a nondeterministic Turing machine. In fact Cook [1971] and Karp [1972] introduced the polynomial time reducibilities for showing the satisfiability problem for propositional formulas and some fundamental graph problems to be **NP**-complete. Though it is not known that **NP** properly contains **P**, this is widely believed, whence **NP**-hardness can be taken as strong evidence for intractability. In the sequel a great number of interesting problems could be shown to be **NP**-complete (see Garey and Johnson [1978]).

Though the main interest in the polynomial time reducibilities is directed to intractability proofs based on hardness results, the exploration of the algebraic structure of these reducibilities on the recursive sets and their corresponding degree structures attracted a number of researchers. This chapter is devoted to these structural investigations which are mainly of recursion theoretic nature.

Our goal is to summarize the most important results in this area, to state some of the central open problems, and, by sketching three of the most fundamental diagonalization techniques, to give some flavour of the methods used in this area. In Sections 2 and 3 we state some of the fundamental facts and discuss the role of effective finite extension arguments in this setting. Section 4 is devoted to the delayed diagonalization technique, which is the key for density type results. The third important technique, here called iterated look-ahead technique, is introduced in Section 5.

The last two fundamental techniques were introduced in 1975 in the seminal paper of Ladner [1975], which can be viewed as the origin of the theory of the polynomial time degrees. Section 6 discusses some global properties of the degree structures focusing on results about their first-order theory. Finally, in Section 7, we shortly discuss some other reducibilities and some attempts for axiomatic approaches to the subrecursive reducibilities in general.

We conclude this introduction with explaining some of our notation. We identify the set $\omega = \{0, 1, 2, \ldots\}$ of natural numbers with the set $\Sigma^* = \{0, 1\}^*$ of (finite) binary strings by letting x be the $(x + 1)$-th string in the canonical ordering. The length of x is denoted by $|x|$. We use the terms *set* and *problem* synonymously to denote sets of strings, while the term *class* refers to a set of problems. Lower case letters (with the exception of f, g, h, t which are reserved for functions on ω) denote strings, capital letters denote sets, and boldface capital letters denote classes. A set A is identified with its characteristic function, i.e. $A(x) = 1$ iff $x \in A$ and $A(x) = 0$ iff $x \notin A$. $A \restriction x = \{y < x\colon\ y \in A\}$ is the finite initial segment of A below x.

We assume the reader to be familiar with the basic facts on time and space complexity (see, e.g., Hopcroft and Ullman [1979]). Besides the classes **REC**, **PRIM**, **ELEMENTARY** of the recursive, primitive recursive, and elementary recursive sets, respectively, we will focus on the complexity classes

$$\begin{aligned} \mathbf{P} &= \mathbf{DTIME}(poly(n)) \quad \text{(Deterministic Polynomial Time)},\\ \mathbf{NP} &= \mathbf{NTIME}(poly(n)) \quad \text{(Nondeterministic Polynomial Time)},\\ \mathbf{EXP} &= \mathbf{DTIME}(2^{poly(n)}) \quad \text{(Deterministic Exponential Time)}, \end{aligned}$$

and we will use the same notation for the corresponding function classes.

We call a recursive function $f : \omega \to \omega$ *p-constructible* if $f(n)$ can be computed in $poly(f(n))$ steps, and by saying that f is *hyperpolynomial* we abbreviate that f is a nondecreasing function dominating all polynomials. Then a class $\mathbf{C}$ is called a *hyperpolynomial time class* if $\mathbf{DTIME}(f(n)) \subseteq \mathbf{C}$ for some p-constructible hyperpolynomial function f. Note that, for such a class $\mathbf{C}$, $\mathbf{P} \subset \mathbf{C}$. A class $\mathbf{C}$ of recursive sets is called *recursively presentable* (*r.p.*) if $\mathbf{C}$ is empty or if $\mathbf{C}$ consists of the rows of a binary recursive set, i.e. $\mathbf{C} = \{U_n\colon\ n \geqslant 0\}$ for some recursive set $U \subseteq \omega \times \Sigma^*$ where $U_n = \{x\colon\ (n, x) \in U\}$. Finally, we say that A is a subproblem of B if $A = B \cap C$ for some set $C \in \mathbf{P}$.

2. Basic definitions and results

By imposing polynomial time bounds on the reduction procedure, the classical Turing and many-one reducibilities turn into the two most fundamental polynomial time reducibilities.

DEFINITION 2.1. (a) (Cook [1971]) A set A is *Turing reducible* to a set B *in polynomial time* ($A \leqslant_T^P B$) if there is a deterministic polynomial time bounded oracle Turing machine M which accepts A with oracle B, i.e. $A(x) = M^B(x)$ for all strings x.

(b) (Karp [1972], Levin [1973]) A set A is *many-one reducible* to a set B *in polynomial time* ($A \leqslant_m^P B$) if there is a function $f \in \mathbf{P}$ such that $A = f^{-1}(B)$, i.e. $A(x) = B(f(x))$ for all strings x.

If $A \leqslant_T^P B (A \leqslant_m^P B)$ we shortly say that A is p-T-reducible (p-m-reducible) to B. In the following we let r stand for m and T. Note that $A \leqslant_m^P B$ implies $A \leqslant_T^P B$. Also note that any subproblem of a set A is p-m-reducible to A, i.e. $A \cap B \leqslant_m^P A$ for all $B \in \mathbf{P}$. Moreover, the p-reducibilities are reflexive and transitive, whence *p-r-equivalence* defined by

$$A =_r^P B \quad \text{iff} \quad A \leqslant_r^P B \text{ and } B \leqslant_r^P A$$

is an equivalence relation. The equivalence class of a set A is called the *p-r-degree* of A and it is denoted by

$$deg_r^P(A) = \{B\colon A =_r^P B\}.$$

We let boldface lower case letters denote p-r-degrees of recursive sets. For any class $\mathbf{C}$ we let $\mathbf{C}_r$ be the class of p-r-degrees of sets in $\mathbf{C}$,

$$\mathbf{C}_r = \{deg_r^P(A)\colon A \in \mathbf{C}\},$$

and we say that $\mathbf{C}_r$ is recursively presentable if $\mathbf{C}$ is r.p. The partial order induced on the p-r-degrees by $\leqslant_r^P$ is denoted by $\leqslant$:

$$\mathbf{a} \leqslant \mathbf{b} \text{ iff } (\exists A \in \mathbf{a})(\exists B \in \mathbf{b})(A \leqslant_r^P B).$$

Then $(\mathbf{REC}_r, \leqslant)$ is a partial order with least element $\mathbf{0} = \mathbf{P}$. (In case of $\leqslant_m^P$ we ignore the sets $\emptyset$ and $\{0,1\}^*$ to avoid trivialities.) In fact, $(\mathbf{REC}_r, \leqslant)$ is an *upper semilattice* (*u.s.l.*) where the join (supremum) $\mathbf{a} \vee \mathbf{b}$ of two degrees $\mathbf{a}$ and $\mathbf{b}$ is represented by the join $A \oplus B = \{0x\colon x \in A\} \cup \{1x\colon x \in B\}$ of any sets $A \in \mathbf{a}$ and $B \in \mathbf{b}$.

A class $\mathbf{I}$ of recursive p-r-degrees is called an ideal in $(\mathbf{REC}_r, \leqslant)$ if $\mathbf{I}$ is closed under $\vee$ and downwards closed under $\leqslant$. An ideal $\mathbf{I} = [\mathbf{0}, \mathbf{a}]$ with greatest element $\mathbf{a}$ is called a principal ideal. The classes $\mathbf{EXP}_r$, $\mathbf{ELEMENTARY}_r$, $\mathbf{PRIM}_r$ are ideals in $(\mathbf{REC}_r, \leqslant)$ and $\mathbf{NP}_m$ is an ideal in $(\mathbf{REC}_m, \leqslant)$, but it is not known whether $\mathbf{NP}$ is closed under $\leqslant_T^P$. (In fact, Ambos-Spies [1987a] constructed an oracle relative to which $\mathbf{NP}_T$ is not an ideal in $(\mathbf{REC}_T, \leqslant)$.) Moreover, the classes $\mathbf{EXP}$ and $\mathbf{NP}$ possess p-m-complete problems. (Examples of such problems are bounded versions of the halting problem consisting of the coded triples $\langle e, x, \tilde{n}\rangle$ such that the eth deterministic respectively non-deterministic Turing machine accepts input x within n steps, where for $\mathbf{EXP}$ the runtime is presented in binary ($\tilde{n} = n$) and for $\mathbf{NP}$ in unary ($\tilde{n} = 0^n$); see, e.g., Balcazar et al. [1988].) Hence the ideals $\mathbf{EXP}_r$ and $\mathbf{NP}_m$ are principal, whereas $\mathbf{ELEMENTARY}_r$ and $\mathbf{PRIM}_r$ are not principal (see below).

For analyzing the structure of the polynomial degrees of recursive sets, it is crucial to note that there are effective enumerations of the p-r-reductions, i.e. there are uniformly recursive sequences of total functionals defining these reducibilities. In fact, such enumerations can be found in any hyperpolynomial time class. For the sake of simplicity, here we will focus on **EXP** which is the least of the commonly investigated deterministic time complexity classes which properly contain **P** and which are downward closed under $\leqslant_r^P$. Moreover, **EXP** is the least deterministic time class known to contain **NP**. We should mention, however, that most results on the p-r-degrees of **EXP** presented below easily carry over to any sufficiently closed hyperpolynomial time class.

So, for the following, we fix enumerations $\{f_n\colon n \geqslant 0\}$ and $\{M_n\colon n \geqslant 0\}$ of the p-m-reductions and p-T-reductions, respectively, such that $f_n(x)$ is $|x|^n + n$ time bounded and, simultaneously, $2^{|x|} + n$ time bounded uniformly in x and n, and similarly for M_n. Moreover, we let $P_n = M_n^{\emptyset}$ so that $\{P_n\colon n \geqslant 0\}$ is an enumeration of **P** within **EXP**.

The existence of recursive enumerations of the reductions easily implies that $(\mathbf{REC}_r, \leqslant)$ has no greatest, in fact no maximal elements. Namely, given any recursive set A, the set B defined by $B(n) = 1 - M_n^A(n)$ is recursive and $B \not\leqslant_T^P A$ (hence $B \not\leqslant_m^P A$), whence $A <_r^P A \oplus B$. Moreover, if A is chosen to be primitive (elementary) recursive, then B is primitive (elementary) recursive too, whence also $(\mathbf{ELEMENTARY}_r, \leqslant)$ and $(\mathbf{PRIM}_r, \leqslant)$ do not possess greatest elements.

More fundamental facts on the p-r-degrees can be obtained by the *effective finite extension method*, which is the effective version of the Kleene–Post type arguments used in the degrees of unsolvability (Kleene and Post [1954]). Here a recursive set A is constructed to meet an infinite sequence of finitary requirements R_n by effectively enumerating initial segments $A \upharpoonright x_n$ of A of growing length, where the finite extension $A \upharpoonright x_n$ of $A \upharpoonright x_{n-1}$ is chosen so that it forces that the requirement R_n is met. For instance, by such an argument, the observation on the nonexistence of maximal degrees can be extended to show that the partial order $(\mathbf{REC}_r, \leqslant)$ is not total. Given a recursive set $A \notin \mathbf{P}$ we can construct a recursive set B such that $A |_T^P B$ by merging the above strategy for ensuring $B \not\leqslant_T^P A$ with requirements

$$R_n\colon\ A \neq M_n^B$$

ensuring $A \not\leqslant_T^P B$. Namely, given a finite initial segment $B \upharpoonright x$ of B, we can effectively find an extension $B \upharpoonright y$ of $B \upharpoonright x$ ensuring R_n as follows. Since $B \upharpoonright x$ is finite, $M_n^{B \upharpoonright x} \in \mathbf{P}$ whence $A \neq M_n^{B \upharpoonright x}$ and we can effectively find the least z with $A(z) \neq M_n^{B \upharpoonright x}(z)$. Then, by letting $y = 0^m$ for the least m such that $|x|, |z|^n + n < m$ and by letting $B \upharpoonright y = B \upharpoonright x$, this disagreement is preserved if we replace the oracle $B \upharpoonright x$ by the oracle B.

The above argument can be easily modified to show that there is a countable r.p. sequence $\{\mathbf{a}_n\colon n \geqslant 0\}$ of p-r-degrees which is independent, i.e. which satisfies $\mathbf{a}_n \not\leqslant \bigvee_{m \in F} \mathbf{a}_m$ for any $n \in \omega$ and any finite subset $F \subseteq \omega$ such that $n \notin F$. This easily implies that every countable partial order can be embedded in $(\mathbf{REC}_r, \leqslant)$. Moreover,

Ambos-Spies [1984b] used an effective finite extension argument to show that the u.s.l. $(\mathbf{REC}_T, \leqslant)$ of the recursive p-T-degrees is not distributive (as a u.s.l.) whereas $(\mathbf{REC}_m, \leqslant)$ is distributive.

THEOREM 2.2 (Ambos-Spies [1984b]).
(a) *The u.s.l.* $(\mathbf{REC}_m, \leqslant)$ *is distributive, i.e.*

$$(\forall \mathbf{a})\ (\forall \mathbf{b})\ (\forall \mathbf{c})\ (\mathbf{c} \leqslant \mathbf{a} \vee \mathbf{b} \Rightarrow (\exists \mathbf{d}_a \leqslant \mathbf{a})\ (\exists \mathbf{d}_b \leqslant \mathbf{b})\ (\mathbf{c} = \mathbf{d}_a \vee \mathbf{d}_b)).$$

(b) *For any ideal* $\mathbf{I}_T$ *in* $(\mathbf{REC}_T, \leqslant)$ *such that* $\mathbf{I}$ *is a hyperpolynomial time class, the u.s.l.* $(\mathbf{I}_T, \leqslant)$ *is not distributive. In particular, the u.s.l.* $(\mathbf{REC}_T, \leqslant)$ *is not distributive.*

In order to apply finite extension arguments to questions related to the meet operator, we next observe that introducing large gaps in sets is a simple method for destroying p-r-reductions.

3. The meet operator and gap languages

Ladner [1975] has shown that $(\mathbf{REC}_r, \leqslant)$ is not a lattice, i.e. that there are p-r-degrees $\mathbf{a}$ and $\mathbf{b}$ such that the meet (infimum) $\mathbf{a} \wedge \mathbf{b}$ of $\mathbf{a}$ and $\mathbf{b}$ does not exist. He has also shown, however, that there are minimal pairs, i.e. degrees $\mathbf{a}, \mathbf{b} > \mathbf{0}$ such that $\mathbf{a} \wedge \mathbf{b} = \mathbf{0}$. Ladner used a quite sophisticated look-ahead technique (which we will discuss in Section 5 below) in order to prove these results, and the sets he obtained were not elementary recursive. Later Machtey [1976], Chew and Machtey [1981], and Landweber et al. [1981], obtained much stronger results on minimal pairs of low complexity by quite simple arguments based on the following observation on gaps. Call sets A_0 and A_1 *$t(n)$-separated*, if

$$(\forall x < y)(\forall i \in \{0, 1\})(x \in A_i \ \&\ y \in A_{1-i} \Rightarrow t(|x|) < |y|).$$

LEMMA 3.1 (Gap Lemma). *Let* $A \in \mathbf{DTIME}(t(n))$ *and let* $B_0, B_1 \in \mathbf{P}$ *be* $t(n)$-*separated. Then, for any set* C *which is* p-r-*reducible to both* $A \cap B_0$ *and* $A \cap B_1$, $C \in \mathbf{P}$.

PROOF. (For $r = m$.) Fix p-m-reductions g_i from C to $A \cap B_i$ and polynomial time bounds p_i for g_i $(i = 0, 1)$. Then, given x, $C(x)$ can be computed in polynomial time as follows. By assumption,

$$C(x) = A \cap B_0\big(g_0(x)\big) = A \cap B_1\big(g_1(x)\big)$$

and, by symmetry, we may assume that $g_0(x) \leqslant g_1(x)$. Now, if $g_0(x) \notin B_0$ or $g_1(x) \notin B_1$, then $x \notin C$. So in the remainder assume that $g_i(x) \in B_i$ for $i = 0, 1$, whence, by the $t(n)$-separation property of B_0 and B_1, $t(|g_0(x)|) \leqslant |g_1(x)| \leqslant p_1(|x|)$. So $C(x) = A(g_0(x))$ and the latter can be computed in $p_1(|x|)$ steps. □

COROLLARY 3.2 (Machtey [1976]). *There is a minimal pair in* $(\mathbf{EXP}_r, \leqslant)$.

PROOF. Let δ be the iterated exponential function defined by $\delta(0) = 0$ and $\delta(n+1) = 2^{\delta(n)}$, and let $B_i = \{0^{\delta(2n+i)}: n \geqslant 0\}$ $(i = 0, 1)$. Then $B_0, B_1 \in \mathbf{P}$ and B_0 and B_1 are 2^n-separated. Define $A \subseteq B_0 \cup B_1$ by letting $A(0^{\delta(2n+i)}) = 1 - P_n(0^{\delta(2n+i)})$. Then $A \in \mathbf{DTIME}(2^n)$ and $A \cap B_0$, $A \cap B_1 \notin \mathbf{P}$. So, by the Gap Lemma, $deg_r^P(A \cap B_0)$ and $deg_r^P(A \cap B_1)$ form a minimal pair. □

Sets like the set A in the above proof were introduced and investigated by Ambos-Spies [1986b] and were called *super sparse* there. Typical for a super sparse set A is that $A \in \mathbf{DTIME}(t(n)) - \mathbf{P}$ for a strictly increasing p-constructible function t and that A is contained in a $\mathbf{P}$-set $D = \{x_0, x_1, x_2, \ldots\} \subseteq \{0\}^*$, where $t(|x_n|) < |x_{n+1}|$ for all numbers $n \geqslant 0$. In a p-T-reduction to a super sparse set, for any input there is at most one (recognizable) relevant oracle query, and in fact this property is inherited by reductions among sets below a super sparse set. This easily implies (see Ambos-Spies [1986b]) that, for any super sparse set A, the partial orders $(\mathbf{REC}_T(\leqslant A), \leqslant)$ and $(\mathbf{REC}_m(\leqslant A), \leqslant)$ of the p-T, respectively, p-m-degrees below the degree of A are isomorphic, and that these structures are distributive lattices. Moreover, the degrees of the subproblems of a super sparse set A, i.e. $deg_r^P(A \cap B)$ for $B \in \mathbf{P}$, give an embedding of the atomless countable Boolean algebra into $(\mathbf{REC}_r(\leqslant A), \leqslant)$ (as a Boolean algebra). On the other hand, super sparseness can be used to give a simple proof that neither $(\mathbf{REC}_r, \leqslant)$ nor $(\mathbf{EXP}_r, \leqslant)$ is a lattice.

THEOREM 3.3 (Ladner [1975], Ambos-Spies [1992a]). *For* $r = m, T$, $(\mathbf{REC}_r, \leqslant)$ *is not a lattice. In fact, for any class* $\mathbf{C}$ *containing* $\mathbf{EXP}$, $(\mathbf{C}_r, \leqslant)$ *is not a lattice.*

PROOF. It suffices to construct sets A and B in $\mathbf{DTIME}(2^n)$ which are super sparse via δ, i.e. which have domain $\{0^{\delta(n)}: n \geqslant 0\}$, and which satisfy

$$(\forall C)\ (\exists D)\ \big(C \leqslant_m^P A, B \Rightarrow D \leqslant_m^P A, B \ \&\ D \not\leqslant_m^P C\big).$$

By considering the polynomial time sets $D_e = \{0^{\delta(\langle e,n\rangle)}: n \geqslant 0\}$ (for some polynomial time pairing function $\langle \cdot, \cdot \rangle$), the latter can be ensured by guaranteeing

$$\begin{aligned} &f_{e_0}^{-1}(A) = f_{e_1}^{-1}(B) \\ &\quad \Rightarrow A \cap D_e = B \cap D_e \ \&\ A\big(0^{\delta(\langle e,n\rangle)}\big) \neq A\big(f_{e_0}\big(f_n\big(0^{\delta(\langle e,n\rangle)}\big)\big)\big) \end{aligned} \qquad (*)$$

for all numbers $e = \langle e_0, e_1 \rangle$ and n. (Namely, if $C \leqslant_m^P A$ via f_{e_0} and $C \leqslant_m^P B$ via f_{e_1} then the hypothesis of $(*)$ holds, whence, for $D = A \cap D_e = B \cap D_e$, $D \leqslant_m^P A, B$ but $D \not\leqslant_m^P C$ since $D(0^{\delta(\langle e,n\rangle)}) \neq C(f_n(0^{\delta(\langle e,n\rangle)}))$ for all numbers n.)

Now, for the inductive definition of A and B fix $x = 0^{\delta(\langle e,n\rangle)}$, let $e = \langle e_0, e_1 \rangle$, and assume that $A \restriction x$ and $B \restriction x$ have been defined already. For the definition of $A(x)$ and $B(x)$ distinguish the following cases: if $f_{e_0}(f_n(x)) = f_{e_1}(f_n(x)) = x$ then let

$A(x) = 0 \neq 1 = B(x)$ thereby refuting the hypothesis of $(*)$. Otherwise, by symmetry, without loss of generality $f_{e_0}(f_n(x)) \neq x$ and, by choice of δ we may assume that $f_{e_0}(f_n(x)) < 0^{\delta(\langle e,n\rangle+1)}$. So, by super-sparseness of A, $A(f_{e_0}(f_n(x))) = (A \upharpoonright x)(f_{e_0}(f_n(x)))$, whence we can let $A(x) = B(x) = 1 - A(f_{e_0}(f_n(x)))$ thereby locally satisfying the conclusion of $(*)$. □

4. Delayed diagonalization or the looking back technique

Effective finite extension arguments, even in combination with the gap technique, do not suffice to obtain downward density or general density results for the structure of the p-r-degrees. Ladner [1975] introduced the delayed diagonalization or looking back technique to attack this type of questions, thereby also giving access to the structure of the p-r-degrees of **NP**-sets (assuming $\mathbf{P} \neq \mathbf{NP}$). We illustrate this technique by a simple example.

THEOREM 4.1 (Ladner [1975]). *For any recursive set $A \notin \mathbf{P}$ there is a recursive set $B \notin \mathbf{P}$ such that $B \leqslant_m^P A$ but $A \nleqslant_T^P B$.*

PROOF (*Sketch*). We let $B = A \cap C$ be a subproblem of A, whence $B \leqslant_m^P A$, where the set $C \in \mathbf{P}$ is defined by induction. So it suffices to meet the requirements

$$R_{2n}\colon\ B \neq P_n \quad \text{and} \quad R_{2n+1}\colon\ A \neq M_n^B$$

in order to ensure that $B \notin \mathbf{P}$ and $A \nleqslant_T^P B$, respectively. Instead of using a finite extension argument to meet these requirements directly (as done in Section 2), here we observe that, for any finite variant B of A, all requirements R_{2n} are met since $A \notin \mathbf{P}$ and $\mathbf{P}$ is closed under finite variants, and similarly, for any finite variant B of $\emptyset$, all requirements R_{2n+1} are met. Moreover, since the requirements R_n are finitary, for meeting a single requirement it suffices to have the agreement between B and A (or B and $\emptyset$) on a sufficiently long interval. So we can meet all requirements by letting B alternately look like A and $\emptyset$ on appropriately chosen intervals. More formally, we inductively define a strictly increasing function $g : \omega \to \omega$ such that, for $C = \{x\colon\ (\exists n)(g(2n) \leqslant |x| < g(2n+1))\}$ and $B = A \cap C$ all requirements are met: given $g(2n)$, effectively find the least x with $|x| \geqslant g(2n)$ such that $A(x) \neq P_n(x)$, and let $g(2n+1) \geqslant |x| + 1$, thereby ensuring that R_{2n} is met. Similarly, given $g(2n+1)$, effectively find the least string x such that $A(x) \neq M_n^{B \upharpoonright g(2n+1)}(x)$, and let $g(2n+2) > \max(g(2n+1), |x|^n + n)$, thereby ensuring that this disagreement is preserved for oracle B, whence R_{2n+1} is met.

It remains to argue that $C \in \mathbf{P}$. For this it suffices to make $g(n)$ p-constructible. This is achieved by putting down $g(n)$ not *immediately* after the corresponding diagonalization step is completed: the definition of $g(n)$ is *delayed* to a sufficiently large number m such that, by *looking back* from $g(n) = m$, the diagonalization witness can be recovered in time linear in m. □

COROLLARY 4.2 (Ladner [1975]). *If* $\mathbf{P} \neq \mathbf{NP}$ *then there is a set* A *in* $\mathbf{NP} - \mathbf{P}$ *which is not p-r-complete for* **NP**.

By variants of the above argument, Ladner showed that the partial order of the p-r-degrees is dense and that every nonzero p-r-degree splits, i.e. is join-reducible. In fact these two results combine, and Landweber et al. proved the dual fact for branching, i.e. meet-reducible degrees. For the latter, note that the gap technique of Section 3 for creating meets smoothly combines with the delayed diagonalization technique, which is also based on introducing gaps in given sets.

THEOREM 4.3 (Ladner [1975], Landweber et al. [1981]). *Let* $\mathbf{a}, \mathbf{b} \in \mathbf{REC}_r$ *where* $\mathbf{a} < \mathbf{b}$. *There are p-r-degrees* $\mathbf{c}_0, \mathbf{c}_1, \mathbf{d}_0, \mathbf{d}_1$ *such that* $\mathbf{a} < \mathbf{c}_0, \mathbf{c}_1, \mathbf{d}_0, \mathbf{d}_1 < \mathbf{b}$, $\mathbf{b} = \mathbf{c}_0 \vee \mathbf{c}_1$ *and* $\mathbf{a} = \mathbf{d}_0 \wedge \mathbf{d}_1$.

There are several extensions and variations of this result. For example, Breidbart [1977] and Mehlhorn [1976] showed that every countable partial order can be embedded in any interval of the p-r-degrees. Theorem 4.3 and most of its variants can be viewed as lattice embedding results, whence they are subsumed by the following general embeddability theorem for distributive lattices.

THEOREM 4.4 (Ambos-Spies [1987a]). *Let* $\mathcal{L}$ *be any countable distributive lattice, let* $\mathbf{a}, \mathbf{b}$ *be recursive p-r-degrees where* $\mathbf{a} < \mathbf{b}$, *and let* $\mathbf{C}_r$ *be an r.p. class of p-r-degrees such that* $\mathbf{C}_r$ *is contained in* $(\mathbf{a}, \mathbf{b})$. *There are lattice embeddings* $f_i : \mathcal{L} \to [\mathbf{a}, \mathbf{b}]$ *of* $\mathcal{L}$ *into the interval* $[\mathbf{a}, \mathbf{b}]$, *where* f_0 *maps the least element* 0 *of* $\mathcal{L}$ *(if any) to* $\mathbf{a}$ *and* f_1 *maps the greatest element* 1 *of* $\mathcal{L}$ *(if any) to* $\mathbf{b}$. *Moreover, for any* $\mathbf{c} \in \mathbf{C}_r$ *and* $a \in \mathcal{L} - \{0, 1\}$, $f_i(a)$ *is incomparable with* $\mathbf{c}$.

Note that, by taking $\mathcal{L}$ to be the 2-atom Boolean algebra $\mathcal{B}_2 = \{0, a_1, a_2, 1\}$, Theorem 4.4 implies Theorem 4.3 by letting $\mathbf{c}_i = f_1(a_i)$ and $\mathbf{d}_i = f_0(a_i)$. Moreover, by taking the 3-element total order $O_3 = \{0, a, 1\}$ and by letting $\mathbf{C}_r = \{\mathbf{c}\}$ for some degree $\mathbf{c}$ with $\mathbf{a} < \mathbf{c} < \mathbf{b}$ or by letting $\mathbf{C}_r$ be an r.p. anti-chain in $(\mathbf{a}, \mathbf{b})$, Theorem 4.4 shows that for any degree $\mathbf{c}$ in the open interval $(\mathbf{a}, \mathbf{b})$ there is another degree in $(\mathbf{a}, \mathbf{b})$ incomparable with $\mathbf{c}$ (Balcazar and Diaz [1982]) and that no recursive anti-chain in $(\mathbf{a}, \mathbf{b})$ is maximal (Schmidt [1984]).

In general Theorem 4.4 cannot be extended to lattice embeddings which simultaneously preserve the least and greatest elements: Ambos-Spies [1986a] (for p-m) and Downey [1992] (for p-T) have shown that there is a p-r-degree $\mathbf{a} > \mathbf{0}$ which is not the join of a minimal pair. (In fact, in case of p-m, Ambos-Spies, Homer and Soare [1994] have shown that the p-m-complete degree for **EXP** has this property, whereas, for p-T, only non elementary recursive sets with this property are known.) On the other hand, Breidbart [1977] has shown that this observation is optimal in the sense that for any $\mathbf{a} < \mathbf{b}$ there are *three* pairwise incomparable p-r-degrees $\mathbf{c}_0, \mathbf{c}_1, \mathbf{c}_2$ such that $\mathbf{a} = \inf(\mathbf{c}_0, \mathbf{c}_1, \mathbf{c}_2)$ and $\mathbf{b} = \sup(\mathbf{c}_0, \mathbf{c}_1, \mathbf{c}_2)$. By extending Breidbart's result, Ambos-Spies [1987a] has shown that every finite distributive lattice which is nowhere complemented can be embedded into any interval of p-r-degrees by a map which preserves both the least and the greatest elements. By distributivity of the p-m-degrees and by the existence of distributive initial segments of the p-T-degrees (see Section 3), these results yield the following characterization of the finite lattices densely embeddable into $(\mathbf{REC}_r, \leqslant)$ preserving 0 and 1.

THEOREM 4.5. *Let $\mathcal{L}$ be a finite lattice. The following are equivalent.*

(i) *$\mathcal{L}$ is distributive, $\mathcal{L}$ has at least 2 elements, and no element in $\mathcal{L} - \{0, 1\}$ has a complement in $\mathcal{L}$.*

(ii) *$\mathcal{L}$ is lattice embeddable in every interval of $(\mathbf{REC}_m, \leqslant)$ by a map which preserves the least and greatest element.*

(iii) *$\mathcal{L}$ is lattice embeddable in every interval of $(\mathbf{REC}_T, \leqslant)$ by a map which preserves the least and greatest element.*

(iv) *$\mathcal{L}$ is lattice embeddable in $(\mathbf{EXP}_m, \leqslant)$ by a map which preserves the least and greatest element.*

Another extension of Ladner's density theorem is the following general extendability result for partial order embeddings, also proved by delayed diagonalization.

THEOREM 4.6 (Shore and Slaman [1992]). *Let $\mathcal{L}$ be a sublattice of some finite partial order $\mathcal{K}$, where $\mathcal{L}$ and $\mathcal{K}$ have the same least and greatest elements. Then every p.o. embedding of $\mathcal{L}$ into $(\mathbf{REC}_r, \leqslant)$ can be extended to a p.o. embedding of $\mathcal{K}$.*

Though, as pointed out in Section 3, there are principal ideals $[\mathbf{0}, \mathbf{a}]$ of p-r-degrees which are distributive lattices, no interval of the p-r-degrees is a Boolean algebra. This follows from the observation that every nonzero p-r-degree has the strong anti-cupping property [Ambos-Spies, 1991]: given a recursive set $A \notin \mathbf{P}$ and a hyperpolynomial p-constructible function f, the f-shift

$$A_f = \{0^{f(|x|)}1x\colon\ x \in A\}$$

of A has the property that $A_f <_r^P A$ and, for any set B, $A \leqslant_r^P A_f \oplus B$ implies $A \leqslant_r^P B$. So, to make $deg_r^P(A_f)$ an anti-cupping witness for A, it suffices to construct f so such that $A_f \notin \mathbf{P}$, which can be done by a delayed diagonalization.

Further applications of the delayed diagonalization technique include results on upper bounds for r.p. sequences and ideals in $(\mathbf{REC}_r, \leqslant)$. For example, Ambos-Spies [1989a] has shown that no ascending r.p. sequence of p-r-degrees has a minimal upper bound, whence an r.p. ideal has a least or minimal upper bound iff the ideal is principal. Furthermore, Nies [1997] has shown that, for any independent r.p. sequence $\mathbf{A} = \{\mathbf{a}_n\colon\ n \geqslant 0\}$ and any uniform upper bound $\mathbf{a}$ for $\mathbf{A}$, every Σ_2-subsequence of $\mathbf{A}$ can be selected by some degree below $\mathbf{a}$. To be more precise, given a recursive set $A \subseteq \omega \times \Sigma^*$ representing the sequence $\mathbf{A}$ in the sense that $\mathbf{a}_n = deg_r^P(A_n)$ for the nth row A_n of A and given a Σ_2-set $S \subseteq \omega$, there is a subproblem A_S of A such that $A_n \leqslant_r^P A_S$ iff $n \in S$.

In all of the above results obtained by delayed diagonalization, the positive reductions are witnessed by subproblems or (in case of the anti-cupping property) by constructible shifts, whence these reductions are p-m-reductions. By closure of **NP** under p-m-reducibility, it follows that if we started with **NP**-sets then the proofs yield **NP**-sets again. So, assuming $\mathbf{P} \neq \mathbf{NP}$, we obtain results on the structure of $(\mathbf{NP}_r, \leqslant)$. E.g., by Ladner's Theorem 4.1, there are sets in $\mathbf{NP} - \mathbf{P}$ which are not p-r-complete for **NP**, and Theorem 4.3 implies that $(\mathbf{NP}_r, \leqslant)$ is dense, that the **NP**-complete degree splits into two lesser ones, and that there is a minimal pair in $(\mathbf{NP}_r, \leqslant)$.

Following Ladner's paper, the delayed diagonalization technique was not only applied in order to obtain stronger results, but variants of this technique, like its "structural variant" (Landweber et al. [1981]) were introduced which helped to simplify this method. Moreover, for many applications it is useful to note that diagonalizing over p-r-reductions is closely related to diagonalizing over recursively presentable classes which are closed under finite variants (c.f.v.). Obviously $\leqslant_r^P$ is c.f.v. whence, by the existence of effective enumerations of the p-r-reductions, it easily follows that classes like $deg_r^P(A)$, $\leqslant_r(A) = \{X\colon X \leqslant_r^P A\}$ and $[A, B]_r = \{X\colon A \leqslant_r^P X \leqslant_r^P B\}$ are r.p. and c.f.v. (for all recursive sets A, B). So many of the above results can be obtained as direct corollaries of general diagonalization theorems for r.p. and c.f.v. classes. A very elegant example of such a diagonalization lemma is due to Schöning. Intuitively it says that two recursive diagonals D_0, D_1 of r.p. and c.f.v. classes $\mathbf{C}_0, \mathbf{C}_1$, respectively, can be merged to a diagonal D for the union of these classes.

THEOREM 4.7 (Schöning [1982]). *Let $\mathbf{C}_0, \mathbf{C}_1$ be r.p. and c.f.v. classes and let D_0, D_1 be recursive sets such that $D_0 \notin \mathbf{C}_0$ and $D_1 \notin \mathbf{C}_1$. There is a recursive set $D \leqslant_m^P D_0 \oplus D_1$ such that $D \notin \mathbf{C}_0 \cup \mathbf{C}_1$.*

Note that Theorem 4.1 follows from Theorem 4.7 by letting $\mathbf{C}_0 = deg_r^P(A)$, $\mathbf{C}_1 = \mathbf{P}$, $D_0 = \emptyset$ and $D_1 = A$. For a more complete treatment of the delayed diagonalization technique we refer to Balcazar et al. [1988, Chapter VII], or Ambos-Spies [1987a].

5. The iterated look-ahead technique

In this section we present the iterated look-ahead technique which is another fundamental diagonalization technique for constructing polynomial time degrees. This powerful method has the disadvantage, however, that it yields sets which are not elementary recursive. So – in contrast to the delayed diagonalization technique – it cannot be used for investigating the degrees of low complexity classes like **NP** or **EXP**. The iterated look-ahead technique was introduced by Ladner ([1975]) in order to construct minimal pairs and pairs without infima. Since these results were obtained by quite simple arguments in the sequel (see Sections 3 and 4 above), the original proofs were forgotten, however, and in Ambos-Spies [1987b] the technique was rediscovered in order to show that every recursive nonzero degree is half of a minimal pair in $(\mathbf{REC}_r, \leqslant)$.

THEOREM 5.1 (Ambos-Spies [1987b]). *For any recursive set A there is a recursive set $B \notin \mathbf{P}$ such that*

$$(\forall C)\left(C \leqslant_r^P A, B \Rightarrow C \in \mathbf{P}\right).$$

Hence every nonzero recursive p-r-degree is the half of a minimal pair in $(\mathbf{REC}_r, \leqslant)$.

PROOF. We sketch the proof for $r = m$. Given A recursive, it suffices to construct a recursive set B satisfying the requirements

$$R_{2n}\colon\ B \neq P_n$$

and, for $n = \langle n_0, n_1 \rangle$,

$$R_{2n+1}\colon\ f_{n_0}^{-1}(A) = f_{n_1}^{-1}(B) \Rightarrow f_{n_1}^{-1}(B) \in \mathbf{P}.$$

Now, ignoring the task of making B recursive for the moment, these requirements can be met by a standard finite extension argument as follows. Given the finite initial part $B \restriction x$ of B, by letting $B(x) = 1 - P_n(x)$ the extension $B \restriction x + 1$ will obviously force R_{2n}. In order to force R_{2n+1} one attempts to refute the hypothesis of the requirement and one will argue that if this attempt fails then the conclusion of the requirement must hold: given $B \restriction x$ search for the least string y such that $f_{n_1}(y) \geqslant x$. If such a y exists then, by letting $B(f_{n_1}(y)) = 1 - A(f_{n_0}(y))$, the extension $B \restriction f_{n_1}(y) + 1$ will force that $f_{n_0}^{-1}(A) \neq f_{n_1}^{-1}(B)$ thereby guaranteeing R_{2n+1}. On the other hand, if such a string y does not exist then the range of f_{n_1} is finite, whence the polynomial time reduction of $f_{n_1}^{-1}(B)$ to B uses only a finite part of B so that $f_{n_1}^{-1}(B) \in \mathbf{P}$. Here the search for the diagonalization witness y is not bounded, however, whence the construction is not effective and does not yield a *recursive* set B.

The effectivization of the construction is based on the following observation. Assume that we know a time bound $t(n)$ for the set B (under construction) in advance. Then requirement R_{2n+1} can be met as follows: for (almost) every x, search for a string $y \leqslant t(|x|)$ such that $f_{n_1}(y) = x$. If such a y is found, then, by letting $B(x) = 1 - A(f_{n_0}(y))$, $B \restriction x + 1$ will force R_{2n+1}. On the other hand, if (almost) all of these searches fail, then, for (almost) all strings y, $t(|f_{n_1}(y)|) < |y|$, whence, by the assumed time bound for B, $f_{n_1}^{-1}(B)(y) = B(f_{n_1}(y))$ can be computed in $poly(|y|)$ steps. So in this case $f_{n_1}^{-1}(B) \in \mathbf{P}$.

This observation leads to a *slow diagonalization* (*wait-and-see*) construction of B, where $B(x)$ is specified at stage x. At this stage R_m is forced for the least $m < x$ (if any) such that R_m has not yet been satisfied but can be satisfied by appropriately choosing $B(x)$. The set B constructed this way is recursive but, unfortunately, not in $\mathbf{DTIME}(t(n))$: the search for a diagonalization witness $y < t(n)$ for a requirement R_{2m+1} at stage n requires more than $2^{t(n)}$ steps. So we cannot argue that the odd index requirements are met.

This obstacle is overcome by letting each requirement R_{2m+1} work with its own time bound t_m, where $t_m(n)$ measures the time required to construct $B \upharpoonright n$ and to find out the action the requirements $R_{2m+2}, \ldots, R_n$ want to take at stage n. In particular, at stage n, the search for a diagonalization candidate for a requirement R_{2m+1}, $2m + 1 \leqslant n$, which has not yet been satisfied before is restricted to strings y with $|y| \leqslant t_m(n)$. Since every requirement acts at most once, for each m there is a stage n_m such that, for $n > n_m$, the value of $B(n)$ does not depend on the requirements $R_0, \ldots, R_{2m+1}$ anymore, whence each t_m actually is a time bound for B. So now the construction of B is correct. □

Note that in the above proof, for every m, $t_m(n) \geqslant 2^{t_{m+1}(n)}$ almost everywhere. So every time bound t_m for B defined by some requirement R_{2m+1} possesses an exponential speed up t_{m+1}, whence the time bounds for B obtained in the proof are not elementary recursive. So the iterated look-ahead technique in general only yields (primitive) recursive but not elementary recursive sets. In case of Theorem 5.1 this apparent shortcoming of this method, however, is unavoidable. By a simple padding argument, it can be shown that, in general, the set B there cannot be elementary recursive in A.

LEMMA 5.2 (Book; see Ambos-Spies [1987b]). *Let A be p-m-complete for* **EXP** *and let B be elementary recursive. Then $deg_r^P(A)$ and $deg_r^P(B)$ is not a minimal pair.*

Theorem 5.1 can be easily extended to a general exact pair theorem for r.p. ideals. Here a pair **a**, **b** of recursive p-r-degrees is called an exact pair for the ideal $\mathbf{I}_r$ in $(\mathbf{REC}_r, \leqslant)$ if $\mathbf{I}_r = \mathbf{I}[\mathbf{a}, \mathbf{b}]$ for the ideal

$$\mathbf{I}[\mathbf{a}, \mathbf{b}] = \{\mathbf{c}\colon\ \mathbf{c} \leqslant \mathbf{a}\ \&\ \mathbf{c} \leqslant \mathbf{b}\}$$

defined by **a** and **b**.

THEOREM 5.3 (Ambos-Spies [1987b, 1991]). *Let $\mathbf{C}_r$ and $\mathbf{I}_r$ be r.p. classes of p-r-degrees such that $\mathbf{I}_r$ is an ideal and let $\mathbf{a} \in \mathbf{REC}_r$ be an upper bound for $\mathbf{I}_r$. There is a recursive p-r-degree* **b** *such that $\mathbf{b} \notin \mathbf{C}_r$ and* **a**, **b** *is an exact pair for $\mathbf{I}_r$.*

In particular, every r.p. ideal $\mathbf{I}_r$ possesses an exact pair in $(\mathbf{REC}_r, \leqslant)$. Since, conversely, every pair of recursive degrees **a**, **b** defines an r.p. ideal $\mathbf{I}[\mathbf{a}, \mathbf{b}]$, the r.p. ideals are just the ideals which possess an exact pair (see Ambos-Spies and Nies [1992]). This allows to identify r.p. ideals of degrees with pairs of degrees, which makes it possible to talk about these ideals in the first-order language of partial order. This observation was exploited in the first undecidability proofs for the theories of the p-r-degrees (see Section 6 below).

Some further consequences of Theorem 5.3 are: for every p-r-degree **a** there is a p-r-degree **b** such that the infimum $\mathbf{a} \wedge \mathbf{b}$ of **a** and **b** does not exist (Ambos-Spies [1987b]), and every minimal pair **a**, **b** of p-r-degrees is extendible, i.e. there are p-r-degrees $\mathbf{a}' > \mathbf{a}$ and $\mathbf{b}' > \mathbf{b}$ such that $\mathbf{a}'$, $\mathbf{b}'$ is a minimal pair too (Zheng [1994]).

The iterated look-ahead technique was also used to solve the embedding problem for the p-T-degrees. Recall that delayed diagonalizations suffice to settle the embedding problem for distributive lattices (and thus, by distributivity of $(\mathbf{REC}_m, \leqslant)$, the general embedding problem for the p-m-degrees). The question which nondistributive lattices are embeddable into the u.s.l. of the recursive p-T-degrees, however, cannot be solved by this method, since here the gap lemma cannot be applied. As a first step, Ambos-Spies [1984a] could show that the two five-element nondistributive lattices (which are the prime examples of nondistributive lattices in the sense that every nondistributive lattice contains one of these lattices as a sublattice) can be embedded into $(\mathbf{REC}_T, \leqslant)$. By using a representation theorem for finite lattices by finite partition lattices, Shore and Slaman extended this to all finite lattices.

THEOREM 5.4 (Shore and Slaman [1992]). *Every finite lattice $\mathcal{L}$ is lattice embeddable into $(\mathbf{REC}_T, \leqslant)$ by a map which preserves the least element.*

Ambos-Spies [1992b] has shown that in these lattice embeddings the iterated look-ahead technique can be avoided if one does not require that the embeddings preserve the least element, thereby showing that every finite lattice can be lattice embedded into the p-T-degrees of the double exponential time class $\mathbf{DTIME}(2^{2^n})$. The embedding problem for $(\mathbf{EXP}_T, \leqslant)$, however, and, in case of embeddings preserving 0, also for $(\mathbf{ELEMENTARY}_T, \leqslant)$ is still open.

Further important applications of the iterated look-ahead technique are related to the theories of the recursive p-m- and p-T-degrees, which we will discuss in the next section.

6. The theory of the polynomial time degrees

Many global questions on the structure of the recursive p-r-degrees can be expressed by properties of the first-order theory $\mathrm{Th}(\mathbf{REC}_r, \leqslant) = \{\varphi\colon (\mathbf{REC}_r, \leqslant) \models \varphi\}$ of these structures ($r = m, T$). Therefore these theories were intensively investigated where, in particular, their complexity in terms of the number of types and their degree of unsolvability has been analyzed.

The n-type of an n-tuple $\mathbf{a}_1, \ldots, \mathbf{a}_n$ of recursive p-r-degrees is defined as the set of first-order formulas with n free variables $x_1, \ldots, x_n$ which hold for $\mathbf{a}_1, \ldots, \mathbf{a}_n$ in $(\mathbf{REC}_r, \leqslant)$:

$$\mathit{type}(\mathbf{a}_1, \ldots, \mathbf{a}_n) = \big\{\varphi\colon (\mathbf{REC}_r, \leqslant) \models \varphi_{x_1,\ldots,x_n}[\mathbf{a}_1, \ldots, \mathbf{a}_n]\big\}.$$

The number of types of a countable structure is of interest, since, by Ryll-Nardzewski's theorem (see, e.g., Chang and Keisler [1978]), the existence of infinitely many different n-types (for some fixed n) implies that the theory of this structure possesses a countable nonstandard model.

The existence of infinitely many 1-types for $(\mathbf{REC}_m, \leqslant)$ was shown by Ambos-Spies [1984a]. There, by combining the look-ahead-technique of Theorem 5.1 with the construction of a degree which is not the top of a minimal pair, for every $n \geqslant 2$, a recursive p-m-degree $\mathbf{a}_n$ is constructed such that the n-atom Boolean algebra but not the $(n+1)$-atom Boolean algebra can be embedded (as a Boolean algebra, i.e. by a lattice embedding preserving 0 and 1) into the ideal $[\mathbf{0}, \mathbf{a}_n]$.

The Embedding Theorem 5.4 of Shore and Slaman implies the existence of infinitely many 3-types in $(\mathbf{REC}_T, \leqslant)$. Work by Ambos-Spies and Yang [1990] on the honest polynomial time reducibilities can be adapted to the standard p-T-reducibility to show the existence of infinitely many 1-types in $(\mathbf{REC}_T, \leqslant)$ too. This work exploits properties of the k-super sparse sets ($k \geqslant 0$), a concept generalizing the super sparseness notion introduced in Section 3: p-T-reductions to a k-super sparse set collapse to bounded query reductions allowing only k queries on every input. Now, exploiting this fact, Ambos-Spies and Yang have shown that for the so-called 1-n-1-lattices $\mathcal{L}_{1\text{-}n\text{-}1}$ consisting of n mutually incomparable elements which pairwise join to the greatest element and meet to the least element of the lattice, the maximal number n such that the lattice $\mathcal{L}_{1\text{-}n\text{-}1}$ can be embedded in the principal ideal $[\mathbf{0}, \mathbf{a}]$ for some k-super sparse top $\mathbf{a}$ is growing with k.

As the result on 1-types for the p-m-degrees, this result is also based on the iterated look-ahead technique, whence it left open the question of the abundance of 1-types in case of the elementary recursive or even exponential time degrees. Again using k-super sparse sets, however, Ambos-Spies [1992a] answered this question in the affirmative.

THEOREM 6.1 (Ambos-Spies [1992a]). *For any ideal* $\mathbf{I}_r$ *in* $(\mathbf{REC}_r, \leqslant)$ *such that* $\mathbf{I}_r$ *contains a hyperpolynomial time class,* $\mathrm{Th}(\mathbf{I}_r, \leqslant)$ *has infinitely many* 1*-types.*

The proof of Theorem 6.1 is based on nondistributivity properties (in the u.s.l. sense) and meet properties of principal ideals. Call a principal ideal $[\mathbf{0}, \mathbf{a}]$ a *k-lattice* if there are degrees $\mathbf{a}_1, \ldots, \mathbf{a}_k < \mathbf{a}$ such that $[\mathbf{0}, \mathbf{a}_i]$ is a lattice (for $1 \leqslant i \leqslant k$) and $\mathbf{a} = \mathbf{a}_1 \vee \cdots \vee \mathbf{a}_k$; and call $[\mathbf{0}, \mathbf{a}]$ *k-distributive*, if for all degrees $\mathbf{a}_1, \ldots, \mathbf{a}_k$ which pairwise join to $\mathbf{a}$ and for any degree $\mathbf{b} < \mathbf{a}$, $\mathbf{b} = \mathbf{b}_1 \vee \cdots \vee \mathbf{b}_k$ for degrees $\mathbf{b}_i < \mathbf{a}_i$ $(1 \leqslant i \leqslant k)$. Then, by finite extension arguments similar to the ones used in the proofs of Theorems 3.3 and 2.2, it is shown that for principal ideals $[\mathbf{0}, \mathbf{a}]$ with k-super sparse top the maximum number k' such that the k'-lattice property (for p-m) or k'-distributivity (for p-T) can fail for $[\mathbf{0}, \mathbf{a}]$ is growing with k.

The above results on 1-types were obtained by distinguishing lower cones by their elementary properties. It is an interesting open problem whether there are upper cones which can be distinguished by elementary properties or which at least are not isomorphic:

HOMOGENEITY PROBLEM. Let $\mathbf{REC}_r(\geqslant \mathbf{a}) = \{\mathbf{b} \in \mathbf{REC}_r\colon\ \mathbf{a} \leqslant \mathbf{b}\}$. Are there recursive p-r-degrees $\mathbf{a}$ and $\mathbf{b}$ such that $\mathbf{REC}_r(\geqslant \mathbf{a}) \not\cong \mathbf{REC}_r(\geqslant \mathbf{b})$ or such that in fact $\mathrm{Th}(\mathbf{REC}_r(\geqslant \mathbf{a}), \leqslant) \neq \mathrm{Th}(\mathbf{REC}_r(\geqslant \mathbf{b}), \leqslant)$?

The undecidability results for the p-degrees are based on interpretations of structures with hereditarily undecidable theories in the partial order of p-r-degrees.

Hereditary undecidability is preserved by these interpretations. Moreover, these interpretations may use parameters so that for degree classes $\mathbf{C}_r$ and $\mathbf{C}'_r$ such that $\mathbf{C}_r \subseteq \mathbf{C}'_r$ and $\mathbf{C}_r$ is definable with parameters in $\mathbf{C}'_r$, the hereditary undecidability of $\mathrm{Th}(\mathbf{C}_r, \leqslant)$ implies (hereditary) undecidability of $\mathrm{Th}(\mathbf{C}'_r, \leqslant)$.

The first undecidability result for the p-r-degrees was obtained by Shinoda and Slaman. Using a very sophisticated iterated look-ahead argument, they gave an interpretation of first-order arithmetic in the u.s.l. of recursive p-T-degrees, thereby not only showing that $\mathrm{Th}(\mathbf{REC}_T, \leqslant)$ is undecidable but also classifying the degree of unsolvability of this theory as the highest possible one, namely that of first-order arithmetic.

THEOREM 6.2 (Shinoda and Slaman [1990]). $\mathrm{Th}(\mathbf{REC}_T, \leqslant)$ *is undecidable. In fact* $\mathrm{Th}(\mathbf{REC}_T, \leqslant) =_T \mathrm{Th}(\omega, +, \cdot)$.

The coding schema in the proof of Theorem 6.2 is nondistributive, hence does not carry over to the p-m-degrees. Ambos-Spies and Nies [1992], however, showed that the lattice of Σ_2-sets is definable in $(\mathbf{REC}_m, \leqslant)$ with parameters, which by hereditary undecidability of $\mathrm{Th}(\Sigma_2, \subseteq)$ (Herrmann [1984]) implies the (hereditary) undecidability of $\mathrm{Th}(\mathbf{REC}_m, \leqslant)$.

THEOREM 6.3 (Ambos-Spies and Nies [1992]). $\mathrm{Th}(\mathbf{REC}_m, \leqslant)$ *is undecidable.*

The proof of Theorem 6.3 exploits the distributivity of $(\mathbf{REC}_m, \leqslant)$ and the exact pair theorem. It is shown that for any independent r.p. sequence $\mathbf{A} = \{\mathbf{a}_n\colon n \geqslant 0\}$, the partial order $(\mathbf{DI}(\mathbf{A}), \subseteq)$ of the ideals with exact pairs in $\mathbf{A}$ is isomorphic to $(\Sigma_2, \subseteq)$. Hence, by the exact pair theorem, it suffices to give such a sequence $\mathbf{A}$ definable in the ideal $\mathbf{I}(\mathbf{A})$ generated by $\mathbf{A}$. By distributivity, however, any sequence $\mathbf{A}$ of degrees, which are pairwise minimal pairs but not the top of any minimal pair, has this property, since the elements of $\mathbf{A}$ will be the maximal non-top elements of $\mathbf{I}(\mathbf{A})$. This completed the proof, since, in his first proof for the existence of infinitely many 1-types for the p-m-degrees based on Boolean algebras, Ambos-Spies already had constructed such a sequence.

By using similar structural properties of $(\mathbf{REC}_m, \leqslant)$ and a new coding schema for arithmetic, Nies (private communication) extended this result by showing that $\mathrm{Th}(\mathbf{REC}_m, \leqslant)$ has the highest possible degree too.

THEOREM 6.4 (Nies). $\mathrm{Th}(\mathbf{REC}_m, \leqslant) =_T \mathrm{Th}(\omega, +, \cdot)$.

By using the iterated look-ahead technique, the proofs of the above undecidability theorems left open the undecidability of $\mathrm{Th}(\mathbf{ELEMENTARY}_r, \leqslant)$ and that of $\mathrm{Th}(\mathbf{C}_r, \leqslant)$ for still smaller complexity classes $\mathbf{C}$. Ambos-Spies [1992b] made some progress on this question by modifying the proof of Theorem 6.3 in order to obtain undecidability of $\mathrm{Th}(\mathbf{C}_m, \leqslant)$ for any class $\mathbf{C}$ containing the double exponential time class $\mathbf{DTIME}(2^{2^n})$. A break through, however, was obtained only very recently by Downey and Nies:

THEOREM 6.5 (Downey and Nies [1997]). *Let* $\mathbf{C}$ *be a class of recursive sets containing a hyperpolynomial time class. Then* $\mathrm{Th}(\mathbf{C}_m, \leqslant)$ *and* $\mathrm{Th}(\mathbf{C}_T, \leqslant)$ *are undecidable.*

The proof uses a new coding method based on interpretations of ideal lattices of Boolean algebras, and it requires only properties of the p-r-degrees provable by simple finite extension arguments and delayed diagonalizations. Moreover, it heavily depends on the existence of a super sparse set A in any hyperpolynomial time class $\mathbf{DTIME}(h(n))$. The coding is essentially done in the principal ideal $[\mathbf{0}, deg_r^P(A)]$ but it uses additional parameters from $\mathbf{DTIME}(h(n))$. Nies [ta] further extended this method to show that, for any super sparse p-r-degree $\mathbf{a}$, the theory of the p-r-degrees below $\mathbf{a}$ has the highest possible degree:

$$\mathrm{Th}([\mathbf{0}, \mathbf{a}], \leqslant) =_T \mathrm{Th}(\omega, +, \cdot).$$

In contrast to hereditary undecidability, however, this property is not inherited by every super class. In particular it is still not known whether this result extends to the theory of the p-r-degrees of **EXP** and **ELEMENTARY**.

All of the undecidability results above are obtained by defining structures with (hereditarily) undecidable theories in the degree structures with parameters. This naturally leads to the question of definability without parameters in the degree structures. Here, however, essentially nothing is known. Interesting open questions in this direction include: is there a nonzero p-r-degree which is (first-order) definable in $(\mathbf{REC}_r, \leqslant)$? For any natural complexity class $\mathbf{C}$ like $\mathbf{C} \in \{\mathbf{EXP}, \mathbf{ELEMENTARY}, \mathbf{PRIM}\}$, is $\mathbf{C}_r$ definable in $(\mathbf{REC}_r, \leqslant)$? Is $(\mathbf{REC}_r, \leqslant)$ definable in the u.s.l. of all (not necessarily recursive) p-r-degrees?

These definability questions are related to the – also still open – automorphism problem: Is there a nontrivial automorphism of $(\mathbf{REC}_T, \leqslant)$? The construction of p-T-degrees $\mathbf{a} \neq \mathbf{b}$ in Haught and Slaman [1997] such that the initial segments $[\mathbf{0}, \mathbf{a}]$ and $[\mathbf{0}, \mathbf{b}]$ are isomorphic might be a first step for obtaining such a nontrivial automorphism. In case of p-m-reducibility, by the existence of non-selfdual sets, i.e. sets A such that the complement $\overline{A}$ of A is not p-m-reducible to A (see Ladner et al. [1975]), the function mapping a set to its complement induces a nontrivial automorphism of $(\mathbf{REC}_m, \leqslant)$, but no other nontrivial automorphisms are known.

The undecidability of the theory $\mathrm{Th}(\mathbf{REC}_r, \leqslant)$ $(r = m, T)$ leads to the question which fragments of these theories are decidable. The embeddability of all finite partial orders into the u.s.l. $(\mathbf{REC}_r, \leqslant)$ implies decidability of the 1-quantifier theory $\exists^*$-$\mathrm{Th}(\mathbf{REC}_r, \leqslant)$. Shore and Slaman [1992] deduced from Theorems 5.4 and 4.5 the decidability of the 2-quantifier theory $\forall^*\exists^*$-$\mathrm{Th}(\mathbf{REC}_T, \leqslant)$ of the u.s.l. of the p-T-degrees of recursive sets, and subsequently Ambos-Spies et al. [1996] deduced from Theorems 4.4 and 4.5 a decision procedure for $\forall^*\exists^*$-$\mathrm{Th}(\mathbf{REC}_m, \leqslant)$ too. In fact the latter could be extended to decision procedures for $\forall^*\exists^*$-$\mathrm{Th}(\mathbf{ELEMENTARY}_m, \leqslant)$ and $\forall^*\exists^*$-$\mathrm{Th}(\mathbf{EXP}_m, \leqslant)$, whereas for p-T the corresponding questions are still open and require the solution of the lattice embedding problem (for embeddings preserving 0 and 1) for these structures. In the opposite direction Nies (private communication) has announced that the 4-quantifier theory of $(\mathbf{REC}_m, \leqslant)$ is undecidable, but the exact level at which undecidability starts is open for both $(\mathbf{REC}_m, \leqslant)$ and $(\mathbf{REC}_T, \leqslant)$.

7. Other reducibilities and the axiomatic approach

Polynomial time bounded versions of other effective reducibilities besides many-one and Turing reducibility have been studied in the literature. Ladner et al. [1975] introduced the polynomial time bounded versions of one-one (p-1, $\leqslant_1^P$), bounded truth-table (p-btt, $\leqslant_{btt}^P$) – with bounded truth-table reducibility of norm k (p-$btt(k)$, $\leqslant_{btt(k)}^P$) as special cases – and truth-table (p-tt, $\leqslant_{tt}^P$) reducibility. They showed that the implications

$$\leqslant_1^P \Rightarrow \leqslant_m^P \Rightarrow \leqslant_{btt(k)}^P \Rightarrow \leqslant_{btt(k+1)}^P \Rightarrow \leqslant_{btt}^P \Rightarrow \leqslant_{tt}^P \Rightarrow \leqslant_T^P$$

are proper on the exponential time sets. In fact, Ladner et al. showed that, for $r, r' \in R = \{1, m, btt(k)(k \geqslant 1), btt, tt, T\}$ such that $\leqslant_r^P$ is stronger than $\leqslant_{r'}^P$, there are exponential time sets A and B such that $A =_{r'}^P B$ but $A |_r^P B$. Simon and Gill [1977] investigated upward separations for the reducibilities in $R - \{1\}$: they showed that, for any recursive set $A \notin P$ and for $r, r' \in R - \{1, m\}$ such that $\leqslant_r^P$ is stronger than $\leqslant_{r'}^P$ there is a recursive set B such that $A \leqslant_{r'}^P B$ but $A \not\leqslant_r^P B$. For m and $1 - tt$, however, upward separation in general fails. Moreover, by constructing a recursive set $A \notin P$ such that $B \leqslant_1^P A$ iff $B \leqslant_T^P A$ for all sets B, Simon and Gill have shown that downward separation in general completely fails. (Note that, by closure of **EXP** under p-T-reducibility, every p-1-complete set A for **EXP** witnesses this failure too.)

The question of separating the completeness notions and degrees induced by the reducibilities in R has been addressed too. For the class **EXP** of the exponential time sets, Berman [1977] and Homer et al. [1993] showed that p-1-, p-m- and p-$btt(1)$-completeness coincide, while Watanabe [1987] separated the other completeness notions. For **NP**, separation results for the reducibilities and the corresponding completeness notions are only known under stronger hypotheses than $\mathbf{P} \neq \mathbf{NP}$. For details see the recent survey on complete sets by Buhrman and Torenvliet [ta].

For the degrees in general, appropriately defined resource-bounded generic sets give complete separations, i.e. for generic A and for $r, r' \in R$ such that $\leqslant_r^P$ is stronger than $\leqslant_{r'}^P$, $deg_r^P(A) \subset deg_{r'}^P(A)$ (see Ambos-Spies [1996]). On the other hand, $deg_T^P(A)$ collapses to $deg_{btt(1)}^P(A)$ for every super sparse set A (Ambos-Spies [1986b]) and, as already pointed out above, $deg_{btt(1)}^P(A)$ collapses to $deg_1^P(A)$ for any p-m-complete set A for **EXP**. It is not known, however, whether these results can be combined, i.e. whether there is a recursive set $A \notin \mathbf{P}$ such that $deg_r^P(A) = deg_T^P(A)$ for $r \in \{1, m\}$.

With the exception of p-$btt(k)$-reducibility for $k \geqslant 2$, the above reducibilities are transitive. For the transitive reducibilities between $\leqslant_m^P$ and $\leqslant_T^P$ many results on the degree structure carry over. For example, all of the above results proved by delayed diagonalization hold for these reducibilities too. Nondistributivity arises for the reducibilities allowing at least two oracle queries (cf. Theorem 2.2). No structural differences between $(\mathbf{C}_m, \leqslant)$ and $(\mathbf{C}_{btt(1)}, \leqslant)$ on the one side and between $(\mathbf{C}_T, \leqslant)$ and $(\mathbf{C}_r, \leqslant)$ for $r \in \{btt, tt\}$ on the other side are known for $\mathbf{C} \in \{\mathbf{NP}, \mathbf{EXP}, \mathbf{ELEMENTARY}, \mathbf{PRIM}, \mathbf{REC}\}$. Since below a super sparse set the degree structures for the reducibilities between $\leqslant_m^P$ and $\leqslant_T^P$ are all isomorphic, the undecidability theorem of Downey and Nies (Theorem 6.5) extends to all of these reducibilities, and, similarly, the proof of Theorem 6.1 on 1-types works for these reducibilities too.

The structure of the p-1-degrees, however, shows fundamental differences to the other degree structures: $\leqslant_1^P$ is not closed under finite variants and the partial order $(\mathbf{REC}_1, \leqslant)$ is neither a u.s.l. nor dense (see Ambos-Spies [1984a]). This structure has not been studied in detail so far, but Kampen [1993] has adapted the proof of Theorem 6.3 to show that $\mathrm{Th}(\mathbf{REC}_1, \leqslant)$ is undecidable.

Another interesting type of p-reducibility, which has no direct counterpart in the effective case, are the honest reducibilities. Note that in a p-r-reduction, for any input x and any oracle query y on this input, the length $|y|$ of the query is polynomially bounded in the input length $|x|$. If conversely, $|x|$ is polynomially bounded in $|y|$ too, the reduction is called honest. We let $\leqslant_{h-r}^P$ denote the honest counterpart of $\leqslant_r^P$. (Here, in case of many-one reducibility, we allow the reduction function $f(x)$ to assume the values $+$ and $-$ in order to indicate that the input belongs, respectively, does not belong to the set to be reduced and that this fact can be recognized in polynomial time without access to the oracle. This convention helps to avoid some pathological features of $\leqslant_{h-m}^P$, see Ambos-Spies [1989b].)

Honesty is necessary for inverting a p-r-reduction, whence it plays an important role in the investigation of isomorphism problems like the Berman–Hartmanis conjecture, which states that all p-m-complete sets for $\mathbf{NP}$ are p-isomorphic, i.e. isomorphic under polynomial time computable and invertible bijections (see Buhrman and Torenvliet [ta] for details).

Though many constructions of p-degrees in the literature yield honest reductions and therefore many results on the p-reducibilities carry over to the honest variants, there are some notable exceptions. Looking at nonrecursive sets, Homer [1987] showed that no p-m-degree is minimal whereas assuming that $\mathbf{P} = \mathbf{NP}$ there are (necessarily nonrecursive) minimal p-h-m-degrees. An example of a construction based on a nonhonest reduction is the proof of the anti-cupping-property based on shifts (presented in Section 3), and, assuming $\mathbf{P} = \mathbf{NP}$, there is a recursive p-h-r-degree ($r = m, T$) without the anti-cupping-property (Ambos-Spies [1989b]). So assuming $\mathbf{P} = \mathbf{NP}$ the theories $\mathrm{Th}(\mathbf{REC}_r, \leqslant)$ and $\mathrm{Th}(\mathbf{REC}_{h-r}, \leqslant)$ differ, but it is not known whether a difference can be given without this assumption. For the elementary recursive sets, however, such differences are known. Ambos-Spies and Yang [1990] have shown that for the honest reducibilities in general iterated look-ahead constructions can be replaced by simple exponential look-ahead constructions yielding elementary recursive sets. In particular they showed that, in contrast to Lemma 5.2, every nonzero p-h-r-degree is half of a minimal pair in $(\mathbf{ELEMENTARY}_{h-r}, \leqslant)$. Moreover, Ambos-Spies and Yang (unpublished) used the above observation to prove undecidability theorems for the honest reducibilities corresponding to Theorems 6.2 and 6.3 not only for the recursive but also for the elementary recursive sets.

When comparing subrecursive reducibilities, it is also natural to ask which results on the p-reducibilities actually exploit specific properties of **P**. It seems that all results on $(\mathbf{REC}_r^P, \leqslant)$ obtained so far can be obtained for other time bounded (or space bounded) reducibilities for sufficiently closed complexity classes too. So it was quite natural to try some axiomatic approach to the subrecursive reducibilities by proposing axiom systems which are satisfied by the common subrecursive reducibilities (with the exception of the reducibilities of the one-one type which, as pointed out above, show a behaviour quite different from that of the other reducibilities) and which are sufficient for proving interesting results on their structure.

A first, quite powerful axiom system along these lines was proposed by Mehlhorn [1976]. Besides fundamental axioms expressing that the reducibility is given by a uniformly recursive sequence of recursive functionals and that it is closed under finite variants, Mehlhorn's system contains axioms allowing to carry out the delayed diagonalizations of Ladner [1975]. Variants of Mehlhorn's system were introduced by Schmidt [1985] and Mueller [1991]. The system of Schmidt is quite simple and elegant, in its spirit comparable to Schöning's diagonalization theorem (Theorem 4.7), but it seems to be weaker than Mehlhorn's original system. Mehlhorn's system has the defect, however, that it does not allow to handle meets. Recently Merkle [1996] introduced a new axiom system which is satisfied by the standard subrecursive reducibilities in the literature and which also suffices to obtain the basic results on the meet operator. So his axioms suffice to prove the Embedding Theorem 4.4 and the Exact Pair Theorem 5.3. Merkle also discusses additional axioms sufficient for the distinction of the distributive and nondistributive reducibilities.

References

K. Ambos-Spies

[1984a] *On the Structure of the Polynomial Time Degrees of Recursive Sets*, Habilitationsschrift, Universität Dormund.

[1984b] On the structure of polynomial time degrees, in: *Proceedings STACS 84*, Lecture Notes in Comput. Sci., Vol. 166, Springer, pp. 198–208.

[1985] Sublattices of the polynomial time degrees, *Inform. and Control*, 65, pp. 63–84.

[1986a] An inhomogeneity in the structure of Karp degrees, *SIAM J. Comput.*, 15, pp. 958–963.

[1986b] Inhomogeneities in the polynomial time degrees: the degrees of super sparse sets, *Inform. Process. Lett.*, 22, pp. 113–117.

[1987a] Polynomial time degrees of NP-sets, in: *Trends in Theoret. Computer Science*, E. Börger, ed., Computer Science Press, pp. 95–142.

[1987b] Minimal pairs for polynomial time reducibilities, in: *Computation Theory and Logic*, Lecture Notes in Comput. Sci., Vol. 270, Springer, pp. 1–13.

[1989a] On the relative complexity of hard problems for complexity classes without complete problems, *Theoret. Comput. Sci.*, 63, pp. 43–61.

[1989b] Honest polynomial time reducibilities and the $P = NP$ problem, *J. Comput. System Sci.*, 39, pp. 250–281.

[1991] *A note on minimal and exact pairs in the polynomial degrees*, Unpublished research note.

[1992a] On the theory of the polynomial time degrees of exponential time sets, in: *Structure and Complexity*, Dagstuhl-Seminar-Report 30 (Abstract). (Full paper in preparation.)

[1992b] *The theory of the polynomial time many-one degrees of the elementary recursive sets is undecidable*, Unpublished research note.

[1996] Resource-bounded genericity, in: *Computability, Enumerability, Unsolvability*, London Math. Soc. Lecture Notes Series, Vol. 224, Cambridge Univ. Press, pp. 1–59.

K. Ambos-Spies, P. A. Fejer, S. Lempp and M. Lerman

[1996] Decidability of the two-quantifier theory of the recursively enumerable weak truth-table degrees and other distributive upper semi-lattices, *J. Symbolic Logic*, 61, pp. 880–905.

K. Ambos-Spies, S. Homer and R. I. Soare

[1994] On minimal pairs and complete problems, *Theoret. Comput. Sci.*, 132, pp. 229–241.

K. Ambos-Spies and A. Nies

[1992] The theory of the polynomial many-one degrees of the recursive sets is undecidable, in: *Proceedings STACS 92*, Lecture Notes in Comput. Sci., Vol. 577, Springer, pp. 209–218.

K. Ambos-Spies and D. Yang

[1990] Honest polynomial time degrees of elementary recursive sets, in: *Proceedings CSL 89*, Lecture Notes in Comput. Sci., Vol. 440, Springer, pp. 1–15. (Extended version in preparation.)

P. Axt

[1959] On a subrecursive hierarchy and primitive recursive degrees, *Trans. Amer. Math. Soc.*, 92, pp. 85–105.

J. L. Balcázar and J. Díaz

[1982] A note on a theorem by Ladner, *Inform. Process. Lett.*, 15, pp. 84–86.

J. L. Balcázar, J. Díaz and J. Gabarró

[1988] *Structural Complexity*, Vol. I, Springer, Berlin.

L. Berman

[1977] *Polynomial Reducibilities and Complete Sets*, Ph.D. Thesis, Cornell University.

S. I. Breidbart

[1977] *The Structure of Complexity Classes and Degrees*, Ph.D. Thesis, University of California, Santa Barbara.

H. Buhrman and L. Torenvliet

[ta] Complete sets and structure in subrecursive classes, in: *Proceedings of Logic Colloquium 95*, San Sebastian.

C. C. Chang and H. J. Keisler

[1978] *Model Theory*, North-Holland, Amsterdam.

P. Chew and M. Machtey

[1981] A note on structure and looking back applied to the relative complexity of computable functions, *J. Comput. System Sci.*, 22, pp. 53–59.

S. A. Cook

[1971] The complexity of theorem proving procedures, in: *Proceedings 3rd Annual ACM Symp. on Theory of Comp.*, pp. 151–158.

R. DOWNEY

[1992] Nondiamond theorems for polynomial time reducibility, *J. Comput. System Sci.*, 45, pp. 385–395.

R. DOWNEY AND A. NIES

[1997] Undecidability results for low complexity degree structures, in: *Proceedings of the 12th Ann. IEEE Conf. on Computational Complexity*, Ulm, pp. 128–132.

M. R. GAREY AND D. S. JOHNSON

[1978] *Computers and Intractability: A Guide to the Theory of NP-Completeness*, H. Freeman, San Francisco.

C. A. HAUGHT AND T. A. SLAMAN

[1997] Automorphisms in the PTIME-Turing degrees of recursive sets, *Ann. Pure Appl. Logic*, 84, pp. 139–152.

E. HERRMANN

[1984] The undecidability of the elementary theory of the lattice of recursively enumerable sets (abstract), in: *Proceedings of the 2nd Frege Conference*, Akademie-Verlag, Berlin, pp. 66–72.

S. HOMER

[1987] Minimal degrees for polynomial reducibilities, *J. Assoc. Comput. Machin.*, 34, pp. 480–491.

S. HOMER, S. KURTZ AND J. ROYER

[1993] A note on many-one and 1-truth table complete sets, *Theoret. Comput. Sci.*, 115, pp. 383–389.

J. E. HOPCROFT AND J. D. ULLMANN

[1979] *Introduction to Automata Theory, Languages and Computation*, Reading.

J. KAMPEN

[1993] *Über Struktur und Unentscheidbarkeit der Theorie der Polynomiellen* 1*-Grade rekursiver Mengen*, Diplomarbeit, Universität Heidelberg.

R. M. KARP

[1972] Reducibility among combinatorial problems, in: *Complexity of Computer Computations*, Plenum, New York, pp. 85–103.

S. C. KLEENE

[1958] Extensions of an effectively generated class of functions by enumeration, *Colloq. Math.*, 6, pp. 67–78.

S. C. KLEENE AND E. L. POST

[1954] The upper semi-lattice of degrees of recursive unsolvability, *Ann. of Math.*, 59, pp. 379–407.

R. E. LADNER

[1975] On the structure of polynomial time reducibility, *J. Assoc. Comput. Machin.*, 22, pp. 155–171.

R. E. LADNER, N. LYNCH AND A. L. SELMAN

[1975] A comparison of polynomial time reducibilities, *Theoret. Comput. Sci.*, 1, pp. 103–123.

L. H. LANDWEBER, R. J. LIPTON AND E. L. ROBERTSON

[1981] On the structure of sets in *NP* and other complexity classes, *Theoret. Comput. Sci.*, 15, pp. 181–200.

L. LEVIN

[1973] Universal sorting problems, *Problemy Peredachi Informacii*, 9, pp. 115–116 (in Russian).

M. MACHTEY

[1974] The honest subrecursive classes are a lattice, *Inform. Comput.*, 24, pp. 247–263.

[1976] Minimal pairs of polynomial degrees with subexponential complexity, *Theoret. Comput. Sci.*, 2, pp. 73–76.

K. MEHLHORN

[1976] Polynomial and abstract subrecursive classes, *J. Comput. System Sci.*, 12, pp. 147–178.

W. MERKLE

[1996] *A Generalized Account of Resource Bounded Reducibilities*, Ph.D. Thesis, Universität Heidelberg.

W. MUELLER

[1991] *Abstract Degree Structures*, Ph.D. Thesis, Mount Holyoke College, South Hadley.

A. NIES

[1997] A uniformity of degree structures, in: *Complexity, Logic and Recursion Theory*, A. Sorbi, ed., Lecture Notes in Pure and Applied Mathematics, Vol. 187, Marcel Dekker, New York, pp. 261–267.

[ta] Effectively dense Boolean algebras and their applications, in preparation.

D. SCHMIDT

[1984] On the complement of one complexity class in another, in: *Logic and Machines: Decision Problems and Complexity*, Lecture Notes in Comput. Sci., Vol. 171, Springer, pp. 77–87.

[1985] The recursion-theoretic structure of complexity classes, *Theoret. Comput. Sci.*, 38, pp. 143–156.

U. SCHÖNING

[1982] A uniform approach to obtain diagonal sets in complexity classes, *Theoret. Comput. Sci.*, 18, pp. 95–103.

J. SHINODA AND T. A. SLAMAN

[1990] On the theory of the *PTIME* degrees of the recursive sets, *J. Comput. System Sci.*, 41, pp. 321–366.

R. A. SHORE AND T. A. SLAMAN

[1992] The p-t-degrees of the recursive sets: lattice embeddings, extensions of embeddings and the two quantifier theory, *Theoret. Comput. Sci.*, 97, pp. 263–284.

I. SIMON AND J. GILL

[1977] Polynomial reducibilities and upward diagonalizations, in: *Proceedings 9th Ann. ACM Symp. on Theory of Computation*, pp. 183–194.

O. WATANABE

[1987] A comparison of polynomial time completeness notions, *Theoret. Comput. Sci.*, 54, pp. 249–265.

X. ZHENG

[1994] On the maximality of some pairs of p-t-degrees, *Notre Dame J. Formal Logic*, 34, pp. 29–35.

Author Index

Upright numbers refer to pages on which the author (or his/her work) is mentioned in the text of a chapter. Italic numbers refer to reference list pages. (No distinction is made between first and coauthor(s).)

Abel, N.H. 370, *435*
Aberth, O. 429, *435*
Abraham, U. 102, *115*, 157, *166*
Ackermann, W. 9, 28, *33*, 535, 541, *582*, 590, *670*
Aczel, P. *357*, 359
Adams, W.W. 385, 435, *435*
Adian, S.I. 421, *435*
Ahmad, S. 107, *115*, 145, *145*
Ajtai, M. 601, *670*
Allen, B. 591, 625, 641, 654, *670*
Alton, D. 426, *435*
Ambos-Spies, K. 112, *115*, 131, 140, 141, *146*, 175, 177–179, 182, 183, 187, 191, 192, *193*, *194*, 202, *244*, 686, 688, 689, 691–701, *702*, *703*
Anshel, M. 432, *438*
Antonio, G. 428, *444*
Arslanov, M.M. 135, 138, 139, *146*
Artin, E. 370, *435*
Ash, C.J. 425, *435*, 521, 529, *530*
Asser, G. 610, *670*
Asveld, P.R.J. 431, *436*
Atiyah, M. 372, 409, 415, *436*
Avenhaus, J. 432, *436*
Axt, P. 684, *703*
Ayoub, C.W. 385, *436*

Babai, L. 640, *670*
Babbage, C. 370, *436*
Balcázar, J.L. 686, 691, 693, *703*
Barker, E. 529, *530*
Barrington, D.A. 599, 624, *670*
Barwise, K.J. 358, *358*, 432, *436*
Baumslag, G. 411, 421, *436*
Baur, W. 414, 416, 417, *436*
Beame, P.W. 607, *670*
Bean, D. 79, *82*
Becker, H.A. 192, *194*
Becker, T. 385, 435, *436*
Beeson, M.J. 434, *436*
Bel'tyukov, A. 626, 670, *670*
Bellantoni, S. 638, 646, 647, 652–654, 659, *670*
Beller, A. 295, 296, *297*
Benda, M. 521, *530*
Bennett, J.H. 590, 608, 610, 622, 633, 634, *670*
Berger, U. 254, 256, *274*, 535, 544, *582*
Bergstra, J.A. 262, *274*, 427, 428, *436*, *437*
Berman, L. 700, *703*
Bernays, P. *33*
Blanck, J. 434, *437*
Bloch, S. 654–656, *670*, *671*
Blum, L. 316, 338, 344, 347, 348, *358*, 433, *437*
Boerger, E. *358*, 359
Boolos, G. 30, *31*
Boone, W.W. 421, 422, *435*, *437*
Boppana, R. 608, *671*
Borodin, A. 597, 607, *671*
Bourbaki, N. 410, *437*
Breidbart, S.I. 691, *703*
Bridges, D. 371, 434, *437*
Buchholz, W. 534, 542, 576, 578, *582*
Buhrman, H. 700, 701, *703*
Bulitko, V.K. 100, *115*
Burris, S. 243, *245*, 426, *437*
Buss, J. 656, *671*
Buss, S.R. 592, 596, 600, 601, 607, 637, 642, 668, *671*
Byrnes, J. 11, 13, *35*

Cai, J.-Y. 630, *671*
Calhoun, W. 144, *146*
Cannonito, F.B. 411, 421, 431, *436*, *437*
Carnielli, W. *32*
Case, J. 143, *146*

Cenzer, D. 38, 39, 48–50, 61, 62, 64–67, 72–74, 76, 82, *82*, *83*, 431, *437*
Chandler, B. 421, *437*
Chandra, A. 595, 597, 600, 621, *671*
Chang, C.C. 508, 511–513, 516–518, 522, *530*, 696, *703*
Chevalley, C. 423, *437*
Chew, P. 688, *703*
Chisolm, J. 529, *530*
Cholak, P.A. 38, 67, *83*, 109, 114, *115*, 183, 191, *194*, 202, 216, 244, *245*
Chong, C.T. 127, 130, *146*, 289, 290, *298*
Church, A. 5, 7, 9, 10, 14, 15, 19–21, 29, *31*, 36, 43, *83*, 123, *146*, 200, *245*, 590, 592, *671*
Cichon, A. 534, *582*
Clapham, C.R.J. 422, *438*
Clote, P.G. 38, 65, 66, *82*, 535, *582*, 591, 592, 608, 610, 618, 621, 623–625, 627, 630, 637, 639, 640, 642, 646, 653, 660, 663, 666–668, *671*, *672*
Cobham, A. 535, *582*, 591, 608, 610, 622, *672*
Cohen, P.J. 124, *146*
Cohn, P.M. 372, 409, 426, *438*
Collins, D.J. 422, *438*
Constable, R. 638, 639, 657, *672*
Cook, S.A. 536, *582*, 591, 592, 607, 643, 645–647, 652, 656–660, 668, *670*, *672*, *674*, 684, 685, *703*
Cooper, S.B. 91, 107, 111, 113, 114, *115*, *118*, 123, 124, 126, 128–130, 132, 133, 136–145, *146*, *147*, *151*, 162, 165, *166*, 184, 188–190, *194*
Copestake, C.S. 143, 144, *147*
Csillag, P. 638, *672*
Cucker, F. 433, *437*
Cutland, N.J. 17, 27, 30, *31*, 36, 324, 325, 340, 343, *358*, 372, *438*

David, R. 295–297, *298*
Davis, M. 5, 10, 14, 15, 19, 26, 30, *31*, 32, *32*, 35, 36, 245, *245*, 247, 248, 424, *438*
Dedekind, R. 7, 28, 30, *32*, 367, *438*
Degtev, A.N. 97, 100–102, 108–110, *116*, 136, *147*
Dehn, M. 421, *438*
Dekker, J. 45, 48, *83*
Denisov, S.D. 96, 98, 108, *116*, 483, 492, *502*
Deutsch, D. 602, *672*
Díaz, J. 686, 691, 693, *703*
Ding, D. 138, 139, *148*
Dobrica, V.P. *438*
Domanski, B. 432, *438*
Dosch, W. 570, *582*
Downey, R.G. 38, 48–50, 67, 69, 72, 73, *82*, *83*, 108–110, 114, *115*, *116*, 127, 130, 133, 139, *146*, *148*, 176, 178, 183, 191–193, *194*, 202, 216, 217, 244, *245*, 425, *435*, 691, 699, *704*
Drake, F.R. *358*, 359
Dvornikov, S.G. 484, 498, *502*
Dyck, W. 421, *438*

Edalat, A. 434, *438*
Edwards, H.M. 372, *438*
Ehrenfeucht, A. 39, 76, *83*
Enderton, H.B. 372, *438*
Engeler, E. 317, *358*
Epstein, R.L. *32*, 123, 126, 129, 132, 137, *147*, *148*
Ershov, Yu.L. 94, 95, 108, *116*, 119, 124, 137, *148*, 254, *274*, 369, 378, 391, 395, 396, 404, 426, 428, *438*, 474, 478, 481, 482, 484, 491, 495, 496, 498, 499, 502, *502*, 535, 544, *583*

Fagin, R. 591, *672*
Fairtlough, M.V.H. 535, 541, 542, *583*
Fedoryaev, S.T. *439*
Feferman, S. *32*, 98, 105, *116*, 123, 124, *148*, 424, *439*
Fejer, P.A. 105, 110, 112, *115*, *116*, 130, *148*, 178, 187, 192, *193*, *194*, 699, *703*
Fenstad, J.E. *313*, 316, *358*, 432, *439*
Fischer, P.C. 93, *116*
Fitting, M. *32*
Fortnow, L. 640, *670*
Fortune, S. 558, *583*, 602, *672*
Fridman, A.A. 422, *439*
Friedberg, R.M. 105, *117*, 142, *148*, 171, *194*, 201, *245*
Friedman, H.M. 82, *83*, 316, *358*, 541, 542, *583*
Friedman, S.D. 288, 291–297, *298*
Fröhlich, A. 369, 371, 391–394, 401, 403, *439*
Furst, M.L. 598, 601, 621, 630, *671*, *673*

Gabarró, J. 686, 693, *703*
Galois, E. 370, *439*
Gandy, R.O. 8, 10–13, 15, 28, 31, *32*, 36, 305, *313*, 358, *358*, 371, *439*, 542, *583*
Garey, M.R. 684, *704*
Gatterdam, R.W. 431, *437*, *439*
Gill, J. 700, *705*
Girard, J.-Y. 541, *583*, 668, *673*
Gödel, K. 5, 7, 9, 10, 14, 15, 24, 26, 30, 32, *32*, *33*, 92, *117*, 123, *148*, 200, *245*, 271, *274*, 508, *530*, 570, 574, *583*, 590, 591, 608, *673*
Goerdt, A. 669, *673*
Goguen, J.A. 428, *442*
Gold, E.M. 137, *148*
Goldschlager, L. 602, *673*

Goldsmith, J. 656, *671*
Goncharov, S.S. 384, 423, *439*, 493, *502*, 514, 517, 521, 526, 527, 529, *530*
Goodstein, R.L. 371, *439*
Grätzer, G. 426, *439*
Griffor, E.R. 254, *275*, 312, *313*, 434, *445*, 544, *585*
Grilliot, T. 262, *274*, 309, *313*
Groszek, M.J. 57, *83*, 103, *117*, 157, *167*, 290, *298*
Grzegorczyk, A. 450, 457, *470*, 542, *583*, 590, 608, 625–627, 632, 638, 639, *673*
Guichard, D. 425, *439*
Gurevich, Y. 591, 632, 669, *673*
Gutteridge, L. 106, *117*, 144, *149*

Haas, R. 137, *148*
Hájek, P. 636, *673*
Handley, W.G. 624, 627, 637, *673*
Hanf, W. 77, *83*, 98, *117*, 123, *149*
Hardy, G.H. 540, *583*, 590, *673*
Harel, D. 82, *83*
Harizanov, V.S. 508, 529, *530*
Harnik, V. 668, *673*
Harrington, L. *33*, 104, 111, *117*, 136, 138, *147*, *149*, 181, 182, *194*, 201, 202, 205–209, 211, 215–219, 230, 235, 241–244, *245*, *246*, 305, 310, 311, *313*, *439*, 514, 526, 527, *530*
Harrow, K. 632, 633, *673*
Haught, C.A. 104, 111, 112, *117*, 127, *149*, 699, *704*
Hay, L. 495, 496, *503*
Hensel, K. 370, *439*
Hentzel, I.R. 432, *439*
Henzelt, K. 370, *440*
Herken, R. 32, *33*, 34
Herman, G.T. 317, *358*
Hermann, G. 370, 385, 408, 411, 421, *440*
Herrmann, E. 110, *117*, 243, *246*, 698, *704*
Higman, G. 421, 422, *435*, *437*, *440*
Hilbert, D. 7–9, 28, 30, *33*, 590, 667, *674*
Hillis, W.D. 602, *674*
Hingston, P. 411, *440*
Hinman, P.G. 53, 58, *83*, 124, *149*, 317, 331–334, *358*, 539, *583*
Hodges, A. 12, 24, *33*
Homer, S. 691, 700, 701, *703*, *704*
Hoover, H.J. 607, 668, *670*, *674*
Hopcroft, J.E. 685, *704*
Huang, W. 82, *84*
Hugill, D.F. 130, *149*
Hyland, J.M.E. 254, 260, 266, *274*, 542, *583*

Ignjatovic, A. 591, 660, 663, 666–668, *671*
Immerman, N. 591, 596, 604, 607, 609, 621, *670*, *674*
Isard, S.D. 317, *358*
Ishmukametov, Sh.T. 138, *149*

Jacobs, D.P. 432, *439*
Jacobsson, C. 411, 431, *440*
Jeffrey, R. 30, *31*
Jensen, R.B. 278, 280, 290, 291, 294–296, *297*, *298*
Jiang, Z. 139, *149*
Jockusch Jr., C.G. 38, 48–50, 52–54, 56, 57, 60, 67, 72, 73, 77, *82*, *83*, 100, 101, 108–110, *116*, *117*, 126, 127, 129, 130, 134–140, *146*, *149*, *150*, 158, 159, 161, 164, 165, *167*, 179, 183, 191, *193*, *195*
Johannsen, J. 592, *674*
Johnson, D.S. 684, *704*
Jones, J.P. 639, *674*
Jones, N.D. 590, *674*

Kaddah, D. 138, *150*
Kalantari, I. 425, *440*
Kálmar, L. 590, 638, *674*
Kampen, J. 701, *704*
Kannan, R. 622, *674*
Kapron, B.M. 591, 643, 645, 646, 657–660, 663, 666–668, *671*, *672*, *674*
Karass, A. 421, *441*
Karp, R.M. 603, *674*, 684, 686, *704*
Kaye, R. 426, *440*
Kechris, A.S. 192, *195*, 318, 338, 340, 341, 348, 350, 352, *358*
Keisler, H.J. 508, 511–513, 516–518, 522, *530*, 696, *703*
Ker-I Ko 668, *674*
Khisamiev, N.G. *438*, *440*
Khutoretskii, A.B. 491, 492, *503*
Kierstead, H. 82, *84*
Kirby, L.A. 289, *299*
Kleene, S.C. 5–10, 14, 15, 17–25, 27, 29, *31*, *33*, *34*, 36, 38, 43, 45, 52, 56, 74, *83*, *84*, 93, *117*, 124, 134, *150*, 156–158, *167*, *195*, 200, *246*, 252–255, 258, 266, *274*, 279, *298*, 302, 310, *313*, 316, 317, *358*, 371, *440*, 509, 512, 517, *531*, 534, 535, 542, 544, 578, *583*, 590, 608, *674*, 684, 687, *704*
Kline, M. 16, *34*
Knight, J.F. 529, *530*
Kobzev, G.N. 110, 112, *117*

König, G. 370, *440*
Kozen, D. 595, 597, 600, *671*
Krajíček, J. 592, *674*
Kramer, R. 137, *148*
Kreisel, G. 38, 52, 55, 61, *84*, 253–256, 271–273, *274*, 278, 279, *298*, 535, 541, 544, 576, *583*, 668, *674*
Kreitz, C. 432, *440*
Kronecker, L. 370, 400, *440*
Krull, W. 371, 401, 406, *440*
Kucera, A. 57, 59, *84*, 134, 135, *150*
Kudaibergenov, K.Zh. 515, *531*
Kudinov, O.V. 384, 426, *440*
Kumabe, M. 127, 128, 135, *150*
Kuratowski, K. 43, *84*
Kurtz, S. 700, *704*
Kutyłowski, M. 640, 641, *674*
Kuz'mina, T.M. 497, *503*
Kuznetsov, A.V. 422, *440*

Lachlan, A.H. 94, 96–98, 107, 110, 111, 113, *115*, *117*, 126, 128, 131, 138, 139, 144, *147*, *150*, 157, 160, *167*, 173–175, 177, 178, 182, 183, 192, *195*, 201, 203, 205, 207, 216, 220, 241, 242, *246*, 429, *441*, 483, *503*
Lacombe, D. 82, *84*, 450, 457, *471*, *503*, 668, *674*
Ladner, R.E. 112, 685, 688–691, 693, 699, 700, 702, *704*
LaForte, G.L. 139, *150*
Lagrange, J.L. 370, *441*
Lallement, G. 424, *441*
Landweber, L.H. 688, 691, 693, *704*
Lang, S. 372, *441*
Larsen, K.G. 545, *583*
Lavrov, I.A. 483, *502*
Lazard, D. 411, *441*
Lebeuf, R. 126, *150*, 157, *167*
Leivant, D. 558, *583*, 646, 653, 656, 669, *675*
Lempp, S. 112, *115*, *118*, 130, 138, 139, 141, *146–148*, *150*, 176–178, 180, 191, *193–195*, 699, *703*
Lerman, M. 27, *34*, 112, 113, *115*, *118*, 123–126, 130, 132, 140, 141, *149*, *150*, 157, 158, 160, *167*, 175–179, 181, *193*, *195*, *196*, 241, 243, *246*, 282, 286, *298*, 508, 511, *531*, 699, *703*
Levin, L. 684, 686, *704*
Li, A. 130, *150*
Lin, C. 424, 433, *441*
Lind, J.C. 610, 623, *675*
Lindström, I. 254, *275*, 434, *445*, 544, *585*
Lipton, R.J. 432, *441*, 688, 691, 693, *704*
Löb, M.H. 542, *583*
Longo, G. 255, *274*
Loustaunau, P. 385, 435, *435*
Löwenheim, L. 9, *34*
Lynch, N. 699, 700, *704*
Lyndon, R.C. 421, 422, *437*, *441*

Maass, W. 178, *196*, 209, 211, 212, 214, *246*, 280, 292, *298*
MacDonald, I. 372, 409, 415, *436*
Machtey, M. 372, *441*, 684, 688, 689, *703*, *704*
MacIntyre, A. 422, *441*
MacQueen, D. 310, *313*
Madison, E.W. 426, 429, *435*, *441*
Madlener, K. 432, *436*
Magidor, M. 297, *299*
Magnus, W. 421, *437*, *441*
Mal'cev, A.A. 104, *118*
Mal'cev, A.I. 368, 369, 372, 380, 422, 424, 426, 430, *441*, 474, 487, *503*
Manaster, A. 82, *84*
Marchenkov, S.S. 102, *118*, 136, *151*, 202, *247*, 639, *675*
Marion, J.-Y. 669, *675*
Markov, A.A. *441*
Martin, D.A. 69, 78, *84*, 127, 130, 136, *151*, 159, *167*, 201, 241, *247*
Martin, G. 67, 68, *84*
Martin-Löf, P. 59, *84*
Matijasevič, Y. 123, *151*, 639, *674*
McAloon, K. 426, *441*
McEvoy, K. 111, *118*, 143, 144, *151*
McKenzie, R.N. 243, *245*, 426, *442*
McNulty, G.F. 82, *84*, 426, *442*
Medvedev, Y.T. 107, *118*
Mehlhorn, K. 591, 657, 658, *675*, 691, 702, *705*
Meinke, K. 427, *442*
Merkle, W. 702, *705*
Meseguer, J. 428, *442*
Metakides, G. 82, *84*, 391, 393, 403, 404, 406, 408, 424, 425, *442*, *471*
Meyer, A.R. 626, *675*
Michaux, C. 340, 347, 348, *358*
Millar, T.S. 515, 516, 519–521, 523, 525, 526, 528, *530*, *531*
Miller III, C.F. 411, 421, *436*, *442*
Miller, D.P. 107, *118*, 190, *196*, 202, *247*
Miller, W. 136, *151*
Mines, R. 371, 434, *442*
Minsky, M.L. 536, *584*
Mints, G.E. 535, 578, *584*
Mix Barrington, D. 596, 607, 609, 621, *670*
Moggi, E. 255, *274*
Mohrherr, J. 108, *118*

Moldestad, J. 316, 318, 340, 344, *359*, *442*, 544, *584*
Molk, J. 434, *442*
Monien, B. 644, *675*
Morley, M.D. *439*
Moschovakis, Y.N. 55, *84*, 192, *195*, *313*, 317, 318, 334, 336–338, 340, 341, 348, 350, 352, *358*, *359*, 434, *442*
Moss, L. 428, *442*
Mostowski, A. 43, *84*
Mourad, K.J. 289, 290, *298*
Muchnick, S.S. 626, *675*
Muchnik, A.A. 171, *196*, 201, *247*
Mueller, W. 702, *705*
Müller, H. 626, *675*
Myers Jr., J.P. 359, *359*
Myhill, J. 45, 48, *83*, 103, 111, *118*, 133, 136, 142, *151*, 201, *247*, 458, *471*, 499, *503*, 534, 537, *584*
Mytilinaios, M. 290, *298*, *299*

Nagel, E. 358, *359*
Nepomnjascii, V.A. 622, 636, *675*
Nerode, A. 62, 82, *84*, 107, *118*, 125, *151*, 160, 165, *167*, 366, 391, 393, 403, 404, 406, 408, 424–426, 431, *442*, *443*, *471*, 529, *530*, 668, *675*
Neumann, B.H. 422, *443*
Nguyen, A.P. 653, *675*
Nies, A. 91, 97, 98, 102, 103, 105, 112, *115*, *118*, 180–182, 184–186, *195*, *196*, 202, 244, *244*, *245*, *247*, 692, 695, 698, 699, *703–705*
Niggl, K.-H. 536, *584*, 669, *675*
Normann, D. 23, *34*, 254, 260, 262, 269, 270, *275*, 302, 312, *313*, 535, 543, *584*
Novikov, P.S. 421, *443*
Novy, L. 370, *443*
Nurtazin, A.T. *438*, 514, 526, 527, *530*

O'Donnell, M.J. 359, *359*, 558, *583*, 592, *675*
Oberschelp, W. *358*, 359
Odifreddi, P.G. 27, *34*, 41, 60, 67, *85*, 91, 104, 108, 110, 111, *118*, *119*, 123, 124, 126, 128, 135, 136, 144, *151*, 163, *167*, 372, *443*, 474, *503*
Otto, J. 656, *676*
Owings, J. 62, 67, 68, *85*

Pakhomov, S.V. 631, *676*
Paliutin, E. 91, *119*
Paris, J.B. 289, *299*, 624, 627, 633, 637, 640, *673*, *676*
Peacock, G. 370, *443*
Peano, G. 7, 30, *34*
Peirce, C.S. 28, 31, *34*
Penrose, R. 25, *34*, 35
Peretyat'kin, M.G. 519–521, *531*
Péter, R. 7, 28, 30, *34*, 538, *584*, 590, *676*
Phillips, I.C.C. 372, *443*
Pillay, A. 523, *531*
Pitt, F. 642, *676*
Platek, R.A. 7, *34*, 318, *359*, 536, 544, 559, *584*
Plotkin, G.D. 94, 95, *119*, 535, 559, 570, *584*
Posner, D.B. 123, 126, 128–130, 140, *149*, *151*, 164, *167*
Post, E.L. 5, 6, 19–22, 25, 29, 31, *34*, *35*, 36, 93, 105, *117*, *119*, 123, 124, 133, 135, 136, *150*, *151*, 156–158, *167*, 170, *195*, *196*, 200, 201, 203, *247*, *443*, 508, *531*, 687, *704*
Pour-El, M.B. 69, 78, *84*, 371, 429, 434, *443*, 454, 455, 458–461, 466, 468, 470, *471*
Pudlák, P. 602, 636, *673*, *676*
Putnam, H. 25, 30, *35*, 137, *151*

Qian, L. 138, 139, *148*
Qing Zhou 470, *471*

Rabin, M.O. 369, 391, 396, 422, 423, *443*
Ramachandran, V. 603, *674*
Reed, C.R. 521, *531*
Reid, C. 370, *443*
Remmel, J.B. 39, 66, 74, 76, 80, 82, *83*, *85*, 366, 424–426, 431, *437*, *442*, *443*, 668, *675*
Retzlaff, A. 425, *440*, *443*
Rice, H.G. 454, *471*, 495, *503*
Richards, I. 371, 429, 434, *443*, 454, 455, 458–461, 466, 468, 470, *471*
Richman, F. 371, 411, 418, 434, *437*, *442*, *443*
Richter, M.M. *358*, 359
Ritchie, D. 626, *675*
Ritchie, R.W. 591, 608, 628, 630, *676*
Robbin, J.W. 542, *584*
Robertson, E.L. 688, 691, 693, *704*
Robinson, J. 633, *676*
Robinson, R.M. 454, *471*, 610, *676*
Robinson, R.W. 129, *151*, *152*, 183, 187, 189, *196*
Rodgers, H. 491, 495, *503*
Rogers Jr., H. 19, 27, *35*, 63, *85*, 92, 98–100, 105, *117*, *119*, 140, 142, *148*, *152*, 158, *167*, 171, *196*, 202, *247*, 279, *299*, 372, *443*, 491, *503*, 539, *584*
Rose, H.E. 358, *359*, 542, *584*, 609, 622, 670, *676*
Rosenstein, J. 82, *84*
Rosser, B. 75, *85*
Rotman, J. 421, 424, *443*
Routledge, N.A. 534, 537, *584*
Royer, J.S. 668, *676*, 700, *704*

Rozinas, M.G. 143, 144, *152*
Rubin, M. 243, *247*
Ruffini, P. 370, *444*
Ruitenburg, W. 371, 434, *442*
Ruzzo, W.L. 607, *676*
Rybina, T.V. 498, *502*

Sacks, G.E. *35*, 109, *119*, 125, 128, 131, 132, 138, 140, *152*, 157–159, 161, *167*, 171, 172, 179, 180, 192, *196*, 202, *247*, 278–280, 282, 290, 294, 296, *298*, *299*, 302, 306–309, 311, 312, *314*, 433, *444*, 515, 521, 525, *531*, 539, *584*
Sankappanavar, H.P. 426, *437*
Sasso, L.P. 112, 130, *152*
Savitch, W.J. 600, *676*
Saxe, J.B. 598, 601, 621, *673*
Scedrov, A. *439*, 668, *673*, *675*
Schinzel, B. *358*, 359
Schlipf, J. 508, *531*
Schmerl, J.H. 82, *85*, 511, *531*
Schmidt, D. 691, 702, *705*
Schnorr, C.P. 632, *676*
Scholz, H. 590, *676*
Schöning, U. 693, *705*
Schreiber, U. 434, *447*
Schreier, O. 370, *435*
Schröder, E. 9, *35*
Schupp, P.E. 421, 422, *441*
Schütte, K. 578, *584*
Schwichtenberg, H. 359, *359*, 534, 535, 541, 542, 558, 570, 576, 577, 579, *585*, 626, 669, *676*
Scott, D. 60, *85*, 114, *119*, 134, 142, *152*, 544, 545, *585*
Scott, P. 668, *673*
Seetapun, D. 129, 132, *152*
Seidenberg, A. 371, 411, 417, 419, *444*
Selivanov, V.L. 105, *119*, 137, *152*, 492, 497, *503*
Selman, A.L. 142, *152*, 590, *674*, 699, 700, *704*
Seth, A. 659, 668, *677*
Sheard, M. 541, 542, *583*
Shelah, S. *117*, 181, 182, *194*, 297, *299*, 591, 632, *673*
Shepherdson, J.C. 323, 358, *359*, 369, 371, 391–394, 401, 403, 426, 433, *439*, *444*, 536, *585*
Shiloach, Y. 602, *677*
Shinoda, J. 698, *705*
Shoenfield, J.R. 10, 17, 22, 27, *35*, 38, 39, 44, 45, 53, 75, *85*, 100, *119*, 125, 129, 134, *152*, 171, 172, 177, *196*, 241, *247*, 539, *585*, 668, *674*
Shor, P. 602, *677*
Shore, R.A. 38, 48–50, 67, 72, 73, *82*, *83*, 91, 100, 102, 104, 105, 107, 110–113, *115–119*, 124–126, 128–130, 132, 137–139, 141, 142, 144, *146*, *148–152*, 157, 160, 161, 163, 165, *166*, *167*, 175, 177–179, 181–186, 190–193, *193–197*, *246*, 281, 284, 290, *299*, 425, *444*, *471*, 508, *531*, 692, 696, 699, *705*
Shub, M. 316, 338, 344, 347, 348, *358*, 433, *437*
Sieg, W. 8–15, 26, 31, *35*
Silver, J.H. 288, 293, *299*
Simmons, H. *357*, 359, 646, *677*
Simon, H.U. 432, *444*
Simon, I. 700, *705*
Simpson, S.G. 59, 60, 82, *83*, *85*, 126, 135, *149*, *153*, 160, *168*, 282, *299*, *439*, 508, *532*
Sipser, M. 598, 601, 608, 621, *671*, *673*
Skolem, T. 7, 28, 30, *35*, 590, *677*
Slaman, T.A. 57, *83*, 91, 103, 112, 113, *117–119*, 124, 128, 129, 131–133, 135, 138, 140, 141, 143–145, *146*, *152*, *153*, 157, 158, 160, 163–166, *167*, *168*, 177, 178, 180–187, 190, 192, *195–197*, 289, 290, *298*, *299*, 309, 311, 312, *314*, 692, 696, 698, 699, *704*, *705*
Smale, S. 316, 338, 344, 347, 348, *358*, 433, *437*
Smith, R.L. 38, 61, 62, 64–67, 69, 82, *82*, *83*, *85*, 404, 424, 425, *443*, *444*
Smullyan, R.M. 493, *503*, 634, *677*
Soare, R.I. 3, 5–7, 12, 18, 21, 27, 30, *33*, *35*, 38, 49, 50, 52–54, 56, 57, 60, 65, 66, 77, *82*, *83*, *85*, 109, *119*, 123, 124, 131, 133–136, 138, *147*, *149*, *153*, 159, *167*, 171–175, 177–179, 181, 183, 189, 191, *193–197*, 200–209, 211, 212, 214–220, 227, 230, 235, 241–243, *246*, *247*, 372, *444*, *471*, 508, *531*, *532*, 691, *703*
Solitar, D. 421, *441*
Solovay, R.M. *149*
Sorbi, A. 145, *147*
Soublin, J.-P. 410, *445*
Spector, C. 39, *85*, 124, 125, *153*, 157, *168*
Sperschneider, V. 428, *444*
Stavi, J. 297, *299*
Steel, J.R. 129, 133, 135, *153*, 192, *197*
Steele Jr., G.L. 602, *674*
Steinitz, E. 369, 370, *444*
Stephan, F. 106, 107, *119*, 136, *150*
Stewart, I.A. 372, 432, *444*
Stillwell, J. 421, *445*
Stob, M. 73, *83*, 133, *153*, 202, 216, *245*, *246*
Stockmeyer, L.J. 595, 597, 600, 607, 621, 664, *671*, *677*
Stoltenberg-Hansen, V. 254, *275*, 316, 340, 344, *359*, 366, 405, 411, 415, 422, 426, 428, 429, 431, 433, 434, *440*, *442*, *445*, 544, *585*
Straubing, H. 596, 607, 609, 621, *670*

Sturgis, H.E. 323, *359*, 536, *585*
Sturm, C.-F. 370, *445*
Suppes, P. 358, *359*
Suter, G.H. 433, *445*

Tait, W.W. 576–579, *585*
Takeuti, G. 535, 541, *585*, 591, 592, 608, 621, 623–625, 642, 646, *672*, *677*
Tamburrini, G. 16, 31, *36*
Tarski, A. 358, *359*
Taylor, W.F. 426, *442*
Tennenbaum, S. 426, *445*
Thomas, W. *358*, 359
Thompson, D.B. 591, 608, 623, 630, *677*
Thue, A. 424, *445*
Torenvliet, L. 700, 701, *703*
Townsend, M. 658, *677*
Tretkoff, C. 432, *445*
Troelstra, A.S. 434, *445*
Trotter, W. 82, *84*
Tucker, J.V. 316, 340, 344, *359*, 366, 405, 411, 415, 422, 423, 426–429, 431, 433, 434, *436*, *437*, *442*, *445*, *446*
Turing, A.M. 5, 11–15, 19, 21, 22, 24, 26, 31, 36, *36*, 108, *119*, 123, *153*, 170, *197*, 200, 201, *248*, *446*, 590, 592, 596, *677*

Ullmann, J.D. 685, *704*
Urquhart, A. 668, *672*
Uspenskii, V.A. 474, *503*

Valiant, L. 640, *677*
Valiev, M.K. 432, *446*
van Dalen, D. 434, *445*
van de Wiele, J. 303, 313, *314*
van der Waerden, B.L. 370, *446*
van Emde Boas, P. 592, 602, *677*
van Heijenoort, J. 6, 32–35, *36*
Vandiver, H.S. 371, 434, *446*
Vaught, R.L. 514, 520, *532*
Velickovíc 296, *298*
Ventsov, Yu.G. *446*
Vishkin, U. 602, 607, 621, 664, *671*, *677*
Vollmer, H. 591, 640, *677*

Waack, S. 432, *446*
Wagner, K. 591, 592, 608, 609, 631, 638–640, 644, *677*, *678*
Wainer, S.S. 38, 65, 66, *82*, 262, *275*, *357*, *358*, 359, 535, 541, 542, 570, *582*, *583*, *585*
Washihara, M. 470, *471*
Watanabe, O. 700, *705*
Watson, P. 138, 139, *147*
Weber, H. 370, *446*
Wechler, W. 427, *447*
Wechsung, G. 592, 609, 631, 638, 639, 644, *678*
Weiermann, A. 534, 542, *582*, *586*
Weihrauch, K. 429, 432, 434, *440*, *447*, 470, *471*, 474, *503*
Weinstein, B.J. 176, 183, 191, *197*
Weispfenning, V. 385, 435, *436*
Welch, L.V. 128, *153*
Welch, P. 295, 296, *297*
Wilkie, A.J. 624, 627, 633, 636, 637, 640, *673*, *676*, *678*
Winskel, G. 545, *583*
Wirsing, M. 427, 428, *447*
Woodin, W.H. 91, 112, 113, *119*, 124, 128, 129, 140, 141, 143, 145, *153*, 160, 163–166, *168*, 182–184, *197*, 289, *299*
Woods, A. 621, 636, 637, *678*
Wrathall, C. 591, 634, *678*
Wüssing, H. 421, *447*
Wyllie, J. 602, *672*

Yang, D. 697, 701, *703*
Yang, Y. 290, *298*
Yasugi, M. 470, *471*
Yates, C.E.M. 125, 127, 130, 132, *153*, 171, 173, 183, *197*, 358, *358*
Yi, X. 138, 139, 145, *147*, *153*
Young, P.R. 91, 92, *119*, 372, *441*

Zabell, S.L. 11, *36*
Zalcstein, Y. 432, *441*
Zemke, F. 534, *586*
Zheng, X. 695, *705*
Ziegler, M. 422, *447*
Zucker, J.I. 316, *359*, 433, *446*

Subject Index

$\exists$-theory of $\mathcal{R}$, 171, 172
$\exists\forall$-theory of $\mathcal{R}$, 175
$\forall\exists$-theory of $\mathcal{R}$, 177, 179, 180
$\forall\exists\forall$-theory of $\mathcal{R}$, 180
$\{e\}(x)$, 302
${}^2E$, 261, 269
$\mathbf{0}''$ constructions or methods, 181
$\mathbf{0}'''$ constructions or methods, 177, 181
1-3-1, 174, 183, 191
1-degree, 108, 110
1-generic, 124, 129, 130, 141–143
 in e-degrees, 143
1-generic degrees, 126–128
1-genericity, 129
1-section, 270
2-generic, 130

α^*, 281
$\alpha^*(B)$, 281
$\alpha_n^*(B)$, 281
$\aleph_\omega^L$, 280
α-c.e., 137
α-cardinal, 281
α-cosemidecidable, 376
α-decidable, 376
α-degree, 279, 288, 290
 above $\mathbf{0}'$, 288
α-finite, 278, 280, 281, 288
α-finite injury, 282
α-RE, 279, 280, 282, 283
α-RE degrees, 284
α-recursion theory, 279, 281, 289, 290
α-recursive, 278, 279
α-recursive, degree 279
α-recursive, functions 317
 partial, 278
α-recursive, set 281
α-recursively enumerable (α-RE), 278
α-region, 221
α-RE degree, 279, 280
α-semidecidable, 376
α-stability lemma, 283
α-stable, 283
α-states, 222
(α, β)-computable, 376
$\mathfrak{A}$-explicit functionals, 321–323
abstract computability theory, 433
accepted, 595
accepting computation tree, 597
accepts, 594, 597, 606
admissibility spectrum, 293–296
admissible
 ordinal, 282, 317, 332
 recursion theory, 432
 R, 294, 295
 weakly, 292
Alexandrov condition, 550
algebra, history, 369–371
algebraic
 closure, 402
 dependence, 393, 408
 dependence algorithm, 393
algebraically closed fields, 403
algorithm, 8
almost disjoint forcing, 294, 295, 297
almost homogeneity, 521, 522
almost prompt a.p., 217
alternating Turing machine, 597
analytic engine, 8
answer set, 657
anti-cupping property, 692, 693
application, 557
approximable map, 548
approximation, 501
arbitrary fan-in, 606
area thesis, 16
arithmetic, 517, 521
 first-order, 97
 second-order, 97
 structure with, 339–344

arithmetization, 92, 608
Arslanov completeness criterion, 135
Artinian ring, 409
associate, 253, 254, 262, 263, 266
atomic diagram, 510
atomless, 241
automatic machine, 200
automorphism, 95, 96, 107, 112, 114, 140, 145, 158, 164–166, 212, 215
  base, 131, 140, 141, 187, 188
  invariant, 381
  of e-degrees, 145
  of $\mathcal{R}$, 183, 184, 188, 193
  problem, 699
  theorem, 234
  Turing, 140–142
autoreducible, 68
axiomatizable theory, 69, 74–79, 508, 511

β-degree, 291
β-finite, 291
β-RE, 291
β-recursion theory, 280, 290, 291
β-recursive, 291
β-recursively enumerable, 291
  tamely, 291
β-reducibility, 291
(β-RE), 291
btt-degrees, 111, 112
Babbage, 8
back half, 641, 654
bar recursion, 265
basic feasible functionals, 658
basic feasible functionals of higher type, 668
basis, 38, 51–60
Berman–Hartmanis conjecture, 701
bi-immune, 56
biinterpretability, 129, 131, 141, 193 183, 185, 186, 188,
biinterpretable, 163–165
binary successor functions, 609
bitgraph, 602
block size, 617
blocking lemma, 285
Boolean circuits, 606
Boolean formula valuation problem, 600
Boolean functions, 323
bottom, 254
bounded, 41, 55, 56
  2-value recursion, 643
  fan-in, 606
  linear logic, 668
  loop programs, 668
  minimization, 627, 632
  product, 638
  quantifier formulas, 633
  recursion, 625
  recursion on notation, 591, 622
  shift left function, 641
  summation, 638
  truth-table reducibility, 99
branch, 265
branching, 182
branching degree, 182, 187
branching instructions, 627
BSS-computable, 339
BSS-machine, 338

$\mathcal{C}$-consistent, 236
C-B, 44
C-B rank, 47, 48
calculable, 28
calculus ratiocinator, 8
canonical basis, 392
canonical computable extension field, 391, 392, 394, 402, 403
Cantor singleton, 67, 68
Cantor–Bendixson derivative, 39, 43
Cantor–Bendixson rank, 43, 44, 61–69, 71
Cartesian closed category, 556
category, 58
Cauchy–Weierstrass thesis, 16
characteristic function, 608
Church's Theorem, 170
Church's Thesis, 10
Church–Kleene ω_1, 278, 279
Church–Rosser property, 427
Church–Turing thesis, 6, 16, 365, 366
circuit, 605
circuit families, 605
closure ordinal, 327
codability, 218
codable, 218
coding lemma, 160, 162
coding theorem, 296
coherence algorithm, 410
cohesive, 136, 144
cohesive degrees, 136
collapse, 537
coloring, 39, 74
commutation with directed unions, 551
commutator, 599
commutator subgroup, 599
compact, 254, 550
companion theorem, 317

complement, 467
complete, 599
formula, 511
partial orderings, 254
term rewriting systems, 427
type, 511, 512, 515, 521
completeness theorem, 509, 510
composition, 321, 322, 325, 328, 332, 552, 609
computability, 6
Banach space, 462
axioms for, 464
computability structure, 462, 464
heat dissipation, 460
wave propagation, 460
computability for $C[a, b]$
definition, 455
differentiation, 458
integration, 458
intermediate value theorem, 459
max–min theorem, 458
computability for L^p-spaces, 451
computability on higher types, 22
computability structure, Banach space, 450
computability theory, 5
computability, relative, 122–153
computable, 5, 509, 511, 515
algebra, 365, 367
history, 369–371
algebraic closure, history, 369–371
automorphism, 376, 381, 395
coherence, 408, 410, 419
dimension, 383
directed family of rings, 390
extension field, 391
extension theorems, 396, 402
family of rings, 389
field, 365
history, 369–371
functions, 12, 201
421–424,
homomorphism theorem, 387
mapping, 376
module, 409
Noetherian rings, 408
partial functions, 18
real, 451, 453, 454
real numbers, 428
ring, 365, 374
history, 369–371
spectrum, 383
subring, 384
universal algebras, 426
vector spaces, 424
computably
autostable, 382, 395, 403
relative, 382
categorical, 529
enumerable (c.e.), 5, 123, 170, 201, 509
enumerable degrees, 123, 128, 130, 141
enumerable in, definability of 131
inseparable, 511, 516
presentable, 510, 516
presented, 508, 510
stable, 379, 394, 529
relative, 380
stable module, 409
computation, 6, 252, 258, 259
computation tree, 262, 303
computational tree, 595
computations, 268
computer, 11
computor, 11
computorable, 12
concatenation function, 612
concatenation recursion on notation, 610, 665
concurrent random access machine, 603
conditional
parallel, 563
sequential, 563
conditional function, 611
configuration, 593, 615
confluence property, 427
Congo, 312
conjecture, 193
conjunctive reducibility, 99
conservative, 342
consistent, 236, 546
constructive
algebra, 434
history, 370
arithmetic, 591, 633
parus, 501
ring, 375
continuous, 323, 327
continuous functional, 251, 253, 258
continuum hypothesis, 103
control variables, 646
convergent sequence, 260
cosemicomputable ring, 374
countable, 38
countable chain condition, 308
countable functional, 253, 266
countable thin, 38
countably categorical, 181
counter Turing machine, 668

course-of-values recursion, 638, 643
covering theorem, 297
Craig's interpolation theorem, 512
CRE (A), 204
creative, 203, 205
critical triple, 176
cuppable, 191
cuppable degree, 183
cupping for $\mathbf{D}_n$, 139
curve thesis, 16

δ^1_2, 280
Δ^0_2 degrees, 123
Δ^0_3-automorphism method, 220
Δ_1 master code, 291, 292
d-computably enumerable (d.c.e.), 137–140
data types, 427
decidability, 124, 132
  below $\mathbf{0}'$, 124–127, 132
  in e-degrees, 145
  in n.c.e. degrees, 140
decidable, 39, 75, 127, 132, 140, 508, 510, 515, 518–522, 525–527
  ideal, 385
  model, 518
    completion, 516
  subring, 384
decided, 510, 594
decision problem, 170, 405, 421–425
Dedekind cut, 50
deductive closure, 549
deductively closed, 546
definability, 95, 96, 128, 140, 317
  in e-degrees, 143–145
  in $\mathcal{R}$, 183, 187, 188, 193
  in cones, 131
  of $\mathcal{E}$, 131
  of 'computably enumerable in', 131
  Turing, 122, 128–133, 135, 140–142
    in cones, 131–133
    of $\mathcal{E}$, 131
definable, 140, 143
definition by cases, 321, 322, 325, 328, 612
degree, 38, 41, 51, 57, 59, 60, 65, 67, 68, 72, 73, 90, 135, 517
  Δ^0_2, 123–153
  minimal, 57, 74
  r.e., 57, 67, 73
  invariant, 192, 193
  of unsolvability, 22
  PA, 124, 134, 135, 142
  Turing, 122–153
degree theory, local, 121–153
delayed diagonalization, 690, 692, 693, 699
density, 256
  of $\mathcal{D}_e(\leqslant \mathbf{0}'_e)$, 144
density theorem, 172, 177, 189, 255, 256, 258, 280, 284, 290
depth, 606, 660
descriptive complexity theory, 591
deterministic, 319
diagonally noncomputable functions, 134
diamond theorem, 145
  for e-degrees, 145
direct connection language, 607
direct limit, 387–391
direct product, 388
  of numbered sets, 481
direct sum, 387–391
  of numbered sets, 481
directed set, 550
disjunctive reducibility, 99
distributive, 93
distributive uppersemilattice, 93
divergence witnesses, 305
divide and conquer, 638
divide and conquer recursion, 625
divisible group, 424
DNC, 134
Downey diamond theorem, 139
Downey's conjecture, 139

η-hyperhypersimple, 108
η-maximal, 108
$\mathcal{E}$-definable, 241
$\mathcal{E}^*$, 191
E-closed, 321
e-degrees, 106, 107, 142
E-recursion, 321
e-subset, 479
E-closed, 303
E-r.e. degrees, 312
E-recursive in, 304
E-recursively enumerable, 304
effective
  algebra, 365, 367, 368, 421, 435
  bounding, 306
  density lemma, 465
  distributive uppersemilattice, 96
  lattice, 96
  spectrum, 383
  unbounded search, 313
effective omitting types theorem, 528
effectively

calculable, 5
enumerable, 19
enumerable set, 20, 200
immune, 56
inseparable, 60
noncappable, 191
Ehrenfeucht theory, 520–524
eigenvectors theorem, 451, 452, 470
element special, 482
elementarily definable with parameters (e.d.p.), 243
elementary
chains, 513
diagram, 510
functions, 590
induction, 334
embedding theorem, 171
embeddings, 257
Entscheidungsproblem, 9, 10
Entscheidungsverfahren, 9
enumeration degrees, 105–107, 114, 115, 142–145
enumeration reducibility, 105–107
enumeration reducible, 105–107, 142
envelope, 345
epimorphism, 480
equational theories, 427
equivalence of numberings, 477
Ershov hierarchy, 137
essentially undecidable theory, 69, 75
evaluation functional, 319
exact pairs, 125, 143, 173, 695
in e-degrees, 143
existential theory of $\mathcal{D}(\leqslant \mathbf{0}')$, 124
expansion, 321, 328, 658
explicit definition, 556
explicit field, 370
explicit ring, 375
extended rudimentary, 590
extendible, 42
extension of embedding problem, 172, 177, 179
extension of embedding theorem, 177, 190
extension property, 382
extensionality, 255

factorization, 480
family computable, 486
family structurally definable, 493
fan functional, 262, 263
fan-in, 605
fan-out, 605
feasible type 2 functionals, 656
final algebras, 427
fine structure theory, 278, 281, 288
finitary, 522, 524
finitary functional, 318
finite
algorithmic procedure (FAP), 316
basis, 639
property, 526
branching, 40, 52, 55
tree, 52
combinatory process, 19
extension argument, 699
extension method, 687
injury, 177
sequences, coding, 340
types, 338, 667
finitely generated, 550
commutative ring, 375
ring, 387
finitely observable property, 550
finitely related module, 410
finiteness conditions, 374
finitist program, 8
first main theorem, 451, 452, 467
first recursion theorem, 318, 326, 332, 343
first-order arithmetic, 97, 104, 115
fixed point, 326, 327, 561
free, 135
free degrees **FPF**, 134
least, 326, 327
principal least, 328
simultaneous, 327
$\mathcal{F}$LOGSPACE, 623
forcing, 53
with Π^0_1 classes, 53
over an E-closed, 305
formal systems, 91, 98, 105
FPF degrees, 135
fragments of Peano arithmetic, 289
Friedberg–Muchnik theorem (solution of Post's problem, 289
front half, 641, 654
full approximation, 126, 131, 132
fully specified OCRAM, 665
function
constant, 552
length cost, 657, 660
S_3-definable, 287
universal, 483
function algebra, 608
functional
computable, 560

hereditarily total, 542
partial recursive, 542
recursive, 543
recursive in pcond and ∃, 564
complexity theory, 656
composition, 658
interpretation, 271
structure, 319
substitution, 658

Gödel's constructible universe L, 278
Galois group, 406
game, 220
gap language, 688
gates, 605
general recursive, 21
functions, 9
generalised computability theories, 432
generalised high/low hierarchy, 129, 130
generalized discrete upper semilattice of rank k, 497
generalized recursiveness, 316
generated set, 200
generic, 53, 58, 59
generic c.e. set, 214
genetic Turing machine, 602
global read/write, 602
global search, 368, 430
global shared memory, 602
Goncharov's theorem, 384
graph, 39, 74, 602
coloring, 37, 79–81
Grzegorczyk hierarchy
extended, 540, 559
Grzegorczyk's classes, 626
Gutteridge's theorem, 144

H_n, 193
h-simple, 201
half function, 654
halted configuration, 593
halting problem, 170, 517–519, 525
Hardy-functions, 540
Heinermann hierarchy, 626
hereditarily recursive operations, 668
Hermann's theorem, 370, 386, 398, 421
Hessenberg sum, 43, 62
hh-simple, 201
HHM sets, 217
hidden operations, 427
hierarchy, 137
high, 54, 71, 190
high c.e. degrees, 201
high_2, 193
high_n, 186, 189
high/low hierarchy, 124, 128, 141, 143
generalised, 129, 130
in e-degrees, 143
higher type parallel complexity classes, 668
highly recursive, 40, 52
Higman's theorem, 422
Hilbert basis theorem, 370, 385, 386
homogeneity, 95, 96, 103, 106, 181
homogeneous, 514, 518–521
homogeneous models, 517, 518
honest reducibility, 701, 702
hyperarithmetic, 39, 55, 56, 63, 517, 519, 521, 529
hyperelementary, 334–338
hyperfinitary functionals, 270
hyperhyperimmune, 49, 144
hyperhyperimmune, strongly, 136
hyperimmune, 45, 47, 49, 53, 64, 66, 136, 143
free degree, 134, 135
sets, 103
hyperimmune-free, 54, 67, 103
degree, 103
hyperregular, 281
hypersimple, 48, 64
hypersimple set, 99, 170

ideal, 546
ideal membership algorithm, 408, 413, 415, 417, 420
ideal membership problem, 385, 413, 420
identity, 552
immune, 56, 136, 143
immune-free e-degrees, 144
in-degree, 605
inadmissible, strongly, 292
incomplete high RE set, 290
incompleteness, 205
result, 92
theorem, 9, 170
incremental instructions, 627
index, 252, 263
index answer tape, 595
index query tape, 595
index set, 495
indexing, 317, 325, 329, 330, 332, 339
inductive, 334–338
inductive definition, 317
infinite injury, 177
infinite injury priority argument, 284

infinite injury priority method, 290
information content, 21
information system
  coherent, 558
  flat, 546
  object of, 546
initial
  algebras, 427
  computation, 262, 268
  functionals, 321
  segment, 125, 128, 130, 132, 145, 157, 158
    of $\mathbf{D}_e(\leqslant \mathbf{0}'_e)$, 145
    of $\mathcal{D}(\leqslant \mathbf{0}')$, 125, 126
  subset, 39, 45, 46
intrinsically computably enumerable, 529
invariant, 241
isolated, 138
isolated degree, 138, 139
isolating, 138
isolating degree, 138, 139
isomorphism, 480
  invariant, 374
iterable, 327
iterated look-ahead technique, 693, 695
iterating, 326
iteration, 609
  functional, 559

J_α-hierarchy, 290
Jensen's $\diamond$-principle, 292, 293
Jensen's coding theorem, 296, 297
Jensen's covering theorem, 295
Jensen's fine structure theory, 297
jump, 129
  classes, 186–188, 193
  inversion, 123, 125, 133, 134, 143
  of minimal degrees, 129, 130
  of the strongly hyperhyperimmune degrees, 136
  on $\mathcal{D}_e$, 143
  on e-degrees, 143
  operator, 288, 290
  Turing, 123, 128, 129, 137, 138

$\kappa_n(B)$, 281
k-section, 310
k-bounded recursion, 630
k-bounded recursion on notation, 624
k-bounded tree recursion on notation, 642
k-envelope, 311
k-function, 641
k-initial subsets, 46
k-retraceable, 46–49
$k+1$-retraceable, 46
Kleene fixed point theorem, 7
Kleene numbering, 487
Kleene's normal form theorem, 18
Kreisel–Lacombe–Shoenfield, 254

λ-abstraction, 557
λ-calculus, 105
λ-definable functions, 9
λ-term, 557
Λ-sequence m-universal, 485
Λ-sequence co-productive, 499
Λ-sequence creative, 499
L^p-computability, 460
  definition, 461
L_{n+1}, 193
labelling systems, 529
Lachlan, 201
Lachlan game, 220
lattice $\mathcal{E}$, 201
lattice embeddings, 174–179, 181, 183, 191, 691, 692, 696
lattices of r.e. substructures, 425
least fixed point, 318, 326, 327
least significant part function, 614
LC, 179
Leibniz, 8
length, 659
Lie group, 423
limited recursion on notation, 658
linear
  dependence, 425
  groups, 423, 432
  growth, 601
  logic, 668
  reducibility, 100
  time hierarchy, 598, 633
lingua characteristica, 8
local degree theory, 121–153
local ring, 415, 431, 433
logarithmic growth, 601
logtime hierarchy, 598
LOGTIME-uniformity, 607
looking back technique, 690
low, 53, 54, 73, 188
low basis theorem, 134
low_1, 214
low_2, 189–191, 193, 212
low_n, 186, 187, 189

low$_2$ density, 138
  for $\mathbf{D}_n$, 138
lowness properties, 212

$\mathcal{M}$-consistent, 224
$\mathcal{M}$-inconsistent, 224
μ-recursive, 18
m-complete r.e. sets, 92
m-degree, 93–98, 108, 114
m-enumerability, 496
m-jump, 496
m-reducibility by separation (sm-reducibility), 498
m-reducibility of Λ-sequences, 485
M_5, 174
M, 179
many-one reducibility, 92–98, 378, 686
many-one reducible, 92, 686
Marchenkov, 202
marriage problem, 82
master code, Δ_1, 291
matrix groups, 423, 432
maximal, 49, 74, 241, 287
maximal set, 201, 281, 290
maximal, α-RE set, 287
measure, 57–59
mechanistic, 15
Miller, D., 202
minimal, 39, 69–74
  complement, 132
  cover, 144, 159
    in e-degrees, 144
    strong, 128
  degree, 57, 74, 125, 127, 130, 132, 141
    below $\mathbf{0}'$, 129, 130
  Π^0_1 class, 39
  pair, 53, 143, 173, 178–180, 182, 183, 190, 191, 688, 689, 693–695
    of e-degrees, 143
mixed type, 271
model, 44
model completion, 515
modular counting gate, 606
modulus of convergency, 260, 269
monomorphism, 480
monotone, 319, 550
morphism of numbers sets, 479
Moschovakis witness, 305
most significant part function, 614
multiple bounded recursion on notation, 646

ν-computable subset, 479
N_5, 174
n-CEA, 137
n-CEA hierarchy, 137
n-computably enumerable (n.c.e.), 137–140
n-generic, 124, 127, 128, 143
  in e-degrees, 143
n-subset, 479
n-type, 696
(n-)type, 696, 697
NC, 176, 191
n.c.e., 137
nearly linear time, 632
neighborhood condition, 279, 291, 292
next configuration, 593
Nies, 202
Noetherian module, 409
Noetherian ring, 409
non-principal type, 511, 519, 524
nonbranching degree, 182, 187
noncappable degree, 183, 191
NonCapping
  in e-degrees, 145
  theorem, 145
noncomputable real, example, 454
noncuppable degree, 183
NonCupping
  in e-degrees, 145
  theorem, 145
nondensity of $\mathcal{D}_e(\leqslant \mathbf{0}_e^{(2)})$, 144
nondensity theorem for $\mathbf{D}_2$, 138
nondeterministic multitape Turing machine, 595
nondiamond theorem, 174, 178
nonlow sets, 211
nonsplitting theorem, 177, 183, 189
norm, 350
normal, 310, 348–352, 646
normal extension, 394
normal forms, 328, 329
normal system, 20
NP-complete, 684
NP-completeness, 684
number selection theorem, 343, 352
numbered
  ring, 373
  separated, 480
  set, 477
    complete over approximation, 501
    completely numbered (complete), 481
    precompletely numbered (precomplete), 481

numbering, 365, 372, 428
  complete (precomplete), 482
  computable, 374, 484, 486
    principal, 487
  cosemicomputable, 374
  decidable (positive, negative), 478
  effective, 373, 377
  equivalence, 478
  invariance, 378
  ν-computable or computable relative to, 479
  of a set, 477
  recursive equivalence, 378
  recursive reduction, 378
  semicomputable, 374
  standard, 377

ω-c.e., 137
ω-computability theory, 6
ω_1^{CK} Church–Kleene ω_1, 280
omitting types theorem, 511, 524–526
operation, 608
operator, 608
oracle, 319
  answer tape, 657
  concurrent random access machine, 663
  query state, 657
  query tape, 596, 657
  registers, 664
  Turing machine, 596, 657
ordered commutative rings, 338
ordinal recursion theory, 278, 280, 282
out-degree, 605
output node, 605

Π_1^0 and r.e. subsets, 44
Π_1^0 class, 37–85, 133–136
  bounded, 41
  countable, 38
  minimal, 37
  recursively bounded, 38–42
  thin, 37
Π_1^0 set, 44–51, 72
  retraceable, 39
Π_1^0 set, 37
  retraceable, 45–49
p-r-degree, 686
p-constructibility, 685
p-constructible, 685
p-cylindrification, 486
pth root algorithm, 401
PA degrees, 124, 134, 142

pairing function, 615
parallel, 563
parallel conditional, 563
parallel machine model, 602
parallel random access machine, 602
parity gate, 606
partial
  α-recursive, 278
  $\mathfrak{A}$-recursive, 328
  $\mathfrak{A}$-recursive, simply, 328
  continuous functionals, 557
  E-recursive, 302
  κ-recursive, 332
  μ-recursive, 18
  recursive functions, 18
Pascal, 8
path through $\mathcal{O}$, 539
Peano arithmetic, 39, 60, 75, 289
  fragments, 289
pentagon, 174
perfect, 108
perfect field, 401
perfect set, 108
persistent, 163
PFS, 551
POLY, 658
polynomial
  growth, 601
  ring, 370, 374, 377, 381, 417
  time, 685, 686
    algebras, 431
    degree, 686
    groups, 432
    hierarchy, 598
    oracle Turing machine, 657
    reducibility, 683–702
polynomially bounded branching recursion, 641
polynomially bounded recursion on notation, 648
positive elementary, 334–338
positive extended rudimentary, 590
positive reducibility, 99
Post numbering, 488
Post's problem, 171, 192, 193, 201, 282, 288, 292, 293
  above $\emptyset'$, 288
  β-recursion theory, 292
  for all admissible ordinals, 280
  Sacks splitting theorem, 280
  uniform solution to, 131
Post's program, 201

Presburger arithmetic, 633
Presburger definable sets, 633
prime field, 394, 400
prime model, 513, 518–520, 526, 527
prime subfield, 404
primitive recursion, 7, 332, 609
 functional, 320
primitive recursive, 41
 algebra, 430
 tree, 41, 42
principal, 626
 computable relative to $\mathfrak{S}$, 479
 function, 45
 numbered subset, 479
 subset, 479
 types, 511, 526
principle of finite support, 551
priority
 arguments, 384, 408
 method, 99, 171, 201
  infinite injury, 290
 resolution, 603
probabilistic Turing machine, 602
problem P for a pair, 484
product of functions, 554
production system, 20
program, 324
projection, 257, 554
projection functions, 609
projectively finitary, 523
promptly simple, 178, 179, 186, 187, 191, 213–215
PS, 178, 179, 187, 191

$Q(A)$, 205
quantum Turing machine, 602
quasilinear space, 632
quasilinear time, 632
query set, 657

$\mathcal{R}$, 171
$\mathcal{R}_{tt}$, 191
$\mathsf{Reg}_{\mathfrak{A}}$, 325
$\mathcal{R}$-consistent, 226
$\mathcal{R}$-inconsistent, 226
R-admissible, 294, 295
r-maximal, 206
r.b. Π^0_1 class, 38
r.e., 60, 64, 67, 72
r.e. m-degrees, 95–105
r.e. tt-degrees, 104
r.e. degree, 57, 67, 73, 74
r.e. set, 45, 60, 200
ramified recurrence, 653
random, 59
random access, 595
rank, 579, 608
rank (k, ℓ), 656
rank-faithful, 68, 69
REA-operators, 156, 161
rearrangement, 321, 328
recursion, 6, 328, 333, 356
 theorem, 7, 263, 269
 theory, 26, 289
  on fragments of Peano arithmetic, 278
recursive, 18, 52, 267, 278, 344
 autoequivalence, 381
 binary tree, 63
 cofinality, 285
 enumeration, 19
 function, 200
 model theory, 508, 509
 ordinal, 43, 61–69, 71
 presentability, 685
 relative to Φ, 22
 ring, 375
 tree, 40, 52
recursively
 bounded, 41, 55, 56
 bounded (r.b.) Π^0_1, 38
 enumerable, 38, 170, 200
 enumerable (r.e.) set, 44–51, 200
 enumerable degrees, 171
 enumerable set, 19
 inseparable, 45
 presentable, 181, 685
 presentable ring, 375
recursiveness, characterizations, 316
reducibility, 279
 bounded truth-table, 99
 conjunctive, 99
 disjunctive, 99
 linear, 100
 of numberings, 477
 positive, 99
reflexive program call, 7
register, 324
 computability, 347
 computable, 325
 decidable, 344
 machine, 17, 316, 323–326
regular, 280, 598
regularity, 280
rejected, 597

Rekursion, 28
relative computability, 21, 122, 170
repeated squaring, 634
replacement axiom, 279
  Σ_1, 280
replacement functional, 320
representation, 366–368, 429
requirement, 290, 292, 293
retraceable, 45
retraceable Π_1^0 set, 39
reverse recursion theory, 289
RE set, incomplete high, 290
rigid, 140, 158, 164, 166
rigidity, 184
rigidity, Turing, 140, 141
ring, computability over, 316
rings, computation over, 338, 344, 347, 348
Ritchie–Cobham property, 659
Robinson's consistency theorem, 516
Rogers, 202
root algorithm, 399
roots of unity, 405
rudimentary, 591
rudimentary functions, 633
rudimentary sets, 634
Ryll-Nardzewski theorem, 522

Σ_1-admissible ordinals, 278, 279
Σ_1-admissibility, 290
Σ_1-complete, 294, 295
Σ_1-replacement axiom, 280
Σ_n projectum (β), 291
Σ_n-admissibility, 278
Σ_n-admissible, 278, 280
Σ_n-cofinality, 281, 291
Σ_n-projection, 291
Σ_n^B-projectum, 281
$\mathbf{S(T)}$, 511
$S(T)$, 513, 514, 519, 521, 522
$S^r(T)$, 514, 519, 521, 522, 525, 526
S_3-definable, 287
$S_p(T)$, 513, 514, 526
sm-equivalence, 498
$\mathbf{S^r(T)}$, 511
$\mathbf{S_p(T)}$, 511
$\mathbf{SP\bar{H}}$, 191
Sacks preservation, 172
Sacks splitting theorem, 161
Sacks–Simpson lemma, 282
safe, 646
  composition, 647
  divide and conquer recursion, 655
  minimization, 652
  recursion, 646, 653, 669
    on notation, 647
  weak recursion on notation, 654
  storage Turing machines, 630
saturated model, 513, 519, 520
schemata-based definition, 23
scheme, 252, 253, 259, 269
Scott
  basis theorem, 134
  condition, 550
  domains, 669
  rank, 518, 525
  sentences, 518
  set, 60
  topology, 550
search, 325, 328, 332
  functional, 319
second main theorem, 451, 452, 469
  consequences, 469
second order polynomials, 660
second recursion theorem, 317, 333, 352
second-order, 318
  arithmetic, 97, 103, 106
  polynomial, 666
section, 345
selection, 305, 309, 312, 346, 352
  functional, 332
  number, 343
semi-algebraic, 347
semi-characteristic function, 335
semi-low$_{1.5}$, 214
semi-recursive, 337, 345
semi-register decidable, 345
semicomputable
  groups, 421–424
  ideal, 385
  module, 409
  ring, 374
  spectrum, 383
semicomputably stable, 379
semirecursive, 108
separating sets, 39, 45
sequence, 260
sequential conditional, 563
sharply bounded
  maximization, 613
  minimization, 613
  product, 638
  quantifier, 612
  recursion on notation, 623
  summation, 638
Shoenfield conjecture, 172, 173, 180

Shore incompleteness lemma, 284, 285
signature, 259
simple, 201, 206
simple set, 92, 110, 170
simultaneous
    bounded recursion, 642
        on notation, 643
        327,
    recursion, 642
        on notation, 642
size, 606
slow diagonalization, 694
small major subset, 206, 208
small sets, 207
small subset, 208
Soare, 202
solvable, 599
space $S(n)$, 594, 595, 597
Spector class, 317
spectrum, 309
splitting, 145
    algorithm, 370, 391, 399, 403
    in e-degrees, 145
    theorem, 172, 177, 178, 183, 187, 189
        for $\mathbf{D}_n$, 138
stability lemma, 465
stable, 522
stack register machine, 627
stack registers, 627
stage comparison, 350
standard model, 98
Steinitz construction, 369, 391, 404
storage instructions, 627
strong minimal cover, 128, 133
strong reducibility, 700
strongly hyperhyperimmune, 136
strongly inadmissible, 292, 293
strongly terminating, 427
structure, 44, 318
    with arithmetic, 339
subcomputation, 262
subrecursive reducibility, 684, 702
subspace membership problem, 425
substitution, 319, 321, 325, 328, 331, 333, 336, 354
successor function, 609
super sparse set, 689, 690, 697, 699, 701
supremum functional, 320, 332
syntactic monoid, 599

T-degree, 110, 111
tt-complete r.e. sets, 99
tt-cylinder, 102
tt-degree, 102–105, 108–111, 114
tt-reducibility, 98–105
tame approximation, 285
tamely β-RE, 291
Tennenbaum's theorem, 426
term evaluation, 341, 344
terminal information system, 556
terminate, 259
theories, 39
theory, 37, 39, 74–79
    axiomatizable, 69
    decidable, 39, 75–79
    essentially undecidable, 69, 75–79
    of $\mathcal{D}_e(\leqslant \mathbf{0}'_e)$, 145
    of $\mathcal{R}$, 171, 172, 175, 177, 179, 180, 182, 184, 185
thin, 69–74
threshold gate, 606
tiering, 646, 669
time $T(n)$, 594, 595, 597
token, 545
topological algebras, 433
total, 255, 258
total e-degrees, 106
tracking functions, 372
transcendence algorithm, 405
transcendence basis, 393
transcendence degree, 393, 404
transition function, 593, 596
transition relation, 595
tree, 40–43, 265
    finite branching, 40
    highly recursive, 40–42
    of possibilities, 307
    primitive recursive, 41
    recursive, 40–42, 63, 64
true arithmetic, 180, 184, 185
true arithmetic theory of $\mathcal{R}$, 184
true path, 230
truth table degrees, 191
truth-table, 67
truth-table conditions, 98
truth-table degrees, 98–105, 112–114
Turing, 11
    computable, 5, 201
    definability, 122, 135
    degree, 21, 122, 156, 170
    jump, 123, 128, 137, 138, 156, 161
    machine, 12, 200, 592, 593
    reducibility, 91, 685
    reducible, 685

Theorem, 12
thesis, 13
type, 252, 511, 556
type $(k;\boldsymbol{l})$, 318
type 2 functional, 656
type 2 recursion theory, 432
type level, 667
typed while programs, 668
types in $\mathcal{R}$, 181–183

unbounded, 606
undecidability, 99, 145, 697
undecidability of $\mathcal{R}$, 181, 182
undecidability proofs, 91
undecidable problems, 405
undecidable theory, 697, 698, 701, 702
uniform solution to Post's problem, 131
uniformly Σ_1-definable, 312
unit cost, 657
model, 658
universal functionals, 339–343
universal Turing machine, 11
unstable, 524
uppersemilattice, 93

Vaught's theorem, 514
vectorized Grzegorczyk classes, 626
very safe divide and conquer recursion, 655
very tardy, 217
VSDCR, 655

wn-subset, 479
wtt-degree, 112
weak bounded primitive recursion, 631
weak bounded recursion on notation, 624, 666
weak density theorem, 138
for n.c.e. degrees, 138
weakly α-recursive in, 279
weakly admissible, 292
structure, 284
well quasi-ordering, 659
well-resided α-states, 225
wellfounded, 265
winning strategy, 235
word problem for groups, 371, 421–423
work register, 627

x-reflecting, 304

yield, 594